Mixed-Signal Embedded Systems Design

Edward H. Currie

Mixed-Signal Embedded Systems Design

A Hands-on Guide to the Cypress PSoC

Edward H. Currie
Hofstra University
Hempstead, NY, USA

ISBN 978-3-030-70314-1   ISBN 978-3-030-70312-7   (eBook)
https://doi.org/10.1007/978-3-030-70312-7

This Springer imprint is published by the registered company Springer Nature Switzerland AG
The registered company address is: Gewerbestrasse 11, 6330 Cham, Switzerland

Preface

This textbook is intended to provide a unique, in-depth look at programmable-system-on-a-chip (PSoC) technology from the perspective of the world's most advanced PSoC technology, viz., Cypress Semiconductor's PSoC product line. The book introduces a wide variety of topics and information intended to facilitate your use of true visual embedded design techniques and mixed-signal technology. The author has attempted to include sufficient background material and illustrative examples to allow the first-time PSoC user, as well as, advanced users, to quickly "come up to speed." A detailed bibliography, and other sourcing of useful supplementary materials, is also provided.

Readers are encouraged to visit Cypress' website at www.cypress.com and explore the wealth of material available there, for example, user forums, application notes, design examples, product data sheets, the latest versions of development tools (which are provided to users at no cost), datasheets on all Cypress products, detailed information regarding Cypress' University Alliance programs, and tutorials. Cypress has amassed a vast wealth of material related to the subject matter of this text, most of which is accessible to readers online, and the author has exercised a complete lack of restraint in harvesting important concepts, illustrations, examples, and source code from this source.

The presentation style employed in this textbook is based on the author's desire to provide relevant and definitive insight into the material under discussion while circumnavigating a swamp of details in which the reader might otherwise become mired. Thus, mathematical derivations are provided in what, in some cases, may be viewed as excruciating detail to best accomplish the task at hand. But various forms of metaphorical "syntactic sugar," or a reasonable facsimile thereof, as well as other forms of embroidery have been freely and liberally applied, as required, to leave the reader with both a strong intuitive grasp of the material and a solid foundation. However, true mathematical rigor in the sense that purists and theoretical mathematicians prefer has been largely avoided.

To the extent feasible, the author has avoided colloquialisms such as "RAM memory" (which is literally "random access memory, memory") and employed accepted abbreviations and mnemonics, hopefully without sacrificing clarity and getting bogged down by details, which may well be required for completeness, but are often of little real-world applicability or value.

The author has made every effort in the preparation of this textbook to ensure the accuracy of the information. However, the information contained herein is provided and intended for pedagogical purposes only, and without any other warranty, either express or implied. The author will not be held liable for any damages caused, or alleged to be caused, either directly, or indirectly, by this textbook and/or its contents. Camera-ready copy was prepared using the author's LaTeX files.

Errors found in this work are the sole and exclusive property of the author, but much of the material found here is either directly, or indirectly, the result of the efforts of many of the employees of Cypress, and of course, its customers. The author welcomes your suggestions, criticisms, and/or observations and asks that you forward any such comments to edward.currie@hofstra.edu.

The author wishes to express his profound appreciation to the following people who made the publication of this textbook possible: Ata Khan, George Saul, Dennis Sequine, Heather Montag, David Versdahl, Dave Van Ess, Don Parkman, and especially Dr. Patrick Kane.

Hempstead, NY, USA   Edward H. Currie

Contents

List of Figures

List of Tables

Legal Notice

In this textbook the author has attempted to teach the techniques of mixed-signal, embedded design based on examples and data believed to be accurate. However, these examples, data, and other information contained herein are intended solely as teaching aids and should not be used in any particular application without independent testing and verification by the person making the application. Independent testing and verification are especially important in any application in which incorrect functioning could result in personal injury, or damage to property. For these reasons, the author and Cypress Semiconductor Corporation[1] expressly disclaim the implied warranties of merchantability and of fitness for any particular purpose, even if the author and Cypress Semiconductor Corporation have been advised of a particular purpose, and even if a particular purpose is indicated in the textbook. The author and Cypress Semiconductor Corporation also disclaim all liability for direct, indirect, incidental, or consequential damages that result from any use of the examples, exercises, data, or other information contained herein and make no warranties, express or implied, that the examples, data, or other information in this volume are free of error, that they are consistent with industry standards, or that they will meet the requirements for any particular application. The author and Cypress Semiconductor Corporation expressly disclaim the implied warranties of merchantability and of fitness for any particular purpose, even if the author and Cypress Semiconductor Corporation have been advised of a particular purpose, and even if it is indicated in the textbook. The author and Cypress Semiconductor Corporation also disclaim all liability for direct, indirect, incidental, or consequential damages that result from any use of the examples, exercises, references, data, or other information contained herein.

Cypress, PSoC, CapSense, and EZ-USB are trademarks or registered trademarks of Cypress in the United States and other countries. Other names and brands contained herein may be claimed as the sole and exclusive property of their respective owners.

Introduction to Embedded System

*Embedded Systems are
application-domain specific,
information processing systems that
are tightly coupled to their
environment.*

Dr. T. Stefano (2008)

Abstract

This chapter provides a brief review of the history of embedded systems, microprocessors and microcontrollers. Also presented are basic concepts of programmable logic devices, overviews of the 8051 microcontroller, brief descriptions of some of the more popular and currently available microcontrollers that are in widespread use and introductions to a number of subjects related to microcontrollers and embedded systems, e.g., types of feedback systems employed in embedded systems, microcontroller subsystems, microprocessor/microcontroller memory types, embedded system performance criteria, interrupts, introductory sampling topics, etc.

The architectures of the early microprocessors are discussed briefly, as well as, the role of polling, interrupts and interrupt service routines (ISRs), each of which is discussed in more detail in later chapters. DMA controllers and tri-stating are also introduced.

It can be argued that the evolution of the mixed-signal, embedded system began with Autonetic's introduction of the world's first all solid state computer, the VERDAN (Goldstein et al., Calif. Digit Comput Newslett 9(2–9):2– via DTIC, 1957), and similarly the evolution of the modern microcomputer can be said to have begun with the introduction of the Altair, in January of 1975 (This was developed by H. Edward Roberts who caused the phrase "personal computer" to come into widespread use.). While A/D and D/A converters were available as early and 1930s, high-speed converters were not widely available until the 1950s as a result of Bernard Marshall Gordon's introduction of a high-speed/precision A/D design in 1953(High-speed D/A converters follow as a natural consequence of the availability of high-speed A/D converters.). The evolution of a large family of A/D's based on solid state technology at a low price made it practical to create embedded systems based on microprocessors/microcontrollers , A/D's, D/A's, operational amplifiers, etc., which were capable of controlling and monitoring very complex systems for both commercial and military applications.

© Springer Nature Switzerland AG 2021
E. H. Currie, *Mixed-Signal Embedded Systems Design*,
https://doi.org/10.1007/978-3-030-70312-7_1

1

1.1   The Origin of the Embedded System

It is difficult to say when the first, all solid-state, embedded system appeared, particularly when compared to modern embedded systems. However, a very early example occurred as a result of the introduction of a series of inertial navigation systems, developed by Autonetics, a division of North American Rockwell, in the early nineteen fifties (Fig. 1.1). This work included, at least philosophically, a predecessor of the modern-day microcontroller which was called VERDAN (Versatile Digital Analyzer), shown in Fig. 1.2. This system, known by various names (VERDAN, MARDAN, D9) evolved from work in the latter part of the 1940s by Autonetics, into a fully transistorized, flight control system [2] consisting of some 1500 germanium transistors, 10,670 germanium diodes, 3500 resistors and 670 capacitors. [7]. This was a particularly amazing accomplishment in light of the then-known problems with manufacturing germanium transistors and their associated, and for some, infamous thermal "runaway" problems.[1]

The resulting "macro-computer-based" system was able to navigate [9], and control, airframes capable of operating at speeds well in excess of Mach 3, i.e., greater than 2300 mph. The computer portion of the system consisted of three sections: (1) a General Purpose (GP) section based on 24-bit data formats, fifty-six 24-bit instructions[2] and integral multiply/divide[3] hardware, (2) an I/O section capable of handling multiple shaft encoder and resolver[4] I/O channels and (3) 128 digital integrators in the form of a Digital Differential Analyzer (DDA). The equations of motion, a set of coupled, partial differential equations, were solved in the DDA section, based on continual input from a combination a star tracker and inertial guidance platform. The GP Section interacted with the DDA to update various parameters in the equations of motion. The DDA then solved the equations of motion in real-time and

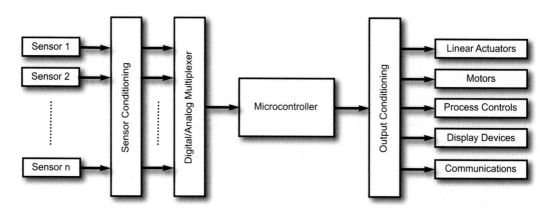

Fig. 1.1 A typical embedded system architecture.

---

[1]Some transistors exhibit a phenomenon known as "thermal runaway" which results from the fact that the number of free electrons present is a function of temperature. Thus, as the temperature rises, the current increases leading to additional Ohmic heating and therefore further temperature increases. This type of problem can lead to the destruction of the transistor, but also means that circuit behavior, based on such devices, can become "unpredictable".

[2]Although the instruction length was 24 bits the word length was 26-bits with one sign bit and on reserved bit.

[3]Later systems, e.g. in the Minutema n missiles, were based on the same computer hardware and architecture, but supported only hardware multiply. Division was carried out by inverting the divisor and multiplying. Functions such as square root, if needed, were handled programmatically. VERDAN multiplication and division results were each 48-bits in length.

[4]Encoders were used to determine shaft rotation position.

the solutions were subsequently passed to the GP section which communicated with the I/O section to output control information and commands to the system's flight control actuators.

VERDAN's architecture was to reappear in several incarnations [4], e.g., VERDAN II (a.k.a MARDAN [11]) part of The US Navy Ships Inertial Navigation System (SINS) and the D17, which was employed in Minute Man Missiles.[5] Each of these consisted of multiple plug-in cards sharing a common 84 pin bus. Autonetics may have also been the first to consider a floppy disk as a rotating memory device, but ultimately decided on, what appears to have been, the first "rotating disk" memory. It had fixed heads and the magnetic disk, the only moving part, rotated at speeds comparable to present-day hard disks, viz., 6000 RPM.[6]

VERDAN was repackaged to include a "front panel" used to facilitate field access to the circuitry and renamed MARDAN (Marine Digital Analyzer) [8]. The bus structure remained the same and most of the gold-plated printed, circuit boards used in VERDAN, of which there were approximately 96, were fully compatible with MARDAN's bus architecture. MARDAN, shown in Fig. 1.3, was to remain in service, from 1959 until 2005, on the Polaris [5], Trident, etc., -class submarines. VERDAN was repackaged for a third time as the navigational computer in the later versions of the Minuteman systems.In addition, some aspects of VERDAN's design were incorporated into the Apollo guidance system and the Apollo simulator.

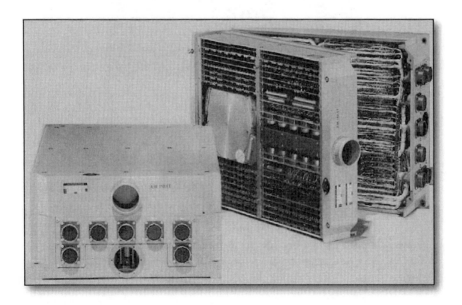

Fig. 1.2 The VERDAN computer.

---

[5]Autonetics also marketed desktop versions of the VERDAN called RECOMP II and III [10] that were intended to be used as general-purpose computers [4, 6].

[6]Modern hard drive disks rotate at 5400–7200 RPM.

Fig. 1.3 The MARDAN computer.

1.2   Evolution of Microprocessors

The first microprocessor (Intel 4004) was introduced by Intel in 1971. It was followed by the Intel 8008 (1972) and the Intel 8080 (1974).[7] All these microprocessors required a number of external chips to implement a "useful" computing system and required multiple operating voltages (-5, +5 and +12vdc), cf. Fig. 1.4. Von Neumann and Harvard architectures were to find their way into specific classes of computer applications, viz., desktop/laptop/workstations and digital signal processors (DSP's)/microcontrollers, respectively.[8]

Intel introduced its first true microcontroller with onboard memory and peripherals, known as the "Intel 8048", an N-channel, silicon-gate, MOS device, which was a member of Intel's MCS-48 family of 8-bit microcontrollers, in 1976. The 8048 had 27 I/O lines,[9] one timer/counter, hardware

---

[7]Dr. H. Edward Roberts, Paul Allan and Bill Gates [1] were all watching the evolution of Intel's microprocessors during the early 1970s. Dr. Roberts designed the "first" true microcomputer based on the Intel 8080 microprocessor and, in addition to coining the phrase "personal computer", and collaborated with Bill Gates and Paul Allan who ported the first high-level language running on an Intel 8080 resulting in the introduction of the Altair computer in January of 1975 which was subsequently available with Microsoft BASIC.

[8]The Intel 4004, 8008 and 8080 did not provide hardware support for multiplication or division, but instead left hat for software implementation. Multiplication and division can of course be carried out on these architectures out by repeated addition and subtraction, respectively.

[9]There were two, eight-bit, "quasi-bidirectional ports; one bidirectional BUS port and 3 test inputs (t0, T1 and INT).

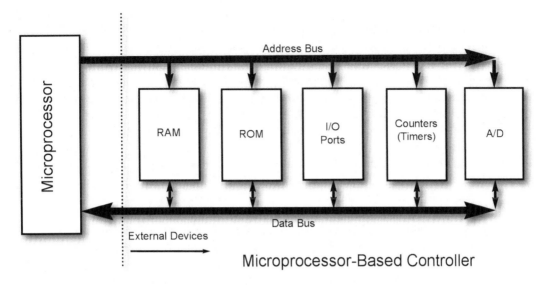

Fig. 1.4 External devices required for Intel's early microprocessor-based (micro)controller.

reset, support for external memory, an 8-bit CPU, interrupt support, hardware single-step support and utilized a crystal-controlled clock.

In 1977, Intel introduced the 8085, which was a modified form of the Intel 8080, but it relied less on external chips and required only +5 volts for operation. As the number of microprocessor-based, embedded system applications grew it soon became apparent that utilizing a simple microprocessor, such as the Intel 8080, with its requirement for multiple, associated support chips and the limitations of available external peripheral devices made microprocessors inadequate for many potential and actual embedded system applications (Fig. 1.5).

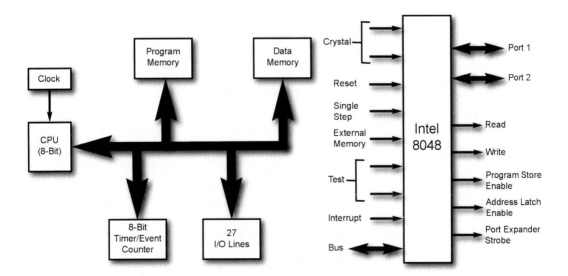

Fig. 1.5 A microcontroller incorporates all the basic functionality on-chip.

Embedded system microcontrollers, and microprocessors, have historically employed either a Von Neumann, or Harvard, memory architectures, as shown in Fig. 1.6. The Von Neumann memory configuration has only one memory address "zero" location for both data and memory, and therefore data and program code must reside in the same memory space. Since data and addresses each have their own bus in the Harvard configuration, the program and the data are stored separately in different regions of memory to allow instructions to be fetched, while data is being processed and stored/accessed. Instruction Address zero is distinct from Data Address zero, in the Harvard architecture, which makes it possible for address and data to have different bit sizes, e.g., 8-bit data versus 32-bit instructions.

A third configuration, sometimes referred to as a "modified Harvard" architecture, allows the CPU to access both data and instructions in separate memory spaces, but allows program memory to be accessed by the CPU, as if it were data, to permit instructions and text to be treated as data, so that they can be moved and/or modified.

Because an 8-bit address bus is limited to addressing a maximum of 256 bytes of memory, some other method must be employed to address additional RAM. Early microprocessors sometimes employed multiplexing of address lines to allow a larger memory space to be employed, e.g., 64K addressable by 16-bits, i.e., two bytes. In such applications, the lower byte of the address was placed on the address bus and latched in an external register. The upper byte was then placed on the address bus and latched in a second external register. The data byte held in the external memory location, and addressed by the two bytes, was then placed on the bus for retrieval by the microcontroller.

Another memory addressing technique, employed by both microprocessors and microcontrollers, is based on the concept of *paging* and utilizes a special register that holds a "page number" which serves as a pointer to a particular page, or segment of memory, of predefined size and location. Thus, a CPU can, for example, in principle at least, address an arbitrary number of pages of 256 bytes, or more, by reading/writing a byte from/to a memory location pointed to by this special register. This

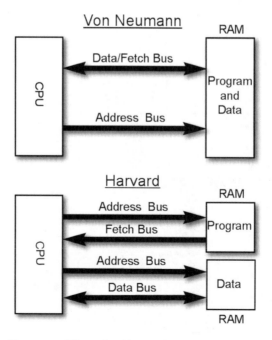

Fig. 1.6 Comparison of Von Neumann and Harvard architectures.

type of memory structure is sometimes referred to as *segmented,* or *paged*. With the advent of the Intel 8080, the address bus became 16-bits wide and therefore the page size became 64K.

However, paged/segmented memory structures imposed additional overhead on the CPU and increased the complexity of writing applications software. Microprocessors such as Motorola's 68000 had a memory structure referred to as linear/sequential/contiguous[10] that made it possible to directly address all available memory, which could be as large as 16 MB (Fig. 1.7).[11]

Fig. 1.7 An INTEL 8749 ultraviolet (UV) erasable microcontroller.

Memory-mapped I/O, which is not to be confused with memory-mapped file I/O,[12] uses the same address bus to address both memory and I/O devices. In such cases, the CPU instructions used to access memory are also used for accessing external devices. In order to accommodate the I/O devices, areas of CPU's addressable memory space must be reserved for I/O. The reservation might be temporary, in some systems, to enable them to bank switch between I/O devices and permanent and/or RAM. Each I/O device monitors the address bus and responds to any access of device-assigned address space, by connecting the data bus to an appropriate device hardware register.

Port-mapped I/O uses a special class of CPU instructions specifically designed for performing I/O. This was generally the case for most Intel microprocessors, specifically the IN and OUT instructions which can read and write a single byte from/to an I/O device. I/O devices have a separate address space from general memory, either accomplished by an extra "I/O" pin on the CPU's physical interface, or an entire bus dedicated to I/O. Thus inputs/outputs are accomplished by reading/writing from/to a predefined memory location.

The 8051 microprocessor was introduced by Intel in 1980 as a "system on a chip" and has subsequently become something of a worldwide standard in the fields of microcontrollers and embedded systems. It was a refinement, and an extension, of the basic design of the Intel 8048 and in its simplest configuration had the following:

- A Harvard memory architecture
- An ALU
- Seven on-chip registers
- A serial port (UART)
- A power-saving mode
- Two 16-bit counter/timers

---

[10]Linear, sequential, contiguous memory architectures use a memory space in which memory is addressable physically and logically in the same manner.

[11]Modern microcontrollers/microprocessors support multiple *direct* , and *indirect*, addressing modes for accessing registers and memory.

[12]Memory-mapped file I/O refers to a technique of treating a portion of memory as if data, contained therein, is organized in a file format.

- Internal memory consisting of GP bit-addressable storage, register banks and special function registers
- Support for 64K of external memory (code)[13]
- Support for 64K external memory (data)
- Four 8-bit, bidirectional I/O ports.[14]
- Two 10 bit-addressable locations.[15]
- 128 bytes of internal RAM
- 4k bytes of internal ROM
- Multiple addressing modes—indirect/direct to memory, register direct via the accumulator
- 12 clock cycles per machine cycle
- Very efficient execution, because most instructions required only one or two machine cycles.
- 0.5–1 MIPS performance at a clock speed of 12 MHz.
- An on-chip clock oscillator
- Six sources, five vector-interrupt handling
- 64K program memory address space
- 64K data memory address space
- Extensive Boolean handling capability

Intel ceased production of the 8051 in 2007, but a significant number of chip manufactures continue to offer 8051-like architectures, many of which have been very substantially enhanced and extended compared to the original design, e.g., Atmel, Cypress Semiconductor, Infineon Technologies, Maxim (Dallas Semiconductor), NXP, ST Microelectronics, Silicon Laboratories, Texas Instruments and Winbond. Although the original 8051 was based on NMOS[16] technology, in recent years, CMOS[17] versions of the 8051, or similar architecture, have become widely available.

In addition to supporting memory-mapped I/O, the 8051's registers were also memory-mapped and the stack resided in RAM which was internal to the Intel 8051.[18] The 8051's ability to access individual bits made it possible to set, clear AND, OR, etc., individual bits utilizing a single 8051 instruction. Register banks were contained in the lowest 32-bytes of the 8051's internal memory. Eight registers were supported, viz, R0–R7, inclusive, and their default locations were at addresses 0x00–0x07, respectively. Register banks could also be used to provide efficient context switching and the active register bank was selected by bits in the Program Status Word (PSW). At the top of the internal RAM there were 21 special registers located at addresses $0x80 - 0xFF$. Some of these registers were bit- and byte-addressable, depending upon the instruction addressing the register.

As designers became increasingly more comfortable with microcontrollers they began to take on more, and more, complex embedded system applications. This resulted in the need for a much wider variety of peripheral devices. PWMs, additional UARTs, A/D converters and D/A converters were some of the first modules to be available "on-chip". The demand continued to grow for more memory, CPU functionality, faster clock speeds, better interrupt handling [12], support for more levels of interrupt, etc., and with the introduction of OpAmps *on-chip,* and all typically *interoperable,* the

---

[13]The 8051 utilizes separate memory for data and code, respectively.

[14]I/O ports in the Intel 8051 are "memory-mapped", i.e., to read/write from/to an I/O port the program being executed must read/write to the corresponding memory location in the same manner that a program would normally read/write to any memory location.

[15]128 of these are at addresses 0x20-0x2F with the remaining 73 being located in special function registers.

[16]Negative channel metal oxide (NMOS).

[17]Complimentary symmetry metal oxide (CMOS).

[18]Intel has licensed a number of manufacturers to produce microprocessors based on the Intel 8051.

demand for analog devices increased, as well. On-chip real estate has always been an extremely valuable asset and while additional digital functionality was also desirable, the need for more analog support was greater.

Thus, a compromise was required and resulted in so-called "mixed-signal"[19] techniques being introduced into embedded system applications space. As a result, there are now a large number of manufacturers of microcontrollers utilizing a variety of cores, many of which are 8051 derivatives, fully integrated with various combinations/permutations of analog and digital "peripherals" on-chip. Some representative types that are currently available are listed below with a brief description:

68HC11 (Motorola) CISC, 8-bit, two 8-bit accumulators, two sixteen-bit index registers, a condition code register, 16-bit stack pointer, 3-5 ports, 768 bytes internal memory, max of 64k external RAM, 8051 8-bit ALU and registers, UART, 16-bit counter/timers, four-byte (bidirectional) I/O port, 4k on-chip ROM, 128-256 bytes of memory for data, 4 kB of memory for the program, 16-bit address bus, 8-bit data bus, timers, on-chip oscillators, bootloader code in ROM, power-saving modes, in-circuit debugging facilities, i2C/SPI//USB interfaces, reset timers with brownout detection, self-programming Flash ROM (program memory), analog comparators, PWM generators, support for LIN/CAN busses, A/D and D/A converters, non-volatile memory (EEPROM) for data, etc.

at91SAM3 (ATMEL at91SAM series) ARM Cortex-M3 revision 2.0, 32 bits, core, max clock rate 64 MHz, Memory Protection Unit (MPU), Thumb-2 instruction set, 64–256 kB embedded Flash, 128-bit wide access memory accelerator (single plane), 16–48 kB embedded SRAM, 16 kB ROM with embedded bootloader routines (UART, USB) and IAP routines, 8-bit static memory controller (SMC): SRAM/PSRAM/NOR and NAND Flash support, embedded voltage regulator for single supply operation, Power-on-Reset (POR), brown-out detector (BOD) and watchdog, quartz or ceramic resonator oscillators: 3 to 20 MHz main power with Failure Detection and optional low power 32.768 kHz for RTC or device clock, high precision 8/12 MHz factory trimmed internal RC oscillator with 4 MHz default frequency for device startup, (in-application trimming access for frequency adjustment),slow clock internal RC oscillator as permanent low-power mode device clock, two PLLs up to 130 MHz for device clock and for USB, temperature sensor, 22 peripheral DMA (PDC) channels, low power modes (sleep and backup modes, down to 3 μA in backup mode), ultra low power RTC, USB 2.0 (12 Mbps, 2668 byte FIFO, up to 8 bidirectional endpoints, on-chip transceiver), 2 USARTs with ISO7816 (IrDA), RS-485, SPI, Manchester and modem Mode), two 2-wire UARTs, 2 two-wire I2C compatible interfaces SPI (1 Serial Synchronous Controller (I2S), 1 high-speed multimedia card interface (SDIO/SD Card/MMC)), 6 three-channel 16-bit timers/counters with capture/waveform/compare and PWM mode, quadrature decoder logic and 2-bit Gray up/down counter for driving a stepper motor, 4-channel 16-bit PWM with complementary output/fault input/12-bit dead time generator counter for motor control, 32-bit real-time timer and RTC with calendar and alarm features, 15-channel 1Msps ADC with differential input mode and programmable gain stage, one 2-channel 12-bit 1Msps DAC, one analog comparator with flexible input selection/window mode and selectable input hysteresis, 32-bit Cyclic Redundancy Check Calculation Unit (CRCCU), 79 I/O lines with external interrupt capability (edge or level sensitivity)/debouncing/glitch filtering and on-die series resistor termination, three 32-bit parallel input/output controllers, and peripheral DMA assisted parallel capture mode.

ST92F124xx (STMicroelectronics ST92F Family) Register-oriented 8/16 bit CORE with RUN, WFI, SLOW, HALT and STOP modes, single Voltage Flash 256 kB (max), 8 kB RAM (max), 1 kB E^3(Emulated EEPROM), In-Application Programming (IAP), 224 general purpose registers

---

[19]Mixed-signal refers to an environment in which both analog and digital signals are present and in many cases being processed individually by analog and digital modules that are interoperable.

(register file) available as RAM, accumulators or index pointers, Clock, reset and supply management, 0–24 MHz operation (Int. Clock), 4.5–5.5 V range, PLL Clock Generator (3–5 MHz crystal), minimum instruction time: 83 ns (24 MHz int. clock), 80 I/O pins, 4 external fast interrupts + 1 NMI, 16 pins programmable as wake-up or additional external interrupt with multi-level interrupt handler, DMA controller for reduced processor overhead, 16-bit timer with 8-bit prescaler, and watchdog timer (activated by software or by hardware), 16-bit standard timer that can be used to generate a time base that is independent of the PLL clock generator, two 16-bit independent Extended Function Timers (EFTs) with prescaler, two input captures and two output compares, two 16-bit multifunction timers with a prescaler, two input captures and two output compares, *Serial Peripheral Interface* (SPI) with selectable master/slave mode, one multi-protocol *Serial Communications Interface* with asynchronous and synchronous capabilities, one asynchronous Serial Communications Interface with 13-bit LIN Synch Break generation capability, J1850 Byte Level Protocol Decoder (JBLPD), two full I2C multiple master/slave interfaces supporting the access bus, two CAN 2.0B active interfaces, 10-bit A/D converter (low current coupling).

ATmega8 (Atmel AVR 8-Bit Family) Low-power AVR 8-bit Microcontroller, RISC Architecture, 130 instructions (most are single-clock cycle execution), 32x8 general purpose registers, fully static operation to 16 MIPS throughput at 16 MHz, on-chip 2-cycle multiplier, High 8 k–32 kB of In-System Self-programmable Flash program memory, 512 Bytes EEPROM, 1 K Byte SRAM, 10,000 Flash/100,000 EEPROM write/erase cycles(20 years data retention, optional boot code section with independent lock bits, in-system programming by on-chip boot program, true Read-While-Write operation, programming lock for software security, two 8-bit timer/counters with separate prescaler and one compare mode, one 16-bit timer/counter with separate prescaler and compare/capture mode, a real-time counter with a separate oscillator, three PWM Channels, 6–8 channel ADC with 10-bit Accuracy, byte-oriented two-wire Serial Interface, programmable Serial USART, master/Slave SPI Serial Interface, programmable watchdog timer with separate on-chip oscillator, analog comparator, power-on Reset and programmable brown-out detection, calibrated RC Oscillator, external and internal interrupt sources, five sleep modes (Idle, ADC noise reduction, power-save, power-down, and standby), 23 programmable I/O lines, operating voltages: 2.7–5.5 V, clock speeds: 0–16 MHz, current drain (4 Mhz, 3 V, 25 C)—active: 3.6 mA/idle: 1.0 mA/power-down: 0.5 μA

80C51 (Atmel) 8051 Core Architecture, 256 Bytes of RAM, 1 kB of SRAM, 32K Bytes of Flash, Data Retention: 10 years at 85C, Erase/Write cycle: 100K, Boot code section with independent lock bits, 2K Bytes Flash for Bootloader, in-system programming by Boot program, CAN, UART and IAP Capability, 2 kB of EEPROM, Erase/Write cycle: 100K, 14-sources 4-level interrupts, three 16-bit timers/counters, full-duplex UART, maximum crystal frequency of 40 MHz (X2 mode)/20 MHz (CPU Core, 20 MHz), five ports: 32 + 2 digital I/O Lines, five-channel 16-bit PCA with: PWM (8-bit)/high-speed output /timer and edge capture, double data pointer, 21-bit watchdog timer (7 Programmable Bits),10-bit analog to digital converter (ADC) with 8 multiplexed inputs, on-chip emulation logic (enhanced hook system),power-saving modes: idle and power-down. Full CAN controller (CAN Rev2.0A and 2.0B), 15 independent message objects: each message object programmable on transmission or reception, individual tag and mask filters up to 29-bit identifier/channel, 8-byte Cyclic Data Register (FIFO)/message object, 16-bit status and control register/message object, 16-bit time-stamping register/message object, CAN specification 2.0 Part A or 2.0 Part B programmable for each message object, access to message object control and data registers via SFR, programmable reception buffer length up to 15 message objects, priority management of reception of hits on several message objects at the same time, priority management for transmission message object, overrun interrupt, support for time-triggered communication autobaud and listening mode, programmable automatic reply mode, 1-Mbit/s maximum transfer rate at 8 MHz crystal frequency in X2 mode, readable error counters, a programmable link to the timer for time-stamping and network synchronization, independent baud rate prescaler, data/remote error and overload frame handling.

PIC (MicroChip PIC 18F Family) Utilizes an 8-bit, Harvard architecture, one or more accumulators, small instruction set, general purpose I/O pins, 8/16/32 bit timers, internal EEPROM, USART/UART, CAN/USB/Ethernet support, internal clock oscillators, hardware stack, capture/-compare/PWM modules, A/D converter, etc., multiple power-managed modes (Run: CPU on, peripherals on, Idle: CPU off, peripherals on, Sleep: CPU off, peripherals off, multiple power consumption modes (PRI_RUN: μA, 1 MHz, 2 V, PRI_IDLE: 37 μA, 1 MHz, 2 V SEC_RUN: 14 μA, 32 kHz, 2 V SEC_IDLE: 5.8 μA, 32 kHz, 2 V RC_RUN: 110 μA, 1 MHz, 2 V RC_IDLE: 52 μ A, 1 MHz, 2 V sleep: 0.1 μ A, 1 MHz, 2 V) timer1 oscillator: 1.1 μA, 32 kHz, 2 V watchdog timer: 2.1 μA two-Speed Oscillator Start-up, four Crystal modes(LP, XT, HS: up to 25 MHz) HSPLL: 4–10 MHz (16–40 MHz internal), two External RC modes, up to 4 MHz, two External Clock modes, up to 40 MHz, internal oscillator block (8 user-selectable frequencies: 31, 125, 250, 500 kHz, 1, 2 , 4, 8 MHz), 125 kHz–8 MHz calibrated to R1%, two modes select one or two I/O pins, OSCTUNE allows user to shift frequency, secondary oscillator using Timer1 @ 32 kHz, fail-safe clock monitor (allows safe shutdown if peripheral clock stops), high current sink/source 25 mA/25 mA, three external interrupts, enhanced Capture/Compare/PWM (ECCP) module (One, two or four PWM outputs, selectable polarity, programmable dead time, auto-Shutdown and Auto-Restart, capture is 16-bit, max resolution 6.25 ns (TCY/16), compare is 16-bit, max resolution 100 ns (TCY)), compatible 10-bit, 13-channel Analog-to-Digital Converter module (A/D) with programmable acquisition time, enhanced USART module: supports RS-485, RS-232 and LIN 1.2 auto-wake-up on start bit, auto-Baud Detect, 100,000 erase/write cycle Enhanced Flash program memory typical 1,000,000 erase/write cycle Data EEPROM memory typical Flash/Data EEPROM Retention: > 40 years self-programmable under software control priority levels for interrupts, 8x8 Single-Cycle Hardware Multiplier, extended Watchdog Timer (WDT)(Programmable period from 41 ms to 131s with 2% stability), single-supply 5 V In-Circuit Serial Programming, (ICSP) via two pins, In-Circuit Debug (ICD) via two pins, wide operating voltage range: 2.0–5.5 V.

MSP430(Texas Instruments MSP 430F Family) RISC[20] set computer instruction set, Von Neumann architecture, 27 instructions, maximum of 25 MIPS, operating voltage: 1.8–3.6 V, internal voltage regulator, constant generator, program storage: 1–55 kB, SRAM: 128–5120 Bytes, I/O: 14-48 pins, multi-channel DMA, 8–16 channel ADC, watchdog, real-time clock (RTC), 16 bit timers, brownout circuit, 12 bit DAC, Flash is bit, byte and word addressable, maximum of twelve 8-bit bidirectional ports (Ports 1 and 2 have interrupt capability), individually configurable pull-up/pull-down resistors, supporting static/2-mux/3-mux and 4-mux LCDs, an integrated charge pump for contrast control, single supply OpAmps, rail-to-rail operation, programmable settling times, OpAmp configuration includes: unity gain mode/comparator mode/inverting PGA/non-inverting PGA, differential and instrumentation modes, hardware multiplier supports 8/16 x 8/16 bit bit, signed and unsigned with optional "multiply and accumulate", DMA accessible, maximum of seven 16-bit sigma-delta A/D converters, each has up to 8 fully differential multiplexed pinouts including a built-in temperature sensor, supply voltage supervisor, asynchronous 16-bit timers with

---

[20]Computers are typically categorized as either Complex Instruction Set Computers (CISC), or as Reduced Instruction Set Computers (RISC). The basic concept is that RISC instructions require fewer machine cycles per instruction than CISC instructions. In RISC machines the instructions are typically of fixed length, each is responsible for a simple operation, general purpose registers are used for data operations and not memory, data is moved via load and store instructions, etc.

up to 7 capture/compare registers, PWM outputs, a USART, SPI, LIN, IrDA and I2C support, programmable baud rates, USB 2.0 support at 12 Mbps, USB suspend/resume and remote wake-up.

PSoC 1 (Cypress CY8C29466) A programmable system on a chip, Harvard architecture, M8C Processor Speeds up to 24 MHz, two 8x8 Multiply with 32-Bit accumulate, 4.75–5.25 V operating voltage, 14-Bit ADCs, 9-Bit DACs, programmable gain amplifiers, programmable filters, programmable comparators, 8–32-Bit timers/counters, PWMs, CRC/PRS, four full-duplex or eight half-duplex UARTs, multiple SPI masters/slaves (connectable to all GPIO pins), Internal ± 4% 24 MHz Oscillator, High Accuracy 24 MHz with Optional 32.768 kHz Crystal and PLL, Optional External Oscillator, up to 24 MHz, Internal Low Speed, Low Power Oscillator for Watchdog and Sleep Functionality, 24 MHz Oscillator high accuracy 24 MHz clock, optional 32.768 kHz crystal and PLL support, external oscillator support to 24 MHz, watchdog and sleep functionality, flexible on-chip memory, 32K Bytes Flash program storage, 100k erase/write cycles 2K Bytes SRAM data storage, In-System Serial Programming (ISSP), partial Flash updates, flexible protection modes, EEPROM emulation in Flash, programmable pin configurations: 25 mA sink, 10 mA drive on all GPIO Pull Up, Pull Down, High Z, Strong, or Open Drain Drive Modes on All GPIO, Up to 12 Analog Inputs on GPIO [1], Four 30 mA Analog Outputs on GPIO Configurable Interrupt on all GPIO, I2C master/slave or multi-master operation to 400 kHz, watchdog/sleep timers, user-configurable low voltage detection, integrated supervisory circuit and precision voltage reference.

PSoC 3 (Cypress CY8C34 Family) A single-cycle 8051 CPU core, DC to 67 MHz operation, multiply/divide instructions, Flash program memory to 64 kB 100,000 write cycles, 20 years retention, multiple security features, max of 8 kB Flash ECC or configuration storage, max of 8 kB SRAM, max of Up to 2 kB EEPROM (1M cycles, 20 years retention), 24 channel DMA with multi-layer AHB bus access, programmable chained descriptors and priorities, high bandwidth 32-bit transfer support, operating voltage range from 0.5 to 5.5 V, high efficiency boost regulator (0.5 V input to 1.8–5.0 V output), current drain 330 µA at 1 MHz, 1.2 mA at 6 MHz, 5.6 mA at 40 MHz, 200 nA hibernate mode with RAM retention and LVD, 1 µA sleep mode with real-time clock and low voltage reset, 28 to 72 I/O channels (62 GPIO, 8 SIO, 2 USBIO, any GPIO to any digital or analog peripheral routability, LCD direct drive from any GPIO (max of 46x16 segments), 1.2–5.5 V I/O interface voltages (max of 4 domains), maskable independent IRQ[21] on any pin or port, Schmitt trigger TTL inputs, all GPIO configurable (open drain high/low, pull up/down, High-Z, or strong output), configurable GPIO pin state at power on reset (POR), 25 mA sink on SIO, 16 to 24 programmable PLD-based Universal Digital Blocks, full CAN 2.0b 16 RX, 8 TX buffers, USB 2.0 (12 Mbps) using an internal oscillator, max of four 16-bit configurable timer, counter, and PWM blocks, 8, 16, 24, and 32-bit timers, counters, and PWMs, SPI, UART, I2C, Cyclic Redundancy Check (CRC), Pseudo Random Sequence (PRS) generator, LIN Bus 2.0, Quadrature decoder, configurable Delta-Sigma ADC with 12-bit resolution (programmable gain stage: x0.25–x16, 12-bit mode, 192 ksps, 70 dB SNR, 1 bit INL/DNL), two 8-bit 8 Msps IDACs or 1 Msps VDACs, four comparators with 75 ns response time, two uncommitted OpAmps with 25 mA drive capability, two configurable multifunction analog blocks. (configurable as PGA, TIA, Mixer, and Sample/Hold), JTAG (4 wire), Serial Wire Debug (SWD) (2 wire), Single Wire Viewer (SWV) interfaces, Bootloader programming supportable through I2C, SPI, UART, USB, and other interfaces, precision, programmable clocking (1–48 MHz (±1% with PLL), 4–33 MHz crystal oscillator for crystal PPM accuracy, PLL clock generation to 48 MHz, a 32.768 kHz watch crystal oscillator, Low power internal oscillator at 1 and 100 kHz.

---

[21]The acronym for an interrupt request is IRQ.

PSoC 4 (Cypress CY8C34 Family) A scalable and reconfigurable platform architecture for a family of mixed-signal programmable embedded system controllers with an ARM Cortex-M0 CPU. It combines programmable and reconfigurable analog and digital blocks with flexible automatic routing. The PSoC 4200 product family, based on this platform, is a combination of a microcontroller with digital programmable logic, high-performance analog-to-digital conversion, OpAmps with Comparator mode, and standard communication and timing peripherals. The PSoC 4200 products will be fully upward compatible with members of the PSoC 4 platform for new applications and design needs. Programmable analog and digital sub-systems allow flexibility and in-field tuning of the design. Features 32-bit MCU Sub-system. 48-MHz ARM Cortex-M0 CPU with single-cycle multiply. Up to 32 kB of flash with Read Accelerator. Up to 4 kB of SRAM Programmable Analog. Two OpAmps with reconfigurable high-drive external and high-bandwidth internal drive, Comparator modes, and ADC input buffering capability. 12-bit, 1-Msps SAR ADC with differential and single-ended modes; Channel Sequencer with signal averaging. Two current DACs (IDACs) for general-purpose or capacitive sensing applications on any pin. Two low-power comparators that operate in Deep Sleep mode Programmable Digital Four programmable logic blocks called universal digital blocks, (UDBs), each with 8 Macrocells and data path. Cypress-provided peripheral component library, user-defined state machines, and Verilog input Low Power 1.71-V to 5.5-V Operation. 20-nA Stop Mode with GPIO pin wakeup. Hibernate and Deep Sleep modes allow wakeup-time versus power trade-offs.

PSoC 4 and Capacitive Sensing Cypress CapSense Sigma-Delta (CSD) provides best-in-class SNR (>5:1) and water tolerance. Cypress-supplied software component makes capacitive sensing design easy. Automatic hardware tuning (SmartSense)Segment LCD Drive.LCD drive supported on all pins (common or segment). Operates in Deep Sleep mode with 4 bits per pin memory. Serial Communication: Two independent run-time reconfigurable serial communication blocks (SCBs) with reconfigurable I2C, SPI, or UART functionality. Timing and Pulse-Width Modulation: Four 16-bit timer/counter pulse-width modulator (TCPWM) blocks. Center-aligned, Edge, and Pseudo-random modes. Comparator-based triggering of Kill signals for motor drive and other high-reliability digital logic applications. Up to 36 Programmable GPIOs. Any GPIO pin can be CapSense, LCD, analog, or digital. Drive modes, strengths, and slew rates are programmable. Five different packages: 48-pin TQFP, 44-pin TQFP, 40-pin QFN, 35-ball WLCSP, and 28-pin SSOP package. 35-ball WLCSP package is shipped with I2C Bootloader in FlashExtended Industrial. Temperature Operation: 40 °C to + 105 °C operation. PSoC Creator Design Environment Integrated Development Environment (IDE) provides schematic design entry and build (with analog and digital automatic routing). Applications Programming Interface: (API) component for all fixed-function and programmable peripherals. Industry-Standard Tool Compatibility: After schematic entry, development can be accomplished with ARM-based industry-standard development tools.

PSoC 5LP (Cypress CY8C53 Family) 32-bit ARM Cortex-M3 CPU core, DC to 80 MHz operation, Flash program memory (max 256 kB, 100,000 write cycles, 20 year retention), multiple security features, 64 kB SRAM (max), 2 kB EEPROM (1 million cycles, 20 years retention), 24 channel of DMA with multi-layered AHB bus access, programmable chained descriptors and priorities, high bandwidth 32-bit transfer support, operating voltage ranges: 0.5–5.5 V, high efficiency boost regulator (0.5 V input to 1.8–5.0 V output, current drain of 2 mA at 6 MHz), 300 nA hibernate mode with RAM retention and LVD, 2 μA sleep mode with real-time clock and low voltage reset, 28 to 72 I/O channels (62 GPIO, 8 SIO, 2 USBIO), any GPIO to any digital or analog peripheral routability, LCD direct drive from any GPIO (max of 46x16 segments), 1.2–5.5 V I/O interface voltages (max of 4 domains), maskable independent IRQ on any pin or port, Schmitt trigger TTL inputs, all GPIO configurable (open drain high/low, pull up/down, High-Z, or strong output), configurable GPIO pin state at power on reset (POR), 25 mA sink on SIO, 20 to 24 programmable PLD based Universal Digital Blocks, full CAN 2.0b 16 RX, 8 TX buffers, full-USB 2.0 (12 Mbps using internal oscillator), max of four 16-bit configurable timer, counter, and PWM blocks, 8, 16, 24, and 32-bit timers, counters, and PWMs SPI, UART, I2C, Cyclic Redundancy Check (CRC), Pseudo Random Sequence (PRS) generator, LIN Bus 2.0, Quadrature decoder, SAR ADC (12-bit at 1 Msps), four 8-bit 8 Msps IDACs or 1 Msps VDACs, four comparators with 75 ns response time, four uncommitted OpAmps with 25 mA drive capability, four configurable multifunction analog blocks (PGA, TIA. Mixer and Sample and hold), JTAG (4 wire), Serial Wire Debug (SWD) (2 wire), Single Wire Viewer (SWV), and TRACEPORT interfaces, Cortex-M3 Flash Patch and Breakpoint (FPB) block, Cortex-M3 Embedded Trace Macrocell (ETM) for generating an instruction trace stream. Cortex-M3 Data Watchpoint and Trace (DWT) for generating data trace information, Cortex-M3 Instrumentation Trace Macrocell (ITM) for printf-style debugging, DWT, ETM, and ITM blocks that can communicate with off-chip debug and trace systems via the SWV or TRACEPORT, Bootloader programming supportable through I2C, SPI, UART, USB, and other interfaces, Precision, programmable clocking from 1 to 72 MHz with PLL, 4 to 33 MHz crystal oscillator for crystal PPM accuracy, PLL clock generation up to 80 MHz, 32.768 kHz watch crystal oscillator support, low power internal oscillator at 1 and 100 kHz.

PSoC 6 (Cypress CY8C53 Family) A scalable, and reconfigurable, platform architecture for a family of programmable embedded system controllers with Arm Cortex CPUs (single and multi-core). The PSoC6 product family, based on an ultra low-power 40-nm platform, is a combination of a dual-core microcontroller with low-power Flash technology and digital programmable logic, high-performance analog-to-digital and digital-to-analog conversion, low-power comparators, and standard communication and timing peripherals. 32-bit Dual Core CPU Subsystem 150-MHz Arm Cortex-M4F CPU with single-cycle multiply(Floating Point and Memory Protection Unit) for user application 100 MHz Cortex M0+ CPU with single-cycle multiply and MPU for System functions (not user-programmable) User-selectable core logic operation at either 1.1 V or 0.9 V. Inter-processor communication supported in hardware. 8 kB 4-way set-associative. Instruction Caches for the M4 and M0+ CPUs respectively. Active CPU power consumption slope with 1.1 V core operation for the Cortex M4 is 40 and 20 A/MHz for the CortexM0+, both at 3.3 V chip supply voltage with the internal buck regulator. Active CPU power consumption slope with 0.9 V core operation for the Cortex M4 is 22 and 15 A/MHz for the CortexM0+, both at 3.3 V chip supply voltage with the internal buck regulator. Two DMA controllers with 16 channels each Flexible Memory Subsystem.1 MB Application Flash with 32 kB EEPROM area and 32 kB Secure Flash.128-bit wide Flash accesses reduce power. SRAM with Selectable Retention Granularity. 288 kB integrated SRAM. 32 kB retention boundaries (can retain 32–288 kB in 32 kB increments). OTP E-Fuse memory for validation and security Low-Power 1.7–3.6 V Operation. Active, Low-power Active, Sleep, . Deep Sleep mode current with 64 kB SRAM retention is 7 A with 3.3 V external supply and internal buck. On-chip Single-In Multiple Out (SIMO) DC-DC Buck converter,<1 A quiescent current. Backup domain with 64 bytes of memory and Real-time Clock(RTC)Flexible Clocking Options. On-chip crystal oscillators (High-speed, 4–33 MHz, and Watch crystal, 32 kHz). Phase-locked Loop (PLL) for multiplying clock frequencies. 8 MHz Internal Main Oscillator (IMO) with 2% accuracy. Ultra low-power 32 kHz Internal Low-speed Oscillator (ILO)with 10% accuracy. Frequency Locked Loop (FLL) for multiplying IMO frequency. Serial Communication. Nine independent run-time reconfigurable serial communication blocks (SCBs), each is software configurable as I2C,SPI, or UART. Timing and Pulse-Width Modulation. Thirty-two Timer/Counter Pulse-Width Modulator (TCPWM)blocks. Center-aligned, Edge, and Pseudo-random modes. Comparator-based triggering of Kill signals Up to 78 Programmable GPIOs. Drive modes, strengths, and slew rates are programmable. Six overvoltage tolerant (OVT) pins.

1.3   Embedded System Applications

Embedded systems can be found in an ever-increasing number of applications including: televisions, cable boxes, satellite boxes, cable modems, routers, printers, microwave ovens, surround-sound systems, computer monitors, digital cameras, zoom lenses, cars and trucks (some vehicles have 100+ such systems), stereos, dishwashers, dryers, washing machines, cell phones, digital multimeters, calculators, air conditioners, mp3 players, heaters, flight-control systems (fly-by-wire), running shoes, tennis rackets, traffic lights, elevators, telecommunications systems, medical equipment, airplanes, automotive cruise controls, ignition systems, personal digital assistants, pleasure boats, motorcycles, children's toys, oscilloscopes, ships, industrial and process control applications, railway systems, laboratory equipment, personal computers, data collection/logging equipment, numerical processing applications, "smart" shoes, robotics, fire/security alarms, biometric systems, proximity detectors, inertial guidance systems, GPS devices, UAVs, etc. The major markets for embedded systems include automotive, medical, avionic, communications, industrial and consumer electronics.

For example, increasing numbers of automobile manufacturers produce products that utilize embedded systems that control their vehicle's major functions, such as power-train management, air conditioning, (heating/cooling systems), seat positioning mechanisms, fuel systems, braking mechanisms, dashboard instrumentation, GPS systems, etc. Automakers must also continue to respond to steadily growing requirements for advanced safety, environmental protection and driver convenience, thus increasing the number of microelectronics components, in a vehicle, all of which continue to require more and more "lines of code". Embedded systems are also the foundation of autonomous vehicles control and monitoring systems.

In the last two decades, the total number of lines of code employed by automobile manufacturers has reportedly grown from approximately one million lines to close to one hundred million lines of proprietary and third-party code. Thus, the ease with which new code can be developed and reusable in future designs becomes of paramount importance. This is particularly true as microcontrollers evolve with increasingly more complex architectures in an attempt to meet market demands.[22]
The need to continually:

- reduce the time to market for new designs,
- introduce less expensive microcontrollers with ever-increasing capability and in some cases more specialization,
- support ever-increasing application complexity,

and,

- support lower power consumption,

has, in turn, increased the demand for a wide variety of generic and specialized microcontrollers and substantially advanced the state of the available microcontroller technology and associated peripherals.

---

[22] In the early days of microcomputers/microcontrollers, software developers would often brag about the number of lines of code they had written for an application, with the implication being, the more lines of code, the more sophisticated the developer who often spoke in terms of k-locs (thousand of lines of code). It soon became apparent that the true measure of a developer's prowess was the inverse.

Automotive Electronics Vehicle manufacturers continue to move aggressively in implementing more, and more, embedded system technology into new vehicles to increase their competitive strengths in meeting the new challenges of their competitors and public demand for more efficient, reliable and feature-rich transportation.

Currently, the number of microprocessors/microcontrollers in automobiles ranges from 10, to more than 100, with current estimates suggesting that as much as 40% of the value of some automobiles is invested in the electronics systems and networking. Some modern vehicles employ three, or more, network protocols, e.g., LIN[23] (10 kbits/sec), CAN[24] (1 Mbits/sec) and FlexRay[25] (10 Mbits/sec) to address the wide range of real-time responses needed in contemporary vehicles.

High-speed networking, utilizing FlexRay and high-speed CAN, is required to handle fuel ignition and exhaust systems, spark/valve timing, fuel injection systems, anti-lock braking systems, cruise control, airbags, active suspension, steer-by-wire, brake-by-wire, and other "x-by-wire" systems. Low-speed networking, utilizing LIN and low-speed CAN are employed to handle less demanding real-time requirements, such as the instrument panel (dashboard), air conditioning, windshield wipers, power windows, mirror adjustments, seat controls, alarm systems, door locks, headlights, internal lighting systems, stop/tail/fog lights and high/low beams, seat temperature controllers, etc.

Avionic Electronics Private, commercial and military avionic systems make extensive use of embedded systems for fly-by-wire , GPS-based and other navigational systems such as inertial navigation systems. Heads-up displays, power plant monitoring and control, instrument displays, in-flight communication systems and related equipment, transponders, instrument panels, internal/external lighting systems, offensive and defensive weapon systems, weather radar, navigational systems, etc., are also increasingly controlled and/or monitored by embedded systems.

Consumer Electronics Since the advent of the microprocessor, consumer electronics have continuously taken increasing advantage of semiconductor technology and most particularly of microcontrollers. Modern homes make extensive use of embedded systems in the form of security, lighting control, stereo, telecommunications, cable TV and Internet systems, personal computers, MP3 players, etc.

Communications Electronics Cell phones, telephone switches, GPS, routers, microwave and satellite-based systems, etc., make extensive use of embedded systems.

Industrial Electronics Process control systems, numerically-controlled milling and drilling machines, 3D printing systems, robotics, automated inspection systems, etc., are heavily dependent on embedded systems particularly for high volume, close tolerance manufacturing processes and systems.

Medical Electronics Blood pressure, young child/adult heart rate, fetal heart rate, pulse oximetry, blood glucose, electrocardiogram, ventilation/respiration, electronic stethoscopes, vital signs, blood sugar and anesthesia monitors all employ embedded systems and are used in homes as well as,

---

[23]Local Interconnect Network (LIN).

[24]Controller Area Network (CAN).

[25]FlexRay is an open, scalable network protocol created by a consortium consisting of Philips Semiconductor, BMW, Daimler Chrysler, Motorola, BMW, Ford Motor Company, General Motors Corporation and Robert Bosch GMBH specifically for automotive applications. It supports both synchronous and asynchronous data transfers and is capable of operating in either a single or double channel mode, if redundancy is required.

clinics, doctors offices and hospitals. Imaging systems, e.g., acoustic (sonograms), X-Ray CT(X-Ray Computed Tomography), MRI(Magnetic Resonance Imaging), SPECT(Single photon emission computed tomography), and PET(Positron Emission Tomography), powered patient beds, monitoring systems, operating room systems, robotic surgery systems are also important embedded system applications.

Each of these relies on one, or more, embedded systems to gather input data from devices called "sensors", and/or other data sources. Based on the information gathered, they then engage in numerical/logic processing of the input data, subject to certain predefined constraints and/or operating modes (states), make decisions based on the input data and subsequently provide outputs to various types of devices, such as other computer systems, display devices, linear/rotary actuators, motors, speakers, data transport channels, etc.

1.4 Embedded System Control

Regardless of the application, type of data input sources whether digital or analog, real-time or on-demand, response time, sequential or parallel operation, etc., embedded systems all share a number of common characteristics.

1.4.1 Types of Embedded System Control

Embedded systems are capable of functioning in a number of different modes, e.g.,

1. **Event-Driven Mode (EDM)**—perhaps the most common type of embedded system which is that which is constrained to responding to previously defined events, and providing predefined responses. The system waits for an event to occur in the form of a key depression, a parameter meeting some threshold level and thus representing an event, or other "triggering" event(s),
2. **Continuous-Time Mode (CTM)**—such systems are continuously monitoring input channels and reacting to various input conditions,
3. **Discrete-Time Mode (DTM)**- these systems *wake-up* at predetermined intervals, sample input data, carry out the appropriate responses and then go back to *sleep*,

or some permutation, thereof.

For example, a system may be required to *wake-up*, respond to some set of input conditions on an event-driven basis and then go back to *sleep*. Some systems employ *watchdog* functions[26] that, in the absence of the system responding within a predetermined period of time, automatically reset the system as a way of avoiding the system becoming "locked-up" because of some anomalous situation, or malfunction, and subsequently failing to function appropriately.

While some embedded systems are primarily involved in control functions, and to a lesser extent data processing, others are predominantly engaged in data processing/collection and some control functions, and still others are heavily involved in both. In such cases, the former are usually described as state machines that move from "state-to-state" as a result of certain events, or input data conditions/values. In such cases, the embedded system remains in a given state until sufficient

---

[26]Watchdog parameters can either be hard-coded or soft-coded.

conditions arise in terms of events, or input data, that meet the criteria for a state transition. Resetting/setting such systems causes the state machine to enter a predetermined "home" state. Whether functioning as:

- a controller designed to maintain certain parameters, or operating conditions, of a system, or process, within predefined ranges or contexts,
- part of a network of embedded systems engaged in making decisions, monitoring activity and/or exchanging information regarding the various systems, or processes, to be monitored or controlled and their respective states,
- an application-specific embedded system for image/video processing, graphics, multimedia processing,
- an embedded system for demanding computational applications and interfacing applications,
- a data logging system for applications such as remote sensing systems,

or,

- a specialized/custom digital communications processing system, such as part of a data link,

each consists of a CPU, memory, registers, address/data busses and various peripheral devices such as analog-to-digital converters, pulse-width modulators, digital-to-analog converters, various types of signal conditioners such as filters, comparators, etc.

Some embedded systems employ real-time operating systems [3], while others are merely subsystems in a real-time, operating system [4] environment.[27] In the latter case, failure of one, or more, of the embedded subsystems might allow some portion of the total system to continue to function. In systems for which the embedded system has primary control, any failure could prove catastrophic, and therefore requires much more attention to failure modes, and how best to address them by employing, e.g., fail-safe modes.[28] However, real-time operating systems add complexity, cost and processing overhead which, for some applications, is undesirable and can significantly degrade the system's performance.

1.4.2   Open-Loop, Closed-Loop and Feedback

Embedded systems may be implemented as open- or closed-loop systems. An "open-loop" system, sometimes referred to as a *feedforward* system, as shown in Fig. 1.8, acquires input information and produces appropriate outputs based on the inputs, without any ability to determine whether, ultimately, the correct action, or actions, have taken place.[29] Furthermore, such a system assumes that the input data is always correct and that there are no disturbances, e.g., noise, or other anomalies, to take into account.

---

[27]It should be borne in mind that an operating system involves both benefits and costs. For example, a significant cost can be the effective number of machine cycles required to support the operating system's overhead. For this reason designers are often reluctant to include an operating system, particularly if there are other less expensive approaches in terms of memory, CPU cycles, I/O transfer rates, etc.resources.

[28]Fail-safe engineering refers to a design methodology that assures that in the event of a software and/or hardware failure of a critical system/subsystem, it defaults to a "safe mode."

[29]In some situations, no action may constitute the correct action.

Fig. 1.8 Schematic view of an open-loop system.

An open-loop, embedded system gathers information, reacts to the input parameter values in a predefined way and produces the appropriate output signals and/or commands, e.g., a thermostat senses temperature, (*TempSensed*), compares the temperature to a preset value, (*TempUpperLimit*), and if

$$TempUpperLimit < TempSensed, \tag{1.1}$$

closes some switch contacts to turn on a fan. However, this simple system does not know whether the fan is operating, or if it is operating at a speed sufficient to return the temperature to an acceptable value, within a required period of time. Furthermore, as long as the (TempSense) exceeds *TempUpperLimit*, the system will continue to attempt to provide cooling, but makes no attempt to take further corrective action. This system is representative of the open-loop system shown in Fig. 1.8.

Should the fan fail to be activated at the proper speed, the controller, in this example, would not initiate any further action, i.e., since there is no indication returned to the controller that cooling is, or is not, taking place. This type of open-loop system is referred to as a *bang-bang* system since it does not provide proportional control of the device that it controls, i.e., the fan is either operating at a constant speed (RPMs), or is inactive. It would of course be possible to program the controller to monitor the input temperature as an explicit function of time so that if it, for example, found that the temperature was not changing it could take other actions, e.g., sounding of an alarm.

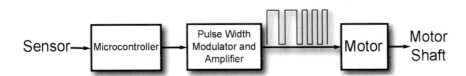

Fig. 1.9 Embedded system motor controller.

A similar example of an open-loop system is shown in Fig. 1.9, in which a motor is controlled by a sensor and an embedded system consisting of a microcontroller and a pulse-width modulator (PWM),[30] in this case under the control of a microcontroller that, in the present example, allows the speed of the motor to be varied over a wide range, driving an amplifier with sufficient output voltage, and current, to drive a motor. In this case, the speed of the motor is determined by the embedded system which controls the duty-cycle[31] of the pulse train produced by the PWM. Therefore, the controller is able to provide proportional control of the fan by controlling the average amount of power provided to it.

Some motors have integral tachometers and/or Hall effect sensors[32] that can be used to produce an analog signal that can be returned directly to the summing junction to produce an error signal to be

---

[30] A Pulse Width Modular (PWM) is a device capable of producing pulses of variable width and frequency.

[31] Duty-cycle is defined as the ratio of time-on to time-off, over some predefined period of time.

[32] Hall effect and other forms of magnetic sensors are discussed in Sect. 3.2.5.

processed by the controller, and confirm that the motor is running at the appropriate speed. In Fig. 1.10 the system returns a signal, e.g., an analog voltage/current, or digital data, that reflects the output state of the system, to the input for comparison.

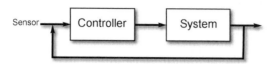

Fig. 1.10 Schematic view of a closed-loop system with direct feedback.

In some cases, the embedded system is provided with input parameters that produce desired state of a system. The controller compares the values that characterize the current state of the system with such predefined state conditions and makes decisions regarding what steps must be taken, if any, to bring the system into compliance with the appropriate conditions. Other embedded systems utilize a sensor, or sensors, to define the parameter values characterize the desired state of a system and additional sensors that represent the actual state of the process, or system. In both cases, the input and output signals are provided to a *summing junction*, or equivalent, to produce an error signal.

Note that in the former case, one sensor is used to establish a comparison between *a setpoint*, i.e, the desired input parameter value, and another sensor is employed to determine the output, i.e., *actual state*, so that the desired versus *actual state* can be determined to allow the controller to establish what error, if any, exists and take such action, or actions, as may be required to minimize the resulting *error signal* at the summing junction. Modern thermostats and cruise controls are examples of such systems. This configurability allows the *equilibrium point* to be set externally, while in the latter case that point is established programmatically within the controller, e.g., as in the case of inertial navigation systems, anti-lock braking systems, etc.

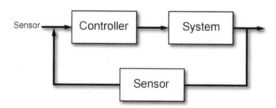

Fig. 1.11 Schematic view of a closed-loop system with "sensed" output feedback.

When designing embedded systems, it is important for characteristics such as latency, phase shift and stability to be taken into account. Figure 1.11 shows a representation of a system with simple feedback represented symbolically. Note that the blocks representing the Controller and System transfer functions can be combined as shown in Fig. 1.12.
This is equivalent to combining the two transfer functions as follows:

$$G_1 = H_{Controller} H_{System} \tag{1.2}$$

where $H_{Controller}$ and H_{System} represent the transfer functions for the Controller and System, respectively.

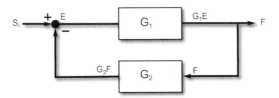

Fig. 1.12 A generalized SISO feedback system.

Thus the embedded system represented in Fig. 1.11 can be represented by the following:

$$E = S_I - G_2 F \tag{1.3}$$

$$F = G_1 E \tag{1.4}$$

$$G_2 F = G_2 G_1 E \tag{1.5}$$

which leads to the result that:

$$\frac{f(t)}{s(t)} = \frac{G_1}{1 + G_1 G_2} \tag{1.6}$$

and assuming that this system is a LTI (linear, time-invariant) system, the corresponding Laplace Transform[33] can be expressed, symbolically as:

$$\frac{F(s)}{S(s)} = \frac{(s - z_1)(s - z_2)(s - z_3) \ldots (s - z_{m-1})(s - z_m)}{(s - p_1)(s - p_2)(s - p_3) \ldots (s - p_{n-1})(s - p_n)} \tag{1.7}$$

where $s = \sigma + j\omega$, the z_m terms are the zeros of the transfer function and the p_n terms are the poles. The stability, of lack thereof, of this system can then be determined by an examination of the location of the system's poles, p_n, in the complex plane, or by other techniques. An embedded system is said to be *Bounded Input, Bounded Output* (BIBO) stable if any bounded input results in a bounded output. An embedded system's stability may be one of several types, e.g., unstable, uniformly stable, marginally stable, conditionally stable, etc.

Although it is also beyond the scope of this textbook, a further refinement of this type of mathematical model for an embedded system would be to include the impact of perturbations, i.e., various types of disturbances, that the embedded system may be subject to such as electromagnetic interference, vibration, frictional effects, variation in the loading of motors and actuators, nonlinear effects, effects of stray magnetic and/or electric fields, etc. Note also that when using sensors in an embedded system, it is sometimes necessary to employ various types of signal conditioning, e.g., a variety of filtering techniques can be employed to maintain signal integrity and thereby assure that the appropriate current/voltages limitations are imposed, etc. Fig. 1.13 represents such a system.

Adaptive embedded systems[34] are employed, when required, to allow them to modify their characteristics to meet, often in real-time, variable "environmental" conditions such as power supply

---

[33] Refer to Chap. 11 for more discussion regarding transfer functions, poles, zeros and Laplace transforms.

[34] An embedded system is considered *adaptive* if it is able to reconfigure its program and hardware resources in real-time to continue to meet the applicable functional and performance specifications. In some cases, degradation in these specifications may be regarded as being acceptable, if the embedded system performance remains within defined boundaries.

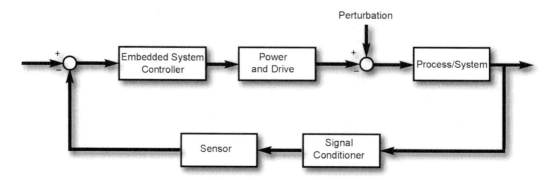

Fig. 1.13 An embedded system that is subject to external perturbations.

fluctuations/degradation and externally variable process and system conditions. Many embedded systems are able to reduce their clock frequencies, enter sleep modes, and operate at lower power levels in response to changes in environmental conditions. Others are able to move tasks between multiple cores to optimize performance and minimize hot spots.

Arguably, any system that employs feedback could be considered *adaptive* since the embedded systems is adapting its responses based on input data. However, adaptability in the present context refers to adaptation within the controller itself in response, e.g., to changing environmental conditions. Adaptive systems may employ fuzzy logic, neural networks, Radial Basis Functions (RBFs),[35] Kalman filtering,[36] etc., which are often used for approximation, interpolation and to overcome limitations imposed by wavelets, polynomial interpolation, least-square and other techniques when multidimensional parameters are involved, as for example, in the case of signal conditioning.

1.5   Embedded System Performance Criteria

Two of the most important considerations for embedded systems are (1) that they perform each task correctly and (2) that all the reactions/responses by an embedded system occur in a timely manner. It is also important that an embedded system be robust.[37] Timeliness, or lack thereof, in the present context can be characterized as Soft, Firm or Hard. [9]

A *hard real-time system* (HRTS) is one in which failure could produce a catastrophic result, e.g. failure of a fire alarm system, or a pacemaker malfunction resulting in death. A *Firm Real-Time System*(FRTS) failure might be an automotive cruise control for which the latest value of the current speed is not available in time for the cruise control algorithm to determine what, if any, corrective action is required. In such cases, the algorithm may be able to use the previously reported speed and still perform the necessary operations to maintain relatively constant speed. A *Soft Real-Time System* error(SRTS) is exemplified by the failure of an ATM which, while perhaps inconvenient, is hardly a firm, or hard failure. Failures of these three types (HRTS, FRTS and SRTS) are usually analyzed in terms of *deadline misses* and their respective impacts on the system. Because such systems are expected to react in real-time, they are, by their nature, typically asynchronous.

---

[35] Sums of radial basis functions can be used to approximate functions.

[36] Kalman filtering, is sometimes referred to as linear quadratic estimation (LQE).

[37] Robustness is defined as exhibiting substantive resistance to perturbations.

Latency, in the case of an embedded system, refers to the delay $(t_1 - t_0)$ between the time (t_0) when a condition exists requiring a response and the time that the response occurs (t_1). Such delays can arise as the result of hardware delays in sensors, microcontrollers, peripheral output devices and/or software delays produced by program-execution overhead, e.g. program execution speed and interrupt servicing.[38]

Embedded systems are often asynchronous and receive input data from multiple sources. In such cases, this data must wait until the embedded system is available to accept it. Some input devices introduce a delay between the time an input parameter is sensed/updated and the time at which the data has been *latched* for input to the microcontroller. Latching is often employed to be sure that when data is available from an external device such as a sensor, the microcontroller has sufficient time to complete, or suspend, its current tasks. Suspension of ongoing tasks occurs when an *interrupt request*, of sufficient priority, is received. For tasks of lower priority, than the currently running task, or when large data sets are involved, various techniques can be employed to *buffer* the inputs from sensors until they can be processed. Flip-flops, which are bistable devices, are commonly used to latch input, or output data, particularly at the byte level, while waiting for a device, such as a microcontroller, to enter a ready state and subsequently accept the data. Various *shared memory* techniques such as Direct Memory Access transfers, can also be employed to provide the needed buffering.

Embedded systems employ various techniques to minimize latency:

- Direct Memory Access (DMA)—this technique, while not normally involving any pre-processing of data, allows I/O to occur relatively transparently without requiring significant CPU overhead. I/O devices can transfer data to/from the embedded system by directly accessing the microcontroller's memory space. A DMA controller, such as that shown in Fig. 1.14, is used to facilitate the transfer, once the CPU has defined where the data is located, or to be stored, within local memory.

  In some cases, a region of memory is predefined as assigned to the DMA controller and therefore the CPU's direct involvement in data transfers under DMA control is obviated. The DMA controller can set a flag, or flags, indicating whether new data is available for processing by the CPU, or has been transferred to one, or more, external devices. This technique addresses both latency and bandwidth overhead by allowing the data to be transferred at a rate most appropriate for the external device(s), whenever it is available.

  The CPU, under software control, can be programmed to initialize the DMA controller and provide the data addresses for both source and destination, and the amount of data to be transferred. Microcontrollers allow the DMA controller, upon request, to take control of the bus and transfer data in a so-called *burst mode*. In doing so, the CPU's access to the address bus is usually *tri-stated*[39] to avoid bus conflicts (cf. Fig. 1.15). When the transfer is complete, the DMA controller returns control of the bus to the CPU.

  A *cycle-stealing* mode can be employed by the DMA controller, in which case, it relinquishes control of the memory bus after each transfer. Note that most DMA controllers have address and length registers that are of different sizes, so that if the address register is larger than the length register it can address a large portion of memory. If the size of the address register is 32-bits, and the length register is 16-bits, then the DMA controller can transfer data in blocks of 64 kB, anywhere within 4 GB of memory. As shown in Fig. 1.14, data may be transferred either to/from memory internal to the microcontroller, or, if necessary to external memory depending on the amount of data to be transferred, latency concerns and the application.

---

[38]Interrupts are discussed in Sect. 1.5.1.

[39]Tri-stating, or tri-stating, refers to placing a device input or output in one of three states, e.g., high (1), low (0) or high impedance, the latter effectively removing it from a circuit, e.g., a bus. This technique prevents bus conflicts and the possibility of two subsystems attempting to "drive the bus", i.e., apply or receive signals, at the same time.

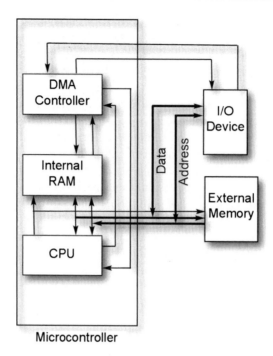

Fig. 1.14 Block diagram of a typical microcontroller/DMA configuration.

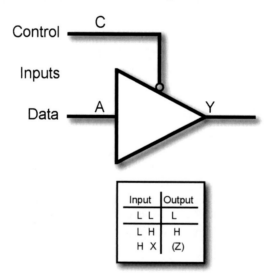

Fig. 1.15 An example of a tri-state device.

- Looping—Since an I/O device can itself be in an active (busy) or waiting (idle) state, an embedded system can remain in a loop waiting for a flag to be set, before continuing with program execution. This has the advantage that having the microcontroller remain in an idle state waiting for data reduces, at least, some latency. But, this technique has the limitation that the CPU is unable to accept data from other sources, while in this mode.

- Polling—Alternatively, an embedded system can *poll* status flags for I/O devices to determine if a device: (1) has data available for transfer to the microcontroller, (2) is available for receipt of data, or (3) is busy. Polling may be periodic, or aperiodic, depending on the application.
- FIFOs—First-In-First-Out buffers, and other forms of buffering, can be used with external devices to buffer I/O until the microcontroller's resources are available. This technique, as in the case of the use of DMA, can be employed for occasions when data is being gathered faster than it can be handled by the microcontroller.
- Interrupts—Interrupt schemes can be employed that interrupt the CPU only when I/O needs to occur.

1.5.1 Polling, Interrupts and ISRs

The computing world was somewhat slow in fully appreciating the importance of interrupts.[40] But interrupts, and interrupt service routines, are obviously integral to virtually any microcomputer/microcontroller application today, particularly in applications in which microprocessors/microcontrollers are operating in complex environments, performing complex tasks, or both. The introduction of interrupt-driven systems was met with mixed emotions by some system developers.[41] The environment in which a CPU found itself had become a lot more complex, and perhaps more importantly nondeterministic, as Dijkstra pointed out (Fig. 1.16)[3].

Interrupts are an important part of any embedded application. They free the CPU from having to continuously poll for the occurrence of a specific event and, instead, notify the CPU only when that event occurs. In system-on-chip (SoC) architectures such as PSoC, interrupts are frequently used to communicate the status of on-chip peripherals to the CPU.

There are 32 interrupt lines int[0]to int[31] in PSoC 3 and PSoC 5LP as shown Fig. 1.17. Each interrupt line can be assigned one of eight priority levels (0 to 7), where 0 is the highest priority. Each interrupt line is assigned an interrupt vector address, which refers to the starting address of the interrupt code. The CPU branches to this address after receiving an interrupt request. The interrupt code is referred to as the Interrupt Service Routine (ISR).The interrupt controller acts as the interface between the interrupt lines and the CPU. It sends the interrupt vector address of an interrupt line to the CPU along with the interrupt request signal. The interrupt controller also receives acknowledgment signals from the CPU on interrupt entry and exit conditions. The interrupt controller resolves interrupt priority in the case of requests from multiple interrupt lines.

An interrupt is a request initiated by a device requesting that the CPU be *interrupted* to service, i.e., *process*, some particular task. If the device initiating the interrupt has a task of sufficiently high

---

[40] 'The interrupt fundamentally changed the nature of computer operation, and therefore also the nature of the software that runs on it. An interrupt not only creates a break in the temporal step-by-step processing of an algorithm, but also creates an opening in its operational space. It breaks the solipsism of the computer as a Turing Machine, enabling the outside world to "touch" and engage with an algorithm.The interrupt acknowledges that software is not sufficient unto itself, but must include actions outside its coded instructions. In a very basic sense, it makes software social, making its performance dependent upon associations with other processes and performances elsewhere. These may be human users, other pieces of software, or numerous forms of phenomena traced by physical sensors such as weather monitors and security alarms. The interrupt connects the data space of software to the sensorium of the world. Polling was often used to pause program flow periodically to determine if any external tasks were waiting to be processed. Interrupt controllers allowed program flow to be "interrupted" whenever a higher priority task needed immediate attention." *Simon Yuill*[12].

[41] "It was a great invention, but also a Box of Pandora. Because the exact moments of the interrupts were unpredictable and outside our control, the interrupt mechanism turned the computer into a nondeterministic machine with a non-reproducible behavior, and could we control such a beast. . ." [3].

Interrupt Vector #	Fixed Function Interrupt Sources		DMA nrq Interrupt Sources	UDB Interrupt Sources
	Interrupt Source	PSoC Creator Component		
0	Low Voltage Detect (LVD)	Global Signal Reference	phub_termout0[0]	udb_intr[0]
1	Cache	Global Signal Reference	phub_termout0[1]	udb_intr[1]
2	Reserved	Not Applicable	phub_termout0[2]	udb_intr[2]
3	Power Manager	RTC, SleepTimer, Global Signal Reference	phub_termout0[3]	udb_intr[3]
4	PICU[0]	Digital Input Pin, Digital Bidirectional Pin	phub_termout0[4]	udb_intr[4]
5	PICU[1]		phub_termout0[5]	udb_intr[5]
6	PICU[2]		phub_termout0[6]	udb_intr[6]
7	PICU[3]		phub_termout0[7]	udb_intr[7]
8	PICU[4]		phub_termout0[8]	udb_intr[8]
9	PICU[5]		phub_termout0[9]	udb_intr[9]
10	PICU[6]		phub_termout0[10]	udb_intr[10]
11	PICU[12]		phub_termout0[11]	udb_intr[11]
12	PICU[15]		phub_termout0[12]	udb_intr[12]
13	Comparators Combined	Unsupported	phub_termout0[13]	udb_intr[13]
14	Switched Caps Combined	Unsupported	phub_termout0[14]	udb_intr[14]
15	I2C	I2C	phub_termout0[15]	udb_intr[15]
16	CAN	CAN	phub_termout1[0]	udb_intr[16]
17	Timer/Counter0	Timer, Counter, PWM	phub_termout1[1]	udb_intr[17]
18	Timer/Counter1	Timer, Counter, PWM	phub_termout1[2]	udb_intr[18]
19	Timer/Counter2	Timer, Counter, PWM	phub_termout1[3]	udb_intr[19]
20	Timer/Counter3	Timer, Counter, PWM	phub_termout1[4]	udb_intr[20]
21	USB SOF Int	USBFS	phub_termout1[5]	udb_intr[21]
22	USB Arb Int		phub_termout1[6]	udb_intr[22]
23	USB Bus Int		phub_termout1[7]	udb_intr[23]
24	USB Endpoint[0]		phub_termout1[8]	udb_intr[24]
25	USB Endpoint Data		phub_termout1[9]	udb_intr[25]
26	Reserved	Not Applicable	phub_termout1[10]	udb_intr[26]
27	LCD	Segment LCD	phub_termout1[11]	udb_intr[27]
28	DFB	Filter	phub_termout1[12]	udb_intr[28]
29	Decimator	Delta Sigma ADC	phub_termout1[13]	udb_intr[29]
30	PHUB Error	Unsupported	phub_termout1[14]	udb_intr[30]
31	EEPROM Fault	Unsupported	phub_termout1[15]	udb_intr[31]

Fig. 1.16 PSoC 3 and PSoC 5LP interrupt sources.

priority, i.e., a higher priority than that of the task being conducted by the CPU at the time the interrupt request was received, and there are no other interrupt requests of higher priority waiting to be processed, then:

- the interrupt request is accepted,
- interrupts of the same, or lower, priority are blocked,

- the current task is suspended (which requires that the state of the CPU,[42] be preserved to allow the interrupted task to be completed at a later time),

and,

- the requested task is processed.

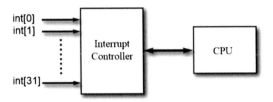

Fig. 1.17 PSoC 3, 5LP support 32 levels of interrupt.

If other interrupt requests of higher priority than that of the original task exist, they will all be processed before the CPU returns to continue the processing its original task. If the microcontroller is engaged in a task and a series of increasingly higher priority interrupt requests occur, before the preceding interrupt has been fully serviced, then the stack will contain state information about each of the interrupted tasks that has been suspended by a higher priority task request and the original task, except for the highest priority interrupt, which will then be serviced by an interrupt service routine (ISR). It is important to fully preserve the state of each lower priority task on the stack, e.g., by storing the contents of the accumulator, program counter, program status word and any other registers involved. The microcontroller can then restore the previous task(s) and continue program execution until the next interrupt occurs.

1.5.2 Latency

The overall latency, of an embedded system, is the time between when the input data is ready (latched) and the time at which the microcontroller is able to input the data, process it and produce the required results. In the case of an output device, for example, a hard disk, UART, or device which has *busy* states, the microcontroller must wait until the device is ready to accept data/commands, i.e., is in a *non-busy* state. Therefore, the total maximum latency (L_{max}) for a system can be defined as:

$$L_{max} = L_{sensors} + L_{microcontroller} + L_{peripherals} + L_{actuators} + \ldots \tag{1.8}$$

---

[42]The accumulator, program status word (PSW) program counter, and any related registers, are typically stored on the stack when a task is suspended to service an interrupt to allow the interrupted task to be fully restored.

where, $L_{microcontroller}$ is a function of program execution times, time required to service interrupts, wake-up time,[43] boot time,[44] etc., and each of the Latency parameters represented in Eq.(1.8) represent the worst-case conditions.

Interrupts introduce delays because of the time required to service a given interrupt and the fact that they are handled in order of priority. In the worst-case, a low order priority interrupt will have to wait until all higher order interrupts have been serviced before it is serviced. If higher-order interrupts occur frequently, it is possible that the lower-order priority interrupts may be blocked for unacceptable periods of time, or they may not be serviced at all. In some applications, as long as the embedded system responds within a predefined period of time, the system is performing satisfactorily. In other cases, different response times are required depending on the state of the processes being monitored/controlled.

Thus, for interrupt-driven I/O, an additional latency factor is priority assignment which determines task precedence. While higher priority tasks, whether input or output, are addressed earlier than lower priority tasks, in some applications, all tasks may be assigned the same priority, so that no task takes precedence, over any other. Alternatively, as discussed previously, the embedded system may poll input/output devices to determine whether such devices are busy, have data available for input to the microcontroller, are available to transfer/receive data, etc. However, polling can result in the significant waste of machine cycles when polling for data that is not available and/or conditions that don't exist very often. Interrupts also make it possible to detect conditions internal to the microcontroller, such as timer/counter overflow, data available in an internal UART, an internal UART being available for character transmission, that a multiplication product is available, etc.

Therefore, a microcontroller responds to interrupts by first determining if more than one interrupt has occurred. If so, the microcontroller then services the interrupts on the basis of priority by halting execution of the current task, storing all the information required to restore that task and then *servicing* the interrupt request,[45] e.g., by collecting the latched input data, taking whatever action maybe required, such as storing the data, subjecting it to numerical processing and/or taking appropriate action such as setting/transmitting output parameters for actuators, data to transmission channels, data to display devices, etc. It should be noted that most microcontrollers have a reserved interrupt referred to as a *None-Maskable Interrupt* (NMI) which has priority over all other interrupts. This interrupt is usually reserved for catastrophic events such as hard disk, or other serious failures (Fig. 1.18).

1.6   Embedded Systems Subsystems

Microcontrollers need a wide range of subsystems, if they are to be the basis for complex embedded system applications, e.g., voltage A/D and current/voltage D/A converters, mixers, pulse-width modulators (PWMs), programmable gain amplifiers (PGAs), instrumentation amplifiers, etc., as shown in Table 1.1. In many applications, embedded systems involve multiple analog/digital data input channels because data is often provided by a wide variety of sensors, communications channels,

---

[43]Some microcontrollers are programmed to go to *sleep* when nothing interesting is occurring, in order to conserve power. They can be awakened periodically, or by the occurrence of an interrupt. In such cases, when the required tasks are completed the microcontroller can then be returned to a sleep state until needed again. It may be necessary, in some applications, to take into account the latency associated with returning from a sleep state to an active state.

[44]In some embedded systems, in the event that the embedded system becomes "locked-up", the embedded system reboots itself after a predetermined period of time.

[45]The routine responsible for responding to the interrupt request is referred to an Interrupt Service Routine (ISR).

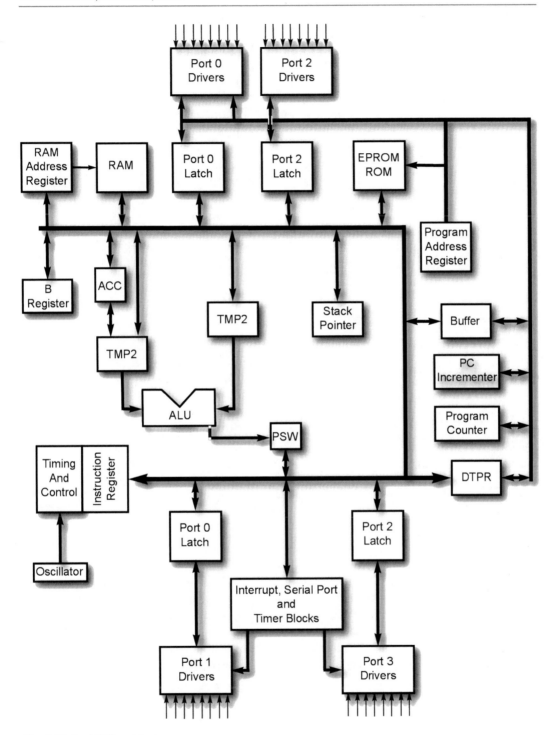

Fig. 1.18 Intel 8051 architecture.

Table 1.1 Some of the types of subsystems available in microcontrollers.

Amplifiers	Prog. gain	Instr	Transconductance	Comparators	OpAmp
A/D converters	Delta-sigma	SAR	Incremental		
D/A converters	Multiplying	Current DAC	Voltage DAC	–	–
Dialer	DTMF	–	–	–	–
Counters	8 Bit	16-Bit	24-Bit	32-Bit	–
Timers	16 Bit Tach/Timer	8 Bit	16-Bit	24-Bit	32 Bit
Random sequence	PRS8	PRS16	PRS24	PRS32	–
PWMs	PWM8	PWM16	PWM24	PWM32	–
Analog muxs	AMUX4	AMUX8	RefMUX	Virtual	Sequencer
Filters	Lowpass	Bandpass	Highpass	Notch	Adaptable
Digital comm	UARTs	USART	CRC generators	–	–
Digital comm	SPI	SPIM	SPIS	CAN	LIN
Digital comm	IDaTX/IrDARX	I2Cm	I2CHW	USBFS	–
Digital comm	One Wire	I2C	FlexRay	I2S	–
MAC	–	–	–	–	–
LCD	Character	Segment static	Segment	–	–
LED	LED 7 segment	–	–	–	–
Sleep Timer	–	–	–	–	–
LVDT	–	–	–	–	–
Logic	AND	OR	XOR	NAND	NOR
Logic	NOT	XNOR	Logic high	Logic low	LUT
Logic	Digital MUX	De-multiplexer	D Flipflop	–	–
Registers	Control	Status	–	–	–
DMA	–	–	–	–	–
Pins/ports	Analog	Digital bi-direct	Digital input	Digital output	–
Logic	AND	OR	XOR	NAND	NOR
Logic	NOT	XNOR	Logic high	Logic low	
Logic	Digital MUX	De-multiplexer	–	–	–
Registers	Control	Status	Shift	–	–
Mixer	–	–	–	–	–

etc. The embedded system employs a microcontroller/microprocessor[46] for numerical computation, and logic functions, that are to be performed on the input data, e.g., numerical processing of input data, and the decision-making based thereon. Output drivers for a variety of devices, e.g., motors, linear/rotary actuators, LCD and other types of display devices, communications devices for I2C, CAN, SPI, RS232,[47] etc., are also required as part of the embedded system and interconnect directly with the microcontroller and the actuators, displays, PCs, networks, etc.

---

[46]The distinction between microcontroller and microprocessor has become somewhat ambiguous in modern parlance and the two terms are sometimes used interchangeability with little regard for their differences. In the discussion that follows the term microcontroller shall refer, at a minimum, to a microprocessor, memory and some form of I/O capability all within the confines of a single chip, that functions as a "system on a chip".

[47]Within the family of RS232 type drivers are RS422 and RS485 protocols which are specific hardware protocols as opposed to data protocols and provide support for master/slave operation, as well as, greatly improved noise immunity and longer transmission paths.

Typically, an embedded system involves a combination of analog and digital devices, under the control of a microcontroller, that collectively serve as a feedback/control system to monitor, and control, a wide variety of electromechanical, electro-optical systems, chemical process, etc. Embedded systems can be as simple as a fan controller, used to control one, or more, fans, to maintain predefined temperature(s) in a server, or consist of a complex network of embedded systems collecting and sharing data, as well as, handling various output/control functions.

In addition, embedded systems may also be required to provide real-time actions in terms of responding within predefined time limits to certain critical input conditions, or lack thereof, with appropriate output responses, as in the case of anti-lock brakes, deployment of air bags, response to the failure of one or more devices, initiating critical shut-down procedures, gathering data at sufficiently high rates to allow for data processing and appropriate control functions, etc.

Modern-day embedded systems must, of necessity, and to the extent feasible, also be adaptable to changing market requirements, avoid steep learning curves for the designer, etc. Furthermore, issues such as low component costs, minimal printed circuit board real estate requirements, ease of manufacture, minimal reliance on external components, in-circuit debugging/programming capability, support for standard communications protocols, and interoperability with other devices and systems are, also important.

The advent of microprocessor/microcontroller technology led early adopters to conclude that embedded systems of the future would consist simply of one, or more, analog-to-digital and digital-to-analog converters conjoined with a microprocessor/microcontroller, as shown in Fig. 1.19. The basic design philosophy was to immediately convert all input signals to their digital counterpart, process the resulting digital form of the inputs and then, if required, convert the digital results back to an analog signal, via digital-to-analog converters, for connection to the external world.

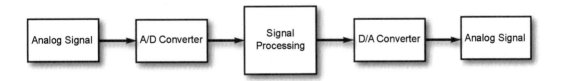

Fig. 1.19 A simple example of analog signal processing.

This view was strengthened by the fact that analog signals can be degraded by component tolerances, undesirable nonlinearities, sensitivity to electrical noise (EMI), changes in environmental conditions such as temperature and humidity, vibration, limited current/voltage dynamic range, storage of analog information in other than digital formats, etc.

However, while it was soon realized that analog signal processing was an important part of many embedded systems, it was also important to minimize power consumption,[48] provide fast response times for converting all data to a digital format so was the ability to deal with a variety of problems, e.g., aliasing, digital filtering overhead, ease of debugging, and so on.

When dealing with digital methods for gathering and processing data, careful consideration must be given to the amount of data gathered per unit time over, a given period, and the rate with which such data is gathered. It is assumed, in the following discussion, that the signal under consideration

---

[48]Which would, inter alia, reduce systemic noise.

is a continuous-time, *well-behaved*[49] signal, and that the goal is to convert the analog signal into its digital equivalent under conditions sufficient to allow the original analog signal to be accurately reconstructed. Sampling at a rate below the highest frequency component of a given signal can give rise to a phenomenon known as *aliasing*, as shown in Fig. 1.20. In this case, a fixed frequency, sinusoidal signal is sampled at a rate of once per second while the implied signal derived by sampling, is seen to have a period of approximately 10 s.

The Nyquist–Shannon Sampling Theorem, also referred to as the Nyquist, or Shannon, criteria, requires that under these conditions the sample rate be equal to, or greater than, twice the highest frequency component of the signal, or equivalently, that if the frequency component is B Hertz, that the sample rate be:

$$f_s = 2\beta B \tag{1.9}$$

where f_s is the sampling frequency, B represents the bandwidth (based on the highest frequency component of the signal) and β is a measure of the amount of oversampling, if any. Oversampling becomes important when attempting to minimize anti-aliasing effects, particularly where A/D conversion is involved.[50]

Sampling, in the present context, refers to the periodic, or aperiodic, collection of data resulting in a *discrete-time series*. The rate of sampling, in terms of samples per sec, is usually determined by the application, the hardware used in the embedded system and Eq. (1.9). The amount of data gathered per sample is obviously determined by the number of bits (bytes) gathered per sample and the rate is determined by how often a sample is taken.

For example, if each sample consists of two bytes, or equivalently 16-bits per sample, the sampling rate is 200 samples second, and the length of time over which samples are gathered at this rate is 24 h, the size of the sampled data set, D, is given by:

$$D = (bits\ per\ sample)(\#\ of\ samples\ per\ second)(total\ sampling\ time) \tag{1.10}$$

and therefore:

$$D = \frac{(16)(200)(24)(3600)}{8} = 34.56\,\text{MB} \tag{1.11}$$

Note that, in this example, it is tacitly assumed that the highest frequency component in the sampled signal is 100 Hz for unit oversampling, i.e., for $\beta = 1$. Furthermore, *sampling* which is often introduced when relying on digital signal processing techniques such as A/D and D/A conversion, can result in the loss of information (aliasing), adds additional CPU overhead and introduces potential quantization issues and round-off errors.

In some applications, the output of a D/A converter is followed by digital filtering, which while capable of providing an excellent filter response, can be an example of obtaining excellent output characteristics at the expense of data processing time, and therefore latency, which can preclude their use in certain types of control systems. In such cases, analog filters may be employed that, while perhaps offering much less sophisticated filtering capability, are often cheaper, faster and characteristically have a larger dynamic range. As in the case of any optimized embedded system design, trade-offs are frequently required in order to provide the best overall solution in terms of

---

[49]For the purposes of the present discussion the phrase *well-behaved* signal shall be assumed to mean any signal that yields gracefully to whatever mathematical technique that is applied to it.

[50]Oversampling is sometimes used to provide an averaging mechanism and to reduce quantization noise.

response time, power consumption, cost, manufacturability, component count, printed circuit board (PCB) real estate, etc.

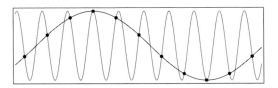

Fig. 1.20 Example of aliasing due to undersampling.

1.7   Recommended Exercises

1-1 Compare the architectures of the Intel 4004, 8008 and 8080 . Explain what the differences are and what made the Intel 8080 a much better choice for the first microcomputer. Include brief discussions of the relative architectures, instruction set differences and clock speeds.

1-2 What is the shortest program that you write that can be executed on the architectures described in 1-1. What does it do?

1-3 In some applications, a filter is placed at the input of an A/D converter and a filter is placed after the output of the D/A converter. Explain the justification for each.

1-4 Contrast,and compare, polling versus the use of interrupt service routines (ISRs)and explain the advantages and disadvantages of each,

1-5 Virtually every microprocessor/microcontroller has a NO-Operation(NOP) instruction. Why?

1-6 Explain how you could create a D/A using an A/D and some additional logic circuits. Can you create an A/D using a D/A and some additional logic?

1-7 Can a processor with a given word length of 4-bits and clock speed of 1 megahertz compete with a processor of word length of 8-bits, say 8-bits and faster clock rate have the same execution speed simply by What would be the resultant limitation of the 4004 in such a case, if any?

1-8 If you want to gather data in "real-time" that has a time-stamp associated with each sample, and the CPU clock period is more than the time between samples what methods could you employ? If the clock period and time between samples are equal is it possible collect time-stamped data? If so, how?

1-9 The VERDAN has a 26-bit word length and a clock speed of 345 kHz. Assum ing that two bits are reserved. If the VERDAN is the flight computer for an aircraft traveling at 2000 miles per hour, could a PC laptop running Windows with a clock speed of 2 GHz be used instead as the guidance computer?

1-10 A sample and gold circuit is typically required when using an A/D converter to collect data but not with a D/A converter which are used in conjunction with a microcontroller or microprocessor. Why?

1-11 Find the closed-loop gain of the block diagram shown below, and assuming that $\beta_1 = 0.23$, $\beta_2 = 1$, $A_1 = A_2 \geqslant 1$, derive an approximate expression for the closed-loop gain.

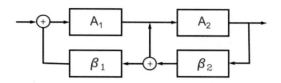

References

1. P. Allen, Idea man: a memoir by the cofounder of microsoft. Portfolio; Reprint edition. (2012)
2. Digital Computers for Aircraft. Flight International. 85 (2867): 288. ISSN 0015-3710 (1964)
3. E.W. Dijkstra, My recollections of operating system design. Oper. Syst. Rev. **39**(2), 4–40 (2005)
4. L. Donglin, H. Xiabo, S. Lemmon, D. Michael, L. Qiang, Firm real-time system scheduling based on a novel QoS constraint. IEEE Trans. Comput. **55**(3), 320–333 (2006)
5. Elliott Bros And Autonetics Fit Verdan Computers To Polaris Submarines. Electronics Weekly, 2 January 2018
6. D. Goldstein, J. Gordon, A. Neumann, COMPUTERS. U. S. A. autonetics, RECOMP, Downey, Calif. Digit. Comput. Newslett. **9**(2–9), 4–12, 2– via DTIC (1957)
7. H. Kreiger, VERDAN technical reference manual. EM-1319-1. Autonetics a division of North American Rockwell Corporation. (1959). Revised 13 June 1962
8. MARDAN Computer - Time and Navigation. https://timeandnavigation.si.edu/multimedia-asset/mardan-computer
9. C.F. O'Donnell, *Inertial Navigation Analysis and Design* (McGraw Hill Company, New York, 1964)
10. Recomp III Service Manual. A3958-501. Autonetics Division of North American Rockwell. 20 August 1959
11. The amazing MARDAN - accelerating vector. https://acceleratingvector.com/2014/06/21/the-amazing-mardan/
12. S. Yuill, M. Uller, *Software Studies: A Lexicon,Cambridge, Massachusetts, London, England* (The MIT Press, Cambridge, 2008)

Microcontroller Subsystems

2

Abstract

In this chapter, the discussion focuses on subsystems using PSoC 3 and PSoC 5LP (PSoC 3 Architecture Technical Reference Manual. Document No. 001-50235 Rev. *M Cypress Semiconductor, 8 April 2020; PSoC 5LP Architecture Technical Reference Manual. Document No. 001-78426 Rev. *G. Cypress Semiconductor, 6 Nov 2019) as illustrative examples of the fundamental aspects of current microcontroller architectures (J.A. Borrie, Modern control systems—a manual of design methods. Prentice Hall, London, 1986). Included is a detailed discussion of the 8051 instruction set, the wrapper concept as employed in PSoC 3 to integrate an 8051 core into the PSoC environment, basic concepts of interrupts and interrupt handling, DMA transfer concepts including using various DMA functions in conjunction with a peripheral hub for transferring data and to/from peripherals, clock sources and clock distribution, internal and external memory use, power management, sleep/hibernate considerations, implementation of a RTC, hardware testing and debugging, etc. In the following chapters, the discussion focuses on microcontrollers, digital and analog peripherals, the development environment and modules such as delta-sigma converters, PWMs, OpAmps, etc. and finally conclude with a detailed implementation of a digital voltmeter. Various subsystems common to microcontrollers, viz., the CPU, interrupt controller, DMA functionality, busses, memories, clocking, general-purpose I/O (GPIO), power management and hardware debugging support. PSoC 3 and PSoC 5LP are also topics throughout this chapter to illustrate the key concepts involved in each of these topics. (It should be noted that the basic architectures of both PSoC 3 and PSoC 5LP are quite similar but because of the dramatic differences in the microprocessor cores employed in each case, implementation details of some aspects of these programmable systems on a chip are quite different. However, such differences are not the primary focus of this chapter and shall be treated, if at all, in detail elsewhere in this textbook.)

2.1 PSoC 3 and PSoC 5LP: Basic Functionality

Before beginning a discussion of microcontroller subsystems it is important to discuss the functionality common to PSoC 3 and PSoC 5LP, e.g., they have:

- the same pin-out configuration, and are therefore pin and peripheral compatible,
- support for a variety of communications protocols, e.g., USB, I2C, CAN, UART, etc.,

- a common development environment, viz., PSoC Creator,
- high precision/performance analog functionality with up to 20-bit ADC and DAC support, in addition to comparators, OpAmps, PGAs, mixers, TIAs, configurable logic arrays, etc.,
- an easily configurable logic array,
- SRAM, Flash and EEPROM memory,
- analog systems [7] that include both switched-capacitance (SW) and continuous-time (CT) blocks, 20-bit sigma-delta converter(s), 8-bit DACs configurable for 12-bit operation, PGAs, etc.,
- digital systems that are based on Universal Digital Blocks (UDB) and specific function peripherals such as CAN and USB,
- programming and debugging support via JTAG, Serial Wire Debug (SWD) and Single Wire Viewer (SWV),
- a nested, vectored interrupt controller,
- a high-performance DMA controller,

and,

- flexible routing to all pins.

However, there are some significant differences between PSoC 3 and PSoC 5LP, e.g.,

- PSoC 3's CPU subsystem (core) [16] is based on a 8-bit 8051 core plus DMA controller and digital filter processor, at up to 67 MHz single-cycle,[1] 8-bit, 8051-based, Harvard architecture processor capable of operating at clock speeds up to 67 MHz, which permits it to outperform standard 8051 incarnations by as much as a factor of ten, or equivalently one order of magnitude.
- PSoC 5LP's CPU subsystem (core) is based on a 32-bit, Harvard architecture, three-stage, pipelined, ARM Cortex-M3 processor capable of operating at clocks speeds up to 80 MHz. Its instruction set is the same as Thumb-2 and supports both 16- and 32-bit instructions. PSoC 5LP has a Flash cache that reduces the number of Flash accesses required and thereby lowers power consumption.

2.2   PSoC 3 Overview

The fundamental approach to the PSoC architecture [1], and philosophy, has remained basically unchanged as it has evolved from being based on the proprietary M8C microprocessor [12] to support for both 8051 and ARM cores [2]. The latter processors, while based on quite different architectures, both control a standard set of analog/digital blocks and the system's I/O ports as shown in Figs. 2.1 and 2.2, respectively.

PSoC 3 integrates a single-cycle-per-instruction 8051 core, a programmable digital system, programmable analog components and configurable digital system resources together with a highly configurable I/O system. Internal communication is primarily based on the Arm Advanced High-Performance Bus (AHB) in conjunction with a multi-spoke bus controller called the Peripheral Hub (PHUB).[2] This allows many of the functional blocks within PSoC 3 to communicate with little

---

[1] Single-cycle refers to instructions being executed in a single machine (clock) cycle.

[2] The PHUB bus is based on Arm's Advanced Microcontroller Bus Architecture (AMBA) AHB protocol, introduced by ARM in 1996, and consists of a central hub and radial spokes that are connected to one or more peripheral blocks.

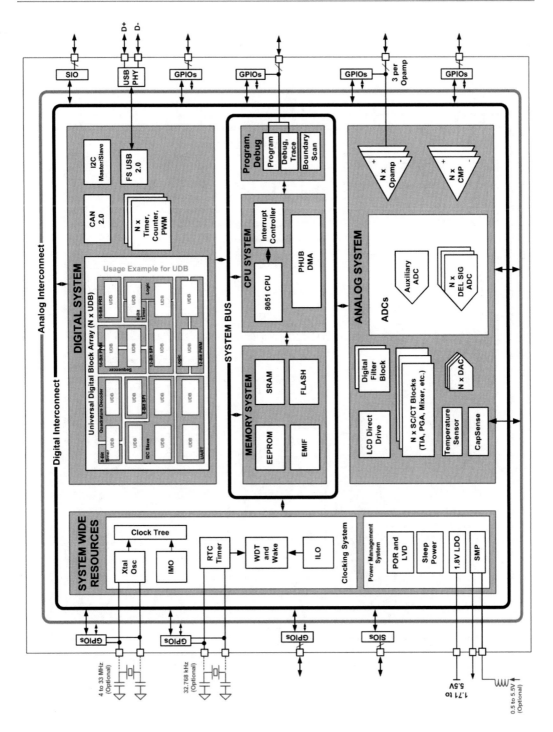

Fig. 2.1 Top level architecture for PSoC 3.

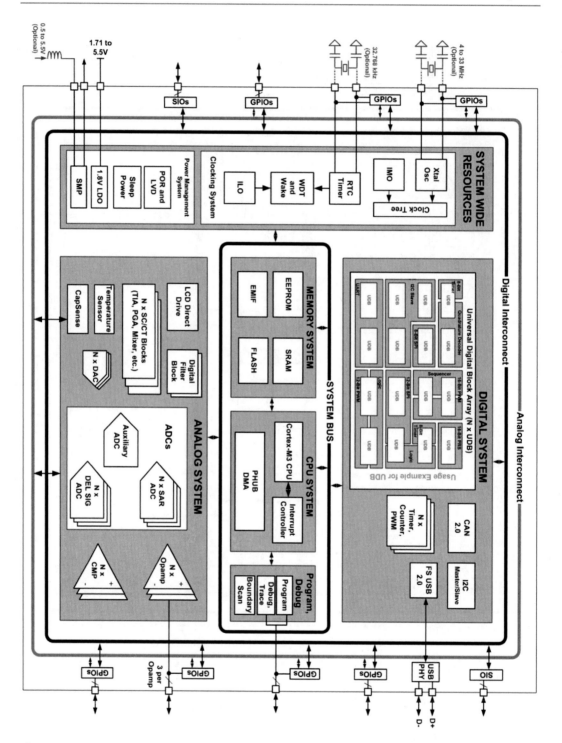

Fig. 2.2 Top level architecture for PSoC 5LP.

or no CPU involvement. In addition, there is an Analog Global Bus (AGB) that can be used to connect to/from the I/O system. A secondary bus structures allows the CPU to communicate directly with the I/O ports. The EEPROM, Flash and SPC blocks are connected this bus to enable SPC programming control. CPU subsystem connections to the cache and interrupt controller allow the CPU to communicate directly with both, thereby minimizing the latency and any requirements for communicating with the peripheral controller.

The 8051 "core"[3] is capable of being clocked from DC to 67 MHz, provides both hardware multiply and divide, 24 channels of Direct Memory Access (DMA), up to 8K each of Flash and SRAM and up to 2K of one million cycles, 20-year retention, EEPROM.

2.2.1 The 8051 CPU (PSoC 3)

The PSoC 3 architecture shown schematically in Fig. 2.3 is based on an 8051 core (Fig. 2.4).[4]

As discussed in Chap. 1, the 8051 microcontroller is something of a classic in the field of microprocessors and microcontrollers dating from 1980 when it was introduced by Intel Corporation (Fig. 2.5).

In its simplest configuration it consisted of:

- An ALU
- Seven on-chip registers
- A serial interface
- Two 16-bit timers
- Internal memory consisting of GP/bit-addressable storage and register banks and special function registers
- Support for 64K of external memory (code)[5]
- Support for 64K external memory (data)
- Four 8-bit I/O ports[6]
- 210 bit-addressable locations[7]

In addition to supporting memory-mapped I/O, the registers are also memory-mapped and the stack resides in RAM that is internal to the 8051. The ability to access individual bits makes it possible to set, clear AND, OR, etc., utilizing a single 8051 instruction. Register banks are contained in the lowest 32 bytes of the 8051's internal memory. Eight registers are supported, viz, R0–R7, inclusive, and their default locations are at addresses 0x00-0x07. Register banks can be used to provide efficient context switching and the active register bank is selected by bits in the Program Status Word (PSW). At the top of the internal RAM there are 21 special registers located at addresses $0x80 - 0xFF$. Some of these registers are bit- and byte-addressable, depending on the instruction addressing the register.

---

[3]This core is fully compatible with the MCS-51 instruction set [11], i.e., it is "upward-compatible".

[4]The CORTEX M3 core is based on an ARM processor which is optimized for "cost-sensitive" microcontroller applications. The CORTEX M3 used in the PSoC 5LP employs a 3-stage, Harvard architecture with pipeline core that supports single-cycle hardware multiply/divide, branch speculation and Thumb-2 instructions.

[5]The 8051 utilizes separate memory for data and code, respectively.

[6]I/O ports in the 8051 are "memory-mapped", i.e. to regions for SFRwrite from/to an I/O port the program being executed must read/write to the corresponding memory location in the same manner that a program would normally read/write to any memory location.

[7]128 of these are at addresses 0x20-0x2F with the remaining 73 being located in special function registers.

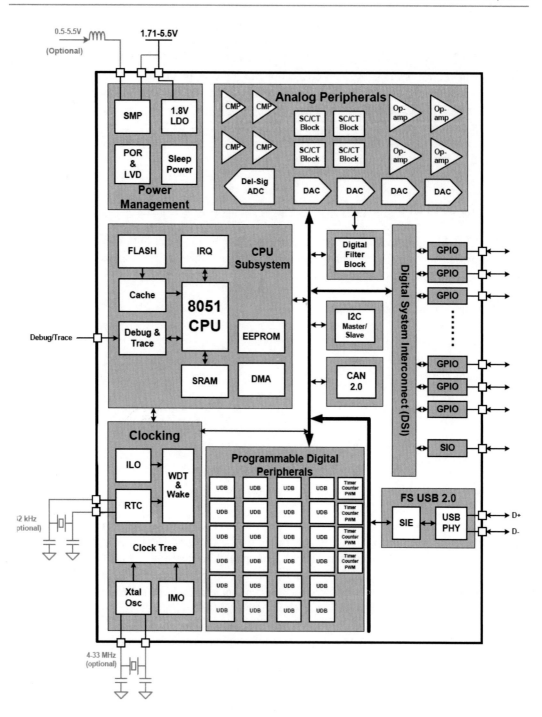

Fig. 2.3 Simplified PSoC 3 architecture.

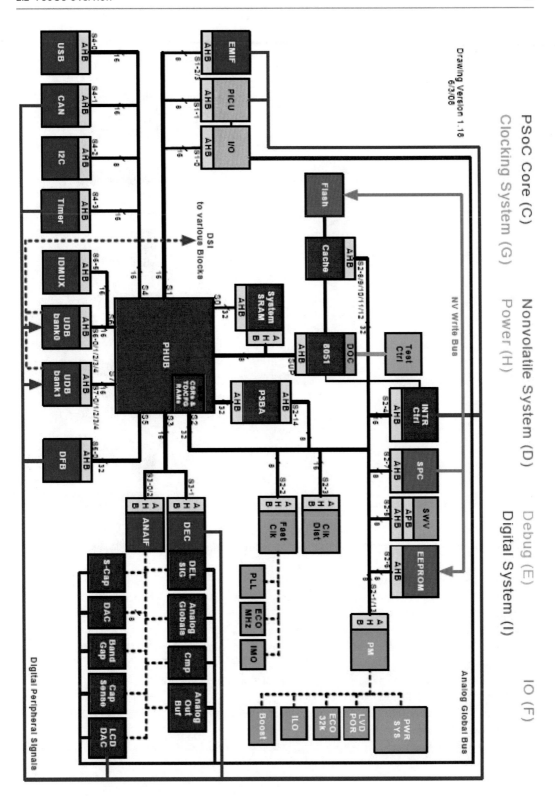

Fig. 2.4 PSoC 3—top level architecture.

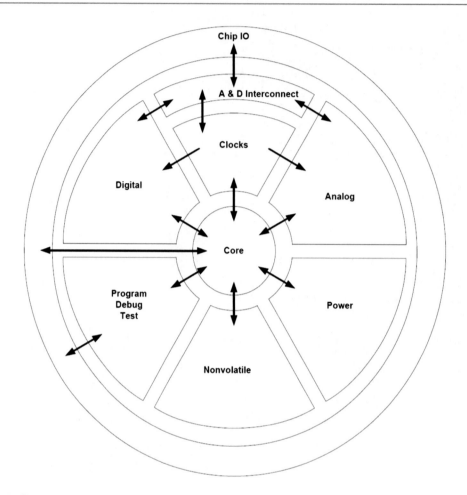

Fig. 2.5 Communications paths within PSoC 3.

The 8051 implementation in PSoC 3 has the following features:

- The architecture is RISC-based and pipe-lined.
- It is 100% binary-compatible with the industry standard 8051 instruction set, i.e., it is upward compatible in terms of executables.
- Most instructions operate in one, or two, machine cycles.[8]
- It supports a 24-bit external data space that enables access to on-chip memory and registers, and off-chip memory.
- A new interrupt interface has been provided that enables direct interrupt vectoring.
- 256 bytes of internal data RAM are available.
- The Dual Data Pointer (DPTR) has been extended from 16-bits in the "standard 8051" architecture to 24-bits to facilitate data block copying.
- Special Function Registers (SFRs) provide fast access to PSoC 3 I/O ports and control of the CPU clock frequency.

---

[8]Instruction cycle refers to the time required to fetch an instruction, interpret the instruction and execute it. Machine or clock cycle is the clock period.

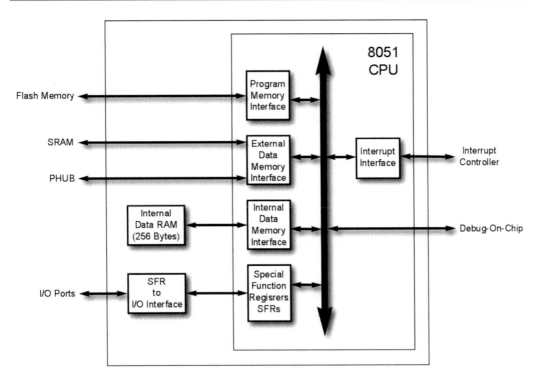

Fig. 2.6 PSoC 3's 8051 wrapper.

2.2.1.1 8051 Wrapper

In order to most efficiently, and effectively, incorporate the 8051 core into PSoC 3, a "wrapper" [1], is provided as shown in Fig. 2.6. This so-called wrapper[9] is in fact simply logic that surrounds the core and provides an interface between the core and the rest of the PSoC 3 system. The 8051 is one of two bus masters, the other being the DMA controller. Two bus slaves are available in the form of the PHUB, discussed in Sect. 2.2.7, and the on-chip SRAM, and are accessible via the 8051's external memory space. This configuration provides access to all of PSoC 3's registers, as well as, external memory. The wrapper also provides an SFR-I/O interface and gives direct access to the I/O port registers via the SFRs. A CPU clock divider is also included in the wrapper. Each port has two interfaces, one of the interfaces is to the PHUB to allow boot configuration and access to all the I/O port registers, and the second interface is to the SFRs in the 8051 which gives faster access to a limited set of I/O port registers. The clock divider makes it possible to operate the CPU at frequencies that are divisors of the bus clock speed, cf. Sect. 2.2.11.

2.2.1.2 8051 Instruction Set

The 8051 instruction set consists of 44 basic instructions as shown in Table 2.1.[10] These basic instructions result in 256 possible instructions, of which 255 (24 3-bytes, instructions, 92 2-byte instructions and 139 1-byte instructions) are documented. The full set of opcodes is shown in Table 2.1 (Figs. 2.7 and 2.8).

---

[9]The term wrapper is used to describe encapsulating code so that its use can be effected via an API, or otherwise simpler interface.

[10]The mnemonics used here for the 8051(8052) are copyrighted by Intel Corporation, 1980.

Table 2.1 The complete set of 8051 opcodes.

	0x00	0x01	0x02	0x03	0x04	0x05	0x06	0x07	0x08	0x09	0x0a	0x0b	0x0c	0x0d	0x0e	0x0f
0x00	AJMP	LJMP	RR	INC	INC	INC	INC	INC	INC	INC	INC	INC	INC	INC	INC	INC
0x10	ACALL	LCALL	RRC	DEC	DEC	DEC	DEC	DEC	DEC	DEC	DEC	DEC	DEC	DEC	DEC	DEC
0x20	AJMP	RET	RL	ADD	ADD	ADD	ADD	ADD	ADD	ADD	ADD	ADD	ADD	ADD	ADD	ADD
0x30	ACALL	RETI	RLC	ADDC	ADDC	ADDC	ADDC	ADDC	ADDC	ADDC	ADDC	ADDC	ADDC	ADDC	ADDC	ADDC
0x40	AJMP	ORL	ORL	ORL	ORL	ORL	ORL	ORL	ORL	ORL	ORL	ORL	ORL	ORL	ORL	ORL
0x50	ACALL	ANL	ANL	ANL	ANL	ANL	ANL	ANL	ANL	ANL	ANL	ANL	ANL	ANL	ANL	ANL
0x60	AJMP	XRL	XRL	XRL	XRL	XRL	XRL	XRL	XRL	XRL	XRL	XRL	XRL	XRL	XRL	XRL
0x70	ACALL	ORL	JMP	MOV	MOV	MOV	MOV	MOV	MOV	MOV	MOV	MOV	MOV	MOV	MOV	MOV
0x80	AJMP	ANL	MOVC	DIV	MOV	MOV	MOV	MOV	MOV	MOV	MOV	MOV	MOV	MOV	MOV	MOV
0x90	ACALL	MOV	MOVC	SUBB	SUBB	SUBB	SUBB	SUBB	SUBB	SUBB	SUBB	SUBB	SUBB	SUBB	SUBB	SUBB
0xa0	AJMP	MOV	INC	MUL	?	MOV	MOV	MOV	MOV	MOV	MOV	MOV	MOV	MOV	MOV	MOV
0xb0	ACALL	CPL	CPL	CJNE	CJNE	CJNE	CJNE	CJNE	CJNE	CJNE	CJNE	CJNE	CJNE	CJNE	CJNE	CJNE
0xc0	AJMP	CLR	SWAP	XCH	XCH	XCH	XCH	XCH	XCH	XCH	XCH	XCH	XCH	XCH	XCH	XCH
0xd0	ACALL	SETB	SETB	DA	DJNZ	XCHD	XCHD	XCHD	XCHD	XCHD	XCHD	XCHD	XCHD	XCHD	XCHD	XCHD
0xe0	AJMP	MOVX	MOVX	CLR	MOV	MOV	MOV	MOV	MOV	MOV	MOV	MOV	MOV	MOV	MOV	MOV
0xf0	ACALL	MOVX	MOVX	MOVX	CPL	MOV	MOV	MOV	MOV	MOV	MOV	MOV	MOV	MOV	MOV	MOV

2.2.1.3 Internal and External Data Space Maps

A diagram of the 8051's XDATA address map[11] and internal data space is shown in Table 2.2 and Fig. 2.9, respectively. This space is divided into five regions as shown. While the Internal Data Memory addresses are in fact only one byte wide, implying that the address space is limited to 256 bytes, direct addresses higher than 7FH access one memory space and indirect addresses higher than 7FH access a different memory space. Thus the upper 128 bytes can be used as SFR space for ports, status bits, etc., if direct addressing is employed. Sixteen of the addresses in the SFR memory space are both bit and byte-addressable.

The lower 128 bytes consists of the lowest 32 bytes grouped as 4 banks of 8 registers referred to as R0–R7. Bank selection is determined by two bits in the PSW. The next 16 bytes, i.e., above the register banks, is a bit-addressable memory space with bit addresses ranging from 00H-7FH, inclusive. The 8051's instruction set includes a number of instructions for manipulating single bits in this area, using direct addressing.

2.2.1.4 Instruction Types

The 8051 has five types of instructions:

1. **Arithmetic**—addition, subtraction, division, incrementing and decrementing.
2. **Boolean**—clearing a bit, complementing a bit, setting a bit, toggle a bit, move a bit to carry, etc.
3. **Data Transfer**—internal data, external data and lookup table data.
4. **Logical**—Boolean operations such as AND, OR, XOR and rotating/swapping of nibbles.

and,

5. **Program branching**—Conditional and unconditional jumps (branches) to modify the program execution flow

2.2.1.5 Data Transfer Instructions

The 8051 is capable of three types of data transfer:

---

[11] Table 2.2 is a tabulation of the external memory addressable by PSoC 3.

Mnemonic	Description	Cycles
MOV A,Rn	Move register to accumulator	1
MOV A,Direct	Move direct byte to accumulator	2
MOV A,@Ri	Move indirect RAM to accumulator	2
MOV A,#data	Move Immediate data to accumulator	2
MOV Rn,A	Move accumulator to register	1
MOV Rn,Direct	Move direct byte to register	3
MOV Rn,#data	Move immediate data to register	2
MOV Direct,A	Move accumulator to direct byte	2
MOV Direct,Rn	Move register to direct byte	2
MOV Direct,Direct	Move direct byte to direct byte	3
MOV Direct@Ri	Move indirect RAM to direct byte	3
MOV Direct,#data	Move immediate data to direct byte	3
MOV @Ri,A	Move accumulator to indirect RAM	2
MOV @Ri,Direct	Move direct byte to indirect RAM	3
MOV @Ri,#data	Move immediate data to indirect RAM	2
MOV DPTR,#data16	Load data pointer with 16-bit constant	3
MOVC A,@A+DPTR	Move code byte relative to DPTR to accumulator	5
MOVC A,@A+PC	Move code byte relative to PC to accumulator	4
MOVX A,@Ri	Move external RAM (8-bit) to accumulator	3
MOVX A,DPTR	Move external RAM (16-bit) to accumulator	2
MOVX @Ri,A	Move accumulator to external RAM (8-bit)	4
MOVX DPTR,A	Move accumulator to external RAM (16-bit)	3
PUSH Direct	Push direct byte onto stack	3
POP Direct	Pop direct byte from stack	2
XCH A,Rn	Exchange direct byte with accumulator	2
XCH A,Direct	Exchange direct with accumulator	3
XCH A<@Rn	Exchange indirect RAM with accumulator	3
XCH A,@Ri	Exchange low order indirect digit RAM with accumulator	3

Fig. 2.7 Data transfer instruction set.

1. **External Data Transfer**—MOVX instructions are used to transfer data between the accumulator and an external memory address.
2. **Internal Data Transfers**—Direct, indirect, register and immediate addressing instructions allow data to be transferred between any two internal RAM locations of SFRs.

0xF8	0/8 (Bit Addressable)	1/9	2/A	3/B	4/C	5/D	6/E	7/F
0xF8	SFRPRT15DR	SFRPRT15PS	SFRPRT15SEL					
0xF0	B		SFRPRT12SEL					
0xE8	SFRPRT12DR	SFRPRT12PS	MXAX					
0xE0	ACC							
0xD8	SFRPR6DR	SFRPRT6PS	SFRPRT15S6					
0xD0	PSW							
0xC8	SFRPRT5DR	SFRPRT5PS	SFRPRT5SEL					
0xC0	SFRPRT4DR	SFRPRT4PS	SFRPRT4SEL					
0xB8								
0xB0	SFRPRT3DR	SFRPRT3PS	SFRPRT3SEL					
0xA8	IE							
0xA0	P2AX	CPUCLK_DIV	SFRPRT1SEL					
0x98	SFRPRT2DR	SFRPRT2PS	SFRPRT2SEL					
0x90	SFRPRT1DR	SFRPRT1PS		DPX0		DPX1		
0x88		SFRPRT0PS	SFRPRT0SEL					
0x80	SFRPRT0DR	SP	DPL0	DPH0	DPL1	DPH1	DPS	

Fig. 2.8 Special function register map.

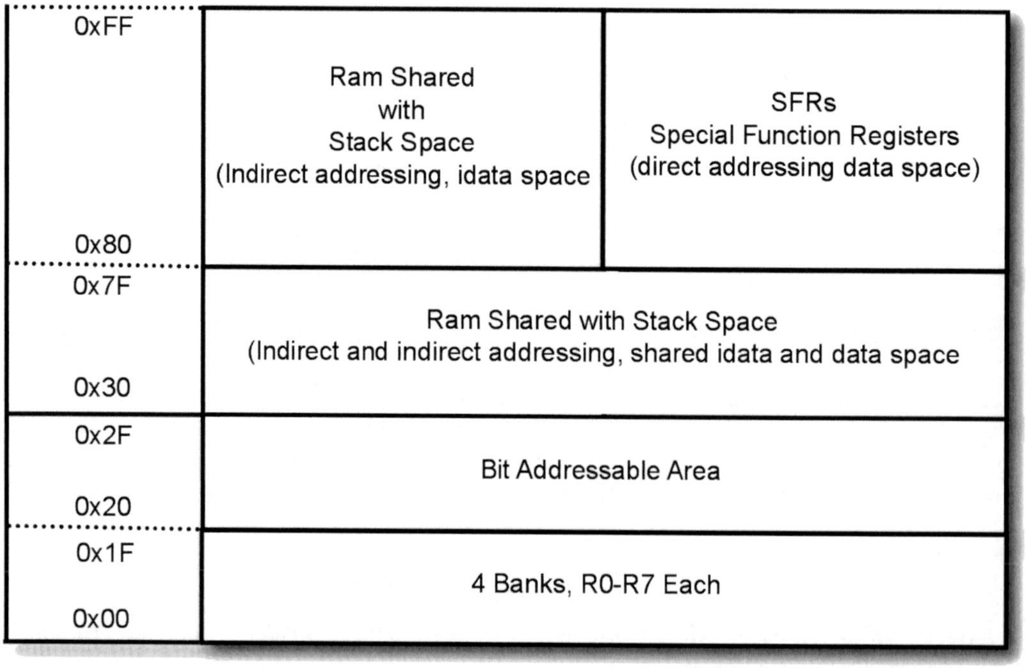

Fig. 2.9 8051 internal data space map.

Table 2.2 XDATA address map.

Address Range	Purpose
0x00 0000 - 0x00 1FFF	SRAM
0x00 2000 - 0x00 2FFF	Trace SRAM
0x00 4000 - 0x00 42FF	Clocking, PLLs and Oscillators
0x00 4300 - 0x00 430F	Power Management
0x00 4400 - 0x00 44FF	Interrupt Controller
0x00 4500 - 0x00 45FF	Ports Interrupt Control
0x00 4600 - 0x00 46FF	Reserved
0x00 4700 - 0x00 47FF	Flash Programming Interface
0x00 4900 - x00 49FF	I2C Controller
0x00 4E00 - 0x00 4EFF	Decimator
0x00 4F00 - 0x00 4FFF	Fixed Timer/Counter/PWMs
0x00 5000 - 0x00 51FF	General Purpose IOs
0x00 5300 - 0x00 530F	Output Port Select Register
0x00 5400 - 0x00 54FF	External Memory Interface (EMIF) Control Registers
0x00 5800 - 0x00 5FFF	Analog Subsystem Interface
0x00 6000 - 0x00 60FF	USB Controller
0x00 6400 - 0x00 6FFF	UDB Configuration
0x00 7000 - x00 7FFF	PHUB Configuration
0x00 8000 - 0x00 8FFF	EEPROM
0x00 A000 - 0x00 A400	CAN
0x00 C000 - 0x00 C800	Digital Filter Block
0x00 0000 - 0x00 FFFF	Digital Interconnect Configuration
0x00 0000 - 0x00 FFFF	Direct Access To Cache Memory
0x00 0000 - 0x00 FFFF	Flash Memory
0x00 0000 - 0x00 FFFF	Direct Access To 8051 IDATA
0x00 0200 - 0x00 021F	Test Controller (Internal)

3. **Lookup Table Transfers**—MOVC instructions are used to transfer data between the accumulator and program memory addresses.

2.2.1.6 Data Pointer

The Data Pointer (DPTR) is located in a 16-bit register at 0x83 (high byte) and 0x82 (low byte), respectively which is used to access up to 64K, inclusive, of external memory.

2.2.1.7 Dual Data Pointer SFRs

In order to facilitate the copying of blocks of data, four Special Function Registers (SFRs), as shown in Table 2.3, are employed to hold two 16-bit pointers, DPTR0 and DPTR1 and INC DPTR can be used to switch between them. The active DPTR register is selected by the SEL bit (0x86) in the SFRs space, e.g., if the SEL bit is equal to zero, then DPTR0 (SFRs 0x83:0x82) is selected, otherwise, DPTR1 is selected.

Table 2.3 Special function registers.

	0/8 (Bit Addressable)	1/9	2/A	3/B	4/C	5/D	6/E	7/F
0xF8	SFRPRT15DR	SFRPRT15PS	SFRPRT15SEL					
0xF0	B		SFRPRT12SEL					
0xE8	SFRPRT12DR	SFRPRT12PS	MXAX					
0xE0	ACC							
0xD8	SFRPR6DR	SFRPRT6PS	SFRPRT15S6					
0xD0	PSW							
0xC8	SFRPRT5DR	SFRPRT5PS	SFRPRT5SEL					
0xC0	SFRPRT4DR	SFRPRT4PS	SFRPRT4SEL					
0xB8								
0xB0	SFRPRT3DR	SFRPRT3PS	SFRPRT3SEL					
0xA8	IE							
0xA0	P2AX	CPUCLK_DIV	SFRPRT1SEL					
0x98	SFRPRT2DR	SFRPRT2PS	SFRPRT2SEL					
0x90	SFRPRT1DR	SFRPRT1PS		DPX0		DPX1		
0x88		SFRPRT0PS	SFRPRT0SEL					
0x80	SFRPRT0DR	SP	DPL0	DPH0	DPL1	DPH1	DPS	

The data pointer select register is used in conjunction with the following instructions:

- INC DPTR
- JMP @A+DPTR
- MOVX @DPTR,A
- MOVX A,@DPTR
- MOVC A,A+DPTR
- MOV DPTR,#data16

2.2.1.8 Boolean Operations

Boolean instructions allow single-bit operations to be performed on the individual bits of registers, memory locations and the CY Flag (the AC, OV and P flags cannot be altered by these instructions). The operations that can be conducted on individual bits are clear, complement, move, set, AND, OR, and tests for conditional jumps.

JC/JNC—jump to a relative address if the CY Flag is set or cleared
JB/JNB—jump to a relative address if the CY FLAG is set or cleared.

JBC—Jump to a relative address if a bit is set or cleared.
Short jump.

2.2.1.9 Stack Pointer

The Stack Pointer (SP) is an 8-bit register located at 0x81 that contains the default value of 0x07, when the system is reset. This causes the first "Push" to the stack to be stored in location 0x80 and therefore Register Bank 1, and potentially Register Banks 2 and 3 may not be accessible. However, initializing the SP pointer will allow all the Register Banks to be used.

2.2.1.10 Program Status Word (PSW)

The Program Status Word is located at 0xD0 and contains information about the 8051's flags as shown in Fig. 2.10. The Carry Flag (CF) can also be used as a 1-bit "Boolean accumulator", i.e., a 1-bit register for Boolean instructions. Flag 0 is a general-purpose flag and RS0/RS1 is used to determine the active register. The Overflow Flag is set after a subtraction or addition if an arithmetic overflow has occurred. The parity bit is used in each machine cycle to maintain even parity of the accumulator byte and B is a bit-addressable register (accumulator) located at 0xF0 for multiply and divide operations.[12]

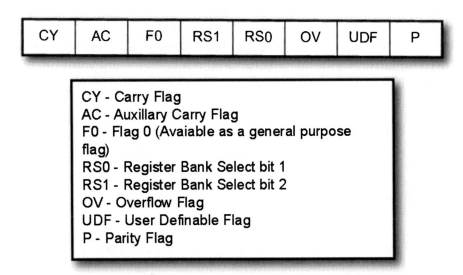

RS1	RS0	Register Bank	Address
0	0	0	0x00-0x07
0	1	1	0x08-0x0F
1	0	2	0x10-0x17
1	1	3	0x18-0x1F

Fig. 2.10 Program status word (PSW) [0xD0].

---

[12]Following an 8-bit by 8-bit multiplication, the resulting 16-bit (two byte) value is stored in A (Low byte) and B (High Byte), respectively.

2.2.1.11 Addressing Modes

The 8051 architecture supports seven addressing modes:

1. **Direct**—the operand is specified by a "direct" 8-bit address, but only internal RAM and special function registers can be addressed by this mode.
2. **Indirect**—a register, either R0 or R1, containing the 8-bit address of the operand is specified by the instruction. In this mode, the Data Pointer (DPTR) is used to specify 16-bit addresses.
3. **Immediate Constants**—Except for the Data Pointer, all 8051 instructions involving immediate addressing utilize 8-bit data values. In the case of the Data Pointer, a 16-bit constant must be used.
4. **Bit Addressing**—In this mode, the operand is specified as one of 256 bits
5. **Indexed Addressing**—Indexed addressing uses the Data Pointer as a base register with an offset stored in the accumulator to point to an address in program memory that is to be read. This addressing mode is intended for reading data from look-up tables. In such cases, a 16-bit "base" register such as the DPTR or PC, is used to point to the base of a table and the accumulator holds a value that points to a particular entry in the table. Thus, the actual address in program memory is the sum of the values held in the accumulator and 16-bit base register.[13]
6. **Register addressing**—eight registers (R0–R7) are used for register addressing. Instructions using these registers utilize the three least significant bits of the instruction opcode to specify a particular register. Since an address byte is not required use of ti mode, where possible, results in improved code efficiency. The bank select bits stored in the PSW determine which bank holds the register.
7. **Register Specific**—some instructions are used only in conjunction with specific registers such as the accumulator or DPTR and therefore an address byte is obviated, e.g., any instruction referencing the Accumulator (A) are both accumulator and therefore register specific.

The following are illustrative examples of some of the most common addressing modes:

- SBB A,2FH (Direct addressing)
- SBB AA,@R0 (Indirect Addressing)
- SBB A,R4 (Register Addressing)
- SBB A,#31 (Immediate Addressing)

- **Absolute**—ACALL and AJMP require the use of absolute addresses and store the 11 least significant bits of the address and the remaining 5 bits are derived form the five most significant bits of the Program Counter.
- **Relative**—relative addressing is used with some of the jump instructions. The relative address serves as a 8-bit (-128-127) offset that is added to the program counter to provide the address of the next instruction to be executed. The use of relative addressing can result in the program code that is relocatable, i.e., does not have any memory location dependence.
- **Long**—LCALL and LJMP require long addressing and consist of a 3-byte instruction that includes the 16-bit destination address as bytes 2 and 3. These opcodes allow the full 64K code space to be used.

---

[13]Another form of indexed addressing is employed by "case jump" instructions, i.e., a jump instruction's destination address is determined by the sum of the base pointer and the value in the accumulator.

2.2.2 The 8051 Instruction Set

The following are brief descriptions of the 8051 instruction set.[14]

- **ACALL LABEL**—unconditionally calls a subroutine located at an address LABEL. When this instruction is invoked the program counter, and stack pointer, are both advanced two bytes so that the next instruction address to be executed, upon return from the subroutine, is stored on the stack.
- **ADD A, <src-byte>**—performs an 8-bit addition of two operands, one of which is stored in the accumulator. The result of the addition is stored in the accumulator and the CY flag is set/reset as required by the results of the addition. <src-byte> can be R_n(a register), Direct (a direct byte), $@R_i$(indirect RAM), or #$data$(immediate data)
- **ADDC A, <src-byte>**—invokes an 8-bit addition of two operands based on the previous value of the CY flag. e.g. when carrying out 16-bit addition operations. <src-byte> can be R_n(a register), Direct (a direct byte), $@R_i$(indirect RAM), or #$data$(immediate data).
- **AJMP addr11**—Absolute jump using an 11-bit address. This is a two-byte instruction that uses the upper 3-bits of the address, combined with a 5-bit opcode, to form the first byte and the lower 8-bits of the address from the second byte. The 11-bit address replaces the 11-bits of the PC to produce the 16-bit address of the target. Therefore, the resulting locations are within the 2K byte memory page containing the **AJMP** instruction.
- **ANL <dest-byte>,<src-byte>**—performs a bitwise logical AND between the dest and src byte and stores the result in dest.
- **ANL, bit**—Performs a logical AND operation between the Carry bit and a bit, placing the result in the Carry bit.
- **ANL, /bit**—Performs a logical AND operation between the Carry bit and the inversion of a bit, placing the result in the Carry bit.
- **CJNE <dest-byte>,<src-byte>, rel**—Compares the magnitude of the first two operands and branches if they are not equal to the address given by adding **rel** (signed relative displacement) to the PC, after it has been incremented to the start of the next instruction. The carry flag is set if the unsigned, integer value of **<dest-byte>** is less than the unsigned integer value of **<src-byte>**; otherwise the carry is cleared. The first two operands allow four addressing mode combinations: the accumulator may be compared with any directly addressed byte or immediate data, and any indirect RAM location or working register can be compared with an immediate constant.
- **CLR A**—Clear the accumulator, i.e., reset all A bits to zero. Flags are not affected.
- **CLR bit**—Clear the indicated bit, i.e., reset to zero. Flags are not affected. CLR can operate on the carry flag, or any directly addressable bit.
- **CPL A**—Complement the accumulator.
- **CPL bit**—Complement a bit. No other flags are affected.
- **DA A**—Decimal adjust the accumulator. This instruction "adjusts" the eight-bit value in the accumulator resulting from the prior addition of two variables, each of which is in packed BCD format, resulting in the production of two four-bit values. Any ADD, or DDC, instruction may have been used for the addition.
- **DEC <src-byte>**—Decrement the operand by one. <src> can be a direct address, an indirect address, the accumulator, or a register. ($00H \Rightarrow 0FFH$).
- **DIV AB**—Divide the unsigned contents of the accumulator (A) by the unsigned contents of the B, placing the resulting integer value of the quotient in A and the integer remainder in B. If B

---

[14]Additional information is available in references [15] and [11].

originally contained 00H then both of the returned values will be undefined and the overflow flag will be set. The carry flag is cleared in all cases.

- **DJNZ <byte>,<rel-addr>**—Decrement the byte value and jump if not zero to the relative address given by adding **rel** (signed relative displacement) to the PC, after it has been incremented to the first byte of the next instruction. No flags are affected and $00H \Rightarrow 0FFH$.
- **INC <src-byte>**—Increment the operand by one. The operand can be a direct address, indirect address, register, accumulator or the data pointer (DPTR).[15] No flags are affected.<src-byte> can be R_n(a register), Direct (a direct byte), $@R_i$(indirect RAM), or #$data$(immediate data).
- **JB bit,rel**—jump if the bit is set to one, otherwise, proceed to the next instruction. The branch destination is computed by adding the signed relative displacement to the PC after incrementing the PC to the first bye of the next instruction. The tested bit is not modified and no flags are affected.
- **JBC bit,rel**—Jump if the bit is set to one and clear the bit. The destination is computed by adding the signed relative displacement to the PC after incrementing the PC to the first bye of the next instruction.
- **JC rel**—If carry is set, then branch to the destination computed by adding the signed relative displacement to the PC, after incrementing the PC to the first bye of the next instruction. No flags are affected.
- **JMP @A+DPTR**—Jump indirect. Add the eight-bit unsigned contents of the accumulator to the sixteen-bit data pointer, and load the resulting sum to the program counter. The resulting sum is the address for the instruction. Sixteen-bit addition is performed (modulo 2^{16}). A carry from the low-order eight bits propagates through the higher-order bits. Neither the Accumulator nor the Data Pointer is altered. No flags are affected.
- **JNB bit**—If the bit is not set, then branch to destination computed by adding the signed relative displacement to the PC after incrementing the PC to the first bye of the next instruction. No flags are affected.
- **JNC rel**—If the carry flag is a zero, branch to the address indicated, otherwise, proceed with the next instruction. The branch destination is computed by adding the relative-displacement to the PC, after incrementing the PC twice to point to the next instruction. The carry flag is not modified.
- **JNZ**—If the accumulator contains a value other than zero, branch to the indicated address, otherwise, proceed with the next instruction. The branch destination is computed by adding the signed relative displacement after incrementing the PC twice. The accumulator is not modified and no flags are affected.
- **JZ**—If the value in the accumulator is zero, branch to the address indicated, otherwise, proceed with the next instruction. The branch destination is computed by adding the signed relative-displacement in the second instruction byte to the PC, after incrementing the PC twice. The accumulator is not modified. No flags are affected (Tables 2.4 and 2.5).
- **LCALL addr16**—calls a subroutine located at the indicated address. The instruction adds three to the program counter to generate the address of the next instruction and then pushes the result onto the stack (low-byte first), incrementing the Stack Pointer by two. The high-order and low-order bytes, of the PC, are then loaded, respectively, with the second and third bytes of the LCALL instruction. Program execution continues with the instruction at this address. The subroutine may therefore begin anywhere in the full 64K-byte program memory address space. No flags are affected.

---

[15]Incrementing the DPTR by 1 causes this 16-bit pointer to be increased by 1, An overflow of the lower byte (DPL), i.e., $0xFF \Rightarrow 0x00$, causes the upper byte (DPH) to be incremented. DPTR is the only PSoC 3, 16-bit register that can be incremented in this manner.

Table 2.4 Jump instructions.

Mnemonic	Description	Cycles
ACALL addr11	Absolute subroutine call	4
LCALL addr16	Long subroutine call	4
RET	Return from a subroutine	4
RETI	Return from an interrupt	4
AJMP addr11	Absolute jump	3
LJMP addr16	Long jump	4
SJMP rel	Short jump (relative address)	3
JMP @A + DPTR	Jump indirect relative to DPTR	5
JZ rel	Jump if accumulator is zero	4
JNZ rel	Jump if accumulator is non zero	4
CJNE A,Direct,rel	Compare immediate data to accumulator	5
CJNE A,#data,rel	Compare immediate data to accumulator	4
CJNE Rn,#data,rel	Compare immediate data to regsiter and jump if not equal	4
CJNE @Ri,#data,re;	Compare immediate data to indirect RAM an jump if not equal	5
DJNZ Rn,rel	Decrement register and jump if non zero	4
DJNZ Direct,rel	Decrement direct byte and jump if non zero	5
NOP	No operation	1

- **LJMP addr16**—Long Jump using a 16-bit address. This is a three-byte unconditional jump to any location in the 64K program space. Address by loading the high-order and low-order bytes of the PC (respectively)with the second and third instruction bytes. The destination may therefore be anywhere in the full 64K program memory address space. No flags are affected.
- **MOV <dest-byte><src-byte>**—The byte variable indicated by the src-byte is copied into the location specified by the first dest-byte. The source byte is not affected. No other register or flag is affected. There are 15 combinations of source and destination addressing modes for this instruction.
- **MOVC A,@A+<base-reg>**—loads the accumulator with a code byte, or constant, from program memory. The address of the byte fetched is the sum of the original unsigned eight-bit accumulator contents and the contents of a sixteen-bit base register, which may be either the Data Pointer or the PC. In the latter case, the PC is incremented to the address of the following instruction before being added to the accumulator, otherwise, the base register is not altered. Sixteen-bit addition is performed so a carry-out from the low-order eight bits may propagate through higher-order bits. No flags are affected.
- **MOVX A,@Ri**—These instructions, tabulated in Table 2.6, transfer data between the accumulator and a byte of external data memory, and is denoted by appending an X to MOV. There are two types of instructions, differing in whether they provide an eight-bit or sixteen-bit indirect address to the external data RAM. In the first type, the contents of R0 or R1, in the current register bank,

Table 2.5 Arithmetic instructions.

Mnemonic	Description	Cycles
ADD A,Rn	Add register to accumulator	1
ADD A,Direct	Add direct byte to accumulator	2
ADD A,@Ri	Add indirect RAM to accumulator	2
ADD A,#data	Add immediate data to accumulator	2
ADDC A,Rn	Add register to accumulator with carry	1
ADDC Direct	Add direct byte to accumulator with carry	2
ADDC A,@Ri	Add indirect RAM to accumulator with carry	2
ADDC A,#data	Add immediate data to accumulator with carry	2
SUBB A,Rn	Subtract register from accumulator with borrow	1
SUBB Direct	Subtract direct byte from accumulator with a borrow	2
SUBB A,@Ri	Subtract indirect RAM from accumulator with borrow	2
SUBB A,#data	Subtract immediate data from accumulator with borrow	2
INC A	Increment Accumulator	1
INC Rn	Increment register	2
INC Direct	Increment direct byte	3
INC @Ri	Increment indirect RAM	3
DEC A	Decrement accumulator	1
DEC Rn	Decrement register	2
DEC Direct	Decrment direct byte	3
DEC @Ri	Decrement indirect RAM	3
INC DPTR	Inrement data pointer	1
MUL AB	Multiply accumulator by B	2
DIV AB	Divide accumulator by B	6
DA A	Decimal adjust accumulator	3

provide an eight-bit address multiplexed with data on P0.[16] Eight bits are sufficient for external 1/0 expansion decoding or for a relatively small RAM array. For somewhat larger arrays, any output port pins can be used to output higher-order address bits. These pins would be controlled by an output instruction preceding the MOVX. In the second type of MOVX instruction, the Data Pointer generates a sixteen-bit address. P2 outputs the high-order eight address bits (the contents of DPH)

---

[16]P0,P1, P2 and P3 are the SFR latches on ports 0, 1, 2 and 3 respectively.

while P0 multiplexes the low order eight bits (DPL) with data. The P2 Special Function Register retains its previous contents while the P2 output buffers are emitting the contents of DPH. This form is faster and more efficient when accessing very large data arrays (up to 64K bytes), because no additional instructions are needed to set up the output ports. It is possible in some situations to mix the two MOVX types. A large R4M array with its high order address lines driven by P2 can be addressed via the Data Pointer, or with code.

Table 2.6 Data transfer instructions.

Mnemonic	Description	Cycles
MOV A,Rn	Move register to accumulator	1
MOV A,Direct	Move direct byte to accumulator	2
MOV A,@Ri	Move indirect RAM to accumulator	2
MOV A,#data	Move Immediate data to accumulator	2
MOV Rn,A	Move accumulator to register	1
MOV Rn,Direct	Move direct byte to register	3
MOV Rn,#data	Move immediate data to register	2
MOV Direct,A	Move accumulator to direct byte	2
MOV Direct,Rn	Move register to direct byte	2
MOV Direct,Direct	Move direct byte to direct byte	3
MOV Direct@Ri	Move indirect RAM to direct byte	3
MOV Direct,#data	Move immediate data to direct byte	3
MOV @Ri,A	Move accumulator to indirect RAM	2
MOV @Ri,Direct	Move direct byte to indirect RAM	3
MOV @Ri,#data	Move immediate data to indirect RAM	2
MOV DPTR,#data16	Load data pointer with 16-bit constant	3
MOVC A,@A+DPTR	Move code byte relative to DPTR to accumulator	5
MOVC A,@A+PC	Move code byte relative to PC to accumulator	4
MOVX A,@Ri	Move external RAM (8-bit) to accumulator	3
MOVX A,DPTR	Move external RAM (16-bit) to accumulator	2
MOVX @Ri,A	Move accumulator to external RAM (8-bit)	4
MOVX DPTR,A	Move accumulator to external RAM (16-bit)	3
PUSH Direct	Push direct byte onto stack	3
POP Direct	Pop direct byte from stack	2
XCH A,Rn	Exchange direct byte with accumulator	2
XCH A,Direct	Exchange direct with accumulator	3
XCH A<@Rn	Exchange indirect RAM with accumulator	3
XCH A,@Ri	Exchange low order indirect digit RAM with accumulator	3

- **MUL AB**—This instruction multiplies the unsigned eight-bit integers in the accumulator and in register B. The low-order byte of the sixteen-bit product is left in the accumulator, and the high-order byte in B. If the product is greater than 255 (OPPH)the overflow flag is set; otherwise, it is cleared. The carry flag is always cleared.
- **NOP**—Execution continues at the following instruction. Other than the PC, no registers or flags are affected.
- **ORL<dest-byte><src-byte>**—performs the bitwise logical-OR operation between the indicated variables, storing the results in the destination byte. No flags are affected. The two operands allow six addressing mode combinations. When the destination is the accumulator, the source can use register, direct, register-indirect, or immediate addressing; when the destination is a direct address, the source can be the accumulator, or immediate data. When this instruction is used to modify an output port, the value used as the original port data will be read from the output data latch, not the input pins.
- **POP direct**—causes the contents of the internal RAM location addressed by the Stack Pointer to be read ("POPed"), and the Stack Pointer is decremented by one. The value read is then transferred to the directly addressed byte indicated. No flags are affected (Table 2.7).
- **PUSH direct**—increments the Stack Pointer by one. The contents of the indicated variable are then copied ("PUSHed") into the internal RAM location addressed by the Stack Pointer. The flags are not affected.
- **RET**—return from a subroutine by "POP"ing the return address from the stack and continue execution from that location. RET pops the high-and low-order bytes of the PC successively from the stack decrementing the Stack Pointer by two. Program execution continues at the resulting address, generally the instruction immediately following an ACALL or LCALL. No flags are affected.
- **RETI**—return from an interrupt service routine by "POP"ing the return address from the stack, restoring the interrupt logic to accept interrupts [3] at the same level of the interrupt as the one just processed and continue execution from the address retrieved from the stack. (Note that the PSW is not automatically restored.) RETI pops the high-and low-order bytes of the PC successively from the stack, and restores the interrupt logic to accept additional interrupts at the same priority level as the one just processed. The Stack Pointer is left decremented by two. No other registers are affected. *Special Note: PSW is not automatically restored to its pre-interrupt status.* Program execution continues at the resulting address, which is generally the instruction immediately after the point at which the interrupt request was detected. If a lower- or same-level interrupt had been pending when the RETI instruction is executed, that instruction will be executed before the pending interrupt is processed.
- **RL A**—rotates the contents of the accumulator A, one bit position to the left. The eight bits in the accumulator are rotated one bit to the left. Bit 7 is rotated into the bit 0 position. No flags are affected.
- **RLC A**—rotates the contents of the accumulator one bit position to the left through the Carry flag. The eight bits in the accumulator are rotated one bit to the left. Bit 7 is rotated into the bit 0 position. No flags are affected.
- **RR A**—rotate the contents of the accumulator one bit position to the right. The eight bits in the Accumulator are rotated one bit to the right. Bit 0 is rotated into the bit 7 position. No flags are affected.
- **RRC A**—rotate the contents of the accumulator one bit position to the right through the Carry flag. The eight bits in the accumulator and the carry flag are rotated together, one bit to the right. Bit 0 moves into the carry flag; the original value of the Carry flag moves into the bit 7 position. No other flags are affected.

Table 2.7 Logical instructions.

Mnemonics	Description	Cycles
ANL A,Rn	AND register to accumulator	1
ANL A,Direct	AND direct byte to accumulator	2
ANL A@Ri	AND indirect RAM to accumulator	2
ANL A,#data	AND immediate data to accumulator	2
ANL Direct,A	AND accumulator to direct byte	3
ANL Direct,#data	AND immediate data to direct byte	3
ORL A,Rn	OR regsiter to accumulator	1
ORL A,Direct	OR direct byte to accumulator	2
ORL A,@Ri	OR indirect RAM to accumulator	2
ORL A,#data	OR immediate data to accumulator	2
ORL Direct,A	OR accumulator to direct byte	3
ORL Direct,#data	OR immediate data to direct byte	3
XRL A,Rn	XOR regsiter to accumulator	1
XRL A,Direct	XOR direct byte to accumulator	2
XRL A@Ri	XOR indirect RAM to accumulator	2
XRL A,#data	XOR immediate data to accumulator	2
XRL Direct,A	XOR accumulator to direct byte	3
XRL Direct,#data	XOR immediate data to direct byte	3
CLR A	Clear the accumulatorComplet	1
CPL A	Complement the accumulator	1
RL A	Rotate accumulator left	1
RLC A	Rotate accumulator left through carry	1
RR A	Rotate accumulator right	1
RRC A	Rotate accumulator right through carry	1
SWAP A	Swap nibbles within accumulator	1

- **SETB**—sets the indicated bit to one. SETB can operate on the Carry flag or any directly addressable bit. No other flags are affected.
- **SJMP rel**—causes a Short Jump using an 8-bit signed offset relative to the first byte of the next instruction. This instruction causes the program to make an unconditional control branch to the address indicated. The branch destination is computed by adding the signed displacement to the PC, after incrementing the PC twice. Therefore, the range of destinations allowed is from 128 bytes preceding this instruction to the 127 bytes following it.

- **SUBB A,<src-byte>**—Subtract with borrow results in the subtraction of an operand and the previous value of the CY flag. (A <= A–<operand>- CY). This instruction subtracts the indicated variable and the Carry flag from the accumulator, leaving the result in the accumulator. SUBB sets the Carry (borrow) flag if a borrow is needed for bit 7, and clears C, otherwise. (If C was set before executing a **SUBB** instruction, this indicates that a borrow was needed for the previous step in a multiple precision subtraction, so the carry is subtracted from the accumulator along with the source operand.) A C is set if a borrow is needed for bit 3, and cleared otherwise. OV is set if a borrow is needed into bit 6, but not into bit 7, or into bit 7, but not bit 6. When subtracting signed integers OV indicates a negative number produced when a negative value is subtracted from a positive value, or a positive result when a positive number is subtracted from a negative number. The source operand allows four addressing modes: register, direct, register-indirect or immediate. <src-byte>can be R_n(a register), Direct (a direct byte), @ R_i(indirect RAM), or #$data$(immediate data).
- **SWAP A**—interchanges (SWAPs) the low- and high-order nibbles (four-bit fields) of the accumulator (bits 3-0 and bits 7-4). The operation is equivalent to a four-bit rotate instruction. No flags are affected.
- **XCH A,<byte>**—loads the accumulator with the value of the byte variable and loads the accumulator contents to the byte variable. The src/dest operand can use register, direct, or register-indirect addressing.
- **XCHD A,@Ri**—exchanges the low-order nibble of the accumulator (bits 3-0), generally representing a hexadecimal or BCD digit, with that of the internal RAM location indirectly addressed by the specified register. The high-order nibbles (bits 7-4) of each register are not affected.
- **XRL <dest-byte><src-byte>**—This instruction performs a bitwise, logical, Exclusive-OR operation between <dest-byte>and <src-byte>, storing the results in <dest>. No flags are affected. The two operands allow six addressing mode combinations. When the destination is the accumulator, the source can use register, direct, register-indirect, or immediate addressing; when the <dest>is a direct address, <src> can be the accumulator or immediate data. (Note When this instruction is used to modify an output port, the value used as the original port data will be read from the output data latch, not the input pins.)
- **XCHD**—exchanges low-order indirect digit RAM with the accumulator.
- **Undefined**—OpCode 0xA5 is an undocumented function.

2.2.3   ARM Cortext M3 (PSoC 5LP)

The ARM CORTEX M3 utilizes a three-stage, pipe-lined, Harvard-based bus architecture to provide a single cycle, hardware-based multiply/divide capability. It also supports the Thumb-2 [2] instruction set.[17]

---

[17]The Thumb instruction set is a subset of the 32-bit instruction set that has been compressed from 32 bits to 16 bits resulting in a reduction in code density of approximately 30%. Because of this reduction it is possible to maintain more instructions in the on-chip memory which further reduces power consumption because of-chip fetches tend to consume more power than on-chip fetches.

Table 2.8 Boolean instructions.

Mnemonic	Description
CLR C	Clear carry
CLR bit	Clear direct bit
SETB C	Set Carry
SETB bit	Set direct bit
CPL C	Complemenmt carry
CPL bit	Complement direct bit
ANL c,bit	AND direct bit to carry
ANL C,/bit	AND indirect bit to carry
ORL C,bit	OR direct but to carry
ORL C,/bit	OR complement of direct
MOV C,bit	Move direct bit to carry
MOV bit,C	Move carry to direct bit
JC rel	Jump if carry is set
JNC rel	Jump if no carry is set
JB bit,rel	Jump if direct bit is set
JNB bit,rel	Jump if direct bit is not set
JBC bit,rel	Jump if direct bit is set and clear bit

2.2.4 Instruction Set [13]

Thumb instructions include arithmetic operations, logical operators, conditional/unconditional branches and store/load data operations [14]. I/O and exception handling typically require the use of 32-bit ARM instructions [10]. It should be noted, however, that there are inherent limitations of the Thumb instruction set. While the use of Thumb instructions, guarantee efficient code execution and power consumption when employing Thumb operators is not guaranteed [9]. The ARM CORTEX M3 provides hardware support that greatly facilitates debugging by providing trace, profiling, breakpoints, watchpoints and code patching. The Advanced RISC Machines (ARMs), have a 32-bit architecture with sixteen registers, one of which is the program counter (PC). Most of the instructions have a 4-bit condition code to facilitate branching (Table 2.8).

2.2.4.1 RISC Versus CISC Systems

Some modern microprocessor architecture can be described as a Complex Instruction Set Computer (CISCs) and are capable of carrying out arbitrarily complex instructions. Another class of microprocessors are referred to as RISC machines, or Reduced Instruction Set Computer(s). RISC instructions are usually executed in a single clock instruction and can result in substantially improved execution times. Such improvement is not without cost because more RISC instructions may be required than CISC instructions to execute a given program, which in turn means increased memory requirements.

Advanced RISC Machines (ARMs) utilize a set of instructions referred to as Thumb Instructions which consist of 16-bit instructions that are "extensions" of the 32-bit ARM instructions. Thumb instructions are fetched as 16-bit instructions[18] and then expanded utilizing dedicated hardware within the microprocessor to 32-bits. Thus, Thumb instructions and their 32-bit counterparts are functionally equivalent.[19] ARM instructions are aligned on a four-byte boundary and Thumb instructions are aligned on a two-byte boundary. Any Thumb instruction that involves data processing operates on 32-bit values. Instruction fetches and data access instructions create 32-bit addresses. In the Thumb state, eight general-purpose, integer registers (R0–R7) are employed.

Thumb instructions fall into the following categories:

- Arithmetic
- Branch
- Extend
- Load
- Logical
- Move
- Process or State Change
- Push and Pop
- Reverse
- Shift and Rotate
- Store

Thumb instructions are a 16-bit subset of ARM 32-bit instructions which can be conditionally executed, while Thumb instructions are always executed. The Thumb-2 instruction set consists of a mixture of 16- and 32-bit instructions.

2.2.5   Interrupts and Interrupt Handling

As discussed in Chap. 1, interrupts and interrupt handling are extremely important aspects of many embedded system applications. Proper treatment of interrupts allows the most efficient response of such systems and ensures that requests are handled in the appropriate order with minimal latency. An interrupt controller provides hardware resources that allow the system to suspend tasks prior to their completion (Fig. 2.11).

The interrupt controller employed in PSoC 3 has a number of enhanced features not available in the original 8051, e.g.,

- eight levels of priority (0–7)
- configurable Interrupt Vector Address
- multiple I/O vectors,
- flexible interrupt sources,
- programmatic interrupts,
- programmatic clearing of interrupts,
- 32 interrupt vectors, and
- dynamic assignment of one of eight priorities,

---

[18]Thus conserving memory space.

[19]It should be noted when handling exceptions, the processor must be in the "ARM state", i.e., exceptions cannot be handled in the Thumb state and therefore cannot be handled by Thumb instructions.

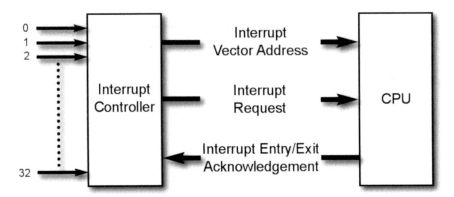

Fig. 2.11 Interrupt controller block diagram.

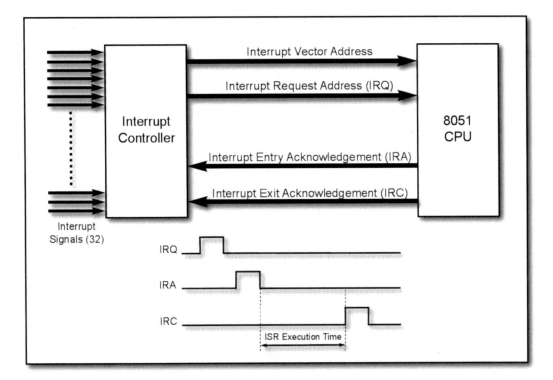

Fig. 2.12 PSoC 3 interrupt controller.

As shown in Fig. 2.12, PSOC 3's integral interrupt controller supports up to 32 interrupt signals (cf. Table 2.9), inclusive, which when active are processed by the interrupt controller. Each of these inputs can be enabled/disabled programmatically and a dedicated interrupt vector table,[20] stores the addresses of the respective interrupt service routines (ISRs). Under program control, the priority assigned to an input signal can be changed as well as the vector address.

When an interrupt occurs, the interrupt controller processes them and assigns them a priority based on the preassigned interrupt priority for each interrupt signal [8]. When an interrupt occurs

---

[20]Such tables are sometimes referred to as a "jump" tables.

all information required to reinstate the interrupted task must be stored on the stack as discussed in Chap. 1. If the program has been written in C, then the C compiler will automatically introduce the necessary code to store the required information on the stack, However, if the program has been written in assembly language the necessary push and pop instructions must be manually included in the assembly source file so that prior to entry into the ISR and following a return and prior to attempting to resume the interrupted task the required information is pushed/popped to/from the stack.

Table 2.9 Interrupt vector table (PSoC 3).

#	Fixed Function	DMA	UDVB
0	LVD	phub_termout0[0]	udb_intr[0]
1	ECC	phub_termout0[1]	udb_intr[1]
2	Reserved	phub_termout0[2]	udb_intr[2]
3	Sleep (Pwr Mgr)	phub_termout0[3]	udb_intr[3]
4	PICU[0]	phub_termout0[4]	udb_intr[4]
5	PICU[1]	phub_termout0[5]	udb_intr[5]
6	PICU[2]	phub_termout0[6]	udb_intr[6]
7	PICU[3]	phub_termout0[7]	udb_intr[7]
8	PICU[4]	phub_termout0[8]	udb_intr[8]
9	PICU[5]	phub_termout0[9]	udb_intr[9]
10	PICU[6]	phub_termout0[10]	udb_intr[10]
11	PICU[12]	phub_termout[11]	udb_intr[11]
12	PICU[15]	phub_termout0[12]	udb_intr[12]
13	Comparator Int	phub_termout0[13]	udb_intr[13]
14	Switched Cap Int	phub_termout0[14]	udb_intr[14]
15	I²C	phub_termout0[15]	udb_intr[15]
16	CAN	phub_termout0[0]	udb_intr[16]
17	Timer/Counter0	phub_termout0[1]	udb_intr[17]
18	Timer/Counter1	phub_termout0[2]	udb_intr[18]
19	Timer/Counter2	phub_termout0[3]	udb_intr[19]
20	Timer/Counter3	phub_termout0[4]	udb_intr[20]
21	USB SOF Int	phub_termout0[5]	udb_intr[21]
22	USB ARB Int	phub_termout0[6]	udb_intr[22]
23	USB Bus Int	phub_termout0[7]	udb_intr[23]
24	USB Endpoint [0]	phub_termout0[8]	udb_intr[24]
25	USB Endpoint Data	phub_termout0[9]	udb_intr[25]
26	Reserved	phub_termout0[10]	udb_intr[26]
27	Reserved	phub_termout0[11]	udb_intr[27]
28	DFB Int	phub_termout012]	udb_intr[28]
29	Decimator Int	phub_termout0[13]	udb_intr[29]
30	PHUB Error Int	phub_termout0[14]	udb_intr[30]
31	EEPROM Fault Int	phub_termout0[15]	udb_intr[31]

Two hardware stacks are maintained by the interrupt controller, one for storage of the interrupt priorities and the other for the related vector addresses. When an interrupt acknowledgment entry (IRA) is received from the CPU, the interrupt controller pushes the current interrupt vector address and priority level to their respective stacks. When an acknowledgment for an interrupt exit (IRC) is received, the interrupt controller pops the previous state information from the stack.

Interrupts can be nested so that a higher priority interrupt can "interrupt" a lower priority interrupt. Interrupts that occur while PSoC 3's microcontroller is shutdown, e.g., while asleep, should be of the type referred to as "sticky" interrupts,[21] i.e., interrupts asserted while PSoC 3 is inactive must be held until PSoC 3 "wakes up".

If an interrupt request occurs which is of higher priority than that assigned to the currently executing task, the current task is suspended, and the higher priority task is invoked. Once completed, the lower priority task resumes. Priorities are assigned numbers in the range of 0–31, with zero being the highest priority and 31 the lowest. If two tasks have been assigned the same priority and their respective interrupt requests occur simultaneously, then the task with the lower vector number has priority.

2.2.5.1 Interrupt Lines

As discussed previously, the interrupt controller has 32 interrupt input lines, numbered as shown in Fig. 2.11, and possible input sources asserted on these lines are defined as:

1. **Fixed Function**—these are asserted by peripherals such as I2C, Sleep, CAN, Port Interrupt Controller Unit (PICU) and the Low Voltage Detector (LVD).
2. **DMA Controller Interrupts**—these signal the completion of a DMA transfer.
3. **UDB**—interrupts initiated by various Universal Device Blocks implemented as timers, counters, etc. (Fig. 2.13)

However, each interrupt line is assigned one of these three types of the interrupt sources as shown in Fig. 2.14, and the designation for each line is determined by the IDMUX control register, IDMUX.IRQ_CTL,

Interrupt lines pass through a multiplexer on their way to the interrupt controller, as shown in Fig. 2.13, which selects one of the following: a Fixed Function interrupt request (IRQ), a Universal Digital Block (UDB) IRQ with level, on a UDB IRQ with Edge or a DMA IRQ. The IDMUX,IRQ_CTL register determines the mux path with respect to IRQ selection.

The interrupt controller supports two types of the interrupt assertions on the lines:

1. **Level Shift**—an interrupt request is initiated by a shift of the level of the interrupt line.
2. **Pulse**—a pulse on a pulse-designated interrupt line creates an IRQ when the low-to-high edge transition occurs, which causes the pending bit for that interrupt line to be set. In the event that a second pulse occurs while the first is still pending, it has no effect. When the CPU acknowledges receipt of the IRQ by transmitting an IRA, the pending bit for that line is reset. If another pulse now occurs, the pending bit is set again, even if the first ISR is still active.

2.2.5.2 Enabling/Disabling Interrupts

The interrupt controller's Enable register (SETEN) and Clear ENABLE registers allow interrupt lines to be enabled and disabled, respectively. Writing a 1 to the SETEN register enables an interrupt, while

---

[21]This is to avoid the possibility of an interrupt being asserted and then cleared while the microprocessor is "sleeping" thereby resulting in a missed interrupt request.

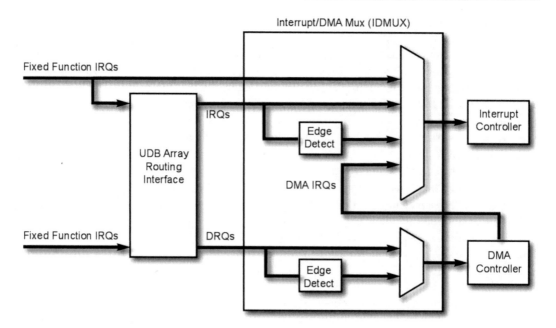

Fig. 2.13 Interrupt processing in the IDMUX.

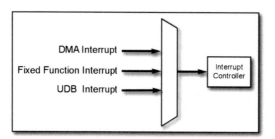

Fig. 2.14 Interrupt signal sources.

a zero has not effect. Reading a one from the SETEN register implies interrupt is enable and a zero implies the interrupt is disabled. Similarly, writing a 1 to CLREN register disables an interrupt and writing a zero has no effect. Reading a one from the CLREN register implies that an interrupt is enabled and reading a zero implies that it is disabled.

2.2.5.3 Pending Interrupts

The "Pending" bit is set when the interrupt controller receives an interrupt signal. This can also be set/cleared programmatically by using the "Set Pending" register (SETPEND) and the "Clear Pending" register (CLRPEND), respectively. Each of the bits in these registers represents the status of one interrupt line. Interrupt requests can be made by either asserting a level shift or a pulse on an interrupt line. In either case, following an IRQ, the pending bit is cleared immediately once an interrupt acknowledgment has been received from the CPU. Should a new pulse be received on the same line after receipt of the CPU's acknowledgment, the pending bit is set. However, when a line-level shift occurs, the interrupt controller checks the status of the line after it receives an acknowledgment that the CPU has exited the interrupt service routine (ISR).

2.2.5.4 Interrupt Priority

The proper handling of priorities is obviously very important and there are two possibilities to be considered in such cases:

1. **An interrupt occurs while an interrupt service routine (ISR) for the previous interrupt is being executed.** If the interrupt is of higher priority, then the ISR is suspended, the information required to reinstate that ISR is placed on the stack and the ISR associated with the higher priory interrupt is invoked. If the priority of the interrupt is lower than that of the prior interrupt, then the ISR continues execution until completed, at which point, if no new interrupt requests have been received, the ISR for the most recent requested is invoked. If the most recent interrupt is of the same priority as that of the currently executing ISR, then the ISR continues execution and, upon completion, the new interrupt ISR is invoked.
2. **Two interrupts occur contemporaneously.** If they are of the same priority, then the interrupt with the lower index number[22] is serviced first. Otherwise, the interrupt with the higher priority is serviced first.

When an interrupt signal occurs, i.e., an IRQ, the pending bit for that line is set, in the pending register, indicating that an IRQ has occurred, and is waiting. The priority for this IRQ is read and a determination is made as to when this request should be serviced. The request and the associated vector address, are then sent to the CPU. At this point, the CPU acknowledges the request and returns an Interrupt Entry Acknowledgment signal (IRA). When the ISR is completed the CPU sends an Interrupt Exit Acknowledgment (IRC) (Tables 2.10 and 2.11).

2.2.5.5 Interrupt Vector Addresses

PSoC 3 allows the ISR starting addresses to be explicitly specified, i.e., the addresses are programmable. Thus calling an ISR does not involve a branch instruction and therefore the ability to make direct calls to an ISR reduces latency. The programmable ISR addresses are stored in the 16-bit Vector Address registers, VECT[0...31].[23] When writing to these registers, the LSB must be written first and followed by the MSB.[24] When an IRQ occurs, the respective address is passed to the CPU for execution of the appropriate ISR.

2.2.5.6 Sleep Mode Behavior

It should be noted that in Sleep Mode, all the status and configuration registers associated with interrupts retain their values. However, the Pending and Interrupt Controller stack registers are set with the "power-on" value at wake-up.

2.2.5.7 Port Interrupt Control Unit

PSoC 3 has a Port Interrupt Control Unit (PICU) which interfaces to the GPIO pins and provides a way to process externally generated interrupts that:

- support 8 pins,
- handle rising/falling/both-edge interrupts,
- interfaces to the PHUB over AHB for reading/writing to its registers,
- does not support Level sensitive interrupts,

---

[22]The interrupt controller's 32 input lines are numbered from 0–31 and referred as "index numbers".

[23]There are 32 of these registers, i.e., one for each of the input lines.

[24]The acronym for most significant byte is MSB.

Table 2.10 Bit status during read and write.

Register	Operation	Bit Value	Comment
SETEN	Write	1	To enable the interrupt
		0	No effect
	Read	1	Interupt is enabled
		0	Interrupt is disabled
CLREN	Write	1	To disable the interrupt
		0	No effect
	Read	1	Interrupt is enabled
		0	Interrupt is disabled

- allows pin interrupts to be individually disabled,
- transmits single interrupt request (PIRQ) to the interrupt controller,

and,

- has pin status bits to allow easy determination of the source of the interrupt, at the pin level.

Thus, each pin of a port can be independently configured by the interrupt "type" register controlling each pin, to detect rising edge/falling edge/both edges interrupts. Based on the mode configured for each pin, when an interrupt occurs, the corresponding status register bit, i.e., the pin's status bit, will be set to "1" and an interrupt request sent to the interrupt controller. Each of the PICU's has a "wakeup_in" input and "wakeup_out" output signal. As shown in Fig. 2.15, all the PICUs are "daisy-changed" together so that a final wake-up signal goes to the power manager.

2.2.5.8 Interrupt Nesting

PSoC 3 supports up to eight levels of "nested" interrupts. Nesting occurs whenever a lower level interrupt service routine is suspended as a result of the receipt of a higher level interrupt. Interrupt nesting involves both the CPU stack and the interrupt controller stack(s) which store the interrupt number and priority. Two upward-growing stacks with a depth of eight levels are maintained by the interrupt controller, viz., STK which stores the interrupt priority and STP_INT_NUM which stores the interrupt number.

Table 2.11 Pending bit status table.

Register	Operation	Bit Value	Comment
SETPEND	Write	1	To put an interrupt to pending
		0	No effect
	Read	1	Interupt is pending
		0	Interrupt is not pending
CLRPEND	Write	1	To clear a pending interrupt
		0	No effect
	Read	1	Interrupt is pending
		0	Interrupt is not pending

The CPU stack is used to store the contents of various registers, e.g., the ACC, B, GPR, PC, PSW and SFR. While the CPU automatically handles pushing and popping of the PC register to/from the stack, the ISR must store any other required register contents.

2.2.5.9 Interrupts Masking and Exception Handling

Exceptions are predefined interrupts designed to handle various, typically serious, fault conditions that can occur such as bus fault, memory management fault, program error, etc. PSoC 5LP provides support for 15 different types of exceptions and Non-Maskable Interrupts (NMIs). NMIs are not programmable in the general sense, but rather predefined ISRs designed to handle serious system faults.

Masking is a technique for blocking an interrupt, or group of interrupts, and includes:

- BASEPRI—Specifying a specific priority level in the BASEPRI register is masked, i.e., blocked.
- FAULTMASK—setting a bit in the FAULTMASK register blocks all interrupts except for NMI.
- PRIMASK—setting a bit in the PRIMASK register blocks all interrupts except Hard Fault (3)and NMI (2).

Exception handling and NMIs are not explicitly supported in PSOC 3.

2.2.5.10 Interrupt "Best Practices"

It's important to exercise care when dealing with interrupts to avoid among other things introducing unnecessary latencies that degrade the embedded system's [4] responsiveness. The following represent

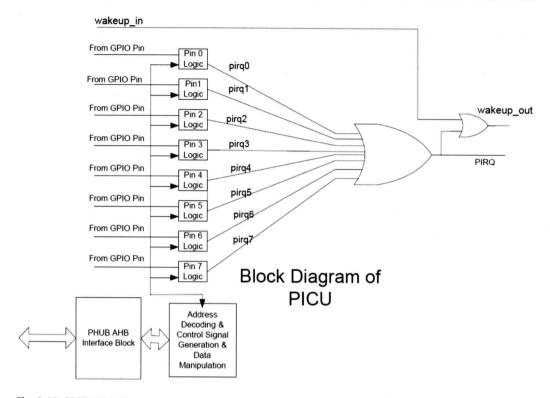

Fig. 2.15 PICU block diagram.

some suggested guidelines that are often overlooked in developing program code that involves interrupts [8]:

1. If a function call occurs in more than one location, e.g., main code and interrupt code, then it should be declared as reentrant.
2. Calling functions from an ISR should be avoided in order to minimize pop/push operations.
3. The status register should be read from within an ISR if the interrupt signal is a level shift.
4. ISR's should involve as little code as possible. (Setting a flag bit in the ISR and then checking its status from main code can significantly reduce latency.)
5. In order for the 8051 (PSoC 3) to service interrupts, the global interrupts enable bit (EA) in the Interrupt-Enable (bit 7 of IE) special function register (SFR 0xA8) must be set.
6. Before enabling an interrupt, the pending bit should be set to avoid unexpected ISR calls. The interrupt should also be disabled before dynamically changing the vector address and priority in software. After the configuration has been completed, the interrupt should then be enabled.

2.2.6 Memory

Up to 8K of Static RAM (SRAM) is employed in PSoC 3 for temporary (volatile) data storage that can be accessed by both the DMA controller and by the 8051. Simultaneous access is also supported provided that there is no attempt to access the same 4K block. Flash is also used as nonvolatile memory for firmware, user configuration data, bulk data storage and optional Error Collecting Codes (ECC) data. Flash memory can be up to 64 kbytes for storage of user program space. An additional 8K of Flash memory space is available for ECCs, but if ECCs are not used then this space is available to store device configuration and bulk user data. However, user code cannot be run from within the ECC memory space.

Current ECC technology is, in general, quite effective at correcting single-bit errors which are the most common form of error. The ECCs used in PSoC 3 are capable of detecting 2-bit errors in every 8 bytes of firmware memory. If an error is detected an interrupt can be generated to allow appropriate action to be taken. Flash output is 9 bytes wide with one byte reserved for ECC data.

2.2.6.1 Memory Security

Maintaining the security of proprietary code is often a major concern and embedded controller such as PSoC 3 and PSoC 5LP have mechanisms for preventing access to, and visibility of, such code and to prevent reverse-engineering, or duplication, of the intellectual property. Flash memory is organized as blocks of 256 bytes of program code, or data, and 32 bytes of ECC, or configuration data. Therefore, up to 256 blocks are provided for 64-kbyte of Flash. There are four levels of protection that can be assigned to each row of Flash as shown in Table 2.12.

Table 2.12 Flash protection modes.

Mode	Description	Read	External Write	Internal Write
00	Unprotect	Yes	Yes	Yes
01	Read Protect	No	Yes	Yes
10	Disable External Write	No	No	Yes
11	Disable Internal Write	No	No	No

Changing these levels of protection can only be accomplished by first imposing a complete Flash erase. The Full Protection mode allows internal reads to occur, but precludes external reads/writes and internal writes, which among other things prevents loading of code to download the internal code. In addition, a fifth option is available called "Device Security" that permanently disables all test, programming and debug ports as a further security measure. While there may be no completely effective method for protecting code, the methods described here do represent the current state-of-the-art.

2.2.6.2 EEPROM

Byte addressable nonvolatile memory, consisting of 128 rows of 16 bytes each, is provided in the form of EEPROM that can be erased and written to at the row level and up to 2 KB of user data can be stored in EEPROM. At the byte level, random access reads can be carried out directly and writes are accomplished by sending write commands to an EEPROM program interface. It is not necessary to suspend CPU activity while EEPROM writes are occurring. However, the CPU cannot execute EEPROM code directly and there is no ECC hardware to secure EEPROM code integrity. In applications requiring ECC protection for EEPROM code, it is necessary to do so at the firmware level.

2.2.6.3 Interfacing External Memory

Many embedded system applications employing microcontrollers rely on external memory to meet a variety of data storage requirements. This usually involves the use of some form of External Memory

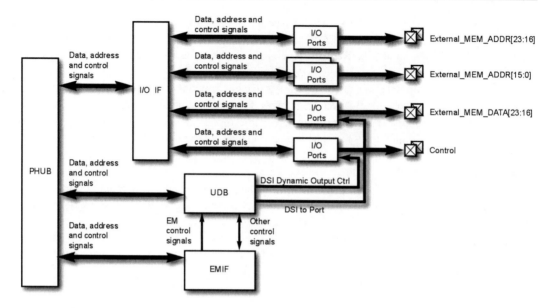

Fig. 2.16 EMIF block Diagram.emi.

Interface (EMIF). This interface makes it possible for the CPU to read from, and write to, external memories. PSoC 3's EMIF functions in conjunction with I/O ports, UDBs and other hardware to provide the necessary external memory control and address signals. Eight or sixteen bit memory can be accessed in a memory space addressable by as much as 24 bits, i.e., 16 MB of memory.

The EMIF is compatible with four types of external memory:

1. Asynchronous SRAM
2. Synchronous SRAM
3. cellular RAM/PSRAM

and,

4. NOR Flash

The EMIF provides external memory control signals for synchronous memory, but not for the other forms of memory. Both 8- and 16-bit external memory can be accessed via either the XDATA memory space (PSoC 3), or the ARM Cortex-M3 external RAM space (PSoC 5LP). EMIF addresses can be one byte, two bytes or three bytes utilizing one, two or three of the ports shown in Fig. 2.16. These ports are selected by configuring the 3-bit *portEmifCfg* field in the PRT*_CTL register which allows the least significant and middle byte or most significant byte of a three-byte address to be assigned to a given port. However, the data transferred via the EMIF is restricted to either two either a single port or two ports. A particular data port can be selected as the path for either the most or least significant byte of the data. The fourth port is used to provide control utilizing 3–6 pins on the fourth I/O port. While unused pins on this port are available for other use, depending on the application, any unused address pins are not to be used for any other purpose. When the system is in Sleep Mode, all the EMIF registers retain their respective configurations.

EMIF clocking is derived from the bus clock which also serves as the clock for the PHUB and the CPU. This signal can be provided, as EM_CLOCK, to eternal memory at frequencies either equal to, one half or one-quarter of the bus clock, i.e., the bus clock divided by a factor of 1, 2 or 4. However, the maximum allowable I/O rate for PSoC 3 and PSoC LP5 GPIO pins is 33 MHz. In addition, the maximum bus clock frequencies are 67 and 80 MHz for PSoC 3 and PSoC 5LP, respectively Therefore, in most cases EM-CLOCK will only be available for external memory at frequencies lower than the bus clock frequency.

2.2.7 Direct Memory Access (DMA)

As discussed in Chap. 1, minimizing latency is often a primary concern in the design of an embedded system because of the need for it to respond within certain time constraints to anticipated events and/or conditions. Furthermore, when dealing with data from, for example, a number of sensors, the rate of data collection may be much faster than the CPU's ability to process it, under some circumstances. Therefore, the ability to move data to/from an embedded system without incurring significant CPU overhead can an important concern when trying to minimize latency.

PSoC 3 and PSoC 5LP have integral Direct Memory Access controllers that are capable of:

- memory-to-memory transfers
- memory-to-peripheral transfers
- peripheral-to-memory transfers
- peripheral-to-peripheral transfers
- supporting up to 24 independent DMA channels,
- handling data transfers that can be initiated, stalled or terminated,
- allowing multiple DMA channels, or transaction descriptors, to be chained, or nested, to perform complex tasks,
- assigning one, or more Transcription Descriptors[25] to each DMA channel for complex operations,
- allowing large data transfers to be split into multiple packets, varying in size from 1 to 127 bytes, which can be transferred in "bursts",
- supporting the triggering of DMA transfers by externally routed, digital signals via GPIO, by another DMA channel, or by the CPU,
- assigning one if eight priority levels (0–7) to each DMA channel.
- supporting up to 128 Transcription Descriptors, inclusive,

and,

- generating an interrupt (nrq) when a data transaction (DMA transfer) has been completed.

One technique for handling such requirements is to employ a combination of a DMA controller and a bus over which peripheral access occurs and bulk data transfers, referred to as DMA transfers, take place and an associated controller. The PHUB is a combination of high-speed bus, arbiter, router, and DMA controller with radiating "spokes" each of which connects to a peripheral, as shown in Fig. 2.17. The bus ports support 16, 24 and 32-bit addressing modes. Since both the CPU and DMA controller

---

[25]Transaction descriptors contain information regarding the transfer of data, e.g., source address, destination address, and number of bytes to be transferred and enable *Termout* signals after the transfer has been completed.

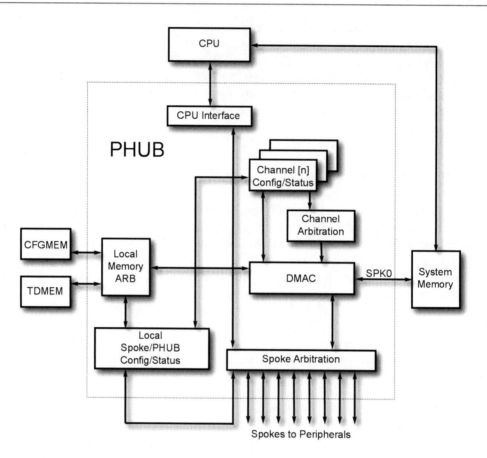

Fig. 2.17   Block diagram of PHUB.

(DMAC) can initiate block transfers, the arbiter determines how such transfers are to be handled.[26] Spoke numbers, the number of peripherals supported by each, the spoke address widths and spoke data width are tabulated in Table 2.13.

Both the CPU and DMAC can act as masters and can initiate transactions on the bus. In the event that multiple requests occur, the arbiter in the central hub determines which DMA channel has the highest priority. If a higher priority transaction request occurs while a lower priority transfer is occurring, the lower priority transfer can be interrupted. The primary configurations of the PHUB are the number of DMA channels and spokes. The PHUB's architecture allows the CPU and DMAC to simultaneously access peripherals located on different spokes.

The PHUB employs two local memories referred to as CFGMEM and TDMEM, respectively. CFGMEM serves as the channel configuration memory which stores information for each record defined as CH[n]_CONF0/1 with one 8 byte set of each per channel with the result that CFGMEM is sized as 8 bytes x the number of DMA channels. CFGMEM is configured as an x64 memory to allow all 8 bytes of a CHn_CONFIG0/1 set to be accessed on a single cycle to maximize DMA processing efficiency. TDMEM stores the transcription descriptor chains for a given channel and contains the

---

[26]The PSoC 5 LP DMA controller cannot directly access SRAM memory locations from 0x1FFF8000 to 0x1FFFFFFF. However, it can access memory in the range from 0x20008000 to 0x2000FFFF. Therefore, it is necessary to add 64k to the starting address to allow the DMA controller to access locations 0x1FFF8000 to 0xFFFFFFFF.

Table 2.13 Spoke parameters.

Spoke Number	Number of Peripherals	Spoke Address Width (Bits)	Spoke Data Width (Bits)
00	1	14	32
01	4	9	16
02	15	19	32
03	3	11	16
04	4	10	16
05	1	11	32
06	6	17	16
07	5	17	16
08-15	0	NA	NA

DMAC instructions required for a DMA transfer via the channel spokes. Such chains are considered as TDs, 8 bytes wide with a maximum of 128 TDs in TDMEM. The allocation of TD chains is based on a given sequence that the DMA channel requires and in a maximal configuration. TDMEM can be a maximum of 8 bytes x 128 TDs, or 1 KB. If multiple bursts are required, the DMAC must keep track of where it was when it completed the last burst, while interleaving other channels' bus access. The intermediate TD states can stored either on top of CH(n)_ORIG_TD0/0 of the TD chain, or in CH[N]_SEP_TD0/1 to allow the chain to be preserved.

The DMA controller has five semi-independent functions that operate, in parallel, in a pipelined manner:

1. **ARB**—arbitrates between the various DMA requests regarding the DMA channels,
2. **DST**—handles data bursting via the Destination Spoke (DST)
3. **Fetch**—causes the transaction description (TD) and configuration (CONFIG) for a channel to be fetched when a channel wins arbitration,
4. **SRC**—causes data bursting on the Source (SRC) spoke for the channel,
5. **WRBAK**—the updated TD and CONFIG information is written back to their respective locations when the burst for a channel has been completed.

2.2.8 Spoke Arbitration

The CPU and DMAC can access all the spokes except for SPK0, which is the SYSMEM spoke, provided that they do not attempt to access the same spoke, contemporaneously. When the CPU and DMA are using different spokes there is no conflict. However, should the CPU and either the DMAC or the DST engines attempt to access the same spoke, the DMAC is required to allow the CPU to access the spoke first. The results of arbitration are a function of (1) the spoke's priority, (2) which attempted access first, (3) did both attempt access at the same time, and (4) whether the spoke in question is a CPU or DMA priority spoke which is determined by SPKxx_CPU_PRI(CFG_15:1). If the DMA engine is waiting for the CPU to finish its use of a spoke, it can still be subject to interruption by a higher priority DMA channel initiating the following sequence of events:

- The interrupted DMA channel completes any data in transit as may be required when it gains access to the spoke(s).
- The state of the channel is then saved by the DMA WRBACK function and the AUTO_RE_REQ bit for than channel is set. Setting of this bit results in the channel being returned to the DMA request pool subject to the normal arbitration rules.
- When that channel again has access to the DMAC, it simply resumes the "burst" where it left off.

While single requests are given immediate access, multiple requests must be subjected to arbitration subject to the following guidelines:

- PRIO is the highest priority and therefore not subject to arbitration (Table 2.14).

2.2.9 Priority Levels and Latency Considerations

As mentioned previously DMA channels of higher priority can interrupt lower priority DMA transfers, i.e., those with a lower priority number, subject to the constraint that the lower priority transfer is allowed to complete its then-current transaction. In cases for which multiple DMA access requests have occurred, a "fairness" algorithm is employed to minimize latency. This algorithm requires that the priority levels 2–7, inclusive, have at least some minimum percentage of the bus's bandwidth. If two requests are tied, then a simple round-robin method is employed to allow each to share half of the allocated bandwidth. However, this technique can be disabled for any of the DMA channels to allow that channel to always have priority. Table 2.15 shows the minimum bus bandwidth allocated for each priority level, once the CPU and DMA transactions of priority 0 and 1 have been completed. If the fairness algorithm has been disabled, then DMA access is based solely on their respective priority levels and without minimum bandwidth constraints.

2.2.10 Supported DMA Transaction Modes

The ability to chain transactions and the flexibility available in configuring each DMA channel; makes it possible to support simple, relatively complex and highly complex transaction modes, e.g.:

- **Auto Repeat DMA**—The same memory contents are repeatedly transferred.
- **Circular DMA**—Multiple buffers and TDs are employed with the last TD chained back to the first TD.
- **Indexed DMA**—This technique allows an external master to access locations on the system bus as if they were in shared memory.
- **Nested DMA**—Since the TD configuration space is memory mapped, one TD can modify another, e.g., a TD loads another TD's configuration and then calls that TD, when the second TD's transaction is complete, it then calls the first, which updates the second TD's configuration, with this cycle repeating as often as required.

Table 2.14 Peripheral interfaces to PHUB.

Spoke Number	Peripheral Number	Short Name
00	00	SYSMEM
01	00	IOIF
01	01	PICU
01	02	EMIF CSR
01	03	EMIF DATA
02	00	PHUB LOCSPK
02	01	PM
02	03	CLKDIST
02	04	IC
02	05	SWV
02	06	EE
02	07	SPC
02	13	PM TRIM
02	14	BIST_ASSIST
03	00	ANALOG I/F
03	01	DECIMATOR
03	02	ANALOG I/F TRIM
04	00	FS USB
04	01	CAN
04	02	I2C
04	03	TIMERS
05	00	DFB
06	00	UDB SET0 8-BIT
06	01	UDB SET0 16-BIT
06	02	UDB SET0 CONFIG
06	03	UDB SET0 DSI
06	04	UDB SET0 CTRL
06	05	UDBIF
07	00	UDB SET1 8-BIT
07	01	UDB SET1 16-BIT
07	02	UDB SET1 CONFIGI
07	03	UDB SET1 DS
07	04	UDB SET1 CTRL

- **Packet Queuing DMA**—packets are employed with specific protocols employing separate configuration, data and status phases for the transmission and receipt of data.
- **Ping-Pong DMA**—Double buffering is used to allow one buffer to be filled while the contents of the other are being transferred.
- **Scatter Gather DMA**—a transaction involving multiple, noncontiguous sources and/or locations for a given DMA transaction.

Table 2.15 Priority level vs. bus bandwidth.

Priority Level	% Bus Bandwidth
0	100
1	100
2	50
3	25
4	12.5
5	6.3
6	3.1
7	1.5

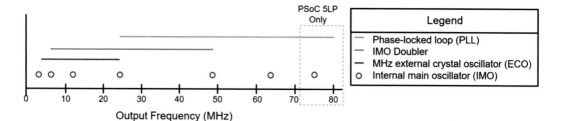

Fig. 2.18 PSoC3/5LP clock source options.

2.2.11 PSoC 3's Clocking System

PSoC 3 and PSoC 5LP contain many clock sources that vary in frequency and accuracy. This section describes each potential clock source in detail. Figure 2.18 shows the clock source options in the MHz range, illustrates the ranges of operation for PLLs, IMO doublers, ECO's and IMOs for both PSoC 3 and PSoC 5LP.

Microprocessor/microcontroller clocks synchronize virtually all internal microprocessor signals and are especially important in imposing error-free communications for both internal and external digital components, and synchronize analog signals conversion from A/D and D/A. There are multiple clock sources in PSoC 3 and PSoC 5L whose frequency and accuracy are variable.

PSoC 3 and PSoC 5LP clocks play a critical part in PSoC operations, e.g., they are used for synchronizing internal signals, ensuring error-free communication with other digital devices, and driving the conversion of signals to, and from, the analog domain. These roles make the configuration of the different clocks used inside of a microcontroller very important.

PSoC 3 and PSoC 5LP contain many clock sources that vary in frequency and accuracy. This section describes each potential clock source in detail. Figure 2.16 shows the ranges of operation for PLLs, IMO doublers, ECO's and IMOs for both PSoC 3 and PSoC 5LP. PSoC 3's clocking generator provides the main/master time bases and allows the designer to make tradeoffs possible between accuracy, power and frequency. A broad range of clock frequencies are available due to the

Table 2.16 Clock naming conventions.

Clock Signal	Description
clk_sync	Synchronizing clock from the Master clock mux used to synchronize the dividers in the distribution.
dsi_clkin	Clocks that are taken as input into the clock distribution from DSI.
clk_bus	Bus clock for all peripherals.
clk_d[0:7]	Output clock from the seven digital dividers.
clk_ad[0:3]	Output clock from the four analog dividers synchronized to the digital domain clock.
clk_a[0:3]	Output clock from the four analog dividers synchronized to the analog synchronization clock.
clk_usb	Clock for USB block.
clk_imo2x	Output of the doubler in the IMO block.
clk_imo	IMO output clock.
clk_ilo1k	1 kHz output from the ILO.
clk_ilo100k	100 kHz from the ILO.
clk_ilo33k	33 kHz output from the ILO.
clk_eco_kHz	32.768 kHz output from the MHz ECO.
clk_eco_kHz	4-33 MHz output from the MHz ECO.
clk_pll	PLL output.
dsi_glb_div	DSI global clock source to USB block.

ability to accommodate multiple clock inputs and employ PSoC 3's highly configurable internal clock distribution system. Table 2.16 provides a summary of the clock naming conventions.

PSoC 3's internal clock generator can use internal/external clock sources,[27] as shown in Fig. 2.19, in the kHz and MHz range and input from Digital System Interconnects (DSI)[28] and an internal Phased-Locked Loop (PLL) can also be used for frequency synthesis.

In addition to support for multiple clock sources, there are eight individually sourced, 16-bit, clock dividers for the digital system peripherals, four individually sourced 16-bit dividers for the analog system peripherals and a dedicated 16-bit divider for the bus clock.

The primary clock sources consist of the following:

---

[27]External clocks sources such as crystal oscillators are often used.

[28]DSI can provide clock signals created in UDBs, of-chip clocks routed through I/O pins and clock signals from the systems clock distribution resources.

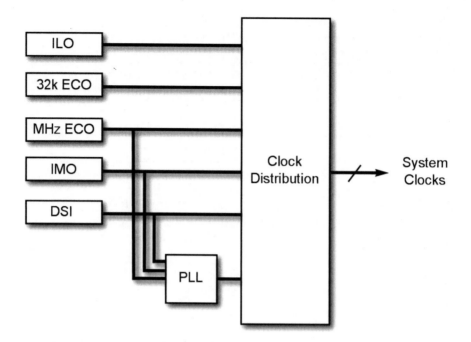

Fig. 2.19 PSoC 3 clocking system.

1. A fixed 36 MHz clock that routes to SPC,
2. A 4–33 MHz crystal oscillator,
3. A 3–67 MHz Internal main oscillator (IMO),
4. 12–67 MHz Doubler output source form from the IMO, MHz external crystal oscillator (MHzECO) or Digital System Interconnect (DSI),
5. 1, 33, and 100 kHz Internal Low speed Oscillator (ILO),

and,

6. Digital System Interconnect from an I/O pin or other logic. 12–67 MHz fractional Phase-Locked-Loop (PLL) driven by the IMO, MHzECO or DSI

If required, the internal PLL[29] can be used to synthesize frequencies in the range from 12–100 MHz. The PLL's input can be from IMO, a MHz crystal oscillator or from a DSI signal. As shown in Fig. 2.20, the Master Clock Mux selects the IMO, DSI, PLL or MHz crystal oscillator as the primary clock source. Note that it is also possible to independently control the phase of the primary clock source for both digital and analog clocks, respectively. This arrangement also makes it possible to change the clock source for the primary clock in multiple systems.

2.2.11.1 The Internal Master Oscillator (IMO)

The IMO produces a stable clock frequency without the use of any external components and contains a doubler circuit that provides an output that is twice the frequency of the IMO frequency, i.e., 6–

---

[29]An integral PLL prescaler(Q) and PLL divider (P) can be used to create clocks that are P/Q times the PLL's input frequency.

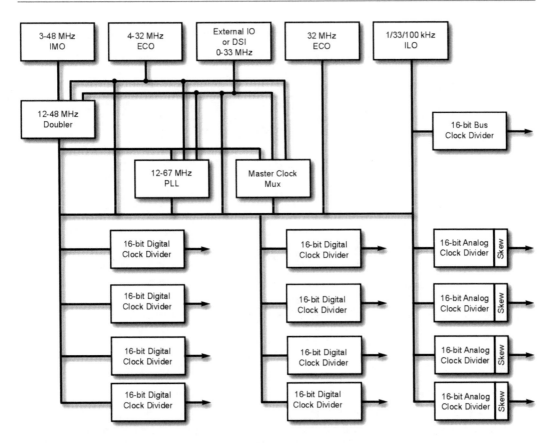

Fig. 2.20 Clock distribution network for PSoC 3 and PSoC 5LP.

24 MHz. However, the IMO output can be either the IMO's frequency or double that frequency, but not both.

Alternatively, other clock sources can be routed through the IMO, as shown in Fig. 2.21, e.g., a DSI or MHz crystal oscillator output and hence through the IMO doubler.[30] As shown, IMO2X_SRC(PSoC 3) selects either DSI or XTAL\CLK as a source for the clock signal. Thus, DSI, XTAL or OSC (3, 6, 12, 24, 48 or 92 MHz) can then be selected by the clock mux (as determined by the (CLKDIST.CR) IMO_OUT register) following which the resulting clock signal (clk_imo) may be IMOCLK, IMOCLK2 or 36 MHz.[31]

The IMO block employs a precision input voltage and current to charge a capacitor from ground to a reference voltage. An integral comparator senses when a predetermined threshold voltage is reached and causes the charging cycle to repeat, between two capacitors, resulting in a pulse on each edge of the input clock and producing a clock frequency which is two times the input clock. (The AHB interface and registers for the IMO are implemented in the FAST Clock interface, i.e., logic for the PLL, IMO and an external oscillator.)

---

[30]To reduce power [5] consumption the doubler can be disabled.

[31]This clock signal (36 MHz) is routed to SPC and available to clock distribution only in test mode, i.e., it is not available in user mode. Its accuracy is approximately 10%.

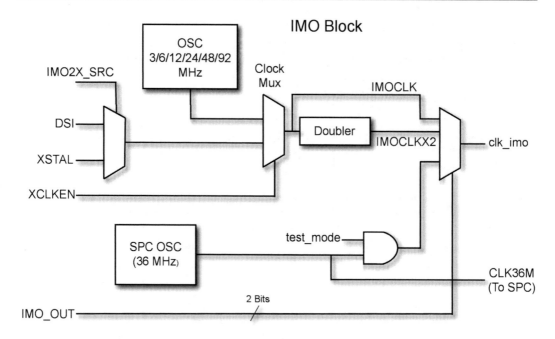

Fig. 2.21 PSoC 3's clock distribution system.

2.2.11.2 Trimming the IMO

The IMO has provisions for "trimming" the clock frequency in terms of both gain and offset. The offset trim step size is determined by the gain setting. Factory trim settings are provided for the proper 24 MHz setting, because this frequency is used for USB operation. For non-USB operation, the gain should be fixed to reduce power requirements. It is recommended that the gain setting employed be the same as the setting for 24 MHz. If it is necessary to change ranges, the offset trim should be loaded first at the lower frequency range, i.e., when moving to a higher frequency range, apply the new offset value and then change the range. Conversely, when moving to a lower frequency range, change the range and then apply the new offset. Note that range and trim values take effect immediately.[32]

The ability to trim the frequency allows automatic "Clock-Frequency Locking" for USB operation to be employed so that small frequency variations of incoming USB signals [6] can be corrected by comparing the incoming USB timing (frame markers) to the IMO clock rate.[33] Alternatively, a crystal controlled clock operating at 24 MHz "doubled" to 48 MHz could be used for Full Speed USB operation, or other crystal-controlled frequencies could be employed in conjunction with the PLL to synthesize 48 MHz.

2.2.11.3 Fast-Start IMO

The IMO can also be operated in the Fast-Start IMO (FIMO) mode which is activated when "waking up" and provides a clock [17] 7 output within 1 μs after exiting the power down mode. In this mode the clock frequency is 48 MHz with an accuracy of about 10% of the primary IMO mode. The FIMO mode is selected by setting the FASCLK_IMO_CR[3] bit in the IMO.CR register which causes the

---

[32]The clock may exhibit one cycle of slight variation.

[33]This will require, however, that the IMO frequency be 24 MHz and that the doubler be used to provided 48 MHz.

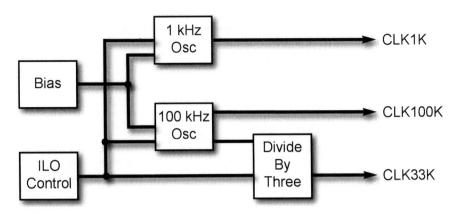

Fig. 2.22 ILO clock block diagram.

IMO clock to be replaced at the next wake-up. The FIMO mode is deselected by clearing the fimo bit resulting in the IMO clock replacing the FIMO in the next wake-up.

2.2.11.4 Internal Low Speed Oscillator

The Internal Low Speed Oscillator (ILO) generates two independent clock frequencies, one at 1 kHz and the other at 100 kHz, respectively, neither of which require external components, as shown in Fig. 2.22.

In addition to operating independently of each other, they are not synchronized with each other and can be enabled/disabled independently or simultaneously. The 1 kHz clock is typically deployed as a "heart beat" timer and for the watchdog timer. The 100 kHz clock serves as a low power system clock and can be used to time sleep mode entry/exit intervals. Finally, a third clock frequency which is derived by applying "divide-by-three" to the 100 kHz clock. The power required, in terms of current, is in the 100 nA to 1 μA range, with an accuracy of 0.20% and 300 μs start-up time.

2.2.11.5 Phase-Locked Loop

The PLL is capable of producing synthesized frequencies in the range of 12–67 MHz. The PLL uses a 4-bit input divider Q (FASTCLK_PLL_Q) to divide the reference clock, selected by the Mux as the IMO, an external crystal oscillator, or the DSI (an external clock signal) and an 8-bit feedback divider P (FASTCLOCK_PLL_P) to divide the output as shown in Fig. 2.23. Outputs of the two dividers are compared by the phase frequency detector (PFD). The PFD compares the phase and frequency difference between the two signals to determine whether the signal fed back from the output is leading or lagging the reference signal F_{ref} defined by Eq. (2.1). The PFD drives the output frequency, via "Up" or "Down" signals, either higher, or lower, as required and then it is "locked" producing an output frequency that is P/Q times the input reference clock. This PLL is capable of locking frequency within 10 μs, and once a lock has been achieved a bit (FASTCLK_PLL_SR[0]) is set and at that point the output frequency is available for distribution to the clock trees.

Thus:

$$F_{ref} = \frac{F_{in}}{Q} \tag{2.1}$$

$$clk\_pll = F_{VCO} = F_{ref}(P) = \left[\frac{F_{in}}{Q}\right]P \qquad (2.2)$$

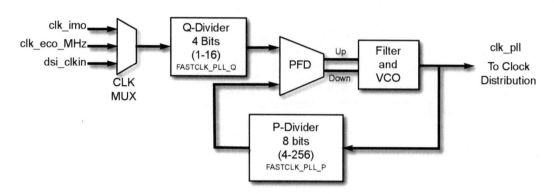

Fig. 2.23 Internal design of the PLL.

In low power operation, during sleep and hibernate modes, the PLL must be disabled to allow "clean entry" into these modes of operation. Following wake-up and lock, the PLL can be enabled so that it can serve as the system clock. PSoC 3/5LP will not enter sleep or hibernate mode as long as the PLL remains enabled.

2.2.11.6 External 4–33 MHz Oscillator

Precision clock signals in the range from 4–33 MHz can be employed by adding an external, fundamental mode, parallel resonance crystal and two capacitors as shown in Fig. 2.24. The pins used for this purpose can also be used with standard I/O functions, e.g., GPIO, LCD and analog global, thus they must be tri-stated when used with an external crystal. The resulting signal is routed to the clock distribution network and may be routed to the IMO doubler, if the crystal frequency is in a valid range for the doubler, i.e., less than, or equal to, 24 MHz. While this configuration is compatible with a wide range of crystals, crystal start-up times are a function of crystal resonant frequency and quality. Oscillator settings can be matched to a given crystal by setting the xcfg bits of the FASTCLK_XMHZ_CFG0[4:0] register. The oscillator is enabled by FASTCLK_XMHZ_CSR[0].

Should the crystal oscillator fail, e.g., due to the adverse effects of moisture, or for some other reason, it is possible to detect this condition by checking the clock error status bit. FAST-CLK_XMHZ_CSR[7]. If the FASTCLK_XMHZ_CSR[6] bit is set and the crystal oscillator fails, then the crystal oscillator output is driven low and the IMO is enabled, assuming that it is not already running, and the output of the IMO is routed through the crystal oscillator output mux. Thus, the system can continue to operate in the event of a crystal fault. When the system is in SLEEP/HIBERNATE mode, it is not necessary to allow the crystal oscillator to continue running and therefore consume power. The 32 kHz oscillator can be kept active if precise timing is required, e.g., for the Real-Time Clock (RTC). However, it is not possible to enter the SLEEP/HIBERNATE mode when the MHz crystal oscillator is running. One approach is to switch clock trees to the IMO source, and then disable the MHz crystal oscillator and the PLL, if it is active. It is then possible to enter a sleep mode. When the system wakes up from a sleep mode, the MHz crystal oscillator, and if necessary the PLL, can be enabled and employed, once stability has been achieved.

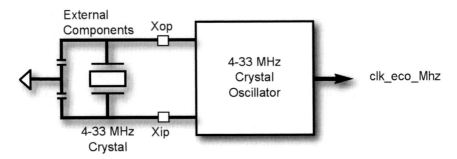

Fig. 2.24 4–33 MHz crystal oscillator.

2.2.11.7 External 32 kHz Crystal Oscillator

The 32 kHz oscillator, kHzECO, utilizes a low cost, external crystal (32.768 kHz) and external capacitors to produces a precision timing signal, and serve as the basis for a real-time clock operating at very low power, i.e., current levels less than 1 µA. The resulting timing signal, clk_eco_Khz, is routed to the clock distribution network within PSoC 3 and serves as a clock source for the clock distribution logic and the Real-Time Clock (RTC) timer. Enabling/disabling of the kHzECO is accomplished by setting/clearing SLOWCLK_X32_CR[0]. This oscillator can operate at one of two power levels, depending on the state of the LPM bit, SLOWCLK_X32_CR[1], and the sleep mode status of the system.

The default mode is "Active" for the kHzECO and a hardware interlock forces the oscillator into its high power mode, which consumes 1–2 µA and minimizes noise sensitivity. Assuming that the LPM bit is set for low power mode, the oscillator only operates at low power when the system when the system is in SLEEP/HIBERNATE mode. However, if LP_ALLOW (SLOWCLK_X32_CFG[7]) is set, the oscillator enters low power mode immediately when the LPM bit is set.

It should be noted that this oscillator is not stable when activated, and therefore some time is required for it to achieve stability. The DIG_STAT status bit, SLOWCLK_X2_CR[4], indicates that oscillation is stable by comparing it to the 33 kHz ILO signal. The ANA_STAT bit, SLOW-CLK_X32_CR[5], use an internal analog monitor to measure the oscillator's amplitude (Table 2.17).[34]

2.2.11.8 Implementing a Real-Time Clock

Many embedded systems require the availability of a real time clock to time events, record time, data logging, etc. The kHzECO oscillator can be used to provide real time clock functionality by dividing the kHzECO signal by 32,768 to produce one pulse per second which in turn can be used to generate interrupts at 1-s intervals, update counters, etc., unless the system is in HIBERNATE mode.

2.2.11.9 Clock Distribution

The clock sources discussed previously produce signals that can be made available to other PSoC resources through the clock distribution logic within PSoC 3 by routing them through analog and digital clock dividers. Some peripherals require specific clocks for their operation, e.g., the Watchdog Timer (WDT) requires the ILO. Clock distribution is facilitated by the use of clock trees. PSoC 3 has four such trees for clock distribution, viz.,

---

[34]To avoid excessively long startup times before stability is achieved it is a good practice to start the oscillator in high power mode.

Table 2.17 Oscillator parameter table.

Source	Frequency	Power	Accuracy	Startup Time	Usage Notes
Internal Main Oscillator (IMO)	3-92 MHz	100-300 μS		5 μS	No external components
Internal Low Speed Oscillator (ILO)	1,33,100 kHz	100 nA - 1μA	20%	300 μS	Watchdog, sleep timer
External 32 kHz Crystal Oscillator		100 nA - 1μA	10 ppm	1 S	Cab be used to provide real time clock
External 4-33 MHz Crystal Oscillator		1-4 mA	10 ppm	100 μS	
Phase-Locked Loop		500 μA	Same as input source	50 μS	Synthesize desired output frequency

1. Analog Clock tree
2. Digital Clock tree
3. System Clock tree

and,

4. USB clock tree

Eight dividers for the digital clock tree and four analog dividers for the analog clock tree are provided as part of the clock distribution system, as shown previously in Fig. 2.20. Clock sources in each case are selected by an eight input mux for connection to the dividers and the outputs of the dividers are synchronized with their respective domain clocks. Distribution of sync clocks is facilitated by the Master Clock Mux and there are options that provide delay for the digital synch clock. All the digital dividers are synchronized to the same digital clock, but the analog dividers can each be synchronized to their respective analog clock with different, or the same, delays.

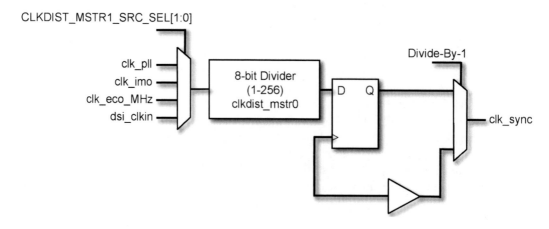

Fig. 2.25 Master clock mux.

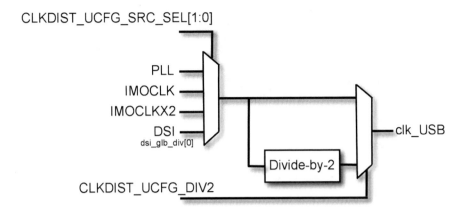

Fig. 2.26 The USB clock mux.

The Master Clock Mux (MCM), shown in Fig. 2.25, selects a clock from the available inputs, viz., PLL, IMO, ECO_MHz or DSI. The output of this mux becomes the source that is supplied to the phase mod circuit to produce skewed clocks selected by the digital and analog phase mux blocks. The MCM provides two re-sync clocks for the system: clk_sync_dig, for the digital clocks, and clk_sync_a for the analog system clocks. The Master clock, which is always the fastest clock in the system, is also the basis for switching the clock source for multiple clock trees simultaneously. Clock trees select the clk_sync_dig or clk_sync_a clock as their input for systems that must maintain known relationships. An 8-bit divider is provided to make it possible to generate lower frequencies clocks, CLKDIST_MSTR0[7:0].

2.2.11.10 USB Clock Support
The advent of the now ubiquitous Universal Serial Bus (USB) has resulted in increasing support for it at the microcontroller level. PSoC 3 provides the USB logic a synchronous bus interface while allowing that logic to operate asynchronously to process USB data.

The USB clock mux, shown in Fig. 2.26, can be used to select the USB clock source as:

- imo1x
  - The 48 MHz DSI clock is subject to the accuracy of the clock.
  - Since the oscillator cannot operate at 48 MHz and therefore imo1x must be multiplied by the PLL to get 48 MHz.
- imo2x
  - 24 MHz crystal with doubler.
  - 24 MHz IMO and doubler with USB lock.
  - 24 MHz DSI with doubler.
- clk_pll
  - Crystal and PLL to generate 48 MHz.
  - IMO and PLL to generate 48 MHz.
  - DSI input and PLL to generate 48 MHz.
- DSI input
  - 48 MHz.

If the internal main oscillator is selected, the oscillator locking function must be used to allow it to develop the required USB accuracy for USB traffic. This automatic clock frequency locking facility allows small frequency adjustments based on the incoming frame marker timing with respect to the IMO frequency. This type of clock frequency locking allows the clock frequency to stay within $\pm0.25\%$, with respect to accuracy, for the USB full speed mode. The locking mode is enabled by setting the FASTCLK_IMO_CR[6]. It is also possible to use a 24 MHz crystal-controlled clock which is subsequently doubled to 48 MHz for full-speed USB operation. Another option is to use other frequencies, e.g., 4 MHz, with the PLL to synthesize 48 MHz. In addition to the clk_imo option, the DSI signal, dsi_glb_div[0], can also be employed.

2.2.12 Clock Dividers

Clock dividers are an integral and important aspect of the clock distribution system. In addition, they also provide some control over the duty cycles. It is possible to generate a single cycle clock pulse. A 50% duty cycle mode produces a clock with approximately a 50% duty cycle. A divider reloads its divide count after it reaches a terminal count of zero. The divider count is set in the CLKDIST_DCFG[0..7]_CFG0/1 register for digital dividers and the CLKDIST_ACFG[0..3]_CFG[0..3]_CFG0/1 register for analog dividers. The counter is driven by a clock source selected by an 8-bit mux controlled by CLKDIST_DCFG[0..7]_CFG2[2:0] for digital dividers and CLKDIST_ACFG[0..7]_CFG2[2:0] for analog dividers, in either a single-cycle pulse mode, or a nominal 50% duty cycle mode.

Regardless of the mode selected, a divide by zero causes the divider to be bypassed, resulting in a divide by one, and the input clock is applied to the output after a resynch, assuming that the sync option has been previously selected. If the loaded value is M, then the total period for the output clock is given by:

$$N = M + 1 \tag{2.3}$$

The CLKDIST_DCFG[x]_CFG2[4] or CLKDIST_ACFG[x]_CFG2[4] bit in the configuration register for each clock output is set high to enable the 50% duty cycle mode. However, it may not be possible to provide a 50% in all cases, because of dependencies on phase and frequency differences between the sync clock and the output clock.

2.2.12.1 Clock Phase

Another important clock parameter is phase. In addition to two duty cycle choices the outputs can be phase-shifted to go high after the terminal count, or at the half-period cycle. The default mode is known as "Standard Phase" and refers to the rising edge of the output, after the terminal count. Alternatively, "Early Phase' refers to the output being effectively shifted to an earlier point in time to an approximate count that is one half of the divide value. Setting the CLKDIST_DCFG_CFG2[5] or CLKDIST_ACFG_CFG2[5] bit in the configuration register for each clock will enable the Early Phase mode and the rising edge will occur near the half count point. While analog clock dividers are architecturally similar to digital dividers, they have an additional resync circuit to synchronize the analog and digital clocks. Synchronizing the digital and analog clocks facilitates communications between the digital and analog domains.

2.2.12.2 Early Phase

The clock outputs can also be phase-shifted by requiring them to "go high" after the terminal count or at the half-period cycle. The phrase "Standard Phrase" refers to the rising edge of the output to occur after the terminal count. The Early Phase option allows the output to be viewed as having been shifted to a point earlier in time for an approximate count that is one-half of the divide value. Early Phase

Mode can be invoked by setting the CLKDIST_DCFG_CFG2[5] bit, so that the rising edge occurs near the half-count point. While analog dividers are similar to digital dividers, they have an additional resynch circuit to synchronize the analog clock to the digital domain clock, thereby synchronizing the output of the analog dividers, called clk_ad, with the digital domain.

2.2.12.3 Clock Synchronization

Each of the clock trees can be set for one of the following options, with respect to the output clocks:

- **Bypassed clock source**—If the divider value is set to zero and the synch bit is reset, The clock tree's selected source is routed to the output without division and results in an asynchronous clock.
- **Phase-Delayed clk_sync**, e.g., as clk_sync_dig—The tree operates at the same frequency as clk_sync but with the appropriate phase. In this case, the input clock source is ignored.
- **Resynchronized clock**—Activating the sync bit causes a clock at clk_sync/2 maximum frequency to be resynchronized by the phase-delayed clk_sync.
- **Unsynchronized divided clock**—This clock is asynchronous and occurs when the synch bit s reset, and the divider has a non-zero value.

Bus Clock (BUS_CLK) is the CPU clock in PSoC 3. However, it can be further divided using 8051 SFR CPUCLK_DIV.

2.2.13 GPIO

Modern microcontrollers make extensive use of buses which are analog or digital transmission paths, typically consisting of multiple conducting paths grouped together to facilitate digital and analog signal transmission, e.g., memory access depends critically upon the availability of high-speed data and address buses to allow program code an data to move quickly and efficiently between, e.g., the CPU and RAM. Similarly, buses are also needed for internal peripherals to allow communication to take place peripheral-to-peripheral, peripheral-to-CPU, CPU-to-peripheral, CPU-to-I/O, etc. Bus design varies, but they are typically a minimum of 8 parallel paths in width. Care must be taken in laying out such paths to assure that the electrical path length is the same for each path in a given bus, particularly as the speed of allowable bus traffic is increased. Since a significant number of devices may have access to the same bus, tri-stating techniques, as described in Chap. 1, are often used to make sure that bus performance is not degraded, to avoid collisions and to simplify bus use.

PSoC 3 and PSoC 5LP make extensive use of buses and particularly of the analog interconnect, digital interconnect and system bus. The system bus allows traffic to be moved between the CPU, Memory and debug facilities and the digital/analog systems. The system bus is also used by the system-wide resources. Routing of data along bus paths is also a common requirement and analog/digital multiplexers, and switches, are used to determine how bus traffic is to be routed.

Switches are functionally quite similar to a multiplexer, because they are both based on analog switches, except for the fact that in the case of multiplexers, while there may be "n" inputs there is only one connected to the output at any given time. However, in the case of a switch, it is possible to have zero to "n" inputs connected to output at any given time. This is an important distinction and it should also be noted that in the case of a multiplexer fewer bits are required to connect an input to an output than is the case for a switch when both have the same number of inputs.[35] PSoC 3 and PSoC 5LP have several analog routing resources, e.g., local analog buses (abus), global analog buses

---

[35] For example, selecting one of 8 inputs requires only three bits for a multiplexer and eight bits for a switch.

(AGs), analog mux buses (AMUXBUS) and an LCD bias bus (LCDBUS). The analog globals and AMUXBUS connect to the GPIO's and provide a method of interconnecting GPIOs and the analog resource blocks (ARBs) such as DACs, comparators, switched-capacitors, CapSense, Delta-Sigma ADC and OpAmps. A voltage reference bus (V_{ref}) that provides precision reference voltages for the ARBs that are created by the precision reference block which is capable of generating precision voltages and currents that are not a function of temperature.

Each GPIO pin can be connected to an analog global path by use of a switch and it is possible to connect two pins on each port to the same global path. The analog global bus provides interconnection options via muxes and switches to the inputs/outputs of the following ARBs for I/O: CapSense (a virtual block), comparator, DAC, Delta Sigma ADC and the Output buffer. Each GPIO pin has two analog switches, one to connect the pin to analog global and the other to connect to the AMXBUS. The control signals required to open, or close, these switches are invoked either by using the PRT[x]_AMUX and PRT[x]_AG registers which is the default option, or dynamically by using the DSI control that is connected to the input of the port pin logic block. However, before using the latter option, it is necessary to set a bit in the Port Bidirection Enable register, i.e., PRT[x]_BIE.

There are nine input/output ports consisting of seven General Purpose I/O (GPIO) ports, one SIO and one mixed-function port. This allows digital input sensing, output drive, pin interrupts, connectivity for analog input/output, LCD and access to internal peripherals to be supported directly via defined ports, or the UDB Individual I/O channels [5] are arranged in groups of eight bits, or pins, and defined as the respective "ports" (Fig. 2.27).

2.3   Power Management

Power management in any embedded system is an important consideration in terms of maintaining the proper power levels, minimizing power consumption, proper distribution of power, minimizing noise and its effects in the supply lines, etc. PSoC 3 and PSoC 5LP maintain separate external analog and digital supply pins for the internal core logic. Two internal 1.8 V, voltage regulators are used to provide V_{ccd} for digital and V_{cca} for the analog circuitry. A sleep regulator is also maintained for operation in the sleep domain, an I^2C regulator for powering I^2C logic and a hibernate regulator for supplying "keep-alive" power to assure state retention when the system is in a hibernate mode. External connection the internal power distribution systems are made via pins labeled V_{dda}, V_{ddd} and V_{ddiox} for the analog, digital and I/O power systems, respectively. Capacitors are required, as shown in Fig. 2.28, preferably placed as physically close to their respective pins as possible. The capacitors are provided to minimize external power supply transients and to minimize adverse load effects. The digital and analog regulators are referred to as "active domain" regulators since they enter low power modes of operation in sleep mode. Sleep and hibernate regulators provide the necessary power when the system enters its lowest power consumption modes.

2.3.1   Internal Regulators

When operating in regions for which external power supplies provide voltages that range from 1.95–5.55 V, the internal regulars draw power from these external supplies via the V_{ddd} and V_{cca} pins. If the external power supply is delivering voltage in the 1.71–1.95 range, the internal regulators remain powered, by default, after power-up. However, register PWR-SYS.CR0 should be used to disable these regulators, after power-up, in order to minimize power consumption.

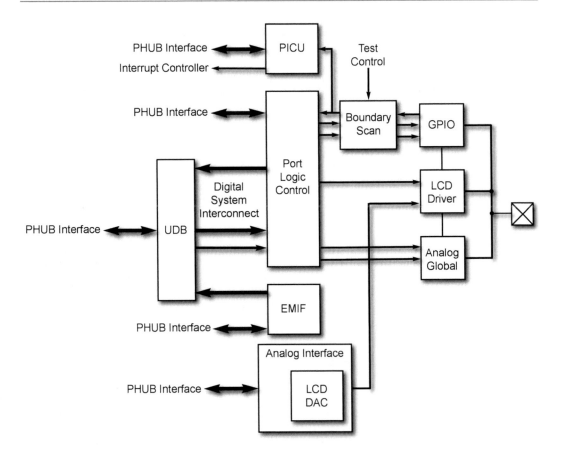

Fig. 2.27 GPIO block diagram.

2.3.1.1 Sleep Regulator

When the system is in sleep mode, a sleep regulator provides a regulated voltage, V_{sleep}, for the 32 kHz ECO, Central Timewheel (CTW), Fast Timewheel (FTW), ILO, RTC Timer and watchdog timer (WDT). The Hibernate regulator supplies keep-alive power, V_{pwrKA}, to those domains responsible for state retention during hibernation.

2.3.1.2 Boost Converter

PSoC 3 and PSoC 5LP are capable of operating from voltage supplies over the range from 1.7 to 5.5 V. However, external supplies may not be able to maintain a constant voltage to the system under some circumstances, e.g., systems using external supplies in the form of batteries can experience a wide variance in supply voltage as the battery system discharges, or as in the case of solar cells, the ambient illumination varies. Therefore, PSoC 3/5LP have an integral boost converter that is capable of accepting input voltages over a wide range, e.g., as low as 0.5 V and producing a constant output voltage at the required power levels. The internal converter requires an input voltage, an external inductor and capacitors, unless the external voltage is greater than 3.6 V, in which case an external

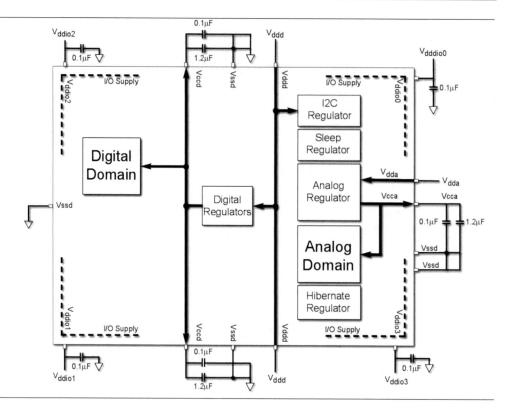

Fig. 2.28 Power domain block diagram.

Schottky diode[36] is also required. In addition to being able to provide voltage for internal use, an external pin, V_{Boost}, is provided for driving voltages for external devices, e.g., an LCD.

PSoC 3/5LP's boost converter can be disabled, or enabled, by setting, or resetting, the BOOST_CR1[3] and the output voltage can be changed by writing to the BOOST_CR0 [4:0] register. At startup the boost converter is enabled and by default the output voltage setting is 1.8 V. If the boost converter is not to be used, then V_{bat} should be "tied to ground" and the IND pin should be left floating.

The following C language code fragment illustrates how to start the boost converter, set its operating frequency at 100 kHz, and then stop it (Fig. 2.29).[37]

```
# include <device.h>
    void main( )
        {
            Boostconv_1_Start( );
            BoostConv_1_SelFrequency(BoostConv_1_SWITCH_FREQ_100KHZ);
            BoostConv_1_Stop( );
```

---

[36]Schottky diodes are named for Walter H. Schottky a German physicist (1886–1976) whose work led to the development of the hot carrier diode, also known as the Schottky diode. It has the important property that when conducting, the voltage drop across the diode is significantly lower than most diodes, viz., 0.15–0.45 V versus 0.7–1.7 V.

[37]Note that the Boost Converter is by default "Active" when the system powers up.

}

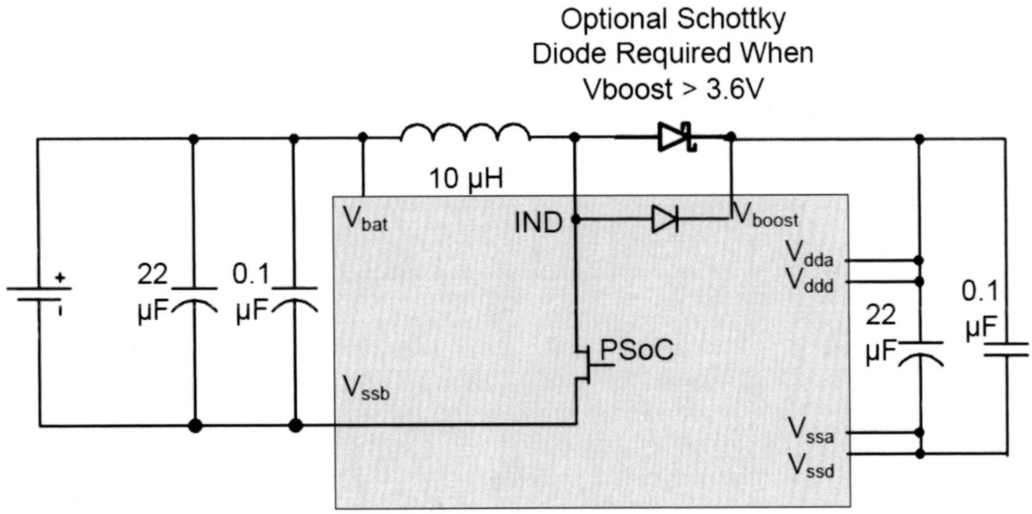

Fig. 2.29 The boost converter.

As shown in Fig. 2.30, when the boost converter's MOSFET is conducting, the voltage across the inductor is:

$$V_{input} = V_L = L\frac{di}{dt} \tag{2.4}$$

and therefore:

$$\frac{di_L}{dt} = \frac{V_{input}}{L} = constant \tag{2.5}$$

The duty cycle is given by:

$$Duty\,Cycle = \frac{T_0}{T} = D \tag{2.6}$$

and therefore:

$$T_0 = DT \tag{2.7}$$

Eq. (2.5) implies that,[38]

$$\frac{di_L}{dt} = \frac{\Delta i_L}{\Delta t} = \frac{\Delta i_L}{DT_2} \tag{2.8}$$

and thus:

---

[38]Some will undoubtedly find the use of Δt and dt in the same expression disturbing, as well they should. However, this transgression does not adversely affect the calculation, or its conclusion.

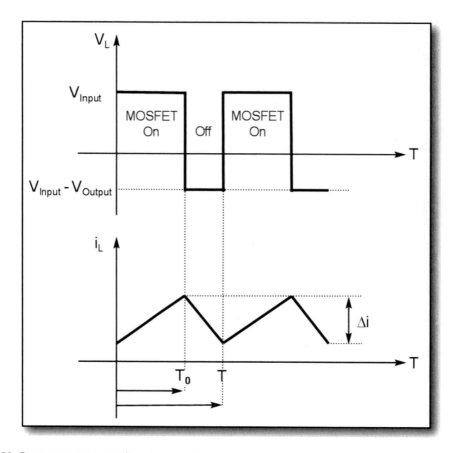

Fig. 2.30 Boost converter current flow characteristics.

$$[\Delta i_L]_{on} = \frac{V_{input}\,DT}{L} \qquad (2.9)$$

When the MOSFET is not conducting (MOSFET-Off state),

$$V_L = V_{input} - V_{output} = L\frac{di_L}{dt} \qquad (2.10)$$

and therefore,

$$\frac{di_L}{dt} = \frac{V_{input} - V_{output}}{L} = \frac{\Delta i_L}{\Delta t} = \frac{\Delta i_L}{(1-D)T} \qquad (2.11)$$

so that:

$$[\Delta i_L]_{off} = \frac{(1-DT)(V_{input} - V_{output})}{L} \qquad (2.12)$$

But,

$$[\Delta i]_{off} + [\Delta i]_{on} = 0 \tag{2.13}$$

and therefore:

$$\frac{V_{input} \, DT}{L} + \frac{(1-D)T(V_{input} - V_{output})}{L} = 0 \tag{2.14}$$

so that:

$$V_{output} = \frac{V_{input}}{(1-D)} \tag{2.15}$$

Therefore, when the MOSFET is conducting, energy is being stored in the inductor's magnetic field and the capacitor is supplying power to the load. Conversely, when the MOSFET is not-conducting, the energy supplied to the inductor, plus the additional input energy, is being supplied to the load and capacitor. It should be noted that in deriving Eq. (2.15), that the inductor's resistance, the diode's resistance and forward conducting voltage were assumed to be negligible and that the MOSFET functioned purely as a switch with insignificant resistance when conducting.

2.3.1.3 Boost Converter Operating Modes

The boost converter can operate in one of three modes that are determined by the BOOST_CR0[6:5] register:

1. **Active**—In this mode, the Boost regulator produces a regulated output voltage from a battery. The switching frequency of the Boost Converter is selected by the BOOST_CR[1:0] register. The available switching frequencies are 100 kHz, 400 kHz and 2 MHz but are not synchronous with any other clock, i.e. these frequencies are "free-running".
2. **Standby**—In the standby mode, only the bandgap and comparators are active and other systems are disabled to reduce the power consumed by the Boost Converter. The Boost Converter's output voltage monitored continuously and supervisory data is available in BOOST_SR[4:0]. The supervisory data is referenced to the selected voltage.
3. **Sleep**—In this mode except for the bandgap, the comparators and other circuits are turned off. In this mode the output of the boost converter is a very high impedance, and the active circuits are powered by the energy stored in the 22 μ capacitor. Over a prolonged period, the voltage across the capacitor will decay. This can be handled in some cases by awakening the system and recharging the capacitor (Fig. 2.31).

2.3.1.4 Monitoring Booster Converter Output

The status register BOOST_SR contains information regarding the input and output voltages of the boost converter referenced to the nominal voltage setting. The BOOST_SR[4:0] register provides the following status information:

1. Bit4: ov—above over-voltage threshold (nominal +50 mV)
2. Bit3: vhi—above the high regulation threshold (nominal +25 mV)
3. Bit 2: vnom—above nominal threshold (nominal)
4. Bit 1: vlo—below low regulation threshold (nominal to 25 mV)
5. Bit 0: uv—below the under-voltage limit (nominal to 50 mV)

Register	Function
PWRSYS_CR0	Regulator control
PWRSYS_CR1	Analog regulator control
Boost_CR0	Boost thump, voltage selection and mode select
Boost_CR1	Boost enable and control
Boost_CR2	Boost control
Boost_CR3	Boost PWM duty cycle
Boost_SR	Boost Status
RESET_CR0	LVI trip value setting
RESET_CR1	Voltage monitoring control
RESET_SR0	Voltage monitoring status
Reset_SR2	Real-time voltage monitoring status

Fig. 2.31 Boost converter register functions.

2.3.1.5 Monitoring Voltages

Two circuits, shown in Fig. 2.32, are provided for monitoring voltages to detect any deviation from the selected thresholds for external analog and digital supplies:

1. **Low Voltage Interrupt** (LVI)—this circuit generates an interrupt when it detects a voltage below the set value. The low voltage monitors default to the off mode. However, the trip level for the LVI is set in the RESET_CR0 register over a range from 1.7–5.45 V, in steps of 250 mV.
2. **High Voltage Interrupt** (HVI)—this circuit generates an interrupt when it detects a voltage above the set value.

2.4   Recommended Exercises

2-1 Explain what function the 8051 listing shown below performs and document each line of the source code:

```
MOV R0,20H;
MOV R1, #30H;
MOV A, @R0;
INCR0;
MOV B,@R0;
MUL AB ;
MOV @R1, B;
INC R1;
MOV @R1, A;
```

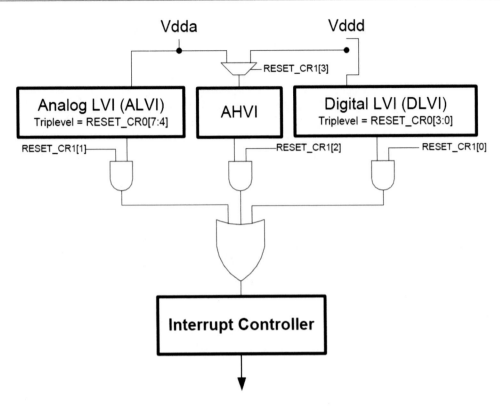

Fig. 2.32 Voltage monitoring block diagram.

HALT: SJMP HALT ;

2-2 Write equivalent assembly language code to that in (2-1) for the Intel 4004 and 8080

2-3 Propose a method for extending the Intel 8051's memory space beyond 64k using some form of paging scheme. Explain the difference between a linear memory space and a paged one. Explain the benefits of each.

2-4 If the CPU clock has temperature dependencies what, if any, are the potential issues resulting from varying ambient temperature? Should disk capacitors and carbon resistors be used to provide a stable CPU clock? If not what other options are available?

2-5 Give examples of how you might use the PSW in various types of control systems?

2-6 Complete the following assembly language program by writing the instruction codes for locations A, B, C, and D. This could be used to rotate through an LED array.

```
org        1000h
mov TMOD,#01h  ; Select  Timer0
mov a,#11111111B
loop: A;Move  Accumulator  to  Port  1
```

```
        call delay
        B ;Rotate Accumulator 1 bit to the left
            jmp loop
        delay:  C ; Timer 0 start counting in the
            TCON register
        again: TL0,#<(65536-1000) ; set TL0's
                    value
                TL0,#<(65536-1000) ; set TL0's
                    value
                D ; Jump to loop3 if TF0(overflow)
                    is 1, clear TF0
        jmp loop2          ; Jump to loop2
```

2-7 Create an 8051 assembly language program that can transfer 16k of data stored at memory location 1000H to memory location 400H. Assume that the start address in each case is stored in a location specified explicitly in your program register.

2-8 Consider the following listing:

```
            mov R1,#02H ;
            mov R2,#04H ;
            mov R3,#00H ;
    Up:   inc 03H           ;
            djnz R2,Up        ;
            djnz R1,UP        ;
    Nm:  simp Nm           ;
```

(a) What is stored in R3 when the program been executed. (b) What is the role of the last instruction in the code?

2-9 Sketch the 8051's interrupt table and explain how to change the interrupt priority under program control. Create an assembly language program to illustrate this ability.

2-10 Create an assembly language program to allow an 8051 to receive serial date at 2400 baud. Assume that the data is transmitted in an 11-bit frame format consisting of one start bit, one data byte, one parity bit and a stop bit. The program must be capable of detecting a parity error and illuminating an LED connected to pin P1.2.

References

1. M. Ainsworth, PSoC 3 8051 Code optimization. Application Note: AN60630. Cypress Semiconductor (2011)
2. J.A. Borrie, *Modern Control Systems - A Manual of Design Methods* (Prentice Hall, London, 1986)
3. V.S. Kanna, PSoC 3 and PSoC 5 interrupts. AN54460. Document No. 001-54460 Rev. *D 1. Cypress Semiconductor (2012)
4. M. Kingsbury, PSoC 3 and PSoC 5LP clocking resources. AN60631. Cypress Semiconductor (2017)
5. E.E. Klingman, *Microprocessor Systems Design* (Prentice Hall, Englewood Cliffs, 1977)
6. J.K. Peckol, *Embedded Systems, A Contemporary Design Tool* (Wiley, Hoboken, 2008)
7. E.W. Kamen, B.S. Heck, *Fundamentals of Signals and Systems*, 3rd edn. (Prentice Hall, Upper Saddle River, 2007)

8. V.S. Kannan, J. Chen, PSoC 3, PSoC 4, and PSoC 5LP temperature measurement with a diode. AN60590. Document No. 001-60590 Rev. *K 1 Cypress Semiconductor (2020)
9. A. Krishswamy, R. Gupta, Profile guided selection of ARM and thumb instructions. LCTES'02-Scopes'02. 19–21 June 2002
10. J. Lemieux, Introduction to ARM thumb. Embedded systems design. Sept 2003
11. MCS-51 Microcontroller Family User's Manual. Chapter 2, pp 28–75. Intel Corporation (1993)
12. F. Nekoogar, G. Moriarty, *Digital Control Using Digital Signal Processing* (Prentice Hall, Upper Saddle River, 1999)
13. R. Phelan, *Improving Arm Code Density and Performance (New Thumb Extensions to the Arm Architecture)* (Arm Limited, 2003)
14. PSoC 3 Architecture Technical Reference Manual. Document No. 001-50235 Rev. *M Cypress Semiconductor, 8 April 2020
15. PSoC 3, PSoC 5 Architecture TRM (Technical Reference Manual). Document No. 001-50234 Rev D, Cypress Semiconductor (2009)
16. PSoC 5LP Architecture Technical Reference Manual. Document No. 001-78426 Rev. *G. Cypress Semiconductor, 6 Nov 2019
17. G. Reynolds, PSoC 3 and PSoC5LP low-power modes and power reduction techniques. AN77900. Document No. 001-77900 Rev.*G1. Cypress Semiconductor (2017)

Sensors and Sensing

3

Abstract

Sensors and sensor types, strain gauge/thermocouple/thermistor sensing and measurement techniques are discussed in this chapter, in some detail. Because sensors are an integral part of virtually every embedded system, considerable discussion is devoted to identifying and characterizing some of the more popular sensor types (Kannan, Chen, PSoC 3, PSoC 4,and PSoC 5LP–temperature measurement with a diode. AN60590. Document No. 001-60590 Rev. *K 1 Cypress Semiconductor, 2020). However, the sensor types discussed in this chapter are only a small sample of the many types currently available. When designing for a specific application and more particularly for a particular environment, set of environments and/or environmental conditions, the reader is encouraged to carefully search the literature for the most appropriate types of sensors for the application and the associated environmental conditions/constraints. It is assumed throughout this chapter that once the selected sensor(s), for a given application, are chosen PSoC 3 and PSoC LP5 are used to access the sensor data via A/D, or other PSoC-related means to allow the data to be processed, the necessary logic applied and appropriate control imposed, e.g., using PSoC D/A components.

3.1 Sensor Fundamentals

Sensors are devices that convert one or more physical parameters into digital, or analog, signals for processing and control applications. Such sensors, often referred to as transducers,[1] typically convert physical parameters such as temperature [4], pressure, linear/curvilinear motion of objects (acceleration, velocity, displacement, etc.), salinity, hydrogen ion concentration (pH) wind speed, ocean currents, vibration, presence of toxic materials, fire, proximity of objects, force (linear, torque), fluid[2] flow (velocity, acceleration, displacement, etc.), radiation measurement (low-frequency

---

[1]Transducers may convert one physical parameter into another, but for the purposes of these discussions, the term *transducer* refers explicitly to a device capable of converting the value(s) of physical parameter into a corresponding voltage, current or digital value.

[2]Both gases and liquids are to be regarded, for the purposes of these discussions, as fluids.

© Springer Nature Switzerland AG 2021
E. H. Currie, *Mixed-Signal Embedded Systems Design*,
https://doi.org/10.1007/978-3-030-70312-7_3

RF, high-frequency RF, microwave, ultraviolet, visible light, infrared, etc.), heat flux, stress/strain, chemical signals (e.g., smells),quasi-static electric and magnetic fields [5] (i.e. non-radiative), altitude, metal, resistance, capacitance, inductance, electrical power, mechanical power, mass flow, volume flow, etc., into an analog voltage or current, or the digital equivalent, thereof. In recent years there have been many advaces in sensors using a variety of sensing techniques, e.g, using Field Effect Transistors (FETs) or light emitting diodes (LEDs)for sensing. [12]

In addition, to *point* measurements utilizing a single sensor, multiple sensors may also be used in groups, in both heterogeneous and homogeneous arrays, to provide fields of data over a given surface, or volume. Regardless of the physical parameter(s) being sensed, i.e. measured or detected, including the lack thereof, the converted parameter is provided as a unique and related voltage, current or digital value, which can then be processed by a microcomputer, or microcontroller, and used for feedback, or feed forward, information in a wide variety of applications.

Some of the more commonly encountered input sensors include microphones, tachometers, thermistors/thermocouples, sonar or other forms of acoustic sensors, pressure sensors, liquid level sensors [11],]infrared sensors (passive and active), ultrasonic sensors, RFID readers, strain gauges, linear/rotary position sensors, mechanical switches of various configurations, distance (altimeters, ranging sensors, etc.), velocity, acceleration, roll, yaw, pitch, GPS, and proximity detectors. Common output devices[3] include speakers, electric motors, linear/rotary positioners, actuators, opto-couplers, solenoids, wireless connections of various types and protocols, liquid crystal (LCD) displays, and PC connections.

Hundreds of types of sensors have been designed to detect acceleration, displacement, force, humidity, spatial position/orientation, temporal parameters, tactile contact or the lack thereof, biometrics (retinal, fingerprint, DNA, facial, etc.), proximity, speed and a host of other parameters. These devices provide resistance, capacitance, inductance, current, voltage, amplitude, frequency, phase, quadrature modulation, and data in binary, or other forms, as output parameters. All these devices are susceptible to electromagnetic noise (EMI), aging, temperature, vibration and other forms of degradation of their output signals. Thus, it is important to carefully calibrate such devices using known references, often against the manufacturers specifications, under a variety of anticipated operating conditions. In some cases, the embedded system can utilize *look-up tables* (LUTs), and other means, to apply corrections to data provided by such devices. Filtering of analog signals can also be used to maintain input data integrity. In the case of wireless sensors [6], additional care must be taken to avoid extraneous signals being imposed by the environment, especially if low-level sensor signals are involved. [8]

In dealing with sensors, it is important to delineate between the *accuracy* of a sensor and its *precision*. The latter refers to the degree to which a sensor is measuring the quantitative value of a parameter and the latter to how close a series of measurements of a parameter, by a given sensor, are grouped, quantitatively. Thus, it is possible for a sensor to make a series of measurements of a parameter and arrive at similar values and yet not represent a particularly accurate assessment of the actual quantitative value. Note that precision is defined in terms of the number of significant figures to which the value is measured, while accuracy is related to how close the value reflects the true value of the parameter being measured. Accuracy, and in some cases, precision is also affected by the conversion of analog data to a digital format. Typical microcontrollers have *data widths* of 8–16 bits, which can affect both accuracy and precision, depending on the application.[4]

Sensors also have inherent limitations with respect to input signal *dynamic range* and *bandwidth* and may introduce noise, suffer from nonlinearities, be affected by offsets, subject to saturation effects,

---

[3]Note these may also be viewed as transducers.

[4]Newer architectures, such as the Cortex M3 that is used in PSoC5, have 32-bit data widths.

and, depending on their input and output impedances, may require proper impedance matching[5] for both input signals and interfacing to the microcontrollers I/O channels.

In the case of analog signals, *dynamic range*, as applied to sensors, is a quantitative figure of merit defined as the ratio of the maximum input signal to the minimum discernible, i.e., detectable, input signal that can be applied to a sensor for which the sensor can produce an output without distortion. Dynamic range, for digital signals, is defined in terms of the *bit error ratio* (BER) which is the ratio defined as:

$$BER = (number\ of\ altered\ bits)/(number\ of\ bits\ transmitted) \tag{3.1}$$

where *altered bits* refers to transmitted bits altered by adverse phenomena such as interference, noise, etc.

Care must be taken to ensure that *impedance mismatches*[6] don't distort measurements and/or adversely affects the sensor values that are input to the microcontroller, and the microcontroller's output to actuators, peripherals, motors, actuators, etc.

3.2 Types of Sensors

Currently, wireless sensors, ultra-low power, plug-and-play, MEMs-based,[7] PWM output and sensor fusion[8] are some of the most popular areas of sensor technology. Communications protocols used in conjunction with sensors include: CAN/CANOpen, Devicenet, Ethernet IP, TCP/IPWireless, USB, RS232, RS485 and various proprietary networks. In terms of zn embedded system designer's priorities, reliability, accuracy, precision, durability (ruggedness/robustness), noise immunity, sensitivity, sensing/dynamic range, resolution, ease of maintenance, ease of setup, and operating environment are the most important concerns, and in descending order. Currently, for modern embedded systems, the most popular types of sensors are vision, wireless [6], rotary position, proximity, linear displacement and photoelectric.

3.2.1 Optical Sensors

Optical sensors are popular because they provide excellent electrical isolation, have good electromagnetic immunity, are capable of both point and distributed configuration, offer a wide dynamic range and can be used in hostile environments. Various types devices are available for measuring of optical parameters and detecting the presence/absence of radiation in the optical portion of the electromagnetic spectrum. These devices can be used to measure temperature, chemical species,

---

[5]Impedance matching considerations are important when employing sensors to avoid adverse perturbations of that being sensed.

[6]*Impedance mismatch* is a phrase often used to refer to differences between the output impedance of one stage, e.g., a sensor, and the input impedance of a second stage, e.g., an amplifier, when power transfer is a consideration. In such cases, the output and input impedances should not be the same, i.e., *matched*, i.e., if it is important that the microprocessor not draw significant power from the sensor to avoid the possibility of distorting the value of the parameter being sensed.

[7]Micro-electromechanical or MEMs sensors can be on the order of 5x5x1 mm in volume, and are available as pressure, acceleration, gyroscopic, gas flow, temperature and other types of sensors.

[8]Sensor fusion is the combining of sensed data, obtained from multiple sensors, to reduce uncertainty in the values being measured.

pressure, flow, force, radiation, liquid level, pH (hydrogen ion concentration), vibration, rotation, magnetic fields, electric fields, acceleration, acoustic fields, electric fields, velocity, strain, humidity, and radiation.

Typically, a beam of radiation, often either visible or infrared, is incident upon a region of interest and one or more of the following are detected optically:

- intensity,
- phase,
- polarization,
- wavelength

and/or

- spectral content

Some sensors employ optical fibers in either intrinsic or extrinsic configurations.[9] Extrinsic sensors employ optical fibers to irradiate the region of interest and a second fiber to collect the radiation emerging from this area and transmit it to a light detector, for example, photomultipliers, pin diodes, photoconductors, or photodiodes.

- Photomultipliers have multiple stages of light amplification that result in a current that is proportional to the intensity of illumination. These devices use low work function materials,[10] e.g., alkali metal-based coatings, to convert photon impacts into electrons and hence currents.
- Pin Diodes/Photodiodes are semiconductors, e.g., PN junctions, which produce a current which is proportional to the intensity of the illumination, i.e., they respond to the amplitude of the incoming radiation. Photovoltaics are a type of photodiode that is based on a PN junction that produces a voltage.
- Photoconductors are devices whose resistance is a function of incident radiation intensity and wavelength. A light-dependent resistor can have a response similar to that of the human eye, but it may respond more slowly than the eye, or other types of light sensors. .

Intrinsic sensors employ fiber optics but are capable of delivering radiation to the region of interest and collecting the resulting radiation without allowing the radiation to leave the optical fiber. Such sensors rely on changes in pressure, temperature [4], curvature, etc., to modify the physical properties of an optical fiber resulting in corresponding changes in polarization, phase spectral content and/or amplitude in the radiation being transmitted by the optical fiber from a radiation source to a detector. Radiation sources include:

- incoherent light-emitting diodes (LEDs) that are available in spectral ranges from ultraviolet (UV) to infrared (IR),
- high power light-emitting diodes,
- coherent laser diodes (LD) that are available in spectral ranges from ultraviolet (UV) to infrared (IR),

---

[9]Intrinsic sensors, employ a fiber optic cable as the sensor, whereas extrinsic sensors, use optical fibers to transmit the light to/from a conventional sensor.

[10]Work functions represent the minimal energy required to eject an electron from a solid and are defined as $W = h\nu$ where ν is the minimum photon frequency required for photoelectric emission to occur from a given solid surface.

and,

- vertical cavity, surface-emitting lasers (VCSEL) that are capable of operating in a multimode configuration.

3.2.2 Potentiometers

Resistive potentiometers are a very popular form of transducer and are frequently used in position sensing applications.This type of sensor can be used to measure both linear and angular positions. Typical potentiometers consist of a length of resistive material and a sliding contact that moves along the material. A voltage is applied across the ends of the material and the voltage on the sliding contact is directly proportional to its position. This type of sensor is relatively inexpensive and is often employed in feedback loops to accurately control the positions of actuators, as well as, sense their position.

3.2.3 Inductive Proximity Sensors

This type of sensor is based on the fact that the inductance of a coil is a function of the proximity of a ferromagnetic object to the coil. Such devices may employ self-inductance, mutual inductance[11] or some combinations of both. The value of an inductive proximity sensor can be measured by a combination of an AC current source and an AC bridge, or by an AC voltmeter.

3.2.4 Capacitive Sensing

In recent years, capacitive sensing has become increasingly more common in applications such as automobiles, mobile phones, a wide range of consumer electronics including home appliances, stereos, televisions, a wide variety of consumer products as well as a broad range of military and industrial applications. Capacitive sensing offers a number of advantages over its mechanical counterpart, e.g., no mechanical parts, completely sealed interface, etc.

Capacitive sensing is based on a very simple relationship between the area of a capacitor, the distance between two conducting surfaces of a capacitor, d, the permittivity of free space, ϵ_0, the relative dielectric constant, ϵ_r and its capacitance, C, as follows:

$$C = \epsilon_0 \epsilon_r \frac{A}{d} \tag{3.2}$$

where C , in MKS units, is the capacitance in Farads, A is the area of the "plates" of the capacitor, d is the distance between the plates, $\epsilon_0 = 8.84 x 10^{-12}$ and ϵ_r is the relative permittivity.

There are two basic types of types of capacitive sensing system (1) based on mutual capacitance in which a charged object, e.g., a finger, or other charge bearing surface, modifies the mutual coupling between two other conducting surfaces, and (2) a charged surface coming into close proximity to another charged surface results in increased, or decreased, capacitive effects between the two, or with

---

[11] Self-inductance is the ratio of magnetic flux of a given circuit to the current that produced it. Mutual inductance is defined as the ratio of the flux produced by one circuit, in a second circuit, to the current that produced it.

respect to ground. In both cases, whether a variation in mutual capacitance, or absolute capacitance, each can be detected and thereby provide a capacitive sensitive mechanism [1].

Current estimates suggest that as many as 2.5 billion buttons and switches have been replaced by this technology. This *non-touch sensing* technology is sufficiently sensitive in some applications to allow it to be employed in applications requiring nanometer resolution. In addition to replacing the traditional buttons, capacitive sensing techniques [11] are also used to function as sliders, proximity detection, LED dimming, volume controls, motor controls, etc. Capacitive sensing is capable of sensing the presence of conductive materials, including fingers and providing proximity sensing for a wide variety of touch-pads and -screens. Many capacitive sensing applications consist of a conducting surface, often protected by glass or plastic, that senses the proximity of one or more fingers.

A typical capacitive sensing arrangement involves two separate conducting surfaces often created by traces on a printed circuit board which represent a capacitance of 10–30 picofarads. Assuming that the traces are protected by an insulating material, perhaps 1 millimeter in thickness, an approaching finger represents a capacitance in the range of 1–2 picofarads.

3.2.5 Magnetic Sensors

There are various forms of magnetic sensors which operate in some cases, by closing or opening switch contacts, e.g., reed relays), utilizing the Hall effect to vary current flow, sensing current flow etc. Magnetic sensors also take advantage of the Curie Point.[12]

They are used in a wide variety of applications including:

- Brushless DC motors
- Pressure sensors
- Rotary encoders
- Tachometers
- Vibration sensors
- Valve position sensors
- Pulse counters
- Position sensors
- Flowmeters
- Shaft position sensors
- Limit switches
- Proximity sensors

Magnetic sensors, unlike other types of sensors do not, for the most part, measure a physical parameter directly. Instead, magnetic sensors [13] react to perturbations in local magnetic fields in terms of strength and direction to determine the state of electric currents, direction, rotation, angular position, etc.

The units of magnetic field are Gauss, Tesla and gamma, and they are related by the following expression:

$$10^5 \, gamma = 10^{-4} \, Tesla = 1 \, Gauss \qquad (3.3)$$

---

[12]The Curie point refers to the temperature at which the magnetic properties of a substance change from ferromagnetic to paramagnetic. If the temperature is subsequently reduced to below the Curie Point, the substance becomes ferromagnetic again.

Magnetic sensors can be classified in terms of the range of magnetic field strength which they sense as follows:

- Low Fields—magnetic fields whose strength is less than 1 Gauss.
- Earth's Field—magnetic fields in the range from 1 microgauss to 10 Gauss
- Bias Magnetic Fields—magnetic fields of strength greater than 10 g.

and, because a magnetic field is a vector field, both magnitude and direction of the field, a magnetic sensor can use of the presence, strength or direction and strength of a magnetic field to measure a particular parameter, e.g.,

- a vector magnetic sensor utilizes both magnitude and direction,
- an omnidirectional magnetic sensor uses the magnetic field in one direction,
- a bidirectional magnetic sensor measures magnetic field in both directions,

and,

- a scalar magnetic sensor utilizes magnetic field strength only.

Finally, magnetic sensors can be further classified as:

- Anisotropic magneto-resistive (AMR) sensors are used for measuring position in terms of angular, linear position and displacement in fields comparable to that of the Earth. They consist of thin-film resistors that are created by depositing nickel-iron on silicon whose resistance can be varied by several percent in the presence of a magnetic field.
- Bias magnetic field sensors use Hall devices,
- Fluxgate sensors[13] are often used in navigation systems.
- Hall effect sensors sense current in a small plate as a result of the Lorentz force $F = q(vxB)$ on electrons. This in turn produces a Hall voltage which is directly proportional to the magnetic field.
- Magneto-inductive sensors utilize a single winding coil which has a ferromagnetic core in the feedback loop of an operational amplifier to form a relaxation oscillator. Changes in the ambient magnetic field alter the frequency of the oscillator by as much as 100%. A shift in frequency can be detected by a microcontroller's *capture/compare*[14] functionality.
- Search coil sensors rely on the fact that a changing magnetic field induces a changing electric field in a coil. However, search coil sensors require that either the magnetic field is varying or the coil is moving.
- Squid sensors are based on the Josephson junction[15] it is the most sensitive and is capable of sensing fields as low as 10^{-15} Gauss and as high as $9x10^4$ Gauss (9 Tesla), which is equivalent to fifteen orders of magnitude.

---

[13]Fluxgate sensors were developed during World War II to detect submarines from low flying aircraft. In its most basic form it consists of a saturable core and a drive coil. It functions as a variable inductance [13] that is a function of the drive current and the external magnetic field.

[14]Capture refers to a microcontroller's ability to time the duration of an event. Compare refers to its ability to compare the values in two registers and subsequently trigger an external event.

[15]In 1962, Josephson, at the age of 22, predicted that a sandwich of superconductor-insulator-superconductor, e.g., niobium-aluminum oxide-niobium, that is 10 Angstroms, or less, in thickness will have the property that current can flow through the junction without a voltage being applied across the junction, up to a critical current, This current depends heavily upon the ambient magnetic field.

3.2.6 Piezoelectric

3.2.7 RF

RF sensors detect fluid viscosity, fluid contamination, fluid flow, linear and rotational speeds, displacement and position in automotive and aeronautical applications. This type of sensor measures the magnetic susceptibility and electric permittivity within a predefined volume of space and can operate with frequencies well in excess of 1 GHz with resolutions down to one part in 10^5 in environments with temperatures from $-170°$ to $1000 °C$. RF sensors are able to discriminate between a wide variety of materials including composites, glasses, plastics, liquids,[16] nonferrous and ferrous materials. It can also be used to measure linear/rotational displacement/position as well as fluid flow, level [1], contamination and viscosity.

3.2.8 Ultraviolet

This type of sensor often relies on the physical characteristics of Zinc Oxide which is transparent when irradiated with visible light and opaque when irradiated with ultraviolet in the 220–400 nanometer range. Silicon photodiodes are also used for UV detection but silicon also absorbs UV which makes it less desirable as a sensor. UV sensors are relatively insensitive to visible and infrared radiation.

3.2.9 Infrared

InfraRed (IR) proximity sensors in applications such as TV remote controls, wireless connections between PCs and printers, utilize light in the range from 600 to 1200 nm, which is not visible to humans.
Various optical techniques are employed including:

- Modulated IR—which modulates an IR beam to control devices remotely. Modulating the carrier provides better signal-to-noise ratios (SNRs) which can be important in IR-noisy environments
- Reflective IR—this technique relies on measuring IR reflected from an object. However, it can be adversely affected by background thermal radiation, but is inexpensive to implement.
- Transmissive—detects objects located between an IR transmitter and receiver.
- Triangulation—offers the best performance for proximity detection using a focused beam and a receiver array to measure the angle of reflection from an object.

3.2.10 Ionizing Sensors

Smoke detectors use a chamber that contains a radioactive source, e.g., Americium-41, to provide alpha particles that ionize the oxygen, and nitrogen, present in the air in the chamber. There is also a set of plates, one positively charged, and the other negatively charged. The negatively charged plate attracts the ionized Oxygen/Nitrogen ions. Similarly, the electrons are attached to the positive plate. The net result is a small, but continuous current flow. However, in the presence of smoke, particulate

---

[16]RF sensors can detect the presence of water in oil down to one part in 10^6.

matter in the smoke binds with the Oxygen/Nitrogen ions thus making the charge of each neutral and reduces the current. This reduction in current is then detected and triggers an alarm.

Photoelectric detectors are also used as detectors in smoke alarms by either monitoring the amount of light reaching a detector which is reduced in the presence of smoke, or by measuring the amount of light scattered from a beam in the presence of smoke.

3.2.11 Other Types of Sensors

While there are obviously a great many types of sensors [10], the three most common forms of sensors are strain gauges, thermistors and thermocouples [11]. The techniques used for making measurements using these three types of sensors are similar and shall be treated briefly.

Since the resistance of a length of wire is a function of length (L), cross-sectional area (A) and a physical quantity that is determined by the type of wire being considered referred to as resistivity (ρ) given by:

$$R = \rho \frac{L}{A} \tag{3.4}$$

and therefore,

$$dR = \rho \left[\frac{A\,dL - L\,dA}{A^2} \right] \tag{3.5}$$

so that,

$$\Delta R = \rho \frac{\Delta L}{A} \tag{3.6}$$

if the change in area is negligible the change in resistance is a linear function of the change in length (dL), assuming that ρ is not an explicit function of length (L), or area (A).

Strain gauges[17] are designed to measure a dimensionless parameter called "strain" which is defined as the deformation of an object when a load is applied, expressed as:

$$Strain = \frac{(\Delta L)}{L} \tag{3.7}$$

where ΔL is the length of deformation and L is the original length. It should be noted that strain can be either compressive or tensile (stretched). Strain gauges are designed to convert mechanical deformation into some form of electronic change, e.g., in resistance, inductance or capacitance, which is proportional to the strain.

Strain gauges measure deformation in only one direction, and therefore, if the deformation is in two or three dimensions, multiple strain gauges should be placed such that they are orthogonal. In addition, most materials tend to be at least somewhat anisotropic[18] so that the same stress applied in orthogonal directions may result in different amounts of strain.
The three basic forms of strain:

---

[17]Strain gauges were invented by Simmons and Ruge in 1938.

[18]If the properties of a material are independent of direction the material is said to isotropic, otherwise, the material is said to be anisotropic.

- Bending strain, sometimes referred to as "moment strain", is defined as the amount of strain resulting from a given force.
- Poisson strain is a measure of the elongation and thinning of an object that occurs as the result of stress applied to an object.
- Shear strain—Shear strain is a strain that is parallel to the face of an object that it is acting upon, as shown in Fig. 3.2.

Stress, illustrated in Fig. 3.1, is defined as:

$$\sigma = \frac{F}{A} \tag{3.8}$$

where F is orthogonal to the area, A, and can exist in five different states:

- Compression—caused by external forces applied to an object which cause adjacent particles within a material to be pushed against each other resulting in "shortening" of the material.
- Flexure—also referred to as bending.
- Tension—caused by external forces applied to an object which causes adjacent particles within a material to be pulled away from each other resulting in "stretching".
- Torsion—occurs when a material is "twisted".

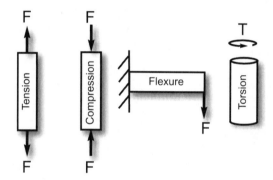

Fig. 3.1 Examples of tension, compression, flexure (bending) and torsion.

- Shear—occurs when adjacent parts of a material "slide" away from each other as shown in Fig. 3.2. Shear may be either vertical or horizontal and in the presence of bending, both occur.

The shear angle is defined as:

$$\theta = arctan\left(\frac{\Delta x}{L}\right) \tag{3.9}$$

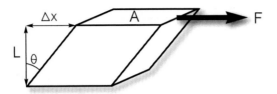

Fig. 3.2 Example of shear force.

Many sensors use stress and strain to measure parameters such as angular displacement, linear displacement, pressure, compression, flexure, torque, force, acceleration, etc. Figs. 3.3 and 3.4 show an application of a strain gauge to a duralumin tensile test specimen.

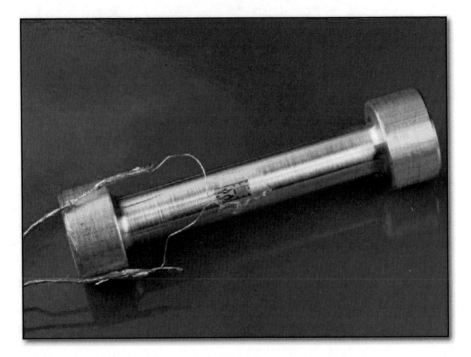

Fig. 3.3 Strain Gauge applied to a duralumin tensile test specimen[19].

[19]This image is taken from Wikipedia [10] and licensed under Creative Commons Attribution 2.5.

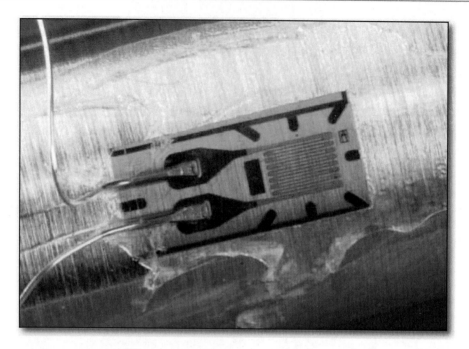

Fig. 3.4 Closeup of a strain gauge[20].

3.2.12 Thermistors

Various techniques are used to sense temperature [3], but the most common thermal sensors are thermistors [14], which are sintered[21] semiconductor materials whose resistance is highly dependent on temperature. Simply stated, a thermistor is a semiconductor, with either a positive or negative temperature coefficient, whose resistance is a function of the ambient temperature. Modern thermistors are based on oxides of cobalt, copper, iron, manganese and nickel. Thermistors are sensitive to static charge and their use is typically restricted to static charge-free applications. The most common types operate temperatures ranging from $0°$ to $100+°$ C. Carbon resistors are often used for extremely low-temperature sensing, e.g., $-250°K \leq -T \leq -150°K$ and have a very linear negative temperature coefficient in this range.

In the simplest case, a thermistor [14] can be characterized by the following relationship:

$$k = \frac{\Delta R}{\Delta T} \tag{3.10}$$

where k is referred to as the temperature coefficient and ΔR is the change in resistance for the corresponding change in temperature, ΔT. Depending on the type of thermistor, k can be either negative or positive.

---

[20]This image is taken from Wikipedia [10] and licensed under Creative Commons Attribution 2.5.

[21]The term sintered refers to the result of heating, without melting, a powdered material into a solid typically porous mass.

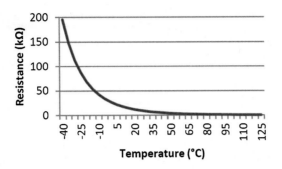

Fig. 3.5 Resistance vs temperature for a NCP18XH103F03RB thermistor.

However, typically thermistors, e.g., of the type shown in Fig. 3.5, do not exhibit a linear relationship, except over a small temperature ranges. The resistance of thermistors typically lies within a range from 1–100 kΩ. Thus, the resistance of leads attached to a thermistor need not be taken into consideration. Metal oxides are used to produce thermistors with negative temperature coefficients. (NTCs) and barium/strontium compounds are used when positive temperature coefficients (PTCs) are required.

A more accurate representation of the change in resistance of a thermistor can be expressed as:

$$\frac{R(T_1)}{R(T_2)} = A^{(T_1 - T_2)} \tag{3.11}$$

where T_1 and T_2 are temperatures, in degrees Kelvin, and A is an empirically derived value that is less than 1. However, an even better approximation to the relationship between resistance and temperature, for a thermistor, is given by:

$$ln(R) \approx a_0 + \frac{a_1}{T} + \frac{a_2}{T^2} + \frac{a_3}{T^3} \cdots + \frac{b_n}{T^n} \tag{3.12}$$

This is often further approximated as:

$$R = exp\left[a_0 + \frac{a_1}{T} + \frac{a_3}{T^3}\right] \tag{3.13}$$

The Steinhart-Hart [5, 15] equation is an empirically-derived relationship with three constants that relate the resistance to a corresponding temperature:

$$\frac{1}{T_c + 273.15} = A + B\ln(R) + C\ln(R)^3 \tag{3.14}$$

The three unknowns can be easily determined by employing three data points: (1) the lowest temperature, (2) the highest temperature and (3) a value midway between the (1) and (2).

It can also be expressed as a 3rd order, logarithmic, polynomial with three constants, i.e.,

$$\frac{1}{T_K} = A + B\ln(R) + C\ln(R)^3 \tag{3.15}$$

where **A**, **B**, and **C** are empirical constants, **R** is the thermistor's resistance in Ohms, and $\mathbf{T}_K$ is the temperature in Kelvins.Generally speaking, the error in the Steinhart–Hart equation is less than 0.02 °C.

A still more useful equation, that provides the temperature in Celsius, is given by:

$$T_C = \frac{1}{A + B \ln(R) + C \ln(R)^3} - 273.15 \tag{3.16}$$

Many thermistors are available with parameters **A**, **B**, and **C** defined. If for a particular thermistor these parameters are not available, their respective values can be calculated by using three points in the conversion table provided by the manufacturer, and solving for these constants. The minimum, maximum, and a middle value for the temperature range of interest are useful points to employ in determining the parameters. The cost of thermistors is primarily determined by the accuracy of their resistance versus temperature characteristics and therefore the exponential nature of thermistors becomes an advantage [2].

For a thermistor with a tolerance of n, the possible temperature error is:

$$(1+n)\, R(T_k) = (1+n)\, A^{T_k} = \left[A^{\frac{\ln(1+n)}{\ln(A)}}\right] A^{T_k} \approx \left[A^{\frac{n}{\ln(A)}}\right] A^{T_k} = A^{\left[T_k + \frac{n}{\ln(A)}\right]} \tag{3.17}$$

which shows that a thermistor's resistance tolerance can be represented as a temperature shift. This shift can be removed by a single point calibration that subjects the thermistor to 25 °C and measuring its resistance, e.g., if its resistance represents a temperature of 26.2 °C, then the embedded system will need to impose an offset of 1.2° in order to determine the actual temperature. In some thermistor applications, the user has access to an offset register via the GUI, and can make the necessary calibration, prior to making a measurement.

A useful heuristic is the fact that a thermistor resistance uncertainty of n% is equivalent to a temperature shift of approximately (n/3)° C and this observation can be used to determine if calibration is necessary. The decision whether to use (3.10), (3.11), (3.13), or (3.16) depends on the application and the required accuracy, cost constraints, available computation time, and other factors.

3.2.12.1 Thermistor Self-Heating

Current flow through a thermistor raises its temperature that in turn introduces an error when measuring the temperature.[22] Manufacturers typically include a dissipation factor with units of mW/°C.

This factor represents the power that will raise the temperature of the thermistor by 1 °C above the ambient temperature, i.e.,

$$Dissipation\,Factor = \frac{T_{Power}}{T_{Error}} \tag{3.18}$$

where T_{Power} supplied to the thermistor and T_{Error} is the difference between the measured temperature value and the ambient temperature.

---

[22]While it may seem that the thermistor's characteristics being a function of temperature is an undesirable feature, there are applications where it can prove useful, e.g., PTC thermistors are often used to limit in-rush current or as current limiting devices because the PTC thermistor's resistance increases with increases in current flow.

3.2.12.2 Thermistor Dissipation Factor

Thermistor self-heating can be an important factor because temperature increases result in more current flowing through the thermistor and that produces an error in the measurement of temperature. Manufacturers refer to self-heating in terms of the "dissipation factor" which is expressed as mW/°C. It is defined as the power required to raise the temperature of the thermistor by 1 °C above the ambient temperature, i.e.,

$$Dissipation Factor = \frac{T_{power}}{T_{error}} \tag{3.19}$$

where T_{power} is defined as the power supplied to the thermistor and T_{error} is defined as the difference between the ambient temperature and the measured temperature

3.2.12.3 Thernistor Gradient Constant "B"

Thermistor characteristics are typically expressed in terms of the operating temperature range, NTC or PTC and a normalized parameter that is a function of the type of ceramic used in the manufacture of the thermistor and the slope of the R versus T curve over a particular temperature range.

It is defined by:

$$B_{(T1/T2)} = \frac{T_2 \times T_1}{T_2 - T_1} \times \ln\left(\frac{R_1}{R_2}\right) \tag{3.20}$$

where T1 is the temperature in degrees Kelvin at the lower temperature point, T2 is the temperature at the upper-temperature point and R^1, R^2 are the resistances at the lower and upper temperatures, respectively.

3.2.13 Thermocouples

Thomas Johan Seebeck[23] discovered that when two dissimilar metals are in contact with each other in the presence of a thermal gradient, a voltage is produced that is a function of the types of the dissimilar metals and the temperatures involved. The *Seebeck effect*[24] is also referred to as the "thermoelectric effect" can be expressed mathematically as:

$$\Delta V = \alpha \Delta T \tag{3.21}$$

where α is referred to as the *Seebeck coefficient* and T is measured in Kelvin. If two wires, e.g., copper and constantan,[25] are joined together the temperature can be determined by measuring the voltage between them, as shown in Fig. 3.6. Copper-Constantan thermocouples, also referred to as type *T* thermocouples, produce approximately $40 \,\mu V$ per ° C ($22 \,\mu V$ per °F). However, in order to measure the voltage, metallic connections must be made to them. If the connections to the device measuring the voltage, e.g., a digital voltmeter, are made of copper, then two additional junctions, viz., copper-to-copper and copper to constantan, are introduced as shown. The copper-to-copper junction will not introduce an additional Seebeck voltage because it does not involve dissimilar metals.

---

[23] Seebeck was a German-Estonian physicist who discovered the thermoelectric effect in 1821.

[24] This effect is also sometimes referred to as the Peltier-Seebeck effect.

[25] Constantan is an alloy of copper (55%) and nickel (45%) with the property that its resistivity remains relatively constant over a wide temperature range. In addition to its use in thermocouples [4], it is also widely used in strain gauge applications.

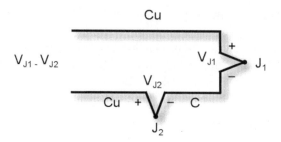

Fig. 3.6 Seebeck potentials.

However, the constantan-to-copper junction will introduce a voltage, V_{J2}. Therefore, the measured voltage is:

$$V_{J1} - V_{J2} = \alpha T_1 - \alpha T_2 = \alpha(T_1 - T_2) \tag{3.22}$$

Because T is in Kelvin:

$$T_2 = t_2 + 273. \tag{3.23}$$

where t_2 is the temperature in degrees Centigrade, Eq. 3.22 becomes:

$$V_{J1} - V_{J2} = \alpha(t_1 + 273 - t_2 - 273) = \alpha(t_1 - t_2) \tag{3.24}$$

In practice, if J_2 is placed in ice, so that $t_2 = 0°C$, Eq. 3.24 reduces to:

$$V_{J1} - V_{J2} = \alpha t_1 \tag{3.25}$$

It should be noted that the National Institute of Technology and Standards uses 0° C, as the reference junction temperature, in this case J_2, in NIST[26] published tables for Type J thermocouples.[27]

3.2.14 Use of Bridges for Temperature Measurement

Sensors such as strain gauges, and thermistors [5] are devices whose resistance is a function of strain and temperature, respectively. In order to collect temperature and strain data it is necessary to read

---

[26]The National Institute of Standards and Technology is a non-regulatory agency of the United States Department of Commerce that serves as a physical science laboratory tasked with measurement techniques and standards in the fields of nanoscale science and technology, engineering, information technology, neutron research, material measurement, and physical measurement. It is sometimes referred to as the "Bureau of Standard" for historical reasons.

[27]Because PSoC 3/PSoC 5LP [2] are capable of measuring voltages as low as $2\,\mu V$, when used in the 20-bit resolution mode, they can be interfaced directly with K-type thermocouples [4] which produce approximately $41\,\mu V/°C$. However, an external IC will be required for cold junction compensation and resistors should be added to compensate for negative thermocouple voltages. The cold junction IC voltage is measured and used to determine the equivalent ambient temperature. If a precision temperature sensor, e.g., Texas Instrument's LM35, is employed for cold junction compensation, at 25° C, the thermocouple will produce 250 mV, i.e., $10\,mV/°C$. The procedure is as follows: Measure the K-type thermocouple differential voltage and add that value to the previously determined cold junction IC voltage.

the resistance of such devices. There are a number of widely employed techniques for making such measurements, two of which are based on Ohms Law [16], e.g.,

- A sensitive current [9] measuring device such as a Wheatstone bridge is used to determine the value of the resistance of a sensor whose resistance is a known function of temperature.
- A known current is passed through a sensor whose resistance is a known function of temperature and the resulting voltage is measured.
- A reference voltage is applied to a reference resistor that is in series with a sensor whose resistance is a known function of temperature, and the resulting voltage is measured.

Once the value of a sensor's resistance has been determined, the data can be compared to available conversion tables[28] that take nonlinearities, and any other transducer-related dependencies, into account.

A common technique for measuring temperature using thermocouples is to employ a Wheatstone bridge,[29] as shown in Fig. 3.7. The value of R_x can be determined by using Ohm's Law,[30] as follows:

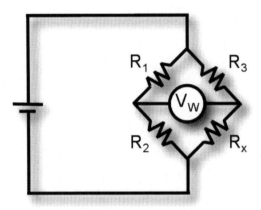

Fig. 3.7 The wheatstone bridge.

$$V_1 = i_1 R_1 \qquad (3.26)$$

$$V_2 = i_2 R_2 \qquad (3.27)$$

$$V_3 = i_3 R_3 \qquad (3.28)$$

$$V_x = i_x R_x \qquad (3.29)$$

where i_1 is the current in R_1, i_2 is the current in R_2, etc.

---

[28]Most sensors are provided with conversion tables, or charts, prepared by the sensor manufacturer.
[29]The Wheatstone bridge was invented by Samuel H. Cristie, in 1833, but later became known as the "Wheatstone Bridge" as a result of the attention drawn to it by Sir Charles Wheatstone, in 1843.
[30]Ohms Law, most commonly expressed as V=IR, states that the potential difference measured across a current carrying resistor is directly proportional to the value of the current through the resistor times the value of the resistance.

If the voltage, V_w, is zero then:

$$i_1 = i_2 \tag{3.30}$$

$$i_3 = i_x \tag{3.31}$$

and therefore:

$$i_1 R_1 = i_3 R_3 \tag{3.32}$$

$$i_2 R_2 = i_x R_x \tag{3.33}$$

$$\frac{i_1 R_1}{i_2 R_2} = \frac{i_3 R_3}{i_x R_x} = \frac{R_1}{R_2} = \frac{R_3}{R_x} \tag{3.34}$$

$$R_x = \frac{R_2 R_3}{R_1} \tag{3.35}$$

This technique while quite sensitive can be replaced by a much more cost-effective, and often desirable, use of microcontrollers such as PSoC 3/5LP [16], to gather data from one or more such sensors by employing a technique that supplies a known current to the sensor, and then measures the resulting voltage across the sensor.

$$i = \frac{(V_1 - V_2)}{R_2} \tag{3.36}$$

$$V_2 = \frac{(V_1 - V_2)}{R_1} R_2 \tag{3.37}$$

$$R_2 = \left[\frac{V_2}{V_1 - V2} \right] R_1 \tag{3.38}$$

A simplified diagram of such an arrangement is shown in Fig. 3.8. A constant current source is used to provide a known value of current to the sensor, and the resulting voltage is measured utilizing an amplifier and analog to digital converter, as shown in Fig. 3.8. If necessary, an amplifier can be employed, that has sufficiently high input impedance to avoid any significant perturbation of the voltage/current to be measured. Obviously, the accuracy of this approach depends critically upon the accuracy of the current source and any errors introduced in measuring the resulting voltage, e.g., gain and offset errors.

A second technique is shown in Fig. 3.9. In this case, sensors resistance can be determined from the following relationship:

$$\frac{V_{ref} - V_{response}}{R_{ref}} = \frac{V_{response}}{R_t} \tag{3.39}$$

and therefore:

$$R_t = \left[\frac{V_{response}}{V_{ref} - V_{response}} \right] R_{ref} \tag{3.40}$$

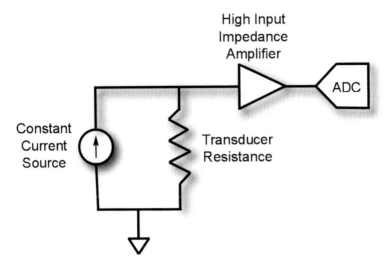

Fig. 3.8 Constant current measurement.

While this technique should be capable of making highly accurate measurements, over a wide range, variances in the values for R_{ref}, amplifier gain and the amplifier's offset voltage can limit its accuracy. Selecting a high-quality resistor for R_{ref}, an OpAmp with good offset characteristics and a very stable reference voltage, as shown in Fig. 3.9, will allow accurate measurements of R_t to be made over a reasonable range for common resistive transducers.

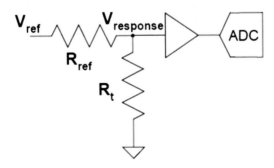

Fig. 3.9 Resistive divider.

3.2.15 Sensors and Microcontroller Interfaces

Embedded systems often employ inputs/outputs in the form of analog voltages, currents and/or digital data in order to be able to obtain information from the widest variety of input devices such as sensors, and to output signals to devices such as motors, actuators, display devices, digital transmission channels, etc. This requires that microcontrollers be able to interface with both analog and digital signals of a wide variety. This is often facilitated, at least in part, by microcontrollers that have configurable I/O pins that allow some sensors to connect directly to the microcontrollers I/O pins.

Input sensors can provide data in the form of analog signals [7] that can be interpreted in terms of phase, amplitude, current, frequency, frequency shift, phase shift, other forms of modulation, or some combination thereof. The input signal may be converted to a digital format by employing an on-chip, analog-to-digital converter for data processing, logging and/or retransmission to external devices. However, it is also possible to employ on-chip peripherals such as analog filters and/or various types of analog amplifiers used in conjunction with on-chip analog-to-digital converters for input, and subsequently output an analog signal without converting the input signal to a digital equivalent.

Handling these types of analog input signals requires the availability of a number of different analog circuits, e.g., analog multiplexers/demultiplexers, analog-to-digital converters, analog comparators, analog demodulators, amplitude/frequency detectors, analog mixers, analog filters, etc. Similarly, digital input signals require digital multiplexers/demultiplexers, the ability to handle serial, parallel or both data formats, support for various protocols such as I2C, RS232 (UART, USART), CAN, SPI, Firewire, USB, etc. as well as, hardware variants of RS232 such as RS422 and RS485. Parallel data can be handled by some microcontroller's ability to simultaneously input data from a group of pins, e.g., P0–P7, for 8-bit parallel input, byte data transfers. In some applications multi-byte input data is transmitted to/from the microcontroller, one byte at a time, using such a technique.

Thus, a microcontroller is essentially a CPU that communicates/interacts in a variety of ways, e.g., by responding to interrupts generated by the peripherals, or to the state of the microcontrollers input pins, with a variety of on-chip analog/digital peripherals that in turn communicate with input devices such as sensors that provide inputs in the form of analog voltages, currents, frequencies, pulses, digital data etc. The microcontroller can apply the applicable numerical algorithms, invoke the appropriate logic and process this data regardless of the original form, apply the appropriate logic sometimes based on the results of numerical processing of the input data and then transmit commands to other on-chip peripheral devices to provide the necessary output signals external devices.

3.3   Recommended Exercises

3-1 What are the key factors to be taken into consideration when choosing one or more sensors to be used in an embedded system?

3-2 A $NCP18XH103F03RB$ thermistor, has a dissipation factor of 1 mW/$°C$. Assume that voltages applied across the thermistor are 3.9 and 1.3 V, 25 and 100 °C, respectively. Given that the thermistor's resistance is 10 kΩ at 25 °C and B = 4538, calculate the temperature error (Terror). How can such errors be reduced? What is the thermistor's resistance at 100 °C?

3-3 Metallic and nonmetallic objects are moving on a conveyor system, The non-metallic objects consist of either ceramic, glass or wood. The metallic objects may be magnetic or nonmagnetic and exhibit resistances that depend on the specific metallic content of each object. Design an array of sensors capable of distinguishing between each type of object.

3-4 A temperature sensor has a highly nonlinear temperature versus resistance characteristic and is sensitive to variations in ambient temperature. Explain how to use such a sensor in conjunction with an 8051 to obtain a very accurate temperature control system.

3-5 The Nyquist criteria suggests that a sensor should be sampled at twice the rate of the highest frequency in the sensor signal in order to create the most accurate data set. In some cases the data is sampled at a much higher rate. Why?

3-6 Cross-sensitivity is defined as a gas sensor's reaction to other stimuli that can alter the sensor's response. Sketch a Wheatstone bridge with one, two and four pressure-sensing resistors and give an equation for the output in each case. Each of the sensors is sensitive to ambient temperature variations. What can be done to minimize cross-sensitivity in this case?

3-7 Explain how a tin-oxide sensor works. What can it measure and what are the design considerations when using it in a microcontroller application?

3-8 The Young's modulus for a piezoelectric disk transducer, charge coefficient in the y-direction, dielectric coefficient and dielectric coefficient in a vacuum are 71 GPa, 0.550 coulombs/meter, 450 and $8.86x10^{-12}$, respectively.

3-9 Gyroscopes can serve as a type sensor which can measure the angular rate of change about any arbitrary axis of an object. Explain a technique employing integration that allows the translation and orientation of an object to be determined/measured. Your explanation must include an 8051 assembly language program that is capable of simple integration.

3-10 The conducting wire in a strain gauge has a length of 3.75 cm and a diameter of 0.0625 mm. It is connected to the top of a cantilever beam. Prior to leading, the beam resistance of the strain gauge is 7.84 Ω. Applying a load to the beam extends the length of the wire to extending the length of the wire by two-tenths of a millimeter so that its electrical resistivity becomes 2×10^{-8}Ω. Poisson's ratio for the wire is 0.443441, find the change in resistance in the wire due to the strain to the nearest hundredth of an Ohm.

3-11 A liquid's temperature is to be measured using an NTC thermistor with a Stein-Hart coefficient of $A = 1.245 \times 10^{-3} K^{-1}$, $B = 2.5 \times 10^{-4} K^{-1}$ and $C = 1 \times 10^{-7} K^{-1}$. The measured resistance when immersed in the liquid is 3.75 kΩ. What is the temperature of the liquid in Centigrade degrees?

3-12 A coil of copper wire can be used as a sensor for measuring magnetic field strength. Assume a 60 turn, 1.8 mH, close-packed coil (inductor), 1 cm in length with a cross-sectional area of $0.75\ cm^2$. If the B field strength is 0.8 Tesla, what is the maximum voltage that can be induced in the coil. Is this sensor linear?

3-13 A strain gauge is attached to a metallic cylinder 2.5 cm in diameter and 2.5 m in length. The cylinder is subjected to strain and extends 25.4 mm. If the initial resistance of the strain gauge was 128 Ω, Estimate what the resulting resistance of the strain gauge will be.

3-14 Ferroelectric Lithium Niobate, (FLN), a man-made material not commonly found in nature, has a number of very useful optical, piezoelectric, electro-optic, elastic, photoelastic, and photorefractive properties. It can be used in pressure sensors employing the piezoelectric effect. If its cross-sectional area is $2.5\ cm^2$ determine the coulomb surface charge and the resulting piezoelectric voltage when the transducer is subjected to 3.14 atmospheres of pressure.

References

1. CE202479 -PSoC 4 capacitive liquid level sensing. Document No. 002-02479 Rev. Cypress Semiconductor (2015)

2. CE210514-PSoC3, PSoC 4, and PSoC 5LP temperature sensing with a thermistor. Document No. 002-10514Rev.*B. Cypress Semiconductor (2018)
3. E. Denton, Tiny Temperature Sensors for Portable Systems. National Semiconductor (2001)
4. T. Dust, PSoC 3/PSoC 5LP - Temperature Measurement with Thermocouples. AN75511. Document No. 001-75511Rev.*F17. Cypress Semiconductor (2017)
5. D.G. Fernandez, A. Blanco, A. Duran, C. Jimenez-Jorquera, Olimpia and Arias de Fuentes. Portable measurement system for type microsensors based on PSoC microcontroller. J. Phys. Conference Series (2013)
6. J. Fraden, *AIP Handbook of Modern Sensors*. Physics, Design and Application. American Institute of Physics (1993)
7. V.S. Kannan, J. Chen, PSoC 3, PSoC 4,and PSoC 5LP–temperature measurement with a diode. AN60590. Document No. 001-60590 Rev. *K 1 Cypress Semiconductor (2020)
8. R. Lossio, PSoC 3, PSoC 4, and PSoC 5LP temperature measurement with a TMP05/TMP06 digital sensor. AN65977 (2016)
9. P. Madaan, Maintaining accuracy with small magnitude signals. EE Times Design (http://www.eetimes.com/design) (2011)
10. T. Matthams, Introduction to mechanical testing. University of Cambridge -DoITPoMS [Online]. [Citação: 2013 de June de 6]
11. R. Ohba, *Intelligent Sensor Technology* (Wiley, Hoboken, 1992)
12. Portable measurement system for FET type microsensors based on PSoC microcontroller. J. Phys. Conf. Ser. **421**(1), 1–5, 2015- · (2013)
13. PSoC 4 Magsense^{TM} inductive sensing. Document Number: 002-24878 Rev.**. Cypress Semiconductor. Revised August 20, 2018
14. G. Singh, S. More, S. Shetty, R. Pednekar, Implementation of a wireless sensor node using PSoC and CC2500 RF module, in *2014 International Conference on Advances in Communication and Computing Technologies (ICACACT)* (2014)
15. J.S. Steinhart, S.R. Hart, Calibration curves for thermistors. Deep Sea Res. Oceanogr. Abstr. **15**(4), 497–503 (1968)
16. D. Van Ess, PSoC 1 temperature measurement with thermistor. PSoC 1 temperature measurement with thermistor. Document No. 001-40882 Rev. *E. Cypress Semiconductor (2002)

Embedded System Processing and I/O Protocols

4

Abstract

Embedded systems are capable of providing different types of functionality, including, but not limited to, data collection/processing/transmission. Because many embedded systems carry out data processing on data that either began as digital data, or was subsequently translated into the digital equivalent of an analog signal, or signals, microcontrollers must be capable of carrying out a number of different low-level computational tasks such as addition, subtraction, multiplication and division, as well as, bit manipulations, shift operations, bit testing. overflow and underflow handling, array manipulations, together with various loop and nesting functions.

4.1 Processing Input/Output

Embedded systems are capable of providing different types of functionality, including, but not limited to, data collection/processing/transmission. Because many embedded systems carry out data processing on data that either began as digital data, or was subsequently translated into the digital equivalent of an analog signal, or signals, microcontrollers must be capable of carrying out a number of different low-level computational tasks such as addition, subtraction, multiplication and division, as well as, bit manipulations, shift operations, bit testing. overflow and underflow handling, array manipulations, together with various loop and nesting functions. Computation of algorithms such as Fast Fourier Transforms, digital filtering, etc., are facilitated by special functions, such as those provided by a MAC.[1]

In some cases, if extensive high-speed computation of a large amount of data must be subjected to complex algorithms, where the time to compute is a concern [10], digital signal (DSPs) or other types of specialized processors may be employed, whose architecture is optimized to perform high speed, often complex, computations. The availability of extremely fast ADCs has made it possible to

---

[1]The combination of multiplication followed by addition is a common occurrence in digital computation for computation of dot-products, matrix multiplication, Newton's method, polynomial evaluation, etc. As a result, many modern computer architectures include hardware modules optimized to carry out a multiplication operated that is followed by an addition. This multiply-accumulate capability is referred to as MAC.

© Springer Nature Switzerland AG 2021

E. H. Currie, *Mixed-Signal Embedded Systems Design*,

https://doi.org/10.1007/978-3-030-70312-7_4

apply a variety of complex digital algorithms to radio frequency (RF) applications. The computational tasks are passed to the DSP or other specialized co-processors and the results of the computation are then made available to the microcontroller via shared memory, or other means. DSPs are specialized microprocessors designed to compute complex algorithms used in digital imaging, radar, seismic, sensor array, statistical, communications, biomedical signal processing. Some examples of such algorithms are shown in Table 4.1. Field Programmable Gate Arrays (FPGAs) are also used as co-processors.

FPGAs are also capable of supporting parallel processing by employing multiple CPUs[2] (multi-core), Multipliers-Accumulators(MACs) and other special function devices such as graphics processors, DMA controllers, etc. on a single chip. MACs provide very high speed multiplications and have the capability of adding the products to a *running sum* of previous products. These algorithms, and others are used to select specific input signals to synthesize, compress, enhance, restore, recover and recognize signals, as well as, predict future values and/or interpolate missing values of a signal. In such cases, the microcontroller can act as the prime controller passing data to the DSP for processing,

Table 4.1 Algorithms used in an embedded system.

Discrete Fourier Transforms (DFT)	Spectral Analysis
Bilinear Transform	Digital Signal Processing
Real-Time Convolution	RADAR Systems
Z-Transforms	Circuit Design Applications
Coordinate Rotations/Translations	Computer Generated Imagery
Quadratic & Higher-Order Polynomials	Numerical Analysis
Discrete Fourier Series	Digital Signal Processing
Discrete Wavelet Transform (DWT)	Image Enhancement
Least-Squares Computation (LMS)	Curve Fitting
Speech Processing Algorithms	Speech Recognition
Correlation Algorithms	Signal Detection
Computer Vision Algorithms	Robotics
Ray Tracing Algorithms	Optics Design
Array Processing Algorithms	3D Graphics
Multimedia Algorithms	Image Compression/Decompression
Character Recognition Algorithms	Optical Character Recognition
Speech Recognition Algorithms	Security
Image Processing Algorithms	Image Enhancement
Video Processing Algorithms	Video Transmission Systems
Target Detection Algorithms	Defense Systems
Compression Algorithms	Video Compression/Decompression
Fingerprint Processing Algorithms	Law Enforcement
EEG/EKG Processing Algorithms	Medical Applications
Digital Filters	Signal Processing
	FIR
	IIR
	Allpass
	Adaptive
	Comb

---

[2]Currently available technology is capable of supporting up to 16 CPUs per FPGA. But the goal is 1000+ per FPGA.

and then carry out the required operations on the result of the DSP calculations. Specialized math processors such as the Intel 80387 floating-point co-processor have the ability to fully control the address and data busses and are often highly optimized for performing certain specialized functions, but they may have somewhat restrictive use.

Microcontrollers with integral MACs are very useful for carrying out multiplications for which the product is *accumulated* a common requirement for many digital signal processing algorithms, such as, vector-dot-products (audio, video, images), Finite Impulse Response (FIR) and Infinite Impulse Response (IIR) filters, Fast Fourier Transforms (FFTs), discrete cosine transforms (DCTs), convolution algorithms, etc.

For vector-dot-product calculations, a typical calculation can be represented by:

$$x = \sum a_i * b_i \tag{4.1}$$

and for convolution calculations,

$$y[n] = y[n] + x[i] * h[n - i] \tag{4.2}$$

similarly for matrix multiplication,

$$\begin{bmatrix} x_1 \\ x_2 \\ x_3 \\ x_4 \end{bmatrix} = \begin{bmatrix} a_{11} & a_{12} & a_{13} & a_{14} \\ a_{21} & a_{22} & a_{23} & a_{24} \\ a_{31} & a_{32} & a_{33} & a_{34} \\ a_{41} & a_{42} & a_{43} & a_{44} \end{bmatrix} \begin{bmatrix} b_1 \\ b_2 \\ b_3 \\ b_4 \end{bmatrix} \tag{4.3}$$

where,

$$x_1 = a_{11}b_1 + a_{12}b_2 + a_{13}b_3 + a_{14}b_4 \tag{4.4}$$

$$x_2 = a_{21}b_1 + a_{22}b_2 + a_{23}b_3 + a_{24}b_4 \tag{4.5}$$

$$x_1 = a_{31}b_1 + a_{32}b_2 + a_{33}b_3 + a_{34}b_4 \tag{4.6}$$

$$x_1 = a_{41}b_1 + a_{42}b_2 + a_{43}b_3 + a_{44}b_4 \tag{4.7}$$

which requires 16 multiplications and 9 additions each time the vector **x** is calculated.

While reliance on co-processors for computation of algorithms, particularly for time-critical applications, i.e. applications for which execution times are an important consideration, has proven successful, in recent years embedded systems have begun to be a synthesis of networking, transmission pathways, sensors, data and signal processing, As a result, microcontroller manufacturers have begun to revise some of their standard microcontroller architectures to allow them to perform one, or more, of the functions formally assigned to co-processors, or other external hardware.

System processes and control algorithms, of the type addressed by embedded systems, can often be expressed in terms of one, or more, systems of linear or in some cases differential equations that involve resistance, capacitance, inductance, OpAmps, etc., such as:

$$b_{11}\frac{d^2y(t)}{dt^2} + b_{12}\frac{dy(t)}{dt} + b_{13}y(t) = a_{11}\frac{d^2x(t)}{dt^2} + a_{12}\frac{dx(t)}{dt} + a_{13}x(t) \tag{4.8}$$

$$b_{21}\frac{d^2y(t)}{dt^2} + b_{22}\frac{dy(t)}{dt} + b_{23}y(t) = a_{21}\frac{d^2x(t)}{dt^2} + a_{22}\frac{dx(t)}{dt} + a_{23}x(t) \tag{4.9}$$

$$b_{31}\frac{d^2y(t)}{dt^2} + b_{32}\frac{dy(t)}{dt} + b_{33}y(t) = a_{31}\frac{d^2x(t)}{dt^2} + a_{32}\frac{dx(t)}{dt} + a_{33}x(t) \qquad (4.10)$$

and equivalently in the form of a system of "difference equations" in the digital domain as:

$$b_{11}y[n-2] + b_{12}y[n-1] + b_{13}y[n] = a_{11}x[n-2] + a_{12}x[n-1] + a_{13}x[n] \qquad (4.11)$$

$$b_{21}y[n-2] + b_{22}y[n-1] + b_{23}y[n] = a_{21}x[n-2] + a_{22}x[n-1] + a_{23}x[n] \qquad (4.12)$$

$$b_{31}y[n-2] + b_{32}y[n-1] + b_{33}y[n] = a_{31}x[n-2] + a_{32}x[n-1] + a_{33}x[n] \qquad (4.13)$$

where b_{mn} and a_{mn} are constants. Note that this system of difference equations only involves decrementing of integer values, multiplication and addition, which are all functions easily performed by a CPU that has MAC support.

4.2   Microcontroller Sub-systems

The discussion that follows is restricted to mixed-signal, microcontroller architectures, i.e., microcontrollers consisting of a microprocessor and some number of analog and digital subsystems, often referred to as "modules". In some cases, the functionality of the analog and digital subsystems will be found to be constrained to a limited range and configurability. In other cases, such as that of PSoC, Cypress' family of Programmable System(s) On-Chip, an unusually high degree of variability, configurability and functionality is provided, as shall be demonstrated throughout this textbook [9].

Compared to a typical personal computer, microprocessors are rather limited in terms of memory resources, a number of registers, clock speeds, program and data capacity, instruction sets, multitasking capability (if any), etc. Microcontrollers typically have, at a minimum, on- chip support for analog-to-digital, digital-to-analog and perhaps pulse width modulation depending on the manufacturer. Microcontroller instructions are typically 8–16 bits wide and interrupt support is relatively limited, compared to personal computers.

Microcontrollers typically include the following subsystems:

CPU—a Central Processing Unit , consisting of an Arithmetic Logic Unit (ALU), e.g., 8051- or ARM-based, microprocessor architectures. The ALU performs mathematical operations such as addition, subtraction, multiplication, division and logic operations such as equality, less than, greater than, AND, OR, NOT, shift right, shift left, etc., and access to very fast, local registers that are used to carry out these operations.

The central processing unit, or as it is more commonly known, the CPU, is the heart of the embedded system and responsible for executing a series of predefined and stored instructions, known collectively as "the program". The CPU fetches instructions from memory, decodes them, performs the instruction and stores the results. A program counter (PC) keeps track of the location in memory from which the next instruction is to be fetched. In cases in which the previous instruction has been executed, and the results stored before the next instruction is available, the CPU must then wait for the new instruction to be loaded before it can begin to decode and execute it. This can result from the program residing in a relatively slow memory compared to the CPU's execution speed.

In some systems, instructions are preloaded from slow memory to a small amount of fast memory, called *cache* and to be fetched from the cache as required. The moving of instructions to the cache occurs as a *background task*, i.e., this particular activity does not require CPU involvement and occurs at a rate sufficient to ensure that instructions are available as needed. Alternatively, so-called *pipe-lining* is sometimes used.[3]

Once an instruction has been *fetched* it must be decoded, to determine what actions are to be taken by the CPU. Typically, these contain specific CPU instructions, operands (or their locations) and locations to which the results are to be written. Typical opcodes might be ADD (addition), SUB (subtraction), MOV (move), etc., and the operands might be characters, numerics, addresses of memory locations, registers within the CPU, or in memory, etc.

Memory—utilized by the CPU for program and data storage that may in fact consist of several types of memory such as SRAM,[4] RAM[5] and EEPROM.[6] Some microprocessors/microcontrollers support both on-chip and off-chip memory. But support for off-chip memory is, in some cases, limited in terms of performance, and/or memory size. Microcontrollers utilizing *paged memory*[7] require the designer to keep track, programmatically, of what is stored on each page, accessing the various pages, etc.

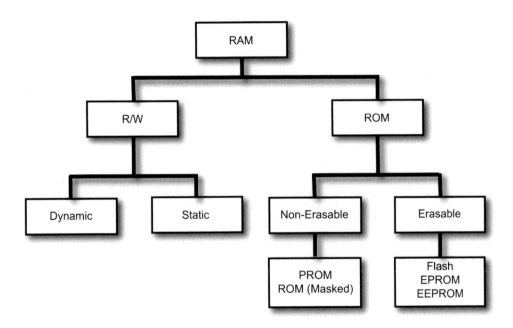

Fig. 4.1 Classification of the types of memory used in/with microcontrollers.

Memory can be classified in terms of its read, write, programmability and erasability characteristics, as shown in Fig. 4.1 viz:

---

[3]Pipe-lining is a technique that fetches the next instructions before the CPU has finished executing the current instruction.

[4]SRAM—Static Random Access Memory (requires no refresh).

[5]Random Access Memory—memory whose storage locations can be arbitrarily accessed.

[6]EEPROM Electrically Erasable Programmable Read-Only Memory that is non-volatile and used in computers and other electronic devices to store small amounts of data that must be saved when power is removed, e.g., calibration tables, or device re-configuration to allow in-field updating.

[7]A system of memory management that causes non-contiguous parts of memory called *pages*, to be treated as contiguous thus creating a virtual, linear memory space.

- ROM—read-only memory
- PROM—programmable read-only memory
- EPROM—Erasable PROM (UV)
- EEPROM —electrically erasable PROM
- FLASH—Mostly read-only, non-volatile
- RAM—read-write, volatile

Memory for microcontrollers and microprocessors falls into the two broad categories: read/write (R/W) and read-only memory (ROM) and as either volatile or non-volatile depending upon whether the program and/or data is to be retained in memory, in the absence of supply voltages. Read/Write memory can be further categorized as either static (SRAM) or dynamic memory (DRAM).

Static RAM (SRAM) consists of large number of so-called "cells" each of which consists of two inverters as shown in Fig. 4.2. This combination of inverters creates a bi-stable device thus making it a viable memory device. A dynamic RAM cell, capable of storing a single bit, consists of a transistor and capacitor combination as shown in Fig. 4.3. While dynamic RAM provides higher density storage than static memory, SRAM is generally much faster than dynamic memory. However, it is also more expensive because it takes as many as four to six transistors (MOSFETs), per bit of storage to implement, but unlike dynamic memory it does not need to be refreshed.

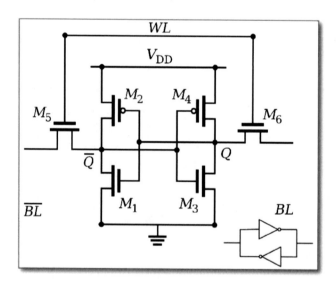

Fig. 4.2 An example of an SRAM cell.

On-chip Flash memory is non-volatile and often used for on-chip program storage while SRAM is employed on-chip to provide fast program execution, for cache RAM[8] and volatile data storage. Depending on the application, DRAM and Flash, SRAM and Flash, or mixtures of SRAM, DRAM and Flash may be used for off-chip storage. Dynamic memory utilizes as little as one transistor per bit but also requires a capacitor for the storage of each bit. Because of capacitive leakage, it is necessary to refresh dynamic memory periodically, e.g., thousands of times per second, in order to maintain memory integrity.

---

[8]Cache memory is defined as temporary memory storage that allows frequently needed information to be accessed much faster than from main memory.

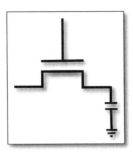

Fig. 4.3 An example of a dynamic cell.

- **Analog Subsystem** —Microcontrollers employ various combinations of analog functions such as those provided by OpAmps, comparators, current/voltage analog-to-digital (A/D) and digital to analog (D/A) converters, mixers, analog multiplexers, programmable gain amplifiers, instrumentation amplifiers, transimpedance amplifiers, filters, etc.
- **Digital Subsystem**—Similarly, digital functions such those provided by as counters, timers, CRC, PRS, PWM, quadrature decoder, shift register, logic (AND, NAND, NOR, NOT, OR, Bufoe, D flip-flop, logic high, logic low, multiplexer, de-multiplexer, virtual multiplexer, look-up table), precision illumination signal modulators (PRISMs), display (LCD) control and status registers, are also to be found in some microcontrollers.
- **Internal Bus Structures**—Obviously, connections between the microprocessor and the various subsystems in a microcontroller are a combination of fixed and variable interconnections and serves as communications pathways between the subsystems, memory, CPU and external world devices and pathways. Microcontrollers that support programmatic changes in internal connections make it possible in some cases to actually reconfigure the internal "wiring" in real-time so that optimal utilization of the microcontrollers internal resources can be achieved and the embedded system can adapt to changing conditions and functional requirements.
- **GPIO System**—A microcontroller's interface (General Purpose I/O system) communicates with external devices and peripherals via its pins. In some cases, the pins are grouped in sets of 8 and referred to as a "port", e.g., for byte I/O transfers. Whether treated as a group of pins, or individually, all GPIO pins are usually configurable either as output or input pins. The impedance characteristics, sourcing and sinking capability of the pins are in some cases configurable, depending on the device and the application. Some GPIO interfaces are also voltage tolerant so that a microcontroller operating at voltages below the voltages applied to one or more pins can operate normally, i.e., without being damaged or malfunctioning.

  Microcontrollers that allows groups of pins to be treated as 8-bit parallel ports so that each of the eight pins assigned to a given port serves as a General Purpose I/O interconnect also allow each pin to have its own input buffer, output driver one-bit register and associated configuration logic. In addition, each pin is programmable with respect to the driving mode required, independent of whether it is part of a multi-bit port configuration.

  Various types of MOSFET-based pin configurations are available, as shown in Fig. 4.4. Configuration (a) is the open drain mode in which both MOSFETs are in an OFF mode causing the output to be in a high impedance state, (b) is referred to as the *strong, slow drive mode* and functions as an inverter, (c) is the high impedance mode, (d) is the *open drain mode* and is compatible with I2C interconnections, (e) is the *pull down mode* (resistive) and provides strong drive capability, (f) functions as an inverter with *strong drive capability* and (g) is the *strong pull up mode*. It should be noted that the use of resistors with the MOSFET configurations can affect the rise and fall times of the various configurations.

Output devices such as motors, actuators and other devices often require more power than can be provided by a microcomputer output channel. Also, motors may react inductively to excitation by a microcontroller, so some form of transient protection may be required, or additional power stages may be required to interface such devices to the microcontroller. Mechanical switches are frequently used for this purpose, as are optically-coupled Darlington pairs in conjunction with protective diodes capable of handling inductive voltages. A typical microcontroller is capable of sourcing 10–25 ma to external devices at nominally five-volt levels. Whether power amplifiers, other solid-state power devices such as Silicon Controlled Rectifiers (SCRs), Thyristors, high power MOSFETs, solid-state relays, optically-coupled Darlington Pairs or other isolated solid-state devices are used to drive motors, actuators, LCDs and other devices requiring significant power, care must be taken to protect the microcontroller and its peripherals, including input devices, from harmful potentials, currents, temperatures, etc.

- **Additional System functionality**—some microcontrollers have an internal boost converter that makes it possible to create voltage levels higher than the available input voltages to provide the "desired system voltage level", advanced microcontrollers can also provide additional clock functionality, the ability to monitor the Die temperature programmatically primarily in cases where it is necessary to write to internal EEPROM, an internal DMA controller,[9] EEPROM (typically

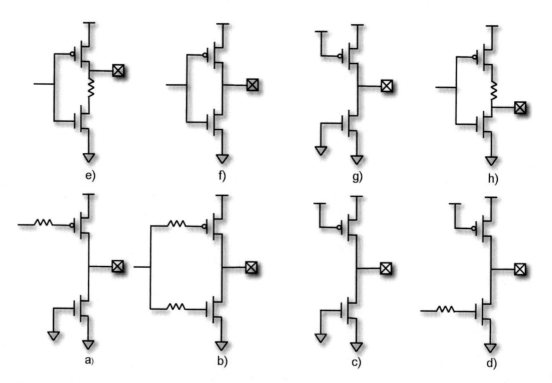

Fig. 4.4 Driving modes for each pin are programmatically selectable.

---

[9]Dynamic memory transfer refers to the ability to transfer data to/from memory without requiring significant CPU overhead.

support is provided for erasing an EEPROM sector, writing to EEPROM, blocking reads while writing, and checking the state of a write), a sophisticated interrupt handler, Real-Time Clocks (RTC), sleep timers, voltage references, etc.

4.3   Software Development Environments

In the early days of computers, programs were written in a language" that was referred to as *machine code*. Each instruction was defined in terms of a unique combination of zeros and ones.[10] The next step was to assign mnemonics, called *OpCodes*, to each such instruction which resulted in the development of *assembly language*.[11] The mnemonics assigned to each machine code usually identifies some aspect of the instruction's function, e.g., NOP for No OPperation[12] or MVI for MoVe an Immediate value, e.g., MVI A, 0, represented the instruction code "00111110 00000000" and resulted in the moving of the "immediate" value zero' to the accumulator.

With the advent of the C language,[13] developers rapidly adopted it for its portability i.e., its ability to produce applications that could run on a significant number of different hardware architectures, and the fact that it provides a somewhat higher level of abstraction than assembly language. Fortunately, C applications do not differ significantly with respect to code size, or execution speed, for most applications, when compared with assembly code. Early development was carried out in Unix-based environments, and at the command level, using a wide array of text editors, preprocessors,[14] compilers,[15] assemblers, linkers, debuggers, profilers and various, preexisting libraries of source and object code.[16] Once graphical user interfaces (GUIs) became ubiquitous, they were soon followed by Integrated Development Environments (IDEs). These environments provided a graphical user interface-based system that supported virtually all the tools required for development. These IDEs could be hosted in a variety of operating system environments including Microsoft Windows, in its various incarnations, and the many *flavors* of UNIX, MAC OS, Linux, etc.

Modern IDEs typically consist of preprocessors, compilers, assemblers, linkers, in some cases profilers, debugging tools of various levels of sophistication and collections of predefined functionality in the form of user-defined modules and/or so-called "standard libraries". The available debuggers usually provide, at a minimum, single-stepping of each line of executable source code and the setting of breakpoints, watchpoints, views of user-defined memory locations, views of registers, etc.

---

[10]Unique, that is, for a particular architecture. There is, in principle aside from any copyright issues, no prohibition on different architectures using the same combination of zeros and ones to represent the same or different instructions.

[11]It should be noted that assembly language does not, in and of itself, offer any new functionality but merely substitutes mnemonics which are related to the instruction's specific function and far more efficient to work with, from a software development standpoint, and much easier to debug.

[12] One might well ask why the need for an instruction that does nothing. While it is true that such an instruction does not result in any action, it does consume CPU cycles and therefore provides a way of introducing delays into the execution of a program.

[13] C, a general-purpose computer programming language, was developed in 1972 by Dennis Ritchie, at Bell Telephone Laboratories, and soon found widespread use for developing portable application software. C as one of the most popular programming languages, especially for embedded system applications development.

[14]Preprocessors load the applicable include files, conditional compilation requirements, and macros prior to the invocation of a compiler.

[15] Some C, and C++, compilers produce assembly language to be processed by an assembler which outputs object code and is subsequently linked to create an executable program for the target system.

[16]The first releases of C++ were in fact translators that translated C++ to C which could then be compiled by a C compiler. This approach made it possible to compile C++ source code on virtually any hardware architecture for which a robust C compiler existed.

Profilers, while less common in such IDEs, are used to determine how much execution time is spent in a particular location, or locations, in a software program. Such knowledge of *hot spots* makes it possible to *tune*, i.e. optimize, the hardware/software performance of an embedded system for efficient program execution. Debugging and profiling are generally the most effective in *single-tasking* environments. Microcontrollers running operating systems can sometimes prove difficult to debug in complex applications.

Typically a designer creates the required source code in an editor-environment, e.g., Notepad, VI, Ultra-Edit, Emacs, or an IDE with an integral editor, and then invokes an assembler, or C compiler, to create an object or assembly language source file. This can result in the generation of warnings, and/or error messages,[17] that may arise due to program inconsistencies, syntax errors, and so on If the compiler produces assembly language output, as opposed to an object file, an assembler is then invoked.

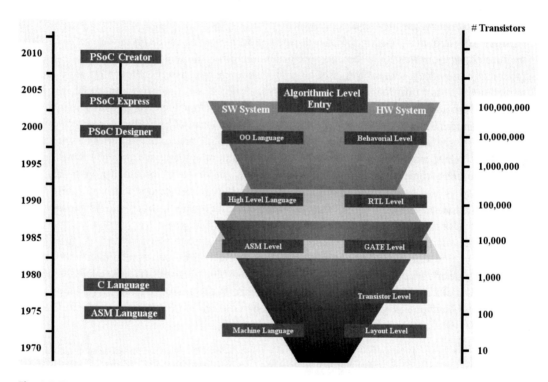

Fig. 4.5 Development tool and hardware evolution.

Then the *relocation process*, sometimes referred to as *Link-Editing*, takes place. This requires that the linker replace the symbolic references, or names, used in each library routine with the appropriate relocatable addresses. In the course of linking all the applicable object files, the linker must also attempt to resolve all unresolved symbols and report the inability to do so, as well as, other errors and/or potential errors. Linkers also provide symbolic information to assist in the debugging of programs. The linker produces a script which contains information for the locator, e.g., the stack size and location and other information that is used to create an absolute file. The linker-locator phase

---

[17] *Warnings* are indications issued by the compiler, or assembler, of potential problems with the program which may or may not, at the sole option of the designer, be ignored, as opposed to *errors* which are serious defects in a program and should be corrected before attempting to use the application.

represents the final stage of the linkage process and, among other things, determines where various aspects of the executable code will reside in the target's physical memory space. It should be noted that libraries are usually designed to be relocatable, i.e. there is no specific memory address dependence. When the linker-locator has determined where each portion of the code is to be physically located, an executable file, referred to as *firmware*, is produced which can then be downloaded to the target hardware.

In some applications, a program may be running in what is, in effect, a "virtual" memory space which appears to the program to be a linear memory space, but is in actually, non-contiguous portions of memory that are mapped into an apparently linear memory space. Debuggers typically provide support for breakpoints, watchpoints and trace buffers so that a program can be interrupted and the last "n" instructions examined, as well as, the ability to monitor/trap registers and memory locations during execution. Also, some IDEs provide simulators, although they are often nothing more than programs that allow the developer to test the program's logic. In a later chapter, a modern IDE will be examined in detail and used to illustrate various aspects of embedded system design.

Linkers perform other tasks, as well, before producing the executable file referred to as *firmware*. Once created, the firmware is downloaded into the *target* microcontroller. Some microcontrollers support real-time debugging via physical handshaking with external hardware and transferring information to external platforms for analysis. In other cases, in-circuit emulators (ICEs) or logic analyzers are employed, as debugging aids.

There are four basic types of problems encountered with an embedded system:

1. **Compile Time Errors** which are errors resulting from syntax and logic errors that are reported by the compiler. These are the most common problems encountered by developers of embedded systems.
2. **Runtime Errors** are encountered only at runtime and therefore can be quite difficult to resolve, and can require careful and often detailed analysis to resolve. Good use can sometimes be made *of isolate and eliminate* techniques. Other cases may require the use of diagnostic hardware, such as logic analyzers that allow the system to operate at full speed while providing the ability to closely track the system's activities.
3. **Hard System Crashes** in such cases the system fails to perform at all. This class of problems can be extremely difficult to resolve because all information leading up to the crash may have been lost. Many newer microcontroller-based systems have some form of hardware debugging that may be of help in debugging this class of problems. Hardware diagnostics that include so-called *deep memory*[18] can sometimes be very effective in diagnosing hard failures by recording the system's history, prior to a crash.
4. **Lock Up** refers to an embedded system getting stuck in some routine, or mode, so that it is unable to continue normal program execution. This can be a coding, or other problem, that only appears when the program is waiting for some hardware condition to be met that will allow normal program execution to continue, or as the result of timing errors, etc.

Whenever possible designers should be relieved of the necessity of dealing with many of the low level implementation details of the hardware involved and be able to engage in the development of designs at a much higher level of abstraction. Unfortunately, manufacturers have been slow, in many cases, to keep pace with hardware technology when it comes to evolving their respective

---

[18]Deep memory refers to the use of arbitrarily large amounts of memory to allow thousands, or tens of thousands, of instructions to be stored, in order to trap causes and effects that are separated by events, e.g., that are hundreds of thousands of instructions apart.

development environments. Examination of Fig. 4.5 shows that from the time of the advent of the first microprocessor in 1972, until 2001, there were virtually no significant advances in software tool development for an embedded system designers other than some modest advances in compiler technology, and minor improvements in debugging functionality, text editors and linkers. Tools such as Source Code Control Systems (SCCSs) have advanced, but they tend to be of primary use to large groups of developers working on the same source code. In recent years, IDE's such as Cypress' PSoC Designer and PSoC Creator has made significant advances in IDE technology and make it possible for application developers and designers to create increasingly more complex embedded system applications that incorporate significantly more sophisticated, mixed-signal, embedded systems.

4.4   Embedded Systems Communications

An important component of an embedded systems are the channels that support input and output [8], particularly as they relate to links between various aspects of a system, or a group or groups of systems, that employ standard communications protocols, such as CAN,[19] I2C,[20] RS232,[21] SPI[22] and an ever-increasing array of communications schemes and protocols.

Communications protocols exist for information exchange within a chip, between chips and for both long and short distances for information transfer to/from an embedded system. They may be state-based, event-based, serial or parallel communication-based, and either point-to-point (data links), or shared media networks (data highways). Master-slave configurations may involve a single master and multiple slaves, or multiple masters and multiple slaves, where point-to-point is a peer configuration and therefore there are neither masters nor slaves.

4.4.1   The RS232 Protocol

Early microcontrollers provided limited communications capability and tended to rely on the RS232 protocol, as shown in Fig. 4.6, operating at baud rates (bits per sec) varying from 60 to 115K baud. This serial data transmission protocol is fundamentally a three-wire system, in which one wire is a dedicated transmission line (Tx) a second is a dedicated receive line (Rx) and the third is maintained as a common ground for both TX and RX. Handshaking, a form of signaling between two systems linked by an RS232 connection is sometimes employed, utilizing additional "control" lines, e.g., Clear to Send, Data Terminal Ready, etc., when implementing RS232 communications in order to avoid collisions and lost data by making sure that when one system is transmitting the other is listening, and vice versa.

Data is commonly transmitted from one location to another in the form of "packets" which may be as little as a single byte. The format of these packets is based on a number of well known, standard

---

[19]Controller Area Network (CAN or CAN-bus)[9] is a message-based, standard protocol that allows microcontrollers and devices to communicate. It has been used in automotive, industrial automation and medical applications.

[20]The Inter IC bus (*I2C*, I^2C or IIC) is a two-wire, bidirectional bus that was developed by Philips originally as a 100 Kb/s bus. Currently, the protocol supports a maximum data rate of 3.4 Mb/s.

[21]RS-232 (Recommended Standard 232) is a standard hardware protocol for serial transmission of binary data signals and is most commonly used in conjunction with personal computer serial ports and external devices.

[22]SPI (Serial Peripheral Interface) [5] is a full-duplex, four-wire, serial bus serving as a synchronous, serial, data link. Communication occurs in a master/slave mode with the master devices initiating the data frame.

protocols. Each packet may include a Cyclic Redundancy Check (CRC)[23] byte or a parity bit[24] which is used to detect the receipt of a packet that was corrupted during transmission. This allows the receiver to ask that the packet be re-sent by the transmitter and provides a simple method for assuring some level of transmitted data integrity. Even if the receiver is not able to request a retransmission of one or more packets, the receiving system is at least aware of the fact that it has received compromised data. While RS232 systems are still in use, they are rapidly being replaced by other protocols such as the Universal Serial Bus (USB).

Additional treatment of this protocol is provided in Chap. 8, along with a discussion of the related RS422 and RS485 protocols.

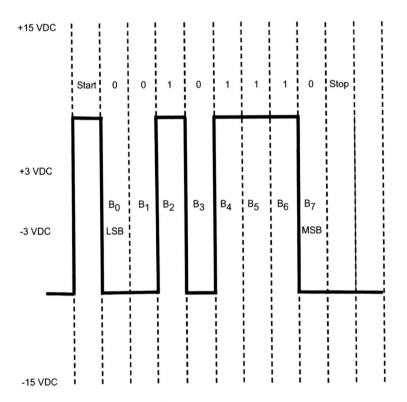

Fig. 4.6 The RS232 protocol (1 start bit, 8 data bit, 1 stop bit).

---

[23] A CRC check is based on division of a packet, referred to as the "message", containing one or more bytes by a polynomial resulting in a CRC value, e.g.,16- or 32-bit, which is then appended to the packet prior to transmission. Upon receipt of a message, it is divided by the same polynomial. Note that this division includes the CRC value in the dividend. If the result is zero then the integrity of the message is confirmed, otherwise, a transmission error has occurred.

[24] A parity bit is a bit added to a group of bits, e.g., a byte, to indicate whether the number of bits having the value '1', exclusive of the parity bit, is an even or odd number. For even/odd parity, the parity bit is set to '1' indicating that the number of bits in the group is odd/even, thus making the total number of bits even/odd, inclusive of the parity bit. Parity is typically used in systems in which single bit errors are the most probable.

4.4.2 USB

USB [2] was originally designed to operate at speeds up to 12 Mbps but currently 480 Mbps implementations[25] are available and in widespread use. It is most commonly used to connect personal computers and a wide variety of peripherals. However, USB does have some significant limitations, e.g., it is limited to a cable length of approximately 5 m, as a result of timing limitations imposed by the USB specification. These limitations can be overcome in some respects, but related protocols such as RS422, and RS485 offer cable length support up to 4800 feet, and Master/Slave support, although at much lower baud rates than USB.

It is a four-wire system consisting of Data Plus (D+), Data (−) (which form a differential pair), V_{bus} a five volt power line and ground. Data is transmitted in packets separated by idle states. Current drain is limited to 500 milliamps. Pullup resistors on D+ and D− enable a host such as PC to determine whether it is connected to a low or full speed. USB 1.0 (low speed mode) USB 1.1 (full speed mode) and USB 2.0 (high speed mode) operate at 1.5 Mb/s, 12 Mb/s and 480 Mb/s, respectively low, full and high speed data voltages are 3.5 volts peak-to-peak, 3.5 volts peak-to-peak and 400 millivolts peak-to-peak, respectively.

Communication takes place asynchronously with error detection/correction, and device detection/configuration occurring automatically USB supports several data flow types: Bulk (aperiodic, burst mode, large packets), Control (aperiodic, burst mode, host initiated response/request), Interrupt (bounded latency, low periodicity) and Isochronous (periodic, continuous data transfer, e.g., audio and video).USB data transfers employ packets and each block of data transferred which begins with the host transferring a token that identifies the type of transfer that will occur. Data is transferred in the direction identified in the token followed by a handshake packet sent to determine whether the data was transferred successfully (Fig. 4.7).

4.4.3 Inter-Integrated Circuit Bus (I2C)

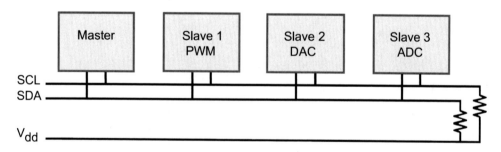

Fig. 4.7 Hardware example of the I^2C network.

The Inter-Integrated Circuit (bus) or I2C is effectively a small area network (SAN) protocol. It was created to facilitate communications between integrated circuits on a printed circuit board and is limited in terms of line distance to ≈4 m. Both I2C and SPI rely on a clock signal (max 100 kHz) on one wire (SCL), data (SDA) on a second wire and a third wire for common ground. I2C is a bidirectional system, with the direction of data determined by the I2C protocol and no limit on the

---

[25]Equivalent to 60 Mb/s.

length of a data transferred. Slave addresses are 7–10 bits and each byte transferred is acknowledged. I2C speeds fall in the range from 100 Kb/s to 3.4 Mb/s. Distinct start and stop conditions are imposed and slaves each have a 7–10 bit address. The master generates the clock, sets the start/stop conditions, transmits a slave address and determines the direction of data transfer.

4.4.4 Serial Peripheral Interface (SPI)

The Serial Peripheral Interface (SPI), originally developed by Motorola, is a protocol that was created primarily for communications with peripheral devices. Data is transferred synchronously but the data is transferred along with the clock signal and therefore the clock rate is variable. SPI is a master-slave protocol, as illustrated in Fig. 4.8 and the master controls the clock signal. There can also be multiple slaves, as shown in Fig. 4.9, but no data transmission unless a clock signal is present. SPI can be operated as either a "single-wire" system, or in full-duplex mode to allow transmission in both directions simultaneously.

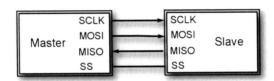

Fig. 4.8 A graphical representation of the simplest form of SPI communication.

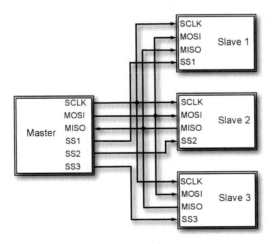

Fig. 4.9 SPI—single master multiple slaves.

4.4.5   Controller Area Network (CAN)[26]

The CAN bus was developed in the 1980s as a low cost, multi-master, serial bus specifically intended to be capable of operating in electrically noisy environments. CAN supports Master-Slave, Peer-to-Peer and Multi-Master operating modes. It was first used in an automotive environment by Mercedes-Benz in 1992. It has since become "the standard" in the automotive industry and is supported by a variety of controllers for controlling/monitoring airbags, door locks, the vehicle's power train, anti-locking brakes, windshield wipers, rain detection sensors, engine timing, the dash panel illuminators/indicators, seat heating, seat belt systems, seat position systems, navigation aids, automotive voice/data communications, cruise control, mirror adjustment, radio/CD player systems, etc.

The physical data path is provided by ribbon cable (RC), shielded twisted pair (STP) or unshielded twisted pair (UTP). Transmission is non-synchronous, so that any node on the bus is able to transmit provided that the bus is not then in use. However, it is also possible for multiple nodes to initiate transmissions contemporaneously, in which case bitwise arbitration is invoked to determine which message has the highest priority.

Four types of messages are supported:

1. Data Frame[27]—this is the frame used to transmit data.
2. Error Frame—this frame alerts the other nodes on the network that the data integrity of the data frame has been compromised and instructs the master to re-send the data frame.
3. Overload Frame—this message occurs when a device is unable to receive data.
4. Remote Frame—is a frame request for data to be transmitted.

Every node on the bus is able to listen to the bus traffic and therefore should a node detect an error in a transmission it is able to request that the transmitter, either a master or another node, resend the message. The transmission of a frame begins with the transmission of a Start Of Frame (SOF) bit which is then followed by the arbitration field, either 11 or 29 bits, which defines the type message and the node from which the message originated. Next, the data is transmitted beginning with 4 bits to define the length of the data, then follows the data. Next the cyclic redundancy field is sent. The transmitter computes the CRC, places it in the data frame prior to transmission and then upon receipt by the receiver, the receiver calculates its own CRC and compares it to that of the transmitter. If they are the same value then the receiver assumes that the data integrity has been preserved, otherwise, the receiver sends back a message indicating that the data frame needs to be re-sent.

4.4.6   Local Interconnect Network (LIN)

The LIN bus[28] [6] is another bus employed by the automotive industry that functions in single-master/multiple-slave modes. It is typically used in conjunction with the CAN protocol, shown in Fig. 4.10, in order to reduce costs. The LIN protocol is much cheaper to implement but has lower

---

[26]Refer to ISO 11898 for a detailed specification.

[27]Frame refers to the format used to package a message for transmission.

[28]Refer to ISO9141 for full details.

performance capability so that it often serves as a subnet to CAN. While LIN is not currently supported by PSoC 3 and PSoC 5LP it is supported by PS4C 4 and is included here simply for completeness.

LIN, a "single-wire"[29] serial communication protocol is based, in part, on the UART[30] and is designed to handle low demand, automotive applications such as power windows, lights, door controls, and other low demand applications. The maximum data rate for a LIN network is 20 Kb/s. The LIN message format is shown in Fig. 4.11. LIN's architecture is self-synchronizing so that nodes do not require crystals or resonators. A master determines message priority and order, controls error handling and provides the systems clock reference. Slaves are limited to a maximum of sixteen and listen for messages with their respective IDs. Although there can be only one master, a slave can function as a master. It should be noted that both CAN and LIN can be interconnected with higher-level networks, if required. The message frames contain a synch byte followed by an ID byte that includes information about the sender, the intended receiver(s), the purpose of the message and the field length of the data.

The frame begins with a break consisting of 13 dominant bits.[31] The next field is the synch which is defined as x55. This field slaves to adjust their baud rates so that they are synchronized with that of the bus. After the synch field has been transmitted, 1 of 64 possible ID fields is transmitted. O through 59 are data frames, 60–61 contain diagnostic data, 62 is reserved for user-defined purposes, and 63 is reserved for future use. The byte representing this field contains two parity bits and the remaining lower six bits are reserved for the ID. Slave response is a field containing from one to eight bytes of data followed by an 8-bit checksum field. Two methods are employed in creating the checksum: (1) the bytes in the data field are summed or by summing the data bytes and the ID. The latter is referred to as the "enhanced checksum;".

Arbitration Field STD ID (11 Bits) EXT ID (29 Bits)	Control Field (6 Bits)	Data Field (6-8 Bits)	CRC Field (16 Bits)	ACK Field (2 Bits)	EOF Field (7 Bits)	INT Field (3 Bits)

Fig. 4.10 CAN frame format.

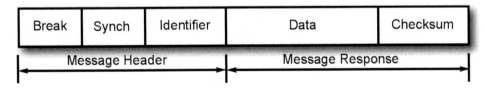

Fig. 4.11 LIN frame format.

<hr />

[29]The phrase "single-wire" is somewhat misleading, but refers to a system in which data transmission takes place using a single-wire referenced to ground.

[30]Universal Asynchronous Receive/Transit (UART) protocol—A start bit is followed by 7–8 data bits which in turn are followed by stop bit(s).

[31]Dominant bits are defined as zeros. Recessive bits are ones.

4.5   Programmable Logic

Because of the inherent cost in designing ICs and the sophistication of the equipment and techniques for manufacturing them, the most efficient way of producing them is in large quantities. However, many IC designs are needed in relatively small quantities and ideally, an IC should be manufacturable in small quantities if required but producible in large quantities if needed. This has given rise to a family of programmable logic devices [3], as shown in Fig. 4.12, which can be economically manufactured in large numbers, but can also be programmed to provide large numbers of various relatively low-volume configurations. Programmable devices are available that are field-programmable, some of which are erasable and reprogrammable to allow field updates and develop prototypes which, if successful, can then b manufactured in high volume as conventional integrated circuits. The permanent form of programmable logic is either mask-programmed, or employs fuses or anti-fuses.[32] The primary manufacturers of such devices have been Actel, Altera, Atmel, Cypress, Lattice, Lucent technologies, QuickLogic and Xilinx.

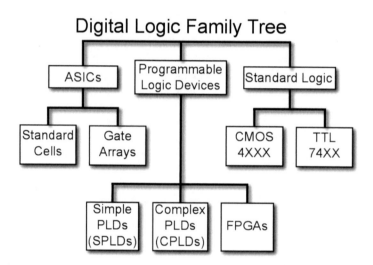

Fig. 4.12 Digital logic family tree.

The first programmable logic device appeared in 1984 and consisted of 320 gates, packaged as a 20-pin device capable of operating at speeds up to 10 MHz. The first field-programmable gate array (FPGA) appeared in 1989 and consisted of 100K gates, represented more than ten million transistors, and was capable of speeds as high as 100 MHz.

There are three basic types of Programmable Logic Devices (PLDs):

---

[32]Fuses are links, i.e., connections that can be opened and anti-fuses are potential connections that can be "linked", i.e. connected.

1. Programmable Read-Only Memory (PROMs)
2. Programmable Array Logic devices (PALs)44
3. Programmable Logic Arrays (PLAs)

The earliest, user-programmable, solid-state device that could be used to implement logic circuits in the field was the Programmable Read-Only Memory (PROM). The address lines were used as the input and the data lines as the output. However, a PROM used for this type of application is inherently more complex from a hardware perspective than is really necessary. PROMs were subsequently followed by the Field-Programmable Logic Array (FPLA), also referred to as a PLA.

A PLA consists of two planes, or levels, one with AND gates and a second with OR gates, as shown in Fig. 4.13. The links in both the AND and OR arrays are programmable which has made PLAs very versatile. However, PALs only allows the AND plane to be programmed, i.e., the OR plane connections are fixed. This makes PLAs less flexible than PALs but has the advantage that ORs switch faster than their programmable link counterparts. Logic expanders can be employed with PALs, but can result in significant propagation delays. In the case of PROMs, the AND array is fixed and the OR array are programmable (Figs. 4.14 and 4.15).

A PLD employs systems of so-called "MacroCells" consisting of simple combinations of gates and a flip-flop. Each MacroCell can be configured to provide various Boolean equations in hardware and it has input and output connections that are used by the Boolean equation. The resulting equation combines the state of an arbitrary number of inputs to produce and output that, if necessary, can be stored in the integral flip-flop until the appropriate clock signal occurs. PLAs and PALs are characterized by the number of AND gates, number of OR gates and the number of inputs [1].

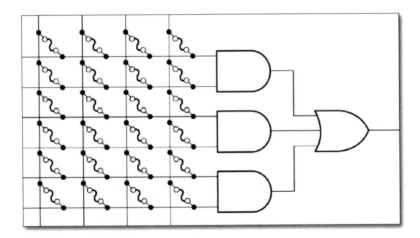

Fig. 4.13 Unprogrammed PAL.

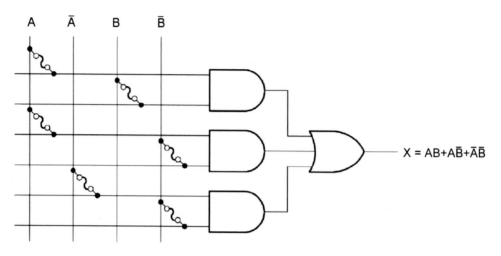

Fig. 4.14 An example of a programmed PAL.

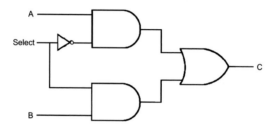

Fig. 4.15 An example of a multiplexer based on a PLD.

4.6 Mixed-Signal Processing

The earliest embedded systems were, for the most part, required to deal with analog signals in the form of voltages and relied on analog-to-digital and digital-to analog converters to interface the embedded system with the real world. In recent years, mixed-signal applications have been required to provide significant increases in digital and analog processing capability involving a wide variety of digital/analog sensors and peripherals, communications protocols that employ both digital techniques and a broad range of the electromagnetic spectrum transmission/receiving and propagation techniques and devices, often in complex combinations. The integration of complex analog and digital devices (peripherals) on a single chip has proved highly successful in addressing many of the emerging embedded system requirements and led ultimately to the concept of a programmable system-on-a-chip, or PSoC.[33]

4.7 PSoC: Programmable System on Chip

Cypress Semiconductor's family of PSoC 1/3/5 devices employ a highly configurable system-on-chip architecture for an embedded control design, providing a Flash-based equivalent of a

---

[33]PSoC is a trademark of Cypress Semiconductor Corporation.

field-programmable ASIC without imposing lead-time or NRE[34] penalties. PSoC devices integrate configurable analog and digital circuits, controlled by an on-chip microcontroller, providing both enhanced design revision capability and component count savings. A single PSoC device can provide as many as 100 peripheral functions, while requiring minimal board space and power consumption, improving system quality and reducing system cost.

All PSoC devices are also dynamically reconfigurable, so that their internal resources can be to "morphed" on-the-fly, utilizing fewer components to perform a given task. This text focuses on two particular PSoC-based architectures that provide excellent performance and unmatched time-to-market, integration, and flexibility across 8-, 16-, and 32-bit applications. These programmable, analog and digital, embedded design platforms are powered by an innovative development environment called the PSoC Creator Integrated Development Environment (PSoC Creator IDE), which has a unique, schematic-based, design-capture functionality and fully tested, libraries of prepackaged analog and digital peripherals that are easily customizable by the use of user-intuitive wizards and APIs. PSoC Creator enables designers to develop new designs in a highly intuitive manner that strongly reflects the manner in which designers think about their designs and dramatically shortens time-to-market.

The programmable analog and digital peripherals in PSoC 3/5LP, high performance 8-bit and 32-bit MCU sub-systems and capabilities such as motor control, intelligent power supply/battery management and support for human interfaces with CapSense touch sensing, LCD segment displays, graphics controls, audio/voice processing, communication protocols, and much more, make it possible for designers to address a wide variety of mixed-signal, embedded applications, including all phases of the industrial, medical, automotive, communications and consumer markets.

The PSoC 3/5LP architectures include high-precision, programmable analog resources that can be configured as ADCs, DACs, TIAs, Mixers, PGAs, OpAmps, etc. and enhanced programmable logic-based digital resources that can be configured as 8-, 16-, 24- and 32-bit timers, counters, and PWMs and advanced digital peripherals such as Cyclic Redundancy Check (CRC), Pseudo-Random Sequence (PRS) generators, and quadrature decoders. These resources allow designers to customize PSoC 3/5LP's general purpose PLD-based logic. These architectures also support a wide range of communications interfaces, including Full-Speed USB, I2C, SPI, UART, CAN, LIN, and I2S.

The new PSoC 3/5LP architectures [4] are powered by high performance, industry-standard processors. The PSoC 3 architecture is based on a new, high-performance, 8-bit 8051 processor that provides up to 33 MIPS. The PSoC 5LP architecture includes a powerful 32-bit ARM Cortex-M3 processor and is capable of providing 100 DMIPS.[35] Both architectures meet the demands of extremely low-power applications based on their availability to operate over a voltage range from 0.5 to 5 volts and hibernate current as low as 200nA. They provide a seamless, programmable design platform from 8- to 32-bit architectures with pin and API compatibility between PSoC 3 and PSoC 5LP, along with programmable routing, allowing any signal, whether analog or digital, to be routed to any general-purpose I/O to ease circuit board layout. This capability includes the ability to route LCD Segment Display and CapSense signals to any GPIO pin.

PSoC 3/5LP architectures serve as scalable platforms with the computing power of high-performance MCUs, the precision of stand-alone analog devices and the flexibility of PLDs, all within

---

[34]Non-Recurring Engineering.

[35]DMIPS refers to the Dhrystone test which is a computer benchmark developed in 1984, and used to assess the relative performance of a processor. Its execution speed for a particular processor is measured in terms of *millions of instructions per second* (MIPS). However, this is solely a test of the performance of the CPU and not of external memory access/read/write performance, or that of any peripheral devices, which are often much slower than the CPU.

the scope of PSoC 3/5LP's powerful, easy-to-use design environment. Thus, designers of 8-, 16- and 32-bit applications able to fully exploit the inherent flexibility and integration of PSoC 3/5LP's true system-level programmability and extend the concept of programmability beyond instructions for the processor to configuring peripherals and customization of digital functions.

PSoC 3/5LP's internal architecture, cf. Fig. 4.16, consists of 14 configurable digital modules (PWM, UART, A/D, etc.), and 10 analog modules (A/D, filter, etc.) and a CPU core (either 8051 or ARM Cortex-M3) with interrupt controller, internal oscillator, digital clocks, Flash, SRAM, I2C and USB controllers, switch-mode pump, decimator, MAC. All these resources are supported by an extensive programmable, interconnect and routing facility that provides virtually unlimited configurations and interconnections of digital and analog modules and resources. External interfacing is provided by 8 ports (0–7).

4.8   Key Features of the PSoC 3/5LP Architectures

- A programmable precision analog sub-system that provides up to 20-bit resolution for the integral Delta-Sigma ADC, sample rates up to 1 msps for the 12-bit SAR ADC, a reference voltage accurate to $+/-0.1$, over typical industrial temperature and voltage ranges, up to four 8-bit, 8 sMsps DACs, 1-50x PGA, general-purpose Op-amps with 25 mA drive capability, up to four comparators with 30 ns response time, DSP-like digital filter implementation for instrumentation and medical signal processing, a large library of pre-characterized, analog peripherals in PSoC Creator and CapSense functionality for all devices.
- A programmable, high-performance, digital array of "Universal Digital Blocks" (UDBs) each consisting of a combination of uncommitted logic (PLD), structured logic (datapath), and flexible routing to other UDBs, I/O or peripherals, a large library of pre-characterized digital peripherals in PSoC Creator Software, e.g., 8-, 16-, 24- and 32-bit timers, counters and PWMs.
- A customizable digital system is made possible by the full-featured general-purpose PLD-based logic provided on-chip.
- PSoC 3/5LP support high-speed connectivity support for full Speed USB, I2C, SPI, UART, CAN [7], LIN and I2S.
- A high-performance CPU sub-systems based on either PSoC3's 8-bit 8051 core with 33 MIPS performance (PSoC 3 or PSoC 5LP's 32-bit ARM Cortex-M3 core with 100 (Dhrystone) MIPS performance, 24-channel, multi-layer, Direct Memory Access (DMA) with simultaneous access to SRAM and CPU on-chip debug and trace functionality with JTAG and Serial Wire Debug (SWD)and the availability of a wide variety of industry-standard compilers and real-time operating systems.
- PSoC3/5's low-power operation modes provide an operating range from 0.5–5.0 volts with no degradation in analog performance. PSoC3/5's active power consumption is 1.2 mA at 6 MHz for PSoC 3 and 2 mA at 6 MHz for PSoC 5LP. Sleep-mode power consumption for PSoC 3 is 1 μA for PSoC 3 and 2 μA for PSoC 5LP. Hibernate-mode power consumption for PSoC 3 is 200 nA and 300 nA for PSoC 5LP.
- PSoC3/5 provide programmable, feature-rich I/O & clocking by providing interconnection of any pin to any analog or digital peripheral, LCD segment display on any pin with up to 16-commons/736 segments, CapSense on any pin for replacing mechanical buttons and sliders.

PSoC3/5 also support 2–5.5V I/O interface voltages, up to 4 domains for easy interface with systems running at different voltage domains, and a 1–66 MHz, internal, $+/-1\%$ oscillator with PLL over the full temperature and voltage range.

In the chapters that follow, more detailed discussions and illustrative software and hardware examples are provided that are related to the topics in this chapter, as well as others, with particular emphasis on the PSoC3/5 family of programmable systems on a chip.

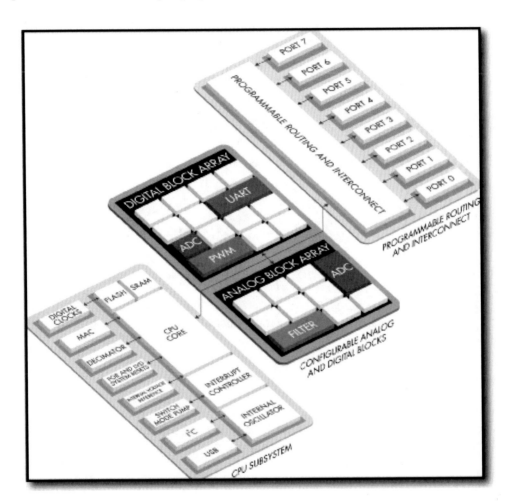

Fig. 4.16 PSoC 1/PSoC 2/PSoC 3 architectures.

4.9 Recommended Exercises

4-1 Based on the truth table shown below:

$$
\begin{array}{ccc|cc}
A & B & C & F_1 & F_2 \\
\hline
1 & 1 & 1 & 1 & 1 \\
1 & 1 & 0 & 0 & 0 \\
1 & 0 & 1 & 1 & 1 \\
1 & 0 & 0 & 1 & 0 \\
0 & 1 & 1 & 0 & 0 \\
0 & 1 & 0 & 0 & 0 \\
0 & 0 & 1 & 0 & 0 \\
0 & 0 & 0 & 0 & 0 \\
\end{array}
$$

(4.14)

create a circuit that employs a PAL with three inputs, two output and three product terms.

4-2 Create a PAL-based combinational circuit based on the following Boolean equations.

$$
\begin{aligned}
w(A, B, C, D) &= \sum(2, 12, 13) \\
x(A, B, C, D) &= \sum(7, 8, 9, 10, 11, 12, 13, 14, 15) \\
y(A, B, C, D) &= \sum(0, 2, 3, 4, 5, 6, 7, 8, 10, 11, 15) \\
z(A, B, C, D) &= \sum(1, 2, 8, 12, 13)
\end{aligned}
$$

(4.15)

4-3 Implement a programmable array with the following logic equations:

$$
\begin{aligned}
X &= AB + AC' \\
Y &= AB' + BC'
\end{aligned}
$$

(4.16)

4-4 Compare and contrast RS432, RS422 and RS485 protocols showing the benefits and the limitations of each. Explain the advantages that the Universal Serial Bus (USB) has over each, if any. How does the USB 2 standard differ from the USB 3 standard?

4-5 Using PSoC Creator design a circuit that would continuously transmit, an RS232 serial port, "Hello World" every second for 100 repetitions.

4-6 A black box has the following I/O table where A_n and B_n represent the input and output, respectively.

A_1	A_2	A_3	B_1	B_2	B_3	B_4	B_5	B_6
0	0	0	0	0	0	0	0	0
0	0	1	0	0	0	0	0	1
0	1	0	0	0	1	0	0	0
0	1	0	0	1	0	0	1	1
1	0	0	0	1	0	0	0	0
1	0	1	0	1	1	0	0	1
1	1	0	1	0	0	1	0	0
1	1	1	1	1	0	0	0	1

Design an 8x4 circuit that stores this tabulated data in ROM.

4-7 Compare and contrast the UART, SPI and I2C protocols showing the benefits and limitations of each. How do they with the RS232, RS422 and RS485?

4-8 Compare and contrast CAN with SPI and I2C protocols showing their benefits and limitations. Give examples of appropriate applications of each.

4-9 A frame of RS232 data is shown below. Identify each time interval in the data frame and identify the character being transmitted.

Data Flow ==>   *Time*

4-10 Explain, in detail, the various differences and similarities between the I2C and SPI protocols. Give several examples for which you would choose one protocol over the other and explain the basis for your choice.

References

1. M.D. Anu, I2C to SPI Bridge. AN49217. Document No. 001-49217 Rev. **. Cypress Semiconductor (2008)
2. A. Bhat, The unlikely origins of USB, the port that changed everything (29.5.19) https://www.fastcompany.com/3060705/an-oral-history-of-the-usb
3. J.M. Birkner, *PAL Programmable Array Logic Handbook* (Monolithic Memories, Santa Clara, 1978)
4. Full Speed USB (USBFS) 3.20. Document Number: 002-19744 Rev. *A. Cypress Semiconductor (2017)
5. I2C Master/Multi-Master/Slave2.0. Document Number: 001-62887 Rev. *B. Cypress Semiconductor (2013)
6. LIN Slave5.0. Document Number: 002-26390 Rev. *A. Cypress Semiconductor (2019)
7. R. Murphy, PSoC®3and PSoC 5LP–Introduction to Implementing USB Data Transfers. AN56377. Document No. 001-56377 Rev.*M. Cypress Semiconductor (2017)
8. R. Murphy, PSoC 3 and PSoC 5LP USB General Data Transfer with Standard HID Drivers. Document No. 001-82072Rev. *. Cypress Semiconductor (2017)
9. M. Ranjith, PSoC 3and PSoC 5LP–Getting Started with Controller Area Network (CAN). AN52701. Document No.001-52701 Rev. *L1. Cypress Semiconductor (2017)
10. D. Van Ess, What's Next For Programmable Devices? Cypress Semiconductor. (6/29/2009) https://www.cypress.com/documentation/technical-articles/whats-next-programmable-devices

System and Software Development

5

Abstract

In this chapter recommended design stages are discussed including signal flow, The Cypress PSoC integrated design environment (IDE), component library, PSoC debugger and Design rule checker are also described in some detail. The creation of custom components is also discussed regarding additions to the standard component library provided by PSoC Creator. Suggestions are also provided for porting applications between the various PSoC incarnations. The Intel Hex Format, which is the protocol employed when downloading the compiled executable to the target PSoC device. Also described are reentrant code, assembly coding, Big Endian vs. small Endian and 8051 code optimization. A brief discussion is provided of the implementation of a real-time operating system which is PSoC compatible and multi-threading, and concurrency.

5.1 Realizing the Embedded System

The development of embedded systems based on hardware platforms such as PSoC 3 and PSoC 5LP requires a design process that begins with a concept and ends with the completed application [1]. While there is not a "best way" to carry out such a process there are a number of widely adopted models for this type of activity, e.g.,

- Waterfall—a series of sequential steps, as shown in Fig. 5.1, viz., development of a specification, creation of a preliminary design (behavioral), development into a detailed design (structural) and full implementation (physical).[1]
- Top-down—the design progresses from an abstract description to a specific design.
- Bottom-up—begins at the lowest level with individual modules or components and evolves as an aggregate to form a larger system.
- Spiral[2]—can be described as a combination of both top-down and bottom-up methodologies. Design begins with a minimal configuration which is then iterated through incorporating an additional feature or features, then tested, evaluated, followed by the addition of more features

---

[1]The transition to the next step requires prior completion of the preceding step.
[2]This model is particularly useful when requirements are changing during the design process.

© Springer Nature Switzerland AG 2021
E. H. Currie, *Mixed-Signal Embedded Systems Design*,
https://doi.org/10.1007/978-3-030-70312-7_5

and the iteration continued, tested and evaluated until the completed design emerges, as shown diagrammatically in Fig. 5.2 [17].

- V-Cycle[3]—allows testing to occur early in the life of the project and affords an opportunity to discover faults in the design earlier in the design process. The left-hand side of the V represents the definition and decomposition process and the right-hand side represents the verification and integration processes, as shown in Fig. 5.3.

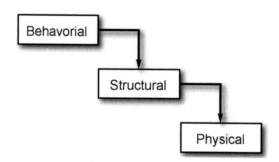

Fig. 5.1 Waterfall design model.

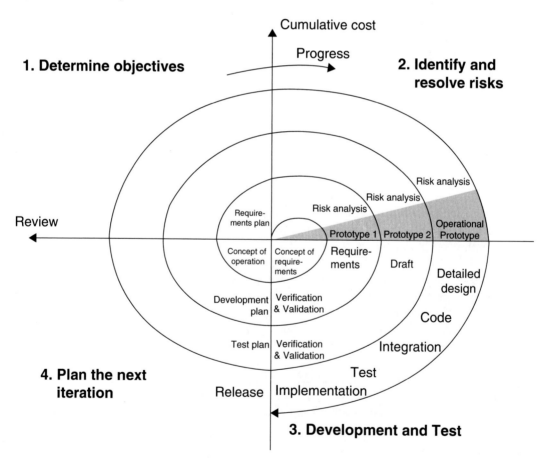

Fig. 5.2 The spiral model.

---

[3] Also referred to as the validation and verification model.

Each of these models is described generically as a *Life-Cycle Model*[4] and each offers certain advantages and disadvantages. Regardless of which a designer may choose, there are an underlying set of principles and steps that serve as the foundation for each and are discussed in the rest of this chapter, in some detail. The models discussed originated largely as software development models, but because the line drawn between hardware and software can often, in some respects at least, be regarded as arbitrary, they are quite applicable to the design process for embedded systems.

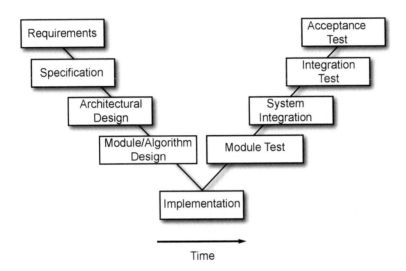

Fig. 5.3 The V-model.

5.2   **Design Stages**

In setting out to design an embedded system, the first step is to define the physical system that is to be controlled and determine from the associated requirements,[5] for the system. It is then possible to draw a functional block diagram of the system and from that point to derive a schematic. The schematic can then be used to create a signal flow diagram, an associated block diagram, or a state-space representation of the system. This results in either an open- or closed-loop system that can be implemented, and tested, to determine whether it conforms to the requirements and specifications. It is common for embedded systems that are intended to control a system, or process, to employ negative feedback loops to assure that the control system remains stable and/or to allow the system to be "self-correcting". Open-loop systems can result in drift away, or in some cases rapid departure from, the desired operating (set) points[6] and conditions for the controlled system, or process. Also, as discussed

---

[4]There is a fourth design methodology referred to as the "big bang" model. In this case, development proceeds for some period of time in relative isolation and is then released in the feverent hope that, with any luck at all, it will prove acceptable.

[5]In the present context, the term *requirements* refers to a description of an explicitly specified set of requirements whereas the term *specification* refers to a description of the system that is in full compliance with the requirements [100].

[6]Set points are the desired, or target, parameter values for a controlled system, or process, that a controller is to maintain, well within acceptable limits, e.g., a process temperature, flow rate, angular velocity, etc.

in a previous chapter, system disturbances must be taken into account, as well, even for well-designed embedded systems.

Parameters to be considered, when designing embedded systems, include:

- variables that are to be controlled,
- variables that are to be manipulated,
- variables that are associated with disturbances,
- controller output variables,
- error signals variables,
- internal setpoints employed by the controller,
- external setpoints associated with the controlled system or process,

and,

- process or system variables that are to be controlled.

Note that in addition to these variables, their respective rates of change may also be important variables, e.g., derivatives of first and higher-order. One of the figures of merit of an embedded system is robustness, which is defined in terms of sensitivity, in the present context, as a measure of an embedded system's ability to perform, as required, in the presence of external disturbances,[7] and the system's ability to maintain setpoints, in spite of external disturbances.

There are various approaches/techniques for modeling systems, e.g.,

- Deterministic versus stochastic
- Linear versus non-linear
- Continuous-time versus discrete-time
- Time-invariant versus time-variant

Systems that employ feedback, and therefore have outputs that depend on previous variable values, are often modeled as a set of differential equations for which the independent variable is time [7]. Such a system can then be mapped into the frequency domain and represented analytically in the form of transfer functions making it relatively easy to study the system's stability. Non-linear systems can be considerably more challenging than linear systems in that, as noted by Poincaré, "...it may happen that small differences in the initial conditions produce very great ones in the final phenomena. A small error in the former will produce an enormous error in the latter. Prediction becomes impossible..." [103].

5.3   Signal Flow and the Schematic View of the System

A signal flow graph is simply a graphical representation of nodes that are interconnected by several directed branches and represent variables such as inputs, outputs, etc. A directed branch illustrates the dependence of one variable on another, the gain associated with each branch and the signal flow direction. The default value for gain is unity, and the allowed direction of signal flow is defined by the direction of the arrow on each branch.

---

[7]It is defined as the ratio of the relative change in steady-state output to the relative change of a system parameter.

A *path* is defined as any branch, or continuous sequence of branches, that can be traversed in moving from one given node to a second given node. Two loops are said to be *non-touching,* if they do not share a common node. Branches that share one or more common nodes are said to be *touching.* Any path that begins and ends on the same node is referred to as a *loop.* A *forward path* is defined as a path from a *source* to a *sink.* The *gain of a path* is defined as the multiplicative product[8] of the gains of each of the branches that are part of the path. Figure 5.4 illustrates examples of some commonly encountered block diagrams, and the associated signal graphs.

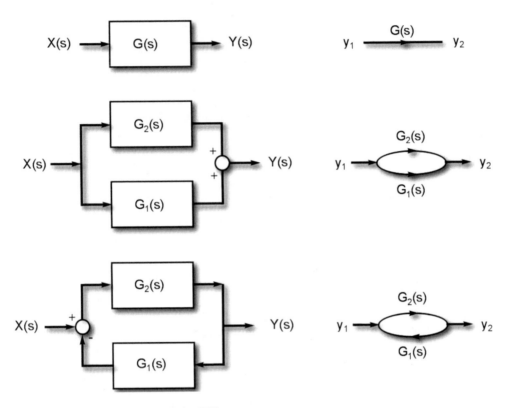

Fig. 5.4 Block diagrams and their respective SFGs.

Thus, a signal graph, in the simplest terms, is just a graphical representation of a set of linear relationships and, as such, is only applicable to linear systems[9] [103]. It is also referred to as a *directed graph* because the direction of signal flow is indicated by the arrows in each branch. The nodes in the signal graph represent the variables of a set of linear equations representing the system, as shown for example in Fig. 5.5. Note that a_{11} and a_{22} are non-touching, *self-loops* and the loop formed by a_{12} and a_{21} is also a self-loop.

---

[8]If the gain of each branch is expressed in terms of dB, then the overall gain, as expressed in terms of dB, is the summation of the dB gain of each branch.

[9]However, if the nonlinear terms of the system can be considered sufficiently small, in some cases the system may be *approximated* by a set of linear equations. If a non-linear system is constrained to operate in a linear region only, it may be possible to employ the techniques described in this chapter that would otherwise be reserved only for truly linear systems. The reader is cautioned, nonetheless, that ignoring non-linear aspects of any system can lead to unintended consequences, e.g., chaotic behavior, as was shown by Poincaré in 1908, who was well aware of the sensitivity of such systems to initial conditions.

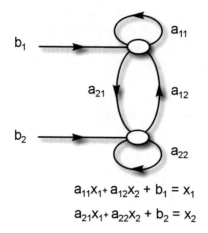

$$a_{11}x_1 + a_{12}x_2 + b_1 = x_1$$
$$a_{21}x_1 + a_{22}x_2 + b_2 = x_2$$

Fig. 5.5 A simple signal path example and the associated linear equations.

The pair of linear equations shown in this figure can be rewritten as

$$(1 - a_{11})x_1 - a_{12}x_2 = b_1 \tag{5.1}$$

$$-a_{21}x_1 + (1 - a_{22})x_2 = b_2 \tag{5.2}$$

and solving for x_1 and x_2 yields

$$x_1 = \frac{(1 - a_{22})}{\Delta b_1} + \frac{a_{12}}{\Delta b_2} \tag{5.3}$$

$$x_2 = \frac{(1 - a_{11})}{\Delta b_2} + \frac{a_{21}}{\Delta b_1} \tag{5.4}$$

where

$$\Delta \overset{def}{=} \begin{vmatrix} (1 - a_{11}) & -a_{12} \\ -a_{21} & a_{22} \end{vmatrix} = determinant \tag{5.5}$$

$$= 1 - a_{11} - a_{22} + a_{22}a_{11} - a_{12}a_{21} \tag{5.6}$$

Using so-called "block rules", it is sometimes possible to substantially simplify a block diagram of a system before attempting to create the signal path graph. Several of these rules are shown in Figs. 5.6, 5.7, and 5.8. Thus, as illustrated by this simple example, it is possible to begin with a graphical representation of signal flow for a particular system and develop therefrom an analytic representation of the system, in a straight forward manner.

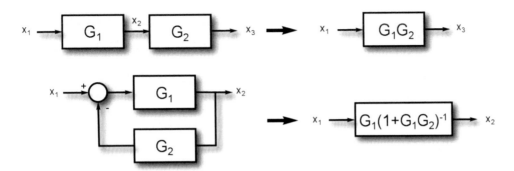

Fig. 5.6 Reduction of two blocks to one.

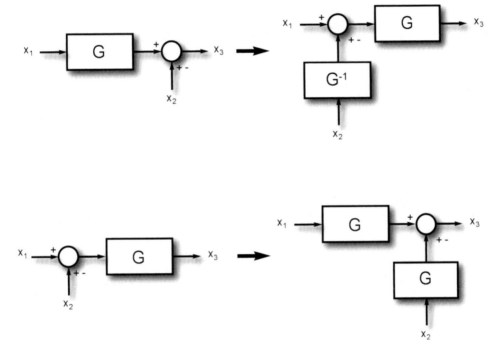

Fig. 5.7 Expansion of single blocks involving summing junctions.

5.3.1 Mason's Rule

[10]An important parameter of a system, such as those discussed in the previous section, is its overall gain [58] which can be expressed, for linear systems, by:

$$H = \frac{y_{out}}{y_{in}} = \sum_{k=1}^{N} \frac{G_k \Delta_k}{\Delta} \qquad (5.7)$$

---

[10]Mason's rule is also referred to as "Mason's gain formula" or more properly as "Masons gain equation".

where y_{in} and y_{out} represent the input and output node parameters, respectively, and H is the *transfer function* that represents the total gain of the system, G_k is the forward gain of the kth forward path and Δ_k is the loop gain of the kth loop. Δ, the determinant is formally defined, in the present context, as:

$$\Delta = 1 - \sum L_i + \sum Li L_j - \sum L_i L_j L_k + \cdots + (-1)^n \sum \cdots + \cdots \tag{5.8}$$

Equation 5.9 can be expressed in words as:

$\Delta = 1 - (sum\ all\ the\ different\ loop\ gains)$

$\quad + (sum\ of\ the\ products\ of\ all\ pairs\ of\ loop\ gains\ for\ non - touching\ loops)$

$\quad - (sum\ of\ products\ of\ all\ the\ triples\ of\ loop\ gains,\ for\ non - touching\ loops)$

$\quad + \cdots \tag{5.9}$

The gain of the circuit shown in Fig. 5.9 can be determined by applying Mason's rule as follows:

$$M_1 = A_2 \tag{5.10}$$

$$\Delta_1 = 1 \tag{5.11}$$

and therefore,

$$\Delta = 1 - L_1 = 1 - A_2 A_1 \tag{5.12}$$

so that

$$H = \frac{\sum M_j \Delta_j}{\Delta} = \frac{A_2}{1 - A_1 A_2} \tag{5.13}$$

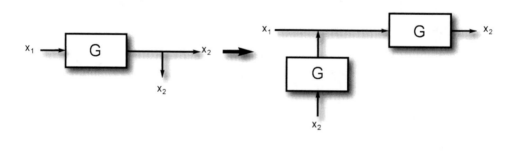

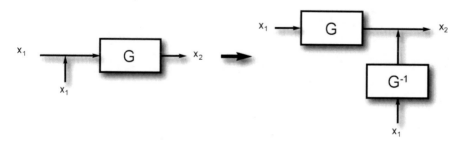

Fig. 5.8 Expansion of single blocks to two equivalent blocks.

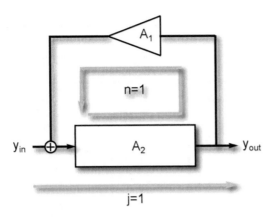

Fig. 5.9 A simple application of Mason's rule.

5.3.2 Finite State Machines

Systems that consist solely of combinational logic, do not have any explicit time dependence and therefore the outputs are not time, or prior history, dependent. Simply stated, the outputs of such systems do not depend on any previous values of the input, or output.[11] However, the outputs, at any point in time, of a finite state machine[4], are dependent on the states that the system passed through in order to reach the current state, the current input values and therefore the time required to produce the current outputs.[12]

Finite state machines (FSMs) are commonly used to implement decision-making algorithms which are a key element of most embedded systems. They are particularly attractive for systems that are highly event-driven and are often employed as an alternative to a system based on a real-time operating system. State machines are used in applications where distinguishable, discrete states exist. Finite state machines are based on the idea, that for a given system that has a finite number of states, there are two types of FSMs (Mealy and Moore) and they are distinguished by their output generation, viz., a Mealy machine has outputs that depend on the state and the input, and a Moore machine has outputs that depend on the state only. FSMs can also be represented by graphical representations in the form of state-charts and hierarchically nested states, as illustrated by the example shown in Fig. 5.10.[13]

5.3.3 Coupling and Cohesion

There are a number of important characteristics of an embedded system, e.g., fault tolerance/ prevention, identification of exceptions, exception handling and module independence in terms of coupling and cohesion. The term *coupling* refers to the relative interdependence of modules and can be broadly characterized in terms of either *tight-* or *loose-coupling*. Examples of loose- and tight-coupling are shown in Figs. 5.11 and 5.12, respectively.

---

[11] This assumes of course that the outputs of such systems are not subject to delays within the system and therefore are an immediate consequence of the inputs to the system.

[12] Chapter 8 treats FSMs in more detail.

[13] State nesting allows new states to be defined in terms of previously defined states and therefore, in terms of differences from previous states, thus fostering reusability. This technique is based on the concept of inheritance.

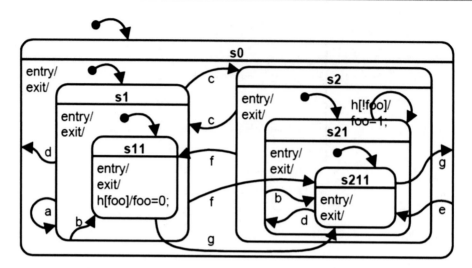

Fig. 5.10 An example of a six-state, nested state chart.

> The terms *tight-* and *loose-coupling* express the degree to which all the elements of a module are directed towards a single task/procedure and all elements directed towards that task/procedure are contained within a single component.

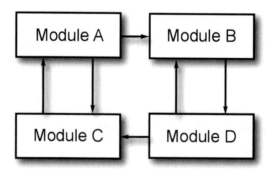

Fig. 5.11 Weak coupling between modules A, B, C and D.

A design consisting of a two, or more, loosely-coupled modules, can provide some immediate benefits because the complexity of a system is often directly proportional to the degree of coupling between modules,[14] i.e., the tighter the coupling the greater the interdependence between modules, and therefore, the greater the complexity of the interaction between them. Thus, the interactions between modules should, as a general rule, be kept to the minimum level required to allow them to interactive effectively. By using loosely-coupled modules in a design, debugging can often be significantly reduced and design modifications/troubleshooting, in the field, can be greatly

---

[14]An analogous situation can arise in applications that employ software modules that rely on global variables, when passing data by value or reference is preferable.

facilitated.[15] There are, of course, systems in which some modules, by necessity, require extremely tight coupling in order to function effectively, e.g., in cases in which error-free communication is required and/or high-speed data transfer rates are involved.

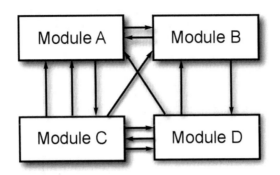

Fig. 5.12 Tight coupling between modules A, B, C and D.

Cohesion is a measure of the degree to which a set of tasks/procedures within a module are related. There are a variety of cohesion types, e.g.,

*   *Coincidental cohesion*—Procedures/tasks just happen to be grouped within a module, but the interdependence between such procedures/tasks is weak.
*   *Temporal cohesion*—Independent tasks are grouped within a module because they have some time dependencies, e.g., they must be completed within some predefined time period, and/or are sequentially ordered with respect to time.
*   *Sequential cohesion*—A given task depends on procedures that must be ordered sequentially.
*   *Functional cohesion*—The module's sole function is to carry out a specific task and the procedures within that module are restricted to those necessary to perform the task.
*   *Communication cohesion*—All the operations within a module are working on a common set of input data and/or produce the same output data.
*   *Logical cohesion*—A set of tasks/procedures that are related logically, and not functionally. Typically, several logically-related tasks reside within a given module and are selected by an external user.
*   *Procedural cohesion*—Related tasks/procedures are contained within a module to ensure a particular order of execution. In such modules, it is control, and not data, that is passed from one procedure/task to another.

5.3.4   Signal Chains

The phrase *signal chain*[16] refers to a signal's path through a series of signal-processing components, of the type used in embedded systems, that acquire data signals and process them serially. A

---

[15]Field troubleshooting is often based on the time-honored practice of "isolating and destroying" techniques. Weakly coupled modules typically make isolating problems much easier.

[16]The phrase *signal processing chain* is sometimes used instead of *signal chain*, but in either case refers to a series of signal-conditioning components [13, 14] involved in analog signal acquisition, processing and control, that are typically encountered in mixed-signal, embedded systems.

programmable signal chain (PSC) is based on programmable analog devices deployed in conjunction with digital logic and a high-performance CPU in the form of a microcontroller, microprocessor or DSP.[17] Such configurations are quite capable of providing embedded systems that are highly adaptive, versatile and effective for addressing a wide variety of mixed-signal applications. This is particularly important when designing systems that can benefit from such a system's ability to reconfigure itself, in real-time, to meet variable operating environments and conditions.[18]

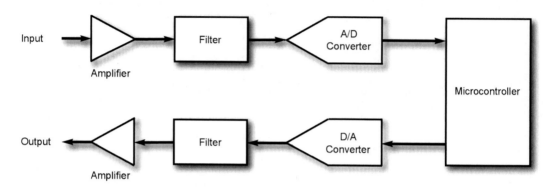

Fig. 5.13 A commonly encountered signal chain.

One of the most commonly encountered signal chains is shown in Fig. 5.13. The input to such a system is often from various types of sensors and the output may be to actuators, data channels, wireless transmission, or other devices. Input amplifiers, in the form of generic OpAmps, instrumentation amplifiers, lock-in amplifiers, radio frequency (RF) amplifiers, etc., are typically used to accept inputs from low-level, high impedance sources and convert them into low impedances,[19] high-level signals. The OpAmps employed are often five-terminal devices, i.e., positive input, negative input, ground and two supply voltages, e.g., $+/-6$ volts. However, there are a variety of special application amplifiers available that are designed for specific types of signal handling capable of:

- demodulating low-level signals,
- preparing analog signals for processing by A/D and D/A converters,
- supplying low output impedance and high-speed output and providing automatic gain control (AGC) or variable gain control (VGC),
- demodulating low-level signals and compressing signals with high dynamic range,[20]
- extracting a low-level, differential signal from a larger common -mode signal, while filtering out transient and/or other unwanted signals,

and,

- accurately reproducing input signals, by minimizing distortion of the input waveform.

---

[17]Digital signal processors are highly specialized microprocessors whose architecture is specifically designed to perform highly optimized signal processing functions, e.g., real-time processing of video signals.

[18]This ability is referred to as *dynamic reconfigurability*.

[19]Low impedance makes it possible to supply sufficient power to successive stages to avoid adversely affecting the signal.

[20]Dynamic range is defined as the ratio of possible high-to-low signal values, either current or voltage, supported by a given device.

Some sensors produce output signals that are in the form of *modulated carriers,* that is, the signal is transmitted as either a variable amplitude, or a variable frequency signal, or some permutation thereof, e.g., a signal embedded in a carrier is shown in Fig. 5.14.

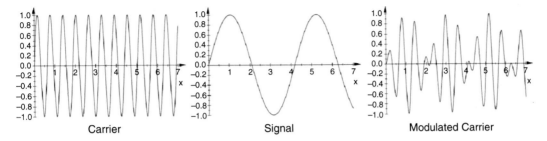

Fig. 5.14 An example of amplitude modulation.

This technique is often used to minimize the effects of amplifier noise, offset and grounding problems. One technique, employed in such cases, is to use a so-called *coupling transformer* to provide DC isolation between the sensor and the input amplifier, as shown in Fig. 5.15.

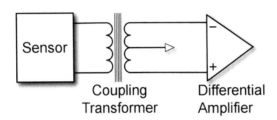

Fig. 5.15 Transformer-coupled input sensor.

In such cases, a center tap on the secondary, or output of the transformer, provides a common ground as a reference for the output from the secondary winding of the transformer. Note that this technique converts a single-ended input into a differential output.

Demodulation of this signal can be accomplished by the use of a variety of techniques, e.g., for extremely low-level signals, lock-in amplifiers capable of detecting a signal as low as 100 dB[21] below the ambient noise level can be employed. Signals may be measured as either single-ended or differential inputs. In the former case, the input signal is measured with respect to signal ground[22] and is sometimes coupled capacitively to the amplifier's input. Differential signals require that both

---

[21]dB is a logarithmic unit of measurement which is based on the ratio of a physical quantity with respect to a reference level of the same physical quantity. In the case of power, it is defined as $10 \log_{10} \left(\frac{P_1}{P_0}\right)$ and for voltage as $20 \log_{10}\left(\frac{V_1}{V_0}\right)$. Thus, 100 dB below ambient is equivalent to -100 dB or 0.0000000001 (10^{10}) of the associated reference value.

[22]Note that signal ground may, or may not, be that of the power supply or a common ground. The ground used can be a source of unwanted signals, i.e., noise, and careful attention must be paid in such cases to employing adequate grounding techniques.

the positive and negative inputs of an amplifier be used,[23] and the differential measurement of both inputs is therefore, not with respect to a common ground.

While at the block level, signal chains can be relatively simple, even the simplest of a signal chains involving perhaps the measurement of an external resistance that is related to some physical parameter of interest such as temperature, pressure, flow rate, etc. can present substantive issues that must be taken into consideration. For example, nonlinearities in resistance versus the value of the physical parameter being measured, temperature versus resistance variations in the sensor and the connections to the sensor,[24] accuracy and precision requirements, ambient interference environment and associated adverse effects on the sensor and connections to the embedded system, variations in resistance of the interconnections between the sensor, and the embedded system, gain variations of the input stage, sensor excitation requirements, crosstalk interactions, and required filtering must also be taken into account.

5.4   Schematic View of the System

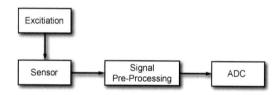

Fig. 5.16 A block diagram of a temperature measuring signal chain.

As an illustrative example, a simple signal chain, related to making a temperature measurement, as shown in Fig. 5.16, can be represented in the form of a block diagram. A schematic representation of this block diagram is shown in Fig. 5.17. It should be noted that in practical applications, the ground connections for the sensor resistor, and ADC, are all made in proximity when implemented in a physical system. This type of differential measurement connection reduces noise problems by virtue of the fact that common-mode measurements tend to cancel out signals that are picked up by the differential input lines. An even simpler signal chain for this type of measurement is shown in Fig. 5.18 in which the digital-to-analog converter has been replaced by a digital-to-current converter, thus eliminating the requirement for the reference resistor.

Temperature measurements using resistive-devices, such as thermistors,[25] whose resistance is a function of the ambient temperature, tend to exhibit non-linear characteristics. As discussed, in Chap. 1, the resistance of a thermistor, as a function of temperature, can be approximated by the equation

---

[23] An OpAmp with both positive and negative inputs can serve as either a single-ended or differential amplifier. Single-ended applications are accomplished by simply grounding one of the amplifiers inputs and applying the input signal to the other input.

[24] Including such considerations as temperature gradients along the wires connected to the sensor and embedded system input.

[25] Thermistor is an acronym for thermal resistor which refers to a resistor whose resistance is a known function of temperature. While usually a non-linear relationship exists between temperature and resistance for such devices, they can sometimes be treated as being quasi-linear, over the range of interest.

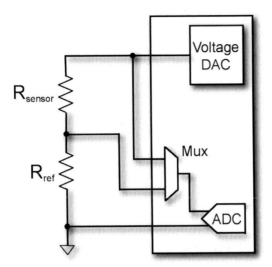

Fig. 5.17 A schematic diagram of the signal chain shown in Fig. 5.16.

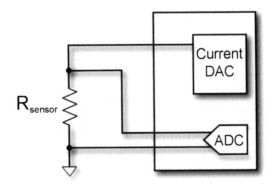

Fig. 5.18 A further simplification resulting from employing a current DAC.

$$\frac{1}{T} = A + B \, ln(R) + C[ln(R)]^3 \tag{5.14}$$

where T is the thermistor's ambient temperature, R is the measured resistance and A, B and C are constants characterizing the particular thermistor involved. Rather than solving this equation explicitly for each temperature measurement, it is simpler, and more efficient, to employ a lookup table that has the discrete values for resistance and temperature for that particular type of thermistor. If necessary, linear interpolation[26] using the coordinates of the known points on the curve can be employed, for additional quantitative detail.

---

[26] Linear interpolation is based on the idea that if two points are known on a given curve, e.g. a graph of temperature versus resistance (T vs. R), then the slope of a line drawn between the two points can be easily determined and the value of the temperature for a given resistance between those two points can be approximated by evaluating the following expression: $T = T_0 + [(T_1 - T_0)/(R_1 - R_0)]$ where (R_0, T_0) and (R_1, T_1) are known values and are chosen sufficiently close to provide the required accuracy. Linear interpolation is believed to have been used by Babylonians as early as 2000-1700 BC [84].

Resistance temperature detectors[27] (RTDs) are a particularly useful type of temperature sensor that has the property that at $0\,^\circ$C the resistance is nominally 100Ω, and the rate of change of resistance with respect to temperature (dR/dT) is 3.85Ω per degree Centigrade.[28] RTDs have very low resistance and therefore the effect of any additional low-wire-resistance paths must be taken into consideration. Typically, the voltage across an RTD is measured by using either 3- or 4-wire methods. The 3-wire method is shown in Fig. 5.19.

The voltage measured by the ADC is a combination of the drop across the RTD and the wire connecting the DAC to the resistor, and the voltage across the resistor. The current through the wire and resistor is known and the resistance of the wire connection can be measured so that the voltage drop across the wire is known.

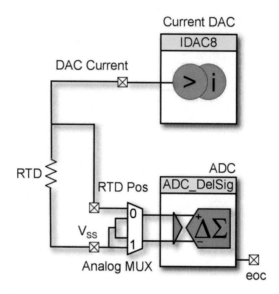

Fig. 5.19 Schematic view of the 3-wire circuit in PSoC Creator.

5.5   Correlated Double Sampling (CDS)

Signal chains often involve small amplitude, and relatively low frequency, signals. The accuracy, as well as precision, of measurements of such signals can be limited by various non-deal characteristics

---

[27]C.H. Meyers [86] first proposed the RTD in the form of a helical platinum coil on a crossed mica web inside a glass tube. However, its thermal response time was too slow for many applications and it was subsequently supplanted by a design by Evans and Burns [40] that instead used an unsupported platinum coil which allowed it to move freely as a result of thermal expansion and contraction.

[28]Platinum RTDs are extremely accurate and stable compared to thermistors and other commonly encountered temperature sensors.

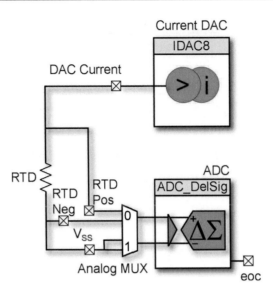

Fig. 5.20 Schematic view of the 4-wire circuit in PSoC Creator.

such as offset and noise. Offset potentials,[29] offset potential drift[30] and low frequency noise,[31] all of which are functions of temperature, arise frequently in systems such as those shown in Figs. 5.19 and 5.20. Obviously, these are highly undesirable effects to have present when making sensitive measurements. Some OpAmps employ chopper stabilizing to minimize drift by periodically grounding the input(s) of the amplifier.[32]

A technique known as *correlated double sampling* [10] (CDS) can be used to minimize these effects [111]. CDS functions as a high-pass filter which allows the (1/f) noise to be reduced and is a signal processing method that reduces unwanted effects that often occur when employing sensitive sensors. This technique is most effective in addressing slow-changing, in terms of frequency and amplitude, signals of the type encountered when using Hall-effect,[33] capacitive or thermocouple sensors.[34]

---

[29]Ideal OpAmps have zero output, when the input voltage differential is zero, as opposed to non-ideal OpAmps that exhibit some output voltage under such conditions. This can be "offset" in some cases by applying a small potential to one of the inputs sufficient to assure that the output is zero volts, under such input conditions. PSoC 3/5 offset voltage is spec'ed as a maximum of 2 mv.

[30]Typical offset voltage drift (TCV_{os}) for a PSoC 3/5 OpAmp is $\approx 6\mu v/°C$ ($12\mu v/°C$ maximum). PSoC 3's delta-sigma analog-to-digital converter (ADC_DelSig) includes a feature known as Vref_Vssa. When an external reference is being supplied, the Vref_Vssa connection can be routed through the analog routing fabric to an external pin on the device. A connection to this pin of an external reference eliminates any offset in the reference as a result of internal IR drops in the Vssa pin and bonding wire.

[31]The noise considered, in this example, is classified as "1/f noise" which is found in any semiconductor device.

[32]Early operational amplifiers used in applications such as integrators employed mechanical switches that were electrically driven. Current devices employ semiconductor switches for this purpose.

[33]The output of a Hall-effect sensor is a function of the ambient magnetic field. A thin piece of conductive material is used which has two connections that are placed perpendicular to the direction of current flow through the device. An external magnetic field will cause a potential to arise between the two connections that is directly proportional to the ambient magnetic field.

[34]The basic techniques described in this section are applicable to PSoC 1, PSoC 3 and PSoC 5LP [116, 139].

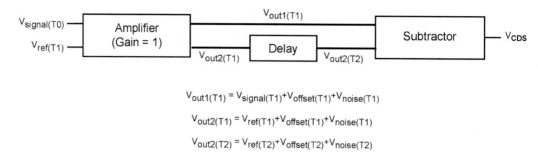

Fig. 5.21 CDS OpAmp block diagram.

As shown in Fig. 5.21,

$$v_{out1(T1)} = v_{signal(T1)} + v_{offset(T1)} + v_{noise(T1)} \qquad (5.15)$$

and,

$$v_{out2(T2)} = v_{signal(T2)} + v_{offset(T2)} + v_{noise(T2)}. \qquad (5.16)$$

assuming that

1. v_{ref} and v_{signal} are constant for values of t such that $t_1 < t < t_2$.
2. v_{offset} has a constant value, i.e., it is not an explicit function of time.

and,

3. system noise is solely a function of time.

Therefore, subtracting (5.15) from (5.16) yields

$$v_{CDS} = (v_{out(T1)} - v_{out(T2)}) \qquad (5.17)$$

$$= (v_{signal} - v_{ref}) + (v_{noise(T1)} - v_{noise(T2)}) + (v_{offset(T1)} - v_{offset(T2)})$$

$$= (v_{noise(T1)} - v_{noise(T2)}) \qquad (5.18)$$

This method is based on making two measurements, one from a sensor with an unknown input and one with a known input. Because the amplifier is assumed to be capable of producing only one output at a time, delaying the output of one signal with respect to the other, allows Eq. (5.17) to be evaluated.[35] Then by subtracting the result of the known input from the unknown input, it is possible to compensate for the offset. This technique is based on first measuring the offset potential across the sensor with both inputs shorted and then measuring

$$v_t = v_{tc} + v_n + v_{ov}, \qquad (5.19)$$

[35] Some applications use a sample-and hold-circuit followed by a subtractor, instead of employing a delay. PSoC Creator's Sample and Hold component is discussed in detail in Sect. 11.6.2.

where v_t is the zero referenced voltage, v_{tc} is the actual thermocouple voltage, v_n is the noise voltage and v_{ov} is the offset voltage.

Thus, for the previous zero-referenced sample

$$v_{zref} = v_n + v_{ov} \tag{5.20}$$

and,

$$v_{zref\_prev} = (v_n + v_{ov})Z^{-1} \tag{5.21}$$

in the continuous-time domain.

It can be transformed into the discrete-time domain by employing Tustin's method,[36] i.e.,

$$v_{signal} = (v_{tc} + v_n + v_{offset}) - (v_n + v_{offset})Z^{-1} \tag{5.22}$$

$$v_{signal} = v_{tc} + (v_n + v_{offset})(1 - Z^{-1}) \tag{5.23}$$

where Z is the bilinear transform, i.e.,

$$Z = \frac{(1 + \frac{sT}{2})}{(1 - sT)} \tag{5.24}$$

and,

$$T = \frac{1}{f_{sample}} \tag{5.25}$$

and therefore, using Eqs. (5.24) and (5.25),

$$\frac{1}{Z} = \frac{1 - \frac{sT}{2}}{1 + \frac{sT}{2}} = \frac{\left[1 - \frac{s}{2f_s}\right]}{\left[1 + \frac{s}{2f_s}\right]} \tag{5.26}$$

so that

$$1 - \frac{1}{Z} = 1 - \frac{\left[1 - \frac{s}{2f_{sample}}\right]}{\left[1 + \frac{s}{2f_{sample}}\right]} = \frac{\left[1 + \frac{s}{2f_{sample}}\right] - \left[1 - \frac{s}{2f_{sample}}\right]}{\left[1 + \frac{s}{2f_{sample}}\right]} = \frac{2s}{(s + 2f_{sample})} \tag{5.27}$$

and therefore [116],

$$v_{signal} = v_{tc} + v_n \left[\frac{2s}{(s + 2f_{sample})}\right] = v_{tc} + v_n \left[\frac{2}{1 + (\frac{2f_{sample}}{s})}\right] \tag{5.28}$$

---

[36]This method, also known as the *bilinear transform*, is actuality a conformal mapping and in the present case represents a mapping of a linear, time-invariant function in the time domain, to a linear shift-invariant transfer function in the discrete-time domain. Conformal mappings preserve certain key aspects of the functions being mapped, e.g., in the present case preservation of characteristics in the frequency domain. The Tustin method is often used to provide good matching in the frequency domain between the discrete and continuous time domains, and in cases where a system's dynamics [10] near the Nyquist frequency [8, 10, 11] are of interest.

assuming that the offset is not a function of time. Equation (5.28) is clearly a high pass response, as shown in Fig. 5.22.

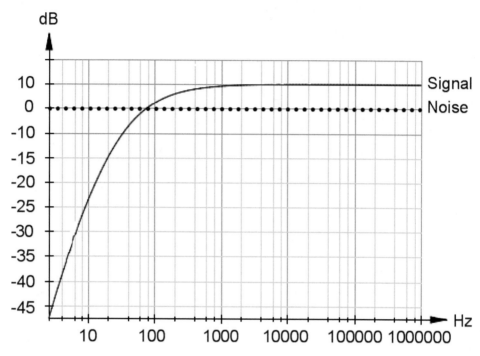

Fig. 5.22 CDS frequency response.

However, because this configuration does not reduce higher frequency noise, additional filtering may be required, e.g., an infinite impulse response (IIR) filter[37] can be used in such cases to reduce unwanted high-frequency components.

A similar technique can be employed in using PSoC 3/5LP's delta-sigma ADC in a CDS configuration as shown in Fig. 5.23. In this case, the input signals, V_{signal} and V_{ref} are alternately passed to a buffer stage,[38] which can introduce unwanted offset and noise,[39] before being supplied to the ADC. The resulting digital forms of these two signals are then subtracted in firmware. Note that this delta-sigma ADC can be configured in either a single, or differential, input mode,[40] as shown in Fig. 5.24a, b. Single and differential input modes can also be implemented using an analog multiplexer as shown in Fig. 5.25.

---

[37]These filters can be implemented as $y[n] = \sum_{k=0}^{M} x[n-k] + \sum_{k=1}^{N} y[n-k]$ where the b_k are the feed-forward coefficients and the a_k are the feedback coefficients.

[38]The sampling time between the two signals acts as the delay employed in the previous OpAmp example.

[39]Offset and noise may also be introduced by other devices in the signal path. In order to minimize such effects it is important that the reference signal and input signal both follow the same *signal path* to the extent feasible/possible.

[40]While in principle connecting an input to signal ground is equivalent to a zero input, in reality signal ground can introduce noise, so that in practical applications the differential mode is often preferable.

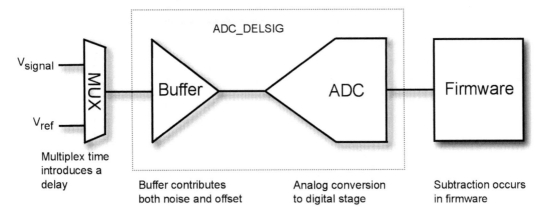

Fig. 5.23 CDS implementation for PSoC 3/5 *ADC_DELSIG*.

The following is an example of the PSoC Creator source code for this application:

```
/*Get the first sample Vout1 */
    AMux_1_Select(0);
    ADC_DelSig_1_StartConvert();
    ADC_DelSig_1_IsEndConversion(
        ADC_DelSig_1_WAIT_FOR_RESULT);
    iVout1 = ADC_DelSig_1_GetResult32();
    ADC_DelSig_1_StopConvert();

/*Get the second sample Vout2 */
    AMux_1_Select(1);
    ADC_DelSig_1_StartConvert();
    ADC_DelSig_1_IsEndConversion(
        ADC_DelSig_1_WAIT_FOR_RESULT);
    iVout2 = ADC_DelSig_1_GetResult32();
    ADC_DelSig_1_StopConvert();

/*perform CDS*/
    iVcds = iVout1 - iVout2;
```

In the system shown in Fig. 5.20, it is clear that its accuracy is solely a function of the IDAC's accuracy. Undesirable variations, i.e., deviations, in the output of the IDAC and ADC gain errors, can result from temperature dependencies. IDAC and ADC errors of the type found in this particular type of application can be reduced by introducing an additional, more accurate, resistance[41] as shown in Fig. 5.26.

When making such measurements it is important to:

- Select the most appropriate sensor for the application.
- Employ a technique such as CDS to avoid offset errors[42]

---

[41]Commercial resistors are available whose variances are less than 0.1%, as a function of temperature.

[42]A filter can be used to remove noise when employing a thermocouple.

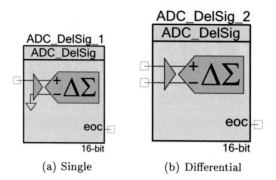

(a) Single   (b) Differential

Fig. 5.24 Single, versus differential, input mode for ADC_DelSig. (**a**) Single. (**b**) Differential.

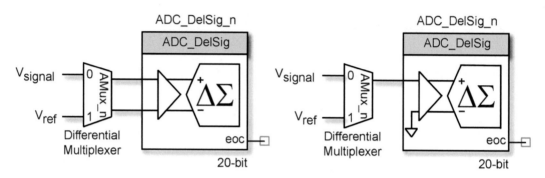

Fig. 5.25 Single/differential input using an analog multiplexer.

- Use current excitation to avoid inaccurate reference resistance.[43]
- Use a Delta-Sigma ADC with high accuracy and resolution to assure the highest possible overall accuracy.[44]

5.6   Using Components with Configurable Properties

The ADC and current DAC of the previous section are basic components supported by PSoC Creator, and as such, they are highly configurable, as are most of the PSoC Creator components. The current source, *IDAC*, may be controlled by hardware, software or some combination of the two and can function as either a source, or a sink (Fig. 5.27).

Similarly, the Delta-Sigma, analog-to-digital converter, *ADC_DelSig*, is also highly configurable. PSoC Creator provides tabulated dialog boxes, an example of which is shown in Fig. 5.28a, b, to allow user-defined configuration for a given application.

The conversion modes (0 – single sample, 1 – multiple samples, 2 – continuous samples or 3 – multiple samples (Turbo)), resolution (8–20 bits inclusive), conversion rate (2000–48,000 samples

---

[43] If voltage excitation is employed, 4-wire measurement techniques should be used.

[44] PSoC 3/5 are excellent platforms for such measurements in that functions such as a very high-resolution Delta-Sigma ADC, current source, voltage source, multi-pole filter and high resolution/speed digital processing are all tightly integrated within a single chip.

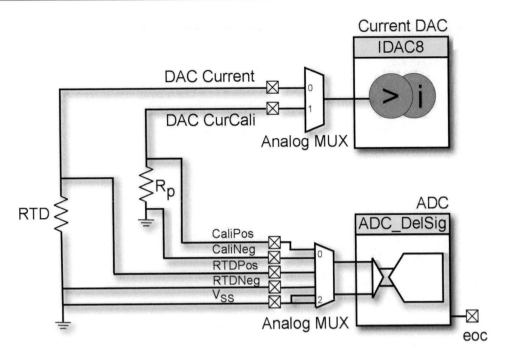

Fig. 5.26 4-wire RTD with gain error compensation.

per second), clock frequency[45] for each of the components in its integrated *Component Catalog* that allow the designer to specify key requirements for a given design, e.g., parameters such as *power* (low, medium or high), *conversion mode* (fast filter, continuous, fast FIR), resolution (8, 9, 10, 11, 12, 13, 14, 15, 16, 17, 18, 19, 20), *conversion rate*, *clock frequency*, *input buffer gain* (1, 2, 4, 8 and disabled), *reference* (various forms of internal V_{ref} [$1.024 volts$] and external reference connections), *clock frequency*, external/internal *clock source*, etc., give the designer the ability to adapt the PSoC 3/5 embedded system to fit the relevant specifications and requirements of each application (Table 5.1).

5.7   Types of Resets

PSoC 3/5LP support power-on resets (POR),[46] hibernate resets (*HRES*), watchdog resets (*WRES*),[47] The software resets (*SRES*) and external resets (*XRES_N*)[48] via the reset module as shown in Fig. 5.29. When a reset occurs, regardless of the type, all registers are restored[49] to their default

---

[45]The clock frequency is a function of the resolution and changes programmatically as a function of the conversion rate selected.

[46]When PSoC 3/5 is powered up, it is held in reset until all the VDDx and VCCx supplies reach the appropriate levels for correct operation.

[47]Watchdog reset is used to recover from errors that would otherwise keep the system from functioning properly and that may be recoverable, if the system is "reset", i.e., "re-booted". The Watchdog Timer (WDT) circuit automatically reboots the system if the WDT is not continuously reset within a user-defined period of time.

[48]If a reset pin is not required then this pin can be reprogrammed to be a GPIO.

[49]DMA is much faster than CPU intervention at populating device registers.

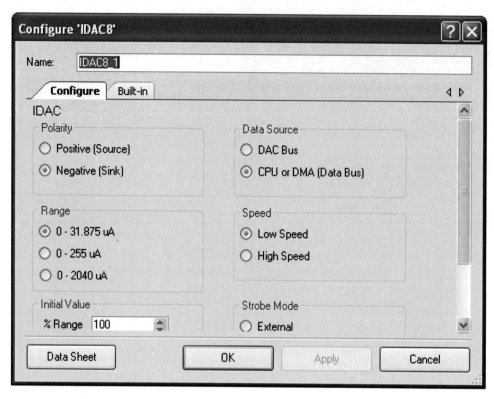

Fig. 5.27 The *Configure IDAC8* dialog box in PSoC Creator.

states except for so-called *persistent* registers.[50] Figure 5.30 demonstrates various reset responses as a function of time with respect to the change in V_{dd}/V_{cc} as well as the time dependencies for reset during normal power-up (POR). In some designs, a low startup time is essential. In these designs, there are a number of steps that can be taken to reduce PSoC 3's startup time. Gains depend heavily upon the configuration of the target device, but switching from CPU to DMA population may save on the order of 1–20 ms. Running the partially trimmed IMO at 48 MHz instead of 12 MHz will speed up most portions of startup by a factor of 4, but a fully trimmed IMO at a higher frequency will also improve startup time.

Because most startup occurs under partially trimmed IMO, the benefits will not be as significant as changing the partially trimmed IMO frequency. As with increasing the partially trimmed IMO frequency, this change will increase current consumption. Much of startup [3] is CPU or DMA limited, and these two resources will operate at the speed of the IMO. The downside to increasing the speed of the partially trimmed IMO is that device current consumption will increase. While the power supply ramp is not normally considered part of microcontroller startup, it does block the beginning of the startup procedure and performance can be improved in some cases by increasing the speed of the VDD ramps, to further minimize startup time, (startup).

---

[50]Both the $RESET\_SR0$ and $RESET\_SR1$ registers contain "persistent status bits" which can only be reset under particular circumstances, e.g., in the case of a POR.

Specific bits in these registers are set for each type of reset and remain set until the *tsrst_en* bit is cleared and either a POR or a user/application induced reset occurs. However, these bits are only accessible if the *tstrst_en* bit (bit 4) in the test controller's *TC_TST_CR2* register is set.

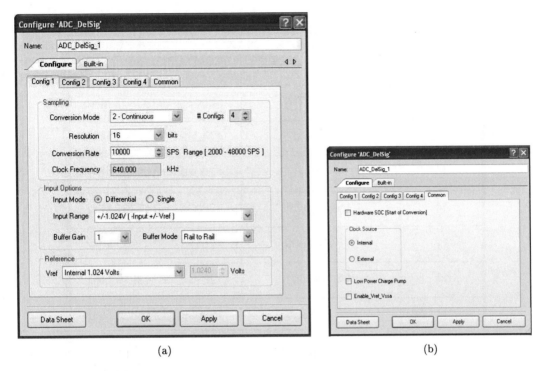

Fig. 5.28 PSoC Creator *Configure ADC_Del_Sig_n* dialog boxes.

5.8   PSoC 3 Startup Procedure

Application software designed to realize an embedded system relies on operating in a known hardware environment and under the constraints imposed by a specific set of initialization parameters and conditions [6], when power is applied to the microcontroller. Thus, the microcontroller must be provided with code designed to cause it to enter a known state with the appropriate initialization. This is accomplished by a combination of two firmware components known, respectively, as the bootloader [2] and a bootloadable project [5].

Following *powerup*, or alternatively, a *reset* caused by the XRES pin, watchdog timer, low voltage detection circuit, power-on-reset or other source, the PSoC 3/5 hardware is configured by initiating the appropriate hardware startup procedures.[51] *Power-on-reset* (*POR*) occurs during the ramp-up of the supply voltage and is not released until all associated power supplies have reached their appropriate operating values. Once the POR has been released, the device enters the boot phase in which a hardware state-machine controls the basic configuration and trim of the target device, using direct memory access (DMA).

Startup begins after the reset of a reset source, or following the end of a power supply ramp.[52] There are two primary startup segments: hardware and firmware as shown in Fig. 5.31.

---

[51] The device's I/O pins are placed in the high-Z drive mode while the reset is asserted and until pin behavior has been loaded.

[52] Because the power supply ramp blocks the beginning of startup, V_{dd} is referred to as $Svdd$ in Cypress Semiconductor Corporation datasheets, and should be taken into consideration as part of the design process.

Table 5.1 Resolution vs. conversion rate and clock frequency.

Resolution	Conversion Rate (sps) (Single Sample)	Clock Frequency
8	1911- 91,701	.128037 - .6143967 MHz
9	1,543 - 74,024	.128069 -.6143992 MHz
10	1,348 - 64,673	.128060 - .6143935 MHz
11	1,154 - 55,351	.128094 - .6143961 MHz
12	978 - 46,900	.128118 - .6143900 MHz
13	806 - 38,641	.128154 - .6143919 MHz
14	685 - 32855	.128095 - .6143885 MHz
15	585 - 28054	.128115 - .6143826 MHz
16	495 - 11861	.128205 - .3071999 MHz
17	124 - 2965	.128464 - .3071740 MHz
18	31 - 741	.128464 - .3070704 MHz
19	4 - 93	.131840 - .3065280 MHz
20	2 - 46	.263680 - .3032320 MHz

Once the hardware startup phase has been completed the system begins the firmware startup. The firmware loads the configuration registers, subject to the requirements set by the application and PSoC Creator, e.g., configuring the analog and digital peripherals, clocks, routing, etc. In addition, the debugging, bootloader and DMA resources are also configured.[53] Upon completion of the firmware startup, the CPU begins executing the user-authored code beginning at memory address location zero.

Register RESET_SR0 (0x46FA)[54] contains information about the status of the software reset, watchdog reset, analog HVI detector, analog LVI detector and digital LVI detector and RESET_SR1

---

[53]Not all the PSoC Creator components will be fully configured following the firmware startup phase. In some cases, additional code will be required to fully activate them.

[54]Reset and voltage status register 0 (RESET_SR0).

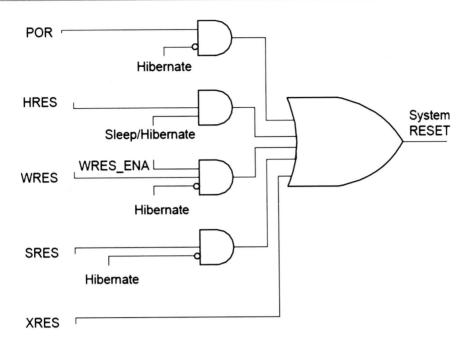

Fig. 5.29 Reset module logic diagram.

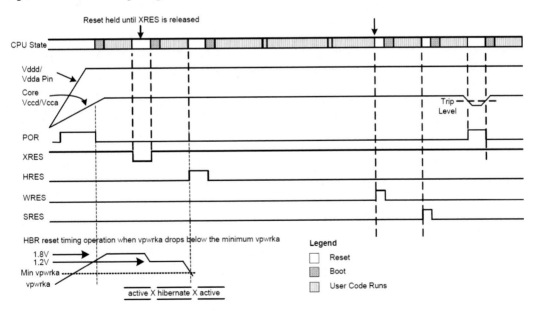

Fig. 5.30 Resets resulting from various reset sources.

$(0x46F8)$[55] contains information about the status of the analog PRES, digital PRES,[56] analog LPCOMP[57] and digital LPCOMP.

---

[55]Reset and voltage detection status register 1 (RESET_SR1).

[56]Precision POR (PRES) refers to a reset that occurs based on a precision trip point. An imprecise POR (IPOR) refers to a reset that occurs during power-up that keeps the target device in reset until Vdda, Vcca, Vddd and Vccd are at the values specified in the deices datasheet.

[57]LPCOMP refers to PSoC 3/5's low-power compare circuit.

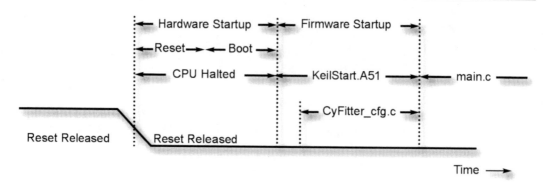

Fig. 5.31 Overview of PSoC 3 startup procedure [64].

RESET_CR2 (0x46F6) controls the software initiated reset (SRES). Setting bit 0 of this register (*swr*) to 1 will cause a system reset that can be initiated by software, firmware, or DMA which will result in the setting of RESET_SR [8]. It will remain set until reset by the user, or until a POR/HRES[58] reset occurs.

KeilStart.A51[59] contains 8051 assembly code that is executed at the beginning of the firmware startup to configure some of the basic components in PSoC 3, e.g., debugging, bootloaders, DMA endpoints and, if required, the clearing of SRAM. KeilStart. A51 code begins at memory address 0 in Flash which contains an unconditional jump to STARTUP. KeilStart also calls *CyFitter_cfg()* which can be used by the designer to handle certain clock startup errors, e.g., bad MHz crystal, loss of PLL lock, etc., and to configure some analog device default settings. The "clear IDATA" step, shown in Fig. 5.32, writes zeros to program memory allocated for IDATA.[60] The "DMAC configuration" step configures the DMA resources subject to the specification of PSoC Creator for the particular application.

The function *CyFitter_cfg()* is invoked by *CyFitter_cfg.c* and results in the population of a significant number of registers, as illustrated in Fig. 5.33, the largest group of which are those associated with analog and digital resources. This step may be carried out under either CPU control, or via DMA.[61] A somewhat smaller group of registers are configured as a result of the *ClockSetup()* API call which results in the configuration of PSoC3's clock tree and clock resources. The specific configuration of the project's clocks is determined by PSoC Creator.[62]

After the target device has been reset, it is clocked by the fast output of the internal main oscillator (IMO) which is based on a fast reference. Once the normal reference becomes stable, the normal IMO becomes valid. The IMO begins to source the normal reference during the reset phase. The IMO is then running nominally at either 12 or 48 MHz, as configured by the device's non-volatile latches

---

[58]POR/HIB refers to power-on reset and/or hibernate-reset.

[59]KeilStart.A51 is proprietary, 8051- based source code owned by Cypress Semiconductor for incorporation in PSoC 3 applications developed with Keil development tools.

[60]This memory allocation is usually for variables.

[61]The function $cfg\_write\_bytes_{c}ode()$ loads this group of registers by utilizing the CPU. The function $cfg\_dma\_dma\_init()$ loads the same group of registers via DMA.

[62]The clocks tab in PSoC Creator can be accessed by double-clicking the *.cydwr* file for a project.

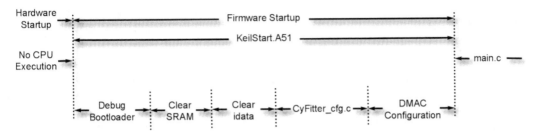

Fig. 5.32 KeilStart.A51 execution steps [64].

(NVLs).[63] T_{IO_init} is the delay, as specified in the target device's datasheet, that determines the delay after which the pins, and other resources, begin to behave as required by the application.

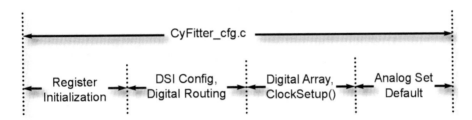

Fig. 5.33 $CyFitter\_cfg.c$ execution steps [64].

The population of registers is determined by the Mode selection options in the *.cydwr* tab, in PSoC Creator, as shown in Fig. 5.34. The compressed mode option causes the CPU to populate the configuration registers and store data in Flash, optimizing Flash usage, rather than startup time. The DMA mode populates the registers under DMA control blocking the CPU execution until the DMA configuration of the registers has been completed. As expected, DMA population is significantly faster than CPU population. "Clear SRAM" determines whether SRAM is to be cleared after a reset[64] for an IMO speed of 12 MHz. "Enable Fast IMO" selects the IMO speed as either 2 or 48 MHz, partially trimmed, i.e., slow or fast boot mode, respectively.

It should be noted that startup code is regenerated each time a change is made to a PSoC Creator schematic, or design, resources.

Thus, if the designer has made any changes to the *KeilStart.A51*, and/or *CyFitter_cfg.c*, files can be lost. In order to avoid any such loss, these files must only be edited when there is no need to perform a "generate" operation to ensure that the configuration in the automatically generated source files matches the application's design resources and schematic. The design wide resources (DWR) and schematic changes are followed by a "clean and build"[65] and then editing of the source files. The

---

[63] A Nonvolatile Latch (NVL or NV latch) is an array of programmable, nonvolatile memory elements whose outputs are stable at low voltage. It is used to configure the device at Power-on-Reset. Each bit in the array consists of a volatile latch paired with a nonvolatile cell. On POR release, nonvolatile cell outputs are loaded to volatile latches and the volatile latch drives the output of the NVL.

[64] Clearing of 8k of SRAM requires approximately 4500 CPU clock cycles, at 12 MHz. However, if SRAM is not cleared but variables are initialized properly, not clearing SRAM will have no adverse effect on firmware operation.

[65] PSoC Creator's *Clean and Build Project* command causes the intermediate and output files of any previous build to be deleted prior to initiating a new build.

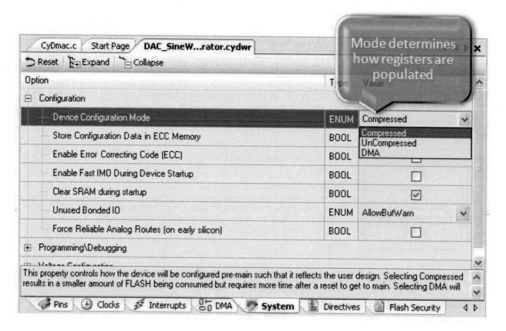

Fig. 5.34 Device register mode selection.

project can then be subjected to a "build" and the resulting firmware will reflect the respective edits. However, any subsequent "clean and build" actions will result in modifications to the generated source code.

5.8.1 PSoC 3/5LP Bootloaders

Once the source code for an embedded system application has been compiled, linked with the appropriate libraries, and debugged,[66] it is ready to be downloaded to the target device, e.g., a PSoC 3 or PSoC 5LP. This is accomplished, in part, by employing a *bootloader*.[67] A PSoC3/5 bootloader reads data from a communications port and writes it to internal Flash. In addition to downloading an application to a target during the design and manufacturing phase, the ability to download firmware upgrades and bug fixes in the field, to a target, non-invasively, are often very important when employing embedded system applications.

Communication ports in common use in such cases include USB, I2C, UART, JTAG and SWD. However, USB, I2C and UART are often preferred for loading software into a system in the field, rather than SWD and JTAG.[68] In addition, many systems utilize a USB, I2C or UART communication channels, used by the embedded system, to meet other application requirements.

---

[66]Debugging can also be carried out after the executable has been downloaded to the target.

[67]PSoC Creator provides a programmer which employs a default bootloader in the target device. However, in some cases use of this bootloader in the field is undesirable, in which case the designer must provide an appropriate bootloader for field programming/updating.

[68]USB, I2C and UART ports in an embedded system can be used for multiple purposes since they are generic communications protocols whereas SWD and JTAG are somewhat application specific.

A *bootloader project* is application software loaded by the bootloader into the target's Flash memory.[69] The functions that can be implemented when creating a bootloader are restricted to:

- CyBtldrCommRead—read function
- CyBtldrCommWrite—write function
- CyBtldrCommStart—initiate communication
- CyBtldrCommStop—halt communication
- CyBtldrCommReset—reset the communication channel

The application code, and associated data, are transferred to the target's Flash memory. The file type created by PSoC Creator for boot loadable files is **.cyacd* and consists of a five-byte header, followed by the data records where the header record format consists of a

- [Four byte SiliconID][one-byte SiliconRev]

followed by data records in the format given by:

- [One-byte ArrayID][Two-byte RowNumber][Two-byte DataLength][N-byte Data][One-byte Checksum]

where the checksum's value is computed by summing all the bytes, other than the checksum, and then taking the 2's complement of the resulting sum. The SiliconID is a value that identifies the target's package type and the SiliconRev is a value identifying the associated revision number.

The bootloader is responsible for accepting/executing commands, and passing responses to those commands back to a communications component. The bootloader collects/arranges the received data and manages the actual writing of Flash through a simple command/status register interface. A bootloader component is not presented in PSoC Creator as a typical component, i.e., it is not available in the *Component Catalog*. The *communications component* manages the communication protocol used to receive commands from an external system, and passes those commands to the bootloader. It also passes command responses from the bootloader back to the off-chip system.[70]

In order to create a bootloader component, and the associated code, it is necessary to create both a *bootloader* and a *Bootloadable* project in PSoC Creator. When a bootloader project is created, a *bootloader Component* is automatically created by PSoC Creator. The design typically requires dragging a communications component onto the schematic, routing I/O to pins, setting up clocks, etc.

While a standard project resides in Flash starting at address zero, a bootloader project occupies memory at an address above zero and the associated bootloader begins at memory address zero, as shown in Fig. 5.35.

The bootloader project code transfers a boot loadable project, or new code, to the Flash via the bootloader project's communications component. After the transfer has been completed, the processor is always reset, causing the execution of the code to begin at memory address zero. The bootloader project is also responsible, at reset time, for testing for certain conditions and possibly auto-initiating a transfer, if the boot loadable project is non-existent, or is corrupt. At startup, the bootloader code

---

[69]There can be only one bootloadable project in use in a PSoC3/5 at a time.

[70]I2C is the only supported communication method for the bootloader and the hardware I2C must be selected and not the UDB-based I2C.

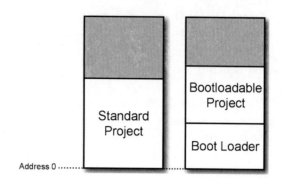

Fig. 5.35 Comparison of memory maps for a standard project and a bootloader project.

loads its respective configuration bytes. It must also initialize the stack and other resources/peripherals involved in the transfer. When the transfer is complete, control is passed to the bootloadable project via a software reset. The bootloadable project then loads configuration bytes for its own configuration; and re-initializes the stack, other resources and the peripherals for its functions. The boot loadable project can call the *CyBtldr_Load( )* function in the bootloader project to initiate a transfer.[71]

Whether a bootloader or bootloadable project is built, an output file is produced that contains both the bootloader and the bootloadable project. It is used to facilitate downloading both projects via either JTAG, or SWD, to Flash memory in the target device. The configuration bytes for a bootloader project are always stored in main Flash, but not in ECC Flash. However, the configuration bytes for a bootloadable project may be stored in either main Flash, or in ECC Flash. The format of the Bootloadable project output file is such that when the device has ECC bytes that are disabled, transfer operations are executed in less time. This is done by interleaving records in the Bootloadable main Flash address space with records in the ECC Flash address space. The bootloader takes advantage of this interleaved structure by programming the associated Flash row once the row contains bytes for both main Flash and ECC Flash. Each project has its own checksum, which is included in the output files at project build time.

Avoiding unintended overwriting of the bootloader can be accomplished by setting the Flash protection settings for the bootloader section of Flash. When the bootloader is built in PSoC Creator, the *Output* window displays the amount of Flash memory required for the bootloader, e.g.,

```
Flash used: 6859 of 65536 bytes (10. 5%).
```

in which case the bootloader occupies 27 rows (ceiling 6859/256) of Flash, i.e., Flash locations 0x000 to 0x1B00. It is protected by highlighting this part of Flash in the Flash Security tab and setting Flash protection as *W-Full protection* (Fig. 5.36).
The bootloader project always occupies the bottom N, 256-byte blocks of Flash, where N is large enough to provide sufficient memory for the

- vector table for the project, starting at address 0 (PSoC 5LP only),
- bootloader project configuration bytes,
- bootloader project code/data,

and,

- checksum for the bootloader portion of Flash.

---

[71] This results in another software reset.

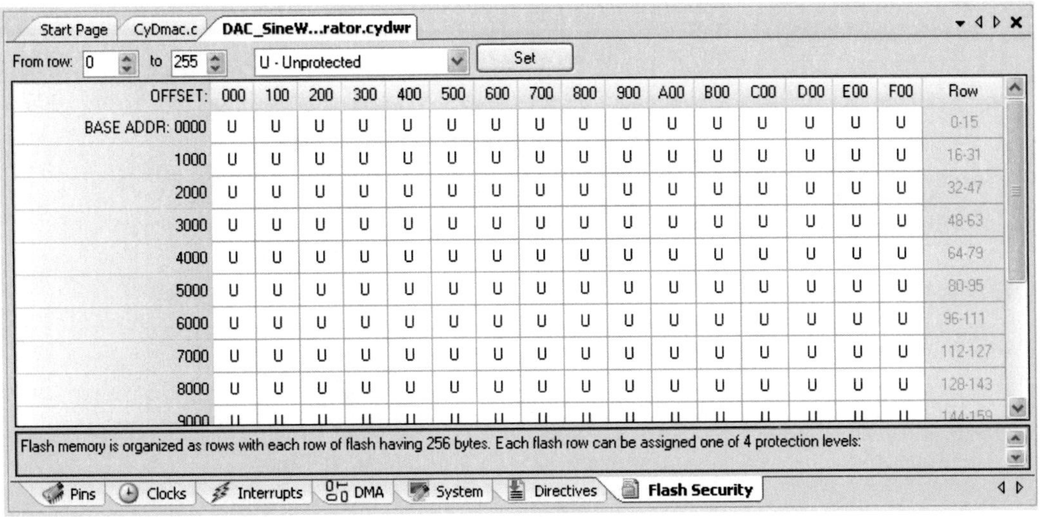

Fig. 5.36 Flash security tab.

The relevant option is removed from the project's *.cydwr* file. The bootloader portion of Flash is protected and can only be overwritten by downloading via JTAG/SWD.

The highest 64-byte block of Flash is used as a common area for both projects. Various parameters are saved in this block, which may include the:

- entry in Flash of the Bootloadable project (4-byte address)
- amount of Flash occupied by the Bootloadable project (Number of Flash rows)
- checksum for the Bootloadable portion of Flash (a single byte)

and

- size of the Bootloadable portion of Flash (4-bytes)

The bootloadable project occupies Flash starting at the first 256-byte boundary after the bootloader, and includes the vector table for the project (PSoC 5LP only), and the bootloadable project code and data. Storage of the bootloadable project's configuration bytes, in either main Flash or in ECC Flash, is determined by settings in the project's *.cydwr* file. The highest 64-byte block of Flash is used as a common area for both projects. Various parameters are saved in this block, e.g., the entry point in Flash of the bootloadable project (a 4 byte address) the amount of Flash occupied by the bootloadable project (the number of Flash rows) the checksum for the bootloadable portion of Flash (a single byte) and/or the size of the bootloadable portion of Flash (4 bytes).

The only *exception vector* supported by PSoC 3 is the 3-byte instruction at address 0, which is executed at processor reset.[72] Therefore, at reset the 8051 bootloader code simply starts executing

---

[72]The interrupt vectors are not in Flash. They are supplied by the Interrupt Controller (IC).

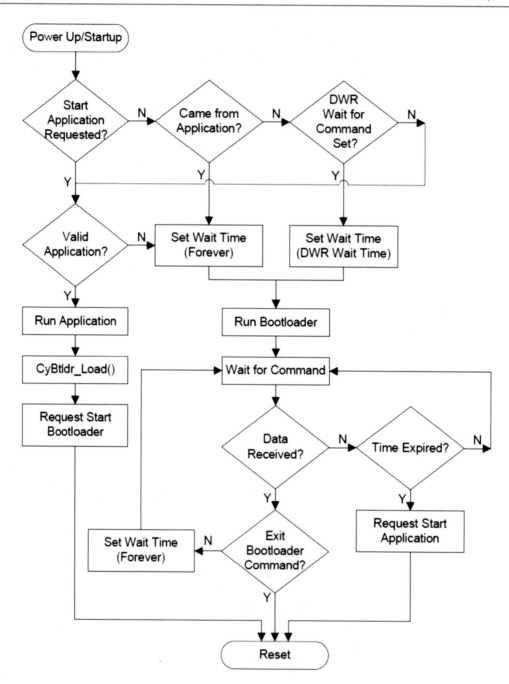

Fig. 5.37 Bootloader flow chart.

from Flash address 0. In the PSoC 5LP, a table[73] of exception vectors exists at address 0 and the bootloader code starts immediately after the table. The table contains the initial stack pointer (SP) value for the bootloader project, the address of the start of the bootloader project code and vectors for

---

[73] This table is pointed to by the *Vector Table Offset Register.*, at address 0xE000ED08, whose value is set to 0 at reset.

the exceptions/interrupts to be used by the bootloader. The bootloadable project also has its own vector table, which contains that project's starting stack pointer (SP) value and first instruction address. When the transfer is complete, as part of passing control to the bootloadable project, the value in the *Vector Table Offset Register* is changed to the address of the bootloadable project's table.

- **Wait for Command**—At reset, if the bootloader detects that the checksum in bootloadable project Flash is valid, then it may optionally wait for a command to start a transfer operation before jumping to the Bootloadable project code. If the selection is "yes", then the *Wait for Command Time* parameter is editable. If the selection is "no", then that parameter is grayed out. In that case an external system typically is not able to initiate a transfer. However, the Bootloadable project code can still launch a transfer operation by calling *Bootloader_Start().*[74] [9].
- **Wait for Command Time**—At reset, if the bootloader detects that the checksum in bootloadable project Flash is valid, then it may optionally wait for a command to start a transfer operation before jumping to the Bootloadable project code. This parameter is the wait timeout period. Allowable settings are 1–255 (inclusive), in units of 10 ms.[75]
- **I/O Component**—This is the communications component that the bootloader uses to receive commands and send responses. One, and only one, communications component must be selected. Only two-way communications components are used, e.g. a UART must have both RX and TX enabled, and an infrared (IrDA) component could not be used. A *design rule check* (DRC) exists for the case where no two-way communications component has been placed onto the bootloader project schematic. This property is a list of the available I/O communications protocols, on the schematic, that have bootloader support. There is typically only one communications *Component* on a bootloader project schematic, but there may be more in the case where the bootloader must also perform a custom function during the transfer.[76]

The bootloader has a public API that can only be used to launch a transfer operation from a bootloadable project. When called, a software reset occurs followed by the bootloader taking control of the CPU. Bootloadable code containing interrupts is not executed in this case. When the transfer begins, resources and peripherals are reconfigured as required and all other resources/peripherals are disabled. When the transfer has been completed, the CPU is automatically reset. *void CyBtldr_Load(void)* starts a transfer and reconfigures the device per the bootloader project. Although the CPU is reset upon completion of the transfer, there is no return value. Figure 5.37 shows the flowchart for the bootloader.

5.9   Recommended Exercises

5-1 Given the finite state machine shown below, find this 2-bit counter's truth table, logic equations and a logic circuit consisting of two D FF's and any required gates.

---

[74]The default value is "yes"

[75]This parameter is editable only if the *Wait for Command* parameter is set to *yes*, otherwise it is grayed out.

[76]If there is only one communications component on the schematic, then it is the only one available in the DWR dropdown.

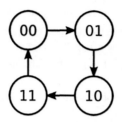

5-2 A sequence detector (FSM) must identify the sequence 1010011 by setting $y = 1$ when the sequence is detected. Sketch the state table, state diagram and excitation tables assuming that the input is x and the output is y. Simplify the equivalent logic circuit using k-map techniques. Is this a Mealy or Moore machine?

5-3 Draw a FSM state diagram (Mealy/Moore)for a serial comparator that utilizes n-bit unsigned numbers x and y as inputs. The FSM takes two bits x_i and y_i every clock cycle. State the bit-ordering that you have chosen, e.g. x_{n-1} and y_{n-1} arrive at time unit1 and x_{n-2} and y_{n-2} at time unit2. The output of the FSM must be 00, if the two values are equal, 10, if x has a greater value, and 01 if y has a greater value than x.

5-4 A finite state machine (FSM) functions as a downward, 3-bit, prime number counter. In the initial state, the prime value is 010. Create the state transition table and the transition diagram. The state must be stored in three D-FFs.[77]

5-5 The Nyquist criteria requires that in order to avoid aliasing, sampling [12] should occur at twice the highest frequency component in the signal. Some systems sample at much higher rates. Explain the reason for exceeding the Nyquist frequency[78]

5-6 Under what circumstances, is it advisable to employ sample-and-hold techniques when engaged in data acquisition of the type described in this chapter. List the key characteristics of the sample and hold phase of data collection hardware and software.

5-7 An analog signal is represented by:

$$x(t) = 2 \sin(240\pi t) + 2.75 \sin(360\pi t)$$

and sampled at a rate of $600sps$. What are the Nyquist sampling rate and folding frequency? Determine the Nyquist sampling rate for $x(t)$ and the folding frequency? What frequencies are to be found in the discrete time signal? Assuming that x[n] is processed by an ideal D/A converter, what the reconstructed signal?

5-8 Given that

$$H_p(s) = \frac{a}{s+a}$$

---

[77]In this case, the set of allowable prime numbers consists of 2, 3, 5 and 7.

[78]The Nyquist criterion must be used with some caution. It assumed that the sample to be measured at discrete intervals is in fact truly band limited. In order for this to be true the signal would have to extend to infinity. Sampling at a rate 0f 10–20 times the highest frequency component in the signal may be required, in order to achieve the best results.

show that

$$H(z) = \frac{\left(1 - e^{-aT}\right) z}{z - e^{-aT}} = \frac{\left(1 - e^{-aT}\right)}{1 - e^{-aT} z^{-1}}$$

5-9 A bandpass filter has a transfer function given by:

$$H(s) = \frac{s}{s^2 + 2\zeta \omega_n s + \omega_n^2}$$

Show that the z-domain transfer function is given by:

$$H(z) = 2T \frac{z^2 - 1}{\left(T^2\omega_n^2 + 4\zeta T\omega_n + 4\right) z^2 + \left(2T^2\omega_n^2 - 8\right) z + \left(T^2\omega_n^2 - 4\zeta T\omega_n + 4\right)}$$

Assuming that $\zeta = 0.25$, $\omega_n = 3.47$ and $T = 0.063$, sketch the z-plane graph and the bandpass characteristics.

5-10 Given a 5-bit counter that has an active high pin for load and increment input, design a logic circuit that generates and repeats the following sequence: $\rightarrow$ 00000 $\rightarrow$ 00010 $\rightarrow$ 00100 $\rightarrow$ 01010 $\rightarrow$ 01100 $\rightarrow$ 01110 $\rightarrow$ 00000 $\rightarrow$ $\cdots$

References

1. M. Ainsworth, PSoC 3 to PSoC 5LP Migration Guide. AN77835. Document No. 001-77835Rev.*D1. Cypress Semiconductor (2020)
2. M.D. Anu, T. Rastogi, PSoC 3 and PSoC 5LP I2C Bootloader. Document No. 001-60317 Rev. *L1. AN60317. Cypress Semiconductor (2020)
3. T. Dust, G. Reynolds, Designing PSoC Creator Components with UDB Datapaths. AN82156. Document No. 001-82156Rev. *I. Cypress Semiconductor (2020)
4. J. Kathuria, C. Keeser, Implementing State Machines with PSoC 3, PSoC 4, and PSoC 5LP. AN62510. Document No. 001-62510 Rev. *F. Cypress Semiconductor (2017)
5. C. Keeser, PSoC Designer Boot Process, from Reset to Main. AN73617. Document No. 001-73617Rev. *C1. Cypress Semiconductor (2017)
6. M. Kingsbury, PSoC 3 and PSoC 5LP Startup Procedure. AN60616. Document No. 001-60616Rev. *G. Cypress Semiconductor (2017)
7. A. Megretsk, Multivariable Control Systems. Massachusetts Institute of Technology. Department of Electrical Engineering and Computer Science. April 3, 2004
8. H. Nyquist, Certain topics in telegraph transmission theory. Trans. AIEE. **47**(2), 617–644 (1928)
9. P. Phalguna, PSoC 3 and PSoC 5LP SPI Bootloader. AN84401. PSoC 3 and PSoC 5LP SPI Bootloader. Document No. 001-84401Rev.*D. Cypress Semiconductor (2017)
10. C.E. Shannon, A mathematical theory of communication. Bell Syst. Tech. J. **27**(3), 379–423 (1948)
11. C.E. Shannon, Communication in the presence of noise. Proc. Inst. Radio Eng. **37**(1), 10–21 (1949). https://doi.org/10.1109/jrproc.1949.232969. S2CID 52873253
12. A. Yarlagadda, PSoC 3 and PSoC 5LP Correlated Double Sampling to Reduce Offset, Drift, and Low-Frequency Noise. AN66444. Document No. 001-66444 Rev. *D. Cypress Semiconductor (2017)
13. https://www.planetanalog.com/design-considerations-the-analog-signal-chain-part-1-of-2/. January 30, 2011
14. https://www.planetanalog.com/design-considerations-the-analog-signal-chain-part-2-of-2/#. February 4, 2011

Hardware Description Languages

6

Abstract

VLSI digital circuits often involve hundreds of logic cells and perhaps thousands of interconnections. Therefore, the associated difficulty of developing such PDL applications capable of performing complex functions has given rise to what is termed hardware description languages which allow the designer to model digital systems. These HDL languages are supported by development environments that typically have the ability to provide schematic design, simulations/verification and the ability to transform the design into a "configuration file and then download it to the targeted device. The HDL form of a design is a temporal and spatial description of the design and includes expressions that formally describe the design's digital logic circuits. The descriptions are text-based, include explicit time dependencies, and take into account interconnections between blocks that are expressed in a hierarchical order. Converting from the description of a logic circuit to an implementation-defined in terms of gates is referred to as *synthesis*. The output of the synthesis process is a *netlist*.[1]

6.1 Hardware Description Languages (HDL)

An ideal HDL should be capable of supporting designs involving tens of thousands of gates, provide high-level constructs for describing complex logic, support modular design methodologies and multiple levels of hierarchy, support both design and simulation, be capable of producing device-independent designs, support schematic capture, as well as, HDL descriptions, etc. In the discussion that follows several HDLs are discussed, each of which, to a greater or lesser degree, meets these criteria [10].

6.2 Design Flow

Various modeling techniques are employed in designing embedded systems, e.g., dataflow for systems involving parallelism and signal analysis, discrete-event for systems involving explicit time dependencies, state machines [7] for systems based on sequential decision logic, time-driven for

---

[1] A netlist is a text-based description of the gates used in the design and their interconnections.

© Springer Nature Switzerland AG 2021
E. H. Currie, *Mixed-Signal Embedded Systems Design*,
https://doi.org/10.1007/978-3-030-70312-7_6

systems that involve periodic and/or time-dependent action, continuous-time for systems involving dynamics, etc. HDLs are capable of facilitating each of these approaches in a wide variety of cases, but may have some limitations with respect to certain types of applications [1].

A design can begin with a graphical representation of the logic circuits, i.e., a schematic, or a purely text-based description of the design. Available tools often include both schematic and HDL editors. When the HDL, or schematic design, phase has been completed, it is then possible to simulate the design for the purposes of behavioral evaluation, conducting preliminary performance evaluations, creation of test vectors, etc. Test bench waveforms can also be introduced as part of the simulation process. Following the synthesis phase of the design, it is possible to carry out additional simulation tests to verify the performance of the logic design including timing, which although limited in scope in some cases, can still provide important information about the design.

Once the descriptions, constraints and netlists have been created, they can then be merged into a database so that the *place-and-route process* can be invoked. In modern development environments the floorplan and routing are displayed graphically providing the designer with additional control over the design by instituting manual changes in the design.

So detailed and complete are these descriptions that they can be used in conjunction with simulators that use the descriptions as input to study the corresponding circuit's behavior and performance in complete detail. Furthermore, once the modeling is completed using simulators, the descriptions can be employed as input to CAD[2] tools to synthesize hardware designs. HDLs are used to create formal descriptions of digital circuits. Some simulators can actually interact with hardware implementations of the design to further optimize the system under design.

VHSP/*VHDL* was originally designed to ease the documentation challenges of *ASIC* designs but was soon found to be an important tool for facilitating the design of very high-speed integrated circuits and is referred to as *VHSIC*. VHDL is a VHSIC hardware description language based on ADA and developed by the Department of Defense beginning in 1983.

6.2.1   VHDL

While some may choose to characterize VHDL as simply "yet another programming language" it is in reality much more. It came into being as a result of a government initiative in 1980 by the Department of Defense (DoD) and was originally intended to be a formal methodology for describing digital circuits. However, it soon became apparent that its scope could be extended to allow it to serve not only as a language standard for digital circuit descriptions, but for simulation of digital circuitry as well. VHDL has gone through a number of reincarnations[3] and its notation is defined, in each case, by a language reference manual (LRM). It is regulated by the IEEE and is maintained as an international standard. VHDL supports both top-down/bottom-up design and as some have suggested even "middle-out" [1].

The VHSIC[4] hardware description language (VHDL) offers a designer a number of important benefits in developing new designs, particularly those that involve tens of thousands of logic gates. For example, VHDL supports very sophisticated and powerful constructs for describing complex logic, a modular design methodology, multiple levels of hierarchy and a VHDL description can be used

---

[2]Computer-Aided-Design.

[3]While there are a number of versions of VHDL extant e.g., VHDL'87, VHDL'93, VHDL'2000, VHDL'2008, etc. version.VHDL'93 remains the most widely used version.

[4]Very-High-Speed integrated circuit.

for both design and simulation. The resulting designs are device-independent and therefore highly portable so that the designer can select the optimum vendor, device and synthesis.

A designer can begin with a very high-level abstraction for a design, use VHDL to develop an architecture for the design and then decompose that structure into subsystems, sometimes referred to as "sub-designs". These subsystems can, in turn, often be decomposed further into subsystems of subsystems until one finally arrives, if required, at the equivalent of a standard set of "basis" modules,[5] e.g., commonly available integrated circuits.

A simple module at the lowest level of this hierarchy might well consist of a device, or devices, with two-inputs and one output, e.g., a gate. Such a module, referred to as an instance of an *entity*, need not be decomposed any further, and can be treated strictly in terms of its characteristics, i.e. the relationship(s) between the input and output signal levels, propagation delays, etc. [2]. Modules at the base of such a hierarchy are typically described in *behavioral*, or *functional* terms. However, if the basis modules employ feedback, the behavioral/functional module descriptions become complex. Fortunately, VHDL is designed to address this situation, as well.

VHDL is based on constructs such as *architectures, configurations entities, packages* and the corresponding *package bodies*. Architectures are functional descriptions of modules, configurations that define the architecture and entities required to build a model. Entities define interfaces and often involve a port list. Packages contain the definitions of data types such as constants, various data types and *subprograms*.[6] In addition, process code, which is a sequence of statements executed in a defined order, is employed as a concurrent object.

VHDL supports both sequential and concurrent statements. Sequential statements are contained within functions, procedures or process statements. *If-then-else*, *case* and *loop* statements are examples of sequential statements. Concurrent statements occur within the architecture in the form of statements containing concurrent procedure calls, signal assignments and component instantiations.[7]

Signals are used to transmit information between design statements, specifically between entities and processes. They are treated as globals in architectures and blocks. When declared in a PORT, signals must be assigned a direction. However, if they are declared in an architecture, block or package, no direction is required. Signal assignments typically include specifications, i.e., time expressions of the delay time that must occur before the signal is allowed to assume a new value. If the time expression is not specified the default value is zero femtoseconds.[8] It should be noted that variable updates occur immediately in VHDL while signal updates occur after a delay, or at the end of a process.

The design flow diagram, shown in Fig. 6.1, illustrates the various stages in the development of a VHDL-based design. The basic steps in the design flow are as follows:

1. *Design entry*—this is often done within the context of a computer-aided design (CAD) tool, and results in the design being available in a machine-readable format.
2. *Functional simulation*—step 1 produces the design description and this step simulates the design to confirm that the design does, in fact, meet the requirements specification. This is often referred

---

[5]The phrase basis modules refers to a set of modules which form a basis in the linear algebra sense.

[6]Subprograms in VHDL are the analog of functions in C.

[7]There is a distinction to be made between concurrency and parallelism in that concurrency refers to parts of a program that at the conceptual level are to be executed simultaneously, i.e. logical concurrency. Parallelism implies that certain parts of a program are in fact executed simultaneously at the hardware level. Programs that execute instructions sequentially are therefore non-concurrent. Some compilers are capable of carrying out a dataflow analysis and outputting parallel code for hardware capable of supporting parallelism.

[8]fs is an acronym for a femtosecond and represents 1×10^{-15} of a second.

to as a behavioral, or functional simulation[9] and its purpose is to verify that the behavior is correct from a logic perspective.

3. *Synthesis*—a CAD tool is employed for this step to interpret the VHDL description and employ a sufficient set of standard building blocks, e.g., LUTs, multiplexers, registers, adders, etc., to implement the design. The step produces a netlist[10] which will be used by the next step, viz.,*Implementation*.

4. *Implementation*—This step involves the invoking of a translate phase (TRANSLATE) which translates the netlist produced by the previous step into a format consistent with the targeted device. A mapping process (MAP) maps the standard set of blocks used by the synthesis process onto the available devices in the target hardware. This is followed by the allocation of the target resources and routing of all the required interconnections between these resources (PLACING and ROUTING). At this juncture, because the actual propagation and other types of delays have been taken into account, it is possible to carry out what is referred to as the POST-PLACE and ROUTE simulation which represents a model of the actual behavior of the physical design.

5. *Programmer Download*—At this stage, the design has been verified and is ready for down-loading to the target device. This is accomplished by the creation of a file that is downloaded in serial fashion to a programmer for the target device.[11]

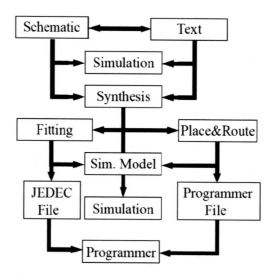

Fig. 6.1 VHDL design flow diagram. Fitting and Place&Route are similar operations with the former producing a JEDEC (Joint Electron Device Engineering Council.) file and the latter producing an arbitrary file in a format acceptable to the programmer for the targeted device.

---

[9] It is important to bear in mind that this type of simulation does not take into account any of the physical implementation details, e.g., actual propagation and other types of timing delays introduced by the various physical components involved are not taken into account by this level of simulation. Therefore, while the simulation may be regarded as a necessary step, it is hardly sufficient, in and of itself. It is, however, possible to include, at this level, statements that assign values for various delays, but these are not synthesizable and are included merely to reflect the impact of the delays on behavior.

[10] In the case of PLDs and CPLDs, a sum-of-products equations may be produced instead of a netlist.

[11] This step should be regarded as nonlinear in the sense that system performance can depend in a nonlinear manner on the components selected by this process.

As discussed previously, VHDL designs are descriptions, also referred to as "design entities", consisting of an *ENTITY* declaration that describes the design's I/O and an *ARCHITECTURE* body that describes the content of the design, e.g., a two-input AND function can be expressed in VHDL as

```
ENTITY and2 IS PORT (
        a,b : IN std_logic;
            f:OUT std_logic);
END and2;
ARCHITECTURE behavioral OFand2IS
BEGIN
        f <= a ANDb;
END behavioral;
```

The ENTITY declaration is formally defined as

```
ENTITY entity_name IS PORT (
    -- optional generics
    name : mode type
        ...
);
END entity_name;
```

where *entity_name* is an arbitrary name, *generics* are used for defining parameterized components, *name* is the signal/port identifier,[12] *mode* describes the direction of data flow and *type* defines the set of values a port name may be assigned. PORTS are points of communications, often associated with a device's pins, that are a special class of SIGNAL with an associated *name, mode* and *type*.

MODE represents the direction of data flow and can be:

- IN—data enters the entity but does not exit from it,
- OUT—data leaves the entity but does not enter and is not used internally,
- INOUT—Data goes in and out of the entity, i.e., it is bidirectional,

or,

- BUFFER—data exits the entity and is also fed-back internally.

6.2.2 VHDL Abstraction Levels

VHDL allows the designer to approach a design at varying levels of abstraction, viz., at the algorithm level which is merely a set of instructions to be carried out without regard for the clock, except perhaps loosely in terms of the ordering of execution of instructions, or delay issues. Alternatively, a register level approach can be employed that is referred to as register transfer level (RTL). At this level of abstraction, the description includes clock dependence which gates all operations. However, propagation and the various forms of temporal delay are not supported. And finally, at the lowest level of abstraction, the description is expressed in terms of a network of registers and gates that are instantiated from standard libraries.

---

[12]This can be a separate list for ports of identical modes and types.

VHDL has available five very fundamental constructs:

1. Entity declarations that specify NAME and PORTS.[13]
2. Architecture bodies that model the circuits within entity Configuration declarations that define which architecture is to be used with which entity.[14]
3. Package[15] declaration which is similar in function to that of a header file in a C program.
4. Package bodies that are similar to implementation files in C programs.

In addition to supporting multiple architectures within a given entity, VHDL allows the designer to determine which architecture is to be employed during the synthesis phase. The order of these constructs within a VHDL file is: entity, architecture and configuration. The IEEE has defined certain standard VHDL libraries, e.g., IEEE 1164[16] which establishes both standard signals and data types.

Consider the case of a four input, single output logic function. An entity description can be of the form:

```
library IEEE;
use IEEE>std_logic_1164.all;
entity LogicFunction is
        port (
        a: in std_logic;
        b: in std_logic;
        c: in std_logic;
        d: in std_logic;
        e: out std\_logic;
        );
end entity LogicFunction;
```

The architecture body defines the internal functionality of the entity, i.e., the circuitry, based on one of the following four modalities:

1. Behavioral modeling in the form of a set of sequential assignment statements referred to as a *process*.
2. Dataflow modeling expressed in terms of a set of "concurrent" signal assignment statements.
3. Structural modeling in terms of a set of interconnected components,

or, as

4. Some permutation of the behavioral, dataflow and/or structural modeling.[17]

As an example, consider a single bit full adder as shown in Fig. 6.2. The VHDL description for this circuit is

---

[13]Entities are analogous to the classic "Black Box" for which the internal functionality is hidden but the input and output ports, that is the interfaces, are specified. The entity is the functional equivalent of a software "wrapper".

[14]The architecture contains a detailed description of the entity's internal functionality/behavior.

[15]Packages are libraries of procedures, functions, overloaded operators, type declarations and components that consist of a BODY section and a declarative section. The constituents of a package can be used by more than the entity in a design.

[16]Standard Multivalue Logic System for VHDL Model Interoperability(1993).

[17]Such combinations are often referred to as "mixed-models".

```
entity full_add_1 is
    port(
          a1: in bit; addend in
          a2: in bit; addend
          c1: in bit; carry in
          sum: out bit; sum
          c2: out bit); carry out
          end full_add_1
```

where the port is defined in terms such as input, output or bidirectional, i.e., a port is defined in terms of the signal associated with the port, its direction and type. The signal type can be either the single bits, 0 or 1, or in the form of a bit vector representing an array of bits. User-defined data types are also supported, such as bytes or mnemonics.

The previous example of a single bit adder can be extended as in the case of an 8-bit adder represented by the following code fragment:

```
entity: full_add_8 is
    port(
          a1: in bit_vector(7 downto 0);
          a2: in bit_vector(7 downto 0);
          c1: in bit; carry
          sum: out bit_vector(7 downto 0); sum
          c2: out bit);
        end full_add_8;
```

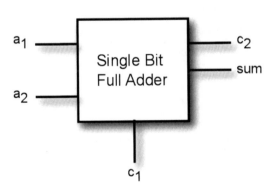

Fig. 6.2 A single bit full adder.

It is also possible, using VHDL, to describe the behavior of a module at a higher-level of abstraction for a single-bit adder, as illustrated by the following code fragment:

```
architecture dataflow of full_add_1 is
  begin
    sum <= a1 xor a2 xor c1 after 3 ns;
    c2 <= (a1 and a2) or (a1 and c1) or (a2 and c1) after 3 ns;
  end;
```

Identifiers are not case-sensitive and must not contain any keywords. Underscores may be used in identifiers, but they must not occur at the beginning, or end, of an identifier and two or more underscores cannot occur in succession.[18] *Extended identifiers* are formally defined as:

```
extended_identifier ::= \ graphic_character {graphic_character}
```

An extended identifier is case-sensitive and may contain spaces, consecutive underscores and/or keywords. Supported delimiters[19] are shown in Table 6.1.

The supported character set for VHDL'93, consisting of 256 characters, that includes uppercase letters, lowercase letters, digits, and a collection of non-alphanumeric characters.

Character strings are delineated by double quotes, e.g.,

"This is a string'"

becomes

""This is a string""

Bit strings are expressed as arrays of type bit, e.g.,

literal_bit_string ::= base-specifier "bit_value"

Table 6.1 Delimiters supported by VHDL.

Single Char Delimiters	Description	Double Char Delimiters	Description
#	Literal Base	?>	Conditional
'	Single Quote	?<	Conditional
"	Double Quote	**	Exponentiation
&	Concatenation	/=	Inequality
(	Left Parenthesis	??	Conditional
)	Right Parenthesis	?=	Conditional
+	"Addition" or "Positive"	?>=	Conditional
-	"Substraction" or "Minus"	?<=	Conditional
*	Multiplication	?/=	Conditional
:	Data :Type Separator	<>	Box
;	Instruction Terminator	>=	Greater Than Or Equal
/	Division	<=	Signal Assignment
?	Conditional	:=	Variable Assigment
|	OR Operator	=>	""Gets" or "Then"

---

[18]The 1993 standard for VHDL (VHDL'93) permits the use of identifiers that are case-sensitive, begin or end with a backslash, consist of graphical characters in any order and of arbitrary length.

[19]Delimiters are defined as separators which have predefined meanings.

6.2.3 VHDL Literals

VHDL supports five types of literals: bit strings,[20] enumeration, numerical, strings and NULL. Bit string literals begin and end with ", may include underscores, e.g., B"0101_0101", and are treated as a one-dimensional array that comply with VHDL'93 256 character specifications. Supported bases include binary(B), octal(O) and hexadecimal(X). While the bit string may contain underscores, the length of the string does not include the underscores. Enumeration literals may be bit, or character.

Numerical literals can contain underscores, the letters "E", or "e", to denote the inclusion of an exponent and "#" to define a base within the range of 2–16 inclusive. Physical types must have a space between the numerical value and the physical type's unit of measure. The NULL literal's use is restricted to pointers in cases for which the pointer is "empty". The "based" literals are formally defined as

$$\text{based\_literal::= base\#based\_integer\{based\_integer\}\#\{exponent\}}$$

and

$$\text{based\_integer::=extd\_digit\{[underlined]extd\_digit\}}$$

where

$$\text{extd\_digit::=digit|letters\_A-F}$$

and the base and exponent must be expressed in decimal form. Numeric literals default to decimal, may contain underscores to enhance readability but not spaces and should not have a base point or negative exponents. The use of scientific notation is restricted to integer exponents

---

Examples of VHDL Literals:

a) VHDL Bit String Literals: B"10101010" - - decimal 170
                                        B"1010_1010" - - decimal 170
                                            O"252" - - decimal 170
                                            X"AA" - - decimal 170
b) VHDL Numeric Literals:

6.2.4 VHDL Data Types

Each of VHDL's data objects has associated data types, cf. Fig. 6.3, that defines the allowable set of values for the data type.

VHDL is a *strongly-typed* language,[21] i.e., each data object is a predefined type. This constraint means that a data object of one type cannot be assigned to an object of another data type.

---

[20]Bit strings are frequently used to in initialize registers.

[21]The phrase 'strongly-typed" is somewhat ambiguous but implies, inter alia, that the type of each variable must be declared prior to use and includes strict rules with respect to any variable manipulations. One advantage of such languages is that their respective compilers are able to catch many bugs prior to runtime. C++, C# and Java are regarded as strongly-typed whilst C is regarded as weakly- or loosely-typed. It is perhaps more accurate to state that the former is more strongly typed languages than the latter.

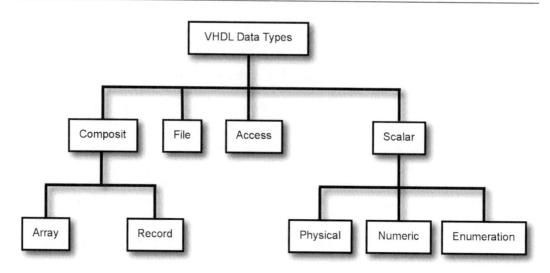

Fig. 6.3 VHDL data types.

VHDL supports the following data types:

1. Integers—allowable values −2147483647 to 214748347
2. Floating point (Real)—allowable values −1.0 E 38 to 1.0 E 38. Precision is a minimum of six decimals.
3. Physical—a numerical type that represents physical quantities such as time, mass, length, voltage, resistance, etc. The base unit must be specified in the declaration, e.g.,

```
type  resistance  is  range  0  to  1E8
       units
          ohms ;
          kohms =1000  ohms ;
          Mohms=1E6  ohms ;
       end  units ;
```

4. Arrays—no limit on array dimensions, can be indexed by any discrete type, logical and shift operations only applicable to arrays with bit or elements.
5. Signals—within a block or architecture signals are global. Signals are assigned values through the use of '<=' and receive default values through the use of ':=' Signals associated with a port must have a direction, but not within a block or architecture.
6. Signal attributes[22] -

    signal_name'event (returns Boolean TRUE for signal event occurrence, else returns Boolean FALSE)

    signal_name'active (returns Boolean TRUE if a transaction on a signal occurs, else returns Boolean FALSE)

    signal_name'transaction (returns a signal of type bit that toggles for each subsequent transaction on the signal)

    signal_name'last_event (returns Boolean TRUE if a signal event occurs else returns Boolean FALSE)

---

[22] Attributes are functions that return a value type or the range of a data type.

signal_name'last_active (returns Boolean TRUE if a signal event occurs else returns Boolean FALSE)

signal_name'delayed(T) (returns Boolean TRUE if a signal event occurs else returns Boolean FALSE)

signal_name'stable[T] (returns Boolean TRUE if no event on signal has occurred during time T, else returns Boolean FALSE. Default value of T is zero.)

signal_name'quiet[T] (returns Boolean TRUE if no transaction on signal has occurred during time T, else returns Boolean FALSE. Default value of T is zero.)

7. Scalar attributes: *scalar_type'left* (returns the first or left-most value)
scalar_type'left *(returns the last or right-most value)*
scalar_type'low (returns the lowest value)
scalar_type'high (returns highest value)
scalar_type'ascending (returns TRUE if T is ascending, else FALSE)
scalar-type'value(s) (returns value T represented by the string value *s*)

and,

8. Array attributes:
MATRIX'left(N) (left-most element index)[23]
MATRIX'right(N) (right-most element index)
MATRIX'high'(N) (upper-bound)
MATRIX'low(N) (lower-bound)
MATRIX'length(N) (number of elements)
MATRIX'range(N) (range)
MATRIX'reverse_range(N) (reverse range)
MATRIX'ascending(N) (TRUE if index is an ascending range, else FALSE)

In VHDL, enumeration, numeric and physical data types are scalar. Numeric data types can be either real or integer [−2147482647, 2147482647]. Physical data types are scalar numeric values associated with a system of units and/or physical measurements. Time is supported as a predefined physical type but other physical types must be defined by the user. Time values may range from 0 to 1E20 with units in femtoseconds. Arrays, consisting of multiple elements of the same type are supported by VHDL with strings, bit_vectors (Bit_vector) and standard logic vectors (Std_logic_vector) being predefined arrays.

6.2.5 Predefined Data Types and Subtypes

VHDL supports a number of predefined data types, viz, integers, reals, time (fs), bit (0,1), (true, false), bit_vector (an unconstrained array of bits), character (128 chars in VHDL'87 and 256 in VHDL'93), severity_level (note, warning, error, failure), file_open_kind (read_mode, write_mode, append_mode)file_open_status (open_ok, status_error, name_error, mode_error), string (an unconstrained array of characters). The predefined subtypes are natural (0-2147483647), positive (1-2147483647) and delay_length 90 fs - 2147483647).

The precedence for:

---

[23]N is optional for any array attribute for which the matrix is a one-dimensional array.

- integers, reals and time is abs, **, *, /, mod, rem, +(sign), -(sign), + (addition), - (subtraction), =, /=, <, <=, >, and >=,
- bit_vector(s) is NOT, &, sll, srl, sla, sra, rol, ror, =, /=, <, <=, >, AND, NAND, OR, NOR, XOR, and XNOR. The precedence for bit and is NOT, =, /=, <, <=, >, >=, AND, NAND, OR, NOR, XOR, and XNOR,
- natural and positive is the same as that for integers,

and

- delay_length is the same as that for time.

6.2.6 Operator Overloading

VHDL allows the user to assign new definitions to existing operators such as +, −, *, NAND, etc. when creating user-defined data types.[24] This object-oriented approach relies on VHDL determining what the appropriate operator action is for a given data type (argument). Objects can be overloaded by overloading, operators, parameters or subprogram names. Overloaded functions can have the same name but a different number of arguments or different argument types. In such cases VHDL uses the number, or type, of arguments to determine the appropriate action(s). Function names can also be in the form of an operator, so that a function can be called by symbols such as +, −, >, <, etc. For example the + operator could be "overloaded" to support vector addition, addition of two strings, etc.

6.2.7 VHDL Data Objects

VHDL data objects include variables, signals and the associated signal attributes. Variables must be declared before they can be used and only with the context of a subprogram or process. Variable declarations must also include a specification of the data type. Initializing variables at the time of declaration is optional but if left unspecified, then the default value is the leftmost element of the declared data type declared.

Signals are declared, in a similar manner, to that of variables and are subject to the to following conditions and constraints:

- They are declared as *either intermediate nodes* (architecture) or *ports* (entity)
- Entities can have port access.
- Assignment of values to input ports is supported by VHDL.
- "Signals" refer to nodes that may have voltage dependencies which are a function of time.
- Signal assignments employ the "<=" delimiter.
- A *transaction* is the scheduling of a value to a signal.
- An *"event"* is said to have occurred when a signal's value changes.
- Signals that are assigned a value without the specification of a delay will, during simulation, only change the value after the simulation's sub-interval has transpired.
- Signals have the following attributes:
  1. *X'active* returns TRUE if a transaction has occurred during the current simulation time, otherwise, it returns FALSE.

---

[24] The assignment of a new function to an existing operator is referred to "operator overloading".

2. *X'quiet(n)* returns TRUE if no transaction has occurred during the previous 'n' seconds.
3. *X'event* returns TRUE if the value of X changed during the current simulation time.
4. *X'stable(n)* returns TRUE if X did not experience an event during the past 'n' seconds.
5. *X'delayed* delays the signal X for n seconds.
6. *X'last_active* returns the time that has elapsed since the last transaction.
7. *X'last_event* returns the time that has elapsed since the last event.
8. *X'last_value* returns the previous value of X.

6.2.8   VHDL Operators

VHDL expressions consist of operators and so-called primaries.[25] Logical operators such as AND, *NAND, OR, ROR*, NOR and *NOT* can operate on arrays as well as can be applied to one-dimensional arrays, s or values of type bit.

Operator types include:

1. Logical operators:[26] *AND, NAND, XOR, OR, NOR, XNOR* and *NOT*
2. Unary sign operators: plus $(+)$ and minus$(-)$
3. Addition operators: plus $(+)$,
4. Addition operators: Plus $(+)$, minus $(-)$ and concatenation[27] $(\&)$,
5. Shift operators:[28] Shift right logical[29] (srl), shift left logical (sll), shift left arithmetic[30] (sla), shift right arithmetic (sra), rotate left (rol) and rotate right (ror,)
6. Multiplication operators: multiply (*), divide [0, modulus[31] (mod) and remainder (rem)],
7. Exponentiation (**) is subject to the constraint that the left-hand operand must be an integer, or floating-point, value and the right-hand operand must be integer only.
8. Absolute value (abs) This operator can be applied to any numeric type within an expression.
9. NOT—the inversion operator.

The order of precedence for these operators from the highest to the lowest is: exponentiation, absolute value and not (inversion) followed by multiplication, addition, shifting, relational and finally logical operators. If two operators of equal precedence are encountered then the left-hand operator is evaluated followed by the right-hand operator. All these rules are applied beginning with the most deeply nested parentheses in the expression.

Arithmetic operations such as division and multiplication can be applied to floating-point and integer values. If the right-hand operand is negative, the left-hand operand must be a floating-point value floating-point numbers or physical types. While the exponential operator can be applied to either floating-point or integer values, the right-handed operand must be an integer. Relational operators such as =, /=, >, >=, <and <= produce results but both the right-hand and left-hand operands must be the

---

[25] The term *primaries* refers to function calls, object names, literals and parenthetical expressions.

[26] These operators are not subject to any precedence order so liberal applications of parentheses is recommended.

[27] The concatenation operator combines the bits on either side of the concatenation operator.

[28] Shift operators have two operands. The left-hand operand is the bit_vector to be shifted, or rotated, and the right-hand operand is an integer value representing to the number of shifts, or rotations. A negative value for the latter results in the inverse operation being invoked.

[29] The *fill value* for sll and srl is '0'.

[30] The fill value for sla is the right-hand bit and the left-hand bit for sra.

[31] This mod and rem operators are only applicable to integer types.

same type. Two values are treated as equal provided that the corresponding elements of each are equal. The concatenation operator, typically used to join strings, can be applied to two one-dimensional arrays with as little as one element each. Single elements can be concatenated with multi-element arrays.

6.2.9   Conditional Statements

VHDL supports both *if-then-else* and *case* statements. Execution is subject to user-defined conditions in the form of expressions which evaluate to values. If such an expression evaluates as true, then the corresponding statements are executed otherwise the else statements are executed.

If statements are of the general form[32]

```
if <condition> then
        statements
            ...
    [
    elsif <condition> then
            statements
                ...
    else
            statements
                ...
    ]
    end if;
```

and case statements are of the form

```
case <expression> is

when <choice(s)> =>

<expression>;
        ...

when ...

[when others => ... ]

end case;
```

Case statements allow execution to depend on the value of a selection statement. Case statements must include all possible values of the expression to be evaluated, however, values that are not to be treated by the case statement can be included as "OTHERS' in conjunction with the reserved word "NULL" resulting in no action with for those values. Supported expression types include integer, enumerated and one-dimensional character arrays.

---

[32]Note that "conditions" are restricted to type, "elsif" contains one "e" and "end if" is two words.

6.2.10 FOR, WHILE, LOOP, END and EXIT

FOR and WHILE loops are supported in VHDL together with EXIT which is used to leave a loop[33] and END LOOP to end a loop. LOOP can be used to repeat a loop indefinitely, e.g.,

```
loop
     some_activity;
end loop;
```

The formal syntax for both WHILE and FOR is

```
loop statement ::=
     [loop label :] [ while -expression | for ]
loop
     { statements }
end loop
```

The *while* reserved word evaluates a test condition prior to each iteration and if the expression evaluates as true the next iteration in invoked, otherwise the loop terminates. The *for* iteration loops for a predefined number of iterations, and a loop parameter keeps track of the number of iterations that have occurred. A *next* statement can be used to terminate the current iteration and an *exit* statement terminates the loop thereby passing control to the next statement to be executed. The *null* statement is typically used to indicate no action is to take place.

The *assert* statement provides exception handling and is of the following formal form:

```
assertion_statement::=[label : ]assertion;
```

where

```
assertion::=
          Assert condition
               [Report expression]
               [Severity expression];
```

If the status is not consistent with this condition and the *report* clause is present, a message, e.g., "assertion violation" occurs. The *severity* clause assigns a severity level, viz., *note, warning, error* or *failure*. If the severity clause is not present, then the security level defaults to *error*. The assert statement can be used to halt the execution of a simulation.

6.2.11 Object Declarations

Three types of objects are supported in VHDL, viz., variables, constants and signals. Constants are initialized with a specific value that may not be modified thereafter[34] unlike variables whose values may be modified after initialization. A *deferred constant* declaration occurs in a corresponding package declaration, but is assigned a value in the *package body*. Non-shared variables are treated as local variables in subprograms and processes and shared variables are treated as global variables. Variables are assigned values by the use of " := ". If an object is merely declared without initialization, its value defaults to that of the first value occurring in the package body.

---

[33] Also referred to as "jumping out of a loop".

[34] In effect, constants in VHDL are read-only.

6.2.12 FSMs and VHDL

Finite state machines [7] are often described in VHDL by using processes *sensitive* only to the clock and asynchronous resets for state transitions. Outputs, in such cases, are expressed as concurrent statements external to the process. State machines can be viewed simply as black boxes and therefore a behavioral model utilizing an entity/architecture pair is sufficient to describe it.[35] The internal states can be defined in terms of enumerated types.

A combinatorial process can be used to provide the next-state conditioning logic, a synchronous process can be used to provide the current state variables and a third process can provide the output logic. Each of the processes operates concurrently in VHDL, and therefore the combination would function as an FSM. The next-state conditioning logic determines the next state as a function of the current state and the inputs. In VHDL the selection of the next state can be managed by the use of a *case* statement.[36] The synchronous process can handle handle the registers with respect to the current state condition and to reset the state machine to a predefined state. The output logic can be implemented in VHDL as a set of *if-then-else* statements.[37]

6.3   Verilog

Verilog [5] is a hardware description language, similar in some respects to VHDL, that is used to model electronic circuits primarily at the register level.[38] Originally introduced in 1984, it provided designers with a description language that for most designs was much easier to learn and use than VHDL. With fewer data types than VHDL, limited casting allowed, no user-defined types supported and relying on primitive types, the language could employ a fast, memory-efficient, simpler compiler than that required for VHDL. While originally a proprietary language, in 1990, Open Verilog International (OVI) was formed and a joint effort was undertaken to create a Verilog standard reference manual that ultimately led to the establishment of an IEEE standard for the language [9].

The Verilog reserved word list (aka keywords list) is shown in Table 6.2. Both line and block comments are supported in Verilog with two forward slashes representing the start of the comment which is assumed to extend to the end of that line. Block comments begin with /* and end with */ and cannot be nested.

Identifiers that begin with a backslash (\) and end with a white space, e.g., newline, space or tab are treated as "escaped" identifiers. However, the leading backslash and ending white space are treated as part of the identifier.

In Verilog [8], a bit can take of one of four values: 0, 1, X or Z corresponding to logic zero, logic one, an unknown logic value[39] or high impedance (floating).

---

[35]The entity defines the interface and the architecture defines the internal behavior.

[36]It is important to specify all the possibilities for a given case statement, even if some possibilities are not used in order to avoid exceptions.

[37]It's a good practice to include an else statement for each if statement, because VHDL signals have "implicit memory".

[38]Verilog [3,6] takes into account signal timing, propagation delays and edge transitions in the description and modeling.

[39]Unknown logic values are restricted to 0, 1 or Z or a transition from one of three to another of the three allowed values. X represents either a "don't know" or "don't care" state, or both.

6.3.1 Constants

Constant values can be expressed as either simple decimals, i.e., as a sequence of digits employing values 0–9 or as a sized constant[40] representing a based number.[41] String constants are treated as unsigned integer constants represented by a sequence of 8-bit ASCII values, with each such value representing a particular character.

6.3.2 Data Types

Verilog supports data types belonging to one of three classes, viz.,

- *nets*—The net data type represents a physical connection between hardware blocks and can be driven by a continuous assignment statement or the output of a module or gate. However, a net data type does not store its value. A net data type can be a wire,[42] tri, supply0 or supply1 type. Wire and tri data types are identical in terms of syntax and functionality and are supported to delineate between wire nets that are driven by a single gate and tri nets that are driven by multiple drivers, supply0 and supply1 are the nets representing logic 0 (ground) and logic1 (power) when modeling power supplies. Declaration of a scalar net must include the range of bits, e.g.,

```
wire [7:0] dataA; /*dataA where bit0 is the LSB and bit7 is the MSB.

wire [0:7] dataA; /*dataA where bit0 is the MSB and bit7 is the LSB.
```

   A net can be either a scalar[43] or vector, the former representing individual signals and the latter representing bus signals. The *strength* of a net is specified by drive strength and charge.[44,45]
- *Registers*—Register data types store their values, until changed by an assignment in *always* blocks functions r tasks, and are used as variables. To declare a reg type the reserved word *reg.* is employed and is declared by using the keyword *reg.*[46] The integer register data type is used for values that are not to be treated as registers. Register variables may be declared as either scalars or vectors. Vector register declarations include a specification of the range of the bits after *reg* or *integer*. The left- and right-hand values in this range specify the most significant and least significant bits, respectively.
- *Parameters*—Parameter values are used in parameterized models are treated as constants during runtime and are declared as follows:[47]

```
parameter_assignment {,parameter_assignments}
parameter_assignment ::= parameter_identifier=constant_expression
```

---

[40]Sized constants consist of three tokens, viz., an optional size, a single quote followed by a base formed character and a sequence of digits representing the value.

[41]For example: hexadecimal, octal, binary or decimal.

[42]A wire represents a 1-bit interconnection between modules.

[43]By default all nets are treated as scalars.

[44]Specified as weak0, weak1, highz0, highz1, pull0, pull1, pullup or pulldown.

[45]Specified as small, medium or large.

[46]*reg* is not necessarily a hardware register, or a flip-flop, but rather simply denotes the fact that the value is retained. Uninitialized *reg* values are treated as X, i.e., undefined.

[47]If a given parameter is not intended to be modified by a higher-level module, the compiler directive *define* should be used.

Examples of parameter use are given by:

```
parameter lsb = 0, msb =3; // lsb and msb are parameters
reg [msb,lsb] x ;                      // x is a vector with range 3:0
parameter tPD = 7;                     // parameter tPD is used to represent
                                       propagation delay
```

Parameters are treated as strings of arbitrary length unless constrained by the user, e.g.,

```
parameter unconst_param = 12 /* unconstrained
                             // (size is determined by usage) */
parameter [3:0] const_param = 12; //constrained to 4 bits
```

6.3.3   Modules

A Verilog module encapsulates the description of a design that can be either:

1. a behavioral (algorithmic) description that defines the behavior of a circuit in abstract, high-level algorithms, or expressed in terms of low-level Boolean equations,

or,

2. a structural description that defines the structure of the circuit in terms of components and resembles a net-list that describes a schematic equivalent of the design and supports concurrency.[48]

Table 6.2   Verilog reserved words.

always	and	assign	automatic	begin
buf	bufifo	bufif1	case	casex
casez	cell	cmos	config	deassign
default	defparam	design	disable	edge
else	end	endcase	endconfig	endfunction
endgenerate	endmodule	endprimitive	endspecify	endtable
endtask	event	for	force	forever
fork	function	generate	genvar	highz0
highz1	if	ifnone	incdir	include
initial	inout	input	instance	integer
join	large	liblist	library	localparam
nmos	no	noshowcancelled	not	notif0
notif1	or	output	parameter	pmos
posedge	primitive	pull0	pull1	pulldown
pullup	pulsestyle_oneevent	pulsestyle_ondetect	nrcmos	real
realtime	reg	release	repeat	mmos
rpmos	rtran	rtranif0	rtranif1	scalared
showcancelled	signed	small	specify	specparameter
strong0	strong1	supply0	supply1	table
task	time	tran	tranif0	tranif1
tri	tri0	tri1	triand	trior
trireg	unsigned	use	vectored	wait
wand	weak0	weak1	while	wire
wor	xnor	xor		

---

[48]Structural descriptions contain a hierarchy in which the components are defined at different levels and the logic is defined in terms of gate primitives.

6.3.3.1 Module Syntax

A Verilog design consists of one or more modules[49] interconnected by ports. Each port has an associated name and mode, viz., *input*, *output* and *inout*. Module definitions cannot be nested. A module is defined using the following syntax:

```
module <name> (interface_list) ;{ module_item }
endmodule
interface_list ::= port_reference
| {port\_reference {, port_reference}}
port\_reference ::= port_identifier
| port_identifier [ constant_expression ]
| port_identifier [ msb_constant_expression : lsb_constant_expression ]
module_item ::= module_item_declaration
| continuous_assignment
| gate_instantiation
| module_instantiation
| always_statement
module_item_declaration ::= parameter\_declaration
| input_declaration
| output_declaration
| inout_declaration
| net_declaration
| reg_declaration
| integer_declaration
| task_declaration
| function\_declaration
```

Example:

```
// a module definition for a d flip-flop
module
my\_dff (clk, d, q);
input clk, d;
output q;
wire clk, d;
reg q ;
always @(posedge clk)
begin
q = d ;
end
endmodule
// a module definition for module
my\_dff (clk, d, q);
input clk, d;
output q;
wire clk, d;
reg q ;
        always @(posedge clk)
                begin
                    q = d ;
                end
        endmodule
```

---

[49]Warp [12] treats the keywords *macromodule* and *module* as synonyms.

6.3.4   Operators

Verilog supports a variety of operations including *arithmetic, bit-wise, concatenation, conditional, equality, logical, reduction, relational, replication,* and *shift*:

- *arithmetic operators* Binary operators for addition, subtraction, multiplication divisions and modulus and unary operations for specifying the sign of a value, i.e., plus or minus. Integer division is supported but truncates the fractional part. Register data types are treated as unsigned values and negative values are expressed in a twos-complement format. The modulus operator assigns the result the same sign as that of the first operand.
- *bit-wise operators* perform bit-wise operations on the respective bits of the two operands. If the two operands are of different bit lengths the shorter value, bitwise, will be padded with zeros sufficient to match the bit length of its counterpart. Supported bit-wise operations include and (&), inclusive or (|), exclusive or (^) and exclusive nor (^~or ~^), i.e., equivalence.
- *concatenation operator* uses braces to encapsulate the values to be concatenated. Each such value is delimited by a comma, e.g., {a, eb[4:0], c, 5'b11011}.
- *conditional operators* have the following syntactic form:

$$\mathrm{condition} \ ? \ \mathrm{expression1};$$
$$\mathrm{expression2}$$

   If *condition* evaluates as false, i.e., zero, then *expression2* is evaluated, otherwise *expression1* is evaluated. If *condition* evaluates as either z or v, then both *expression1* and *expression2* are evaluated and the resulting value is determined by a bit by bit examination based on Fig. 6.4. If *expression1*, or *expression2*, are of type real the value of the whole expression is zero. If *expression1* and *expression2* are of different lengths, then the length of the entire expression is assigned the length of the longer-expression and trailing zeros are added to the shorter-expression as required.

	0	1	x	z
0	0	x	x	x
1	x	1	x	x
x	x	x	x	x
z	x	x	x	x

Fig. 6.4 Value of conditional expressions containing x,z,1 and/or 0.

- *equality operators*—there are two types of equality operators, viz., case and logical. For case equality the operators are a==b (a is equal to be for 0, 1, z and x) and a!==b (a is not equal to be for 0, 1, z and x). For logical equality the operators are a==b and a!=b and in some cases the result may be undefined. These operators are compared bit-by-bit with zeros being added to make the two operands the same length. If either operand contains a z or an x, then the result is x for $a == b$ and $a! = b$. If either operand contains an x or a z, then a==b and a!==b can only be true if the respective bits in a and b have the same values of x and z.

- *logical operators*—are logical *negation* (!), logical *or* (||) and logical *and* (&&). Logical *negation* and logical *and* are evaluated from left-to-right.
- *reduction operators*—are the unary operators *and*, *or*, *xor*, *nand*, *nor* and *xnor* that are bit-wise operations on a single operand and produce a single-bit result, e.g.,
  $$\&(4'b0101) = 0 \& 1 \& 0 \& 1 = 1'b0.$$
- *relational operators*—are the less than, greater than, less than or equal to and greater than or equal to operators that produce a scalar value of zero, the relation is false, 1 if the relation is true and x if any of the operands contains unknown x bits. If any operand is x or z, the result is treated as false.
- *replication operators*—replicates a group of bits *n* times, e.g., {1, 1,{3{1,0}}}= 11101010.
- *shift operators*—performs right- or left-shifts on the right-hand operand where the number of shifts, right (») or left («), is determined by the value of the right-hand operand.[50]

The following is an illustrative example of a Verilog program designed to compute square root:

```
module sqrt32(clk, rdy, reset, x, .y(acc));
input   clk;
output rdy;
input   reset;
input [31:0] x;
output [15:0] acc;
// acc = accumulated result, and acc2 = accumulated acc^2
reg [15:0] acc;
reg [31:0] acc2;
// Track bit being worked on.
reg [4:0]  bitl;
wire [15:0] bit = 1 << bitl;
wire [31:0] bit2 = 1 << (bitl << 1);
// Output ready when bitl counter underflows.
wire rdy = bitl[4];
// guess h=next values for acc. guess2=square of that guess h.
// guess2 = (acc + bit) * (acc + bit)
//        = (acc * acc) + 2*acc*bit + bit*bit
//        = acc2 + 2*acc*bit + bit2
//        = acc2 + 2 * (acc<<bitl) + bit
// Note: bit and bit2 have only a single bit in them.
wire [15:0] guess  = acc | bit;
wire [31:0] guess2 = acc2 + bit2 + ((acc << bitl) << 1);
(* ivl_synthesis_on *)
always @(posedge clk or posedge reset)
if (reset) begin
        acc = 0;
        acc2 = 0;
        bitl = 15;
   end else begin
                    if (guess2 <= x) begin
                    acc  <= guess;
```

---

[50]Vacated bits are replaced by zeros.

```
                              end
                              bit1 <= bit1 - 5'd1;
                  end
                  endmodule
```

6.3.5 Blocking Versus Nonblocking Assignments

In Verilog/Warp a *blocking statement*, which is part of a sequential block, must be executed prior to the execution of those statements that follow it. In the case of nonblocking statements, assignments occur without blocking the procedural flow. Blocking assignments employ the symbol "=" and nonblocking statements employ the symbol "<=" for assignment. Nonblocking statements allow events to be scheduled for a later time.

6.3.6 *wire* Versus *reg* Elements

wire elements are used in Verilog applications to connect the input and output ports of a module instantiation with other elements within a design. However, unlike their counterpart, *reg*, they are not able to store values and must be driven. In effect wires serve as "state-less" connection mechanisms. *wire* elements are only used in cases for which the model is based on combinatorial logic. *reg* elements perform a function similar to that of wires but have the ability to store values in a manner analogous to that of registers. These elements are used in both sequential and combinatorial logic models. While reg cannot be used on the left-hand side of an assign statement, it can be used to create registers in conjunction with always@(posedge clock) statements/blocks. *reg* can also be used on the left-hand side of always@block = or <= symbol. It can also be used as the input to a module, or within a module declaration, but not to connect to the output port of a module.

6.3.7 *always* and *initial* Blocks

In modeling combinatorial and sequential elements the *initial* and *always* blocks play important roles. *initial* blocks[51] are procedural blocks consisting of sequential statements that are executed only once, typically at the start of execution of a simulation, whereas an *always* block is always available for as long as the program is executing. An *always* statement contains a *sensitivity list*[52] that determines when the block of code associated with the *always* block is to be executed. Any change in the signals contained in the sensitivity list will cause the *always* block to be executed.

The standard format for an always statement is defined as:

```
always@(event_expression_1 [or event_expression_2]{or event expression_3}])
```

*event_expression*s can contain timing controls which are either *posedge* or *negedge* for positive or negative-edge triggering, respectively. If sequential triggers are employed in the sensitivity list, sequential logic is synthesized. Asynchronous, or synchronous, triggers may be used in the *sensitivity list*, but not both.

```
            // Always block with asynchronous triggers:
                  always @(x or y)
```

---

[51] Warp ignores *initial* constructs.

[52] Sometimes referred to as the *sensitive list*.

```
        begin
            . . .
        end
        /* Always block which realizes sequential logic with
        rising edge of a clock: */
        always @(posedge clock)
        begin
            . . .
        end
        /* Always block which realizes a sequential logic with
        falling edge of clock and an asynchronous preload */
        always @(negedge clock or posedge load)
        begin
            . . .
        end
```

The following are equivalent syntactically:

```
        always@(signal_1 or signal_2 or  signal_3 or signal_4)
        always@(signal_1,signal_2, signal_3, signal_4)
        always@(*)
        always@*
```

where * refers to all the signals within the *always* block.

6.3.8   Tri-State Synthesis

Warp does not synthesize tri-state logic. In order to include tri-state logic in a Verilog module the cy_bufoe [11] must be instantiated.[53] The tri-state output of this module, y, must then be connected to an inout port on the Verilog module. That port can then be connected directly to a bidirectional pin on the device. The feedback signal of the cy_bufoe, yfb, can be used to implement a fully bidirectional interface, or can be left floating to implement just a tri-state output.

```
        module ex\_tri\_state (out1, en, in1);
                inout out1;
                input en;
                input in1;
                cy\_bufoe buf\_bidi (
                        .x(in1), // (input) Value to send out
                        .oe(en), // (input) Output Enable
                        .y(out1), // (inout) Connect to the bidirectional pin
                        .yfb()); // (output) Value on the pin brought back in
        endmodule
```

---

[53]cy_bufoe is a tri-state, non-inverting buffer [11] with an active high output and a enable input.

6.3.9 Latch Synthesis

Warp synthesizes a latch whenever a variable inside an always block with asynchronous trigger, has to hold its previous value. The following code fragment synthesizes a latch.

```
// example: latch synthesis with if statement
always @ (signal1 or signal2)
begin
        if( signal1 )
                begin
                        out\_sig = signal2 ;
                end
end
```

6.3.10 Register Synthesis

A register is typically a set of flip-flops that share a common clock input and is used to store a group of bits. The register is updated when the next clock edge occurs. Most registers employ both a reset and load input controls. In the case of a shift register, the flip-flops are connected in a chain in which the output of one flip-flop becomes the input to the next flip-flop in the chain. This interconnection scheme allows data to be *shifted* to the next flip-flop each time a clock edge occurs. Shift registers can be serial-in-serial-out (SISO), parallel-in-parallel-out (PIPO), serial-in-parallel-out (SIPO) or parallel-in-serial-out (PISO). Thus, in order to synthesize registers, it is necessary to be able to synthesize flip-flops.

6.3.10.1 Edge-Sensitive Flip-Flop Synthesis

Warp uses the following templates to synthesize synchronous flip-flops. The template for the positive edge sensitive flip-flop is:

```
always @ (posedge clock\_signal)
synchronous\_signal\_assignments
```

and the template for the negative edge sensitive flip flop is:

```
always @ (negedge clock\_signal)
synchronous\_signal\_assignments
```

6.3.10.2 Asynchronous Flip-Flop Synthesis

Warp uses the following format to synthesize asynchronous flip-flops with reset, or preset.

```
always @ (edge\_of clock\_signal or
        edge\_of preset\_signal or
        edge\_of reset\_signal)
if (reset\_signal)
        reset\_signal\_assignments
else if (preset\_signal)
        preset\_signal\_assignments
else
        synchronous\_signal\_assignments
```

The *posedge* construct is used to specify an active high condition and the *negedge* construct to specify an active low condition. The variables in the sensitivity list can appear in any order. Subsequent reset, or preset, conditions can appear in the else-if statements. The last else block represents the synchronous logic. The polarity of the *reset/preset* signal condition used in the sensitivity list and the polarity of the *reset/preset* condition in the *if/else-if* statements should be the same.

Example A *posedge* reset_signal condition in the sensitivity list is required when the reset condition is one of the following forms:

```
if( reset\_signal)
if( reset\_signal == constant\_one\_expression)
```

A *negedge* reset_signal condition in the sensitivity list is required when the reset condition is one of the following forms:

```
if( !reset\_signal)
if( ~reset\_signal)
if( reset\_signal == constant\_zero\_expression)
```

Warp generates an error if the polarity restriction mentioned above is violated. Warp allows more than two asynchronous if/else-if statements before the synchronous else statement as shown in the following example.

```
// An example of two different preset signals:
module asynch\_rpp(in1, clk, reset, preset, preset2, out1);
input in1, clk, reset, preset, preset2;
output out1;
reg out1;
always @ (posedge clk or posedge reset or posedge preset or
posedge preset2)
          if (reset)
             out1 = 1'b0;
          else if (preset)
             out1 = 1'b1;
          else if (preset2)
             out1 = 1'b1;
          else
             out1 = in1;
   endmodule
```

The *posedge* and *negedge* keywords are used to specify active high and low conditions, respectively. Variables in the sensitivity list can occur in any order. The polarity of reset/preset signal conditions in a sensitivity list and the polarity of the reset/preset conditions in corresponding if/else-if statements must be the same.

A *posedge* reset_signal condition in the sensitivity list is required when the reset condition is one of the following forms:

```
if(reset\_signal)
if(reset\_signal == constant_one_expression)
```

A *negedge* reset_signal condition in the sensitivity list is required when the reset condition is one of the following forms:

```
if(!reset\_signal)
if(~reset\_signal)
if(reset\_signal == constant\_zero\_expression)
```

Warp generates an error if the polarity restriction mentioned above is violated. Warp allows more than two asynchronous if/else-if statements before the synchronous else statement as shown in the following example.

```
// An example of two different preset signals:
module asynch\_rpp(in1, clk, reset, preset, preset2, out1);
input in1, clk, reset, preset, preset2;
output out1;
reg out1;
```

6.3.11 Verilog Modules

Verilog modules[54] are used to encapsulate the description of a design expressed either as a behavioral or structural description. A behavioral description defines the behavior of a circuit in terms of abstract, high-level algorithms or in terms of low-level equations.

A structural description defines the circuit's structure in terms of components and resembles a net-list that describes a schematic equivalent of the design. Structural descriptions contain the hierarchy in which components are defined at different levels.

A Verilog design consists of one or more modules[55] connected with each other by means of ports that provide a means of connecting various hardware elements. Each port has an associated name and mode (*input*, *output* and *inout*). A module is defined using the following syntax:

```
module <name>(interface\_list) ;{ module\_item }
endmodule
interface_list ::= port_reference
| {port\_reference {, port\_reference}}
port\_reference ::= port\_identifier
| port\_identifier [ constant\_expression ]
| port_identifier [ msb_constant_expression : lsb\_constant\_expression
module_item ::= module_item_declaration
| continuous_assignment
| gate_instantiation
| module_instantiation
| always_statement
module_item_declaration ::= parameter_declaration
| input_declaration
| output_declaration
| inout_declaration
| net_declaration
| reg_declaration
| integer_declaration
```

---

[54] A Verilog module may represent a single gate, flip–flop, register or other much more sophisticated circuits.
[55] Module definitions cannot be nested.

```
| task_declaration
| function_declaration
```

In Verilog, hierarchical designs are specified by instantiating one, or more, modules in a top-level module not instantiated by any other module. The syntax of the module instantiation statement is as follows:

```
<module_name> [parameter_value_assignment]
<instance_name>
module_instance {, module_instance} ;
module_instance ::= instance_identifier
([list_of_module_connections])
list_of_module_connections ::= ordered_port_connection {,
ordered\_port\_connection }
| named\_port_connection {,named_port_connection }
```

One or more instantiations of the same module can also be specified in a single module instantiation statement. The four instantiation statements in the above example can be combined into one instantiation statement as follows:

```
my\_dff inst\_3(clk, d, q0),
        inst\_2(clk, q0, q1),
        inst\_1(clk, q1, q2),
        inst\_0(clk, q2, q) ;
```

A module connection describes the connection between the signals listed in the module instantiation statement and the ports in the module definition. This connection can be specified in two ways: ordered port association and named port association. In the case of ordered port association, the signals in the instantiation statement should be in the same order as the ports listed in the module definition. In the case of a named port association, the port names of instantiated modules are also included in the connection list.

```
my_dff inst_3(clk, d, q0)              ; // ordered connection list.
my_dff inst_3(.d(d), .q(q0), .clk(clk)) ; /* named association: q0 is
                                          connected to the port q of
                                          my_dff module. */
```

The port expression in the module connection list can be one of the following: a simple identifier, a bit-select of a vector declared within the module or a part-select of a vector declared within the module or some concatenation thereof.

The following describes the behavior of a counter that increments the count by 1 on the rising edge of a clock (trigger). It also contains an asynchronous reset signal that resets the counter to zero.

```
module counter (trigger, reset, count);
    parameter counter_size = 4;
    input trigger;
    input reset;
    inout [counter_size:0] count;
    reg [counter_size:0] tmp_count;
    always @(posedge reset or posedge trigger)
    begin
```

```
if (reset == 1'b 1)
      tmp_count <= { (counter_size + 1){1b 0}};
else
      tmp_count <= count + 1;
end
assign count = tmp_count;
endmodule
```

6.3.12 Verilog Tasks

Tasks are sequences of declarations and statements that can be invoked repeatedly from different parts of a Verilog description. They also provide the ability to break up a large behavioral description into smaller ones for easy readability and code maintenance. A task can return zero or other values.

A *task* declaration has the following syntax:

```
task \textless task\_name\textgreater\ ;{ task\_item\_declaration}
statement\_or\_null endtask
task\_item\_declaration ::= parameter\_declaration
reg\_declaration
integer\_declaration
input\_declaration
output\_declaration
inout_declaration
```

Warp ignores any timing controls present inside a task. The order of variables in the task enable statement calling a task must be the same as the order in which the I/Os are declared inside a task definition. Only *reg* variables can receive output values from a task, i.e., wire variables cannot. Note that Datapath operator inferencing is not supported inside tasks. When datapath operators $(+, -, *)$ are used inside tasks, at least one of the operands must be a constant or an input.

The following is an example of a module task:

```
module task_example(a,b,c,d,sum);
    output sum;
    input a,b,c,d;
    reg sum;
    always @(a or b or c or d)
    begin
        t_sum(a,b,c,d,sum);
    end
    task t_sum;
        input i1,i2,i3,i4;
        output sum ;
        begin
            sum = i1+i2+i3+i4;
        end
    endtask
endmodule
```

6.3.13 System Tasks

Verilog supports a number of *system tasks* that support I/O and measurement functions. These tasks are all prefixed by the symbol "$" and include the following:

- $display—writes text to the screen

$$\texttt{\$display(<parameter\_1>, <parameter\_2>, <parameter\_3>)}$$

- $dumpfile—declare the output file name (VCD format)
- $dumpports—dump the variables (extended VCD format)
- $dumpvars—dump the variables.
- $fdisplay—print to the screen and add a newline.
- $fclose—close and release an open filehandle.
- $fopen—open a handle to a file for either a read or write.
- $fscanf—read a format-specified string from a variable.
- $fwrite—write to a file without a newline.
- $monitor—print the listed variables when any of them change value.
- $random—return a random value.
- $readmemb—read the binary file content into a memory array.
- $readmemh—read the hex file content into a memory array.
- $sscan—read a format specified string from a variable
- $swrite—print a line without the newline to a variable.
- $time—the value of the current simulation time.
- $write—write a line to the screen without a newline.

6.4   Verilog Functions

Similar to tasks, *functions* are also sequences of declarations and statements that can be invoked repeatedly from different parts of a Verilog design. As is the case with *tasks*, *functions* provide the ability to break up a large behavioral description into smaller ones for readability and maintenance. Verilog functions are formally defined as:

```
function [range_or_type] <function_name>
        function_item_declaration {function_item_declaration}
        statement endfunction
function_item_declaration ::= parameter_declaration
        | reg_declaration
        | integer_declaration
        | input_declaration
```

Unlike a *task*, a *function* returns only one value. The *function* declaration will implicitly declare an internal register which has the same type as the type specified in the *function* declaration. The return value of the function is the value of this implicit register. A *function* must have at least one *input* type argument. It cannot have an *output* or *inout* type argument.

A *function* declaration can consist of the following types of declarations: *input, reg, integer*, or *parameter*. The order in which the inputs are declared should match the order in which the arguments are used in the function call. Timing controls and *nonblocking* assignment statements are not allowed inside a function definition. Datapath operator inferencing is not supported inside *functions*. When datapath operators $(+, -, *)$ are used inside *functions*, at least one of the operands must be a constant or an input. The function inputs cannot be assigned to any value, inside the function. All system task functions are ignored by Warp.

Example

```
module func_example(a,b,c,d,sum);
        output[2:0] sum;
        input a,b,c,d;
        reg[2:0] sum;
        always @(a or b or c or d)
        begin
            sum = func_sum(a,b,c,d);
        end
        function[2:0] func_sum;
            input i1,i2,i3,i4;
            begin
                func_sum = i1+i2+i3+i4;
            end
        endfunction
endmodule
```

6.5   Warp™

Cypress Semiconductor supports a subset of Verilog known as Warp™[13]. However, there are a number of significant differences between Verilog and Warp, viz.,

- Warp requires that the first character in an identifier must be a letter.
- Warp renames identifiers beginning with an underscore by adding the prefix 'warp'.
- Warp does not support "escaped" identifiers.
- If an underscore is used in a constant, it is ignored by Warp.
- Warp allows parameters to appear on the right-hand side of another parameter definition.
- Warp ignores the delay expressions, i.e., minimum, typical, and maximum.
- Warp treats the keywords macromodule and module as synonyms.
- Warp ignores the charge strength, drive strength and delay specified in the continuous assignment statements.
- Warp ignores all system tasks and system function identifiers.
- The following Verilog net types are not supported by Warp.
    1. tri0
    2. tri1
    3. wand tri
    4. and
    5. wor
    6. trior
    7. trireg

- Warp ignores the strengths associated with any net. Warp treats integers as 32-bit signed quantities and *reg* datatypes as unsigned quantities, by default, unless specified to be signed quantities.
- Warp does not support multiple drivers for register and integer variables.
- The time, real and real-time declarations are not supported in Warp.
- Ranges and arrays for integers are not supported by Warp. Arrays of register data types (memories) are also not supported in Warp.
- Warp does not automatically handle the size or the signed/unsigned nature of parameters.
- Warp uses the default values, if a parameter does not have a size constraint or a type (signed/unsigned/integer/etc.) designation.
- Warp allows only *defparam* to be used to modify the parameters of immediate instances.
- Parameter values in a module can also be re-defined by using the *defparam* construct. At any level of the design, Warp allows the re-definition of parameters of the modules instantiated at that level only. More than one levels of hierarchical path names are not currently supported.
- Verilog operators supported by Warp include arithmetic, shift, relational, equality, bit-wise, reduction, logical, conditional and concatenation.
- Warp does not support the case equal operators === and !==.
- Although concatenation can be repeated using a repetition multiplier in Verilog, Warp requires that the repetition operator be a constant.
- Warp does not support range specifications in module instantiations (array of instances).
- Warp supports the following primitive gates: *and, nand, or, nor, xor, xnor, buf, not, bufif0, bufif1, notif0, notif1*.
- Warp does not allow assigning a value to a register variable using either blocking or nonblocking assignment.
- Nonblocking assignment statements within a function/task are not supported by Warp.
- Warp does not support parallel block.
- Warp partially supports *casex* and *casez* statements. For the *casex* statement, ?, x, z are allowed in a case-item expression but not allowed in a case expression. Similarly, for the *casez* statement ?, z are allowed in a case-item expression, but not allowed in a case expression.
- When Warp synthesizes any of the case statements, it synthesizes a memory element for each output assigned to it in the case statement, in order to maintain any outputs at their previous values, unless one of the following conditions occurs: (1) All outputs within the body of the case statement are previously assigned a default value within the always block, (2) The case statement completely specifies the design's behavior following any possible result of the conditional test.[56]
- Warp supports two kinds of loop statements: *for* and *while*.[57] The while loop template supported in Warp is written as *while (<comparison><number>)*.
- Warp ignores the intra-assignment timing controls, delay-based timing controls and wait timing controls.[58]
- In structured procedures, Warp ignores the initial construct.
- Warp requires that an *always* statement have a sensitivity list.
- Warp ignores any timing controls present inside a task.
- Warp does not support the disabling of named blocks and tasks using the disable construct.

---

[56]The best way to ensure complete specification of design behavior is to include a default clause within the case statement. Therefore, to use the fewest possible resources during synthesis, either assign default values to outputs in the *always* block or make sure all case statements include a default clause.

[57]In Warp, the loop variable must be initialized to a constant value and the step assignment must be "+" or "−."

[58]Event timing controls are partially supported (only *posedge* and *negedge* event timing controls are supported when used with an *always@*).

- Warp ignores all system tasks and system task functions.
- When an *ifdef* compiler directive is used, Warp compiles only the code within the *'ifdef* Warp block.
- Warp issues a warning when it encounters any of the unsupported compiler directives.
- Warp does not synthesize tri-state logic.[59]
- Warp synthesizes a latch whenever a variable inside an *always* block, with an asynchronous trigger, has to hold its previous value.
- Warp uses the following templates to synthesize synchronous flip-flops. For a positive edge-sensitive flip-flop:

```
always @ (posedge clock\_signal)
synchronous\_signal\_assignments
```

and,

```
always @ (negedge clock\_signal)
synchronous\_signal\_assignments
```

```
is the template for the negative edge-sensitive flip-flop.
```

- Warp uses the following format to synthesize asynchronous flip-flops with reset or preset:

```
always @ (edge\_of clock\_signal or
            edge\_of preset\_signal or
            edge\_of reset\_signal)
    if (reset\_signal)
        reset\_signal\_assignments
    else if (preset\_signal)
        preset\_signal\_assignments
    else
        synchronous\_signal\_assignments
```

The *posedge* construct is used to specify an active high condition and the *negedge* construct is used to specify an active low condition. The variables in the sensitivity list can appear in any order. Subsequent reset or preset conditions can appear in the else-if statements. The last *else* block represents the synchronous logic. The polarity of the reset/preset signal condition used in the sensitivity list and the polarity of the reset/preset condition in the if/else-if statements must be the same.

- Warp allows more than two asynchronous *if/else-if* statements before a synchronous *else* statement.
- Warp allows the user to specify a particular case block to be implemented, e.g., a multiplexer (parallel case) rather than a priority encoder (full case). A parallel case or a full case is specified by including the directives *warp parallel_case* and *warp full_case* before a case statement. These

---

[59] In order to include tri-state logic in a module, the *cy_bufoe* must be instantiated. The tri-state output of this module, y, must then be connected to an *inout* port on the Verilog module. That port can then be connected directly to a bidirectional pin on the device. The feedback signal of the *cy_bufoe*, *yfb*, can be used to implement a fully bidirectional interface or can be left floating to implement just a tri-state output.

directives can be specified within the Verilog comment section (line comment or block comment). The directive must follow the word "warp".

6.6 Verilog/Warp Component Examples

A common use for Verilog/Warp is the creation of special components, e.g., a divide by N, four-bit counter can be easily created using the Verilog/Warp support provided by PSoC Creator, Cypress Semiconductor's development environment. After loading PSoC Creator and starting a new project, e.g., *CountByN*, navigate to the components tab in the Workplace Explorer and right-click on Project 'CountByN'. This will bring up a menu from which you can select *Add Component Item*. The *Add Component* window will then appear at which point you must select *Symbol Wizard* and optionally provide a name for your component, e.g., *DivideByNCounter*. Next, click on the *Create New* button.

This will load the Symbol Creation Wizard whose window allows you to select the name, type and direction of the *DivideByN_Counter*'s terminals. Note that the counter output is labeled count[3:0] indicating that the output is four parallel bits. This window also displays a preview of the *DivideByN_Counter*'s symbol. In the current example, reset and clock are input terminals and the count is a 4-terminal output as shown. Clicking OK will load the *DivideByN_Counter.cysym* page. Right click in area within this window away from the counter symbol. This will load a small menu from which you can select *Symbol Parameter...* At this point is necessary to define the parameter N in terms of its type and value. Select *int* and set the value to '1'.

Example 1 The source code template produced by PSoC Creator will be of the form:

```
//================================================================
                module CountByN (
                        count;
                        clock;
                        reset;
                );
                        output [3:0]
                        input clock;
                        input reset;
                parameter N=1;
        // `#start` body -- edit after this line, do not edit this line
                reg [3:0] count;
                always@(posedge clock or posedge reset)
                        begin
                                if (reset) count \textless= 4'b0;
                                else count \textless= count + N;
                        end
        // `#end` -- edit above this line, do not edit this line
        endmodule
```

Example 2 Similarly the Verilog/Warp code for a four-bit counter with an enable terminal that count from 0 to some defined limit can be expressed as:

```
        module Count4Enable   (
                count;
                clock;
                enable;
        );
```

```
            output [3:0] count;
            input clock;
            input enable;
            Parameter Limit= 15;
//`#start body -- edit after this line, do not edit this line
            reg [3:0] count;
            always@(posedge clock)
            begin
                    if (enable) begin
                            if (count == Limit) count = 4b'0;
                            else count <=count +1;
                            end
            end
//`#end` -- edit above this line, do not edit this line
//endmodule
```

Example 3 A Clocked register equivalent to can be expressed as:

```
        module DFF (
                clk;
                D;
                Q;
        )
                input clk;
                input D;
                output Q;
// `#start` body -- edit after this line,
                do not edit this line
                reg Q;
                always@(posedge clk)
                begin
                        Q <= D;
                end
//`#end` -- edit above this line, do not edit this line
            endmodule
```

Example 4 A clocked register with an asynchronous reset can be implemented by the following:

```
        module DFFR (
                clk;
                D;
                R;
                Q;

        )
                input clk;
                input D;
                input R;
                Output q;
```

```
                        reg q;
                        always @ (posedge clk or posedge R)
                        begin
                                if (R) Q <= 1'b0;
                                else Q <= D;
                        end
                endmodule
```

Example 5 A clocked register with an asynchronous "Set" can be implemented as:

```
                module DFFS (
                        clk;
                        D;
                        S;
                        Q;
                )
                        input clk;
                        input D;
                        input S;
                        output Q;
                        reg Q;
                        always @ (posedge clk or posedge S)
                        begin
                                if (S) Q <= 1'b1;
                                else Q <= D;
                        end
                endmodule
```

Example 6 A 2-input, 1-output mux can be implemented by the following:

```
                module muxA (
                        sel;
                        A;
                        B;
                        Z;
                )
                        reg Z;
                        always@(sel or A or B)
                        begin
                                if (sel) Z = A;
                                else Z = B;
                                end
                endmodule
```

Note that assignment in this example uses the "=" symbol since the assignments are combinatorial, i.e., there is no storage of values. This module is a representation of the Boolean expression

$$Z = sel \cdot A + \overline{sel} \cdot B \tag{6.1}$$

6.7   Comparison of VHDL, VERILOG and Other HDLs

The decision as to which is the best approach to modeling a circuit or system depends heavily on the application, the technology to be employed, the sophistication of the designer, ease of associated tools use, steepness of the associated learning curves, compatibility with other tools, etc. Some designs are not appropriate for HDLs, e.g., simple designs, or designs that cannot take advantage of the benefits of HDLs.

Verilog is based on a simple language syntax and structure allowing a designer to learn Verilog quickly, model both digital and analog circuits[60] Verilog also allows a model's code to be monitored to identify errors at early stages in the design process. Verilog models typically require less memory and therefore often run significantly faster during simulations than is available from similar VHDL models.

VHDL does offer better reusability capability by allowing procedures and functions to be encapsulated in *packages*. VHDL supports libraries as stores for configurations, architectures and packages but no similar concept exists for Verilog.[61] Unlike Verilog, VHDL has functionality intended to facilitate the management of large designs, e.g., *generate* (structure replication), *generic* (generic models), *package*(model reuse) and *configuration* (design structure). Verilog supports reduction operators but VHDL does not. Verilog's support for system tasks and functions allows a designer to incorporate control commands into a description to facilitate debugging, this debugging technique is not supported in VHDL. However, concepts such as user-defined types are supported in VHDL but not in Verilog and there is much more support in VHDL for high-level modeling.

VHDL is often described as "verbose" when compared to other languages in that it offers more than one way of expressing things. VHDL is "strongly-typed" and Verilog is "weakly-typed". VHDL provides a "rich" set of data types and Verilog is a smaller language and typically much easier to use. Verilog and VHDL are syntactically similar but there is no guarantee that Verilog models will behave the same in different tools. Verilog is generally regarded as much easier to learn than VHDL in part because Verilog is more "C-like" than VHDL.

SystemC[62] is sometimes used as an HDL to provide "VHDL-like" capability, but its use can be challenging, when modeling complex circuits. It allows the concept of time and concurrency to be employed in C++ applications, as for example when modeling synchronous hardware. Because it is C++ -based it is supported on a wide range of C++ platforms. SystemC has support for modules that communicate via ports, concurrent processes, channels,[63] events, and fixed-point/logic/extended standard data types.

The following is an example of a simple adder written in SystemC.

```
include "systemc.h"
#define WIDTH   4
SC\_MODULE(adder) {
    sc_in<sc_uint\textless <WIDTH> >   a, b;
    sc_out sc_uint<WIDTH> > sum;
    void do_add() {
        sum.write(a.read() + b.read());
```

---

[60]Verilog-AMS supports both analog and mixed-signal, in part, by supporting a continuous-time simulator capable of solving differential equations in the analog domain and providing the ability to cross-couple the digital and analog domains.

[61]Verilog began life as an interpreter and therefore libraries were not supported.

[62]SystemC is a collection of open-source, C++ classes and macros that function as an event-driven simulation kernel that can be used to model concurrent processes capable of communicating in a simulated, real-time environment.

[63]Channels may be wires, bus channels, FIFO's, signals, buffers, semaphores, etc.

```
        }
    SC_CTOR(adder)          {
    SC_METHOD(do\_add);
    sensitive << a <<  b;
        }
    };
```

6.8   Recommended Exercises

6-1 Express the function given by Eq. (8.22) in the form of a truth table. Use this table to sketch the associated logic diagram. Repeat this process for Eq. (8.26) and write a brief comparison of the two logic diagrams listing the number and types of gates used in both cases.

6-2 Show that

$$F = A \cdot B \cdot \overline{D} + A \cdot \overline{B} \cdot C + \overline{A} \cdot \overline{B} \cdot \overline{D} + A \cdot D + \overline{B} \cdot \overline{C} \cdot \overline{D}$$

can be reduced to

$$F = A + \overline{B} \cdot \overline{D}$$

6-3 Simplify the function F and show that F=1 [4].

$$F = A \cdot B \cdot C \cdot D + \overline{A} \cdot \overline{B} \cdot \overline{C} \cdot \overline{D} + A \cdot \overline{D} + \overline{A} \cdot B \cdot \overline{C} + A \cdot B \cdot \overline{C} \cdot D + B \cdot C \cdot D + \overline{A} \cdot C \cdot \overline{D}$$

6-4 Which of the following is a sum of products and which is a product of sums:

$$\overline{ABC} + A\overline{B}C$$

$$A\overline{B}\,\overline{C} + \overline{A}BC$$

$$(\overline{A} + B)(B + \overline{C} + D)$$

$$(A + \overline{B + C})(\overline{B + C})$$

$$\overline{A}\,\overline{B}\,\overline{C} + AB\overline{C} + A\overline{B}C$$

6-5 Sketch the logic circuit for: $(A + B + C)(\overline{A} + B + \overline{C})(A + B + \overline{C})$.

6-6 Show how to implement a NOT, OR and AND gate using 1, 2 and 3 NOR gates respectively.

6-7 Express each of the following expressions as a sum-of-products:

(a) $(A + B) \cdot (\overline{A} + \overline{B})$
(b) $A \cdot (B + C)$
(c) $-(A + B \cdot C)$

6-8 Write a VHDL entity declaration with the following characteristics:

• Port A is a 12 bit output bus
• Port AD is a 12-bit, three-state bidirectional bus

- Port INT is a three-state output
- Port AS is an output that is used internally
- Port OE is an input bit
- Port CLK is an input bit

6-9 Given the following entity declaration for a comparator:

```
LIBRARY ieee;
USE ieee.std_logic_1164.ALL;
ENTITY compare IS PORT (
a, b: IN std\_logic\_vector(o TO 3);
aeqb: OUT std\_logic);
END compare;
```

Write the VHDL code for an architecture that causes aebq to be asserted when a is equal to b using (a) conditional assignment, (b) Boolean equations and (c) a *process* with *sequential* statements.

6-10 Simplify $A \cdot B \cdot C + A \cdot \overline{B} \cdot \overline{C} + \overline{A} \cdot \overline{B} \cdot \overline{C}$ using a Karnaugh map.

6-11 The truth table for binary addition has three inputs: addedin, augend and carry in. The output consist of a sum and a carry. What is the truth table for the summing portion of binary addition? Simplify the expression representing this table by using a Karnaugh map.[64]

6-12 Draw the state diagram for a 3-bit binary counter as a state machine include any associated truth tables. Show how this counter can be implemented using combinatorial logic and D flip-flops.

6-13 Write an entity/architecture pair for the following circuit:

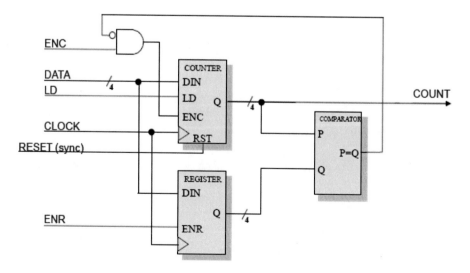

---

[64]Exercises 10 and 11 were suggested by Bob Harbort and Bob Brown, Computer Science Department, Southern Polytechnic State University.

References

1. P.J. Ashenden, *The Designers Guide to VHDL*, 3rd edn. (Elsevier, New York, 2008)
2. P.J. Ashenden, *The VHDL Cookbook*, 1st edn. (1990) http://tams-www.informatik.uni-hamburg.de/vhdl/doc/cookbook/VHDL-Cookbook.pdf
3. Creating A Verilog-Based Component. http://www.ue.eti.pg.gda.pl/~bpa/pusoc/kit_files/Creating a Verilog-based Component.pdf
4. J. Crenshaw, A primer on Karnaugh Maps. Programmers Toolbox. EE Times Design (2003)
5. T. Dust, G. Reynolds, Designing PSoC Creator Components with UDB Datapath. AN82156. Document No. 001-82156Rev. *I1. Appendix D. Auto-generated Verilog Code. Cypress Semiconductor (2017)
6. Just Enough Verilog for PSoC. https://www.cypress.com/file/42161/download
7. J. Kathuria, C. Keeser, Implementing State Machines with PSoC 3, PSoC 4, and PSoC 5LP. AN62510. Document No. 001-62510Rev. *F1. Cypress Semiconductor
8. Y. Magda, Designing PSoC Embedded Systems Using Verilog: A Practical Guide (2020)
9. V.K. Marrivagu, A.R. De Lima Fernandes, PSoC Creator – Implementing Programmable Logic Designs with Verilog. AN82250. Document NUMBER:001-82250Rev.*J1. Cypress Semiconductor (2018)
10. J.A. Peter, The VHDL Cookbook, 1st edn. http://tams-www.informatik.uni-hamburg.de/vhdl/doc/cookbook/VHDL-Cookbook.pdf Cypress Semiconductor (1990)
11. Tri-State Buffer (Bufoe) 1.10. Document Number: 001-50451 Rev. *F. Cypress Semmiconductor (2017)
12. $WARP^{TM}$ Verilog Reference Guide. Synthesis Tool for PSoC Creator. Document 001-48352 Rev.*D. Cypress Semiconductor (2014) http://www.ue.eti.pg.gda.pl/~bpa/pusoc/kit_files/Creating a Verilog-based Component.pdf
13. Warp™ Verilog Reference Guide. Document #001-483-52 Rev. *A. Cypress Semiconductor, San Jose (2009)

Abstract

PSoC Creator (Kannan, PSoC 3 and PSoC 5LP interrupts. AN54460. Document No. 001-54460 Rev. *K. Cypress Semiconductor, 2020) is an Integrated Design Environment (IDE) that enables concurrent hardware and firmware editing, compiling and debugging of PSoC systems. This IDE allows applications to be developed created utilizing more than 150 pre-verified, production-ready peripheral Components (PSoC Creator also provides components that use PSoC 3 and PSoC 5LP Universal Digital Block (UDB) datapath modules that can be used to implement common functions, e.g., UARTs, counters, PWMs, etc., as well as, handling data management tasks that would otherwise consume CPU cycles (Dust and Reynolds, Designing PSoC creator components with UDB datapaths. AN82156. Document No. 001-82156 Rev. *I. Cypress Semiconductor, 2018).) together with a fully integrated schematic capture system. The components supported by PSoC creator are robust analog/digital peripherals and include user-defined components to allow the designer to create custom components (The designer can employ state machine diagrams (Kathuria and Keeser, Implementing state machines with PSoC 3, PSoC 4, and PSoC 5LP. AN62510. Document No. 001-62510 Rev. *F1. Cypress Semiconductor, 2017) or Verilog to further optimize hardware and power consumption.). The user merely drags and drops the appropriate components into the schematic area of PSoC Creator. The user can then set the various parameters associated with such components to meet the design requirements of a broad range of application requirements. Each Component, in the rich mixed-signal Cypress Component Catalog is provided with a full set of dynamically generated API libraries and a customizer dialog. After configuring all the peripherals, firmware can be written, compiled, and debugged within PSoC Creator, or exported to leading 3rd party IDEs such as IAR Embedded Workbench®, Arm®Microcontroller Development Kit, and Eclipse™. The suite of PSoC components provided by the IDE is energy-optimized so that only the required functionality is deployed and thus minimize the design's power requirements.

© Springer Nature Switzerland AG 2021
E. H. Currie, *Mixed-Signal Embedded Systems Design*,
https://doi.org/10.1007/978-3-030-70312-7_7

7.1 PSoC's Integrated Development Environment

PSoC Creator is a free Windows-based IDE that includes:

- Hardware design with complete schematic capture and easy-to-use wiring tool
- Over 150 pre-verified, production-ready Components
- Full communications library including I2C, USB, UART, SPI, CAN, LIN, and Bluetooth Low Energy
- Digital peripherals with powerful graphical configuration tools
- Extensive analog signal chain support with amplifiers, filters, ADC and DAC
- Dynamically generated API libraries
- Free C source code compiler with no code size limitations
- Integrated source editor with inline diagnostics, auto-complete and code snippets and a
- Built-in debugger

7.2 Development Tools

The advent of powerful tools such as the integrated development environment (IDE) has allowed designers to create relatively sophisticated designs utilizing little more than a desktop, or portable computer and a so-called "evaluation board", e.g., of the type shown in Fig. 7.1, that is based on the target device.[1] Early IDE's consisted of a rather simple text editor, an assembler and linker supported, in some respects, by relatively primitive debugging capability. In time, these systems evolved to include various compilers, primitive simulators whose capabilities were generally limited to rather restrictive abilities to check a design's logic, but little else and improved debugging capability.

Debugging, a process which can be the most time-consuming aspect of developing a new design, was initially limited to the post examination of a region of memory after executing a program that had been downloaded to the target, single-stepping through a program one statement at a time and a rather limited capability to set breakpoints. Later IDE debuggers allowed regions of a program, arbitrary memory locations, registers, etc., to be monitored during, and post, execution to determine whether unanticipated, consequential conditions had occurred, as one way to isolate/trap errant code. Some IDEs allowed program variables and expressions to be "watched"[2] and evaluated during program execution. This form of debugging sometimes led to merely "moving the problem to a different place" since such techniques could substantially alter the operating conditions of the executing program, e.g., by introducing too much debugging overhead and adversely affecting the systems responsiveness and execution speed.

---

[1]Evaluation, or eval, boards are provided by microprocessor/microcontroller manufacturers, often at a nominal cost, to allow designers to become familiar with a device, or family of devices, and in some cases to actually incorporate the eval board into a prototype for testing and proof of concept purposes. Such boards generally include several types of I/O connections, LEDs, various types of switches, display devices such as LED/LCD displays and additional hardware to support whatever is required for on-board programming of the target device.

[2]A *watch window* can be used to evaluate and display variables, registers and/or expressions that involve simple variables, array variables, struct variables, registers and assignments. This window is updated immediately following each halt event and displays the name, value, address, type and radix of the parameter being watched. Memory locations can also be watched.

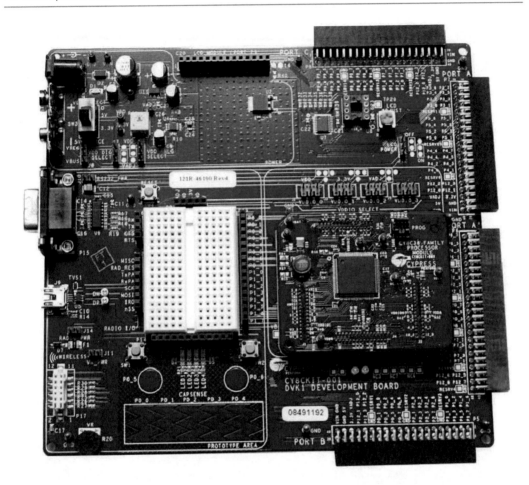

Fig. 7.1 The cypress PSoC 1/3/5LP evaluation board.

Current IDE's for microcontrollers, and microprocessors, tend to primarily support assembly and C language development. However, there are a few notable IDE exceptions that support languages such as BASIC,[3] FORTH,[4] Pascal, etc.[5] Typically, for applications utilizing C language, the associated compiler produces assembly source code as its output. The resulting assembly source code is then processed by the IDE's assembler and subsequently passed to an integral linker.[6] However, debugging, within the context of an IDE, may be restricted to single-stepping and setting of a limited number of breakpoints.

---

[3] Beginners All-Purpose Symbolic Instruction Code (BASIC) is an interpreter originally developed by Thomas Kurtz and George Kemeny, in 1964, at Dartmouth College and placed in the public domain. Subsequently, various incarnations were developed, as interpreters or compilers, some of which are still used to develop applications for microcontrollers, e.g., BASCOM by MCS for Atmel and 8051 architectures.

[4] VFX FORTH for Windows.

[5] IDEs exist for Ada, C/C++, C#, Eiffel, Fortran, Java and JavaScript, Pascal and Object Pascal, Perl, PHP, Python, Ruby, Smalltalk, etc., but not all are either designed or suitable for embedded system development.

[6] A linker, sometimes referred to as a linkage editor, is used to link-edit various object files into a single file that can be used to produce the resulting executable.

7.3 The PSoC Creator IDE

PSoC Creator's user-interface is shown in Fig. 7.2.

It is a combination of a highly intuitive and innovative graphical design editor and a set of sophisticated tools that are well integrated to provide rapid testing of new design ideas, quick response to hardware changes, error-free software interaction with the target's on-chip peripherals, and full access to all aspects of the design. It offers a unique combination of hardware configuration and software development in a single, unified tool. This design frees embedded designers from the innovation-killing division between hardware design and software development characteristic of other IDE systems.

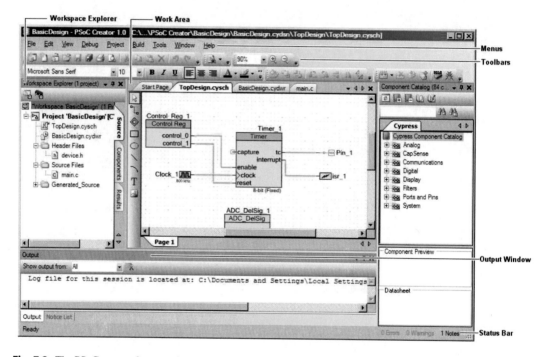

Fig. 7.2 The PSoC creator framework.

PSoC Creator includes an

- integrated schematic capture for device configuration,
- extensive component catalog,
- integrated source editor,
- built-in debugger,
- C/C++/EC++/Ada compiler support,
- support for component creation (affording design reuse),
- a PSoC 3 compiler—Keil PK51 (no code size limit),
- a PSoC 5LP compiler—Sourcery G ®Lite Edition from CodeSourcery,
- sophisticated and reliable bootloading,
- parameter dialogs for comparators, OpAmps, IDACs, VDACs, etc.,

- a static timing checker,
- PSoC 3 instruction cache support,
- a *Generate Application* command/button,

and,

- automated, support case reporting.

When PSoC Creator is opened, it displays the *Start Page* and provides the user with access to recent projects, new project initiation and information on available updates. Links to tutorials, help files, forums, application notes and the reference-design, build projects available online from Cypress' website (www.cypress.com) are also displayed. If a hardware development kit is attached to the designer's PC it is detected by a scan initiated by PSoC Creator to determine the development kit present, and customizes PSoC Creator's *View* for that hardware kit.

7.3.1 Workspace Explorer

PSoC Creator employs a number of dockable windows and allows such windows to be hidden, at the designer's option, via a toggable, pushpin icon option located in the upper right side of the window. When the window is hidden, a small tab remains that upon the occurrence of a mouse-over causes the respective window to reappear. The *Workspace Explorer* window, shown in part in Fig. 7.3, has three tabs: Source, Components and Results. The *Source* tab displays the source and header files for a project in terms of a tree-like structure. Source files displayed in this mode consist of the files generated by PSoC Creator and those introduced by the designer. The *Components* tab displays the components belonging to each project. The *Results* tab is a dynamic listing of files resulting from the most recent build, e.g., programming file, debugging file (if different form the programming file) and in some cases a device file, code generation report, list files and/or map files.

7.3.2 PSoC Creator's Component Library

The *Component Library* includes a wide variety of analog, CapSense. communication, digital, display, filter, port/pin and system components. The designer simply drags each component from the component library to PSoC Creator's work canvas, as shown in Fig. 7.4 and connects the various components as required. Double clicking on a component on the canvas causes a dialog box to appear with the available user-selectable options for the device and access to the component's datasheet. When the components have been selected and interconnected as required for a particular design, a build can be initiated. *Warnings*, *Errors* and *Note* are then displayed in the *Notice List* window.

7.3.3 PSoC Creator's *Notice List* and *Build Output* Windows

The *Notice List* window, shown in Fig. 7.5 combines notices (errors, warnings, and notes) from many sources into one centralized list. If a file and/or error location is shown, double-clicking the entry will display the error, or warning. There are also buttons to *Go To Error* or *View Details*. This window is

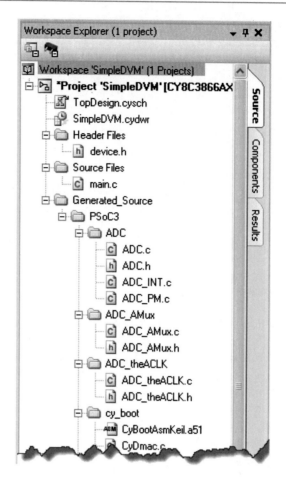

Fig. 7.3 Workspace explorer.

usually located at the bottom of the PSoC Creator framework and often in the same window group as the Output window.[7]

The *Notice List* contains the following columns:

- *Icon*—Displays the icons for the error, warning, or note. A specific row may also contain a tree control containing individual parts of the overall message.
- *Description*—Displays a brief description of the notice.
- *File*—Displays the file name where the notice originated.
- *Error Location*—Displays the specific line number or other location of the message, when applicable.

The number or errors, warnings, and notes also displays on the PSoC Creator Status Bar

- *Errors* indicate there is at least one problem that must be addressed before a successful build can occur. Typical errors include: compiler build errors, dynamic connectivity errors in schematics, and

---

[7]It is possible for a build to fail for no apparent reason and should there be no indication of the cause of failure in the Notice List, the designer should check the Build Output window to determine the cause.

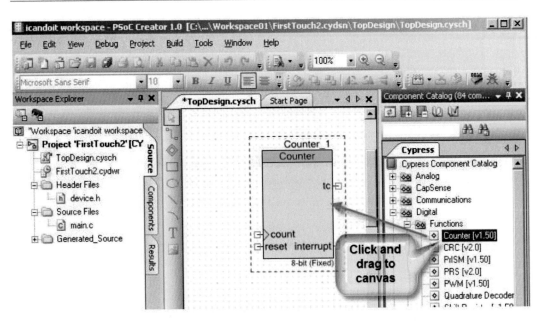

Fig. 7.4 Adding a component to a design.

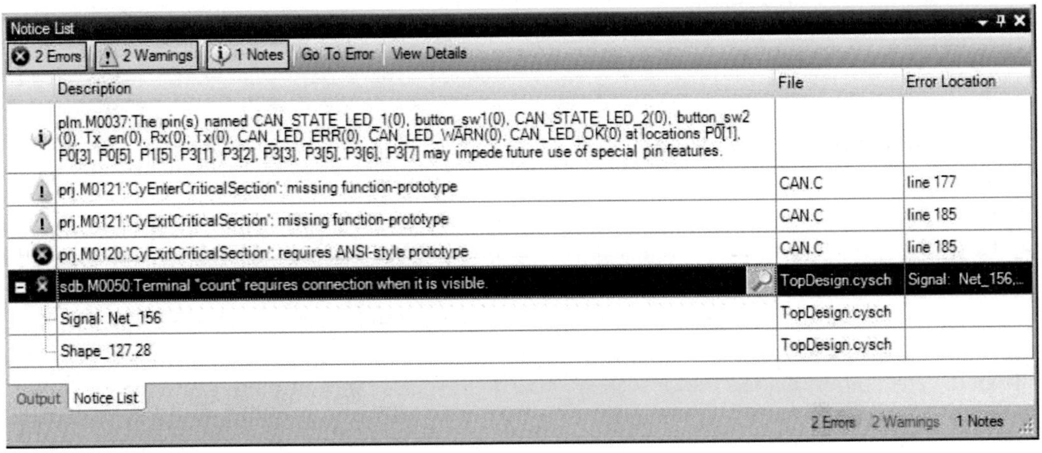

Fig. 7.5 PSoC Creator's *Notice List* window.

Design Rule Checker (DRC) errors. Errors from the build process remain in the list until the next build.

- *Warnings* report unusual conditions that might indicate a problem, although they may not preclude a successful build.
- *Notes* are informational messages regarding the latest build attempt.

The *File* and *Error Location* columns indicate the file in which an error/warning occurred and its location within that file. The three buttons above the *Notice List* labeled *Errors*, *Warnings* and *Notes* can be used to hide/display items in the notice list for each of the three categories.[8] Double-clicking

---

[8]These buttons are labeled with the number of errors, warnings and notes, respectively.

on an error/warning in the *Notice List* opens the associated window and highlights the error. Selecting an error, or a warning, by double-clicking on it in the *Notice List* will cause the associated file/screen to open. The *View Details* button will open a window with additional information about the selected warning/error. As design wide resource and schematic errors are fixed, the Dynamic Rules Checker runs and removes the error/warning from the *Notice List*. Other types of errors will not be removed from the *Notice List*, until the next build occurs. Clicking the *Output* tab causes the window to display the various build, debugger, status, log and other messages, as shown in Fig. 7.6.

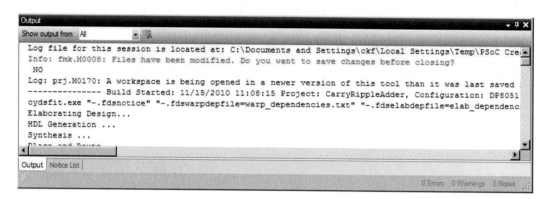

Fig. 7.6 PSoC Creator's *Output* window.

7.3.4   Design-Wide Resources

PSoC Creator provides a design-wide resource (DWR) system that allows the designer to manage all the resources included in a particular design from one location, as shown in Fig. 7.8. Supported resources include clocks, DMA, interrupts, pins, system and directive. Each design has its own default DWR file, with its file type specified as *.cydwr*, and its filename is the same as the project's name. If the *.cydwr* file is deleted for any reason the default values will be used. The pin editor, shown in Fig. 7.8, allows the pins to be assigned and/or locked[9] prior to the build process' place and route operations. Double clicking the *.cydwr* file in Workspace Explorer's *Source* tab causes the DWR window to open and display the pin editor by default (Fig. 7.7).

A signal table is presented in the pin editor that shows the name of each signal, any alias assigned to an individual logic pin or logical port, user-forced pin assignments[10] and an indication of whether a particular pin is locked. Pin assignment is illustrated in Fig. 7.8.

---

[9]Locked pins are constrained to previously specified pin locations. All others are assigned during the build process.

[10]These assignments will not be changed by a build. This column can also be used to make a pin assignment by selecting the desired physical pin from an integral drop-down list. A "-" indicates no assignment has been made for a given signal as does the white background color.

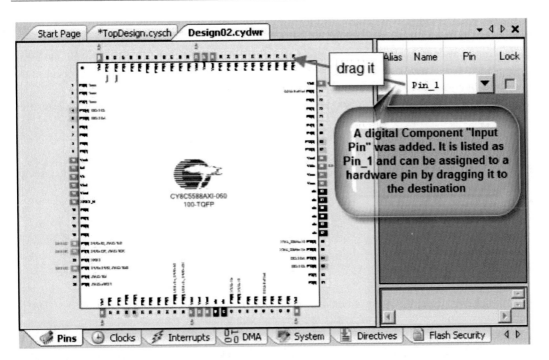

Fig. 7.7 Pin assignment.

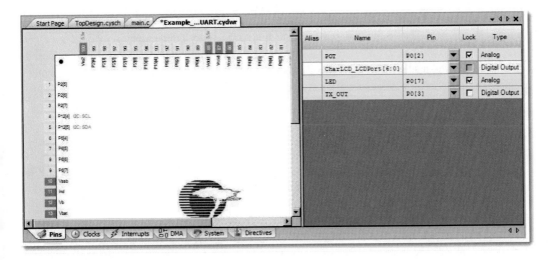

Fig. 7.8 The pin assignment table in the DWR window.

7.3.5 PSoC Debugger

PSoC Creator's built-in debugger supports the following commands:

- *Execute Code/Continue* is used to start/continue a build if the project is out-of-date, update the status bar's message to indicate that the debugger is starting, program the selected target with the latest version of the project's code and start the debugging session.

- *Halt Execution* halts the target.
- *Stop Debugging* ends the debugging session.
- *Step Into* is used to execute a single line of the source code. If the line is a function call, the break in execution will occur at the first instruction in the function, otherwise, a break will occur at the next instruction.
- *Step Over* is used to execute the next line of source code. If the next line is a function call, execution of the function will not occur.
- *Step Out* completes execution of the current function and halts at line of source occurring immediately after the function call.
- *Rebuild and Run* halts the debugging session, recompiles the project, programs the target device and reinitiates the debugger.
- *Restart* resets the program counter (PC) to zero and causes the processor to enter a run state.
- *Enable/Disable All Breakpoints* toggles all the breakpoints in the workspace.

7.3.6   Creating Components

Although PSoC Creator has an extensive catalog of components, e.g., OpAmps, ADCs, DACs, comparators, a mixer, UARTS, etc., it is to add new components. Components can be implemented using several methods, via a schematic, C code or by using Verilog. A *schematic macro* is a mini-schematic that consists of existing components such as clocks, pins, etc. Components created in this manner can consist of multiple macros and macros can have instances, including the component for which the macro is being defined. PSoC Creator's *Component Update Tool* is used to update instances of components on schematics. When a macro is placed on a schematic, the "macroness" of the placed elements disappears. The individual parts of the macro become independent schematic elements. Since there are no "instances" of macros on a schematic, the *Component Update Tool* has nothing to update. However, a schematic macro itself is defined as a schematic. That schematic may contain instances of other components that can be updated via the Component Update Tool.

7.4   Creating a PSoC 3 Design

The following example presents the basic steps required in building PSoC 3/5 designs [8] using PSoC creator. Following installation and opening of PSoC Creator, navigate to the *New Project* window, shown in Fig. 7.9,

This design will involve only three components: a Delta Sigma ADC,[11] LCD display and analog pin, as shown in Fig. 7.10. These components are dragged from the Component Catalog to the *Workspace Canvas*. The wire tool[12] can then be used to connect the analog pin to the positive input terminal of ADC_DelSig. The *Configure 'ADCDelsig'* dialog box, shown in Fig. 7.11 is used in this example to select the *Resolution* as 20-bits, the *Conversion Rate* as 100 samples per second (SPS),[13] the input mode as *single* and the input range as V_{ssa} to 1.024 V(0.0 to V_{ref}).

---

[11]This ADC has 8–20 bits of resolution that can be defined in PSoC Creator menus and/or under software control.

[12]This tool can be activated by using the keyboard's W key.

[13]This selection automatically causes the sampling to be restricted to a range of 8–187 samples per second (SPS).

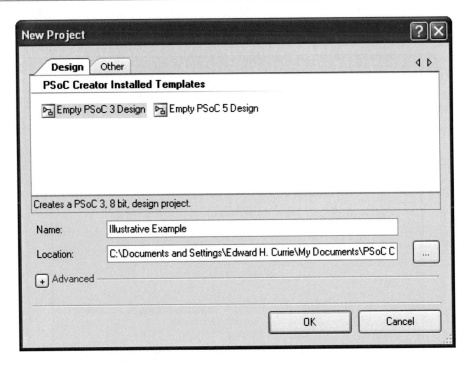

Fig. 7.9 PSoC creator *New Project* window.

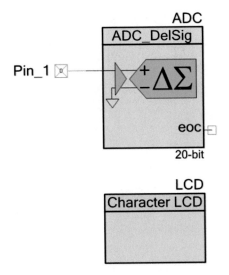

Fig. 7.10 PSoC Creator's Analog Pin, LCD and ADC_DelSig components.

The designer can either select the target manually, or use the Start Auto Select button, shown in Fig. 7.14 to programmatically select the appropriate target device, assuming that the target is connected to PSoC Creator. However, in either case, the designer must manually select the associated *Device Revisions* type for the target as *Production, ES2, ES3*.[14]

---

[14]The default type is *Production*.

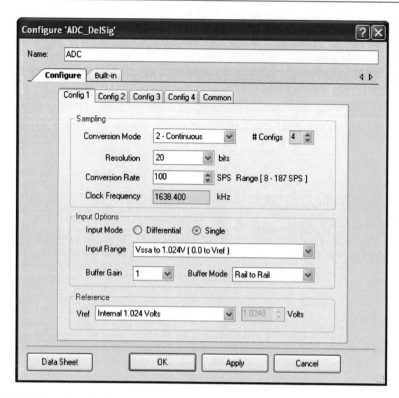

Fig. 7.11 ADC_DelSig settings for the simple voltmeter example.

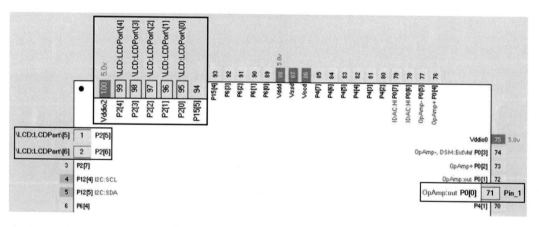

Fig. 7.12 Pin connections for the target device.

Selecting an analog input pin, and connecting it to the ADC_DelSig's input, completes the physical connections for this design. Double-clicking on the associated *.cydwr* tab causes the pin layout for the target device to be displayed as shown in Fig. 7.12.

A *Build* command[15] is then invoked and main.c clicked on opens the tab *main.c* tab as shown in Fig. 7.13.

---

[15]Or alternatively, a *Clean and Build Project* command.

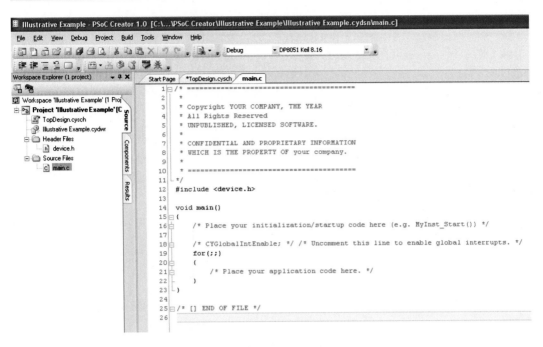

Fig. 7.13 PSoC Creator's main.c tab.

and the screen is shown in Fig. 7.13. Once the source code has been entered as shown below, and following successful compilation and linking, the resulting executable code can be downloaded to the target by invoking the *Program* option in the *Debug* menu.

The source code for this example is shown below and consists of requiring that the result be expressed as 32-bits and be displayed as a floating-point value on the LCD screen. The ADC and LCD must be *started* which requires power to be applied and that they both be initialized. The cursor position is set at (0, 0) and the following message "PSoC Voltmeter" will be displayed on the LCD. A start conversion command will be sent and the system will then enter an infinite loop gathering input readings converting and scaling each input value creating a formatted string and then displaying the result on the 2nd line of the LCD, at which point the process repeats, ad infinitum.

```
/* ========================================
 * PSoC3_Voltmeter
 *
 * Simple project to read a voltage between
 * 0 and 1 volts and display it on an LCD.
 *
 * ========================================
*/
#include <device.h>
#include <stdio.h> /* printf is needed for printing
 output */

void main()
```

```
{
    int32  adcResult;   /* Result to be 32-bit */
    float adcVolts;     /* Result will be displayed as a
     floating point value */
    char   tmpStr[25]; /* 25 character temporary
     string */

    ADC_Start();    /* Initialize and start the ADC */
    LCD_Start();    /* Initialize and start the LCD */
    LCD_Position(0,0);      /* Display message beginning
     at location (0,0) */
    LCD_PrintString("PSoC VoltMeter");
    ADC_StartConvert();     /* Start ADC conversions */

    for(;;)   /* Loop forever */
    {

      if(ADC_IsEndConversion(ADC_RETURN_STATUS) != 0)
       /* Data available?
     */
        {
            adcResult = ADC_GetResult32() ;
/* Get Reading (32-bit) */
            adcVolts = ADC_CountsTo_Volts(adcResult);
/* Convert to volts & scale */
            sprintf(tmpStr,"%+1.3f volts", adcVolts);
/* Create formatted string */

            LCD_Position(1,0);
/* 2nd line of the LCD */
            LCD_PrintString(tmpStr);
/* Display the result */
        }
    }
}
```

Note that *sprintf* is used, in this example, to store the resulting string in a buffer named *tmpStr*, as opposed to *printf* which would result in the string being written to the output stream. *LCD_PrintString* subsequently outputs *tmpStr* to the LCD. Many of the components provided by PSoC Creator must be initialized by a start-device instruction. Following entry of the source code for the application into PSoC Creator's editor[16] the *Device Selector* is used to select the target device as shown in Fig. 7.14.

---

[16] Alternatively, the source can be created by other editors.

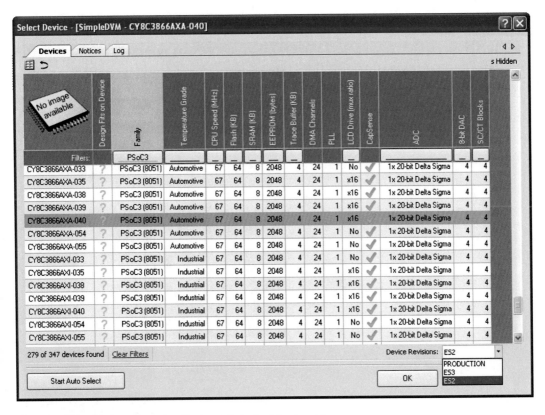

Fig. 7.14 This table is used to select the target device/revision type(s).

The pin assignment for the target device is under the control of the designer. The LCD and input pins for this particular design are set as shown in Fig. 7.15.

7.4.1 Design Rule Checker

PSoC Creator's *Design Rule Checker* (DRC) evaluates the design based on a collection of predetermined rules in the project database. The DRC points to potential errors ,or "rule" violations, in the project that might pose problems and displays the related messages in the *Notice List* window. Some connectivity and dynamic errors update as soon as changes are made to the design, while other errors update following load and save operations.

7.5 The Software Tool Chain

PSoC Creator's integrated development environment (IDE) includes an editor, compiler, assembler, linker, debugger and programmer. The editor is used to enter and/or modify a text file referred to as a *source file*.[17]

---

[17]PSoC Creator-compatible source code can also be created by external, third-party text editors.

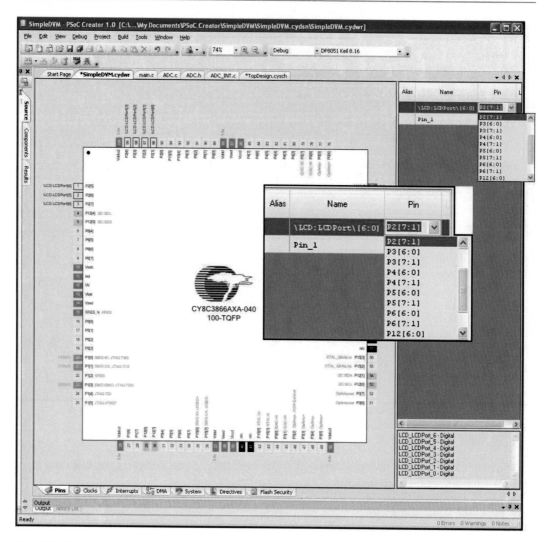

Fig. 7.15 Pin assignment for the target device.

After navigating to *Tools>Options>Text Editor*, various options can be set to include line numbers, set the tab size, enable soft tabs/column guides, highlight the current line and set highlight colors for saved/unsaved changes. The editor's *Find and Replace* command options can be set to display informational messages and to automatically populate *Find What* with text from the editor. In addition to supporting C language development, assembly language programming is also supported, either as a separate assembly source file,[18] or as assembly language instructions within a C source file.[19]

---

[18]To create a separate assembly file, right-click on the project name in the *Workspace Explorer* and select *Add New Item*. Select *8051 Keil Assembly File* and provide a name for the file.This will create an assembly source file, with the extension .a51, in the S*source Files* in the project.

[19]Inline assembly code is placed between the two directives, #pragma asm and #pragma endasm in the C source file. Right click on the C source file in *Workplace Explorer* and select *Build Settings*. Select the *General* option under *Compiler* and set the *Inline Assembly* option to *True*. The compiler will process the assembly language portion of the source file during compilation.

The page background can be set by navigating to *Tools>Options>Design Entry>General* and selecting an appropriate *Canvas Background Color*. Terminal options include *Always Enable Terminal Name Dialog, Always Show Terminals, Schematic Analog Terminal Color, Schematic Digital Color, Symbol Analog Terminal Color, Symbol Digital Terminal Color, Terminal Connector Indicator Color, Terminal Contact Color, Terminal Font* and *Terminal Font Color*.

The colors of the major, and minor, grid lines can also be set as can the *Show Grid* and *Show Grid as Lines* options. Analog and digital *Wire Colors, Wire Bus Size, Wire Dot Size, Wire Font, Wire Font Color* and *Wire Size* can be chosen by the user. It is also possible to add user-defined sheet templates, *Show Hidden Components* and *Enable Param Edit Views*.

Project management options include setting the *Project location, Always Show the Error List window if a build has errors, Always display the workspace in the Workspace Explorer, Display the Output window when a build starts, Reload open documents when a workspace is opened* and *Reload the last workspace on startup*.

The *Programmer/Debugger* options[20] include: *Ask before deleting all breakpoints, Require source files to exactly match the original version* and *Evaluate xx*[21] *children upon expanding in variable view*, The *Default Radix* can be set as *Hexadecimal Display, Octal Display, Decimal Display* or *Binary Display.* options include *On Run/Reset run to Reset Vector, Main* or *First Breakpoint* and *When inserting software breakpoints, warn: Never, On First* or *On Each, Disable Clear-On-Read, Automatically reset device after programming, Automatically show disassembly, after programming if no source is available* and *Allow debugging even if build failed.*

Specific debugger options include *Show Settings for: Breakpoint Windows, Call Stack Window, Debug Intellipoint, Disassembly Window, Locals Window, Memory Window, Registers Window* or *Watch Window.* A wide variety of fonts is provided for debugging and the font size is variable from 6–24 points. *Item foreground* and *Item background* are also user-selectable for *Display items: Plain text, Changed Text, Changing Text* and *Address Text.*

MiniProg3[22] options include *Applied Voltage: 5.0 V, 3.3 V, 2.5 V, 1.8 V* or *Supply Vtarg; Transfer Mode JTAG, SWD, SWD/SWV or Idle; Active Port 10 Pin or 5 Pin; Acquire Mode: Reset, Power Cycle* or *Voltage Sense. Debug Clock Speed* is selectable as *200 Hz, 400 Hz, 800 Hz, 1.5 MHz, 1.6 MHz, 3.0 MHz, 3.2 MHz* or *4 MHz. Acquire Retries* is also user selectable. The Environment options include: *Detect when files are changed outside this environment, Auto-load changes, if saved, At Startup Show Start Page, external Application Extensions File Extensions* and *Require Components Update Dialog Check for up-to-date components when a project is loaded.*

7.6   Opening or Creating a Project

A *project* in PSoC Creator contains all the information about a given design. When PSoC Creator is invoked, it displays a *Start Page,* as shown in Fig. 7.16, that allows the user to either open a previous project or begin a new one. It scans the system for installed development kits and even if none are installed, it will still, if possible, try to configure a device and generate code. However, debugging a

---

[20]Navigate to *Tools>Options.*

[21]"xx" is a integer value provided by the user.

[22]The PSoC MiniProg3 is an all-in-one programmer for PSoC 1, PSoC 3 and PSoC 5LP architectures, that also functions as a debug tool for PSoC 3 and PSoC 5LP architectures [12], and a USB-I2C Bridge for debugging I2C serial connections and communicating to PSoC devices. It supports the following protocols: SWD, JTAG , ISSP and USB-I2C.

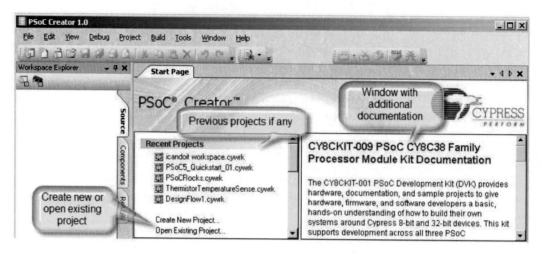

Fig. 7.16 *Start Page* in PSoC creator.

project does require the presence of hardware. The basic steps involved in creating any application in the PSoC Creator development environment consists of the following:

- Creation or opening of an existing project—a project consists of a group of files, e.g.,
  1. *TopDesign.cysch*—a schematic layout of the project
  2. main.c—a file con
- Selection of components to be used in the project—components are selected from the Cypress Component Catalog and dragged to the schematic (.cysch) window.
- Configuration of each of these components—clicking on each of the components will cause the respective dialog box to appear that contains various user options for the component.[23]
- Completion of the schematic—once the required components have been placed in the schematic window the designer can then proceed to incorporate the various interconnections between components that are required.
- Assignment of all resources modification of main.c to allow access to all components used in the design.
- Addition of firmware to main.c
- Building the project
- Downloading of the compiled project
- Debugging the project

by clicking **File>New>Project.**

7.7   Assembly Language and PSoC 3

PSoC Creator supports both C and assembly language application development. An assembler translates symbolic instruction code into object code. Assembly language operation codes are

---

[23]The configuration, and performance characteristics, for a given component are defined by values placed in PSoC resource registers associated with the component.

Fig. 7.17 The linking process.

incorporated in the source in the form of easily remembered mnemonics, e.g., MOV, ADD, SUB, etc.

Assembler source files consist of:

- Directives that define the program's structure and symbols.
- Assembler controls that set the assembly modes and direct flow.
- Machine instructions are the codes that are actually executed by the microprocessor.

A Linker/Locator links (joins) relocatable object modules created by the assembler or compiler, resolves public and external symbols, and produces absolute object modules, as shown in Fig. 7.17. It is also capable of producing a listing file containing a cross-reference of external/public symbol names, program symbols and other information.

PSoC Creator's integral assembler, AX51, is a multi-pass, macro assembler that translates x51 assembly code source files into object files that can then be combined or linked using PSoC Creator's integral linker/locator, LX51, to produce an executable in the form of an absolute object module in an Intel hex file format. The object module generated by the LX51 Linker is an absolute object module that includes all the information required for initializing global variables, zero-initializing global variables, program code and constants, as well as symbolic information, line number information, and other debugging details and the relocatable sections assigned and located at fixed addresses.

7.8 Writing Assembly Code in PSoC Creator

There are two options available for using assembly code in PSoC Creator projects, viz., create a separate assembly source file, or place inline assembly in a C source file. To create separate assembly code source file: Right click on *Project name* in the Project Explorer and then select *Add new item*, select *8051 Keil Assembly File* and provide a name for the file. This will create an assembly source file with a .a51 file extension in the Source Files folder in the project. Assembly code can then be added to this file using standard 8051 instruction codes.[24]

Inline assembly code can be used by placing the assembly code inside the directive *#pragma asm* and *#pragma endasm*,[25] e.g.,

```
extern void test( );
void main (void)  {
    test( );
```

---

[24]See PSoC Creator: *Help > Documentation > Keil > Ax51 Assembler User Guide* which provides instructions, templates, etc., for additional information on assembly language programming.

[25]Pragmas are used in the source code to provide special instructions for the compiler.

```
#pragma asm
        JMP    \$   ; endless loop
# pragma endasm
        }
```

In the *Project* Explorer, right-click on the source file that has the inline assembly and select *Build Settings*. Select the Compiler option and set the value for *Inline Assembly* parameter to *True*. The inline assembly code will then be processed during compilation.

Assembly language source files consist of lines of instructions of the following general form:

$$label:\quad mnemonic\ operand\,,\ operand$$

```
$ITLE(Example  Assembly  Program)
        CSEG    AT  00000h
        JMP  $
        END
```

where $TITLE is a directive[26] and CSEG and END are control statements. The assembler supports symbols that consist of up to 31 characters, inclusive. Supported characters include A-Z, a-z, 0-9, underscore and ?.

Symbols can be defined in the following ways:

```
NUMBER_ONE    EQU    1
TRUE_FLAG     SET    1
FALSE_FLAG    SET    0
```

Labels can be used in an assembly language program to define a place, i.e., address, in a program or data space. Labels must begin in the first text field in a line and be terminated by a colon (:). No more than one label may occur per line, and once defined they must not be redefined. Labels can be used the same way a program offset is used within an instruction. Labels can refer to program code, to variable space in internal or external data memory, or can refer to constant data stored in the program or code space. Labels can also be used to transfer program execution to another location.

Labels are defined as follows (Fig. 7.18):

$$ALABEL:\quad DJNZ\quad R0\,,\ ALABEL$$

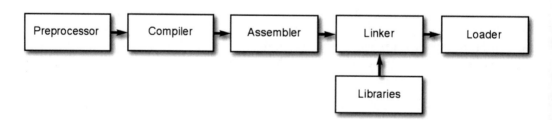

Fig. 7.18 The preprocessor, compiler, assembler, linker and loader chain.

---

[26]There are two types of directives: primary and general. Primary directives occur in the first few lines of the source file and affect the entire source file. General directives can occur anywhere within the source file and may be changed during assembly.

PSoC3/5's assembler supports the following directives:

- CASE—Enable case-sensitive symbol names. (Primary)
- COND—Include conditional source lines skipped by the preprocessor. (General)
- DATE—Specify the appropriate date in the listing file. (Primary)
- DEBUG—Include debugging information in the listing file. (Primary)
- DEFINE—Defines C preprocessor symbols (command line). (Primary)
- EJECT—Insert a form feed into the listing file. (General)
- ELSE—Assemble the current block, if the condition of a previous *IF* is false. (General)
- ELSEIF—Assemble the current block, if the condition is true and a previous IF is false. (General)
- ENDIF—Ends an IF block. (General)
- ERRORPRINT—Specify the file name for error messages. (Primary)
- GEN—Include all macro expansions in the listing file. (General)
- IF—Assemble block, if the condition is true. (General)
- INCDIR—Set additional include file paths. (Primary)
- INCLUDE—Include the contents of another file. (General)
- LIST—Include the assembly source text in the listing file. (General)
- MACRO—Enable preprocessor expansion of standard macros. (Primary)
- MOD51—Enable code generation and define SFRs for classic 8051 devices. (Primary)
- NOAMAKE—Exclude build information from the object file. (Primary)
- NOCASE—Disable case-sensitive symbol names. (All symbols are converted to uppercase.) Primary)
- NOCOND—Exclude conditional source lines skipped by the preprocessor from the listing file. (Primary)
- NODEBUG—Exclude debugging information from the listing file. (Primary)
- NOERRORPRINT—Disable error messages output to the screen. (Primary)
- NOGEN—Exclude macro expansions from the listing file. (General)
- NOLINES—Exclude line number information from the generated object module. (Primary)
- NOLIST—Exclude assembly source text from the listing file. (General)
- NOMACRO—Disable preprocessor expansion of standard macros. (Primary)
- NOMOD51—Suppress SFRs definitions for an 8051 device. (Primary)
- NOOBJECT—Disable object file generation. (Primary)
- NOPRINT—Disable listing file generation. (Primary)
- NOREGISTERBANK—Disables memory space reservation for register banks. (Primary)
- NOSYMBOLS—Exclude the symbol tables from the listing file. (Primary)
- NOSYMLIST—Exclude subsequently defined symbols from the symbol table. (Primary)
- NOXREF—Exclude the cross-reference table from the listing file. (Primary)
- OBJECT—Specifies the name for an object file. (Primary)
- PAGELENGTH—Specifies the number of lines on a page in the listing file. (Primary)
- PAGEWIDTH—Specifies the number of characters on a line in the listing file. (Primary)
- PRINT—Specifies the name for the print file. (Primary)
- REGISTERBANK—reserves memory space for register banks. (Primary)
- REGUSE—Specifies registers modified for a specific function. (General)
- RESET—Set symbols to false that can be tested by IF or ELSEIF.
- RESTORE—Restore settings for the LIST and GEN directives. (General)
- SAVE—Save settings for the LIST and GEN directives. (General)
- SET—Sets symbols, that may be tested by IF or ELSEIF, to true or a specified value. (General)
- SYMBOLS—Include the symbol table in the listing file. (Primary)

- SYMLIST—Include subsequently defined symbols in the symbol table.
- TITLE—Specifies the page header title in the listing file. (Primary)
- XREF—Include the cross-reference table in the listing file. (Primary)

7.9   Big Endian vs. Little Endian

Little-endian and big-endian refer to the ordering of bytes for a particular data format, e.g., big-endian refers to situations in which the most significant byte (MSB) occurs first and the least significant byte occurs last.[27]Conversely, little-endian implies that the least significant byte occurs first and the most significant byte occurs last.[28]

The PSoC 3 Keil Compiler uses the big-endian format for both 16-bit and 32-bit variables. However, the PSoC 3 device uses the little-endian format for multi-byte registers (16-bit and 32-bit register). When the source and destination data are organized in different "endian-ness", the DMA transaction descriptor can be programmed to have the bytes endian-swapped while in transit. The *SWAP_EN* bit of the *PHUB.TDMEM[0..127].ORIG_TD0* register specifies whether an endian swap should occur. If *SWAP_EN* is 1 then an endian swap will occur and the size of the swap is determined by the *SWAP_SIZE* bit of *PHUB.TDMEM[0..127].ORIG_TD0* register. If *SWAP_SIZE = 0* then the swap size is 2 bytes, meaning that every 2 bytes are endian-swapped during the DMA transfer. The code snippet of the TD configuration API to enable byte-swapping for 2 bytes of data is given below.

```
CyDmaTdSetConfiguration(myTd, 2, myTd, TD_TERMOUT0\_EN |
    TD\_SWAP_EN);
```

If SWAP_SIZE = 1 then the swap size is 4 bytes, meaning that every 4 bytes are endian—swapped during the DMA transfer. The code snippet of the TD configuration API to enable byte-swapping for 4bytes data is given below.[29]

```
(myTd, 4, myTd, TD\_TERMOUT0\_EN | TD\_SWAP\_EN |
    TD\_SWAP\_SIZE4);
```

7.10   Reentrant Code

Reentrant code is defined as code that can be shared by multiple processes contemporaneously. In the handling of interrupts, it is fairly common to interrupt a function and allow another process to access the function, e.g., as in the case of a function being called from both the main code and from an interrupt service routine. Declaring a function re-entrant preserves the local variables used in the function, when the function is invoked multiple times. In embedded systems, RAM and stack space is often limited and performance is a major concern which mitigates against calling the same function multiple times concurrently.

Functions, including Component APIs, written using the C51 compiler are typically *NOT* re-entrant. The reason for this limitation is that function as arguments and local variables are stored in fixed memory locations due to limited size of the 8051 stack. Recursive calls to the function use the same memory locations, so that arguments and locals could get corrupted.

---

[27]Jonathan Swift allegedly originated the concept of "ended-ness" in Gulliver's Travels. It arose as a result of the royal edict regarding which end of an egg should be cracked open.

[28]Some architectures allow the ended-ness to be changed programmatically, e.g, ARM.

[29]Unlike the PSoC 3 KEIL compiler, the PSoC 5LP compiler uses little-endian . Hence, the DMA byte-swapping must be disabled when the code is ported to a PSoC 5LP device.

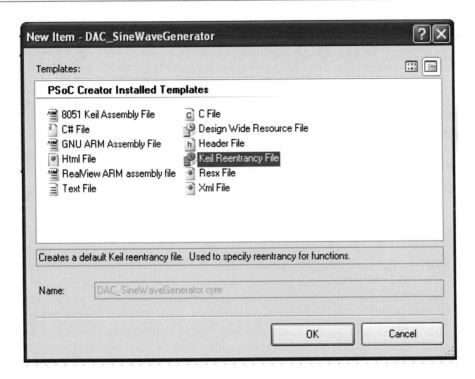

Fig. 7.19 New item dialog in PSoC Creator.

Re-entrant functions [3] can be called recursively, and simultaneously, by two or more processes and are often required in real-time applications, or in situations where interrupt code, and non-interrupt code, must share a function. In spite of the fact that functions are not re-entrant by default in PSoC Creator, functions can be declared re-entrant by creating a *reentrancy file* (*.cyre*) that specifies which functions are to be treated as re-entrant [9]. Specifically, each line of this file must be a single function name.

To create a *.cyre* file for a project the following steps are required:

1. Right click on a project in the Workplace Explorer and select *Add >New Item*
2. Select the *Keil Reentrancy File* to open the file in the editor, as shown in Fig. 7.19.
3. This opens a blank page in the code editor with the filetype *.cyre*. Enter the name of each function to be treated as reentrant, e.g., *ADC_Start*, *PWM_Start*, etc., as a single function name per line.

While the *cyre* file cannot be used for re-entrant functions that are user-defined, it can be used for PSoC Creator generated APIs. In the case of user-defined functions that are to be treated as re-entrant, it is necessary to specify the *CYREENTRANT #define* from *cytypes.h* as part of the function prototype. This will evaluate to the re-entrant keyword, e.g.,

```
void Foo(void) CYREENTRANT;
```

If a custom component requires re-entrancy, the function is declared re-entrant by using the *ReentrantKeil* build expression,[30] e.g.,

---

[30]By default, the function will be a standard function, unless it is listed in the re-entrancy file.

```
void INSTANCE_NAME{Foo(void)} = ReentrantKeil(INSTANCE_NAMR_Foo);
```

The Keil compiler can be used to determine which functions should be re-entrant, if the optimization [2] level is set to 2, or higher and assuming that the functions had not been declared as re-entrant in the source file(s). A build will result in the Keil compiler issuing a warning for any such functions that can be called simultaneously. A function should only be marked as re-entrant when the compiler allocates RAM space for the function, in addition to it being called concurrently. An example of a typical warning output by the Keil linker is:

> Warning: L15 MULTIPLE CALL TO FUNCTION
> NAME: _MYFUNC/MAIN CALLER1: ?C_51STARTUP
> CALLER2: ISR_1_INTERRUPT/ISR_1

which results from the function *MyFunc* being called from both *main()* and the interrupt service routine isr_1.[31]

The Keil compiler establishes a special stack for storage of the re-entrant functions arguments and local variables. The associated stack pointer is used to handle multiple calls to the function in a way that assures that each of the calls is handled correctly.[32] The re-entrant stack is created in *xdata*, *pdata*, or *idata* space,[33] depending on the function's memory model type,[34] i.e., *small*,[35] *compact*,[36] or *large*.[37] Unlike the 8051's hardware stack, the re-entrant stack grows downward and therefore it should be initialized at a high address in memory that ensures that variables located in lower, fixed-memory locations are not overwritten, as the stack grows. The re-entrant stack that corresponds to the memory model being used must be enabled for initialization and the top-of-the-stack address specified in the *KeilStart.A51* file. In this file, the large model, re-entrant stack is enabled, In turn. PSoC Creator initializes the large model, re-entrant stack pointer to point to the top of SRAM, by default.[38]

IBSTACK	EQU 0
XBPSTACK	EQU 1
XBPSTACKTOP	EQU CYDEV_SRAM_SIZE
PBPSTACK	EQU 0

Depending on the application, similar changes can be made to the re-entrant stack pointer for other memory models. It should be borne in mind that using reentrant code techniques in PSoC 3 applications may offer some definite advantages, e.g., a significant reduction in function overhead. However, these same techniques do introduce the additional overhead required to support reentrancy, can result in overwriting other variables in lower memory, and should not be used in firmware, if it requires a significant use of SRAM.

---

[31] If the function in question is an API function that is to be added to the *.re file, then the function name used in the file should not begin with an underscore and should be expressed as the original case sensitive name for the function.

[32] The stack pointer associated with reentrant functions [3] should not be confused with the 8051's hardware stack pointer (SP) SFR whose value is stored in the SFR register in the 8051 CPU.

[33] *idata*, *pdata* and *xdata* refer to memory located on the chip (RAM), memory addressed with an 8-bit address on an external memory page and external memory (RAM) addressed with a 16-bit address, respectively.

[34] The default memory model is large for stack space (xdata).

[35] The PSoC 3 small memory model places function variables and local data segments in internal memory. Although this model imposes a small memory space, it does provide very efficient access to data objects.

[36] The PSoC 3 compact model results in all function/procedure variables and local data segments residing in an external memory page (256 bytes) that is addressable via @$R0$/$R1$.

[37] The large memory model (8051) causes all variables, local data segments to reside in external memory.

[38] The use of the large memory model (8051) requires more instruction cycles to access the "large" external (xdata) memory space.

7.11   Building an Executable: Linking, Libraries and Macros

Once the source code has been completed, the project can be compiled by either invoking *Build All Projects* in the *Build* menu, or by pressing the *F6* function key, on the keyboard. As the build progresses, any associated errors and/or warnings will be displayed in the *Notice List* window. If the build was successful the message *Build Succeeded* will be displayed (Fig. 7.20).[39]

If the compilation and linking are successful the message *Build Succeeded:* followed by the date and time. The linking phase, among other things, binds symbolic to absolute addresses and shared libraries[40] to specific addresses. All the code in PSoC Creator's *Generated Source* tree is compiled into a single library as part of the build process and the compiled library is linked with the user code.[41]

PSoC Creator provides a comprehensive library of components that can be incorporated into an embedded design. The designer can create additional libraries by using the PSoC Library template. []9] The process begins with the selection of a *Symbol* that represents the new component. After right-clicking on the library project,[42] *Add Component Item* is selected and a dialog box will appear ows the designer to select the *Symbol Wizard*. Selecting *Add New Terminals* allows the input and output pins to be defined. It is also possible to specify where in the *Component Catalog* the new component is to be displayed. The functionality of the new component can then be implemented in the form of a schematic, schematic macro[43] or Verilog file.

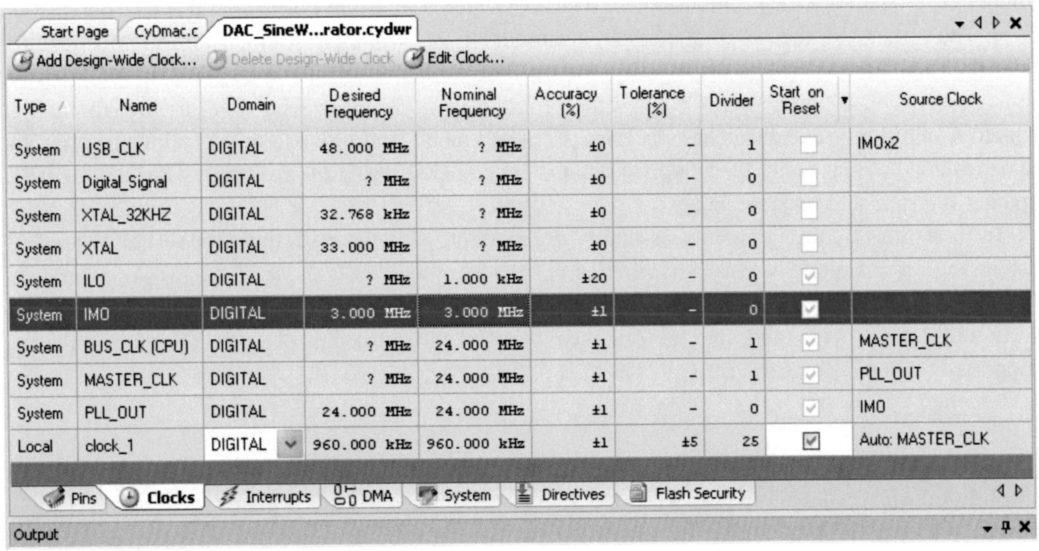

Type	Name	Domain	Desired Frequency	Nominal Frequency	Accuracy (%)	Tolerance (%)	Divider	Start on Reset	Source Clock
System	USB_CLK	DIGITAL	48.000 MHz	? MHz	±0	–	1	☐	IMOx2
System	Digital_Signal	DIGITAL	? MHz	? MHz	±0	–	0	☐	
System	XTAL_32KHZ	DIGITAL	32.768 kHz	? MHz	±0	–	0	☐	
System	XTAL	DIGITAL	33.000 MHz	? MHz	±0	–	0	☐	
System	ILO	DIGITAL	? MHz	1.000 kHz	±20	–	0	☑	
System	IMO	DIGITAL	3.000 MHz	3.000 MHz	±1	–	0	☑	
System	BUS_CLK (CPU)	DIGITAL	? MHz	24.000 MHz	±1	–	1	☑	MASTER_CLK
System	MASTER_CLK	DIGITAL	? MHz	24.000 MHz	±1	–	1	☑	PLL_OUT
System	PLL_OUT	DIGITAL	24.000 MHz	24.000 MHz	±1	–	0	☑	IMO
Local	clock_1	DIGITAL	960.000 kHz	960.000 kHz	±1	±5	25	☑	Auto: MASTER_CLK

Fig. 7.20 PSoC Creator Clocks tab.

[39]If the *Notice List* window is not visible, the number of errors, warnings and notes are displayed on the status bar.

[40]Binding may be either static or dynamic. In the former case, the binding takes place at link time and the latter occurs at runtime. Shared libraries improve runtime and conserves memory.

[41]The GCC Implementation for PSoC 5LP uses all the standard GCC libraries, i.e., libcs3, libc, libcs3unhosted, libgcc, which are linked in by default.

[42]This can be located by selecting the *Component* tab of *Workplace Explorer*.

[43]A schematic macro is a mini-schematic that allows a new component to be created that can be multiple macros with multiple elements, e.g., existing components, pins, clocks, etc. Macros can have instances (including the component for which the macro is being defined), terminals, and wires. Schematic macros are typically created to simplify usage of the components.

In addition to the functionality provided for pins as part of the Pins component, a library of pin macros is provided in PSoC Creator's *cypins.h* file. These macros make use of the port pin configuration register that is available for every pin on the device. Macros for read and write access to the registers of the device are also provided. These macros are used with the defined values made available in the generated cydevice.h, $cydevice_trm.h$ and cyfitter.h files.

7.12   Running/Fixing a Program (Debugger Environment)

Debugging is an important aspect of any embedded system development project. PSoC Creator's debugging capability includes real-time, full speed, in-circuit emulation using an in-circuit emulator (ICE). This capability allows the designer to monitor an application at the source code level, on a line-by-line-of-source-code basis for both C and assembly language applications. In addition to support for breakpoints, watch variables and *dynamic event points*,[44] PSoC Creator also supports the ability to view CPU registers, Flash, RAM and registers and has a 128 kB trace buffer.[45] Traces can be turned on, or off, during program execution, via the use of dynamic event points. Trace display options include saving the trace buffer as a file and viewing, saving and/or printing the trace display in the form of an HTML file. This allows the designer to produce a report that can be external to PSoC Creator.

A status bar at the bottom of the PSoC Creator screen displays ICE-related status information. Single-stepping allows the program execution to be executed on a line-by-line basis at the source code level. The contents of the program counter, accumulator, stack pointer or time stamps corresponding to each "step" are stored and displayed in the trace buffer. User-selectable locations in the program source code, called "breakpoints", allow the program to run until it encounters a "breakpoint", at which point program execution halts.[46] PSoC Creator's menu, and/or icon options, allow the program to be restarted at that point. Once a breakpoint is encountered, program execution halts and the CPU updates registers and variable values.

Because the C compiler emits assembly code, PSoC Creator also supports assembly-level C debugging. In this mode, *single-step instruction, step-over-a-procedure, step-out-of-a-procedure* and *step-into-assembly* and C-level breakpoint capability are also supported.

In order to provide addition performance benefits, three modes of code optimization[47] are supported by PSoC Creator, viz.,

1. Code compression
2. Elimination of unused User Module (area) APIs
3. Multiply/Accumulate at the hardware level

---

[44]Dynamic event points are a type of complex breakpoint that allows multiple events to be monitored, sequenced and logically unified.

[45]A trace buffer of this type maintains a record of the most recent instructions that were executed in a time-sequenced 128k buffer. This allows the designer to follow the precise order of execution of instructions while the system was operating in real-time and at full speed.

[46]It should be noted that while the program is halted at user-specific locations in the program code, the code at that location is not executed until execution resumes.

[47]Optimization, in the present context, refers primarily ro reduction in code size and execution speed. Many compilers such as the Keil compilers offer various levels of optimization.

The types of errors most commonly encountered in developing embedded systems fall into the following categories:

- Corruption of memory by errant code
- Improper use of pointers,
- Hardware design errors,
- Inadequate interrupt handling

and,

- So-called "off-by-one" errors

While PSoC Creator provides excellent facilities for identifying, isolating and locating errors, good coding practices should be employed to minimize debugging time.

7.13 PSoC 3 Debugging

The architectures for PSoC 3 and PSoC 5LP include a Test Controller (TC) that provides access to pins for boundary scanning and to memory/registers via either PSoC 3's Debug On-Chip module [13, 14], or PSoC 5LP's Debug Access Port [15] which supports functional testing, programming and program debugging. Connection to the PSoC 3 debugging is facilitated by the availability of Debug-On_Chip (DOC) and the Single Wire Viewer (SWV). The DOC serves as the interface between the CPU and the Test Controller (TC) and is used to debug, trace code execution and for troubleshooting device configuration.[48]

The test controller serves as a physical interface between a debugging host, and PSoC 3 and PSoC 5LP debug modules and connects to the host via either JTAG or SWD. JTAG support for PSoC 3 and PSoC 5LP exceed the IEEE 149 standard in terms of the access provided to instructions and registers. In the case of PSoC 3, the test controller translates JTAG instructions/registers or SWD accesses to register accesses in the DOC module as indicated schematically in Fig. 7.21.

The DOC has a number of important features:

- It can take control of PSoC 3's CPU (8051) and access any address accessible by the CPU via the PHUB interface. This capability includes the CPU's internal memory, SFRs and PC.
- The DOC can HALT the CPU and single-step through instructions.
- Breakpoint capabilities of the DOC include setting as many as 8 program address breakpoints, setting one memory access breakpoint and setting a Watchdog trigger breakpoint.
- Trace capability includes: tracing the PC, ACC and a single byte from the CPU's internal memory or SFRs; 2048 instruction trace buffer for the PC; 1024 instruction trace buffer for PC, ACC and a single SFR/memory byte; operating in a triggered, continuous or windowed mode; CPU halt or overwrite of the oldest trace when the trace buffer is full and when not tracing the trace buffer is available for other use.

The SWV provides:

---

[48]DOC is used for PSoC 3 and Cypress' Semiconductor's CY8C38 family of devices. Debugging for PSoC 5LP is accomplished by using ARM's Coresight components for debug and traces. SWV targets resident code to provide diagnostic info through a single pin.

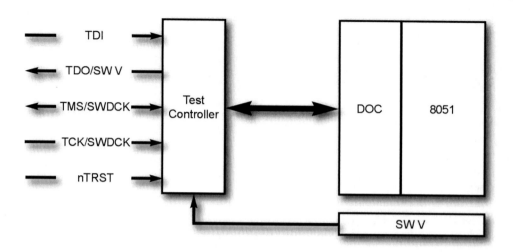

Fig. 7.21 Test Controller for PSoC 3 (8051) block diagram.

- either Manchester or UART for output,
- a simple, efficient packing and serializing protocol,

and,

- Thirty-two stimulus port registers

    PSoC 3 supports three debugging/testing protocols for communicating with PSoC 3:

1. JTAG[49]
2. Parallel test mode (PTM) and,
3. Serial Wire Debug (SWD)—this protocol allows a designer to debug using only two pins of the PSoC 3 device (Fig. 7.22).[50]

    DOC functionality is controlled by accessing registers within DOC. However, these registers are only accessible through the TC interface and not through PHUB. A debugging session utilizing DOC, requires that the CPU enable debugging. Debugging commands are sent to the TC by JTAG or SWD and from there to DOC as shown in Fig. 7.23. Addresses transmitted in this manner are used to access TC registers, DOC registers or alternatively, these addresses are sent on to the DOC memory interface. Within the DOC are a number of memory interfaces and an incoming address is decoded and forwarded to the correct memory interface address output.The DOC waits until the memory access has been completed the DOC transmits a signal to the TC that either the write is complete, or that data from a read command is available.

    The DOC is able to take over control of the CPU memory interfaces and carry-out reads and writes to memory as if the actions were CPU-based. Flash, CPU internal memory, CPU SFRs and the CPU's external memory and registers and the PC can all be accessed by the DOC. Reading and writing to

---

[49]PSoC 3 complies with IEEE 1149.1 (JTAG Specification).

[50]These two pins, once designated for SWD debugging, must be reserved for debug use only and may not be used for any other purposes.

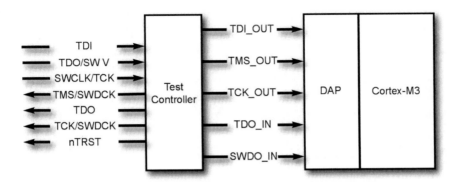

Fig. 7.22 Test controller configuration for PSoC 5LP.

Table 7.1 PSoC memory and registers.

Address Range	Description
0x050000 - 0x0500FF	CPU Internal Memory
0x050000 - 0x0500FF	CPU PC (16-bit register)
0x050000 - 0x0500FF	CPU SFR Space
0x050000 - 0x0500FF	TC and DOC Registers
All Other Addresses	CPU External Memory/Registers

these resources is based on the addresses shown in Table 7.1. In the case of reading or writing to the PC it is necessary to first halt the CPU.

7.13.1 Breakpoints

Breakpoints are a useful tool in analyzing and diagnosing program execution issues, particularly in light of the fact that it is possible to allow the program to operate at normal execution speed before being halted at a breakpoint. PSoC 3 has support for eight, program address breakpoints, a memory access breakpoint and a watchdog trigger breakpoint. Program address breakpoints employ eight registers, DOC_PA_BKPT0 - DOC_PA_BKPT7. Setting an address breakpoint requires that the address for the breakpoint must be stored in bits[15:0].

7.13.2 The JTAG Interface

One of the most popular interfaces for testing integrated circuits (ICs)such as PSoC 3 and PSoC 5LP was developed by the Joint Test and Action Group (JTAG) as a method for controlling and reading an IC's pin values. . The JTAG interface includes the following signals: Test Data In (TDI), Test Data Out (TDO), Test Mode Select (TMS) and a clock signal(TCK). This configuration makes it possible to test multiple ICs on a given board in a daisy-chain manner.)

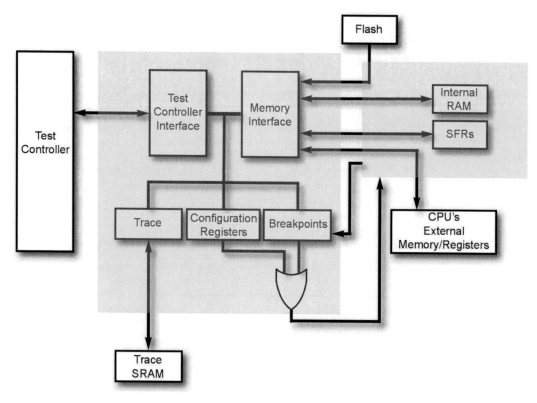

Fig. 7.23 DOC, CPU and TC block diagram.

7.14 Programming the Target Device

PSoC3/5 can be programmed by using the Cypress MiniProg3. This device supports ISSP,[51] SWD[52] and JTAG[53] programming protocols, as well as I2C,[54] SWV,[55] ISSP,[56] SWD[57] and JTAG interfaces

---

[51]*In-System Serial Programming* (ISSP) is a Cypress legacy interface used to program the PSoC 1 family of microcontrollers.

[52]SWD uses fewer pins of the device than JTAG. MiniProg3 supports programming and debugging PSoC 3/5 devices, using SWD.

[53]*JTAG* is supported by many high-end microcontrollers, including the PSoC 3/5 families. This interface allows multiple JTAG devices to be daisy-chained.

[54]A common serial interface standard is the *Inter-IC Communication* (I^2C) standard by Philips. It is mainly used for communication between microcontrollers and other ICs on the same board, but can also be used for intersystem communications. MiniProg3 implements an I2C multimaster host controller that allows the exchange of data with I2C-enabled devices, on the target board. This feature may be used to tune CapSense designs.

[55]The Single Wire Viewer (SWV) interface, is used for program and data monitoring, where the firmware may output data in a method similar to "printf" debugging on PCs, using a single pin. MiniProg3 supports monitoring of PSoC 3/LP firmware, using SWV, through the 10-pin connector and in conjunction with SWD only.

[56]In-System Serial Programming (ISSP) is a Cypress legacy interface used to program the PSoC 1 family of microcontrollers. MiniProg3 supports programming PSoC 1 devices through the 5-pin connector only.

[57]*Serial Wire Debug* (SWD) provides the same programming and debug functions as JTAG, except for boundary scanning and daisy-chaining.

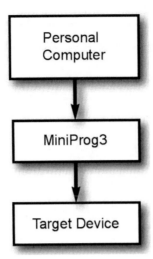

Fig. 7.24 MiniProg3 handles protocol translation between a PC and the target device.

therefore it serves as a protocol converter between a PC and the target device, when connected as shown in Fig. 7.24.

It should be noted that it is designed to communicate with target devices that use only I/O voltages in the range from 1.5 to 5.5 V.

MiniProg3 has five LEDs that indicate status, and are labeled as follows:

- Busy (Red)—indicates that an operation, such as programming or debugging, is in progress.
- Status (Green)—indicates that the MiniProg3 has been "enumerated" on the USB bus and will flash when it receives USB traffic.
- Target Power (Red)—indicates that the MiniProg3 is supplying power to the target's connectors.
- Aux (Yellow)—Reserved.
- Unlabeled (Yellow)—indicates the configuration of the MiniProg3. It flashes during the initial configuration of the MiniProg3 and illuminates continuously when a configuration error has occurred. Following a configuration error, the MiniProg3 must be disconnected from the USB port and reconnected.

7.15   Intel Hex Format

The Intel Hexadecimal 8-bit object format is a representation of an absolute binary object file in an ASCII format. Firmware that is produced by PSoC Creator for the PSoC3/5 microcontrollers, is downloaded in the Intel Hex format, i.e., the downloaded file consists of six parts as shown in Fig. 7.25 and defined as:

1. *Start Code*—a single character, viz., an ASCII colon (:)[58]
2. *Byte Count*—two hex digits representing a single byte that specifies the number of bytes in the data field.

---

[58]The American Standard Code for Information Interchange (ASCII) is a numerical representation of alphabetic and special characters originally developed as a seven-bit telegraph code. It consists of numeric definitions for 128 characters, 94 of which are "printable' and 33 are non-printable such as control characters (line feed, carriage return, etc.) and the "space". The numeric value for colon in the ASCII code is 58 (03AH).

3. *Address*—four hex digits, represented by two bytes, that specify the starting address of the memory location for the data.
4. *Record Type*—two hex digits with values between 00 and 05, inclusive, that define the data field. PSoC Creator generates the following record types:
   - 00 indicating a data record containing data and a 16-bit address
   - 01 indicating an end of a file record referred to as a *file termination record*. This record contains no data and can only occur once for each file.
   - 04 indicates an extended linear address record for full 32-bit addressing. The address field is defined as 0000 and the byte count as 2 bytes. The two data bytes represent the upper 16 bytes of a 32-bit address, when combined with the lower 16-bit address of the 00 record.
5. *Data*—a sequence of N data bytes and represented by 2N hex digits.
6. *Checksum*—two hex digits representing a single byte which is the least significant byte of the two's complement of the sum of the values of all the fields, except for the first and last fields, i.e. except for the start code and the checksum.

Examples of the various record types used by PSoC Creator are:

- :0200000490006A—an extended linear address record, as indicated by the value in the record type field (04). The associated address field is 0000 representing two data bytes. The upper 16-bit portion of the 32-bit address is given by 9000, therefore the base address is 0x90000000 and 6A is the value of the checksum.
- :0420000000000005F7—this represents a data record, as indicated by the value 900 in the record data type field, and the byte count is 04, i.e. there are four data bytes in this record (00000005). The lower 16-bit value of the 32-bit address, as specified in the address field of this record, is 2000 and F7 is the *checksum* for this record.
- :00000001FF—this is the last record and is therefore an end-of-file record, as defined by the value 01 in the record type field.

7.15.1 Organization of Hex File Data

The hex file generated by PSoC Creator contains different types of data,[59] e.g., the main Flash data, ECC data, Flash protection data, customer nonvolatile latch data, write-once latch data and metadata.[60] All this information, including meta data as shown in Table 7.2, is stored at specific

Start Code (Colon Character)	Byte Count (1 Byte)	Address (2 Bytes)	Record Type (1 Byte)	Data (N Bytes)	Checksum (1 Byte)

Fig. 7.25 The Intel hex file format.

---

[59]The data records are in big-endian format (MSB byte in lower address), e.g., the checksum data is address 0x90300000 of the hex file, and the metadata at address 0x9050 0000 of hex. The data records in the rest of the multi-byte regions in a hex file are all in little-endian format (LSB byte in lower address).

[60]Metadata is information contained in the hex file that is not used for programming. It is used to maintain the data integrity of the hex file and store silicon revision and JTAG ID information of the device.

addresses, as shown for example in Fig. 7.26 for PSoC 5LP. This allows the designer to identify which data is meant for what purpose.

- 0x0000 0000—Flash Row Data: The main *Flash data* starts at address 0x0000 0000 of the hex file. Each record in the hex file contains 64-bytes of actual data arranged into rows of 256-bytes. This is because each Flash row of the device is of length 256-bytes. The last address of this section depends on the Flash memory size of the device for which the hex file is intended.[61]
- x8000 0000—*Configuration Data (ECC)*: LP devices have an error-correcting code (ECC) feature, which is used to correct, and detect, bit errors in main Flash data. There is one ECC byte for every eight bytes of Flash data. Thus, there are 32-bytes of ECC data for each row of flash. There is an option to use the ECC memory to store configuration data, if the error-correcting feature is not required. The ECC enable bit, in the device configuration *NV latch* (bit 3 of byte 3), can be used to determine if ECC is enabled. The *NV latch* data byte is located at address 0x90000003. PSoC Creator generates this section of the hex file only if the ECC option is disabled. If this section is present in the hex file, the data needs to be appended with the main Flash data during the Flash programming step. For every 256 bytes in Program Flash, 32 bytes from this section are appended. The last address of this section depends on the device Flash memory capacity. A device with 256 KB of Flash memory has 32 KB of ECC memory. In which case, the last address is 0x80007FFF.
- x9000 0000—*Device Configuration NV Latch Data*: There is a 4-byte device configuration, nonvolatile latch that is used to configure the device, even before the reset is released. These four bytes are stored in addresses starting from 0x9000 0000. One important bit in this NV latch data is the ECC enable bit (bit 3 of byte 3 located at address 0x9000 0003). This bit determines the number of bytes to be written during a Flash row, write process.
- 0x9010 0000—*Secured Device Mode Configuration Data*: This section contains four bytes of the write-once nonvolatile latch data that is to be used for enabling device security.[62] PSoC Creator generates all four bytes as zero, if the device security feature has not been enabled, to ensure that there is no accidental programming of the latch with the correct key. Failure analysis support may be lost on units after this step is performed with the correct key.
- 0x9030 0000 *Checksum Data*: This 2-byte checksum data is the checksum computed from the entire Flash memory of the device (main code and configuration data, if ECC is disabled). This 2-byte checksum is compared with the checksum value read from the device to check if correct data has been programmed. Though the *CHECKSUM* command, sent to the device, returns a 4-byte value, only the lower two bytes of the returned value are compared with the checksum data in the hex file. The 2-byte checksum in the data record is in big-endian format (MSB byte is the first byte).
- 0x9040 0000 *Flash protection data*: This section contains data to be programmed to configure the protection settings of Flash memory. Data in this section should be arranged in a single row, to match the internal Flash memory architecture. Because there are two bits of protection data for each main Flash row, a 256 KB Flash (which has 1024 rows, 256 rows in each of four 64K Flash arrays) has 256 bytes of protection data.
- *Hex file version*: This 2-byte data (big-endian format) is used to differentiate between different hex file versions, e.g., if new metadata information or EEPROM data is added to the hex file generated by PSoC Creator, there is a need to distinguish between the different versions of hex files. By

---

[61]Reference should be made to the respective device datasheet, or the Device Selector menu in PSoC Creator to determine the specific FLASH memory size for different part numbers.

[62]Programming the write-once NV latch with the correct 32-bit key locks the device. This step should only be performed, if all prior steps passed without errors.

Table 7.2 LP hex file metadata organization.

Starting Address	Data Type	Number of Bytes
0x9050 0000	Hex file version	2 (big-endian)
0x9050 0002	JTAG ID	4 (big-endian)
0x9050 0006	Silicon Revison	1
0x9050 0007	Debug Enable	1
0x9050 0008	Internal Use	4

reading these two bytes it is possible to ascertain which version of the hex file is going to be programmed.

- *JTAG ID*: This field has the 4-byte JTAG ID (big-endian format), which is unique for each part number. The JTAG ID read from the device should be compared with the JTAG ID present in this field to make sure the correct device for which the hex file is intended is programmed.
- *Silicon Revision*: This 1-byte value is for the different revisions of the silicon that may exist for a given part number. The byte stored in the hex file should match the value in the chip's *MFGCFG.MLOGIC.REV_ID* register.
- *Debug Enable*: This byte stores a Boolean value indicating, whether debugging is enabled for the program code. (0/1 implies that debugging disabled/enabled.)
- *Internal Use*: This 4-byte data is used internally by PSoC Programmer software. It is not related to actual device programming and need not be used by third-party hardware programmers.

7.16   Porting PSoC 3 Applications to PSoC 5LP

As discussed previously, PSoC 3 and PSoC 5LP are both powerful microcontrollers, the former being based on an 8-bit, 8051 class of microprocessor architecture (33 MIPS) and the latter on a 32-bit, ARM Cortex-M3 (100 DMIPS).[63] Some of the more important differences between the two architectures are shown in Table 7.3.

There may be cases in which it would be of interest to port a PSoC 3 application [7] to a PSoC 5LP environment, perhaps to gain additional performance benefits, take advantage of the 32-bit architecture, deploy components unique to PSoC 5LP, etc.[64] Although the memory maps for the two devices are quite different, in part, as a result of the differences in their respective CPU architectures, the initial phase of such a port can be accomplished by simply using PSoC Creator and navigating to

---

[63] MIPS refers to the millions of CPU instructions executed per second and is not as quantitative a measure of the speed of a processor as DMIPS, which refers to the millions of Dhrystones executed per second. The Dhrystone benchmark is a small integer-based program that is an established benchmark for processors of all types. While both are of some use in comparing processors, they are not the final arbiter of a processor's potential performance in a specific application.

[64] 8051 assembly source code cannot, as a practical matter, be ported directly to the Cortex-M3 space.

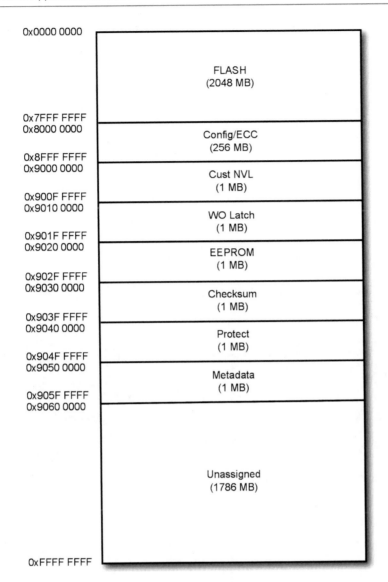

Fig. 7.26 Hex file memory locations for PSoC 5LP.

Project >Device Selector, selecting the targeted PSoC 5LP device from the table shown in Fig. 7.27 and then rebuilding the project.[65]

However, if the port is to be optimized in the new target environment, a number of factors need to be taken into consideration, e.g., PSoC3's memory map consists of three different code spaces as shown in Fig. 7.28.

1. The 8051 internal data space, which is part of the 8051 core, contains 256-bytes of RAM and 128-bytes of special function registers (SFRs). This space is accessed by the *fast* registers and bit instructions and the location of the 8051 hardware stack ($\geq$256-bytes).

---

[65] PSoC Creator supports three compilers and this procedure is based on the assumption that a compatible target compiler is used.

Table 7.3 Comparison of the key differences between PSoC 3 and PSoC 5LP.

Feature	PSoC3	PSoC5
Processor	8-bit 8051	32-bit ARM Cortex-M3
ADC	One DelSig ADC	One DelSig ADC Two SAR ADCs
Flash	Up to 64 KB (inclusive)	Up to 256 KB (inclusive)
RAM	Up to 8 KB (inclusive)	Up to 64 KB (inclusive)
Cache	No	Instruction Cache
EMIF	Data Memory	Data/Code Memory

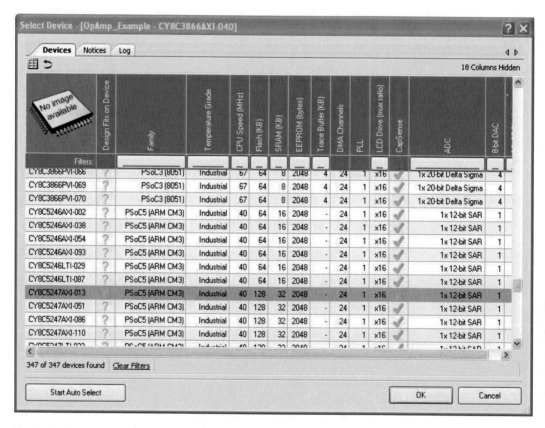

Fig. 7.27 Selection of a new PSoC 5LP target device.

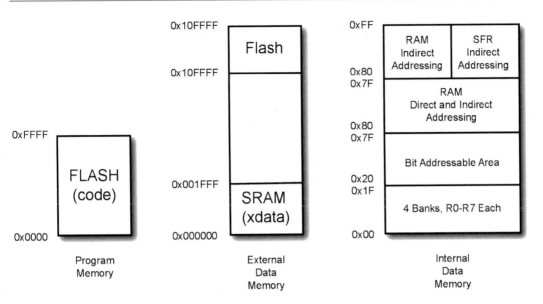

Fig. 7.28 The PSoC3 (8051) memory map.

2. The external data space, while internal to the PSoC 3, is external to the 8051 core. All SRAM, Flash,[66] registers and EMIF[67] addresses are mapped into this space. The assembly instruction MOVX is used to access the external data space which is 16 Mb in size and requires a 24-bit access address.
3. The code space is 64K bytes of Flash memory and it is here that the 8051 instructions reside.

The PSoC 5LP memory space is based on a 32-bit, linear memory map, as shown in Fig. 7.29.

PSoC 5LP's SRAM is located in the memory space defined by [0x1FFF8000, 0x20007FFF] and is centered on the boundary between the code and SRAM memory spaces. The rest of the code space is occupied by Flash beginning at memory address 0. PSoC 5LP registers are located in the peripheral space(s) and the EMIF addresses are located in the external RAM space. The Keil compiler uses the big-endian format for 16- and 32-bit variables for PSoC 3 and little-endian for PSoC 5LP multi-byte variables.[68]

7.16.1 CPU Access

PSoC Creator supports the following macros to allow register access with byte-swapping. These macros are for accessing registers mapped in the first 64K bytes of the 8051 external data space:

$$CY\_GET\_REG8(addr)$$
$$CY\_SET\_REG8(addr, value)$$
$$CY\_GET\_REG16(addr)$$

---

[66] Flash is mapped into this space primarily for DMA data access.

[67] External memory interface (EMIF).

[68] Unlike the PSoC 3 Keil 8051 compiler, all PSoC 5LP compilers use little-endian format.

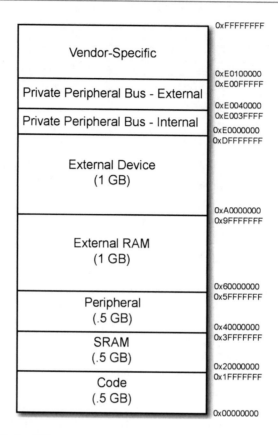

Fig. 7.29 The PSoC 5LP (Cortex-M3) memory map.

CY_SET_REG16(addr , value)
CY_GET_REG24(addr)
CY_SET_REG24(addr , value)
CY_GET_REG32(addr)
CY_SET_REG32(addr , value)

The following macros can be used to access registers mapped above the first 64K bytes of 8051 external data space

CY_GET_XTND_REG8(addr)
CY_SET_XTND_REG8(addr , value)
CY_GET_XTND_REG16(addr)
CY_SET_XTND_REG16(addr , value)
CY_GET_XTND_REG24(addr)
CY_SET_XTND_REG24(addr , value)
CY_GET_XTND_REG32(addr)
CY_SET_XTND_REG32(addr , value)

and they handle endian format translation correctly and can be ported directly to PSoC 5LP compilers.

7.16.2 Keil C 8051 Compiler Keywords (Extensions)

Keil has added a number of important extensions to the set of keywords provided by standard C, e.g.,

- _at_—variables can be located at absolute memory locations[69] using:

```
<< memory_type >> type variable_name_at_constant;
```

where *memory_type* is the variable's memory type, *type* is the variable type, *variable_name* is the name of the variable and *constant* is the variable's address.
- *alien*—used to invoke PL/M-51 routines from C functions by first declaring them external with the alien function type specifier, e.g.,

```
extern alien char plm_func (int, char);
char c_func (void)  {
int i;
char c;
for (i = 0; i < 100; i++) {
c = plm_func (i, c);               /* call PL/M func */
}
return (c);
}
```

To create C functions that may be invoked from PL/M-51 routines, the alien function type specifier must be used in the C function declaration. For example:

```
alien char c_func (char a,  int b)  {
        return (a * b);
}
```

Parameters, and return values of PL/M-51 functions, may be bit, char, unsigned char, int, and an unsigned int. Other types, including long, float, and all types of pointers, can be declared in C functions with the *alien* type specifier. However, these types must be used with care because PL/M-51 does not directly support 32-bit binary integers, or floating-point numbers.

Public variables declared in a PL/M-51 module are available to C programs by declaring them external, like any C variable.
- *bdata*—bit-addressable objects can be addressed as bits or words. Only data objects that occupy the bit-addressable area of the 8051 internal memory fall into this category.
- *bit*—defines a single bit variable,[70] e.g.,

```
bit name << = value >>
```

---

[69]The absolute address following the _at_ keyword must conform to the physical boundaries of the memory space for the variable. The Cx51 Compiler checks for and reports invalid address specifications.

[70]All bit variables are stored in a bit segment located in the internal memory area of the 8051, which is 16 bytes long. Therefore, a maximum of 128 bit variables may be declared within any one scope.

where *name* is the name of the bit variable and *value* is the value to be assigned to the bit.

Bit variables are stored in a segment in the 8051 internal memory space. In most cases, bit variables can be defined, and accessed, in the same manner as any other variable type:

```
bit doneFlag = 0;
```

To port bit variables to PSoC 5LP the bit type can be redefined as

```
#define bit uint8
```

The following is an example of the use of the bit-type:

```
static bit done_flag = 0;      /* bit variable */

bit testfunc (                 /* bit function */
bit flag1,                     /* bit arguments */
bit flag2)
{
        .
        .
        .
    return (0);                 /* bit return value */
}
```

- *code*—Program (CODE) memory is read-only. Program memory may reside within the 8051 MCU, be external, or both. Although 8051 architecture supports up to 64K Bytes of program memory, program space can be expanded using code banking. Program code, including all functions and library routines, is stored in program memory. Constant variables may also be stored in program memory. The 8051 executes programs stored in program memory only. Program memory may be accessed from C programs using the code memory type specifier.
- *compact*—A function's arguments and local variables are stored in the default memory space specified by the memory model. It is possible to specify which memory model to use for a single function by including the *small*, *compact*, or *large* function attribute in the function declaration. For example:

```
#pragma small          /* Default to small model */

extern int calc (char i, int b) large reentrant;
extern int func (int i, float f) large;
extern void *tcp (char xdata *xp, int ndx) compact;

int mtest (int i, int y)                /* Small model */
  {
    return (i * y + y * i + func(-1, 4.75));
  }
    int large_func (int i, int k) large /* Large model */
      {
        return (mtest (i, k) + 2);
```

```
}
```

The advantage of functions using the SMALL memory model is that the local data and function argument parameters are stored in the internal 8051 RAM. Therefore, data access is very efficient. Because internal memory is limited, the small model may not satisfy the requirements of a very large program, in which case, other memory models must be used. In such cases, a function can be declared that uses a different memory model, as shown above.

By specifying the function model attribute in the function declaration, it is possible to specify which of the three possible re-entrant stacks, and the associated frame pointers, are to be used.[71]

- *data*—this memory specifier always refers to the first 128 bytes of internal data memory.[72]
- *far*—this keyword allows variables and constants to be accessed in external memory using 24-bit addresses. For variables, *far* memory is limited to 16 megabytes. Objects are limited to 64K, and cannot cross a 64K boundary. Constants (ROM variables) are limited to 16 megabytes.
- *idata*—this memory specifier refers to all 256 bytes of internal data memory, but it requires indirect addressing which is slower than direct addressing.
- *interrupt*—interrupts can be used for counting, timing, detecting external events and sending/receiving data via a serial interface.[73]
- *large*—selects the large memory model in which all variables and local data segments procedures/-functions are maintained in external memory.
- *pdata*—this memory type is used only for declaring variables and is indirectly accessed by 8-bit addresses of one page of a 256-byte page of external data 8051 RAM.
- *_priority_*—this keyword specifies a task's priority, e.g.,

```
void func (void) _task_ num _priority_ pri
```

where *num* is a task ID number and *pri* is the tasks priority.
- *reentrant*—allows functions to be declared reentrant and therefore called recursively, e.g.,

```
int calc (char i, int b) reentrant  {
        int  x;
        x = table [i];
        return (x * b);
}
```

Small, *compact* and *large* model reentrant functions simulate the reentrant stack in *idata*, *pdata* memory and *xdata* memory, respectively. Bit-type function arguments may not be used and local bit scalars are also not available. The re-entrant capability does not support bit-addressable variables. Re-entrant functions must not be called from *alien* functions and cannot use the alien attribute specifier to enable PL/M-51 argument passing conventions. A re-entrant function may simultaneously have other attributes, such as using an interrupt, and may include an explicit memory model attribute (*small, compact, large*).

Return addresses are stored in the 8051 hardware stack. Any other required PUSH and POP operations also affect the 8051 hardware stack. Although re-entrant functions using different memory models may be intermixed, each re-entrant function must be properly prototyped and include its memory model attribute in the prototype. This is necessary for calling routines to place

---

[71] Note that stack access in the SMALL model is more efficient than in the LARGE model.

[72] Variables stored at that location are accessed using direct addressing.

[73] Thirty-two interrupts are located in the jump table from address 0003h–00FBh, inclusive.

the function arguments in the proper re-entrant stack.[74] For example, if *small* and *large* re-entrant functions are declared in a module, both *small* and *large* re-entrant stacks are created along with two associated stack pointers (one for *small* and one for *large*).

- *sbit*—defines a bit within a special function register (SFR). It is used in one of the following ways:

```
sbit name = sfr-name ^ bit-position;
sbit name = sfr-address ^ bit-position;
sbit name = sbit-address;
```

where *name* is the name of the SFR bit, *sfr-name* is the name of a previously-defined SFR, *bit-position* is the position of the bit within the SFR, *sfr-address* is the address of an SFR and *sbit-address* is the address of the SFR bit, e.g.,

```
/* define the sbit */
sbit PIN1_6 = SFRPRT1DR^6;
/* access the sbit */
PIN1_6 = 1;
```

The *sbit* keyword is used in PSoC 3 for faster access to bits in certain registers, but they cannot be used in PSoC 5LP. Instead the C bit manipulation operators and macros should be used, e.g.,

```
CY_SET_REG8(CYDEV_IO_PRT_PRT1_DR,
CY_GET_REG8(CYDEV_IO_PRT_PRT1_DR) |
0x40;
```

It is often necessary to access individual bits within an SFR and the *sbit* type provides access to bit-addressable SFRs and other bit-addressable objects, e.g.,

```
sbit EA = 0xAF;
```

This declaration defines *EA* as the SFR bit at address *0xAF*, which is the *enable all* bit in the *interrupt enable* register.

Storage of objects accessed using *sbit* is assumed to be little-endian (LSB first). This is the storage format of the *sfr16* type, but it is opposite to the storage of int and long data types. Care must be taken when using *sbit* to access bits within standard data types. Any symbolic name can be used in an *sbit* declaration. The expression to the right of the equal sign specifies an absolute bit address for the symbolic name. There are three variants for specifying the address:

```
sbit name = sfr-name ^ bit-position;
```

The previously declared SFR (*sfr-name*) is the base address for the *sbit* and it must be evenly divisible by 8. The bit-position, which must be a number from 0–7, follows the carat symbol (^) and specifies the bit position to access, e.g.,

```
sfr  PSW = 0xD0;
sfr  IE = 0xA8;
sbit OV = PSW^2;
sbit CY = PSW^7;
```

---

[74]Each of the three possible re-entrant models contains its own re-entrant stack area and pointer.

```
sbit EA = IE^7;
```

```
sbit name = sfr-address ^ bit-position;
```

A character constant (*sfr-address*) specifies the base address for the *sbit* and must be evenly divisible by 8. The bit-position (which must be a number from 0–7) follows the carat symbol (^) and specifies the bit position to access, e.g.,

```
sbit OV = 0xD0^2;
sbit CY = 0xD0^7;
sbit EA = 0xA8^7;
```

```
sbit name = sbit-address;
```

A character constant (sbit-address) specifies the address of the *sbit*. It must be a value from 0x80-0xFF, e.g.,

```
sbit OV = 0xD2;
sbit CY = 0xD7;
sbit EA = 0xAF;
```

Only SFRs whose address is evenly divisible by 8 are bit-addressable and the lower nibble of the SFR's address must be 0 or 8. For example, SFRs at 0xA8 and 0xD0 are bit-addressable, whereas SFRs at 0xC7 and 0xEB are not. To calculate an SFR bit address, add the bit position to the SFR byte address, e.g., to access bit 6 in the SFR at 0xC8, the SFR bit address would be 0xCE (0xC8 + 6). Special function bits represent an independent declaration class that may not be interchangeable with other bit declarations, or bit fields. The *sbit* data type declaration may be used to access individual bits of variables declared with the *bdata* memory type specifier. *sbit* variables must be declared outside the function body.

- *sfr*—defines a special function register (SFR). It is used as follows:

```
sfr name = address;
```

where *name* is the name of the SFR and *address* is the address of the SFR. SFRs are declared in the same fashion as other C variables, except that the type specified is sfr rather than char or int, e.g.,

```
sfr P0 = 0x80;    /* Port-0, address 80h */
sfr P1 = 0x90;    /* Port-1, address 90h */
sfr P2 = 0xA0;    /* Port-2, address 0A0h */
sfr P3 = 0xB0;    /* Port-3, address 0B0h */
```

P0, P1, P2, and *P3* are the SFR name declarations.[75] The address specification after the equal sign must be a numeric constant. *sfr* variables must be declared outside the function body.

- *sfr16*—defines a 16-bit special function register (SFR) and is implemented as follows:

```
sfr16 name = address;
```

---

[75]Names for *sfr* variables are defined just like other C variable declarations and any symbolic name may be used in an *sfr* declaration.

where *name* is the name of the 16-bit SFR and *address* is the address of the 16-bit SFR. The Cx51 Compiler provides the *sfr16* data type to access two 8-bit SFRs as a single 16-bit SFR.

Access to 16-bit SFRs using *sfr16* is possible only when the low byte immediately precedes the high byte (little Big-) and when the low byte is written last. The low byte is used as the address in the sfr16 declaration, e.g.,

```
sfr16 T2 = 0xCC;      /* Timer 2: T2L 0CCh, T2H 0CDh */
sfr16 RCAP2 = 0xCA;   /* RCAP2L 0CAh, RCAP2H 0CBh    */
```

In this example, *T2* and *RCAP2* are declared as 16-bit special function registers. The *sfr16* declarations follow the same rules as outlined for *sfr* declarations. Any symbolic name can be used in a *sfr16* declaration. The address specification after the equal sign must be a numeric constant. Expressions with operators are not allowed. The address must be the low byte of the *SFR* low-byte, high-byte pair. When writing to *srf16*, the code generated by the Keil Cx51 Compiler writes to the high byte first and then the low byte. In many cases, this is not the desired order, and therefore, if the order in which the bytes are written is important, the sfr keyword must be used to define and access the SFRs one byte at a time to assure the order in which the SFRs are accessed. *sfr16* variables may not be declared inside a function, but instead, must be declared outside the function body.

- *small*—A function's arguments and local variables are stored in the default memory space specified by the memory model. However, it is possible to specify which memory model to use for a single function by including the small, compact, or large function attribute in the function declaration, e.g.,

```
#pragma small              /* Default to small model */

extern int calc (char i, int b) large reentrant;
extern int func (int i, float f) large;
extern void *tcp (char xdata *xp, int ndx) compact;
int mtest (int i, int y)                /* Small model */
{
   return (i * y + y * i + func(-1, 4.75));
   }
int large_func (int i, int k) large /* Large model */
{
return (mtest (i, k) + 2);
}
```

The advantage of functions using the *SMALL* memory model is that the local data and function argument parameters are stored in the internal 8051 RAM. Therefore, data access is very efficient. Occasionally, because the internal memory is limited, the small model cannot satisfy the requirements of a very large program and other memory models must be used. In that case, a function must be declared that uses a different memory model. By specifying the function model attribute in the function declaration, it becomes possible to select which of the three possible reentrant stacks and frame pointers to use.[76]

---

[76]Stack access in the SMALL model is more efficient than in the LARGE model.

- *_task_*—this keyword specifies a function as a real-time task when using a real-time multitasking operating system.[77]
- *using*—The first 32 bytes of *DATA* memory (*0x00-0x1F*) are grouped into 4 banks of 8 registers each. Programs access these registers as *R0-R7*. The register bank is selected by two bits of the program status word, *PSW*. Register banks are useful when processing interrupts, or when using a real-time operating system because the MCU can switch to a different register bank for a task, or interrupt, rather than saving all 8 registers on the stack. The MCU can then switch back to the original register bank before returning. The *using* function attribute specifies the register bank a function uses, e.g.,

```
void rb_function (void) using 3
{
.
.
.
}
```

The argument for the *using* attribute is an integer constant from 0-3. The *using* attribute is not allowed in function prototypes and expressions with operators are not allowed. The *using* attribute affects the object code of the function as follows:
- The currently selected register bank is saved on the stack at function entry.
- The specified register bank is set.
- The former register bank is restored before the function is exited.

The following example shows how to specify the *using* function attribute and what the generated assembly code for the function entry and exit looks like.

```
stmt level  source
       1
       2              extern bit alarm;
       3              int alarm_count;
       4              extern void alfunc (bit b0);
       5
       6              void falarm (void) using 3  {
       7    1              alarm_count++;
       8    1              alfunc (alarm = 1);
       9    1          }


        ASSEMBLY LISTING OF GENERATED OBJECT CODE

; FUNCTION falarm (BEGIN)
    0000 C0D0        PUSH   PSW
        0002 75D018    MOV    PSW,#018H
                             ; SOURCE LINE # 6
                             ; SOURCE LINE # 7
```

---

[77]Keil's RTX51 Full and RTX51 Tiny kernels [6] support both real-time control and multitasking to provide several operations to be executed simultaneously and carry out operations that must occur within a predefined period of time.

```
0005 0500    R    INC    alarm_count+01H
0007 E500    R    MOV    A,alarm_count+01H
0009 7002         JNZ    ?C0002
000B 0500    R    INC    alarm_count
000D  ?C0002:
          ; SOURCE LINE # 8
000D D3          SETB   C
000E 9200    E    MOV    alarm,C
0010 9200    E    MOV    ?alfunc?BIT,C
0012 120000 E    LCALL  alfunc
          ; SOURCE LINE # 9
0015 D0D0         POP    PSW
0017 22           RET
          ; FUNCTION falarm (END)
```

In the previous example, the code starting at offset *0000h* saves the initial *PSW* on the stack and sets the new register bank. The code starting at offset *0015h* restores the original register bank by *popping* the original *PSW* from the stack.

The *using* attribute may not be used in functions that return a value in registers. Extreme care should be exercised to ensure that register bank switches are performed only in carefully controlled areas. Failure to do so may yield incorrect function results. Even when the same register bank is used, functions declared with the using attribute cannot return a bit value. The *using* attribute is most useful in implementing interrupt functions. Usually a different register bank is specified for each interrupt priority level. Therefore, one register bank can be employed for all non-interrupt code, a second register bank for the high-level interrupt, and a third register bank for the low-level interrupt.

- *xdata*—External data memory is read/write. Since external data memory is indirectly accessed through a data pointer register (which must be loaded with an address), it is slower than access to internal data memory. *XRAM* space is accessed with the same instructions as the traditional external data space enabled via dedicated chip configuration *SFR* registers and overlaps the external memory space.

  While there may be up to 64K Bytes of external data memory, this address space does not have to be used as memory. A hardware design can map peripheral devices into the memory space so that the program accesses, what appears to be, external data memory to program and control the peripheral.[78] The C51 Compiler offers two memory types that access external data: *xdata* and *pdata*. The *xdata* memory specifier refers to any location in the 64K Byte address space of external data memory. The large memory model locates variables in this memory space. The *pdata* memory type specifier refers to exactly one (1) page (256 bytes) of external data memory. The compact memory model locates variables in this memory space.

---

[78]This is referred to as memory-mapped I/O, in some cases.

7.16.3 DMA Access

DMA transaction descriptors can be programmed to have bytes swapped while transferring data.[79] The swap size can be set to 2 bytes for 16-bit transfers, or 4 bytes for 32-bit transfers. The following examples handle 2- and 4-byte swaps:

```
CyDmaTdSetConfiguration(myTd, 2, myTd,
    TD_TERMOUT0_EN | TD_SWAP_EN);
```

and,

```
CyDmaTdSetConfiguration(myTd, 4, myTd,
    TD_TERMOUT0_EN | TD_SWAP_EN |
                     TD_SWAP_SIZE4);
```

respectively.

7.16.3.1 DMA Source and Destination Addresses

PSoC 3 and PSoC 5LP have the same type of DMA controller (DMAC) which stores 32-bit addresses for both source and destination, in two 16-bit registers. The upper half of the addresses for each DMA channel are specified by the following:

```
DMA_DmaInitalize(..., uppersrcAddr, upperDestAddr)
```

and similarly, the lower half of the addresses are specified for each transaction descriptor (TD) within a DMA channel as:

```
CyDmaTdSetAddress(..., lowerSrcAddr, lowerDestAddr)
```

The contents of a pointer variable cannot be used to provide source or destination address values, because the Keil 8051 compiler uses a 3-byte pointer, i.e., two bytes representing a 16-bit absolute address and a third byte for the memory space being used.

Source in Flash can be accessed by:

```
upperSrcAddr = (CYDEV_FLS_BASE) >> 16
SRAM for source or destination:
upperSrcAddr = 0;
upperDestAddr = 0;
```

and for a peripheral register, for source or destination:

```
upperSrcAddr = 0;
upperDestAddr = 0;
```

The upper half of the PSoC 5LP address for SRAM or peripheral register for source or destination is expressed as:

```
upperSrcAddr = HI16(srcArray);
upperDestAddr = HI16(destArray);
```

and the lower half of the address by using the LO16 macro defined in "the *cytypes.h*" file:

```
lowerSrcAddr = LO16(srcArray);
lowerDestAddr = LO16(destArray);
```

---

[79]DMA byte-swapping must be disabled when the code is ported to PSoC 5LP.

Addresses can also be found by using conditional compilation:

```
#if (defined(__C51__))
  upperSrcAddr = 0;
#else /* PSoC 5 LP*/
  upperSrcAddr = HI16(srcArray);
#endif
```

7.16.4 Time Delays

The *CyDelay* function, defined in *CyLib.c*, is used to generate absolute time delays. It selects the number of loop iterations based on processor type and CPU speed. The supported system function calls include:

- *void CyDelay(uint32 milliseconds)* produces a delay specified by *uint32 milliseconds*.[80] If the clock configuration is changed at run-time, then the function *CyDelayFreq* is used to indicate the new *Bus Clock* frequency. *Cy delay* is used by several components, so changing the clock frequency without updating the frequency setting for the delay can cause those components to fail. *CyDelay* has been implemented with the instruction cache assumed enabled. When the PSoC 5LP instruction cache is disabled, *CyDelay* will be two times larger.[81]
- *void CyDelayUs(uint16 microseconds)* produces a delay specified by *uint16 microseconds*.
- *void CyDelayFreq(uint32 freq)* sets the *Bus Clock* frequency used to calculate the number of cycles required for implementing the delay specified by *CyDelay*. The frequency used is based on the value determined by PSoC Creator at build time, by default.[82]
- *void CyDelayCycles(uint32 cycles)* results in a delay for the specified number of cycles using a software delay loop.

It should be borne in mind that software delays can be affected by interrupts so care must be exercised in their use. If more accurate delays are required a timer or PWM can be used. A simple assembly language delay can be implemented by loading a value into the accumulator and decrementing it until the value becomes zero. If multiple delays are needed, the value to be decremented for a given delay can be loaded from a LUT.

7.17   Re-entrant Code

The Keil compiler assumes that functions are not reentrant by default, and therefore, fixed memory locations in RAM are used to store the function's local variables. If the function must be called from different threads (like main and interrupt handler), or recursively, then it must be specifically defined as a reentrant function:

```
/* reentrant function declaration */
void delay (uint32) reentrant;
```

---

[80]The delay is based on the clock configuration entered in PSoC Creator by default.

[81]*CyDelay* functions implement simple software-based delay loops that are designed to compensate for the bus clock frequency and other factors, e.g., function entry and exit when the delay time is relatively mall.

[82]0: Use the default value, non-0: Set frequency value.

```
* reentrant function definition */
void delay (uint32 x) reentrant
{
. . .
}
```

PSoC 5 compilers define functions as re-entrant and do not support the keyword *re-entrant*. To port functions with this keyword to PSoC 5LP *re-entrant* can be ignored by redefining it as

```
#define reentrant /**/
```

The PSoC 3 Keil compiler provides the various keywords to place variables in different 8051 memory spaces, as shown in Fig. 7.30.

Keywords such as *code*, *idata*, *bdata* and *xdata*, that locate variables in different 8051 memory spaces, can also be ignoring when porting from PSoC 3 to PSoC 5LP by redefining them in a similar manner [4].

Memory Space	Keyword
Internal RAM	bit, data, idata, bdata
Internal SFRs	sfr, sbit
External Memory	xdata
Code (Flash)	code

Fig. 7.30 Keil keywords and related memory spaces.

7.18 Code Optimization

Execution speed and code size are often two paramount concerns when designing an embedded system. Historically, designers have typically tried to resort to departing from C, and higher-level languages when seeking additional optimization and resort to assembly language, particularly in cases which involve microcontrollers with 8051 class microprocessors. However, advances in compiler technology have made it possible to write highly efficient C code, in terms of memory requirements and speed. The Keil compiler has a number of Keil-specific keywords that have been added to support optimization, so that assembly language code may be obviated. However, these keywords are not necessarily supported by compilers for other processors such as the Cortex-M3 in PSoC 5LP.[83] The Keil Compiler supports several levels of optimization with Level 2 being the default level in PSoC Creator. Level 3 optimizes the compiled code with respect to code size by deleting redundant MOV operations, which in some cases have a significant impact on both code size and speed.

The 8051 core is a 256-byte address space that contains 256-bytes of SRAM plus a large set of special function registers (SFRs), as shown in Fig. 7.31, and the 8051 is most efficient when it utilizes this memory. As shown in the figure, the lower 128-bytes is SRAM, and accessible both directly

---

[83]PSoC Creator supports a number of equivalent macros to facilitate porting code from PSoC 3 to PSoC 5LP.

FF
80
7F
00
SRAM
Indirect Address
SRAM
Direct Address
SFRs

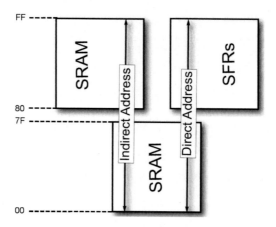

Fig. 7.31 8051 internal space layout.

and indirectly. The upper 128-bytes contain another 128-bytes of SRAM that can only be accessed indirectly. The same upper address space also contains a set of SFRs that can only be accessed directly. Table 7.4 details bytes in the lower address space that can be accessed in other modes. The memory map for the first 256 bytes in the 8051's memory space is shown in Fig. 7.32.

Table 7.4 8051 keyword memory space.

Memory Space	Keyword
Internal RAM	bit, data, idata, bdata
Internal SFRs	sfr, sbit
External Memory	xdata
Code (Flash)	code

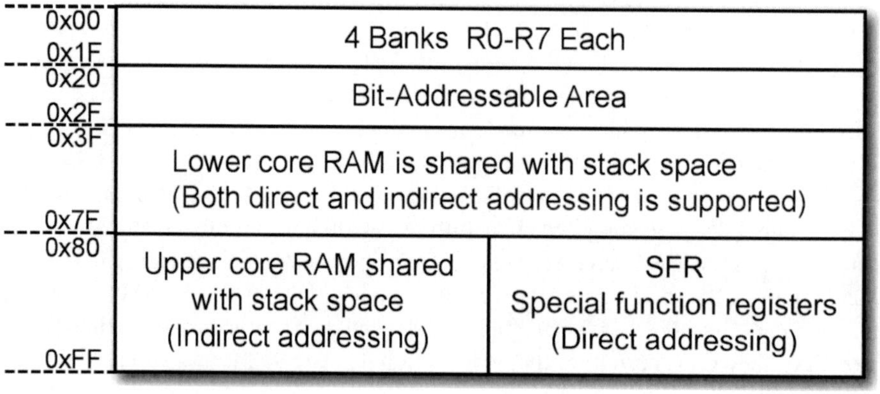

0x00
0x1F
4 Banks R0-R7 Each
0x20
0x2F
Bit-Addressable Area
0x3F
Lower core RAM is shared with stack space
(Both direct and indirect addressing is supported)
0x7F
0x80
Upper core RAM shared
with stack space
(Indirect addressing)
SFR
Special function registers
(Direct addressing)
0xFF

Fig. 7.32 8051 internal memory allocation.

7.18.1 Techniques for Optimizing 8051 Code[84]

Whenever feasible, it is advisable [1] to use bit variables for any variables that will have only binary values, i.e., 0 and not 0, and define then as being of type *bit*, i.e.

$$bit \ myvar;$$

The use of bit variable allows the compiler to draw upon the complete set of 8051 bit-level assembler instructions to create very fast and compact code, .e.g,

```
myvar = ~myVar;
if (!myVar)
{
        . . .
}
```

that causes the compiler to produce the following two lines of assembly code:

```
B200      CPL     myVar
200006    JB      myVar,?C0002
```

which requires 5 bytes of Flash and 8 cpu cycles.

Whenever possible, calling functions from interrupt handlers written in C should be avoided. The Keil C compiler pushes any register contents that it assumes may be changed by the ISR, which can result in a substantial amount of additional code as illustrated by the following example:

```
CY_ISR(myISR)
{
UART_1_ReadRxStatus():
}
```

that is a simple ISR, that when compiled can result in the following assembly language code:

```
C0F0    PUSH  B
C083    PUSH  DPH
C082    PUSH  DPL
C085    PUSH  DPH1
C084    PUSH  DPL1
C086    PUSH  DPS
58600   MOV   DPS,#00H
C000    PUSH  ?C?XPAGE1SFR
750000  MOV   ?C?XPAGE1SFR,#?C?XPAGE1RST
C0D0    PUSH  PSW
75D000  MOV   PSW,#00H
C000    PUSH  AR0
C001    PUSH  AR1
C002    PUSH  AR2
C003    PUSH  AR3
```

---

[84]In this section frequent reference is made to assembly code. However, the reader is asked to recall that any such code discussed herein is presumed, unless stated otherwise, to have been the output of PSoC's C compiler and, as such, is not hand-coded, assembly language, source code.

```
C004  PUSH  AR4
C005  PUSH  AR5
C006  PUSH  AR6
C007  PUSH  AR7
120000  LCALL  UART_1_ReadRxStatus
D007  POP  AR7
D006  POP  AR6
D005  POP  AR5
D004  POP  AR4
D003  POP  AR3
D002  POP  AR2
D001  POP  AR1
D000  POP  AR0
D0D0  POP  PSW
D000  POP  ?C?XPAGE1SFR
D086  POP  DPS
D084  POP  DPL1
D085  POP  DPH1
D082  POP  DPL
D083  POP  DPH
D0F0  POP  B
D0E0  POP  ACC
32    RETI
```

A better approach would be to use a flag in the ISR in the form of a global variable. The flag is simply a single bit that is read by background code accessing the register that contains the flag bit. The following is an example using a flag in the form of a global variable of type *bit* which is subsequently read by background code:

```
CYBIT flag;
CY_ISR(myISR)
{
    flag  =1;
}

void  main()
{
if  (flag)
    {
       flag = 0;
       UART_1+_ReadRxStatus();
           . . .
    }
}
```

The ISR portion of this code results in the following assembly code:

```
D200 SETB flag
32    RETI
```

which is less than 10% of the assembly code produced by the previous example. However, it should be noted that using a flag in this manner, assumes that the status register containing the flag will be checked often enough to result in the desired operation.

Placing variables in the 8051's internal memory can produce substantial benefits. The location of variables in memory should be based on the relative frequency of access, e.g., the most frequently accessed should be of type *data*, the next most frequently accessed as type *idata*, and so on, for *pdata* and *xdata*. As noted previously, because stack space is limited, the Keil compiler stores local variables in fixed memory locations and shares these locations among local variables in functions that don't call each other. Therefore, when possible, variables within functions should be local variables that allows the Keil compiler to store such variables in registers R0-R7. Loop decrementing is more efficient because it is easier to test for zero than for a non-zero value, as shown by the following examples:

```
void main ()
{
    data uint8 i;
    /* loop 10 times */
    for (i = 10; i != 0; i--)
        {

                ...

        }
    }

}
```

is compiled as

```
75000A      MOV i,#0AH ; i = 10
            ?C0002:
E500        MOV A,i ; i != 0
6006        JZ ?C0003
                    ...
1500        DEC i ; i--
80EF        SJMP ?C0002
            ?C0003:
```

as opposed to

```
void main ()
{
    data uint8 i;
    /* loop 10 times */
    for (i = 0; i < 10; i++)
    {

            ...

    }
}
```

that compiles as

```
E4          CLR A ; i = 0
F500        MOV i,A
            ?C0002:
E500        MOV A,i ; i < 10
C3          CLR C
```

```
940A      SUBB  A,  #0AH
5006      JNC  ?C0003
                    . . .
0500      INC  i  ;  i++
80EF      SJMP  ?C0002
          ?C0003 :
```

Bit variables can be used to dramatically improve efficiency, and bit-level assembler instructions can also be employed to implement bit-wise C operations. Some examples of setting bit variables are given by the following:

```
uint8  x;
x  |=  0x10;  /*  set  bit  4  */
x  &=  ~0x10;  /*  clear  bit  4  */
x  ^=  0x10;  /*  toggle  bit  4  */
if  (x  &  0x10)  /*  test  bit  4  */
  {
            . . .
  }
```

8051 bit-level assembly instruction can be used to implement C bitwise operations by using the keyword *sbit* and the $\wedge$ operator.[85]

One method is given by:

```
/*myVar  is  located  in  idata  at  202F  */
          bdata  uint8  myVar;
/*  this  is  bit  4  of  myVar  */
          sbit  mybit4  =  myVar^4;
/*  set  bit  4*/
          mybit4  =  1;
/*  clear  bit  4  */
          mybit4  =  0;
/*  toggle  bit  4
          mybit4  =  ~mybit4 ;
/*  test  bit  4  */
          if  (mybit4)
  {
            . . .
  }
```

which can also be used for variables that are larger than 8-bits, e.g., uint16, uint32, etc. It should be noted that *sbit* and *bdata* definitions global and not local within a function. PSoC Creator provides support for *sbit* and *sfr* keywords as follows:

```
sfr  PSW  =  0xD0;
sbit  P  =  PSW^0;
sbit  F1  =  PSW^1;
sbit  OV  =  PSW^2;
```

---

[85] In this discussion the $\wedge$ operator is not the standard C language exclusive or (XOR).

```
sbit  RS0 = PSW^3;
sbit  RS1 = PSW^4;
sbit  F0  = PSW^5;
sbit  AC  = PSW^6;
sbit  CY  = PSW^7;
```

Alternatively, a bit-addressable SFR can be used given that SFR PSW contains the program status word at D0 and is therefore directly accessible. The *sbit* keyword can be used to access each of the PSW's bits using the same technique discussed in this section, e.g.,

$$F0 = \sim F0;$$

(PSW's F0 and F1 bits are available for general purpose use.) The accumulator (*ACC*) and the *B* register can be used as temporary SFRs. However, the individual bits of each must be specifically defined, e.g.,

```
/* bit 4 of ACC SFR */
sbit  A4 = ACC^4;
/* bit 3 of B SFR */
sbit  B3 = B^3;
```

in which case, faster bit testing can be achieved by using

```
/* assume return value is 8 bits */
ACC = UART_1_ReadRxStatus();
if (A4) /* test bit 4 */
{
        . . .
}
```

The auxiliary B resister can be used for storage to facilitate instructions, such as MUL and DIV, or to switch two 8-bit variables

```
uint8  x, y;
B = x;
x = y;
y = B;
```

Pointers are commonly used in embedded systems and their size is a function of the address space being employed, e.g., a 64k address space will require two-byte pointers, while larger spaces such as those addressed by PSoC 5LP require 4-byte pointers to span the address space. However, PSoC 3's 8051 employs several memory spaces ranging from 256–64k bytes and therefore the Keil C compiler utilizes memory-specific and generic pointers.[86] The use of memory-specific pointers is more efficient than generic pointers and therefore the latter should be used only when the memory type is unknown.[87] An 8051 generic pointer can be used to access data regardless of the memory in which it is stored. It uses 3 bytes—the first is the memory type, the second is the high-order byte of the address, and the third is the low-order byte of the address. A memory-specific pointer uses only one or two bytes depending on the specified memory type.

---

[86]A generic pointer can be used to access data regardless of the memory in which it is stored. It uses 3 bytes—the first of which is the memory type, the second is the high-order byte of the address, and the third is the low-order byte of the address. A memory-specific pointer uses only one or two bytes depending on the specified memory type.

[87]The majority of Keil library functions take generic pointers as arguments and memory-specific pointers are automatically cast to generic pointers.

The C keyword *const*, which be added to an array declaration or variable, is used to require that the variable not be changed but does not control where the variable is stored, e.g.,

```
const  char  testvar  =  37;
void  main ()
{
char  testvar2  =  testvar;
}
```

is compiled as

```
900000   MOV DPTR,#testvar
E0       MOVX A,@DPTR ; MOVX accesses xdata space
900000   MOV DPTR,#testvar2
F0       MOVX @DPTR,A
```

and shows that the const variable *testvar* is stored in Flash, and copied to an SRAM location initialized in the startup code. If there is insufficient SRAM to store all the const variables, the keyword *code* (or CYCODE) must be used in the declaration, i.e.,

```
code  const  char  testvar  =  37;
void  main ()
{
         char  testvar2  =  testvar;
}
```

and then the corresponding assembly code is given by

```
900000 MOV DPTR,#testvar
E4      CLR A
93      MOVC A,@A+DPTR ; MOVC accesses code space
900000 MOV DPTR,#testvar2
F0      MOVX @DPTR,A
```

so that the const variable *testvar* is stored in Flash.

Arrays and strings can be kept in FLASH as illustrated by the following example:

```
const  float  code  array[512]  =  {  ...  };
code  const  char  hello[]  =  "Hello  World";
```

The arguments for C functions are typically passed on the CPU's hardware stack. However, the Keil compiler uses either registers *R0–R7*, or fixed memory locations for passing such arguments and does not pass arguments via the stack. The use of registers is employed because it is faster and uses fewer code bytes. The latter can be important because of the limitation of the 8051 hardware stack to 256 bytes. However, this method has some limitations, as shown in Table 7.5.

If other types of arguments are involved, they can be passed in fixed memory locations. To the extent possible, no more than three function arguments should be employed. However, there is no guarantee that the compiler will pass three arguments in registers.

Arguments of type *bit* are always passed in a fixed memory location in the 8051's bit space (internal memory) and cannot be passed in a register. Bit variables should be declared at the end of a function's argument list, to keep the other arguments consistent with Table 7.5. Function return

Table 7.5 Argument passing via registers.

Argument Number	Char, 1 byte Pointer	Int, 2-byte Pointer	Long, Float	Generic Pointer
1	R7	R7, R6 (MSB)	R7-R4 (MSB)	R3 (mem type) R2 (MSB) R1
2	R5	R5, R4 (MSB)	R7-R4 (MSB)	R3 (mem type) R2 (MSB) R1
3	R3	R3, R2 (MSB)	-	R3 (mem type) R2 (MSB) R1

values are handled as described in Table 7.6. Return values of type bit are always passed via registers. If a function argument is the return value of another function that argument should be the first in the argument list whenever possible.

Table 7.6 Function return values via registers.

Return Type	Register
Bit	Carry Flag
char, 1-byte pointer	R7
int, 2 byte pointer R7, R6 (MSB)	R7, R6 (MSB)
long, float	R7-R4 (MSB)
Generic Pointer	R3 (mem type), R2 (mem type), R1

7.19 Real-Time Operating Systems

In a typical embedded system, there are often multiple tasks[88] involved with a requirement to share and exchange data between such tasks. Scheduling of tasks,[89] and sharing of resources in these cases, can sometimes be greatly facilitated by introducing a real-time operating system[90] so that tasks are processed subject to specific, predefined time constraints. This type of operating system is referred to

---

[88]Some tasks may have to be handled in parallel, others in serial fashion and these activities are referred to collectively as multi-tasking.

[89]Tasks are also referred to as processes and in the present context, the two terms are considered equivalent.

[90]A real-time operating system is a type of operating system that provides one, or more, responses within predefined time periods.

as a real-time operating system, or RTOS.[91] The majority of popular microcontrollers lack the memory space, execution speed, and/or other resources, to adequately support an RTOS. However, the ARM architecture of PSoC 5LP, its clock speed (max of 67 mega Hertz), RAM space (4 gigabytes) and other resources are sufficient to support a real-time operating system (RTOS), such as FreeRTOS.[92]

The role of the real-time operating system is to provide an environment capable of managing the available resources, and provide a variety of services for tasks, e.g.,

- management of system resources and CPU,
- assuring that tasks are handled in a predefined manner and within the imposed time constraints,
- handling data movement and communications between tasks,
- efficiently managing RAM allocation and use,
- determining which resources can be shared and which are allocated exclusively,
- responding to events,
- assigning priorities to tasks,
- coordinating internal and external events,
- synchronizing tasks,

and,

- handling compute and I/O bound tasks.

7.19.1 Tasks, Processes, Multi-Threading and Concurrency

Tasks can be in various states, e.g., running, ready (pended or suspended), blocked (delayed, dormant or waiting). If the embedded system is to employ multitasking, in an optimized fashion, compute-bound and I/O-bound tasks must be assigned priorities such that the executive has a basis for assigning the ordering of execution of tasks. This approach allows lower priority tasks to be preempted by higher-order tasks, by the scheduler. In the case of round-robin scheduling of tasks, tasks of the same priority are executed in a predefined order. Preemptive scheduling assigns the order of task execution based on the concept that the highest priority process, in a group of waiting tasks, is executed first, i.e., it *preempts* other tasks.

Each task is assigned the necessary resources, e.g., RAM space, a task stack, program counter, I/O ports, file descriptions, registers, etc. These resources may be shared with other tasks. The state of a process, at any given time, is determined by the then current program counter value, data values in the task's allocated memory space(s) and/or registers. CPU time is allocated to each task by the operating system, and if tasks are to effectively/efficiently run simultaneously, the CPU must switch from one task to another,[93] often whether be a given task is completed or not, and then return at a later time to the incomplete tasks until each process has been completed.[94] If the CPU switches tasks fast enough, the tasks are said to be running *concurrently*, or alternatively, as *concurrent processes*. In some operating systems the CPU switches task execution at fixed intervals, a practice referred to

---

[91] Real-time operating systems are also referred to as real-time executives and kernels.

[92] $OpenRTOS^{TM}$ is a commercially licensed and supported version of FreeRTOS that includes fully-featured professional-grade USB, file system and TCP/IP components. OpenRTOS is a commercial version of FreeRTOS and provided under license.

[93] This is typically referred to as *context switching*.

[94] It is assumed in this discussion that the CPU is operating at a sufficiently fast clock rate to be able to switch among tasks while assuring that the overall system response meets the system's performance criteria. Some tasks may never be completed, while others have various lifetimes.

as *time slicing*. Tasks waiting for their turn to be executed are said to be in a *waiting state*. A task can be terminated either upon completion, or as a result of being *killed*.[95] Typically, a terminated process, whether completed or killed, is removed from memory and the associated resources are deallocated.

A *thread* is a set of instructions that has access to stack space and registers, and the associated *resources*, needed to carry out a task (process). Tasks can be *ready, blocked, running* or *terminated*. Multiple threads are used when tasks need to occur contemporaneously and are referred to as *parallel processes*. A scheduler is used to control which task is to be run, and when it is to run. A *dispatcher* starts each task, initiates 4 inter-task communications, or any inter-process communication required to exchange information between tasks. Multiple threads may be running in a single- or multi-processor environment.[96]

In the single-processor case, the processor is switched from one thread to another in a mode known as multi-threading based on *time-division multiplexing*. If multiple processors are involved, each may be running a single thread. *Multi-threading* refers to the existence of multiple threads within a given process that, although executing independently, share the resources allocated to the process.

In a multi-thread environment semaphores[97] are sometimes employed to avoid collisions when data is being modified It should be noted however, that threads *are not* synonymous with processes, or tasks, e.g.,

- Context switching from one thread to another, within a given process, is generally substantially faster than process context switching.
- Threads share address space, while processes have independent address spaces.
- Processes rely solely on inter-process communications to exchange data and information.
- Processes are typically independent tasks that may or may share data and./or resources.

While concurrent processing offers a number of attractive benefits that are not available from sequential code execution, it is not a panacea. As noted by Sutter and Larus [], developing concurrent systems is not an easy task, even though, as observed by Lee [11], as suggested by the world is "highly concurrent" and humans are rather adept at analyzing concurrent systems.

7.19.2 Task Scheduling and Dispatching

The RTOS contains both a scheduler and dispatcher within the RTOS' *kernel*. The operating system is responsible for the management of memory, I/O, tasks, file system the file system, networking and interpretation of commands. Task control blocks (TCBs), either static or dynamic,[98] are used to encapsulate the important information associated with a given task, e.g.,

- associated CPU registers
- contents of the program counter

---

[95]The *killing* of a task typically involves sending a signal (message) to a process to terminate.

[96]"Although threads seem to be a small step from sequential computation, in fact, they represent a huge step. They discard the most essential and appealing properties of sequential computation: understandability, predictability, and determinism. Threads, as a model of computation, are wildly nondeterministic, and the job of the programmer becomes one of pruning that nondeterminism." [11].

[97]The simplest form of a semaphore is a Boolean variable or integer that signals that access to a critical section of code, or a critical variable, has occurred.

[98]Static TCB allocation implies that TCBs are created and remain whereas dynamic TCBs are typically deleted once a task has been completed, or terminated.

- state of a process and an associated ID
- list of open files
- a pointer to a function

A typical RTOS employs a set of classes that support kernel services invocable by the applications tasks and include support for

- *Intertask communications*—passing of information between tasks is accomplished by classes such as event flags, mailboxes,[99] messages, queues,[100] pipes, timers, mutexes[101] and semaphores[102]
- *Tasks* manage program execution. While each task is independent of other tasks, tasks can interoperate via data structures, I/O and other constructs. Inter-task communications employs semaphores, messages queues, pipes, shared memory signals, mail slots and sockets.
- *Kernel service routines* process *kernel service requests* initiated by an application to provide operating system functions needed by the application.
- *Interrupts* are an important aspect of a RTOS particularly with respect to prioritization of tasks. However, prioritization is not sufficient to assure that tasks are handled in a timely fashion.

Scheduling can be either clock- or priority-driven. Scheduling variables such as arrival time, computation, deadline, finish time, lateness, period and start time are used to guarantee responsiveness and minimize latency.

- *Arrival time* is defined as the point in time when a task is ready to run.
- *Computation time* is defined as the processor time required to complete the execution of a task, in the absence of interruption.
- *Deadline* is defined as the latest time at which a task is completed.
- *Lateness* is defined as the length of time after the deadline has been passed required to complete a task.
- *Period* is defined as the minimum time that elapses between release of the CPU.[103]
- *Start time* is defined as the time at which the task begins execution.

Each task has a deadline, execution time and period associated with it, as shown in Fig. 7.33.

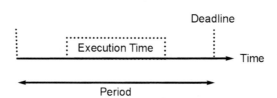

Fig. 7.33   Task timing parameters.

---

[99]Messages and mailboxes are employed to transmit data between a sender and a receiver.

[100]Queues are used to pass data between a producer and a consumer.

[101]Mutexes are binary flags that assure that mutually shared code is also mutually exclusive. Thus, multiple tasks can use a resource, but only one task at a time.

[102]Semaphores are can also be constructs used to synchronize tasks and events. The concept of semaphores was introduced by Edsger Dijkstra, a Dutch computer scientist ,who also, contributed to the deprecation of the *GOTO* statement, created reverse polish notation (RPN) and a multitasking operating system known as "THE".

[103]In some RTOS environments, the tasks with the shortest periods are given the highest priority.

In most cases, the deadline and the period are quantitatively equal. However, a task can start at any time within the period.

7.19.3 PSoC Compatible Real-Time Operating Systems

There are a number of RTOS sources available as commercial, or freeware, implementations that support either PSoC 3 or PSoC 5LP. A brief description of some of these is provided in this section. In some cases the source code for the RTOS is also available, as noted.

Micriμm[104] offers a commercial version of μC/OS III for PSoC 5LP. It has the following features/benefits:

- Relatively small footprint.[105]
- Is written in ANSI C.
- Supports a variety of user-selectable features.
- Employs *round-robin scheduling*.
- Protects critical regions by disabling interrupts while minimizing overhead and providing deterministic interrupt response.
- Supports an arbitrary, although user-selectable, number of priorities.[106]
- Is designed specifically for embedded system applications.
- Blocks NULL pointers ensure that arguments are within allowable ranges.
- Supports user-allocation of kernel objects at run-time.
- Execution times are not a function of the number of executing tasks.
- Places no constraints on maximum task size.[107]
- Allows multiple tasks to run, at the same priority level, in a user-specified, time-slice mode.
- Places no limitations on the number of tasks, semaphores, mutexes, event flags, message queues, timer or memory partitions.
- Supports monitoring of stack growth of tasks

FreeRTOS[108] (PSoC 5) is available as freeware,[109] and has the following features/functionality:

- Minimal ROM, RAM and processing overhead.
- Small footprint.[110]
- Is relatively simple.[111]
- Very scalable,
- Offers a smaller/easier, real-time processing alternative for applications for which eCOS, embedded Linux (or Real-Time Linux) and uCLinux are too large, not appropriate, or not available.

---

[104]http://micrium.com. The source code is available.

[105]The footprint size is determined in part by the user-selectable features chosen.

[106]Typical embedded systems use from 32–256 levels, inclusive, of priority.

[107]Minimum task sizes are imposed.

[108]http://www.freertos.org. The source code is available.

[109]Documentation is available for a nominal sum.

[110]Typical kernel binary image size ranges from 4–9 kB.

[111]The kernel core is contained in three C language files.

RTX51 Tiny[112] (PSoC3) is Keil's real-time operating system that provides an RTOS environment for programs based on standard C constructs and compiled with the Keil C51 C Compiler. Keil additions to the C language allow task functions to be declared without the need for a complex stack and variable frame configuration.

RTX51 (PSoC3) provides the following features:

- Code banking Explicit task switching
- Task ready flag
- Support of CPU idle mode
- User code support in timer mode
- Interval adjustment support
- Scalability

The footprint can be minimized by disabling round-robin switching,[113] stack checking and avoiding unnecessary use of system functions.

The supported functions include:

- *isr_send_signal* causes a signal to be sent to the task's *task_id*. If the task is already waiting for a signal, it is prepared for execution without starting it. Otherwise, the signal is stored in the task's signal flag.
- *isr_set_ready* places the task specified by task_id into the ready state. This function can only be called from interrupt functions. The isr_send_signal function returns a value of 0 if successful and −1 if the specified task does not exist.
- *os_clear_signal* clears the signal flag of the task specified by the *task_id*.
- *os_create_task* causes a task to be marked as ready and executed at the next available opportunity.
- *os_delete_task* stops the task identified by the *task_id* and removes it from the task list.
- *os_reset_interval* is used to correct timer problems.
- *os_running_task_id* determines the *task_id* for the task that is currently running.
- *os_wait* halts the current task and waits for an event such as a time interval, a time-out, or a signal from another task or interrupt.
- *os_switch_task* allows a task to halt execution and allow another task to run. If calling task is the only task ready for execution it resumes running immediately.

When using PSoC 5LP with a real-time operating system, it should be noted that the Cortex M3 core uses numerically low priority numbers to represent HIGH priority interrupts. When assigning an interrupt a low priority it must not be assigned a priority of 0 (or other low numeric value) because it can result in the interrupt actually having the highest priority in the system and could result in a system crash if this priority is above *configMAX_SYSCALL_INTERRUPT_PRIORITY*.

The lowest priority on a Cortex M3 core is 255.[114]

If a PSoC 5LP application provides its own implementation of an interrupt service routine, which accesses the Kernel API, the priority must be equal to, or numerically greater than, the *config-MAX_SYSCALL_INTERRUPT_PRIORITY* so in effect it has a lower priority. To install a customize interrupt service routine, call the *Peripheral_StartEx(vCustomISR)* function (where 'Peripheral' is

---

[112]http://www.keil.com.

[113]Reducing round-robin task switching also reduces the data space requirements.

[114]Different Cortex M3 vendors implement a different number of priority bits and supply library functions that expect priorities to be specified in different ways.

the name of the peripheral to which the ISR relates) passing the interrupt service routine function which has its prototype declared as *C__ISR_PROTO(vCustomISR)* and the function declared with *CY_ISR(vCustomISR)*.

In the function *vInitialiseTimerForIntQTests( )* in *IntQueueTimer.c*, the ISR is installed using a call to *isr_ High_ Frequency_ 2001Hz_ StartEx( )*. Each port # defines '*portBASE_TYPE*' to equal the most efficient data type for that processor. This port defines *portBASE_ TYPE* to be of type long.

7.20   Additional Reference Materials

There a number of valuable resources available, via www.cypress.com, that include training documents/videos, device datasheets, a technical reference manual (TRM), component data sheets, system reference guides, component author guide (CAG), application notes, example projects knowledge base forums and various forums, to assist the designer with the development of PSoC 3/PSoC 5LP embedded systems.

The PSoC 3/PSoC 5LP device datasheets provide a summary of the features, device-level specifications, pin-outs and fixed functional peripheral electrical specifications. The technical reference manual describes the functionality of all the peripherals in detail and includes the associated register descriptions. The component datasheets contain the information required to select and use a component and its functional description, API documentation, assembly language and C example source code, and the relevant electrical characteristics of the component.

The system reference guide (SRG) describes the PSoC Creator *cy_boot* component. This component is automatically included in every project by PSoC Creator[115] and includes an API that can be accessed by firmware for tasks associated with

- Clocking—PSoC 3/5 has flexible clocking capabilities that are controlled in PSoC Creator by selections within the Design-Wide Resources (DWR) settings, connectivity of clocking signals on the design schematic, and API calls that can modify the clocking at runtime.
- DMA—The DMAC files provide the API functions for the DMA controller, DMA channels and Transfer Descriptors. This API is the library version not the auto-generated code that is generated when the user places a DMA component on the schematic. The auto-generated code would use the APIs in this module.
- Flash Linker scripts[116]
- Power management[117]
- Startup code—the *cy_boot* functionality includes a reset vector, setting up the processor to begin execution, setup of interrupts/stacks, configuration of the target device, preservation of the reset status and calling the main() C entry point.[118]

---

[115]Only a single instance can be included in a project, does not include symbolic representation and is not included in the component catalog.

[116] cf. System Reference Guide, *cyboot* Component Document.

[117]ibid.

[118]Initialization of static/global variables and the clearing of all remaining static/global variables is also handled by *cy_boot*.

and,

- Various library functions:
  1. unit8 CyEnterCriticalSection(void)—disables interrupts and returns a value indicating whether interrupts were previously enabled (the actual value depends on whether the device s PSoC 3 or PSoC 5LP).
  2. unit8 CyExitCriticalSection(void)—re-enables interrupts if they were enabled before CyEnter-CriticalSection was called. The argument should be the value returned from CyEnterCritical-Section.
  3. void CYASSERT(uint32 expr)—macro evaluation of an expression and if it is false, i.e., evaluates to 0, then the processor is halted. This macro is evaluated unless NDEBUG is defined, if not, then the code for this macro is not generated. NDEBUG is defined by default for a Release build setting and not defined for a Debug build setting.
  4. void CySoftwareReset(void)—forces a software reset of the device during which the startup code will detect that the reset was the result of a software reset and the SRAM memory area, indicated by corresponding arguments will not be cleared. If any of this area has initialization assignments that initialization will still occur.
  5. void CyDelay(uint32 milliseconds)—invokes a delay[119] by the specified number of milliseconds. By default, the number of cycles to delay is based on the clock configuration. If the clock configuration is changed at run-time, then the function *CyDelayFreq* is used to indicate the new Bus Clock frequency. CyDelay is used by several components, so changing the clock frequency without updating the frequency setting for the delay can cause those components to fail.
  6. void CyDelayUs(uint16 microseconds)—the number of cycles to delay is, by default, based on the clock configuration. If the clock configuration is changed at run-time, then the function *CyDelayFreq* is used to indicate the new Bus Clock frequency. *CyDelayUs* is used by several components, therefore changing the clock frequency without updating the frequency setting for the delay can cause those components to fail.
  7. void CyDelayFreq(uint32 freq)—sets the Bus Clock frequency used to calculate the number of cycles needed to implement a delay with *CyDelay*.The frequency used is based, by default, on the value determined by PSoC Creator at build time.
  8. void CyDelayCycles(uint32 cycles)—the delay, determined by the specified number of cycles, is created by a software delay loop.

7.21   Recommended Exercises

7-1 Using Mason's rule, find the gain for signal flow graph shown below.

7-2 Explain the distinctions between and benefits of multithreading, multitasking and sequential tasking. Describe a physical system that employs all three. Why are threads said to be nondeterministic?

---

[119]The *CyDelay* functions implement simple software-based delay loops that are designed to compensate for bus clock frequency and other factors. Additional factors may also influence the actual time spent in the loop, e.g., function entry and exit, and other overhead factors, may also affect the total amount of time spent executing the function. This may be especially apparent when the nominal delay time is small.

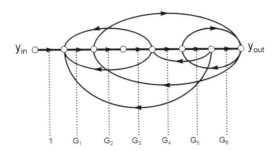

7-3 Draw a block diagram of an embedded system that controls a traffic signal, pedestrian signals and activation buttons at a four-way intersection. Draw the signal graph(s) for such a system and discuss how the design would have to be modified to allow emergency vehicles the right of way.

7-4 Write two callable PSoC 3 routines, one in C and one in assembly, that will produce a delay that can be altered programmatically to provide variable length delays. Comment on the relative speed and overhead requirements of each.

7-5 Sketch an example of frequency modulation of the sensor signal shown in Fig. 5.14.

7-6 Sketch Fig. 5.10 in the form of a signal graph.

7-7 Create a state diagram for a clock that displays minutes hours and seconds.

7-8 Give an example of how to use the RTX51 function calls to facilitate a C program designed to control a traffic light. Assume that the traffic control system is capable of handling emergency traffic such as fire trucks, ambulances, police vehicles, etc., on a priority basis based on the type of emergency vehicle.

7-9 Describe the design for a temperature measuring system that utilizes a temperature dependent resistance, such as a thermistor, whose resistance is a function solely of temperature that can be stored in a look-up table. Provide a requirements description, specification, signal flow graph, and block diagram for the design. How can such a system be implemented if the function itself has other dependencies as well, e.g., ambient pressure?

7-10 Assuming that the design developed in response to Exercise 9 was created for a PSoC 3 and in C. What changes, if any, would be necessary in order to base the design on a PSoC 5LP, i.e., what changes would be required to port it to a PSoC 5LP?

7-11 Create a simple 8-bit down-counter as a datapath-based Component using PSoC Creator. A simple down-counter can be represented by a state machine with two states, as shown Fig. 7.34.

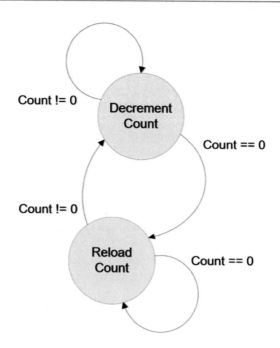

Fig. 7.34 Simple counter state diagram.

References

1. M. Ainsworth, PSoC 3 8051 Code Optimization. Application Note: AN60630. Cypress Semiconductor Corporation (2011)
2. M. Ainsworth, PSoC®3 - 8051 Code And Memory Optimization. AN60630. Document No. 001-60630 Rev.*H. Cypress Semiconductor (2017)
3. M. Ainsworth, PSoC®3 to PSoC 5LP Migration Guide. Document No. 001-77835 Rev.*D1A. AN77835. Cypress Semiconductor (2017)
4. M. Ainsworth, A. Ganesan, M. Balan, K. Mikoleit, B. McAndrews, PSoC Arm Cortex Code Optimization. AN89610. Document Number:001-89610 Rev.*F. Cypress Semiconductor (2020)
5. T. Dust, G. Reynolds, Designing PSoC Creator Components with UDB Datapaths. AN82156. Document No. 001-82156 Rev. *I. Cypress Semiconductor (2018)
6. K. Edlow, PSoC 3 and KEIL RTX51 Tiny. AN64429. Document No. 001-64429 Rev. ** 1. Cypress Semiconductor. November 1, 2010
7. S. Gupta, L. Ntarajan, Migrating from PSoC 3 to PSoC 5. Application Note: AN62083. Cypress Semiconductor Corporation (2011)
8. V.S. Kannan, PSoC 3 and PSoC 5LP Interrupts. AN54460. Document No. 001-54460 Rev. *K. Cypress Semiconductor (2020)
9. V.S. Kannan, J. Chen, PSoC 3, PSoC 4,and PSoC 5LP Temperature Measurement with a Diode. AN60590. Document No. 001-60590 Rev. *K 1 Cypress Semiconductor (2020)
10. J. Kathuria, C. Keeser, Implementing State Machines with PSoC 3, PSoC 4, and PSoC 5LP. AN62510. Document No. 001-62510 Rev. *F1. Cypress Semiconductor (2017)
11. E.A. Lee, The Problem with Threads. Technical Report UCB/EECS-2006-1 (2006). http://www.eecs.berkeley.edu/Pubs/TechRpts/2006/EECS-2006-1.html
12. M.S. Nidhin, Getting Started with PSoC 5LP. AN77759. Document No. 001-77759 Rev.*G. Cypress Semiconductor (2018)
13. PSoC 3 Architecture TRM (Technical Reference Manual), Document No. 001-50235 Rev.*M. Cypress Semiconductor. April 8, 2020
14. PSoC 3 Registers TRM (Technical Reference Manual), Document No. 001-50581 Rev.*. Cypress Semiconductor. March 25, 2020
15. PSoC 5LP Architecture TRM(Technical Reference Manual), Document No. 001-78426 Rev. *G. Cypress Semiconductor. November 6, 2019

Programmable Logic

8

Abstract

In this chapter attention focused on the virtues of programmable logic devices, boundary-scanning techniques for testing programmable devices, Boolean functions and their simplification using Karnaugh maps. Macrocells and logic arrays are shown to form the basis for UDBs are discussed in some detail and the steps required to simplify Boolean expressions have been outlined in detail. Programmable logic devices that are based on combinations of macrocells and logic arrays and discussed in some detail and their use in one incarnation in the form of universal digital blocks. An integral part of using such devices is the ability to form and simply Boolean expressions derived from truth tables or Karnaugh maps. that represent the required logic.

A simple, but straightforward technique has been presented for evaluating Karnaugh maps, suggested by Mendelson (*Schaum's Outline of Theory and Practice of Boolean Algebra*, McGraw-Hill, 1970), Harbort and Brown (https://www.slideshare.net/hangkhong/karnaugh, 2001), et al., that simplifies Boolean expression to minimize hardware requirements in the subsequent implementations. PSoC 3/5LP's universal digital block is discussed with respect to its internal architecture and relationship/interaction with the datapath. The Backus-Naur notation is introduced within the context of a discussion of HDLs and the basic constructs of VHDL, Verilog and WARP are discussed and illustrated by an example. Additionally, finite state machines are introduced and an example of a state machine implementation of a UART using Verilog was presented. PSoC 3/5LP architecture details and functionality were used throughout this chapter to illustrate key aspects of the material presented.

8.1 Programmable Logic Devices

Embedded systems have become truly ubiquitous, outnumbering their PC counterparts by at least one to two orders of magnitude. In the most common incarnation, embedded systems perform a function, or typically a limited set of functions, to which they are tightly constrained. They are expected to be fast, inexpensive, highly responsive, require minimal power, etc. In addition many embedded systems are often required to be aware of changes in their operating environment and to make the necessary

© Springer Nature Switzerland AG 2021
E. H. Currie, *Mixed-Signal Embedded Systems Design*,
https://doi.org/10.1007/978-3-030-70312-7_8

adjustments, if any, in order to maintain high performance. Computations and decisions made by embedded systems are to occur in real-time and not introduce any degradations in overall system performance.

The designer is expected to produce a design capable of meeting, or exceeding, the design specifications while operating well within the constraints imposed by cost-effectiveness, small size, low power consumption, etc. This inevitably results in the designer having to optimize multiple facets of the design in order to produce a system that is overall highly optimized with respect to the key design criteria. Typical metrics include, materials cost, size, robustness, power consumption, manufacturing costs, critical component availability, time-to-market, development cost, etc. In addition to creating an optimized design, the designer must also take into account the various design metrics and arrive at a system that represents the best set of trade-offs with respect to these metrics.

A further complication is introduced by the fact that an embedded system is a synthesis of hardware *and* software. Although an embedded system can be subjected to rigorous testing, the software component is often difficult, if not impossible, to thoroughly test prior to release to the market. An additional complication is that often the designer must have significant expertise in both software and hardware design and implementation in order to effectively optimize the hardware and software aspects of the system. Finally, the designer also needs to have technical competence in a variety of technologies, e.g., optics, analog/digital subsystems, sensors, microcontrollers, ADC/DAC technology, communications protocols, etc.

As discussed briefly in Chap. 1, programmable logic devices allow designers to employ existing generic devices that include the ability to either create internal connections, or destroy existing connections, as may be required, to implement the required functionality. One of the most common of such devices is the field-programmable gate array (FPGA) [3]. Although PLDs can reduce non-recurring engineering (NRE[1]) costs and have the additional advantage that custom devices can be made available almost instantly, they have the disadvantages of often requiring more power, being larger devices, potentially slower than their production counterpart would be, and can be significantly more expensive. While they can be mask-programmed, i.e., factory-programmed, it is usually not practical to use this type of technology, except when large volumes of devices are required. Field programmable devices can be rapidly produced in small quantities and allow the designer to make "mid-course" corrections in the field. FPGA-based designs can also be arbitrarily complex and are typically highly scalable.[2]

The hierarchy of programmable, solid-state, logic devices is shown in Fig. 8.1.

Fortunately, and as a practical matter, it often matters little whether some aspect of an embedded system is implemented in hardware, software or some synthesis of both. As a rule the decision as to what is to be implemented in software and what is to be implemented in hardware is based on any one of a number of trade-offs in terms of cost, power requirements, size, ability to rapidly adapt to changing market needs, etc. In cases in which the anticipated production may represent relatively low volume in terms of the number of units, time-to-prototype, time-to-market, NRE and/or the ability to adapt to changing market needs are major concerns, PLDs can provide an excellent alternative to customized ICs.

---

[1]Non-recurring engineering costs are one-time costs related to the development, engineering, testing and design of a system or product.

[2]As an example of complexity and scalability capability of FPGAs, designs exist that have the functionality of 1000 distinct cores for use in extremely high-speed image processing.

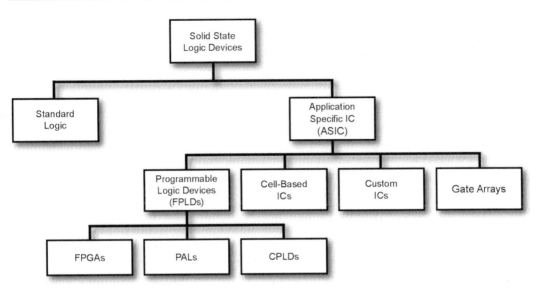

Fig. 8.1 Hierarchy of programmable logic devices.

PLDs typically consist of large numbers of flip-flops and gates[3] that can be connected under software control in arbitrarily complex configurations to provide specific logic functionality. As discussed briefly in Chap. 1, there are three basic types of PLDs, viz., simple PLDs (SPLDs), complex PLDs (CPLDs) [4] and field-programmable PLDs typically referred to as field-programmable gate arrays (FPGAs). The programmable logic arrays, which are one type of PLD, employ fuses that can be permanently rendered in an open state by special types of hardware/software programmers. The generic array logic (GAL) is a PLD similar to the PAL [2] except that it is reprogrammable, and they are relatively high-speed devices that are compatible with both 3.3 and 5 volt logic. In addition to array logic, both GALs and PALs include output logic, e.g., tristate controls and/or gates that allow the combination of logic arrays and output logic referred to as a "MacroCell" to be implemented. PALs and GALS typically have multiple inputs and outputs which further increases their versatility and usefulness.

Programmable logic devices [9] are programmed under software control and typically require that the target device's functionality be provided in the form of state equations, truth tables, Boolean expressions, etc. The programming software can then use these descriptions to produce an industry-standard binary file known as a JEDEC file[4] which is subsequently loaded into a hardware programmer capable of erasing, copying, verifying and/or programming PLDs.

---

[3]PLDs typically have hundreds or thousands of AND, OR and NOT gates and in some cases flip-flops that can be programmatically interconnected to provide a wide variety of devices.

[4]The Joint Electron Devices Engineering Council (JEDEC) has defined standard object file transfer formats for file transport to PLD programmers, e.g., JESD3-C.

8.2   Boundary-Scanning

While it is sometimes highly desirable for a designer to be able to incorporate custom devices into a design or application to facilitate its optimization, it is imperative that the designer be assured that in doing so a new level of complexity has not been introduced. An important aspect of incorporating programmable devices particularly into sophisticated designs is the ability to confirm that each such device is, in and of itself, capable of meeting the relevant specifications and expectations of the designer. Complex systems are generally challenging enough without introducing additional challenges in the form of anomalous or unintended behavior/consequences of a programmable device that is a subcomponent of the system.

Boundary scanning[5] is a technique that allows programmable devices to be tested externally, i.e., without access to the internal logic. Internal registers are provided by the device's manufacturer that allow testing of the internal logic and interconnections. However, the device is not aware that such scanning is taking place and therefore the tests can be carried out while the device under test (DUT) is operating in an unperturbed state, or states. PSoC 3/5LP include, within their respective architectures, a test controller that can be used to access the device's I/O pins for boundary testing by employing an internal, serial, shift register routed across all their pins and hence the name "boundary-scan".

The circuitry at each PSoC 3/5LP pin is supplemented with a multipurpose element called a boundary-scan cell and most GPIO and SIO port pins have a boundary-scan cell associated with them. The interface used to control the values in the boundary scan cells is called the Test Access Port (TAP) and is commonly known as the JTAG interface. It consists of three signals: (1) Test Data In (TDI), (2) Test Data Out (TDO), and (3) Test Mode Select (TMS). Also included is a clock signal (TCK) that clocks the other signals. TDI, TMS, and TCK are all inputs to the device, and TDO is output from the device as shown in Fig. 8.2.

This interface enables testing multiple ICs on a circuit board, in a daisy-chain fashion.

The TMS signal controls a state machine in the TAP. The state machine controls which register (including the boundary scan path) is in the TDI-to-TDO shift path, as shown in Fig. 8.3 for which:

- *ir* refers to the instruction register,
- *dr* refers to one of the other registers (including the boundary scan path), as determined by the contents of the instruction register,
- *capture* refers to the transfer of the contents of a *dr* to a shift register, to be shifted out on TDO (read the dr)

and,

- *update* refers to the transfer of the contents of a shift register, shifted in from TDI, to a dr (write the dr)

---

[5]PSoC 3/5LP support boundary-scanning in accordance with the JTAG IEEE Standard 1149.1 – 2001 Test Access Port and Boundary-Scan Architecture.

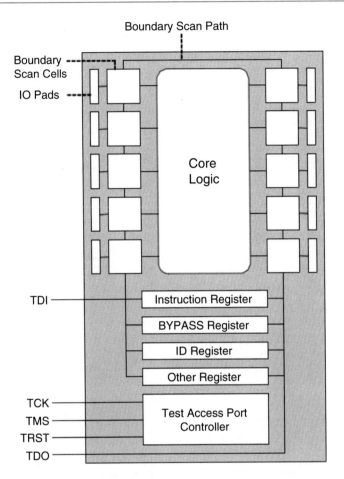

Fig. 8.2 PSoC 3/5LP JTAG interface architecture.

The registers in the TAP are:

- *Instruction*—Typically 2 to 4 bits wide and holds the current instruction that defines which data register is placed in the TDI-to-TDO shift path.
- *Bypass*—1 bit wide, directly connects TDI with TDO, causing the device to be bypassed for JTAG purposes.
- *ID*—32 bits wide and used to read the JTAG manufacturer/part number ID of the device.
- *Boundary Scan Path (BSR)*—Its width equals the number of I/O pins that have boundary-scan cells, used to set or read the states of those I/O pins.

Other registers may be included in accordance with device manufacturer specifications. The standard set of instructions (values that can be shifted into the instruction register), as specified in IEEE 1149, are:

- *EXTEST*—Causes TDI and TDO to be connected to the *boundary-scan path* (BSR). The device is changed from its normal operating mode to a test mode. Then, the device's pin states can be

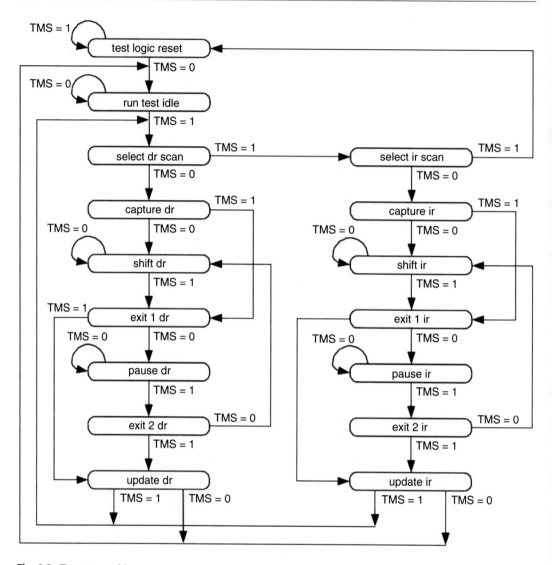

Fig. 8.3 Tap state machine.

sampled using the capture *dr* JTAG state, and new values can be applied to the pins of the device using the update dr state.

- *SAMPLE*—Causes TDI and TDO to be connected to the BSR, but the device is left in its normal operating mode. During this instruction, the BSR can be read by the capture *dr* JTAG state, to take a sample of the functional data entering and leaving the device.
- *PRELOAD*—Causes TDI and TDO to be connected to the BSR, but the device is left in its normal operating mode. The instruction is used to preload test data into the BSR prior to loading an EXTEST instruction. Optional, but commonly available, instructions are:
- *IDCODE*—Causes TDI and TDO to be connected to an IDCODE register.
- *INTEST*—Causes TDI and TDO to be connected to the BSR. While the EXTEST instruction allows access to the device pins, INTEST enables similar access to the core logic signals of a device.

8.3   Macrocells, Logic Arrays and UDBs

Combining gate arrays and macrocells[6] provides a significantly higher level of functionality, particularly when the macrocells include registers, ALUs, flip-flops, etc., than that obtainable through the use of gates alone. One of the simplest configurations of such a combination consists of a *sum-of-products* (SoP) combinatorial logic function and a flip-flop. An example of a device employing multiple such macrocells, as first defined, cells, Logic arrays is shown in Fig. 8.4.

Gates and macrocells serve as fundamental building blocks of PLDs. Combinations of macrocells and gates can be configured in arbitrarily large arrays to provide very complex sequential and combinatorial logic. Some configurations are driven by lookup tables and programmable memory. The information stored in the LUTs can be loaded into memory as required to provide the required logic functions. While in principle the number of inputs to a macrocell are unlimited and therefore should make it possible to have arbitrarily complex functions, a linear increase in fan-ins[7] results in a geometric increase in the number of bits to be stored in the LUT.

PSoC Creator allows the designer to employ universal digital blocks based on macrocell-gate array combinations that are not only configurable but are specifically designed to serve as customizable blocks within PSoC 3/5LP for a broad range of embedded system applications that incorporate a microcontroller and associated peripherals. These blocks, referred to as UDBs consist of a combination of uncommitted logic similar to programmable logic devices, structured logic (databases) and a flexible routing scheme. The UDBs can be further enhanced and supplemented by Boolean elements constructed from basic logic functions supported by PSoC Creator such as AND, NAND, OR, NOR, NOT, XOR, XNOR, D flip-flops, etc.[8] Boolean functions can be created using these basic logic functions to provide the additional functionality that is required for specific applications. Thus designers can create sophisticated systems using the standard set of PSoC 3/5LP blocks or create combinations of Boolean elements and UDBs to provide the required functionality by employing Verilog/Warp.[9]

PSoC 3 has 24 UDBs and in the case of pulse width modulators (PWMs), PSoC Creator will allow the creation of as many as 24 PWMs each of which has two independent outputs. Thus it is possible to have 48 PWM outputs.[10] It is also possible to use the 24 UDBs to configure 12 UARTs in a single PSoC 3/5LP device.

In addition to UDBs, that can be used to provide programmable peripheral functions, PSoC 3/5LP also includes a suite of user-configurable blocks that provide a wide range of additional capability, e.g., analog, CapSense, communications, digital logic, displays, filters, ports/pins and system blocks as shown in Fig. 8.5.

---

[6]In the present discussion, the term macrocell refers solely to a combination of flip-flops and I/O devices exclusive of logic arrays and OR-gates. However, some definitions of the term macrocell represent a broader definition and include all the logic required to provide the Boolean functionality, flip-flops and I/O other than that provided by the logic array. The latter definition is intended to completely encapsulate all of a particular functionality and is often referred to as a "block". In the case of PSoC Creator, a comprehensive set of such "building blocks" are provided, and they are referred to as UDBs, or universal digital blocks.

[7]Fan-in is defined as the number of inputs to a gate, or other device.

[8]Throughout this textbook both upper and lower case will be used when referring to logical operators primarily as a notational convenience.

[9]The discussion of Verilog/Warp begins in Sect. 6.

[10]An additional four, single-output PWMs are also available by using PSoC 3's fixed-function counter/timer/PWMs.

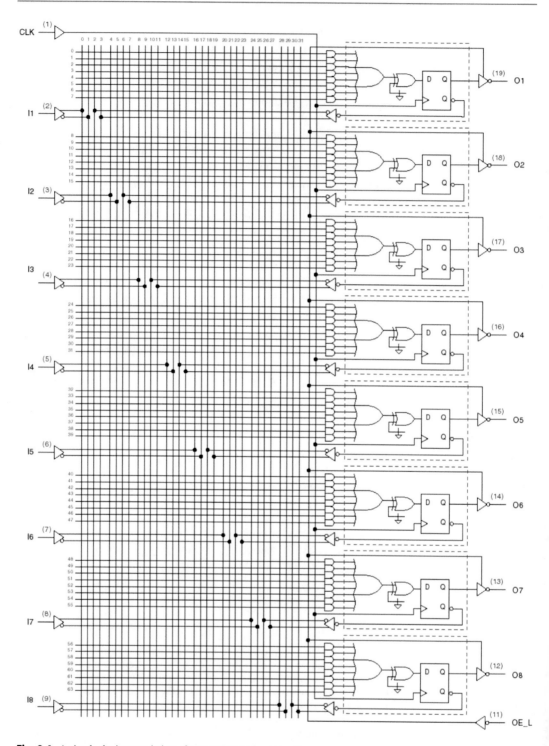

Fig. 8.4 A simple device consisting of gate arrays and macrocells.

All these blocks, i.e., digital, analog and UDBs, are interoperable and, in addition, external components such as resistors and capacitors can be used to further extend PSoC 3/5LP's capability as is shown in Chap. 11.

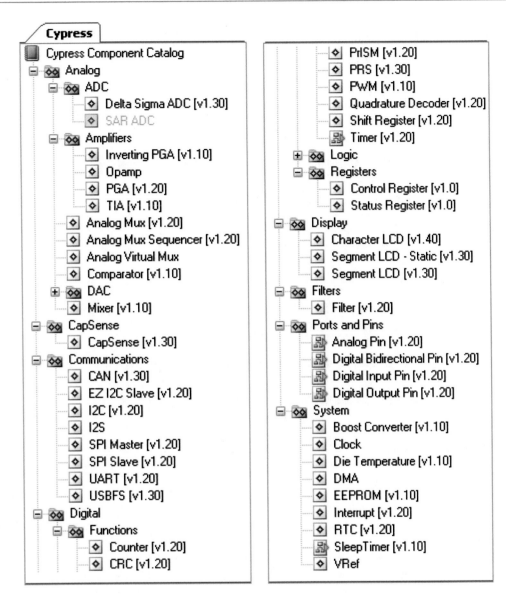

Fig. 8.5 Basic building blocks supported by PSoC creator.

UDB blocks support the following:

– Universal digital block arrays as large as 64 UDBs.
– Portions of UDBs can be either chained, or shared, to enable larger functions.
– Multiple digital functions supported by the UDBs include timers, counters, PWMs (with dead-band generator), UART, SPI, and CRC generation/checking.
– Each UDB includes:
    * an ALU-based, 8-bit datapath

* Two fine-grained PLDs[11]
* A control and status module
* A clock and reset module

as shown in Fig. 8.6.

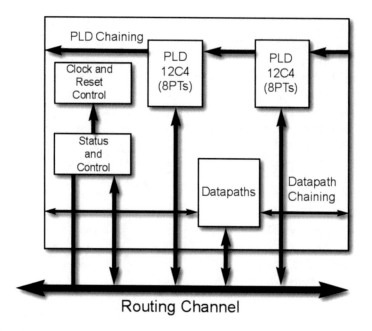

Fig. 8.6 UDB block diagram.

The UDB consists of a pair of PLDs, a datapath and control, status clock and reset functions. The PLDs accept input from the routing and form registered, or combinational sum-of-products logic, to implement state machines,[12] control datapath operations, condition inputs, and drive outputs. The datapath block contains a dynamically programmable ALU, two FIFOs, comparators, and condition generation. The control and status registers provide a way for the CPU firmware to interact, and synchronize, with UDB operations. Control registers drive internal routing, and status registers read internal routing. The reset and clock control block provides clock selection/enabling, and reset selection, for the individual blocks in the UDB.

The PLDs and datapath have chaining signals that enable neighboring blocks to be linked to create higher precision functions. UDB I/Os are connected to the routing channel through a programmable switch matrix for connections between blocks in one UDB, and to all other UDBs in the array. All registers and RAM in each UDB are mapped into the system address space and are accessible as both 8- and 16-bit data.

---

[11] Fine-grained in the present context refers to the implementation of relatively large numbers of simple logic modules, as opposed to coarse-grained, which implies relatively fewer but larger logic modules, often each with two or more sequential logic elements.

[12] State machines are discussed in Sect. 5.3.2.

In addition to a UDB's datapath, status register, control register and 2 PLDs, there is also a *count7* down-counter available that uses certain resources in the UDB, i.e., the control register, the status register's mask register and if a routed load or enable is used, the status register's inputs. In the latter case if the inputs are not used by the count7, the status register remains available for use.[13] In PSoC Creator *count7* can be implemented as shown in Fig. 8.7.

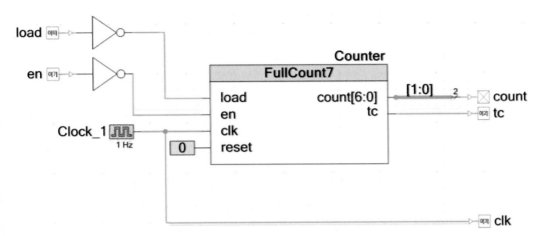

Fig. 8.7 Implementation of a *count7* down-counter in PSoC creator.

Figure 8.8 shows the internal structure of the PLDs. They can be used to implement state machines, perform input or output data conditioning, and to create lookup tables (LUTs). The PLDs may also be configured to perform arithmetic functions, sequence the datapath, and generate status. General-purpose RTL[14] can be synthesized and mapped to the PLD blocks. Each has 12 inputs that feed across eight product terms (PT) in the AND array. In a given product term, the true (T) or complement (C) of the input can be selected. The output of the PTs are inputs into the OR array. The letter C in 12C4 indicates that the OR terms are constant across all inputs, and each OR input can programmatically access any, or all, of the PTs. This structure gives maximum flexibility and ensures that all inputs and outputs are permutable.[15]

PSoC 3/5LP's macrocell architecture is shown in Fig. 8.9. The output drives the routing array, and can be registered or combinational. The registered modes are D Flip-Flop with true or inverted input, and Toggle Flip-Flop on input high or low. The output register can either be set or reset for purposes of initialization, or asynchronously during operation under the control of a routed signal. Outputs of the two PLDs are mapped into the address space as an 8-bit, read-only, UDB working register, that is directly addressable by the CPU's firmware, as shown in Fig. 8.10. The PLDs are chained together (the PLD carry chain) in UDB address order. The carry chain input is routed from the previous UDB in the chain, through each macrocell in both of the PLDs, and then to the next UDB as the carry chain out. To support the efficient mapping of arithmetic functions, special product terms are generated and used in the macrocell in conjunction with the carry chain.

---

[13] However, the status register's interrupt capability (status) is not available for use under these circumstances.

[14] RTL refers to register-level-transfer with respect to Verilog [7] code that describes the transformation of data as it is passed from register-to-register.

[15] Note that there are four outputs: OUT0, OUT1, OUT2 and OUT3.

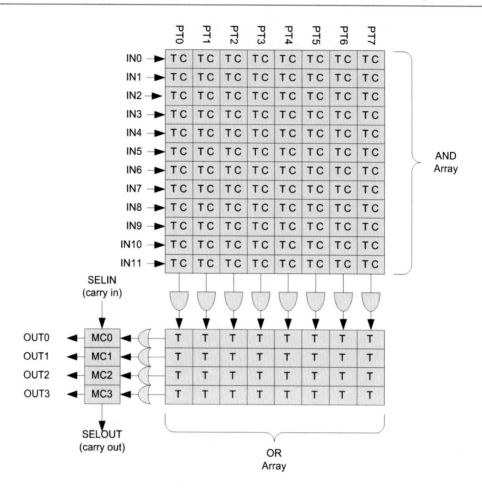

Fig. 8.8 PLD 12C4 structure.

8.4   The Datapath

The datapath, shown in Fig. 8.11, contains an 8-bit single-cycle ALU,[16] with associated compare and condition generation circuits. A datapath may be chained with databases in neighboring UDBs to achieve higher precision functions. The datapath includes a small, RAM-based control store,[17] that can dynamically select the operation and configuration to perform in a given cycle. The datapath is optimized to implement typical embedded functions, such as timers, counters, PWMs, PRS, CRC, shifters and dead-band generators. The add and subtract functions allow support for digital Delta Sigma operations.

---

[16]The single-cycle, arithmetic logic units (ALUs) fetch, execute and store results in a single clock-cycle.

[17]The control store holds the micro-instructions that are used to implement the ALU's instruction set. Some ALUs have been implemented with writable control stores that allow the instruction set to be altered in real-time.

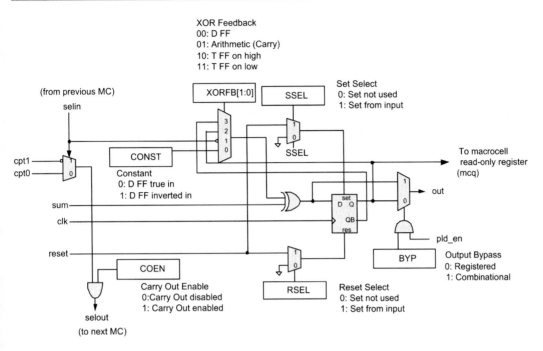

Fig. 8.9 PSoC 3/5LP macrocell architecture.

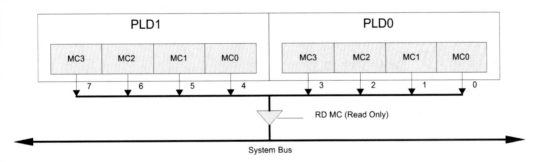

Fig. 8.10 Macrocell architecture read-only register.

Dynamic configuration, or perhaps more appropriately "dynamic reconfiguration", refers to the ability to change the datapath functions and interconnections, on a cycle-by-cycle basis under sequencer control. This is implemented using the configuration RAM, which stores eight 16-bit wide configurations. The address input to this RAM can be routed from any block connected to the digital peripheral fabric, most typically PLD logic, I/O pins, or other databases.

The ALU can perform eight general-purpose functions: increment, decrement, add, subtract, AND, OR, XOR, and PASS. Function selection is controlled by the configuration RAM on a cycle-by-cycle basis. Independent shift (left, right, nibble swap) and masking operations are available at the output of the ALU.

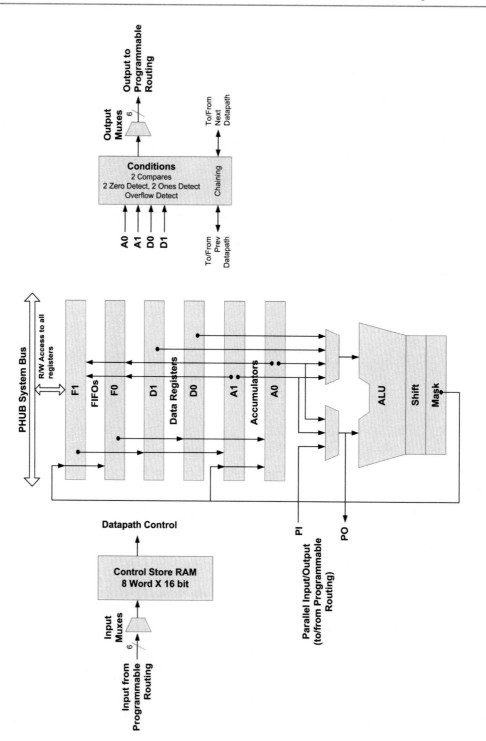

Fig. 8.11 Datapath (top level).

Each datapath has two comparators, with bit-masking options, that can be configured to select a variety of datapath register inputs for comparison. Other detectable conditions include all zeros, all ones, and overflow. These conditions form the primary datapath output selects to be routed to the digital peripheral fabric as outputs, or inputs, to other functions.

The datapath has built-in support for single-cycle Cyclic Redundancy Check (CRC) computation and PseudoRandom Sequence (PRS)[18] generation of arbitrary width and polynomial specification. To achieve CRC/PRS widths greater than 8 bits, signals may be chained between databases. This feature is controlled dynamically, and therefore can be interleaved with other functions. The most significant bit of an arithmetic and shift function can be programmatically specified (variable MSB). This supports variable width CRC/PRS functions and, in conjunction with ALU output masking, can implement arbitrary width timers, counters, and shift blocks.

8.4.1 Input/Output FIFOs

Each datapath contains two 4-byte FIFOs, that can be individually configured for direction as an input buffer (system bus writes to the FIFO, datapath internals read the FIFO), or an output buffer (datapath internals write to the FIFO, the system bus reads from the FIFO). These FIFOs generate status that can be routed to interact with sequencers, interrupt, or DMA requests.

8.4.2 Chaining

The datapath can be configured to chain conditions and signals with neighboring databases. Shift, carry, capture, and other conditional signals can be chained to form higher precision arithmetic, shift, and CRC/PRS functions.

In applications that are oversampled, or do not need the highest clock rates, the single ALU block in the datapath can be efficiently shared with two sets of registers and condition generators. ALU and shift outputs are registered and can be used as inputs in subsequent cycles. Usage examples include support for 16-bit functions in one 8-bit datapath, or interleaving a CRC generation operation with a data shift operation.

8.4.3 Datapath Inputs and Outputs

The datapath has three types of inputs: configuration, control, and serial/parallel data. The configuration inputs select the control store RAM address. The control inputs load the data registers from the FIFOs and capture accumulator outputs into the FIFOs. Serial data inputs include *shift-in* and *carry-in*. A parallel data input port allows up to eight bits of data to be brought in from routing.

There are a total of 16 signals generated in the datapath. Some of these signals are conditional signals, e.g., compares, some are status signals, e.g., FIFO status, and the rest are data signals, e.g., shift out. These 16 signals are multiplexed into the six datapath outputs and then driven to the routing matrix. By default, the outputs are single-synchronized (pipelined). A combinational output option is also available for these outputs.

---

[18]Pseudorandom refers to the fact that sequence is deterministic and at some point repeats itself. Section 12.9.11 presents a discussion of PRS generation techniques.

8.4.4   Datapath Working Registers

Each datapath module has six, 8-bit working registers all of which are readable and writable by the CPU and DMA:

- **Accumulator (A0,A1)**—The accumulators may be both a source and a destination for the ALU. They may also be loaded from a Data register, or a FIFO. The accumulators typically contain the current value of a function, such as a count, CRC, or shift.
- **Data (D0,D1)**—The Data registers typically contain constant data for a function, such as a PWM compare the value, timer period, or CRC polynomial.
- **FIFOs (F0,F1)**—The two 4-byte FIFOs provide both a source and a destination for buffered data. The FIFOs can be configured as one input buffer and one output buffer, two input or two output buffers. Status signals indicate the read and write status of these registers. The FIFOs can be used to buffer TX and RX data in the SPI or UART and PWM compare and timer period data.

Each FIFO has a variety of possible operational modes and configurations:

- **Input/Output**—In input mode, the system bus writes to the FIFO and the data is read and consumed by the datapath internals. In output mode, the FIFO is written to by the datapath internals and is read, and consumed, by the system bus.
- **Single Buffer**—The FIFO operates as a single buffer with no status. Data written to the FIFO is immediately available for reading, and can be overwritten at any time.
- **Level/Edge**—The control to load the FIFO from the datapath internals can be either level or edge-triggered.
- **Normal/Fast**—The control to load the datapath is sampled on the currently selected datapath clock (normal), or the bus clock (fast). This allows captures to occur at the highest rate in the system (bus clock), independent of the datapath clock.
- **Software Capture**—When this mode is enabled, and the FIFO is in output mode, a read by the CPU/DMA of the associated accumulator (A0 for F0, A1 for F1) initiates a synchronous transfer of the accumulator value into the FIFO. The captured value may then be immediately read from the FIFO by the datapath internals. If chaining is enabled, the operation follows the chain to the MS block for atomic reads by databases of multi-byte values.
- **Asynch**—When the datapath is being clocked asynchronously to the system clocks, the FIFO status for use by the datapath state machine (*blk_stat*) is resynchronized to the current DP clock.
- **Independent Clock Polarity**—Each FIFO has a control bit to invert the polarity of the FIFO clock with respect to the datapath clock.

The configurations controlled by the FIFO direction bit are shown in Fig. 8.12.

The TX/RX mode has one FIFO in input mode and the other in output mode. The primary use for this configuration is a serial peripheral interface (SPI) bus communications. The dual capture configuration provides independent capture of A0 and A1, or two separately controlled captures of either A0 or A1. The dual buffer mode provides buffered periods and compares, or two independent periods/compares.

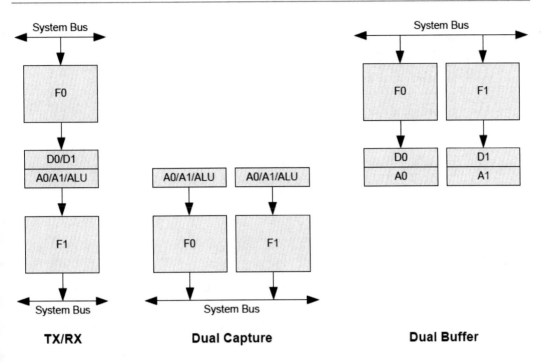

Fig. 8.12 FIFO configurations.

8.5 Datapath ALU

The Datapath block's ALU consists of three, independent, 8-bit, programmable functions that employ an arithmetic/logic, a shifter unit and a mask unit. The ALU functions shown in Table 8.1 are configured dynamically by the RAM control store.[19]

Table 8.1 ALU functions.

Func[2:0]	Function	Operation
000	PASS	srca
001	INC	++srca
010	DEC	--srca
011	ADD	scra + srcb
100	SUB	srca - srcb
101	XOR	srca ^ srcb
110	AND	srca & srcb
111	OR	srca | srcb

---

[19] "srca" and "srcb" refer to the ALU a and b inputs, respectively.

Table 8.2 Carry in functions.

Function	Operation	Default Carry In Implementation
INC	++srca	srca + ooh + ci　　(ci = 1)
DEC	--srca	srca + ffh + ci　　(ci=0)
ADD	srca + srcb	srca + srcb + ci　　(ci=0)
SUB	srca -srcb	srca + -srcb + ci　　(ci=1)

8.5.1　Carry Functions

The *carry-in* option is used in arithmetic operations. There is a default *carry-in* value for each function as shown in Table 8.2.

In addition to the default arithmetic mode for carry operation, there are three additional carry options, as shown in Table 8.3. The CI SELA and CI SELB configuration bits determine the *carry-in* for a given cycle. Dynamic configuration RAM selects either the A or B configuration on a cycle-by-cycle basis. When a routed carry is used, the meaning with respect to each arithmetic function is shown in Table 8.4.[20]

Table 8.3 Additional carry in functions.

Function	Carry Out Polarity	Carry Out Active	Carry Out Inactive
Inc	True	++srca == 0	srca
Dec	Inverted	--srca == -1	srca
ADD	True	(srca + srcb) > 255	srca+srcb
Sub	Inverted	(srca + srcb) < 0	scra-scrb

As shown in Figs. 8.8–8.13, the *carry out* option is a selectable datapath output and is derived from the currently defined MSB position, which is statically programmable. This value is also chained to the next most significant block as an optional *carry-in*. Note that in the case of decrement and subtract functions, the carry-out is inverted. Options for *carry-in*, and for MSB selection for carry-out generation, are shown in Fig. 8.13.

The registered carry-out value may be selected as the *carry-in* for a subsequent arithmetic operation. This feature can be used to implement higher precision functions in multiple cycles.

---

[20]Note that in the case of the decrement and subtract functions, the carry is active low (inverted).

Table 8.4 Routed carry in functions.

Function	Carry In Polarity	Carry In Active	Carry In Inactive
Inc	True	++srca	srca
Dec	Inverted	--srca	srca
ADD	True	(srca + srcb) + 1	srca+srcb
Sub	Inverted	(srca + srcb) - 1	scra-scrb

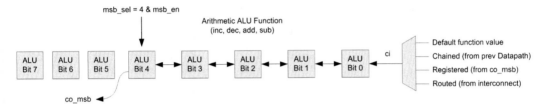

Fig. 8.13 Carry operation block diagram.

Additional *carry-in* functions are provided by CI SEL A and CI SEL B. The A value of 00 imposes the default carry mode. A value of '01' sets the carry mode as *registered* so that add with carry and subtract with borrow operations can be implemented. In this mode, the carry flag represents the result of the previous cycle. A value of '10' sets the *routed carry* mode for cases in which the carry is generated somewhere else and routed to the input allowing controllable counters to be implemented. Finally, the value '11' sets the *chainable* carry mode allowing the carry to be *chained* from the previous datapath and used to implement single-cycle operations of higher precision involving two or more databases.

8.5.2 ALU Masking Operations

An 8-bit mask register in the UDB static configuration register space defines the masking operation. In this operation, the output of the ALU is masked (ANDed) with the value in the mask register. A typical use for the ALU mask function is to implement free-running timers and counters in *powers-of-two* resolutions.

8.5.3 All Zeros and Ones Detection

Each accumulator has dedicated *all zeros* and *all ones* detect capability. These conditions are statically chainable as specified in UDB configuration registers. In addition, the requirement to chain, or not chain, these conditions are statically specified in UDB configuration registers. Chaining of zero detect is the same concept as the compare equal. Successive chained data is ANDed, if the chaining is enabled.

8.5.4   Overflow

Overflow is defined as the XOR of the carry into the MSB and the *carry-out* of the MSB. The computation is done on the currently defined MSB as specified by the MSB_SEL bits. Although this condition is not chainable, the computation is valid when done in the most significant datapath of a multi-precision function, as long as the carry is chained between blocks.

8.5.5   Shift Operations

Table 8.5 Shift functions.

Shift[1:0]	Function
00	Pass
01	Shift Left
10	Shift Right
11	Nibble Swap

Shift operations, shown in Table 8.5, can occur independently from those of the ALU. A *shift out* value is available as a datapath output. Both *shift out right* (sor) and *shift out left* (sol_msb) share that output selection. A static configuration bit (SHIFT_OUT in register CFG15) determines which shift output is used as a datapath output. In the absence of a shift, the sor and sol_msb signal is defined as the LSB[21] or MSB of the ALU function, respectively.

The SI SELA and SI SELB configuration bits determine the shift in data for a given operation. Dynamic configuration RAM selects the A or B configuration on a cycle-by-cycle basis. Shift in data is only valid for left- and right-shift; it is not used for pass and nibble swap. The selections and usage apply to both left and right-shift directions, and if for either SI SEL A or SI SEL B the bit values are 00, the shift in source is default/arithmetic, i.e., the default input is the value of the DEF SI configuration bit (fixed 0 or 1). However, if the MSB SI bit is set, then the default input is the currently defined MSB, but for right-shift only.

If the bit values are 01, then the shift in source is registered and the shift value is driven by the current registered shift-out value from the previous cycle. The shift-left operation uses the last shift-out left value. The right-shift operation uses the last shift-out right value. If the bit values are 10, then the shift-in source is *routed*. Shift is selected from the routing channel, i.e., the SI input. Finally, if the bit values are 11, the shift-in source is chained and shift-in left is routed from the right datapath neighbor.

The shift-out data comes from the currently defined MSB position and the data that is shifted in from the left (*shift-in right*) goes into the currently defined MSB position. Both shift-out data (left or right) are registered and can be used in a subsequent cycle. This feature can be used to implement a higher precision shift in multiple cycles. The bits that are isolated by the MSB selection are still shifted.

---

[21] The acronym for least significant byte is LSB.

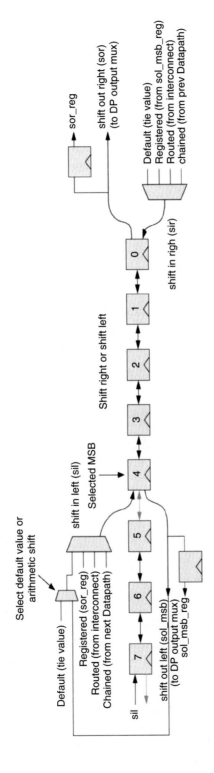

Fig. 8.14 Shift operation.

In the example shown in Fig. 8.14, bit 7 still shifts in the *sil* value on a right-shift and bit 5 shifts in bit 4, on a left shift. The *shift out,* either right or left, from the isolated bits is lost.

8.5.6 Datapath Chaining

As discussed previously, each datapath block contains an 8-bit ALU, which is designed to chain carries, shifted data, capture triggers, and conditional signals to the nearest neighbor databases to create higher precision arithmetic functions and shifters. These chaining signals, which are dedicated signals, allow single-cycle 16-, 24- and 32-bit functions to be efficiently implemented without the timing uncertainty of channel routing resources. In addition, the capture chaining makes possible an atomic read of the accumulators in chained blocks. As shown in Fig. 8.15, all generated conditional and capture signals chain in the direction of least to most significant blocks. Shift left also chains from the least-to-most significant block and shift right chains from the most-to-least significant block. The CRC/PRS chaining signal for feedback chains from the least-to-most significant block; the MSB output chains from the most-to-least significant block.

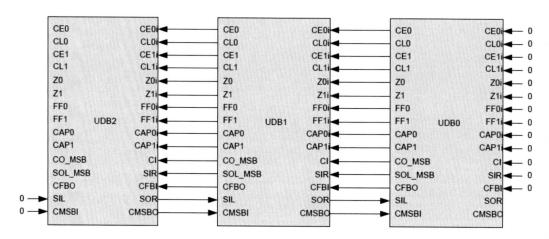

Fig. 8.15 Datapath chaining flow.

8.5.7 Datapath and CRC/PRS

The datapath has special connectivity to allow cyclic redundancy checking (CRC) and pseudo random sequence (PRS) generation. Chaining signals are routed between datapath blocks to support CRC/PRS bit lengths of more than 8 bits. The most significant bit (MSb) of the most significant block in the CRC/PRS computation is selected and routed, while chained across blocks, to the least significant block. The MSB is then XORed with the data input (SI data) to provide the feedback (FB) signal. The FB signal is then routed and chained across blocks to the most significant block. This feedback value is used in all blocks to gate the XOR of the polynomial from the Data0 or Data1 register with the current accumulator value.

Figure 8.16 shows the structural configuration for the CRC operation. The PRS configuration is identical except that the *shift in* (SI) is tied to '0'.

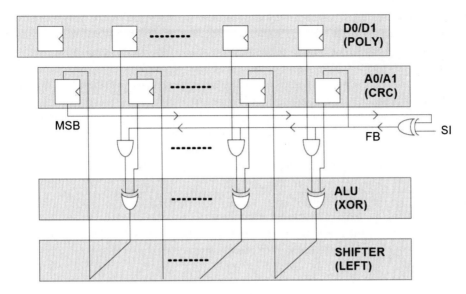

Fig. 8.16 CRC functional structure.

In the PRS configuration, D0 or D1 contains the polynomial value, while A0 or A1 contains the initial or seed[22] value and the CRC residual value at the end of the computation. To enable CRC operation, the CFB_EN bit in the dynamic configuration RAM must be set to '1'. This enables the AND of SRCB ALU input with the CRC feedback signal. When set to zero, the feedback signal is driven to '1', which allows for normal arithmetic operation. Dynamic control of this bit on a cycle-by-cycle basis gives the capability to interleave a CRC/PRS operation with other arithmetic operations.

8.5.8 CRC/PRS Chaining

Figure 8.17 illustrates an example of CRC/PRS chaining across three UDBs. This arrangement is capable of supporting a 17- to 24-bit operation. The chaining control bits are set according to the position of the datapath in the chain, as shown.

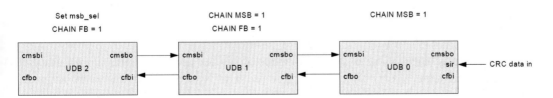

Fig. 8.17 CRC/PRS chaining configuration.

The CRC/PRS MSB signal (cmsbo, cmsbi) is chained based on the following:

---

[22]The phrase *seed value* is used throughout this textbook in various contexts and refers to an initial value with which to begin a process.

- If a given block is the most significant block, the MSb bit (according to the polynomial selected) is configured using the MSB_SEL configuration bits. If a given block is not the most significant block, the CHAIN MSB configuration bit must be set and the MSb signal is chained from the next block in the chain. If a given block is the least significant block, then the feedback signal is generated in that block from the built-in logic that takes the shift in from the right (*sir*) and XORs it with the MSb signal. (For PRS, the *sir* signal is tied to '0'.)
- If a given block is not the least significant block, the CHAIN FB configuration bit must be set and the feedback is chained from the previous block in the chain.

The CRC/PRS MSb signal (*cmsbo*, *cmsbi*) is chained based on the following:

- If a given block is the most significant block, the MSB bit (according to the polynomial selected) is configured using the MSB_SEL configuration bits. If
- If a given block is not the most significant block, the CHAIN MSB configuration bit must be set and the MSB signal is chained from the next block in the chain.

8.5.9　CRC/Polynomial Specification

The following is an illustrative example of how to configure the polynomial for programming into the associated D0/D1 register. Consider the CCITT[23] CRC-16 polynomial, which is defined as $x16 + x12 + x5 + 1$. The method for deriving the data format from the polynomial is shown in Fig. 8.18.

X^{16}	X^{15}	X^{14}	X^{13}	X^{12}	X^{11}	X^{10}	X^9	X^8	X^7	X^6	X^5	X^4	X^3	X^2	X^1	X^0
X^{16}	+			X^{12}			+				X^5			+		1
0	0	0	0	1	0	0	0	0	0	0	1	0	0	0	0	

CCITT 16-Bit Polynomial is 0x0810

Fig. 8.18 CCITT CRC 16 polynomial.

The X0 term is inherently always '1' and therefore does not need to be programmed. For each of the remaining terms in the polynomial, a '1' is set in the appropriate position in the alignment shown.[24]

---

[23]CCITT is an abbreviation for Comité Consultatif International Téléphonique et Télégraphhique which is an international standards organization involved in the development of communications standards.

[24]This polynomial format is slightly different from the format normally specified in HEX. For example, the CCITT CRC16 polynomial is typically denoted as 1021H. To convert it to the format required for datapath operation, shift right by one and add a '1' in the MSb location. In this case, the correct polynomial value to load into the D0 or D1 register is 1810H.

Assuming that D0 contains the polynomial and A0 is used to compute CRC/PRS a suitable polynomial has to be selected and written into D0. Next, a seed value is selected and written into A0.

8.5.10 External CRC/PRS Mode

A static configuration bit may be set (EXT CRCPRS) to enable support for the external computation of a CRC or PRS. As shown in Fig. 8.19, computation of the CRC feedback.

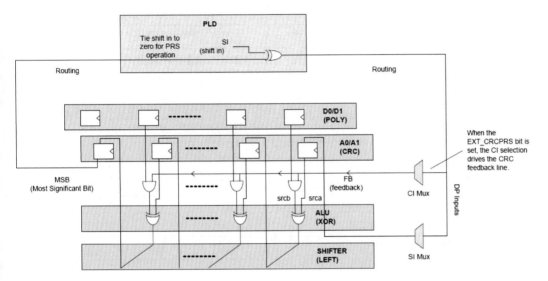

Fig. 8.19 External CRC/PRS mode.

This is done in a PLD block. When the bit is set, the CRC feedback signal is driven directly from the CI (*Carry In*) datapath input selection mux, bypassing the internal computation. The figure shows a simple configuration that supports up to an 8-bit CRC or PRS, inclusive. Normally the built-in circuitry is used, but this feature allows more elaborate configurations, such as up to a 16-bit, inclusive, CRC/PRS function in one UDB, using time-division multiplexing. In this mode, the dynamic configuration RAM bit CFB_EN still controls whether the CRC feedback signal is ANDed with the SRCB ALU input. Therefore, as with the built-in CRC/PRS operation, the function can be interleaved with other functions, if desired.

8.5.11 Datapath Outputs and Multiplexing

Datapath outputs and multiplexing conditions are generated from the registered accumulator values, ALU outputs, and FIFO status. These conditions can be driven to the UDB channel routing for use in other UDB blocks as interrupts, DMA requests, or applied to globals and I/O pins. The 16 possible conditions are shown in Table 8.6. Conditions are generated from the registered accumulator values, ALU outputs, and FIFO status. These conditions can be driven to the UDB channel routing for use in other UDB blocks, use as interrupts or DMA requests, or to globals and I/O pins. The 16 possible conditions are shown in Table 8.6.

Table 8.6 Datapath condition generation.

Condition	Chain ?	Description
Compare Equal	Y	A0==D0
Compare Less Than	Y	A0<D0
Zero Detect	Y	A0==OOh
Ones Detect	Y	A0==FFh
Compare Equal	Y	A1 or A0 == D1
Compare Less Than	Y	A1 or A0 < D1
Zero Detect	Y	A1 == 00h
Ones Detect	Y	A1 == FFh
Overflow	N	Carry(msb) ^ Carry(msb-1)
Carry Out	Y	Carry out of MSB defined bit
CRC MSB	Y	MSB of CRC/PRS function
Shift Out	Y	Selection of shift Output
FIFOO Block Status	N	Depends on FIFO Config
FIFO1 Block Status	N	Depends on FIFO Config
FIFOO Block Status	N	Depends on FIFO Config
FIFO1 Block Status	N	Depends on FIFO Config

There are a total of six datapath outputs. Each output has a 16–1 multiplexer that allows any of these 16 signals to be routed to any of the datapath outputs.

8.5.12 Compares

There are two compares, one of which has fixed sources (Compare 0) and the other has dynamically selectable sources (Compare 1). Each compare has an 8-bit, statically-programmed, mask register, which enables the compare to occur in a specified bit field. By default, the masking is off (all bits are compared) and must be enabled. Comparator 1 inputs are dynamically configurable. As shown in Table 8.7, there are four options for Comparator 1, which applies to both the *less than* and *equal* conditions.

The CMP SELA and CMP SELB configuration bits determine the possible compare configurations. A dynamic RAM bit selects one of the A or B configurations on a cycle-by-cycle basis.

Compare 0 and Compare 1 are independently chainable to the conditions generated in the previous datapath (in addressing order). Whether to chain compares, or not, is statically specified in the UDB configuration registers. Figure 8.20 illustrates *compare equal* chaining, which is just an ANDing of the *compare equal* in this block with the chained input from the previous block.

Table 8.7 Compare configurations.

CMP SEL A CMP SEL B	Compare 1 Compare Configuration
00	A1 compare to D1
01	A1 compare to A0
10	A0 compare to D1
11	A0 compare to A0

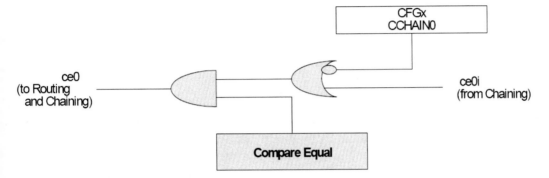

Fig. 8.20 Compare *Equal* chaining.

Figure 8.21 illustrates *compare less than* chaining. In this case, the *less than* is formed by the *compare less than* output in this block, which is unconditional. This is ORed with the condition where this block is equal, and the chained input from the previous block is asserted as less than.

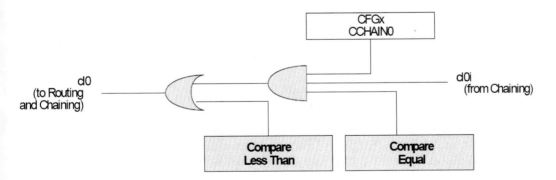

Fig. 8.21 Compare *Less Than* chaining.

8.6   Dynamic Configuration RAM (DPARAM)

Each datapath contains a 16 bit-by-8 word dynamic configuration RAM, which is shown in Fig. 8.22.

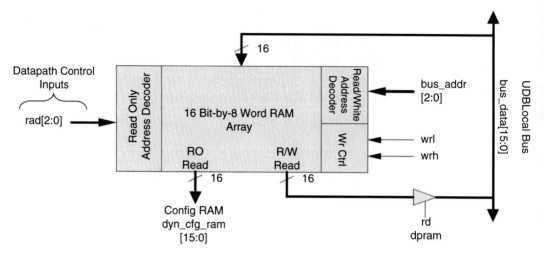

Fig. 8.22 Configuration RAM I/O.

The purpose of this RAM is to control the datapath configuration bits on a cycle-by-cycle basis, based on the clock selected for that datapath. This RAM has synchronous read and write ports for purpose of loading the configuration via the system bus. An additional asynchronous read port is provided as a fast path to output these 16-bit words as control bits to the datapath. The asynchronous address inputs are selected from datapath inputs and can be generated from any of the possible signals on the channel routing, including I/O pins, PLD outputs, control block outputs, or other datapath outputs. The primary purpose of the asynchronous read path is to provide a fast single-cycle decode of datapath control bits.

8.7   Status and Control Mode

When operating in status and control mode, this module functions as a status register, interrupt mask register, and control register in the configuration shown in Fig. 8.23.

8.7.1   Status Register Operation

One 8-bit, read-only status register is available for each UDB and inputs to this register come from any signal in the digital routing fabric. The status register is non-retentive, i.e., it loses its state during sleep intervals and is reset to 0x00 upon awakening. Each bit can be independently programmed to operate in one of two ways, 1) for STAT MD = 0, a read returns the current value of the routed signal (transparent) and 2) for STAT MD = 1, a high on the internal net is sampled and captured (sticky, clear on read).[25]

An important feature of the status register clearing operation is that clearing of status is only applied to the bits that are set. This allows other bits that are not set to continue to capture status, and a coherent view of the process can be maintained.

---

[25] It is cleared when the register is read.

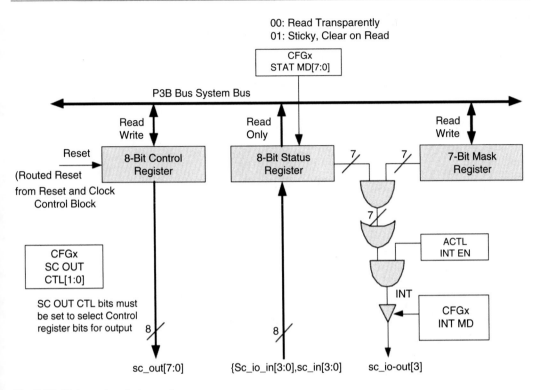

Fig. 8.23 Status and control operation.

8.7.2 Status Latch During Read

Figure 8.24 shows the structure of the status read logic. The sticky status register is followed by a latch, which latches the status register data and holds it stable during the duration of the read cycle, regardless of the number of wait states in a given read.

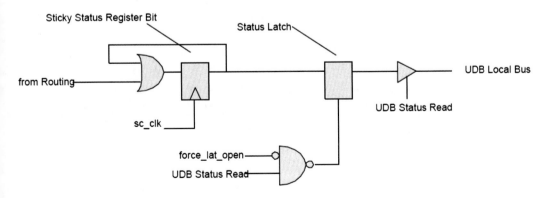

Fig. 8.24 Status read logic.

8.7.3 Transparent Status Read

By default, a CPU read of this register transparently reads the state of the associated routing net. This mode can be used for a transient state that is computed and registered internally in the UDB.

8.7.4 Sticky Status, with Clear on Read

In this mode, the associated routing net is sampled on each cycle of the status and control clock. If the signal is high in a given sample, it is captured in the status bit and remains high, regardless of the subsequent state of the associated route. When CPU firmware reads the status register, the bit is cleared. The status register clearing is independent of mode and occurs even if the block clock is disabled; it is based on the bus clock and occurs as part of the read operation.

8.8 Counter Mode

When a UDB is in counter mode, the control register operates as a 7-bit down-counter with programmable period and automatic reload that can be used for UDB internal operations or firmware applications. 7-bit down-counter. Routing inputs can be configured to control both the enable and reload of the counter. When enabled, control register operation is not available.
The counter has the following features:

- a 7-bit read/write period register, a 7-bit count register that is read/write but can only be accessed when the counter is disabled,
- automatic reload of the period to the count register on terminal count (0),
- a firmware control bit in the Auxiliary Control working register called CNT START, to start and stop the counter,[26]
- selectable bits from the routing for dynamic control of the counter enable and load functions: EN, routed enable to start or stop counting and LD, routed load signal to force the reload of the period,[27]
- it is level sensitive and continues to load the period while asserted. The 7-bit count may be driven to the routing fabric as sc_out[6:0],
- the terminal count may be driven to the routing fabric as sc_out[7].

To enable this mode, the SC_OUT_CTl[1:0] bits must be set to counter output. In this mode the normal operation of the control register is not available. The status register can still be used for read operations, but should not be used to generate an interrupt because the mask register is reused as the counter period register. The use of SYNC mode depends on whether the dynamic control inputs (LD/EN) are used. If they are not used, SYNC mode is unaffected. If they are used, SYNC mode is unavailable.

---

[26]This is an overriding enable and must be set for optional routed enable to be operational.
[27]When this signal is asserted, it overrides a pending terminal count.

8.8.1 Sync Mode

As shown in Fig. 8.25, the status register can operate as a 4-bit double synchronizer, clocked by the current SC_CLK, when the SYNC MD bit is set. This mode may be used to implement local synchronization of asynchronous signals, such as GPIO inputs. When enabled, the signals to be synchronized are selected from UDB pins SC_IN[3:0], the outputs are driven to the SC_IO_OUT[3:0] pins, and SYNC MD automatically puts the SC_IO pins into output mode. When in this mode, the normal operation of the status register is not available, and the status sticky bit mode is forced off, regardless of the control settings for this mode. The control register is not affected by the mode. The counter can still be used with limitations. No dynamic inputs (LD/EN) to the counter can be enabled in this mode.

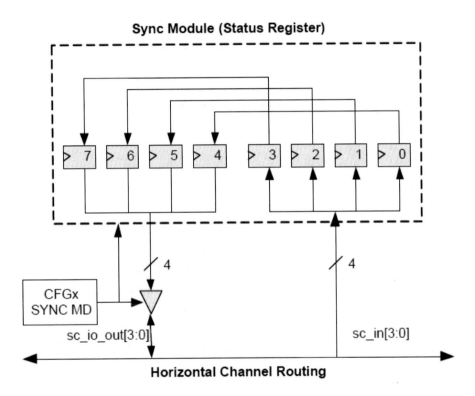

Fig. 8.25 Sync mode.

8.8.2 Status and Control Clocking

The status and control registers require a clock selection for any of the following operating modes:

- Control register in counter mode
- Status register with any bit set to *sticky*[28]

---

[28] Sticky bits are defined as bits that retain their current value until they are reset, e.g., by the CPU.

- Sync mode

The clock for this block is allocated in the reset and clock control module.

8.8.3   Auxiliary Control Register

An Auxiliary Control is a read-write register that controls hardware in the UDB and permits the CPU/DMA firmware to dynamically control the interrupt, FIFO, and counter fixed-function hardware. The CPU and/or the DMA are used to dynamically control the interrupt, counter and FIFO functions. The description of the individual control bits is described in Fig. 8.26.

Auxilliary Control Register							
7	6	5	4	3	2	1	0
		START CNT	INT EN	FIFO1 LVL	FIFO0 LVL	FIFO1 CLR	FIFO0 CLR

Fig. 8.26   Auxiliary control register bit designations.

8.8.3.1 FIFO0 Clear, FIFO1 Clear

The FIFO0 CLR and FIFO1 CLR bits reset the state of the associated FIFO. A '1' written to these bits, clears the state of the associated FIFO. However, these bits must be reset, i.e., zeroed, for FIFO operation to continue. When these bits are left asserted, the FIFOs operate as simple one-byte buffers, without status. FIFO0 Level, FIFO1 Level The FIFO0 LVL and FIFO1 LVL bits control the level at which the 4-byte FIFO asserts bus status (when the bus is either reading or writing to the FIFO) to be asserted. The FIFO bus status depends on the configured direction The register bits and descriptions are shown in Fig. 8.27.

FIFO x LVL	Input Mode (Bus is Writing FIFO)	Output Mode (Bus is Reading FIFO)
0	**Not Full** At least one byte can be written	**Not Empty** At least one byte can be read
1	**At Least Half Empty** At least 2 bytes can be written	**At Least Half Full** At least 2 bytes can be read

Fig. 8.27   FIFO level control bits.

8.8.3.2 Interrupt Enable

If the status register's generation logic is enabled, the INT EN bit gates the resulting interrupt signal.The Count Start The CNT START bit enables and disables the counter provided that the SC_OUT_CTL[1:0] bits are configured for counter output mode.

8.8.3.3 Counter Control

CNTSTART bit is used to enable/disable the counter provided that the SC_OUT_CTL[1:0] configured appropriately, i.e., set for counter output mode.

8.9 Boolean Functions

George Boole[29] published a seminal work in 1847, entitled "The Mathematical Analysis of Logic" that was followed by a second equally important work in 1854 entitled "An Investigation of the Laws of Thought, on Which Are Founded the Mathematical Theories of Logic and Probabilities". His approach was to develop a type of linguistic algebra[30] based on the three constructs AND ($A \cdot B$), OR ($A + B$) and NOT ($\overline{A}$).[31] He was then able to show that they could be used to carry out basic mathematical functions and comparisons. Thus, it became possible to express logical statements in terms of algebraic equations. His work ultimately formed the basis for much of modern computer technology. Claude Elwood Shannon[32] was the first to use Boolean algebra in describing digital circuits.

Simply stated, Boolean Algebra allows any computable algorithm, or realizable digital circuit, to be expressed as a system of Boolean equations. AND, OR and NOT can be easily constructed from NAND gates which is equivalent to an AND gate followed by a NOT gate, as shown in Fig. 8.28.

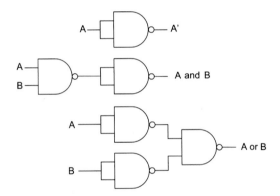

Fig. 8.28 The NAND gate as the basic building block for AND, OR and NOT (inverter) gates.

Having a single type of building block, i.e., NAND gates, as the basis for these three functions allows very complex circuits to be created from the same building block [8]. Boolean functions operate on Boolean variables and the resulting value of a Boolean function is either one or zero. The formal definition of a Boolean function is given by:

---

[29]George Boole (1815–1864), a mathematician, introduced not only a seminal theory of symbolic logic which was ultimately to be known as Boolean logic, but also two important treatises on differential equations and the calculus of finite differences.

[30]Which ultimately became universally referred to as "Boolean Algebra".

[31]Note that although AND and OR are binary operators, NOT is a unary operator since it operates on only one operand.

[32]Claude E. Shannon is regarded by many as the founding father of the electronic communications age. Both a mathematician and an engineer, he applied Boole's logical algebra to telephone switching circuits and authored a classic paper entitled "A Symbolic Analysis of Relay and Switching Circuits". His work on information theory beginning with his two-part paper entitled "A Mathematical Theory of Communication" continues to be widely studied and has contributed much to the evolution of modern computer technology.

A Boolean function is a mapping from the Cartesian product $x^n\{0, 1\}$ to $\{0, 1\}$, i.e., a function $F : x^n\{0, 1\} \Rightarrow set\,B = \{0, 1\}$ where $x^n\{0, 1\}$ is the set of all n-tuples $\{x_1, x_2 \cdots .x_n\}$ and the x_n are either one or zero.

The set B = $\{0,1\}$ is arguably one of the most used sets in the world. Boolean algebra provides the operations and rules for working with this set and forms the foundation for the development and use of digital circuits and for VLSI design. A Boolean algebra consists of a set of operators and a set of axioms. The operators for the Boolean algebra to be discussed here are $+$, $\cdot$ and $'$ for OR, AND and the complement,[33] respectively. The order of precedence for these operators is complement, product and then sum.

The set of postulates includes:

* closure,[34]
* the existence of identity elements for AND (1), and, OR (0) but not for NOT,
    $(A + 0 = A, A \cdot 1 = A)$
* associative: A + (B+C) = (A+B) +C,
* commutative: $A + B = B + A$ and $A \cdot B = B \cdot A$,
* distributive: $A + (B \cdot C) = (A + B) \cdot (A + C)$ and $A \cdot (B + C) = (A \cdot B) + (A \cdot C)$,

and,

* inverse: $A + A' = 1$ and $A \cdot A' = 0$.

Boolean expressions are given either in terms of *minterms* or *maxterms* which are defined respectively as the *product* of N literals, each of which occurs only once, and the *sum* of N literals, each of which occurs only once. A *literal* is a variable within a term of the expression that may be complemented. A Boolean function is a mapping from a domain consisting of n-tuples of zeros and ones to a range consisting of an element of B. Boolean functions can be expressed as a *sum-of-products* (SoP)

$$F = (\overline{A} \cdot B) + (A \cdot \overline{B})\tag{8.1}$$

or as a *product-of-sums* (POS)

$$F = (A + B) \cdot (\overline{A} + \overline{B})\tag{8.2}$$

It is possible to derive a sum-of-products expression for any digital logic circuit, no matter how complex, provided that a description of it in the form of a truth table exists. However, using sum-of-products does not guarantee that the end result will be an optimal design. This is of concern because as a practical matter minimizing the number of gates required can result in very significant reductions in cost, better performance and often increased speed. What may appear to the casual observer, in what follows, addition and multiplication operations is in fact the operations of OR and AND, respectively. Various notations have been adopted for these operations, e.g.,

$$A \cdot B = AB = A\ \textbf{OR}\ B = A \vee B\tag{8.3}$$

---

[33]The complement of one is zero and the complement of zero is one.

[34]x is a Boolean variable if, and only if (iff), its values are restricted to elements of B under AND, OR and NOT.

$$A + B = A \textbf{ AND } B = A \wedge B \tag{8.4}$$

The AND and OR operators are associative,

$$(A \cdot B) \cdot C = A \cdot (B \cdot C) \tag{8.5}$$

$$(A + B) + C = A + (B + C) \tag{8.6}$$

commutative,

$$A \cdot B = B \cdot A \tag{8.7}$$

$$A + B = B + A \tag{8.8}$$

and distributive,

$$A \cdot (B + C) = (A \cdot B) + (A \cdot C) \tag{8.9}$$

$$A + (B \cdot C) = (A + B) \cdot (A + C) \tag{8.10}$$

In addition, for any value A there exists and A' such that $A + A' = 1$ and $A \cdot A' = 0$. All of which leads to some very important and useful results, e.g.,

$$A \cdot B = B \cdot A$$
$$A + B = B + A$$
$$A \cdot (B \cdot C) = (A \cdot B) \cdot C$$
$$A + (B + C) = (A + B) + C$$
$$A \cdot (B + C) = (A \cdot B) + (A \cdot C)$$
$$A + (B \cdot C) = (A + B) \cdot (A + B)$$
$$A \cdot A = A$$
$$A + A = A$$
$$A \cdot (A + B) = A$$
$$A + (A \cdot B) = A$$
$$A \cdot A' = 0$$
$$A + A' = 1$$
$$(A')' = A$$
$$(A \cdot B)' = A' + B'$$
$$(A + B)' = A' \cdot B'$$
$$A + 1 = 1$$
$$A \cdot 1 = A$$
$$A \cdot 0 = 0$$

$$A + 0 = A$$

$$\cdots$$

It should be noted that for any valid Boolean expression, if the + operators in the expression are replaced by · operators, the · operators by + operators and 0s for 1s, and 1s for 0s, the result is also a valid Boolean expression, although the values of the two expressions may not be the same. This property is referred to as *duality*.

DeMorgan's Theorem[35] states that the complement of the product of variables is equal to the sum of the complements of the variables and conversely the complement of the sum of variables is equal to the product of the complements of the variables,[36] i.e.,

$$\overline{A + B} = \overline{A} \cdot \overline{B} \tag{8.11}$$

and

$$\overline{A \cdot B} = \overline{A} + \overline{B} \tag{8.12}$$

which often makes it possible to simplify Boolean expressions and thereby simplify the logic. Some of the most important algebraic rules for Boolean functions are shown in Table 8.8.

Table 8.8 Algebraic rules for Boolean functions.

Associative	$(A \cdot B) \cdot C = A \cdot (B \cdot C)$	$(A + B) + C = A + (B + C$
Distributive	$A \cdot (B + C) = (A \cdot B) + (A \cdot C)$	$A \cdot (B + C) = (A \cdot B) + (A \cdot C)$
Idempotent	$A \cdot A = A$	$A + A = A$
Double Negation	$\overline{(\overline{A})} = A$	—
DeMorgan's	$\overline{A \cdot B} = \overline{A} + \overline{B}$	$\overline{A + B} = \overline{A} + \overline{B}$
Commutative	$A \cdot B = B \cdot A$	$A + B = B + A$
Absorption	$A + (A \cdot B) = A$	$A \cdot (A + B) = A$
Bound	$A \cdot 0 = 0$	a
Negation	$A \cdot (\overline{A}) = 0$	$A + \overline{A} = 1$

8.9.1 Simplifying/Constructing Functions

A function can be expressed as a logic (circuit) diagram, truth table or expression. Logic diagrams show how the individual gates are interconnected. Examples of truth tables are shown in Tables 8.9 and 8.10. The number of possible functions given n inputs and m outputs can be expressed as

$$N = 2^{m2^n} \tag{8.13}$$

---

[35] Augustus De Morgan (1806–1871). DeMorgan was a British mathematician and logician, born in India, who was a contemporary of Charles Babbage and William Hamilton He introduced the phrase *mathematical induction* and served as a significant reformer of mathematical logic. He is best remembered for his work on purely symbolic algebras, De Morgan's laws and symbolic logic.

[36] Note that in general $\overline{A \cdot B \cdot C \cdots} = \overline{A} + \overline{B} + \overline{C} + \cdots$ and $\overline{A + B + C + \cdots} = \overline{A} \cdot \overline{B} \cdot \overline{C} \cdots$.

so that for 2 inputs and one output there are $2^2 = 4$ functions, for two inputs and two output there are $2^8 = 256$ functions, for 3 inputs and two outputs there are $2^{16} = 65,536$ functions, etc.

In order to optimize a logic diagram for which there can be many implementations possible, the designer can begin with a truth table, for example consider the truth table, shown in Fig. 8.9, for the function F as defined in Eq. 8.14.

Table 8.9 A simple truth table.

A	B	$A \cdot B$	$A + (A \cdot B)$
0	0	0	0
0	1	0	0
1	0	0	1
1	1	1	1

$$F = A + A \cdot B = A \cdot 1 + A \cdot B \tag{8.14}$$

$$= A \cdot (1 + B) \tag{8.15}$$

$$= A \cdot 1 \tag{8.16}$$

$$= A \tag{8.17}$$

Note that the initial expression for F with a requirement for two gates could be simplified resulting in an implementation requiring no gates. A more complex case whose truth table, shown in Table 8.10, is illustrated next.

Table 8.10 A more complex example.

A	B	C	$A \cdot B$	$A \cdot \overline{B}$	$B \cdot C$	$A \cdot \overline{B} + A \cdot B + B \cdot C$	$A + B \cdot C$
0	0	0	0	0	0	0	0
0	1	0	0	0	0	0	0
1	0	0	0	1	0	0	0
1	1	0	0	0	0	1	1
0	0	1	0	0	0	0	0
0	1	1	0	0	1	1	1
1	0	1	0	1	0	1	1
1	1	1	1	0	1	1	1

Consider the following:

$$F = A \cdot \overline{B} + A \cdot B + B \cdot C \tag{8.18}$$

By employing the distributive, inverse, and identity properties together with DeMorgan's theorem the function can be significantly simplified, e.g., Eq. 8.18 can be expressed as

$$F = A \cdot (\overline{B} + B) + B \cdot C \tag{8.19}$$

$$= A \cdot 1 + B \cdot C \tag{8.20}$$

$$= A + B \cdot C \tag{8.21}$$

which reduces the number of gates required to implement this function by 50%.
And finally, a still more complex example is given by

$$F = \overline{A} \cdot B \cdot C + A \cdot \overline{B} \cdot C + A \cdot B \cdot \overline{C} + A \cdot B \cdot C \tag{8.22}$$

$$= \overline{A} \cdot B \cdot C + A \cdot \overline{B} \cdot C + A \cdot B \cdot \overline{C} + A \cdot B \cdot C + A \cdot B \cdot C + A \cdot B \cdot C \tag{8.23}$$

$$= \overline{(A \cdot B \cdot C + A \cdot B \cdot C)} + (A \cdot \overline{B} \cdot \overline{C} + A \cdot B \cdot C) + (A \cdot B \cdot \overline{C} + A \cdot B \cdot C) \tag{8.24}$$

$$= (\overline{A} + A) \cdot B \cdot B \cdot C + (\overline{B} + B) \cdot C \cdot A + (\overline{C} + C) \cdot A \cdot B \tag{8.25}$$

$$= B \cdot C + C \cdot A + A \cdot B \tag{8.26}$$

which reduces the number of gates from 14 to 5.

8.9.2 Karnaugh Maps

Karnaugh maps[37] can be used to convert truth tables and logic equations into logic diagrams and as a substitute for both. In addition, Karnaugh maps make it possible to simplify logic diagrams. Consider the truth table shown in Table 8.11. A Boolean function can be expressed as a sum-of-products derived from the corresponding Karnaugh map (K-Map) shown in Table 8.12 and by inspection is found to be

$$F = \overline{A}BCD + A\overline{B}CD + AB\overline{C}D + \overline{ABCD} \tag{8.27}$$

where each square of the Karnaugh map represents one row of the truth table. Note that the Karnaugh map is configured with respect to variables in a manner that allows only one variable to change as you move from one cell to another, whether horizontally or vertically, i.e., $\overline{AB}$, $\overline{A}B$, AB, $A\overline{B}$ and not $\overline{AB}$, $A\overline{B}$, $\overline{A}B$, AB, because $A\overline{B} \Rightarrow \overline{A}B$ is a change in two variables.[38] Any *cell* of the K-Map containing a one represents what is referred to as a *minterm*, i.e., a product term of N variables.

The process involved is largely a mechanical one as opposed to manipulating Boolean expressions, and is considerably simpler. As a practical matter this technique is useful for expressions of six, or less, variables. If more than six variables are involved, the Quine-McCluskey (Q-M) methodology is preferable.[39]
The Q-M algorithm offers a number of advantages:

- There is no limitation on the number of input variables

---

[37]Some material in this section is based, in part, on examples provided by Bob Harbort and Bob Brown, Computer Science Department, Southern Polytechnic State University and reproduced here with their permission.

[38]This type of arrangement is sometimes referred to as Gray coding for a binary system in which any two successive values, e.g., bytes, differ by only one bit.

[39]The Quine-McCluskey algorithm is based on two fundamental properties of Boolean expressions: (1) $A \cdot \overline{A} = 1$ and (2) the distributive law. In addition to being implementable as an efficient computer algorithm, it provides a way to confirm that the resulting Boolean function is in *minimal* form.

Table 8.11 Example truth table.

A	B	C	D	F
0	0	0	0	0
0	0	0	1	1
0	0	1	0	0
0	0	1	1	0
0	1	0	0	0
0	1	0	1	0
0	1	1	0	0
0	1	1	1	1
1	0	0	0	0
1	0	0	1	0
1	0	1	0	0
1	0	1	1	0
1	1	0	0	1
1	1	0	1	0
1	1	1	0	0
1	1	1	1	0

Table 8.12 Corresponding K-map.

	$\overline{C}\overline{D}$	$\overline{C}D$	CD	$C\overline{D}$
$\overline{A}\overline{B}$	0	1	0	0
$\overline{A}B$	0	0	1	0
AB	0	1	0	0
$A\overline{B}$	0	0	1	0

- It always finds the prime *implicants*[40]
- The algorithm can be applied in the form of a computer program.

Both K-Map and Q-M rely on a very simple expression, viz.,

$$A \cdot B + A \cdot \overline{B}$$

and it follows that

$$A \cdot B + A \cdot \overline{B} = A \cdot (B + \overline{B}) = A \cdot 1 = 1 \cdot A = A \qquad (8.28)$$

It is this simple relationship that forms the basis for the Karnaugh map algorithm.
The following steps allow a K-map to be used to simplify a Boolean expression:

1. Draw a *map* in the form of a table with each product term represented by a cell in the table. The cells must be arranged so that moving from one cell to another either horizontally or vertically changes one and only one variable.

---

[40]The product term of a Boolean function F is a *prime implicant* iff the function's value is 1 for all minterms of the product term. In terms of K-maps, a prime implicant is any loop that is fully expanded. An *essential prime implicant* is any loop that does not intersect any other loop.

2. Place a check mark in each box whose labels are product terms and their respective complements.
3. Draw loops around each horizontally, or vertically, adjacent pairs of check marks.[41]
4. For each loop, form an unduplicated list of the terms. Multiple instances of a literal should be reduced to one instance, and a literal and its complement in the list should be deleted from the list.
5. Form the Boolean product of the terms remaining after step 4.
6. Form the Boolean sum of the products resulting from step 5.

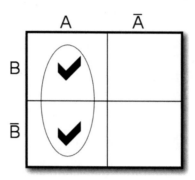

Fig. 8.29 Karnaugh map for $A \cdot B + A \cdot \overline{B}$.

The following simple example will illustrate the procedure outlined by steps 1–6. Assume that the expression to be simplified is $A \cdot B + A \cdot \overline{B}$ which is an expression with two variables, A and B. A table is drawn as shown in Fig. 8.29. Check marks have been placed in the cells representing the AB and $A\overline{B}$ product terms. The loop drawn around these two cells contains A, B, A and $\overline{B}$. The B and $\overline{B}$ cancel and the duplicated A's are reduced to a single A. The end result is: $AB + A\overline{B} = A$. Next, consider the expression $A \cdot \overline{B} + \overline{A} \cdot B + \overline{A} \cdot \overline{B}$. IN this case check marks are placed as shown in Fig. 8.30. Two loops imply that there will be two terms in the simplified expression. The vertical loop yields $\overline{A}, B, \overline{A}$ and $\overline{B}$ and the horizontal loop contains $A, \overline{B}, \overline{A}, \overline{B}$ which are reduced to $\overline{A}$ and $\overline{B}$, respectively. After removing duplicates and invoking the inverse law the expression reduces to

$$A \cdot \overline{B} + \overline{A} \cdot B + \overline{A} \cdot \overline{B} = \overline{A} + \overline{B} \tag{8.29}$$

A third example illustrates the simplification of a three variable, Boolean expression, viz.,

$$\overline{A} \cdot \overline{B} \cdot \overline{C} + \overline{A} \cdot B \cdot C + A \cdot B \cdot C$$

The K-map for this example is shown in Fig. 8.31.

This example involves a *toroidal* loop encompassing cells ABC and $\overline{A}BC$. The simplified expression in this case is given by:

$$F = B \cdot C + \overline{A} \cdot \overline{B} \cdot \overline{C} \tag{8.30}$$

Next, consider a SoP with five product terms each consisting of three variables, viz.,

$$F = \overline{A} \cdot B \cdot \overline{C} + \overline{A} \cdot B \cdot C + A \cdot \overline{B} \cdot C + A \cdot B \cdot \overline{C} + A \cdot B \cdot C \tag{8.31}$$

The K-map for this example is shown in Fig. 8.32, and the equivalent logic circuit is shown in Fig. 8.33. The truth table representing this configuration is shown in Table 8.13.

---

[41] A cell may be in more than one loop and a single loop can span multiple rows, columns or both, provided that the number of enclosed check marks is a multiple of 2, e.g., 1, 2, 4, 8, 16,…

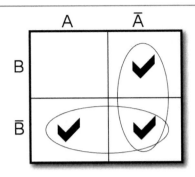

Fig. 8.30 Karnaugh map for $A \cdot \overline{B} + \overline{A} \cdot B + \overline{A} \cdot \overline{B}$.

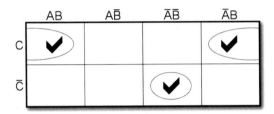

Fig. 8.31 Karnaugh map for $\overline{A} \cdot \overline{B} \cdot \overline{C} + \overline{A} \cdot B \cdot C + A \cdot B \cdot C$.

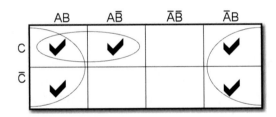

Fig. 8.32 A five product K-map.

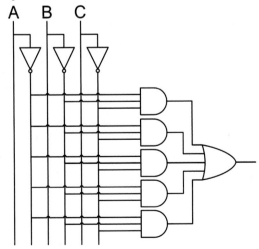

Fig. 8.33 A logic circuit for a five term SoP expression.

Table 8.13 Truth   table   for
Eq. 8.31.

A	B	C	F
0	0	0	0
0	0	1	0
0	1	0	1
0	1	1	1
1	0	0	0
1	0	1	1
1	1	0	1
1	1	1	1

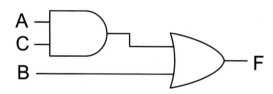

Fig. 8.34 Simplified version of the logic circuit in Fig. 8.33.

After removing redundant instances of variables and their complements have been removed, the expression is reduced to $A \cdot C + B$ and can be implemented in discrete logic as shown in Fig. 8.34.

The loops in a Karnaugh map should be made as large as possible subject to the constraint that the number of check marks within any given loop must be an integer multiple of two. When a Karnaugh map consists of more than two rows, it represents more than three variables and the top and bottom edges are treated as adjacent. A loop that is within loops is not considered because all its terms have been accounted for in the other loops.

8.10   Combinatorial Circuits

A combinatorial circuit is defined as any combination of the basic operations AND, OR and NOT that includes both inputs and outputs. Each of the outputs is related to a unique function. A classic example of a combinatorial circuit is the *half-adder*, which is capable of producing a 1-bit sum, and carry, based on the following functions:

$$\text{Sum} = A' \cdot B + A \cdot B' \tag{8.32}$$

and the resulting carry bit, if any, by

$$\text{Carry} = A \cdot B \tag{8.33}$$

However, although a half-adder can generate a carry it does not have the ability to add a *carry-in* (C_i) to the sum. This capability is embodied in what is referred to as a full-adder that can be represented in terms as

$$\text{Sum} = ABC_i + AB'C_i' + A'BC_i' + A'B'C_i \tag{8.34}$$

and *carry out* (C_o) by

$$C_o = AC_i + BC_i + AB \tag{8.35}$$

Thus a half-adder cascaded with an $n - 1$ number of full-adders can be used to add n bits. However, because combinatorial circuits are employed, some form of memory is needed. This is because for combinatorial circuits, any change in an input results in a change in the outputs[42] and therefore the circuits are *memoryless*.

Fortunately, it is possible to create a very simple memory device from the same basic building block as the logic functions, i.e., from a NAND gate as shown in Fig. 8.35. This configuration is a two-state, or *bistable* configuration for which R and S are normally both set to 1. If input R or S is toggled momentarily, then Q and Q' are forced into opposite states and will remain there until one of the inputs is toggled again. However, if both inputs are set to zero contemporaneously, Q and Q' are forced into the 1 state.

A simple modification of this circuit resolves this potential problem and requires the flip-flop to operate in a synchronous manner when changing state. Figure 8.36 shows the modification which involves the addition of three NAND gates, one of which is configured as an inverter. This configuration is known as a data or D flip-flop and has a clock input (Clk) which allows the flip-flops operation to be synchronous. A clock input, 0–1–0, will cause the data input to be copied to the Q output where it will be *latched,* i.e., retained, until the next clock pulse. Flip-flops can be combined in either parallel configurations to function as memory into which bits may be stored and retrieved in parallel, e.g. as in the case of conventional registers or in a daisy-chain to function as shift registers.

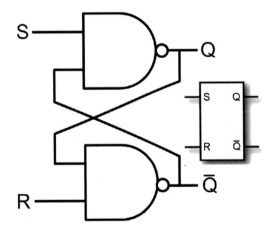

Fig. 8.35 A NAND gate implementation of an RS flip-flop.

[42]There is, of course, some finite amount of propagation delay through a logic circuit, but for the purposes of the present discussion such delays will be ignored. However, it should be noted that propagation delays are often cumulative and, in such cases, it may not be appropriate to ignore them.

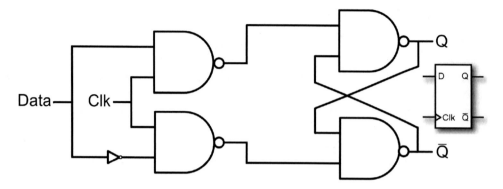

Fig. 8.36 A D flip-flop.

8.11 Sequential Logic

Unlike combinational logic, which has no internal state and whose output depends solely on the state of the input, at any given time, the sequential logic output is a function of its internal state at any given time and inputs. Sequential logic is *clock-based* and relies on a combination of combinational logic and one, or more, flip-flops as shown in Fig. 8.37.

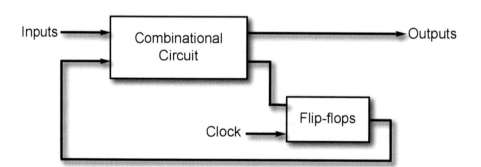

Fig. 8.37 Sequential circuit block diagram.

Flip-flops provide a mechanism for remembering a state and thus for storing data, a capability not achievable by using combinational logic alone. The simplest example of sequential logic is the flip-flop, that can be connected in various configurations, e.g., counters, timers, registers, RAM, etc. Counters are sequential circuits that have a clock signal as their input. The electronic flip-flop was invented by F.W. Jordan and William Eccles in 1919 as a bistable device consisting of two vacuum tubes.[43] The simplest from of flip-flop is the SR flip-flop, sometimes referred to as the SR latch. Its inputs consisted of S(et) and R(eset). The truth table for this device is shown in Table 8.14. A JK flip-flop, shown in Fig. 8.38, is similar to a D flip-flop except that indeterminate states are avoided for

---

[43] Sometimes referred to as a bistable multivibrator. Monostable (pulse) and astable (oscillator) functions using two vacuum tubes in similar configurations were also possible.

cases in which both inputs are being held high by requiring that in such cases, the output *toggles*[44] with the clock, cf. Table 8.15.

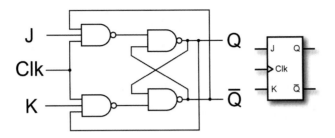

Fig. 8.38 A JK flip-flop.

In some cases, additional control pins are provided for JK flip-flops which allow asynchronous clearing and presetting. The D, or data flip-flop, has the same value as the input when a *clock edge* occurs,[45] as shown in Table 8.16.

Table 8.14 Truth table for an RS flip-flop.

S	R	Q	$\overline{Q}$
0	0	Unchanged	Unchanged
0	1	0	1
1	0	1	0
1	1	Indeterminate	Indeterminate

Table 8.15 Truth table for a JK flip-flop.

J	K	Q	$\overline{Q}$
0	0	Unchanged	Unchanged
0	1	0	1
1	0	1	0
1	1	Toggle	Toggle

Table 8.16 Truth table for a D flip-flop.

Clock	D	Q	$\overline{Q}$
Edge	0	0	1
Edge	1	1	0
Non-edge	Unchanged	Hold	Hold

---

[44] Toggle refers to changing the state of a two-state device, i.e., causing it to change to the other state, by some event, or action.

[45] Some D flip-flops are positive- and some are negative-edge triggered.

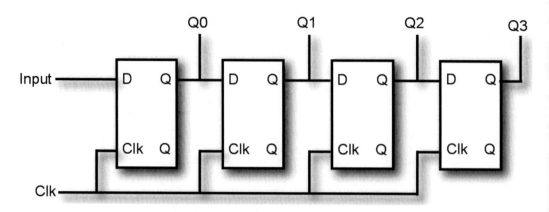

Fig. 8.39 A serial shift register using D flip-flops (SIPO).

Flip-flops can be configured in a variety of ways to provide very useful functionality, e.g., four flip-flops can be configured, as shown in Fig. 8.39, to provide a register that accepts serial data as input and makes the data available in a parallel output format. Each time a clock pulse occurs data stored in each of the flip-flops will be moved one bit position to the right. The bit asserted at the input will move to the first flip-flop and be accessible at output Q0, the data formerly stored in the first flip-flop will move to the second flip-flop and be accessible at the Q1 output and so on. Thus, the bit stored originally stored in the fourth flip-flop is lost. Obviously, the number of clock pulses applied must correspond to the number of bits being added to the shift register for storage. This type of register may be used as a method of converting the serial input to parallel output (SIPO) and/or as a storage location for four bits.

Alternatively, the four flip-flops can also be configured, as shown in Fig. 8.40, to provide parallel input and parallel output (PIPO). Each time a clock pulse occurs the four input bits are stored in their respective flip-flop locations and appear in parallel on Q0, Q1, Q2 and Q3 respectively. If the input data is from a 4-bit parallel data bus, this configuration provides a way to retain bus data in a 4-bit register. This is a common technique for temporarily storing data in a microprocessor.

An important consideration when configuring groups of flip-flops that share a common clock signal is related to the phenomenon known as *clock skew*. This occurs when a device such as a flip-flop is edge-triggered and the clock edge does not arrive at the clock input for each of the flip-flops at the same time. Clock skew can arise as a result in differences in the paths that the clock signal must traverse to reach the clock inputs of the flip-flops. It can also be compounded by the use of a group of flip-flops that are triggered on different edges of the clock signal.

Also in some configurations of flip-flops, gates are required in order to achieve the required sequential logic. In such cases the gates introduce delays if they are in the clock path. In addition to the spatial variations in clock edges i.e., clock skew, temporal variations can occur as well, the latter being referred to as *clock jitter*.[46] Obviously, in a given situation, clock skew is not subject to variation from one clock signal transition to another for any given device, whereas, clock jitter can and often does vary as a function of time on a cycle-by-cycle basis. Clock skew can be introduced not only by path length differences but also by the power supply, temperature and clock driver variations.

Jitter, in the simplest terms, is an undesirable variation in a signal's timing and can arise as a result of variations in the clock source, power supply/temperature variations, capacitive loading and/or

---

[46]All clocks exhibit some degree of jitter. In most cases, the amount of jitter is maintained at a level that does not render it unusable, e.g., by employing a phase-locked loop.

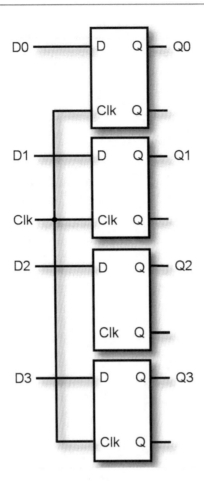

Fig. 8.40 A parallel input/output register using D flip-flops.

coupling, etc. The various types of jitter are depicted in Fig. 8.41. Random jitter, sometimes referred to as background or thermal noise, and manifests itself as a stochastic, Gaussian timing perturbation and because it is present in all such systems it is often referred to as an intrinsic type of jitter [1].

Deterministic jitter[47] can be narrow band, is sometimes periodic and can be bifurcated into data dependent and periodic jitter. The latter can be introduced by crosstalk and other means. Data dependent jitter can be found in some serial data streams and arises as a result of dynamically chaining duty cycles and irregular clock edges. Unlike random jitter, deterministic jitter can often be minimized by identifying the sources and seeking to minimize their undesired effects.

Flip-flops are often used as *static* memory devices in that they are bistable devices which are capable of storing a given state for as long as power is maintained as opposed to dynamic memory devices that store a state in the form of charge on parasitic capacitors and require continuous refreshing. While simpler and cheaper to manufacture, dynamic memories are subject to noise that adversely affects their ability to retain a given state.

---

[47]Deterministic jitter is defined the temporal difference between the time at which a transition occurs and the time the transition should have occurred.

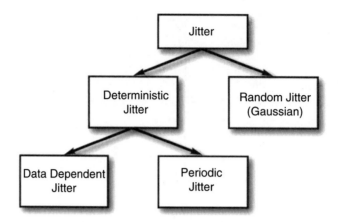

Fig. 8.41 Types of jitter.

Some logic devices are limited to a maximum of two inputs. This restriction can be overcome by combining several devices as shown illustratively in Figs. 8.42 and 8.43, respectively. PSoC Creator allows the designer to configure logic devices such as those shown in Fig. 8.44.

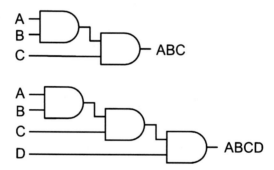

Fig. 8.42 Three and four input configurations for AND gates.

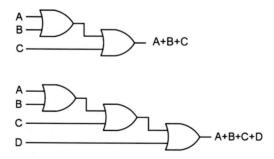

Fig. 8.43 Three and four input configurations for OR gates.

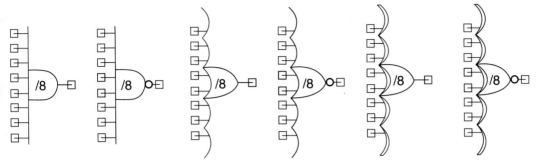

Fig. 8.44 Examples of multiple input gates available in PSoC creator.

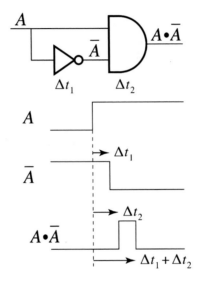

Fig. 8.45 A simple example of a race condition resulting in a "glitch".

However, this type of expansion of inputs obviously increases propagation time and therefore latency and can result in race[48] problems, an example of which is shown in Fig. 8.45.

Assume that input A was asserted previously at a point in time, $t < 0$, sufficient to allow the logic shown in the figure to reach a steady-state condition. If at $t = 0$, $A = 0$ is asserted then after a propagation delay of Δt_1, introduced by the inverter, $\overline{A} = 1$, as shown. This means that for the duration of Δt_1, both A and $\overline{A} = 1$. This produces a pulse of width Δt_1 displaced in time by Δt_2. It is this pulse that is referred to as a *glitch* and said to be caused as a result of *race conditions* in the signal path.

---

[48]Race conditions can occur in logic circuits as a result of propagation time differences that result in an output changing to an inappropriate state, often referred to as a "glitch", caused by a delay in one or more input signals with respect to other inputs to the circuit.

8.12   Finite State Machines

The concept of a finite-state machine (FSM) is an abstraction of a system whose allowed states are restricted to only one of a finite number of states at a time and the "transitions"[49] between those states.[50] A transition between states occurs only as a result of inputs, sometimes referred to as being *event-driven*. While a given state machine may produce an output, or outputs, some state machines do not. It may be the case that the result of a state transition is simply to place the system in a different state. Some machines may have an *error state* to handle unanticipated and/or unexpected inputs. Once a state machine enters an *error state* it remains there, even in the presence of subsequent inputs. Transitions are governed by so-called"rules" or "conditions" and typically these are expressed in the form of case statements. Transitions are triggered by "events" which may be either external or internal. Switch statements and state tables are commonly used to implement FSMs.

FSMs are frequently used in natural language processing, text processing, cellular automata, natural computing, electronic design automation, communications, artificial intelligence, video games, vending machines, traffic control, speech recognition, speech synthesis, parsing, web applications, neurological system modeling, protocol design, process control, vending machines and many other applications.

FSMs are defined in terms of the:

- allowed states
- input signals,
- output signals,
- next-state function,
- output function,

and,

- an initial state.

and, as a result, FSMs are *sequential* machines.

The Moore state machine, shown schematically in Fig. 8.46, has the characteristic that outputs are independent of inputs, i.e., outputs are created within a given state and can only change when a change of state occurs.[51] State assignments may be either arbitrary, or specified. Arbitrary state assignments depend on either combinatorial, or registered, decoded state bits. Specified state assignments are based on either state bits, or on so-called "one-hot" encoding, e.g. as shown in Table 8.17.[52]

For state machines using one-hot encoding, n flip-flops, often referred to as the *state memory*, can be used to represent the n states of the FSM.

The *state vector* is the value currently stored in the *state memory*. Moore FSM outputs are a function of the state vector, but the outputs of a Mealy FSM, shown in Fig. 8.47, are a function of the inputs and the *state vector*. While this method does require n flip-flops to encode and decode the FSM's current state, decoding is simplified by virtue of the fact that no other logic is required

---

[49]"Transition" in the present context refers simply to the change from one state to another.

[50]In theory, any system utilizing memory can be treated as a state machine.

[51]Although inputs can cause a change of state, they do not determine the state to which the FSM moves.

[52]"One-hot" refers to the case in which given a string of bits, only one can be non-zero, e.g., 00010000. The inverse situation is referred to as one-cold, e.g., 11101111. On-hot code is often used decoder, ring counter and some state machine implementations.

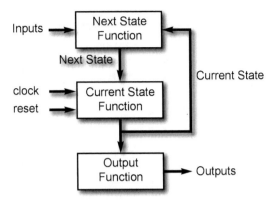

Fig. 8.46 Moore state machine.

Table 8.17 One-hot versus binary encoding.

State	Binary Encoding	One-Hot Encoding
S0	00000001	00000001
S1	00000010	00000010
S2	00000011	00000100
S3	00000100	00001000
S4	000001001	00010000
S5	000001010	00100000
S6	00001011	01000000
S7	00001111	10000000

to determine the current state of the machine. The outputs of Mealy machine are determined either by the present state, or by a combination of the current state and the then-current inputs.[53] Some applications employ both Moore and Mealy FSMs and as a practical matter similar FSMs can be functional equivalents.

State machines are often represented by *state graphs* as shown in Figs. 8.48 and 8.49. By convention, arcs and/or straight lines are used to represent state transitions and each node represents a specific state. In the case of *self-transitions*, the source and target states are the same states. Moore outputs are given within the circle or "bubble" representing the state. Mealy outputs are shown on the associated arc or line.

State machines can also be represented as algorithmic state machines (ASMs) in which case the graphical representation is in the form of a flow chart with the state, decision and conditional boxes. ASMs can be recast as state graphs and vice versa.

Typically, Mealy machines:

- typically have fewer states than that of their Moore counterparts,

---

[53] Mealy FSM outputs can change asynchronously.

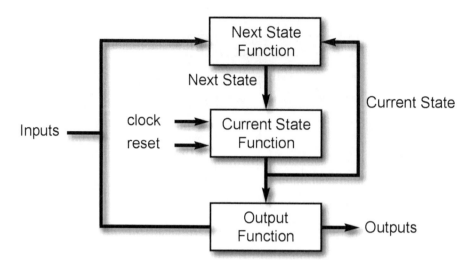

Fig. 8.47 Mealy state machine.

- react faster to inputs,[54]
- has outputs that are a function of both current state and inputs that can change asynchronously,[55]
- outputs can change asynchronously,
- can have fewer states than a Moore FSM,

and,

- can sometimes introduce delays in critical paths.

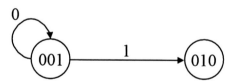

Fig. 8.48 A very simple finite state machine[56].

PSoC 3/5LP are quite capable of implementing state machines and this is facilitated, in part, by the fact that PSoC Creator supports look-up tables (LUTs). LUTs have the characteristic that a particular combination of input values results in the output of a specific combination of outputs. This allows a LUT to provide virtually any logic function and in the case of PSoC Creator, each LUT component is configurable to have as few as one input inputs and one output, e.g., as *in0* and *out0* or as many as five inputs and eight outputs as *in1, in1, in2, in3, in4* and *out0, out1, out2, out3, out4, out5, out6, out7*, respectively. The default configuration is two inputs and two outputs.

---

[54]Moore machines have to wait for the next clock-cycle before changing state.

[55]This can give rise to "glitches". Moore FSMs do not produce glitches.

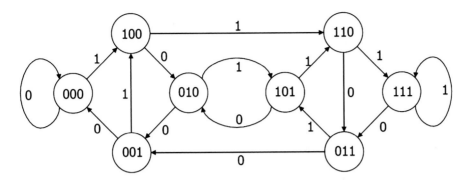

Fig. 8.49 A more complex FSM representing a shift register.

Registering the outputs is accomplished by simply clicking on a check box in PSoC Creator's *Configure "LUT"* dialog box, cf. Fig. 8.50. Registering the outputs and routing some outputs back to the inputs allows state machines to be implemented. The actual implementation of LUTs is based on logic equations stored in the PLDs. LUTs save the designer the trouble of having to create them using combinatorial logic components and by registering a LUT it can be used to implement sequential logic. The registering of the outputs causes the LUT to register the output on the rising edge of the LUT's clock input.

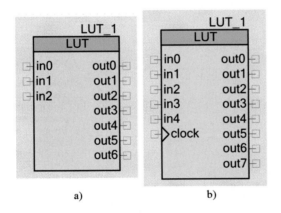

Fig. 8.50 PSoC 3/5LP LUT. (**a**) Unregistered versus (**b**) registered.

The clock speed should not exceed 33 MHz if any of the LUT's outputs are connected to I/O. It should be noted that the LUT is implemented as a hardware-only design and therefore there is no LUT API.

As an example, consider a rising edge detector implemented as a Moore state machine, as shown in Fig. 8.51, that produces a pulse each time a rising edge is detected [6]. The creation of a LUT-based state machine begins with the creation of a table that contains each possible *state* and all possible combinations of inputs. Next, consideration turns to defining the *next state* for each state and the associated inputs. It is then possible to create the LUT. Once this table has been completed, its entries can be entered into the *Configure 'LUT'* dialog box in PSoC Creator. The implementation of the edge detector LUT is shown in Fig. 8.52.

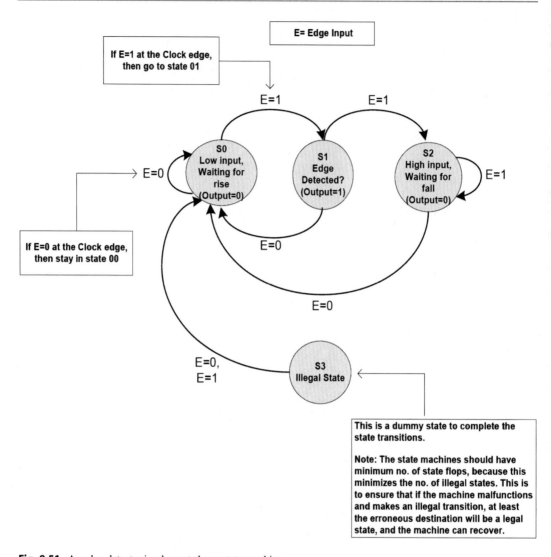

Fig. 8.51 An edge detector implemented as a state machine.

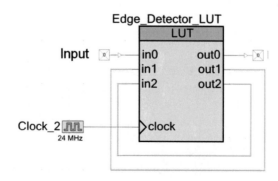

Fig. 8.52 A PSoC 3/5LP implementation of an edge detector as a FSM.

8.13   Recommended Exercises

8-12 Using as a basic building block the PAL shown in Figure 8–13, create a logic array that is capable of subtracting two 4-bit binary values.

8-13 Assuming a reasonable latency for each of the devices in the logic circuit you created in Exercise 8-12, compute the overall latency. List all your assumptions.

8-14 The state machine shown in Fig. 8.53 represents a simple down-counter. The counter decrements from the initial value until it reaches zero,and the period is reloaded. Show how to implement such a counter using Datapath.Use A0 to store the count value and D0 to hold the value that is used when resetting the counter [5].

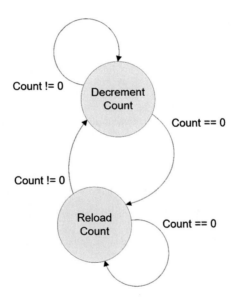

Fig. 8.53 Eight bit down-counter.

8-15 By adding a compare it is possible to turn the simple counter into a PWM. To create a PWM, compare the value in A0 with another fixed value. Use a register, e.g. D1, to hold the reference value, and set the compare block to check if A0 is less than D1.

8-16 Which of the following is a good reason to use a FIFO when transferring data to I/O devices?

(i) FIFOs can be used as a permanent data store.
(ii) FIFOs provide backup and sharing.
(iii) It allows software and hardware to operate at different speeds.
(iv) A FIFO can store an arbitrarily large amount of data.
(v) None of the above.
(vi) All of the above

8-17 Given the message 1110101 and polynomial 110011, compute the CRC at the transmission site and confirm that there is no error at the receiver site by showing that the CRC is zero.

8-18 Derive a truth table and SOPs expression for $F = A \oplus B \oplus C$. Show the logic diagram for this expression.

8-19 An eight-input mux is controlled by the three logic inputs A, B and C. If the inputs are I_n for $n = 0, 1, 2, 3, 4, 5, 6, 7$ and the output is Z, derive the output expression for Z.

8-20 Show the Mealy and Moore state machine diagrams for a sequence detector that outputs a 1 when it detects the final bit in the serial data stream 1101.

8-21 Using the state diagram shown in Fig. 8.49, find its state table. Draw the associated logic circuit and explain how you might implement it in PSoC Creator.

References

1. A Primer on Jitter, Jitter Measurement and Phase-Locked Loops. AN687. Silicon Laboratories. Rev.0.1. (2012)
2. J.M. Birkner, *PAL Programmable Array Logic Handbook* (Monolithic Memories, Santa Clara, 1978)
3. S. Brown, J. Rose, FPGA and CPLD architectures: A tutorial. IEEE Des. Test (Summer 1996)
4. Complex Programmable Logic Devices (CPLD) Information on GlobalSpec., N.p., n.d. Web. 6 Apr. 2013. www.globalspec.com/learnmore/analog_digital_ics/programmable_logic/complex_programmable_logic_devices_cpld
5. T. Dust, G. Reynolds, AN82156. Designing PSoC Creator Components with UDB Datapaths. www.cypress.com. Document No. 001-82156Rev. *I. Cypress Semiconductor (2018)
6. J. Kathuria, C. Keeser, Implementing State Machines with PSoC 3, PSoC 4, and PSoC 5LP. AN62510. Document No. 001-62510 Rev. *F. Cypress Semiconductor (2017)
7. V.K. Marrivagu, A.R. De Lima Fernandes, PSoC creator - Implementing Programmable Logic Designs with Verilog. AN82250. Document NUMBER:001-82250Rev.*J1. Cypress Semiconductor (2018)
8. D. Van Ess, Learn Digital Design with PSoC. A bit at a time. CreateSpace Independent Publishing Platform (August 9, 2014)
9. Xilinx, *Programmable Logic Design – Quick Start Hand Book*, 2nd edn. (Jan 2002)

Communication Peripherals

9

Abstract

This chapter provides a discussion of a number of the more important communications protocols that are used in conjunction with embedded systems. It has not been possible to treat each of these protocols in great detail, but an effort has been made to provide the reader with a broad overview of such protocols and to a lesser extent provide some relative comparisons between them. Each of the protocols discussed offers certain benefits over the others and all are currently in widespread use. The reader is encouraged to review the standards referenced in this chapter for these protocols in order to gain additional insight into their respective architectures and implementation details. Issues such as error detection/recovery, cost of implementation, number of communications paths required, supported transmission speeds, coding complexity, supported master/slave configurations, transmission modes are discussed in some detail.

9.1 Communications Protocols

Embedded systems are often required to communicate with other systems [2] and present data on visual display devices such as LED discrete character displays and LCD screens. Whether communicating with display devices, or other local/remote systems, a wide variety of communications protocols are in common use, e.g., I2C, UART, SPI, USB, RS232, RS485, etc. Many of these protocols have certain features in common and other features that are unique to a particular protocol. Both PSoC 3 and PSoC 5LP are capable of supporting a wide variety of such protocols.[1]

One might well ask why there is a need for so many communication protocols,[2] particularly as they relate to microprocessors and microcomputers. The simple answer is that a typical embedded system communicates with a number of different devices each of which can have its own preferred communications interface. Data transmission from one device, and/or one location to another,

---

[1] In some applications, the support for a particular protocol is part of the PSoC 3/5 architecture, while in other cases, external hardware may be required to interface PSoC 3/5 with external communications channels.

[2] A digital communication's protocol is a formal statement of the governing rules and formats for communications between two or more devices. In addition to setting forth the data formats and syntax involved, the protocol typically defines the parameters used for authenticating a received message and in some cases defines the error detection, and correction, algorithms to be used.

© Springer Nature Switzerland AG 2021
E. H. Currie, *Mixed-Signal Embedded Systems Design*,
https://doi.org/10.1007/978-3-030-70312-7_9

typically relies upon a preferred speed of transmission, support for buffering of data,[3] retention of data integrity and, if possible, error correction.

As a result, many of the extant protocols have either evolved over time into various incarnations, or been replaced by newer protocols, in an attempt to address complexity concerns, speed, data transfer rates, cost, noise immunity, operating levels, interoperability challenges, networking considerations, transfer distance/times, data security/integrity and a myriad of other issues. In some applications multiple protocols are employed in the same application. In other situations, older protocols are still employed to address interfacing requirements imposed by legacy hardware ad software systems. Some protocols address peer-to-peer transmission, others are applicable to master-slave configurations and still others, various networking configurations.

Error detection schemes are often based on the transmission of additional data[4] with each data block that makes it possible to determine whether data integrity has been maintained. When a data block, or frame, of data is received, the redundant data is used to determine if the data that has been changed during transmission. In some applications, when an error is detected, algorithms are applied to the data to correct such errors. Often a trade-off has to be made regarding simply retransmitting data from one location to another and the time required to apply an error correction algorithm.

9.2   I2C

The Inter-integrated Circuit Bus[5] (*I2C*) [7] was originally developed by Phillips Semiconductor to support multi-master inter-communications between devices, such as integrated circuits on a printed circuit board. This makes the I2C component ideal when networking multiple devices whether all on a single board, or as part of a small system. Such systems can employ a single master and multiple slaves, multiple masters, or an arbitrary combination of masters and slaves. Such implementations can employ either fixed hardware I2C blocks, or universal digital blocks (UDBs).

I2C utilizes a two-wire, serial, bidirectional bus connected in a master-slave configuration, as shown in Fig. 9.1.[6]

Although originally limited to a maximum transfer rate of 100 kbits/s, it is currently capable of operating at speeds up to and including 3.4 Mbits/s.[7] Any device on the bus that initiates a data transfer,[8] generates the clock for that transfer and is defined as the then current *master*. The corresponding device receiving the data is the *slave*. Each of the slaves connected to the bus is assigned a unique address[9] that is used by the master to specify which slave is being addressed, to receive a given data transmission. The maximum number of slaves that can be attached to SDA/SCL is determined by the bus capacitance, e.g., 400 pF. Bit transfers are level-triggered, with one bit per clock pulse as the data rate and data changes can only take place during low clocks.

---

[3]Buffering of data is used as a method of holding data until the communication channel is available for transmission.

[4]Referred to as redundancy data.

[5]The abbreviations *I2C* and *I²C* are both in common usage, when referring to the inter-integrated circuit bus.

[6]A slave is defined as any device connected to the bus that is capable of receiving data, e.g., an LCD driver, memory, keyboard driver, microcontroller, etc. In some cases, a device is capable of both receiving and transmitting data and therefore may alternately function as both a master and a slave to allow data transfer on the bus to be in either direction. A device capable of serving as a master can also request data from another device, in which case, that master generates the clock and terminates the transfer.

[7]PSoC's EZI2C components is capable of operating at a maximum data rate of 1000 kbps. Supported "standard rates" are 50, 100 (default, 400 and 1000 kbps.)

[8]Including transmitting the address of the device (slave) that is to receive the data.

[9]Addressing for each device can be based on either 7- or 10-bit addressing.

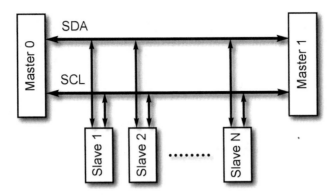

Fig. 9.1 The I2C master-slave configuration.

The serial data (SDL) and serial clock (SCL) signal lines are used in combination to transmit data. If both SCL and SDL are high, no data is transmitted. A high-to-low transition of the SDA line, while the SCL line is high, indicates a *START* condition, which is often designated by *S*. A low-to-high transition of the SDA line, while the SCL line is high, defines a *STOP* condition designated by *P*. Only the current master is capable of generating a *START* condition and, once initiated, the bus enters a busy state until a corresponding *STOP* condition has occurred. If the device addressed by the master is busy, the master may generate a series of *START*s to maintain the bus in a busy state, until the addressed slave becomes available (Fig. 9.2).

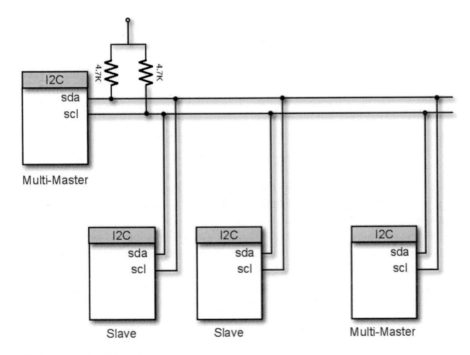

Fig. 9.2 I2C master and multiple slave configuration.

A single bit of each 8-bit data byte is transferred for each clock pulse and there is no inherent limit to the number of bytes that can be transmitted in a given transmission.[10] At the end of the transmission of each byte, an acknowledgment from the receiver is required.[11] The master generates an acknowledge clock pulse and releases the SDA line which then goes high for the duration of the acknowledge clock pulse (Fig. 9.3).

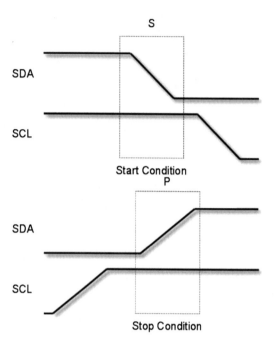

Fig. 9.3 Start and Stop conditions.

The receiving device pulls the SDA line low during the acknowledge clock pulse. If the slave does not acknowledge, the SDA line remains high and the master can then either generate a *STOP*, to terminate the transmission, or a repeated *START* condition to start a new transfer, as shown in Fig. 9.4.

A slave can temporarily suspend further data transmission by placing the SCL line in a low state which results in the master entering a wait state. This capability allows the slave to perform other functions, e.g., servicing an interrupt. When the slave subsequently releases the SCL line, the next byte can then be transmitted. In an actual implementation, the two lines, SDL and SCL, are connected to pullup resistors as shown in Fig. 9.2.

In order for a master to transmit data, the bus must be free. If multiple masters are used in an I2C configuration, a method must be provided to avoid two, or more, masters attempting to transmit data at the same time. This is accomplished by the use of an *arbitration technique*. Each master generates its own clock signals during data transfers on the bus. These signals can be *stretched* by either another master, as a result of arbitration, or by the slow response slave that holds down the clock line. If two, or more, masters attempt to use the bus contemporaneously, the first to introduce a one, while the other introduces a zero, will lose the arbitration and control passes to the latter master. This arbitration may continue in force for multiple bit transfers.

---

[10]Bytes are transferred with the *Most Significant bit* (MSb) being transferred first.

[11]This restriction is relaxed, if one or more of the devices involved is a CBUS receiver. In such cases, a third bus line is required.

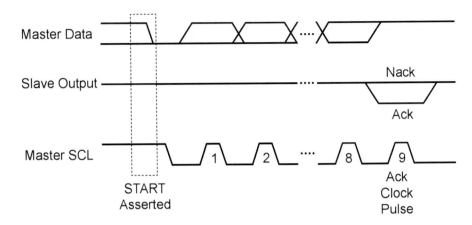

Fig. 9.4 Slave ACK/NACK of the single-byte received.

When multiple masters are used, it is possible for two, or more, of them to generate a *START* condition within the minimum hold time of the *START* condition. Therefore, for each byte to be transmitted, the bus must first be checked to determine whether it is in a *busy state*. An error is returned to the master who loses arbitration.

A PSoC 3/5 master can be operated in either manual or automatic mode. In automatic mode, a buffer is employed that holds the entire transfer. If a write operation is to occur, the buffer is pre-filled with the data to be transmitted. If data is to be read from a slave, a buffer of at least the size of a packet has to be allocated. In the automatic mode, the following function[12] writes an array of bytes to a slave

```
uint8 I2C_MasterWriteBuf(uint8 SlaveAddr, unit8 * wrData, uint8
        cnt, uint8 mode)
```

where *SlaveAddr* is a right-justified, 7-bit slave address; *wrData* is a pointer to the array of data; *cnt* is the number of bytes to be transferred and *mode* determines how the transfer starts and stops.

Similarly, a read operation is initiated by

```
uint8 I2C_MasterReadBuf(uint8 SlaveAddr, unit8 * wrData, uint8
        cnt, uint8 mode)
```

Both of these functions return status information, as shown in Table 9.1.

9.2.1 Application Programming Interface

PSoC Creator provides a set of I2C application programming interface routines (APIs) to allow dynamic configuration of the I2C component during runtime. By default, PSoC Creator assigns the instance name *I2C_1* to the first instance of an *I2C* component in a given design. This instance can be renamed to any unique value that follows the syntactic rules for identifiers. The instance name becomes the prefix of every global function name, variable, and constant symbol. For readability, the instance name used in the following is *I2C*. All API functions assume that the data direction is from

---

[12]The use of the term *function*, in the present context and throughout this text, is a generic reference to methods, function members or member functions.

Table 9.1 Master status info returned by *unit8 I2C_MasterStatus(void)*.

Master Status Constants	Descriptions
I2C_MSTAT_RD_CMPLT	Read transfer complete
I2C_MSTAT_WR_CMPLT	Write transfer complete
I2C_MSTAT_XFER_INP	Transfer in progress
I2C_MSTAT_XFER_HALT	Tranfer has been halted
I2C_MSTAT_ERR_SHORT_XFER	Transfer completed before all of the bytes were transferred
I2C_MSTAT_ADDR_NAK	Slave did not acknowledge address
I2C_MSTAT_ERR_ARB_LOST	Master lost abitration during communications with slave
I2C_MSTAT_ERR_XFER	Error occurred during transfer

the perspective of the I2C master. A write event occurs when data is written from the master to the slave, and a read event occurs when the master reads data from the slave.

PSoC Creator supports a number of function calls that are generic for I2C slave or master operation including:

- *uint8 I2C_MasterClearStatus(void)* clears all status flags and returns the master status and returns the current status of the master.
- *uint8 I2C]_MasterWriteBuf(uint8 slaveAddress, uint8 * wrData, uint8 cnt, uint8 mode)* automatically writes an entire buffer of data to a slave device. Once the data transfer is initiated by this function, further data transfer is handled by the included ISR in byte-by-byte mode and it enables the I2C interrupt.
- *uint8 I2C_MasterReadBuf(uint8 slaveAddress, uint8 * rdData, uint8 cnt, uint8 mode)* automatically reads an entire buffer of data from a slave device. Once the data transfer is initiated by this function, further data transfer is handled by the included ISR in a byte-by-byte mode and it enables the I2C interrupt.
- *uint8 I2C_MasterSendStart(uint8 slaveAddress, uint8 R_nW)* generates a start condition and sends the slave address with a read/write bit. It also disables the I2C interrupt.
- *uint8 I2C_MasterSendRestart(uint8 slaveAddress, uint8 R_nW)* generates a restart condition and sends the slave address with a read/write bit.
- *uint8 I2C_MasterSendStop(void)* generates an I2C stop condition on the bus. If the start, or restart, conditions failed before this function was called, this function does nothing.
- *uint8 I2C_MasterWriteByte(uint8 theByte)* sends one byte to a slave. A valid start, or restart, the condition must be generated before calling this function. This function does nothing, if start, or restart, the conditions failed before this function was called.
- *uint8 I2C_MasterReadByte(uint8 acknNak)* reads one byte from a slave and ACKs, or NAKs, the transfer. A valid start, or restart, condition must be generated before calling this function. This

function does nothing and returns a zero value, if a start or restart condition has failed before this function was called.

- *uint8 I2C_MasterGetReadBufSize(void)*returns the number of bytes that has been transferred by the *I2C_MasterReadBuf()* function. If the transfer is not yet complete, it returns the byte count transferred so far.
- *uint8 I2C_MasterGetWriteBufSize(void)* returns the number of bytes that have been transferred by the *I2C_MasterWriteBuf()*function. If the transfer is not yet complete, it returns the byte count transferred so far.
- *void I2C_MasterClearReadBufSize(void)* resets the read buffer pointer back to the first byte in the buffer.
- *void I2C_MasterClearWriteBufSize(void)* resets the write buffer pointer back to the first byte in the buffer.

9.2.2 PSoC 3/5 I^2C Slave-Specific Functions

The supported slave functions are as follows:

- *uint8 I2C_SlaveClearReadStatus(void)* clears the read status flags and returns their values. No other status flags are affected.
- *uint8 I2C_SlaveClearWriteStatus(void)* clears the write status flags and returns their values. No other status flags are affected.
- *void I2C_SlaveSetAddress(uint8 address)* sets the I2C slave address.
- *void I2C_SlaveInitReadBuf(uint8 * rdBuf, uint8 bufSize)* sets the buffer pointer and size of the read buffer and resets the transfer count returned by the *I2C_SlaveGetReadBufSize()* function.
- *void I2C_SlaveInitWriteBuf(uint8 * wrBuf, uint8 bufSize)* sets the buffer pointer and size of the write buffer. This function also resets the transfer count returned by the *I2C_SlaveGetWriteBufSize()* function.
- *uint8 I2C_SlaveGetReadBufSize(void)* returns the number of bytes read by the I2C master after an *I2C_SlaveInitReadBuf()*, or *I2C_SlaveClearReadBuf()* function was executed.
- *uint8 I2C_SlaveGetWriteBufSize(void)* returns the number of bytes written by the I2C master since an *I2C_SlaveInitWriteBuf()* or *I2C_SlaveClearWriteBuf()* function was executed. The maximum return value is the size of the write buffer.
- *void I2C_SlaveClearReadBuf(void)* resets the read pointer to the first byte in the read buffer. The next byte read by the master will be the first byte in the read buffer.
- *uint8 I2C_SlaveGetWriteBufSize(void)* returns the number of bytes written by the I2C master since an *I2C_SlaveInitWriteBuf()* or *I2C_SlaveClearWriteBuf()* function was executed. The maximum return value is the size of the write buffer.
- *void I2C_SlaveClearReadBuf(void)* resets the read pointer to the first byte in the read buffer. The next byte read by the master will be the first byte in the read buffer.

9.2.3 PSoC 3/5 I^2C Master/Multi-Master Slave

PSoC Creator includes a number of I2C components that support master, multi-master and slave configurations with clocks rates up to 1 megabit per second, inclusive. A typical configuration is shown in Fig. 9.1 with two pullup resistors whose value depends on the applicable supply voltage, clock speed and bus capacitance.

This component has four[13] I/O connections:

- *Clock*—is used to clock the transmission of data on the I2C bus and is derived from the bus as shown in Table 9.2.

Table 9.2 Bus frequencies required for a 16X oversampling clock.

Bus	Clock
50 kbps	800 kHz
100 kbps	1.6 MHz
400 kbps	6.4 MHz
1000 kbps	16 MHz

- *Reset*—maintains the I2C block in a hardware reset state, thereby halting I2C communications. A software reset can be invoked by using the *I2C_Stop()* and *I2C_Start()* APIs.[14]
- *sda*—the serial data i/o channel used to transmit/receive I2C bus data.
- *scl*—the master-generated I2C clock. The slave cannot generate a clock signal, but it can hold the clock low, suspending all bus activity until the slave is ready to send data, or ACK/NAK[15] the latest data, or address.[16]

Address decoding can be based on either hardware, which is the default case, or on software.If only a single slave is involved in the design, hardware decoding is preferable. If hardware address decoding is enabled , the I2C component will automatically NAK addresses other than its own, unless CPU intervention occurs. Each slave recognizes its unique address which is between 00x00 and 0x7F, with a default address of 0x04. A 10-bit address can be used by employing software address decoding, but requires that the second byte of the address be decoded, as well.

Signal connections for the SDA and SCL lines can be one of three possible types:

- I2C0—SCL = SIO pin P12[0], SDA = SIO pin P12 [1].
- I2C1—SCL = SIO pin P12 [14], SDA = SIO pin P12 [16].
- Any (Default)—Any GPIO or SIO pins via schematic routing.

PSoC Creator supports four modes of operation:

- slave-only operation,
- master-only operation,

---

[13]The clock and reset pins are only visible in PSoC Creator when the *Implementation* parameter is set to UDB.

[14]The reset input may be left floating by default, which is equivalent to asserting a logic zero signal to the reset pin.

[15]ACK (acknowledged), NAK (not acknowledged) or NACK (not acknowledged) are handshaking signals.

[16]The pin that is connected to *scl* should be configured as *Open-Drain-Drives-Low*.

- multi-master which supports more than one master,

and,

- multi-master-slave which supports simultaneous multi-master and slave operation.

A slave employs two memory buffers, viz., one for data received from the master and one for the master to read data transmissions from the slave. The I2C slave read and write buffers are set by the initialization commands,

```
void I2C_SlaveInitReadBuf(uint8 * rdBuf, uint8 bufSize)
void I2C_SlaveInitWriteBuf(uint8 * wrBuf, uint8 bufSize)
```

However, these commands do not allocate memory, but instead copy the array pointer and size to the internal component variables. The arrays used for the buffers must be set programmatically because they are not automatically generated by the component. Using these functions sets a pointer and byte count for the read and write buffers. The bufSize for these functions may be less than, or equal to, the actual array size, but it should never be larger than the available memory pointed to by the *rdBuf* or *wrBuf* pointers (Fig. 9.5).

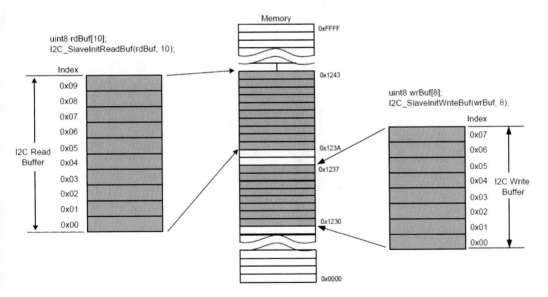

Fig. 9.5 Slave buffer structure.

When the *I2C_SlaveInitReadBuf()*, or the *I2C_SlaveInitWriteBuf()* functions are called, the internal index is set to the first value in the array pointed to by *rdBuf* and *wrBuf*, respectively. As bytes are read/written by the I2C master, the index is incremented until the offset is one less than the *byteCount*. The number of bytes transferred may be determined by calling either *I2C_SlaveGetReadBufSize()* or *I2C_SlaveGetWriteBufSize()* for the read/write buffers, respectively. However, reading/writing, more bytes than are in the buffers causes an overflow error which results in the slave status byte being set.[17]

To reset the index back to the beginning of the array, i.e., zero, use the following commands:

---

[17]This byte can be read via the *I2C_SlaveStatus()* API.

```
void I2C\_SlaveClearReadBuf(void)
void I2C\_SlaveClearWriteBuf(void)
```

The next byte read/write to/by the I2C/SPI [12] is the first byte in the array.[18] Multiple reads, or writes, by the I2C master, continue to increment the array index until a clear buffer command occurs, or the array index exceeds the array size (Fig. 9.6).

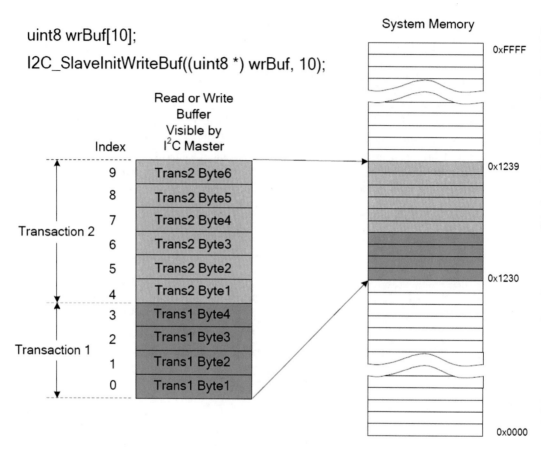

Fig. 9.6 I2C write transaction.

The first write was four bytes and the second write was six bytes. The sixth byte in the second transaction was *NAK*ed by the slave to signal that the end of the buffer had occurred. If the master tries to write a seventh byte for the second transaction, or starts to write more bytes with a third transaction, each subsequent byte will be *NAK*ed and discarded until the buffer is reset. Using the *I2C_SlaveClearWriteBuf( )* function, after the first transaction, resets the index to zero and causes the second transaction to overwrite the data from the first transaction.[19]

---

[18] Before these clear buffer commands are used, the data in the arrays should be read or updated.

[19]The data in the buffer should be processed by the slave before resetting the buffer index.

9.2.4   Master and Multimaster Functions

PSoC 3/5 Master and Multi-Master[20] operations are basically the same, with two exceptions. When operating in Multi-Master mode, the bus should always be checked to see if it is busy. Another master may already be communicating with a slave. In this case, the program must wait until the current operation is complete, before issuing a start transaction. This is accomplished by checking the appropriate return value, to determine whether an error condition has been set, thereby indicating that another master has control of the bus. The second difference is that, in Multi-Master mode, two masters can start at the exact same time.

If this happens, one of the two masters must yield control of the bus and this is accomplished by arbitration. arbitration. A check for this condition must be made after each byte is transferred. The I2C component automatically checks for this condition and responds with an error, if arbitration is lost. Two options are available when operating the I2C master, viz., manual and automatic. In the automatic mode, a buffer is created to hold the entire transfer. In the case of a write operation, the buffer is pre-filled with the data to be sent. If data is to be read from the slave, a buffer of at least the size of the packet to be transmitted needs to be allocated. The following function will write an array of bytes to a slave in automatic mode.

```
uint8 I2C\_MasterWriteBuf(uint8 slaveAddress, uint8 * xferData,
uint8 cnt, uint8 mode)
```

The *slaveAddress* variable is a right-justified, 7-bit slave address ranging from 0 to 127, inclusive. The component's API automatically appends the write flag to the LSb of the address byte. The array of data to transfer is pointed to by the second parameter, *xferData* and the *cnt* parameter is the number of bytes to transfer. The last parameter, *mode*, determines how the transfer starts and stops. A transaction may begin with a restart instead of a start, or halt before the stop sequence. These options allow *back-to-back transfers* where the last transfer does not send a stop, and the next transfer issues a restart, instead of a start.

A read operation is almost identical to the write operation and the same parameters with the same constants are used.

```
uint8 I2C\_MasterReadBuf(uint8 slaveAddress, uint8 * xferData,
uint8 cnt, uint8 mode);
```

Both of these functions return status. See the Table 9.1 for the *I2C_MasterStatus()* function return values. Because the read and write transfers complete in the background, during the I2C interrupt code, the *I2C_MasterStatus()* function can be used to determine when the transfer has been completed.

The following code snippet shows a typical write to a slave.

```
I2C_MasterClearStatus(); /* Clear any previous status */
I2C_MasterWriteBuf(4, (uint8 *) wrData,10,I2C_MODE_COMPLETE_XFER);
for(;;)
{
    if(0u != (I2C_MasterStatus() & I2C_MSTAT_WR_CMPLT))
    {
```

---

[20]In a fixed-function implementation, which does not support undefined bus conditions, for PSoC 3 ES2 and PSoC 5LP, and Master or Multi-Master mode, if the *STOP* condition is set by the software immediately after the *START* condition, the module will generate the *STOP* condition. This occurs after the address field sends 0xFF, if a data write, and the clock line remains low. To avoid this condition, the *STOP* condition should not be set immediately after *START*. At least one byte should be transferred followed by setting *STOP* condition and after *a NAK* or *ACK*.

```
        /* Transfer complete. Check Master status to make sure that
           transfer is completed without errors. */
        break;
    }
}
```

The I2C master can also be operated manually. In this mode, each part of the write transaction is performed with individual commands.

```
status = I2C\_MasterSendStart(4, I2C_WRITE_XFER_MODE);
/* Check if transfer is completed without errors */
if(status == I2C\_MSTR\_NO\_ERROR)
{
    /* Send array of 5 bytes */
    for(i=0; i<5; i++)
    {
        status = I2C\_MasterWriteByte(userArray[i]);
        /* Check if transfer completed without errors */
        if(status != I2C\_MSTR\_NO\_ERROR)
            {
                break;
            }
    }
}
I2C_MasterSendStop(); /* Send Stop */
```

A manual read transaction is similar to the write transaction, except that the last byte should be *NAK*ed. The example below represents a typical manual read transaction.

```
status = I2C\_MasterSendStart(4, I2C_READ_XFER_MODE);
if(status == I2C\_MSTR\_NO\_ERROR)
{
    /* Read array of 5 bytes */
    for(i=0; i<5; i++)
    {
        if(i < 4)
        {
            userArray[i] = I2C\_MasterReadByte(I2C_ACK_DATA);
        }
        else
        {
            userArray[i] = I2C_MasterReadByte(I2C_NAK_DATA);
        }
    }
}
I2C_MasterSendStop(); /* Send Stop */
```

9.2.5 Multi-Master-Slave Mode

In this mode of operation both the Multi-Master and Slave are operational. Although the component can be addressed as a slave, firmware must initiate any master-mode transfers. Enabling *Hardware Address Match* introduces some limitations with respect to arbitrage and address bytes. In the event that the master loses arbitration during an address byte, the hardware reverts to Slave-mode and the byte received generates a slave address interrupt, provided that the slave is addressed. Otherwise, the *lost arbitrage status* will no longer be available to interrupt-based functions.[21] However, the manual-based function, *I2C_MasterSendStart()*,[22] does return correct status information, as shown in Table 9.3, for this particular case.

Table 9.3 *I2C_MasterSendStart()* return values.

Mode Constants	Description
I2C_MSTR_NO_ERROR	Function completed without error
I2C_MSTR_BUS_BUSY	Bus is busy occurred. START condition generation not started.
I2C_MSTR_SLAVE_BUSY	Slave operation in progress.
I2C_MSTR_ERR_LB_NAK	Last byte was NAKed.
I2C_MSTR_ERR_ARB_LOST	Master lost arbitration while generating START.

9.2.6 Multi-Master-Slave Mode Operation

Both Multi-Master and Slave are operational in this mode. The component may be addressed as a slave, but firmware may also initiate master mode transfers. In this mode, when a master loses arbitration, during an address byte, the hardware reverts to Slave mode and the received byte generates a slave address interrupt.

9.2.7 Arbitrage on Address Byte Limitations (*Hardware Address Match* Enabled)

When a master loses arbitration during an address byte, the slave address interrupt is only generated if the slave is addressed. In other cases, the lost arbitrage status is no longer available to interrupt-based functions. The software address detect eliminates this possibility, but excludes the *Wakeup on Hardware Address Match* feature. The manual function, *I2C_MasterSendStart()*, provides correct status information in the case described above.

---

[21] Using software address detection prevents this status from being lost, but excludes the *Wakeup on Hardware Address Match* feature.

[22] This function generates a *START* condition and sends the slave address with a read/write bit.

9.2.8 Start of Multi-Master-Slave Transfer

When using Multi-Master-Slave, the Slave can be addressed at any time. The Multi-Master must have time to prepare to generate a start condition when the bus is free. During this time, the Slave can be addressed and, in this case, the Multi-Master transaction is lost and Slave operation proceeds. Care must be exercised not to break the Slave operation. The I2C interrupt must be disabled before generating a start condition to prevent the transaction from passing the address stage. This action allows a Multi-Master transaction to be aborted and to start a Slave operation correctly.

The following cases are possible when disabling the I2C interrupt:

- The bus is busy (Slave operation is in progress or other traffic is on the bus) before the start generation occurs. The Multi-Master does not try to generate a start condition. Slave operation proceeds when the I2C interrupt is enabled. The *I2C_MasterWriteBuf()*, *I2C_MasterReadBuf()*, or *I2C_MasterSendStart()* call returns the status *I2C_MSTR_BUS_BUSY*. The bus is free before start generation occurs. The Multi-Master generates a start condition on the bus and proceeds with operation when I2C interrupt is enabled. The *I2C_MasterWriteBuf()*, *I2C_MasterReadBuf()*, or *I2C_MasterSendStart()* call returns the status *I2C_MSTR_NO_ERROR*.
- The bus is free before start generation occurs. The Multi-Master tries to generate a start but another Multi-Master addresses the Slave before this and the bus becomes busy. The start condition generation is queued. The Slave operation stops at the address stage because of a disabled I2C interrupt. When I2C interrupt is enabled, the Multi-Master transaction is aborted from queue and Slave operation proceeds. The *I2C_MasterWriteBuf()* or
  - I2C_MasterReadBuf() call does not notice this and returns *I2C_MSTR_NO_ERROR*. The *I2C_MasterStatus()* returns *I2C_MSTAT_WR_CMPLT* or
  - *I2C_MSTAT_RD_CMPLT* with *I2C_MSTAT_ERR_XFER* (all other error condition bits are cleared) after the Multi-Master transaction is aborted. The *I2C_MasterSendStart()* call returns the error status. *I2C_MSTR_ABORT_XFER*.

9.2.9 Interrupt Function Operation

It is possible to assign a priority to a master or slave transaction utilizing interrupts as shown by the following coding example:

- I2C_MasterWriteBuf();

- I2C_MasterReadBuf();

```
I2C\_MasterClearStatus(); /* Clear any previous status */
I2C\_DisableInt(); /* Disable interrupt */
status = I2C\_MasterWriteBuf(4, (uint8 *) wrData, 10, I2C\_MODE\_COMPLETE\_XFER);
/* Try to generate, start. The disabled I2C interrupt halts the transaction in the
address stage, if a Slave is addressed or the Master generates a start condition */
I2C\_EnableInt(); /* Enable interrupt and proceed with the Master or Slave
transaction */
for(;;)
{
if(0u != (I2C\_MasterStatus() & I2\_MSTAT_WR_CMPLT))
{
/* Transfer complete.
```

```
Check Master status to make sure that transfer
completed without errors. */
break;
}
}
if (0u != (I2C_MasterStatus() & I2C_MSTAT_ERR_XFER))
{
        /* Error occurred while transfer, clean up Master status and
           retry the transfer */
}
```

9.2.10 Manual Function Operation

Manual Multi-Master operation assumes that I2C interrupt is disabled, but it is advisable to take the following precaution:

```
I2C_DisableInt(); /* Disable interrupt */
status = I2C_MasterSendStart(4, I2C_WRITE_XFER_MODE);; /* Try to generate start
condition */
if (status == I2C_MSTR_NO_ERROR) /* Check if start generation has completed without
errors */
{
/* Proceed the write operation */
/* Send an array of 5 bytes */
for(i=0; i<5; i++)
{
status = I2C_MasterWriteByte(userArray[i]);
if(status != I2C_MSTR_NO_ERROR)
{
break;
}
}
I2C_MasterSendStop(); /* Send Stop */
}
I2C_EnableInt(); /* Enable interrupt, if it was enabled before */
```

9.2.11 Wakeup and Clock Stretching

The I2C block responds to transactions on the I2C bus, during sleep mode. If the incoming address matches with the slave address, the I2C wakes the system. Once the address matches, a wakeup interrupt is asserted to wake up the system and SCL is pulled low. An ACK is sent out after the system wakes up, and the CPU determines the next action in the transaction.

The I2C slave stretches the clock while exiting sleep mode, as shown in Fig. 9.7.

All clocks in the system must be restored before continuing the I2C transactions. The I2C interrupt is disabled before going to sleep and only enabled after the *I2C_Wakeup()* function is called. During the time between wakeup and end of calling *I2C_Wakeup()*, SCL line is pulled low.

```
...
        I2C_Sleep();           /* Go to Sleep and disable I2C interrupt */
        CyPmSaveClocks();      /* Save clocks settings */
        CyPmSleep(PM_SLEEP_TIME_NONE, PM_SLEEP_SRC_I2C);
        CyPmRestoreClocks(); /* Restore clocks */
        I2C_Wakeup(); /* Wakeup, enable I2C interrupt and ACK the address, until
```

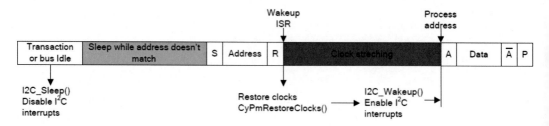

Fig. 9.7 Wakeup and clock stretching.

```
                    the end of this call the SCL is pulled low */
     ...
```

9.2.12 Slave Operation

The slave interface consists of two memory buffers, one for data written to the slave by a master and a second for data read by a master from the slave.[23] The I2C slave read and write buffers are set by the initialization commands discussed below. These commands do not allocate memory, but instead copy the array pointer and array size to the internal component variables. The arrays used for the buffers must be instantiated because they are not automatically generated by the component. The same buffer can be used for both read and write buffers, but care must be exercised to manage the data properly.

The following functions set a pointer and byte count for the read and write buffers.

```
        void I2C_SlaveInitReadBuf(uint8 * rdBuf, uint8 bufSize)
        void I2C_SlaveInitWriteBuf(uint8 * wrBuf, uint8 bufSize)
```

The *bufSize* for these functions may be less than, or equal to, the actual array size, but it should never be larger than the available memory pointed to by the *rdBuf*, or *wrBuf*, pointers. When the *I2C_SlaveInitReadBuf()* or *I2C_SlaveInitWriteBuf()* functions are called, the internal index is set to the first value in the array pointed to by *rdBuf* and *wrBuf*, respectively. As the I2C master reads, or writes the bytes, the index is incremented until the offset is one less than the *byteCount*. At any time, the number of bytes transferred can be queried by calling either I2C_SlaveGetReadBufSize() or I2C_SlaveGetWriteBufSize() for the read and write buffers, respectively. Reading or writing more bytes than are in the buffers causes an overflow error. The error is set in the slave status byte and can be read with the *I2C_SlaveStatus()* API. To reset the index back to the beginning of the array, use the following commands.

```
        void I2C\_SlaveClearReadBuf(void)
        void I2C\_SlaveClearWriteBuf(void)
```

This resets the index back to zero. The next byte the I2C master reads or writes to is the first byte in the array. Before using these clear buffer commands, the data in the arrays should be read or updated.

Multiple reads or writes, by the I2C master, continue to increment the array index until the clear buffer commands are used or the array index tries to grow beyond the array size. Figure 9.6 shows example where an I2C master has executed two write transactions. The first write was four bytes and the second write was six bytes. The sixth byte in the second transaction was NAKed by the slave to signal that the end of the buffer had occurred. If the master tried to write a seventh byte for the

---

[23]Reads and writes are from the perspective of the I2C master.

second transaction or started to write more bytes with a third transaction, each byte would be NAKed and discarded until the buffer is reset. Using the *I2C_SlaveClearWriteBuf()* function after the first transaction resets the index back to zero and causes the second transaction to overwrite the data from the first transaction. Make sure data is not lost by overflowing the buffer. The data in the buffer should be processed by the slave before resetting the buffer index.

Table 9.4 I2C slave status constants.

Slave Status Constants	Value	Description
I2C_SSTAT_RD_CMPLT	0x01	Slave read transfer complete
I2C_SSTAT_RD_BUSY	0x02	Slave read transfer in progress (busy)
I2C_SSTAT_RD_OVFL	0x04	Master attempted to read more bytes than are in the buffer
I2C_SSTAT_WR_CMPLT	0x10	Slave write transfer complete
I2C_SSTAT_WR_CMPLT	0x20	Slave write transfer in progress (busy)
I2C_SSTAT_WR_CMPLT	0x40	Master attempted to read more bytes than are in the buffer

Both the read and write buffers have four status bits to signal that a transfer is complete, a transfer is in progress, and buffer overflow. Starting a transfer sets the busy flag and when the transfer has been completed, the transfer complete flag is set and the busy flag is cleared. If a second transfer is started, both the busy and transfer complete flags can be set at the same time. The values for the slave status constants are shown in Fig. 9.4.

9.2.13 Start of Multi-Master-Slave Transfer

When using a multi-master-slave, the slave can be addressed at any time. The multi-master must take time to prepare to generate a *Start* condition when the bus is free. During this time, the slave could be addressed and, if so, the multi-master transaction is lost and the slave operation proceeds. Care must be exercised to avoid breaking the slave operation and the I2C interrupt must be disabled before generating a *Start* condition to prevent the transaction from passing the address stage. This action allows a multi-master transaction to be aborted and a slave operation to be started correctly.

The following cases are possible when disabling the I2C interrupt:

- The bus is busy, e.g., slave operation is in progress or other traffic is on the bus, before *Start* generation. The multi-master does not try to generate a *Start* condition. Slave operation proceeds when the I2C interrupt is enabled. The *I2C_MasterWriteBuf()*, *I2C_MasterReadBuf()*, or *I2C_MasterSendStart()* call returns the status *I2C_MSTR_BUS_BUSY*.

- The bus is free, before *Start* generation. The multi-master generates a Start condition on the bus and proceeds with operation when the I2C interrupt is enabled. The *I2C_MasterWriteBuf( )*, *I2C_MasterReadBuf( )*, or *I2C_MasterSendStart( )* call returns the status *I2C_MSTR_NO_ERROR*.

- The bus is free before Start generation. The multi-master tries to generate a Start, but another multi-master addresses the slave before this, and the bus becomes busy. The Start condition generation is queued. The slave operation stops at the address stage because of a disabled I2C interrupt. When the I2C interrupt is enabled, the multi-master transaction is aborted from the queue, and the slave operation proceeds. The *I2C_MasterWriteBuf( )* or I2C_MasterReadBuf() call does not notice this and returns *I2C_MSTR_NO_ERROR*. The *I2C_MasterStatus( )* returns *I2C_MSTAT_WR_CMPLT* or *I2C_MSTAT_RD_CMPLT* with *I2C_MSTAT_ERR_XFER* (all other error condition bits are cleared) after the multi-master transaction is aborted. The *I2C_MasterSendStart( )* call returns the error status *I2C_MSTR_ABORT_XFER*.

9.3 Universal Asynchronous Rx/Tx (UART)

PSoC Creator's UART [22] component provides asynchronous communications is often employed to implement the RS232, or RS485[24] protocols.[25] The UART component can be configured for Full Duplex[26] [21], Half Duplex,[27] RX-only, or TX-only versions.[28] However, all four transmission modes have the same basic functionality differing only in the amount of resources used. Two configurable buffers, each of independent size, serve as circular receive and transit buffers that are assigned in SRAM and hardware FIFOs to ensure data integrity.

This arrangement allows the CPU to spend more time on critical, real-time tasks than servicing the UART [4]. In most cases, the UART is configured by choosing the baud rate,[29] parity,[30] number of data bits, and number of start bits. The most common configuration for RS232 is eight data bits, no parity, and one stop bit and designated as *8N1* and is the default configuration. A second common use for UARTs is in multidrop[31] RS485 networks.

The UART component supports a 9-bit addressing mode with hardware address detect, as well as, a TX output enable signal to enable the TX transceiver during transmissions. There are a number of

---

[24]The RS485, also referred to as TIA-485 or EIA-485, protocol is similar to RS232 but differs in that it is a more noise-immune protocol than RS232 and it allows as many as 32 devices to share a common, 3-wire bus and communicate over distances as long as 4000 ft (1200 m). The transmission path is differential (balanced) and consists of a twisted pair and a third wire which serves as ground (there is also a four-wire configuration) that provides very high noise immunity.

[25]The UART can also be employed in a TTL-compatible mode.

[26]A full-duplex system allows simultaneous transmissions in both directions over the communication path.

[27]A half duplex system allows transmissions in both directions over the communication path but not simultaneously.

[28]The UART can also be configured for more advanced protocols such as DMX512, LIN and IrDA or custom protocols.

[29]Baud rate refers to the rate at which bits are transmitted per second.

[30]Parity, in the present context, refers to the use of an optional bit associated with each transmitted byte which has the value 1 if the number of 1s in the byte is even and 0, otherwise. This provides a mechanism to determine, whether the integrity of a byte has been compromised during transmission.

[31]Multidrop implies multiple slaves.

physical-layer and protocol-layer variations of UARTs in common use including RS423,[32] DMX512, MIDI, LIN[33] bus, legacy terminal protocols, and IrDA.[34]

To support the more commonly used variations, the number of data bits, stop bits, parity, hardware flow control, and parity generation and detection is configurable from within PSoC Creator and under software control. As a hardware-compiled option, a clock and serial data stream can be used that transmits the UART data bits only on the clock's rising edge. An independent clock and data output can also be employed for both TX and RX. These outputs allow automatic calculation of the data CRC [6] by connecting a CRC component to the UART.

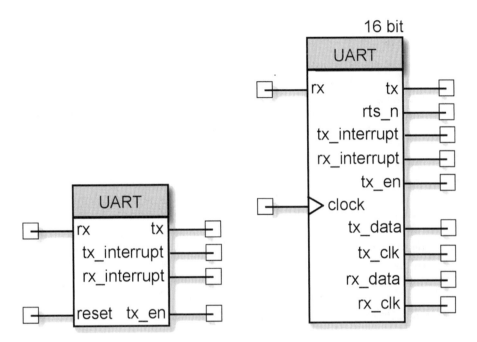

Fig. 9.8 PSoC Creator UART configurations.

The PSoC Creator UART component, shown in Fig. 9.8, has the following features:

- 8x and 16x oversampling,
- 9-bit address mode with hardware address detection,
- Baud rates from 110 to 921,600 bits per second (bps) or arbitrary up to 4 Mbps
- Break signal detection and generation,
- Detection of framing, parity and overrun errors,
- Full duplex, half-duplex, TX only, RX only, optimized hardware,

---

[32]RS423, also referred to as EIA-423 and TIA-423, is an unbalanced (single-ended) interface, that is RS232-like, and employs a single, unidirectional, driver, that is capable of supporting up to 10 slaves. It is normally implemented by using integrated circuit technology and can also be employed for the interchange of serial binary signals between DTE & DCE.

[33]The LIN bus is an inexpensive single wire bus capable of operating at baud rates up to 19.2 kbits/s, employed in a master-slave configuration having a single master and one or more slaves.

[34]The Infrared Data Association (IrDA) has established a standard protocol for IR modulation/demodulation methods and other physical parameters associated with infrared transceivers.

- Rx and Tx buffers = 4 to 65,535 bytes,

and,

- Two out of three voting, per bit.

The UARTs *clock* input determines the serial communication baud rate (bit-rate) which is one-eighth, or one-sixteenth, of the input clock frequency, depending on the value selected for the *Oversampling Rate* parameter. This input is visible, in PSoC Creator, if the *Clock Selection* parameter is set to *External Clock*. If the internal clock is selected, the desired baud rate must be selected during configuration.[35] Resetting the UART, via the *reset* input,[36] places the state machines, *RX* and *TX*, in the idle state, in which case, any data that was currently being transmitted, or received, is discarded.

The *rx* input carries the input serial data from another device on the serial bus.[37] The *tx_output* connection is visible only if the Mode parameter is set to *TX Only*, *Half Duplex*, or *Full UART (RX + TX)*.[38] The *tx_en* output[39] is used primarily for RS485 communication to show that the component is transmitting on the bus. This output goes high, before a transmit starts, low when transmit is complete and shows a busy bus to the rest of the devices on the bus. The *tx_interrupt* output is the logical OR of the group of possible interrupt sources and goes high when any of the enabled interrupt sources are true.[40] The *cts_n* input,[41] (_n), an active-low input, shows that another device is ready to receive data.

The *rx_interrupt* output[42] is the logical OR of the group of possible interrupt sources and goes high while any of the enabled interrupt sources are true. The *tx_data* output is used to shift out the TX data to a CRC component, or other logic.[43] The *tx_clk* output[44] provides the clock edge used to shift out the TX data to a CRC component, or other logic. The *rx_data* output[45] is used to shift out the RX data to a CRC component, or other logic. The *rx_clk* output[46] provides the clock edge used to shift out the RX data to a CRC component, or other logic.[47]

9.3.1   UART Application Programming Interface

The API routines for the UART allow the component to be configured programmatically. The following describes the interface for each function.

---

[35]In such cases, PSoC Creator determines the necessary clock frequency for the required baud rate.

[36]This input is a synchronous reset that requires at least one rising edge of the clock, but can be left floating and the component will assign it a constant logic 0.

[37]This input is visible and must be connected if the Mode parameter is set to *RX Only*, *Half Duplex*, or *Full UART (RX + TX)*.

[38]An external pull-up resistor should be used to protect the receiver from unexpected low impulses during active *System Reset*.

[39]This output is visible when the *Hardware TX Enable* parameter is selected.

[40]This output is visible if the *Mode* parameter is set to *TX Only* or *Full UART (RX + TX)*.

[41]This input is visible if the Flow Control parameter is set to Hardware.

[42]This output is visible if the *Mode* parameter is set to *RX Only*, *Half Duplex*, or *Full UART (RX + TX)*.

[43]This output is visible when the Enable CRC outputs parameter is selected.

[44]Ibid.

[45]Ibid.

[46]Ibid.

[47]Ibid.

- *void UART_Start(void)* is the preferred method to begin component operation. *UART_Start()* sets the *initVar* variable, calls the *UART_Init()* function, and then calls the *UART_Enable()* function.
- *void UART_Stop(void)* disables the UART operation.
- *uint8 UART_ReadControlRegister(void)* returns the current value of the control register.
- void UART_WriteControlRegister(uint8 control) writes an 8-bit value into the control register.
- *void UART_EnableRxInt(void)* enables the internal receiver interrupt.
- void UART_DisableRxInt(void) disables the internal receiver interrupt.
- void UART_SetRxInterruptMode(uint8 intSrc) configures the RX interrupt sources enabled.
- *uint8 UART_ReadRxData(void)* returns the next byte received without checking the status. The status must be checked separately.
- *uint8 UART_ReadRxStatus(void)* returns the current state of the receiver status register and the software buffer overflow status.
- *uint8 UART_GetChar(void)* returns the last received byte of data and is designed for ASCII characters. It returns a uint8 where 1 to 255 are values for valid characters and 0 indicates an error occurred, or that there is no data present.
- *uint16 UART_GetByte(void)* reads the UART RX buffer immediately and returns the received character and an error condition.
- *uint8/uint16 UART_GetRxBufferSize(void)* returns the number of received bytes remaining in the RX buffer.
- *void UART_ClearRxBuffer(void)* clears the receiver memory buffer and hardware RX FIFO of all received data.
- *void UART_SetRxAddressMode(uint8 addressMode)* sets the software-controlled Addressing mode used by the RX portion of the UART.
- *void UART_SetRxAddress1(uint8 address)* sets the first of two hardware-detectable receiver addresses.
- *void UART_SetRxAddress2(uint8 address)* sets the second of two hardware-detectable receiver addresses.
- void UART_EnableTxInt(void) enables the internal transmitter interrupt.
- *void UART_DisableTxInt(void)* disables the internal transmitter interrupt.
- *void UART_SetTxInterruptMode(uint8 intSrc)* configures the TX interrupt sources to be enabled (but does not enable the interrupt).
- *void UART_WriteTxData(uint8 txDataByte)* places a byte of data into the transmit buffer to be sent when the bus is available without checking the TX status register. Status must be checked separately.
- *uint8 UART_ReadTxStatus(void)* reads the status register for the TX portion of the UART.
- *void UART_PutChar(uint8 txDataByte)* places a byte of data into the transmit buffer to be sent when the bus is available. This is a blocking API that waits until the TX buffer has room to hold the data.
- *void UART_PutString(char* string)* sends a NULL terminated string to the TX buffer for transmission.
- *void UART_PutArray(uint8* string, uint8/uint16 byteCount)* places N bytes of data from a memory array into the TX buffer for transmission.
- *void UART_PutCRLF(uint8 txDataByte)* writes a byte of data followed by a carriage return (0x0D) and line feed (0x0A) to the transmit buffer.
- *uint8/uint16 UART_GetTxBufferSize(void)* determines the number of bytes used in the TX buffer. An empty buffer returns 0.
- *void UART_ClearTxBuffer(void)* clears all data from the TX buffer and hardware TX FIFO.
- *void UART_SendBreak(uint8 retMode)* transmits a break signal on the bus.

- *void UART_SetTxAddressMode(uint8 addressMode)* configures the transmitter to signal the next bytes is address, or data.
- *void UART_LoadRxConfig(void)*loads the receiver configuration in half-duplex mode. After calling this function, the UART is ready to receive data.
- *void UART_LoadTxConfig(void)* loads the transmitter configuration in half duplex mode. After calling this function, the UART is ready to transmit data.
- *void UART_Sleep(void)* is the preferred API to prepare the component for sleep. The *UART_Sleep()* API saves the current component state. Then it calls the *UART_Stop()* function and calls *UART_SaveConfig()* to save the hardware configuration. Call the *UART_Sleep()* function before calling the *CyPmSleep()* or the *CyPmHibernate()*function.
- *void UART_Wakeup(void)* is the preferred API to restore the component to the state when *UART_Sleep()*was called. The *UART_Wakeup()* function calls the *UART_RestoreConfig()* function to restore the configuration. If the component was enabled before the *UART_Sleep()* function was called, the *UART_Wakeup()* function will also re-enable the component.
- *void UART_Init(void)*initializes, or restores, the component according to the customizer *Configure* dialog settings. It is not necessary to call *UART_Init()* because the *UART_Start()* API calls this function and is the preferred method to begin component operation.
- *void UART_Enable(void)* activates the hardware and begins component operation. It is not necessary to call *UART_Enable()* because the *UART_Start()* API calls this function, which is the preferred method to begin component operation.
- *void UART_SaveConfig(void)* saves the component configuration and nonretention registers. It also saves the current component parameter values, as defined in the *Configure* dialog or as modified by appropriate APIs. This function is called by the UART_Sleep() function.
- *void UART_RestoreConfig(void)* restores the user configuration of non-retention registers.

9.3.2 Interrupts

The *Interrupt On* parameters allow the interrupt sources to be configured. These values are ORed with any of the other *Interrupt On* parameter to give a final group of events that can trigger an interrupt. The software can reconfigure these modes at any time, and these parameters define an initial configuration.

- RX—On Byte Received *(UART_RX_STS_FIFO_NOTEMPTY)*
- TX—On TX Complete *(UART_TX_STS_COMPLETE)*
- RX—On Parity Error *(UART_RX_STS_PAR_ERROR)*
- TX—On FIFO Empty *(UART_TX_STS_FIFO_EMPTY)*
- RX—On Stop Error *(UART_RX_STS_STOP_ERROR)*
- TX—On FIFO Full *(UART_TX_STS_FIFO_FULL)*
- RX—On Break *(UART_RX_STS_BREAK)*
- TX—On FIFO Not Full *(UART_TX_STS_FIFO_NOT_FULL)*
- RX—On Overrun Error *(UART_RX_STS_OVERRUN)*
- RX—On Address Match *(UART_RX_STS_ADDR_MATCH)*
- RX—On Address Detect *(UART_RX_STS_MRKSPC)*

An ISR can be handled by an external interrupt component connected to the *tx_interrupt* or *rx_interrupt* output. The interrupt output pin is visible depending on the selected *Mode* parameter. It outputs the same signal to the internal interrupt based on the selected status interrupts.

```
#include <device.h>

#define START_CHAR_VALUE    0x20
#define END_CHAR_VALUE      0x7E

uint8 trigger = 0;

void main()
{
    uint8 ch;              /* Data sent on the serial port */
    uint8 count = 0;       /* Initializing the count value */
    uint8 pos = 0;

    CyGlobalIntEnable;

    isr_1_Start();         /* Initializing the ISR */
    UART_1_Start();        /* Enabling the UART */

    for(ch = START_CHAR_VALUE; ch <= END_CHAR_VALUE; ch++)
    {
        UART_1_WriteTxData(ch); /* Sending the data */
        CyDelay(200);
    }

    for(;;) {}
}

void main()
{
  char8 ch;          /* Data received from the Serial port */

  CyGlobalIntEnable; /* Enable all interrupts by the processor. */

        UART_1_Start();

    while(1)
    {
        /* Check the UART status */
        ch = UART_1_GetChar();

        /* If byte received */
        if(ch > 0)
        {
            // Place character
            // Handling code here
        }
    }
```

9.3.3   UART Config Tab

The Mode dialog box determines the mode of operation of the UART, e.g., as a bidirectional *Full UART (TX + RX)*,[48] *Half Duplex UART*, requiring half of the resources, *RX Only* (RS232 Receiver) or *TX Only* (Transmitter). The *Bits-Per-Second* parameter determines the baud-rate, or bit-width, configuration of the hardware for clock generation.[49] The *Data bits* parameter determines the number of data bits transmitted between the start, and stop, of a single UART transaction (Fig. 9.9).[50]

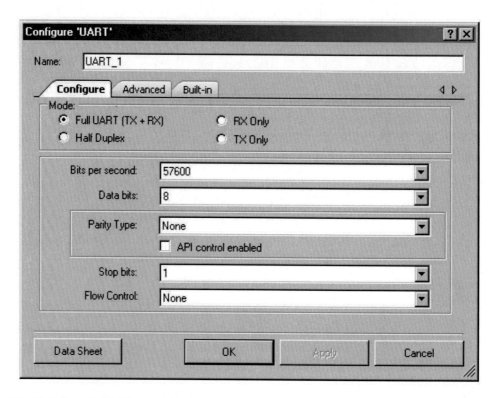

Fig. 9.9 PSoC Creator's UART configuration tab.

9.3.4   Parity

Parity refers to the appending of an extra bit to each byte for the purpose of detecting an error that occurs during serial byte transmission. The parity bit is set to one, if the number of 1 bits[51] is either even or odd, depending on the parity mode selected. Parity can be set as *Even, Odd, None* or

---

[48]This is the default mode.

[49]The default setting for bits per second is 57,600. If the internal clock is used, by setting the Clock Selection parameter, PSoC Creator generates the necessary clock for 57,600 bps.

[50]Options are 5, 6, 7, 8, or 9 data bits. The default setting of 8 bits, results in a transmission of a single byte per transmission. The 9 bit setting utilizes a ninth bit as a parity bit as an indication of either even or odd parity of the eight bits.

[51]Exclusive of the parity bit setting.

Mark/Space. None implies that the ninth bit is not to be employed, i.e. parity is not used, Even/Odd implies that the number of 1s, exclusive of the parity bit, in the byte is even/odd. Upon receipt of each byte, the parity bit can be checked to determine if a change has occurred.[52]

9.3.5 Simplex, Half and Full Duplex

Serial transmission[53] can occur in various modes including simplex, half-duplex and full-duplex as illustrated in Fig. 9.10. A duplex communications system allows communications between two points, in either direction, contemporaneously. Half-duplex systems also allow communications between two points, in both directions, but only in only one direction at a time. Simplex systems allow communication between any two points in one direction only.

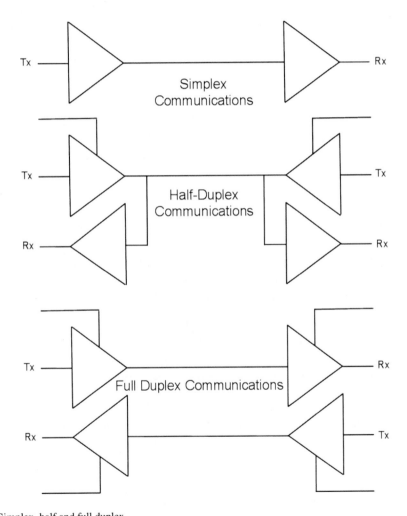

Fig. 9.10 Simplex, half and full duplex.

---

[52] Single bit errors are the most common type of errors that occur during byte transmission.

[53] Serial communication refers to the transfer of information, in a sequential fashion, from one location to another.

9.3.6 RS232, RS422 and RS485 Protocols

The serial communication protocols that have been most prevalent for applications employing UARTs have historically been RS232, RS422[54] and RS485.[55] The RS232 protocol is a point-to-point bidirectional communications link as opposed to RS485, a single channel bus. The RS232 signal path employs a single wire and symmetric voltages about a common ground The RS485 protocol is an EIA[56] standard interface that employs a balanced transmission path[57] and is able to communicate with multiple nodes. It is particularly useful when communication is to occur over relatively long distances and can be employed at distances of up to 1200 m and at rates up to 100 kbit/s.

Table 9.5 Comparison of RS232 and RS485 protocols.

Protocol	RS232	RS485
Max Drivers/Receivers	1/1	32/32
Load Impedance	3k-7k	54
Mode	Single-ended	Differential
Max Data rate	115.200 kbaud	100 -3500 kbaud
Max cable length	15 meters	1200 meters
Max slew rate	5 v 30 V/sec	N/A
Max Driver Current	+/- 6 ma @	+/-100a
Output signal levels	+/- 5 to +/- 15 v	+/- 1.5 v
Supported Duplex Modes	Full and Half	Full and Half
Communication Type	Peer	Multi-point
Sync/Async	Async	Async

The RS232 and RS485 protocols are similar but there are some significant differences as shown in Table 9.5.

---

[54] RS422 is a communication protocol based on differential data transmission which was originally intended to support higher data rates than RS232 and over longer distances. However, this protocol is not a true multidrop protocol in that it will only support one driver and a maximum of ten receivers. There is a four wire implementation of RS422 that supports multiple drivers, but typically, in a half duplex mode.

[55] RS485 is a true *multidrop system* in that it will support multiple drivers and receivers .

[56] Electronic Industries Alliance.

[57] The "balanced" path affords noise immunity making it possible for the receiver to reject common-mode signals and shifts in the ground pathway.

9.4   Serial Peripheral Interface (SPI)

The serial peripheral interface bus, or SPI, was developed by Motorola for intercommunications with relatively slow peripheral devices on an intermittent basis, e.g., transfer of data to a microcontroller from an analog to digital converter. Although comparable to I2C in many respects, SPI is capable of higher data rates and its ability to operate in a full-duplex mode.

PSoC 3/5's Serial Peripheral Interface (SPI) component features include:

- 3- to 16-bit data width
- Four SPI operating modes
- Bit rates up to 9 Mbps[58]

9.4.1   SPI Device Configurations

PSoC Creator supports a number of configurations of SPI masters and slaves as shown in Fig. 9.11.

9.4.2   SPI Master

The *SPI Master* component provides an industry-standard, 4-wire, master SPI interface. It can also provide a 3-wire (bidirectional) SPI interface. Both interfaces support all four SPI operating modes, allowing communication with any SPI slave device. In addition to the standard 8-bit word length, the *SPI Master* supports a configurable 3- to 16-bit word length for communicating with nonstandard SPI word lengths. SPI signals include the standard *Serial Clock* (SCLK), *Master In Slave Out* (MISO), *Master Out Slave In* (MOSI), bidirectional *Serial Data* (SDAT), and *Slave Select* (SS). The *SPI Master* component can be used when the PSoC device must interface with one, or more, SPI slave devices. In addition to SPI slave-labeled devices [19], the *SPI Master* can be used with many devices implementing a shift-register-type serial interface. The *SPI Slave* component should be used in instances in which the PSoC device must communicate with an SPI master device [18]. The *Shift Register* component can be used in situations for which its low-level flexibility provides hardware capabilities not available in the *SPI Master* component.

9.4.3   SPI I/O

PSoC Creator's SPI Master component can be configured using the following:

- *void SPIM_Start(void)* calls both *SPIM_Init()* and *SPIM_Enable().*[59]
- *void SPIM_Stop(void)* disables *SPI Master* operation by disabling the internal clock and internal interrupts, if the *SPI Master* is configured that way.
- *void SPIM_Start(void)* calls both *SPIM_Init()* and *SPIM_Enable()* and should be called the first time the component is started.

---

[58]This value is valid only for MOSI+MISO (Full Duplex) interfacing mode and is restricted up to 1 Mbps in the bidirectional mode, because of internal bidirectional pin constraints.

[59]This should be called the first time the component is started.

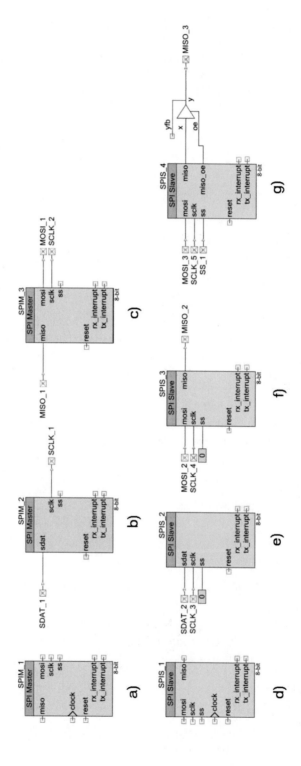

Fig. 9.11 SPI master and slave Schematic Macros supported by PSoC Creator.

- *void SPIM_Stop(void)* disables *SPI Master* operation by disabling the internal clock and internal interrupts.
- *void SPIM_EnableTxInt(void)* enables the internal Tx interrupt irq.
- *void SPIM_EnableRxInt(void)* enables the internal Rx interrupt irq.
- *void SPIM_DisableTxInt(void)* disables the internal Tx interrupt irq.
- *void SPIM_DisableRxInt(void)* disables the internal Rx interrupt irq.
- *void SPIM_SetTxInterruptMode(uint8 intSrc)* configures which status bits trigger an interrupt event.
- *void SPIM_SetRxInterruptMode(uint8 intSrc)* configures which status bits trigger an interrupt event.
- *uint8 SPIM_ReadTxStatus(void)* returns the current state of the Tx status register.
- *uint8 SPIM_ReadRxStatus(void)* returns the current state of the Rx status register.
- *void SPIM_WriteTxData(uint8/uint16 txData)* places a byte/word in the transmit buffer to be sent at the next available SPI bus time. Data may be placed in the memory buffer and will not be transmitted until all other previous data have been transmitted. This function is blocked until there is space in the output memory buffer. It also clears the Tx status register of the component.
- *uint8/uint16 SPIM_ReadRxData(void)*[60] returns the next byte/word of received data available in the receive buffer. It returns invalid data, if the FIFO is empty. Call *SPIM_GetRxBufferSize()*, and if it returns a nonzero value then it is safe to call the *SPIM_ReadRxData()* function.
- *uint8 SPIM_GetRxBufferSize(void)* returns the number of bytes/words of received data currently held in the Rx buffer.
  - If the Rx software buffer is disabled, this function returns $0 = $ FIFO empty or $1 = $ FIFO not empty.
  - If the Rx software buffer is enabled, this function returns the size of data in the Rx software buffer. FIFO data not included in this count
  - *uint8 SPIM_GetTxBufferSize(void)* returns the number of bytes/words of data ready to transmit currently held in the Tx buffer. .
  - If Tx software buffer is disabled, this function returns $0 = $ FIFO empty, $1 = $ FIFO not full, or $4 = $ FIFO full.
  - If the Tx software buffer is enabled, this function returns the size of data in the Tx software buffer.[61]
- *void SPIM_ClearRxBuffer(void)*clears the Rx buffer memory array and Rx hardware FIFO of all received data. It clears the Rx RAM buffer by setting both the read and write pointers to zero. Setting the pointers to zero indicates that there is no data to read. Thus, writing resumes at address 0, overwriting any data that may have remained in the RAM.
- *void SPIM_ClearTxBuffer(void)* clears the Tx buffer memory array of data waiting to transmit. It clears the Tx RAM buffer by setting both the read and write pointers to zero. Setting the pointers to zero indicates that there is no data to transmit. Thus, writing resumes at address 0, overwriting any data that may have remained in the RAM.
- *void SPIM_TxEnable(void)* sets the bidirectional pin to transmit, if the SPI Master is configured to use a single bidirectional pin.
- *void SPIM_TxDisable(void)* sets the bidirectional pin to receive, if the SPI master is configured to use a single bidirectional pin.
- void SPIM_PutArray(uint8/uint16 * buffer, uint8/uint16 byteCount) places an array of data into the transmit buffer.

---

[60]This function returns invalid data, if the FIFO is empty.
[61]The FIFO data not included in this count.

- *void SPIM_ClearFIFO(void)* clears any received data from the Tx and Rx FIFOs.
- *void SPIM_Sleep(void)* prepares the SPI Master for low-power modes by calling the *SPIM_SaveConfig()*and *SPIM_Stop()* functions.
- *void SPIM_Wakeup(void)* prepares the SPI Master to wake up from a low-power mode and calls the SPIM_RestoreConfig() and SPIM_Enable() functions. Also clears all data from the Rx buffer, Tx buffer, and hardware FIFOs.
- *void SPIM_Init(void)* initializes, or restores, the component according to the customizer *Configure* dialog settings. It is not necessary to call *SPIM_Init()* because the *SPIM_Start()* routine calls this function and is the preferred method to begin component operation.
- *void SPIM_Enable(void)* enables the SPI Master for operation. Starts the internal clock, if the SPI Master is configured that way. If it is configured for an external clock, it must be started separately before calling this function. The *SPIM_Enable()* function should be called before SPI Master interrupts are enabled. This is because this function configures the interrupt sources and clears any pending interrupts from device configuration, and then enables the internal interrupts, if there are any. A *SPIM_Init()* function must have been previously called.
- *void SPIM_SaveConfig(void)* saves the SPI Master hardware configuration before entering a low-power mode.
- *void SPIM_RestoreConfig(void)* restores the SPI Master hardware configuration saved by the *SPIM_SaveConfig()* function after waking from a lower-power mode.

9.4.4   Tx Status Register

The Tx status register is a read-only register that contains the various transmit status bits defined for a given instance of the SPI Master component. Assuming that an instance of the SPI Master is named *SPIM*, the value of this register can be obtained by using the *SPIM_ReadTxStatus()* function. The interrupt output signal is generated by ORing the masked bit fields within the Tx status register. The mask can be set by using the *SPIM_SetTxInterruptMode()* function. Upon receiving an interrupt, the interrupt source can be retrieved by reading the Tx status register with the*SPIM_ReadTxStatus()*function. Sticky bits in the Tx status register are cleared on reading, so the interrupt source is held until the *SPIM_ReadTxStatus()* function is called.

All operations on the Tx status register must use the following defines for the bit fields, because these bit fields may be moved within the Tx status register at build time. Sticky bits are used to generate an interrupt or DMA transaction and must be cleared with either a CPU or DMA read to avoid continuously generating the interrupt or DMA. There are several bit fields defined for the Tx status registers. Any combination of these bit fields may be included as an interrupt source. The bit fields indicated with an asterisk (*) in the following list are configured as sticky bits in the Tx status register. All other bits are configured as real-time indicators of status. Sticky bits latch a momentary state so that they may be read at a later time and cleared on a read.

The following # defines are available in the generated header file (for example, SPIM.h):

- *SPIM_STS_SPI_DONE* * Set high as the data-latching edge of SCLK (edge is mode dependent) is output. This happens after the last bit of the configured number of bits in a single SPI word is output onto the MOSI line and the transmit FIFO is empty. Cleared when the SPI Master is transmitting data or the transmit FIFO has pending data. Tells you when the SPI Master is complete with a multi-word transaction.
- *SPIM_STS_TX_FIFO_EMPTY* reads high while the transmit FIFO contains no data pending transmission and reads low, if data is waiting for transmission.

- *SPIM_STS_TX_FIFO_NOT_FULL* reads high while the transmit FIFO is not full and has room to write more data. It reads low, if the FIFO is full of data pending transmit and there is no room for more writes at this time. Tells you when it is safe to pend more data into the transmit FIFO.
- SPIM_STS_BYTE_COMPLETE * set high as the last bit of the configured number of bits in a single SPI word is output onto the MOSI line. Cleared* as the data latching edge of SCLK (edge is mode dependent) is output.
- SPIM_STS_SPI_IDLE * is set high as long as the component state machine is in the SPI IDLE state (component is waiting for Tx data and is not transmitting any data).

9.4.5 RX Status Register

The Rx status register is a read-only register that contains the various receive status bits defined for the SPI Master. The value of this register can be obtained by using the *SPIM_ReadRxStatus()* function. An interrupt output signal is generated by ORing the masked bit fields within the Rx status register. The mask can be set by using the *SPIM_SetRxInterruptMode()* function. Upon receiving an interrupt, the interrupt source can be retrieved by reading the Rx status register with the *SPIM_ReadRxStatus()* function. Sticky bits in the Rx status register are cleared on reading, so the interrupt source is held until the *SPIM_ReadRxStatus()* function is called. All operations on the Rx status register must use the following defines for the bit fields, because these bit fields may be moved within the Rx status register at build time. Sticky bits used to generate an interrupt or DMA transaction must be cleared with either a CPU or DMA read to avoid continuously generating the interrupt or DMA. There are several bit fields defined for the Rx status register. Any combination of these bit fields can be included as an interrupt source. The bit fields indicated with an asterisk (*) in the following list are configured as sticky bits in the Rx status register. All other bits are configured as real-time indicators of status. Sticky bits latch a momentary state so that they may be read at a later time and cleared when read. The following #defines are available in the generated header file (for example, SPIM.h):

- *SPIM_STS_SPI_DONE* * set high as the data-latching edge of SCLK (edge is mode dependent) is output. This happens after the last bit of the configured number of bits in a single SPI word is output onto the MOSI line and the transmit FIFO is empty. Cleared when the SPI Master is transmitting data or the transmit FIFO has pending data. Tells you when the SPI Master is complete with a multi-word transaction.
- *SPIM_STS_TX_FIFO_EMPTY* reads high while the transmit FIFO contains no data pending transmission. Reads low if data is waiting for transmission.
- *SPIM_STS_TX_FIFO_NOT_FULL* reads high while the transmit FIFO is not full and has room to write more data. It reads low, if the FIFO is full of data pending transmit and there is no room for more writes at this time. Indicates when it is safe to send more data to the transmit FIFO.
- *SPIM_STS_BYTE_COMPLETE* * set high as the last bit of the configured number of bits in a single SPI word is output onto the MOSI line. Cleared* as the data latching edge of SCLK (edge is mode dependent) is output.
- *SPIM_STS_SPI_IDLE* * this bit is set high as long as the component state machine is in the SPI IDLE state (component is waiting for Tx data and is not transmitting any data).

9.4.6 Tx Data Register

The Tx data register contains the transmit data value to send and is implemented as a FIFO in the SPI Master. There is an optional higher-level software state machine that controls data from the transmit memory buffer. It handles large amounts of data to be sent that exceed the FIFO's capacity. All APIs that involve transmitting data must go through this register to place the data onto the bus. If there is data in this register and the control state machine indicates that data can be sent, then the data is transmitted on the bus. As soon as this register (FIFO) is empty, no more data will be transmitted on the bus until it is added to the FIFO. DMA can be set up to fill this FIFO when empty, using the *TXDATA_REG* address defined in the header file.

9.4.7 Rx Data Register

The Rx data register contains the received data and is implemented as a FIFO in the SPI Master. There is an optional higher-level software state machine that controls data movement from this receive FIFO into the memory buffer. Typically, the Rx interrupt indicates that data has been received. At that time, that data has several routes to the firmware. DMA can be set up from this register to the memory array, or the firmware can simply call the *SPIM_ReadRxData( )* function. DMA must use the *RXDATA_REG* address defined in the header file.

9.4.8 Conditional Compilation Information

The SPI Master requires only one conditional compile definition to handle the 8- or 16-bit datapath configuration necessary to implement the configured *NumberOfDataBits*. The API must conditionally compile for the data width defined. APIs should never use these parameters directly but should use the following define:

- *SPIM_DATAWIDTH* defines how many data bits will make up a single-byte transfer. The valid range is 3–16 bits.

9.5 Serial Peripheral Interface Slave

PSoC Creator's SPI Slave provides an industry-standard, 4-wire slave SPI interface capable of providing a 3-wire, bidirectional, SPI interface. Both interfaces support all four SPI operating modes, allowing communication with any SPI master device. In addition to the standard 8-bit word length, the SPI Slave supports a configurable 3- to 16-bit word length for communicating with nonstandard SPI word lengths. SPI signals include the standard *Serial Clock* (SCLK), *Master In Slave Out* (MISO), *Master Out Slave In* (MOSI), *bidirectional Serial Data* (SDAT), and *Slave Select* (SS). The SPI Slave component can be used any time a PSoC device is required to interface with an SPI Master device. In addition to being used with SPI Master devices, the SPI Slave can be used with devices implementing a shift register interface. The SPI Master component can be employed in applications requiring that a PSoC device to communicate with an SPI Slave device.

By default, the *PSoC Creator Component Catalog* contains *Schematic Macro* implementations for the SPI Slave component. These macros contain already connected and adjusted input and output pins and clock source. Schematic Macros are available for 3-wire (Bidirectional), 4-wire (Full Duplex) and Full Duplex Multislave SPI interfacing as shown in Fig. 9.11e–g, respectively.

9.5.1 Slave I/O Connections

The slave I/O connections supported by PSoC Creator are as follows:[62]

- *mosi—Input** The *Master Output Slave Input* (MOSI) signal from a master device is applied to the *mosi input*. This input is visible when the *Data Lines* parameter is set to *MOSI + MISO*. If visible, this input must be connected.
- *sdat—Inout** The *Serial Data* (SDAT) signal is applied to the *sdat inout* input which is used when the Data Lines parameter is set to Bidirectional. For both PSoC 3 and PSoC 5LP silicon, an *asynchronous clock crossing* warning will be reported between the component clock and the SCLK signal when timing analysis is performed. The following is an example of such a message: Path(s) exist between clocks *IntClock* and *SCLK(0)_PAD*, but the clocks are not synchronous to each other. This message applies to a path from the register that controls the direction and the sampling of data by SCLK. SCLK should not be running when the direction is being changed. As long as this rule is followed, there is no problem and warning message can be ignored.
- *sclk Input—*The *Serial Clock* (SCLK) signal is applied to the sclk input which provides the slave synchronization clock input to the device. This input is always visible and must be connected.[63]
- *ss Input—*The Slave Select (SS) signal to the device is applied to the *ss input*. This input is always visible and must be connected. The following diagrams show the timing correlation between SCLK and SS signals Generally, 0.5 of the SCLK period is enough delay between the SS negative edge and the first SCLK edge for the SPI Slave to work correctly in all supported bit-rate ranges.
- *reset Input—*resets the SPI Slave and deletes any data that was currently being transmitted, or received. However, it does not clear data from the FIFO that has already been received or is ready to be transmitted. PSoC 3/5 ES2 silicon does not support this reset functionality, so this input is ignored when used with those devices. The use of the reset input results in an asynchronous clock crossing warning being reported between the clock that generates the *Reset input* and the SCLK signal when timing analysis is performed. The following is an example of such a message: Path(s) exist between clocks *BUS_CLK* and *SCLK(0)_PAD*, but the clocks are not synchronous to each other.This message applies to a path from the Reset signal to the operation of the SPI component clocked by SCLK. SCLK should not be running when the Reset signal is changed. As long as this rule is followed, there is no problem and you can ignore this message. The reset input may be left floating with no external connection. If nothing is connected to the reset line the component will assign it a constant logic 0.
- *clock—Input** defines the sampling rate of the status register. All data clocking happens on the*sclk* input, so the clock input does not handle the bit-rate of the SPI Slave. The clock input is visible when the Clock Selection parameter is set to External. If visible, this input must be connected.
- *miso—Output** transmits the *Master In Slave Out* (MISO) signal to the master device on the bus. This output is visible when the Data Lines parameter is set to *MOSI + MISO*.

---

[62]An asterisk (*) in the list of I/Os indicates that the I/O may be hidden for the component symbol under the conditions listed in the description of that I/O.

[63]Some SPI Master devices, e.g., the TotalPhase Aardvark I2C/SPI host adapter, drive the *sclk* output in a specific way. For the *SPI Slave* component to function properly with such devices in modes 1 and 3, when (CPOL = 1), the sclk pin should be set to *resistive pull-up drive mode*. Otherwise, corrupted data is output.

- *interrupt—Output* is the logical OR of the group of possible interrupt sources. This signal goes high while any of the enabled interrupt sources are true.

The PSoC Creator Component Catalog contains *Schematic Macro* implementations for the SPI Slave component that have connected and adjusted input pins, output pins and a clock source. As shown in Fig. 9.11d–g, Schematic Macros are available for 4-wire (Full Duplex), 3-wire (Bidirectional), and Full Duplex Multi-slave SPI interfacing.[64]

9.6   Universal Serial Bus (USB) Basics

The Universal Serial Bus (USB) [12] is an industry standard,[65] serial communication protocol originally designed for communications between computers and peripheral devices such as mice, keyboards modems, external hard drives, etc., as an alternative to larger, and slower, connections that employ serial and parallel ports [10]. In addition to providing faster transfer rates, the intent was to eliminate the various connector configurations used by the different protocols and standardize on a single, physical configuration, connection device. Originally, version 1.0 supported two configurations referred to as low-speed (LS) and full-speed (FS) at 1.5 and 12 Mbits/s. The LS configuration, while significantly slower than its FS counterpart, is much less susceptible to electromagnetic interference. Version 2.0 introduced a higher speed (HS) configuration as part of the specification that supported transmission rates of 480 Mbits/s.

A typical USB application includes a personal computer that serves as a host and several peripheral devices employed as part of a tiered, star topology which can include hubs that provide multiple connection points. The host utilizes at least one host controller and a root hub. Each host controller can support up to 127 connections, inclusive, when used with external USB hubs. The internal root hub is connected to the host controller(s) and provides the first interface layer to the USB. Most PCs are provided with multiple USB ports that are part of the root hub in the PC.

The host controller consists of a hardware chip-set and software driver layer that

- detects the attachment/removal of USB devices
- manages data flow between the host and such devices
- supplies power to the USB connected devices

and,

- monitors the USB bus activity.

Each USB device is assigned an address by the host, a connection pathway referred to as a *pipe* that connects the host and an addressable buffer known as an *endpoint*. The endpoint serves as an addressable buffer that holds data to be transmitted to the host, or that has been received from the host. A USB device can have multiple endpoints each of which has an associated pipe, as illustrated in Fig. 9.12.

---

[64]If schematic macros are not used, the *Pins* component should be configured to deselect the *Input Synchronized* parameter for each of the assigned input pins, i.e., MOSI, SCLK and SS. This parameter is located beneath the Pins>Input tab of the applicable *Pins Config* dialog.

[65]Compaq, DEC, IBM, Intel, Motorola NEC and Nortel collaborated on the development of the specification for the universal serial bus.

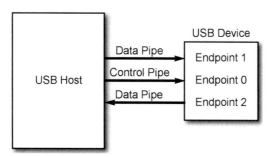

Fig. 9.12 USB pipe model.

The USB specification defines four types of data transfer categories:

- *control transfers* are used for sending commands to a device, to make inquiries and configure a device via the control pipe. Bulk transfers for large data transmissions that exploit all the available USB bandwidth using a data pipe.[66]
- *interrupt transfers* are used for sending small amounts of *bursty* data and to provide a guaranteed minimum latency.
- *isochronous transfers* are used for data that must be transferred at a guaranteed data rate which is based on a fixed bus bandwidth, fixed latency and no error correction.[67]

Every device has a control pipe through which transfers to send, and receive, messages are transmitted. Optionally, a device may have data pipes for transferring data through interrupt, bulk, or isochronous transfers, but the control pipe is the only bidirectional pipe in the USB system. All the data pipes are unidirectional. Each endpoint is accessed with a device address, assigned by the host, and an endpoint number, assigned by the device. When information is sent to the device address and endpoint number are identified with a token packet. The host initiates this token packet before a data transaction. When a USB device is first connected to a host, the USB enumeration process is initiated.

Two files, on the host side, are affiliated with enumeration and the loading of a driver:

- .INF is a text file that contains all the information necessary to install a device, e.g., driver names and locations, windows registry, and driver version information.
- .SYS is the driver needed to communicate effectively with the USB device.

Enumeration is the process of exchanging information between the device and the host that includes learning about the device. Additionally, enumeration includes assigning an address to the device, reading descriptors,[68] and assigning and loading a device driver a process that can occur in seconds. Once this process is complete, the device is ready to transfer data to the host.

The flow chart of the general enumeration process as shown in Fig. 9.13, are:

1. the device is connected to the host,
2. the host resets the device and requests a device descriptor,

---

[66]Bulk transfers cannot be relied on to take place with a specific speed, or latency.

[67]Error correction can introduce variable delays caused by having to delay transmission, while compromised packets are resent.

[68]Descriptors are data structures that provide information about the device.

3. the device responds to the request and the host sets a new address,
4. the host requests a device descriptor using the new address,
5. the host locates and reads the INF file,
6. the INF file specifies the device driver,
7. the driver is loaded on the host,

and

8. the device is configured and ready to use.

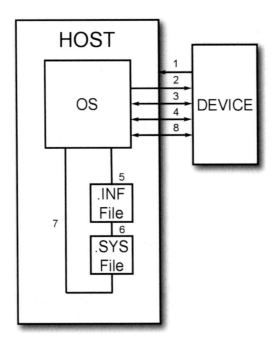

Fig. 9.13 Sequence of enumeration events.

After a device has been enumerated, the host directs all traffic flow, to the devices, on the bus and therefore no device can transfer data without a request from the host controller.

9.6.1 USB Architecture

Only one host can exist in the system and communication with devices is from the host's perspective. A host is an *upstream component*, while a device is a *downstream component*. Data moved from the host to the peripheral is an *OUT transfer*. Data moved to the host from the peripheral is an *IN transfer*. The host, specifically the host controller, controls all traffic and issues commands to devices.

There are three common types of USB host controllers:

- *Universal Host Controller Interface* (UHCI): Produced by Intel for USB 1.0 and USB 1.1. Using UHCI requires a license from Intel. This controller supports both low-speed and full-speed.

- *Open Host Controller Interface* (OHCI): Produced for USB 1.0 and 1.1 by Compaq, Microsoft, and National Semiconductor. Supports low-speed and full-speed and tends to be more efficient than UHCI by performing more functionality in hardware.
- *Extended Host Controller Interface (EHCI)*: Created for USB 2.0 after USB-IF requested that a single host controller specification be created. EHCI is used for high-speed transactions and delegates low-speed and full-speed transactions to an OHCI or UHCI sister controller.

One or more devices are attached to a host. Each device has a unique address and responds only to host commands that are addressed to it. Each is expected to have some form of functionality and not simply be passive. Devices contain one upstream port which serves as the physical USB connection point on the device. A hub is a specialized device that allows the host to communicate with multiple peripheral devices on the bus. Unlike USB peripheral devices, such as a mouse that has actual functionality, a hub device is transparent and is intended to act as a pass-through. A hub also acts as a channel between the host and the device. Hubs have additional attachment points to allow the connection of multiple devices to a single host. A hub repeats traffic to and from downstream devices through one upstream port and up to seven downstream ports. The hub, however, does not have any host capabilities.

As discussed previously, up to 127 devices can be connected to the host controller with the use of hubs. This limitation is based on the USB protocol, which limits the device address to 7 bits. Additionally, a maximum of 5 hubs can be chained together, a limitation imposed by timing considerations. The USB interface can be viewed as being divided into different layers. The *Bus Interface Layer* provides the physical connection, electrical signaling, and packet connectivity. This is the layer that is handled by the hardware in a device. This is accomplished by a physical interface external to the device. The *Device Layer* is used by the USB system software for performing USB operations such as sending and receiving information. This is accomplished with a *Serial Interface Engine*, which is also internal to the device. Finally, the *Function Layer* is the software portion of a USB device that handles the information it receives and gathers data to transfer to the host.

9.6.2 USB Signal Paths

All signals involve a return path, often referred to as the *ground return*.[69] Although ground is typically assumed to be at a potential of 0 V, it is really a reference that may deviate from 0 V as a result of electromagnetic interference, the impedance of the return path, i.e., the ground path, and other phenomena. In the case of long signal paths, there can be a significant difference between the ground at the transmitter (source) and the receiver (sink).

A USB cable consists of multiple conductors that are protected by an insulating jacket. Within this jacket is an outer shield consisting of copper braid. Inside this copper shield are multiple wires: a copper drain wire, a VBUS wire. (red) and a ground wire (black). An inner shield made of aluminum contains a twisted pair of data wires as seen in Fig. 9.14. There is a D+ wire (green) and a D− wire (white). In full-speed and high-speed devices, the maximum cable length is 5-m. To increase the distance between the host and a device, a series of hubs and 5-m cables must be used. While USB extension cables are available, using them to exceed 5 m is not in compliance with the USB protocol. Low-speed devices have slightly different specifications, e.g., their cable length is limited to 3 m and low-speed cables are not required to be a twisted pair, an example of which is shown in Fig. 9.14.

---

[69] In some cases, the return path is simply a ground plane.

The VBUS wire provides a constant 4.40–5.25 V supply to all attached devices. While USB supplies up to 5.25 V to devices, the data lines (D+ and D−) function at 3.3 V. The USB interface uses a differential transmission protocol that is non-return-to-zero inverted (NRZI) encoded with bit stuffing across a twisted pair of conductors.

Fig. 9.14 An example of a twisted pair cable.

NRZI encoding is a method for mapping a binary transmission signal in which a logic 1 is represented by *no change* in voltage level and a logic 0 is represented by a *change* in voltage level as Fig. 9.15 shows. The data that will be transmitted over USB is shown at the top of the figure. The encoded NRZI data is shown in the lower portion of the figure. Bit stuffing occurs by inserting a logic 0 into the data stream, following seven consecutive logic 1s. The purpose of the bit stuffing is for the synchronization of the USB hardware by using a phase-locked loop (PLL). If there are too many logic 1s in the data, then there may not be enough transitions in the NRZI encoded stream to support synchronization. The USB receiver hardware automatically detects this extra bit and disregards it. However, this extra bit stuffing contributes additional USB overhead. Figure 9.15 shows an example of NRZI data with bit stuffing. Although there are eight 1s in the *Data to Send* stream, in the encoded data a logic 0 is inserted after the sixth logic 1. The seventh and eighth logic 1 then follow after the 0 logic bit.

The hardware in USB devices handles all the encoding and bit stuffing upon receiving any data and prior to transmitting any data. The use of differential D+ and D− signals rejects common-mode noise. If noise is coupled into the cable, it will be present on all wires in the cable. With the use of a differential amplifier in the USB hardware internal to the host and device, the common-mode noise can be rejected, as illustrated in Figs. 9.16 and 9.17.

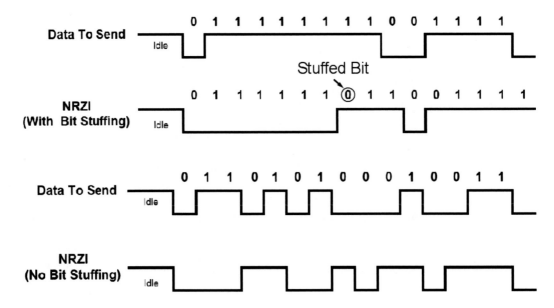

Fig. 9.15 A bit stuffing example.

It should be noted that in the *Data to Send* stream, shown in Fig. 9.15, there are eight 1s. In the encoded data, after the sixth logic 1, a logic 0 is inserted. The seventh and eighth logic 1 then follow after this logic 0. The hardware in USB devices handles all the encoding and bit stuffing upon receiving any data, and prior to transmitting any data. The reason for using the differential D+ and D− signal is for rejecting the common-mode noise. If noise becomes coupled into the cable, it will normally be present on all wires in the cable. With the use of a differential amplifier in the USB hardware internal to the host and device, the common-mode noise can be rejected.

The hardware in USB devices handle all the encoding and bit stuffing upon receiving any data and prior to transmitting any data. The reason for using the differential D+ and D− signal is for rejecting the common-mode noise. If noise becomes coupled into the cable, it will normally be present on all wires in the cable. With the use of a differential amplifier in the USB hardware internal to the host and device, the common-mode noise can be rejected as shown in Fig. 9.17.

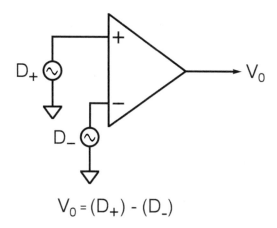

Fig. 9.16 An ideal differential amplifier configuration.

USB communication occurs through many signaling states on the D+ and D− lines. Some of these states transmit the data while others are used as specific signaling conditions. These states are described below with a quick reference list located in Table 9.6 .

- **Differential 0 and Differential 1**: These two states are used in the general data communication across a USB communication path. Differential 1 is when the D+ line is high and the D− line is low. Differential 0 occurs when the D+ line is low, and the D− line is high.
- **J-State and K-State:** In addition to the differential signals, the USB specification defines two additional differential states: J-States and K-States. Their definitions depend on the device speed. On a full-speed and high-speed device, a J-State is a Differential 1 and a K-State is a Differential 0. The opposite is true for a low-speed device.
- **Single-Ended Zero (SE0)** is a condition that occurs when both D+ and D− are driven low, indicating a reset, disconnect, or End of Packet.
- **Single Ended One (SE1)**: Condition that occurs when D+ and D− are both driven high. This condition does not ever occur intentionally and should never occur in a USB design.
- **Idle** is a condition that occurs before and after a packet is sent. An Idle condition is signified by one of the data lines being low and the other line being high. The definition of high vs. low depends on device speed for a full-speed device, an idle condition consists of D+ being high and D− being low. The opposite is true for a low-speed device.

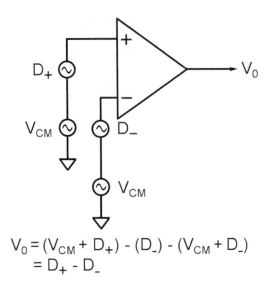

$$V_0 = (V_{CM} + D_+) - (D_-) - (V_{CM} + D_-)$$
$$= D_+ - D_-$$

Fig. 9.17 An example of USB common-mode rejection.

- **Resume** is used to wake a device from a suspend state, by issuing a K-State.
- **Start of Packet (SOP)** occurs before the start of any low-speed or full-speed packet when the D+ and D− lines transition from an idle state to a K-State.
- **End of Packet (EOP)** occurs at the end of any low-speed or full-speed packet. An EOP occurs when an SE0 state occurs for 2-bit times, followed by a J-State for 1-bit time.
- **Reset** occurs when an SE0 state lasts for 10 ms. The device can recognize the reset and begin to enter a reset after an SE0 has occurred for at least 2.5 ms, .
- **Keep Alive** is a signal used in low-speed devices that lack a Start-of-Frame packet that is required to prevent suspend and use an EOP every millisecond to keep the device from entering suspend.

9.6.3   USB Endpoints

In the USB specification, a *device endpoint* is a uniquely addressable portion of a USB device that is the source, or sink, of information in a communication flow between the host and device. The USB enumeration section describes a step in which the device responds to the default address. This occurs before other descriptor information such as the endpoint descriptors are read by the host, later in the enumeration process. During the enumeration sequence, a special set of endpoints are used for communication with the device. These special endpoints, collectively known as the *Control Endpoint* or *Endpoint 0*, are defined as *Endpoint 0 IN* and *Endpoint 0 OUT*. Even though *Endpoint 0 IN* and *Endpoint 0 OUT* are two endpoints, they look and act like one endpoint to the developer. Every USB device must support *Endpoint 0*. For this reason, *Endpoint 0* does not require a separate descriptor.

In addition to *Endpoint 0*, the number of endpoints supported in any particular device is based on its design requirements. A fairly simple design, such as a mouse, may need only a single IN endpoint. More complex designs may need several data endpoints. The USB specification sets a limit on the number of endpoints to 16 for each direction (16 IN/16 OUT = 32 Total) for high and full-speed devices, which does not include the control endpoints *0 IN* and *0 OUT*. Low-speed devices are limited

Table 9.6 USB communications states.

Bus State	Indication
Differential 1	D+ High, D- Low
Differential 0	D+ High, D- Low
Single Ended 0 (SE0)	D+ High, D- Low
Single Ended 1 (SE1)	D+ High, D- Low
J-State: Low Speed Full Speed High Speed	 Differential 0 Differential 1 Differential 1
K-State: Low Speed Full Speed High Speed	 Differential 1 Differential 0 Differential 0
Resume State	K-State
Start of Packet (SOP)	Data lines switch from idle to K-state
End of Packet (EOP)	SE0 for 2 bit times followed by J-statefor 1 bit time

to two endpoints. USB Class devices may set a greater limit on the number of endpoints, e.g., a low-speed HID design may have no more than two data endpoints, typically one IN endpoint and one OUT endpoint. Data endpoints are bidirectional by nature, but it is not until they are configured that they become unidirectional. Endpoint 1, for example, can be either an IN or OUT endpoint. It is in the device descriptors that Endpoint 1 becomes an IN endpoint.

Endpoints use cyclic redundancy checks (CRCs) to detect errors in transactions.[70] Handling of these calculations is taken care of by the USB hardware so that the proper response can be issued. The recipient of a transaction checks the transmitted CRC value against the CRC calculated by the receiver based on the received data. If the two match, then the receiver issues an ACK. If the data and the CRC do not match, then no handshake is sent. This absence of a handshake tells the transmitter to try again.

[70]The CRC is a calculated value used for error checking. The CRC calculation is based on an equation defined in the USB specification.

The USB specification further defines four types of endpoints and sets the maximum packet size, based on both the type and the supported device speed. The endpoint descriptor should be used to identify the type of endpoint requirements.

The four types of *endpoints* and characteristics are:

- *Control Endpoints* support control transfers, which all devices must support. Control transfers send, and receive, device information across the bus. The primary advantages of control transfers are guaranteed accuracy, proper detection of Errors and assurance that the data is resent. Control transfers have a 10% reserved bandwidth on the bus in low and full-speed devices (20% at high-speed) and give the USB system-level control.
- *Interrupt Endpoints* support interrupt transfers which are used on devices that require a highly reliable method to communicate a small amount of data.[71] However, the name of this transfer can be misleading because it is not truly an interrupt based system, but instead employs a polling method. However, it does guarantee that the host check for data at a predictable interval. Interrupt transfers give guaranteed accuracy because errors are properly detected and transactions are retried at the next transaction. Interrupt transfers have a guaranteed bandwidth of 90% on low- and full-speed devices and 80% on high-speed devices. This bandwidth is shared with *isochronous endpoints*. The maximum packet size when employing interrupt endpoints is a function of device speed. High-speed capable devices support a maximum packet size of 1024 bytes. Devices capable of operating at full-speed support a maximum packet size of 64 bytes. Low-speed devices support a maximum packet size of 8 bytes. on low- and full-speed devices and 80% on high-speed devices. This bandwidth is shared with isochronous endpoints. Interrupt endpoint maximum packet size is a function of device speed. High-speed capable devices support a maximum packet size of 1024 bytes. Full-speed capable devices support a maximum packet size of 64 bytes. Low-speed devices support a maximum packet size of 8 bytes.
- *Bulk Endpoints* support bulk transfers, which are commonly used on devices that move relatively large amounts of data at highly variable times where the transfers can use any available bandwidth space.[72] Delivery time for a bulk transfer is variable because there is no predefined bandwidth for the transfer, but instead, varies depending on how much bandwidth on the bus is available, which makes the actual delivery time unpredictable. Bulk transfers give guaranteed accuracy because errors are properly detected, and transactions are resent. Bulk transfers are useful in moving large amounts of data that are not time sensitive. A bulk endpoint maximum packet size is a function of device speed.[73] Devices that support full-speed transfer have a maximum packet size of 64-bytes. Low-speed devices do not support bulk transfer types.
- *Isochronous Endpoints* support isochronous transfers, which are continuous, real-time transfers that have a pre-negotiated bandwidth. Isochronous transfers must support streams of error tolerant data because they do not have an error recovery mechanism, or handshaking. Errors are detected through the CRC field, but not corrected. With isochronous endpoints, a tradeoff must be made between guaranteed delivery and guaranteed accuracy. Streaming music, or video, are examples of an application that uses isochronous endpoints because the occasional missed data is ignored by human ears and eyes. Isochronous transfers have a guaranteed bandwidth of 90% on low and full-speed devices (80% on high-speed devices) that is shared with interrupt endpoints.

---

[71]This is commonly used in *Human Interface Device* (HID) designs.

[72] They are the most common transfer type for USB devices.

[73]High-speed capable devices support a maximum BULK packet size of 512 bytes. Low-speed devices do not support bulk transfer.

High-speed capable devices support a maximum packet size of 1024 bytes, full-speed devices 1023 bytes.[74] There are special considerations with isochronous transfers, e.g., $3\times$ buffering is preferable to ensure data is ready to go by having one actively transmitting buffer, another buffer loaded and ready to transfer, and a third buffer being actively loaded (Table 9.7).

Table 9.7 Endpoint transfer type features.

Transfer	Control	Interrupt	Bulk	Isochronous
Typical User	Device Initialization and Management	Mouse and Keyboard	Printer and Mass Storage	Streaming Audio and Video
Low-speed Support	Yes	Yes	No	No
Error Correction	Yes	Yes	Yes	No
Guaranteed Delivery Rate	No	No	No	Yes
Guaranteed Bandwidth	Yes (10%)	Yes (90%)*	No	Yes (90%)*
Guaranteed Latency	No	Yes	No	Yes

*Shared bandwidth between isochronous and interrupt.

9.6.4 USB Transfer Structure

During the enumeration process, the host requests the device descriptor. The transfer process consists of making the request for the device descriptor, receiving the device descriptor information, and the host acknowledging the successful reception of the data. However, the transfer consists of multiple stages called *transactions*. Each transfer consists of one or more transactions and in the case of the device descriptor request, there are three transactions. The first is the *Setup transaction*, the second is the *Data transaction*, where the descriptor information is sent to the host. The third transaction is the *handshake transaction* where the host acknowledges receiving the packet. Each transaction is made up of multiple packets and contains a token packet at a minimum. Inclusion of a data packet and handshake packet can vary depending on the transfer type.

Each transfer contains one or more transactions, each of which always contains a *token packet.* A *data packet,* and *handshake packet,* may be included depending on the transaction type. Interrupt,

---

[74]Low-speed devices do not support isochronous transfer types.

bulk, and control transfers always include a token, data, and handshake packet with each transaction. Control transfers have three stages: *Setup*, *Data*, and *Status*, and each one of these stages contains a token, data, and handshake packet. Therefore, while an Interrupt and Bulk transfer have a minimum of three packets, a control transfer has nine, or more, with a data stage and six or more without a data stage.

9.6.5 Transfer Composition

A USB packet has the structure as shown in Fig. 9.18. A total of five fields can be populated, four of which are optional, and one is required.

Fig. 9.18 USB Packet contents.

- Packet ID (PID) (8bits: 4 type bits and 4 check bits)
- Optional Device Address (7 bits: Max of 127 devices)
- Optional Endpoint Address (4 bits: Max of 16 endpoints)
- Optional Payload Data (0–1023 bytes)
- Optional CRC (5 or 16 bits)

The *Packet ID* is the only required field in a packet. The *Device Address*, *Endpoint Address*, *Payload Data*, and *CRC* are filled depending on which packet type is sent. *Packet IDs* (PID) are the heart of a USB packet. There are different PIDs depending on which packet is sent (see Table 9.1).

9.6.6 Packet Types

There are four different packet types, as shown in Fig. 9.19 that can potentially represent.

- *Token packets*
  - Initiate a transaction.
  - Identify the device involved in the transaction.
  - Are always sourced by the host.
- *Data packets*
  - Delivers payload data.
  - Are sourced by host or device.
- *Handshake packets*
  - Acknowledge error-free data receipt.
  - Are sourced by the receiver of the data.

Packet Type	PID Name	PID [3..0]	Description
Token	OUT	0001b	Address + endpoint number in host-to-function transaction.
	IN	1001b	Address + endpoint number in function-to-host transaction.
	SOF	0101b	Start-of-Frame marker and frame number.
	SETUP	1101b	Address + endpoint number in host-to-function transaction for SETUP to a control pipe.
Data	DATA0	0011b	Data packet PID even. Data Toggle
	DATA1	1011b	Data packet PID odd. Data Toggle
	DATA2	0111b	Data packet PID high-speed, high bandwidth isochronous Transaction in a microframe. High Speed Only
	MDATA	1111b	Data packet PID high-speed for split and high bandwidth isochronous transactions. High Speed Only
Handshake	ACK	0010b	Receiver accepts error-free data packet.
	NAK	1010b	Receiving device cannot accept data or transmitting device cannot send data.
	STALL	1110b	Endpoint is halted or a control pipe request is not supported.
	NYET	0110b	No response yet from receiver. High Speed Only
Special	PRE	1100b	(Token) Host-issued preamble. Enables downstream bus traffic to low-speed devices.
	ERR	1100b	(Handshake) Split Transaction Error Handshake (reuses PRE value). High Speed Only
	SPLIT	1000b	(Token) High-speed Split Transaction Token. High Speed Only
	PING	0100b	High-speed flow control probe for a bulk/control endpoint. High Speed Only
	Reserved	0000b	Reserved PID.

Fig. 9.19 USB Packet contents.

- *Special packets*
  - Facilitates speed differentials.
  - Are sourced by host-to-hub devices.

Although everything in the packet, except for the PID is optional, token, data, and handshake packets have different combinations of the packet information.

Token packets always come from the host, and are used to direct traffic on the bus. The function of the token packet depends on the activity performed, e.g., *IN tokens* are used to request that devices send data to the host and *OUT tokens* are used to precede data from the host. *SETUP tokens* are used to precede commands from the host and *SOF tokens* are used to mark time frames. With an *IN*, *OUT*, and *SETUP* token packet, there is a 7-bit device address, 4-bit endpoint ID, and 5-bit CRC.

The *SOF* gives a way for devices to identify the beginning of a frame and synchronize with the host. They are also used to prevent a device from entering suspend mode, which it must do if 3 ms pass without an SOF. SOF packets are only found on full and high-speed devices and are sent every millisecond. The SOF packet contains an 8-bit SOF PID, 11-bit frame count value (which rolls over when it reaches maximum value), and a 5-bit CRC. The CRC is the only error check used. A handshake packet does not occur for a SOF packet. High-speed communication goes a step further with microframes. With a high-speed device, a SOF is sent out every 125 μs and frame count is only incremented every 1 ms.

Data packets follow *IN*, *OUT*, and *SETUP* token packets. The size of the payload data ranges from 0 to 1024 bytes, depending on the transfer type. The packet ID toggles between *DATA0* and *DATA1* for each successful data packet transfer, and the packet closes with a 16-bit CRC. The *data toggle* is updated at the host, and the device for each successful data packet transfer. One advantage of the *data toggle* is that it acts as an additional error detection method. If a different packet ID is received than is expected, the device will be able to know there was an error in the transfer and it can be handled appropriately. If an ACK is sent, but not received, the sender updates the data toggle from 1 to 0, but the receiver does not, and the data toggle remains at 1.

Handshake packets conclude each transaction. Each handshake includes an 8-bit packet ID and is sent by the receiver of the transaction. Each USB speed has several options for a handshake response. The handshakes supported depend on the USB Speed:

- *ACK* is an acknowledgment of successful completion. (LS/FS/HS)
- *NAK* is a negative acknowledgment. (LS/FS/HS)
- *STALL* is an error indication sent by a device. (LS/FS/HS)
- *NYET* indicates the device is not ready to receive another data packet. (HS Only)

9.6.7   Transaction Types

Data from the host, and the device, are transferred from point A to B, via *transactions*. IN/Read/Upstream Transactions are terms that refer to a transaction that is sent from the device to the host. These transactions are initiated when the host sends an *IN token packet*. The targeted device responds by sending one or more data packets, and the host responds with a *handshake packet*.

IN/Read/Upstream Special Packets are defined by the USB specification:

- *PRE* is issued to hubs by the host to indicate that the next packet is low speed.
- *SPLIT* precedes a token packet to indicate a split transaction. (HS Only)
- *ERR* is returned by a hub to report an error in a split transaction. (HS Only)

- *PING* checks the status for a Bulk OUT or Control Write after receiving a NYET handshake. (HS Only)

9.6.8 USB Descriptors

As described earlier, when a device is connected to a USB host, the device gives information to the host about its capabilities and power requirements. The device typically gives this information via a *descriptor table* that is part of its firmware. A descriptor table is a structured sequence of values that describe the device, and whose values are defined by the developer.

All descriptor tables contain a standard set of information that describes the device attributes and power requirements. If a design conforms to the requirement of a particular USB device class, additional descriptor information that the class must have is included in the device descriptor structure. When reading or creating descriptors, it is important to assure that the data fields are transmitted with the least significant bit first. Many parameters are 2 bytes long with the low byte occurring first and followed by the high byte.

Device descriptors provide the host with USB specification to which the device conforms, the number of device configurations, and protocols supported by the device, *Vendor Identification*[75] *Product Identification* (Also known as a PID, different from a packet ID), and a *serial number*, if the device has one. The *Device Descriptor* contains the crucial information about the USB device.

Table 9.8 shows the structure for a device descriptor given that:

- *bLength* is the total length, in bytes, of the device descriptor,
- *bcdUSB* reports the USB revision that the device supports, which should be the latest supported revision. This is a binary-coded decimal value that uses a format of 0xAABC, where A is the major version number, B is the minor version number, and C is the sub-minor version number. For example, a USB 2.0 device would have a value of 0x0200 and USB 1.1 would have a value of 0x0110. This is normally used by the host in determining which driver to load,
- *bDeviceClass,bDeviceSubClass*, and*bDeviceProtocol* are used by the operating system to identify a driver for a USB device during the enumeration process. Filling in this field in the device descriptor prevents different interfaces from functioning independently, such as a composite device. Most USB devices define their class(es) in the interface descriptor, and leave these fields as 00h,
- *bMaxPacketSize* reports the maximum number of packets supported by *Endpoint zero*. Depending on the device, the possible sizes are 8 bytes, 16 bytes, 32 bytes, and 64 bytes,
- *iManufacturer*, *iProduct*, and *iSerialNumber* are indexes to string descriptors. String descriptors give details about the manufacturer, product, and serial number. If string descriptors exist, these variables should point to their index location. If no string exists, then the respective field should be assigned a value of zero,

and,

- *bNumConfigurations* defines the total number of configurations the device can support. Multiple configurations allow the device to be configured differently depending on certain conditions, such as being bus-powered, or self-powered (Fig. 9.20).

---

[75] Also known as a VID, which is something that each company gets uniquely from the USB Implementers Forum.

Table 9.8 Device descriptor table.

Offset	Field	Size (Bytes)	Description
0	bLength	1	Descriptor Length = 18 bytes
1	bDescriptior Type	1	Descriptor type = DEVICE (01h)
2	bcdUSB	2	USB Spec Version (BCD)
4	bDeviceClass	1	Device Class
5	bDeviceSubClass	1	Device subclass
6	bDeviceProtocol	1	Device Protocol
7	bMaxPacketSize0	1	Max Packet Size for Endpoint 0
8	idVendor	2	Vendor ID (VID) (Assigned by USB-IF)
10	idProduct	2	Product ID (PID) (Assigned by the manufacturer)
12	bcdDevice	2	Device Release Number (BCD)
14	iManufacturer	1	Manufacturing String Index
15	idProduct	1	Product String Index
16	iSerialNumber	1	Serial Number String Index
17	bNumConfigurations	1	# 0f Supported Configurations

9.6.9   Configuration Descriptor

This descriptor gives information about a specific device configuration, e.g., the number of interfaces, whether the device is bus-powered or self-powered, if the device can start a remote wake-up, and how much power the device needs. Table 9.9 shows the structure for a configuration descriptor.

wTotalLength is the length of the entire hierarchy of this configuration. This value reports the total number of bytes of the configuration, interface, and endpoint descriptors for one configuration.

bNumInterfaces defines the total number of possible interfaces in this particular configuration. This field has a minimum value of 1.

bConfigurationValue defines a value to use as an argument to the *SET_CONFIGURATION* request to select this configuration.

bmAttributes defines parameters for the USB device. If the device is bus-powered, bit 6 is set to 0. If the device is self-powered, then bit 6 is set to 1. If the USB device supports remote wakeup, bit 5 is set to 1. If remote wakeup is not supported, bit 5 is set to 0.

bMaxPower defines the maximum power consumption drawn from the bus when the device is fully operational, expressed in 2 mA units. If a self-powered device becomes detached from its external power source, it may not draw more than the value indicated in this field.

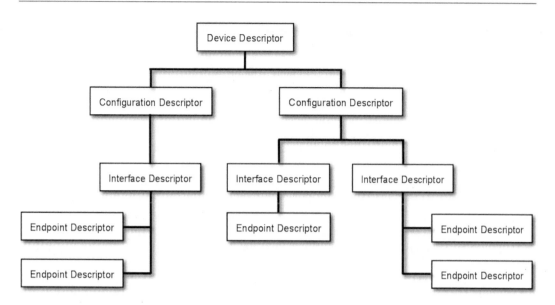

Fig. 9.20 USB descriptor tree.

Table 9.9 Configuration descriptor type.

Offset	Field	Size (Bytes)	Description
0	bLength	1	Descriptor length = 9 bytes
1	bDescription Type	1	Descriptor Type = COMFIGURATION (02h)
2	wTotalLength	2	Total length including interface and endpoint descriptors
4	bNumInterface	1	Number of interfaces in the configuration
5	bConfiguration Value	1	Configuration values used by SET_CONFIGURATION to select the configuration
6	iConfiguration	1	String index that decribes the configuration
7	bmAttributes	1	Bit 7: Reserved (set to 1) Bit 6: Self-powered Bit 5: remote wakeup
8	bMaxPower	1	Max power required for the configuration (in 2 ma units)

9.7   Full Speed USB (USBFS)

PSoC Creator's *USBFS component* provides a USB, full-speed, Chap. 9 compliant device, framework.[76] It provides a low-level driver for the control endpoint that decodes and dispatches requests from the USB host. Additionally, this component provides a USBFS customizer to make it easy to construct the appropriate descriptor. The option of constructing a HID-based device or a generic USB Device is also provided. In PSoC Creator, HID can be selected by setting the *Configuration/Interface* descriptors. The USBFS component can be used to provide an interface that is USB 2.0 compliant.

USB transmissions are based on one of several types of transfer, viz., bulk, control, interrupt and isochronous, depending on the application. While the formal USB specification defines specific commands that may be required for a USB device to receive and respond to USB transmission, it is also possible for the designer to introduce custom commands.[77] Reliable data transmission schemes often rely on data integrity algorithms to detect, and perhaps correct, errors and/or generate an error signal. Handshaking schemes provide feedback to the transmitter to indicate whether or not data integrity has been preserved, and thereby allow retransmission of data to be employed in the event of errors in transmission. The *start-of-frame* (sof) output for the component allows endpoints to identify the start of the frame and synchronize internal endpoint clocks to the host.

9.7.1   Endpoint Memory Management

The USBFS block contains 512 bytes of target memory for the data endpoints to use. However, the architecture supports a *cut-through mode* of operation, referred to as *DMA w/Automatic Memory Management*, that reduces the memory requirement, based on system performance. Some applications can benefit from using Direct Memory Access (DMA) to move data into and out of the endpoint memory buffers.

*   Manual (default) Select this option to use *LoadInEP/ReadOutEP* to load and unload the endpoint buffers.
    -   *Static Allocation*—The memory for the endpoints is allocated immediately after a *SET_CONFIGURATION* request. This takes longest when multiple *Alternate* settings use the same endpoint (EP) number.
    -   *Dynamic Allocation*—The memory for the endpoints is allocated dynamically after each *SET_CONFIGURATION* and *SET_INTERFACE* request. This option is useful when multiple alternate settings are used with mutually exclusive EP settings.

---

[76]SuiteUSB, a set of USB development tools, is available free of charge when used with Cypress silicon. http://www. cypress.com.

[77]Such custom commands, e.g., introduced to provide control of a specific type of device, are often referred to as *vendor commands*.

- *DMA w/Manual Memory Management*[78]—Select this option for manual DMA transactions. The*LoadInEP/ReadOutEP* functions fully support this mode and initialize the DMA automatically.[79]
- *DMA w/Automatic Memory Management*—Select this option for automatic DMA transactions. This is the only configuration that supports combined data endpoint use of more than 512 bytes. *LoadInEP/ReadOutEP* functions should be used for initial DMA configuration.

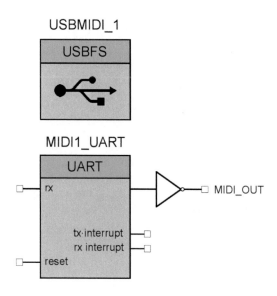

Fig. 9.21 PSoC 3/5's Midi component.

9.7.2 Enabling VBUS Monitoring

USB signals [11] are transmitted via a USB cable consisting of a twisted pair that has a characteristic impedance of 90 Ω, a shield that functions as a ground return and power connections D+ and D−. The protocol assumes that there are no more than 127 devices[80] connected at any one time, in a *tiered-star* topology. The maximum allowable cable length between hubs is 5 m, and no more than six hubs are supported, for a maximum of 30 m. The USB specification requires that no device supplies current on VBUS at its upstream facing port at any time. To meet this requirement, the device must monitor for the presence, or absence, of VBUS and remove power from the D+/D− pull-up resistor, if VBUS is absent. For bus-powered designs, power will obviously be removed when the USB cable is removed from a host but, for self-powered designs, it is imperative for proper operation, and USB certification, that the device comply with this requirement.

---

[78]PSoC 3 [1] does not support DMA transactions directly between USB endpoints and other peripherals. All DMA transactions involving USB endpoints, both in and out, must terminate, or originate, with main system memory. Applications requiring DMA transactions directly between USB endpoints, and other peripherals, must use two DMA transactions to move data to main system memory as an intermediate step between the USB endpoint and the other peripheral.

[79]This option is supported for PSoC 3 [15] production silicon only.

[80]This limitation is a result, in part, of the fact that the address field is 7 bits and that address zero is reserved.

9.7.2.1 USBFS MIDI

The USBFS MIDI component, shown in Fig. 9.21, provides support for communicating with external MIDI equipment and for the USB device class definition for MIDI devices. This component can be used to add MIDI I/O capability to a standalone device, or to implement MIDI capability for a host computer, or mobile device, through a computer's, or mobile device's, USB port. In such cases, it appears to the host computer, or mobile device, as a class-compliant USB MIDI device, and it uses the native MIDI drivers in the host.

The supported features include:

- USB MIDI Class Compliant MIDI input and output.
- Hardware interfacing to external MIDI equipment using UART.
- Adjustable transmit and receive buffers managed using interrupts.
- MIDI running status for both receive and transmit functions.
- Up to 16 input, and output, ports using only two USB endpoints by using virtual cables.

The *PSoC Creator Component* catalog contains a *Schematic Macro* implementation of a MIDI[81] interface. The macro consists of instances of the UART component with the hardware MIDI interface configuration (31.25 kbps, 8 data bits) and a USBFS component with the descriptors configured to support MIDI devices. This allows the user to employ a MIDI-enabled, USBFS component with minimal configuration changes. A *USBMIDI Schematic Macro* labeled *USBMIDI* is available in PSoC Creator that has been previously configured to function as an external mode MIDI device with 1 input and 1 output.

9.7.3 USB Function Calls

PSoC Creator provides an extensive list of USB function calls and, by default, assigns the instance name *USBFS_1* to the first instance of a component in a given design. However, such instance names can be renamed to any unique value that follows the syntactic rules for identifiers. In any event, the instance name becomes the prefix of every global function name, variable, and constant symbol.

For readability, the instance name used in the following is *USBFS*.

- *void USBFS_Start(uint8 device, uint8 mode)* performs all required initialization for the USBFS Component.
- *void USBFS_Init(void)* initializes, or restores, the component according to the customizer *Configure* dialog settings.[82]
- *void USBFS_InitComponent(uint8 device, uint8 mode)* initializes the component's global variables and initiates communication with the host by pulling up the D+ line.
- *void USBFS_Stop(void)* performs all necessary shutdown tasks required for the USBFS component.
- *uint8 USBFS_GetConfiguration(void)* gets the current configuration of the USB device.

---

[81]The Musical Instrument Digital Interface (MIDI), defined by the MIDI Manufacturing Association in 1982, is an industry standard protocol for intercommunication between a wide variety of music related devices. It serves as a software, hardware, communication and instrument categorization standard and is often employed to allow one instrument to control an arbitrary number of other musical instruments, or music-related equipment.

[82] It is not necessary to *call USBFS_Init()* because the *USBFS_Start()* routine calls this function and is the preferred method to begin component operation.

- *uint8 USBFS_IsConfigurationChanged(void)* returns the *clear-on-read* configuration state. It is useful when the PC sends double *SET_CONFIGURATION* requests with the same configuration number.
- *uint8 USBFS_GetInterfacuint8 USBFS_GetEPState(uint8 epNumber)* returns the state of the requested endpoint.
- *uint8 USBFS_GetInterfaceSetting(uint8 interfaceNumber)* gets the current alternate setting for the specified interface.
- *uint8 USBFS_GetEPState(uint8 epNumber)* returns the state of the requested endpoint.
- *uint8 USBFS_GetEPAckState(uint8 epNumber)* determines whether or not an ACK transaction occurred on this endpoint by reading the ACK bit in the control register of the endpoint.[83]
- *uint16 USBFS_GetEPCount(uint8 epNumber)* returns the transfer count for the requested endpoint. The value from the count registers includes two counts for the two-byte checksum of the packet. This function subtracts the two counts.
- *void USBFS_InitEP_DMA(uint8 epNumber, uint8 *pData)*[84] allocates and initializes a DMA channel to be used by the *USBFS_LoadInEP()* or *USBFS_ReadOutEP()* APIs for data transfer. It is available when the Endpoint Memory Management parameter is set to DMA.
- void USBFS_LoadInEP(uint8 epNumber, uint8 *pData, uint16 length) in manual mode: loads and enables the specified USB data endpoint for an IN data transfer. Manual DMA:
  - Configures DMA for a transfer data from data RAM to endpoint RAM.
  - Generates a request for a transfer.
  Automatic DMA:
  - Configures DMA. This is required only once, therefore it is done only when parameter Data is not NULL. When pData pointer is NULL, the function skips this task.
  - Sets Data ready status: This generates the first DMA transfer and prepares data in endpoint RAM memory.
- *uint16 USBFS_ReadOutEP(uint8 epNumber, uint8 *pData, uint16 length)* in manual mode moves the specified number of bytes from endpoint RAM to data RAM. The number of bytes actually transferred from endpoint RAM, to data RAM, is the lesser of the actual number of bytes sent by the host, or the number of bytes requested by the wCount parameter.
  Manual DMA:
  - Configures DMA for a transfer data from endpoint RAM to data RAM.
  - Generates a request for a transfer.
  - After *USB_ReadOutEP()*API and before expected data usage it is required to wait on DMA transfer complete. For example by checking EPstate:
    while *(USBFS_GetEPState(OUT_EP) == USB_OUT_BUFFER_FULL)*;
  Automatic DMA:
  - Configures DMA.[85]
- *void USBFS_EnableOutEP(uint8 epNumber)* enables the specified endpoint for OUT transfers.
- *void USBFS_DisableOutEP(uint8 epNumber)* disables the specified USBFS OUT endpoint.[86]
- *void USBFS_SetPowerStatus(uint8 powerStatus)* sets the current power status. The device replies to *USB GET_STATUS* requests based on this value. This allows the device to properly report its status for *USB Chap. 9* compliance. Devices can change their power source from self-powered to bus-powered, at any time, and report their current power source as part of the device status. This

---

[83]This function does not clear the ACK bit.

[84]This function is automatically called from the *USBFS_LoadInEP()* and *USBFS_ReadOutEP()* APIs.

[85]This is required only once.

[86]Do not call this function for IN endpoints.

function can be called any time the device changes from self-powered to bus-powered, or vice versa, and set the status appropriately.

- *void USBFS_Force(uint8 state)* forces a USB J, K, or SE0 state on the D+/D− lines. This function provides the necessary mechanism for a USB device application to perform a USB Remote Wakeup.[87]
- *void USBFS_SerialNumString(uint8 *snString)* is available only when the *User Call Back* option in the *Serial Number String* descriptor properties is selected. Application firmware can provide the source of the USB device serial number string descriptor during runtime. The default string is used, if the application firmware does not use this function, or sets the wrong string descriptor.
- *void USBFS_TerminateEP(uint8 epNumber)*[88] terminates the specified USBFS endpoint.
- *uint8 USBFS_UpdateHIDTimer(uint8 interface)* updates the HID Report idle timer and returns the status and reloads the timer, if it expires.
- *uint8 USBFS_GetProtocol(uint8 interface)* returns the HID protocol value for the selected interface.

9.8   Controller Area Network (CAN)

The Controller Area Network (CAN) [20] controller implements the CAN2.0A and CAN2.0B specifications as defined in the Bosch specification and conforms to the ISO-11898-1 standard. The CAN protocol was originally designed for automotive applications with a focus on a high level of fault detection thereby ensuring high communication reliability at a low cost. Because of its success in automotive applications, CAN is used as a standard communication protocol for motion-oriented, machine-control networks (*CANOpen*) and factory automation applications (*DeviceNet*). The CAN controller features make it possible to efficiently implement higher-level protocols, without adversely affecting the performance of the microcontroller CPU.

CAN is an *arbitration-free* system in that the highest priority message is always transmitted first. The transmit buffer arbitration scheme employed can be either *round-robin*, the default mode, or *fixed priority*. In the round-robin mode, buffers are served in the following order: $0 - 1 - 2 \cdots 7 - 0 - 1$.[89] In the fixed priority mode, buffer zero is assigned the highest priority which allows it to be the error message buffer thereby assuring that error messages are transmitted first.

9.8.1   PSoC Creator's CAN Component

This component has three standard I/O connections, and a fourth, optional, *interrupt* connection,[90] as shown in Fig. 9.22.

- *rx* is the CAN bus receive (input) signal and is connected to the CAN Rx bus which is external to the transceiver.
- *tx* is the CAN bus transmit signal and is connected to the CAN Tx bus of the external transceiver.
- *tx_en* is the external transceiver enable signal.

---

[87] For more information, refer to the USB 2.0 Specification for details on *Suspend* and *Resume*.

[88] This function should be used before endpoint reconfiguration.

[89] This mode assures that all buffers have the same probability of sending a message.

[90] This output is displayed in PSoC Creator only when the *Add Transceiver Enable Signal* option has been selected in the *Config* dialog.

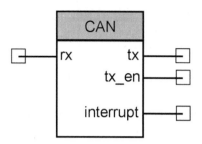

Fig. 9.22 PSoC Creator's CAN component.

The default CAN configuration in the *Component Catalog* is a schematic macro using a CAN component with default settings, and is connected to an *Input* and an *Output Pins* component. The Pins components are also configured with default settings, except that *Input Synchronized* is set to false in the *Input Pin* component.

9.8.2 Interrupt Service Routines

There are several CAN component interrupt sources, all of which have entry points (functions) that allow user code to be placed in them.[91]

- *Acknowledge Error*—The CAN controller detected a CAN message acknowledge an error.
- *Arbitration Lost Detection*—The arbitration was lost while sending a message.
- *Bit Error* —The CAN controller detected a bit error.
- *Bit Stuff Error*—The CAN controller detected a bit stuffing[92] error.
- *Bus Off*—The CAN controller has reached the bus-off state
- *CRC Error*—The CAN controller detected a CAN CRC error.
- *Form Error*—The CAN controller detected a CAN message format error.
- *Message Lost* —A new message arrived, but there was nowhere to put it.
- *Transmit Message*—The queued message was sent.
- *Receive Message*—A message was received.[93]

9.8.3 Hardware Control of Logic on Interrupt Events

The hardware interrupt [20] input can be used to perform simple tasks such as estimating the CAN bus load. By enabling the *Message Transmitted* and *Message Received* interrupts in the CAN component customizer, and connecting the interrupt line to a counter, the number of messages that are on the bus during a specific time interval can be evaluated. Actions can be taken directly in hardware, if the message rate is above a certain value.

---

[91]These functions are conditionally compiled, depending on the customizer.

[92]*Bit stuffing* refers to the introduction of "non-information" bits into frames, buffers, etc., for the purpose of filling them.

[93]The Receive Message interrupt has a special handler that calls appropriate functions for Full and Basic mailboxes.

9.8.4 Interrupt Output Interaction with DMA

PSoC Creator's CAN component does not support DMA operation internally, but the DMA component can be connected to the external interrupt line, if it is enabled and provided that the designer
assumes responsibility for the DMA configuration and operation. However, it is necessary to manage
some housekeeping tasks, e.g., acknowledging the message and clearing the interrupt flags, in code
to handle CAN interrupts properly. With a hardware DMA trigger, registers and data transfers can
be handled when a *Message Received* interrupt occurs, without any firmware executing in the CPU.[94]
The Message Transmitted interrupt can be used to trigger a DMA transfer to reload the message buffer
with new data, without CPU intervention [5].

9.8.5 Custom External Interrupt Service Routine

Custom external ISRs can be used in addition to, or as a replacement for, the internal ISR. When both
external and internal ISRs are used, the Interrupt priority can be set to determine which ISR should
execute first, i.e., internal or external, thus forcing actions before, or after, those coded in the internal
ISR. When the external ISR is used, as a replacement for the internal ISR, the designer is responsible
for the proper handling of CAN registers and events.

The external interrupt line is visible only if it is enabled in the customizer. If an external Interrupt
component is connected, the external Interrupt component is not started as part of the *CAN_Start()*
API, and must be started outside that routine. If an external Interrupt component is connected and the
internal ISR is not disabled or bypassed, two Interrupt components are connected to the same line. In
this case, there will be two separate Interrupt components that will handle the same interrupt events
which in most cases is undesirable.

If the internal ISR is disabled, or bypassed using a customizer option, the internal Interrupt
component will be removed during the build process. If an individual interrupt function call is disabled
in the internal interrupt routine, for an enabled interrupt event by using a customizer option, the CAN
block interrupt triggers, when the relevant event occurs, but no internal function call is executed in
the internal *CAN_ISR* routine. If a specific event needs to be handled, e.g., message received, through
a different path, other than the standard user function call, through DMA. If the internal ISR is to be
customized, using customizer options, the *CAN_ISR* function will not contain any function call other
than the optional PSoC 3 ES1/ES2 ISR patch.

There are several important references that should be consulted when designing systems involving
controller area networks, viz.,

- ISO-11898: Road vehicles—Controller area network (CAN):
  - Part 1: Data link layer and physical signaling
  - Part 2: High-speed medium access unit Controller Area Network (CAN)
  - Part 3: Low-speed, fault-tolerant, medium-dependent interface
  - Part 4: Time-triggered communication
  - Part 5: High-speed medium access unit with low-power mode
- CAN Specification Version 2 BOSCH
- Inicore CANmodule-III-AHB Datasheet

---

[94]This is also useful when handling RTR messages.

9.8.6 Interrupt Output Interaction with the Interrupt Subsystem

The CAN component Interrupt Output settings allow:

- Enabling or disabling of an external interrupt line (customizer option)
- Disabling or bypassing the internal ISR (customizer option)
- Full customization of the internal ISR (customizer option)
- Enabling, or disabling, of specific interrupts handling function calls in the internal ISR, when the relevant event interrupts are enabled using the customizer option. Individual interrupts, e.g., a message transmitted, message received, receive buffer full, bus off state, etc., can be enabled, or disabled, in the CAN component customizer. Once enabled, the relevant function call is executed in the internal *CAN_ISR*. This allows disabling, i.e., removing, of such function calls.

The external interrupt line is visible only if it is enabled in the customizer.

- *uint8 CAN_Start(void)* sets the *initVar* variable, calls the *CAN_Init()* function, and then calls the *CAN_Enable()* function. This function sets the CAN component into run mode and starts the counter, if polling mailboxes available.
- *uint8 CAN_Stop(void)* sets the CAN component into Stop mode and stops the counter, if polling mailboxes available.
- *uint8 CAN_GlobalIntEnable(void)* enables global interrupts from the CAN component.
- *uint8 CAN_GlobalIntDisable(void)* disables global interrupts from the CAN component.
- *uint8 CAN_SetPreScaler(uint16 bitrate)* sets the *prescaler* for generation of the time quanta from the *BUS_CLK*. Values between 0x0, and 0x7FFF, are valid.
- *uint8 CAN_SetArbiter(uint8 arbiter)* sets the arbitration type for transmit buffers. Types of arbiters are Round Robin and Fixed priority. Values 0 and 1 are valid.
- *uint8 CAN_SetTsegSample(uint8 cfgTseg1, uint8 cfgTseg2, uint8 sjw, uint8 sm)* this function configures: *Time segment 1, Time segment 2, Synchronization Jump Width*, and Sampling Mode.
- *uint8 CAN_SetRestartType(uint8 reset)* sets the reset type. Types of reset are *Automatic* and *Manual*. *Manual* reset is the recommended setting. Values 0 and 1 are valid.
- uint8 CAN_SetEdgeMode(uint8 edge) sets *Edge Mode*.error-correcting code Modes are 'R' to 'D' (Recessive to Dominant) and Both edges are used. Values 0 and 1 are valid.
- *uint8 CAN_RXRegisterInit(uint32 *regAddr, uint32 config)* writes CAN receive registers only.
- *uint8 CAN_SetOpMode(uint8 opMode)* sets *Operation Mode*. Operation modes are *Active* or *Listen Only*. Values 0 and 1 are valid.
- uint8 CAN_GetTXErrorflag(void) returns the flag that indicates whether the number of transmit errors exceeds 0x60.
- *uint8 CAN_GetRXErrorflag(void)* returns the flag that indicates whether the number of receive errors has exceeded 0x60.
- *uint8 CAN_GetTXErrorCount(void)* returns the number of transmit errors.
- *uint8 CAN_GetRXErrorCount(void)* returns the number of receive errors.
- *uint8 CAN_GetRXErrorCount(void)* returns the number of receive errors.
- *uint8 CAN_GetErrorState(void)* returns the error status of the CAN component.
- *uint8 CAN_SetIrqMask(uint16 mask)* enables, or disables, particular interrupt sources. *Interrupt Mask* directly writes to the CAN Interrupt Enable register.
- *void CAN_ArbLostIsr(void)* is the entry point to the *Arbitration Lost Interrupt*. It clears the *Arbitration Lost* interrupt flag. It is only generated, if the *Arbitration Lost Interrupt* parameter is enabled.

- *void CAN_OvrLdErrrorIsr(void)* is the entry point to the *Overload Error Interrupt*. It clears the *Overload Error* interrupt flag. It is only generated, if the *Overload Error Interrupt* parameter is enabled.
- *void CAN_BitErrorIsr(void)* is the entry point to the *Bit Error Interrupt*. It clears the *Bit Error Interrupt* flag. It is only generated, if the *Bit Error Interrupt* parameter is enabled.
- *void CAN_BitStuffErrorIsr(void)* is the entry point to the *Bit Stuff Error Interrupt*. It clears the *Bit Stuff Error Interrupt* flag. It is only generated, if the *Bit Stuff Error Interrupt* parameter is enabled.
- *void CAN_AckErrorIsr(void)* is the entry point to the *Acknowledge Error Interrupt*. It clears the *Acknowledge Error* interrupt flag and is only generated, if the *Acknowledge Error Interrupt* parameter is enabled.
- *void CAN_MsgErrorIsr(void)* is the entry point to the *Form Error Interrupt*. It clears the *Form Error* interrupt flag. It is only generated, if the *Form Error Interrupt* parameter is enabled.
- *void CAN_CrcErrorIsr(void)* is the entry point to the *CRC Error Interrupt*. It clears the *CRC Error* interrupt flag. It is only generated, if the *CRC Error Interrupt* parameter is enabled.
- *void CAN_BusOffIsr(void)* is the entry point to the *Bus Off Interrupt*. It puts the CAN component in Stop mode. It is only generated, if the *Bus Off Interrupt* parameter is enabled. Enabling this interrupt is recommended.
- *void CAN_MsgLostIsr(void)* is the entry point to the *Message Lost Interrupt*. It clears the *Message Lost Interrupt* flag. It is only generated, if the *Message Lost Interrupt* parameter is enabled.
- *void CAN_MsgTXIsr(void)* is the entry point to the *Transmit Message Interrupt*. It clears the *Transmit Message Interrupt* flag. It is only generated, if the *Transmit Message Interrupt* parameter is enabled.
- *void CAN_MsgRXIsr(void)* is the entry point to the *Receive Message Interrupt*. It clears the *Receive Message Interrupt* flag and calls the appropriate handlers for *Basic* and *Full* interrupt-based mailboxes. It is only generated, if the *Receive Message Interrupt* parameter is enabled. Enabling this interrupt is recommended.
- *uint8 CAN_RxBufConfig(CAN_RX_CFG *rxConfig)* function configures all receive registers for a particular mailbox. The mailbox number contains *CAN_RX_CFG* structure.
- *uint8 CAN_TxBufConfig(CAN_TX_CFG *txConfig)* configures all transmit registers for a particular mailbox. The mailbox number contains *CAN_TX_CFG* structure.
- *uint8 CAN_SendMsg(CANTXMsg *message)* sends a message from one of the *Basic* mailboxes. The function loops through the transmit message buffer designed as *Basic* CAN mailboxes. It looks for the first free available mailbox and sends it. There can only be three retries.
- *uint8 CAN_SendMsg0-7(void)* are the entry point to *Transmit Message 0–7*. This function checks if mailbox 0–7 already has untransmitted messages waiting for arbitration. If so, it initiates the transmission of the message. It is only generated for Transmit mailboxes designed as *Full*.
- void CAN_TxCancel(uint8 bufferId) cancels transmission of a message that has been queued for transmission. Values between 0 and 15 are valid.
- *void CAN_ReceiveMsg0-15(void)* are the entry point to the *Receive Message* 0-15 Interrupt. They clear Receive Message 0–15 interrupt flags. They are only generated for *Receive* mailboxes designed as *Full* interrupt based.
- *void CAN_ReceiveMsg(uint8 rxMailbox)* is the entry point to the *Receive Message Interrupt* for Basic mailboxes. It clears the *Receive* particular Message interrupt flag. It is only generated, if one of the Receive mailboxes is designed as *Basic*.
- *void CAN_Sleep(void)* is the preferred routine to prepare the component for sleep. The *CAN_Sleep()* routine saves the current component state. Then it calls the *CAN_Stop()* function and calls CAN_SaveConfig() to save the hardware configuration. The *CAN_Sleep()* function must be called before calling the *CyPmSleep()* or the *CyPmHibernate()* function.

- *void CAN_Wakeup(void)* is the preferred routine to restore the component to the state when *CAN_Sleep()* was called. The *CAN_Wakeup()* function calls the *CAN_RestoreConfig()* function to restore the configuration. If the component was enabled before the *CAN_Sleep()* function was called, the *CAN_Wakeup()* function will also re-enable the component.
- *uint8 CAN_Init(void)* initializes, or restores, the component according to the customizer *Configure* dialog settings. It is not necessary to call *CAN_Init()* because the *CAN_Start()* routine calls this function and is the preferred method to begin component operation.
- *uint8 CAN_Enable(void)* activates the hardware and begins component operation. It is not necessary to call CAN_Enable() because the *CAN_Start()* routine calls this function, which is the preferred method to begin component operation.
- *void CAN_SaveConfig(void)* saves the component configuration and non-retention registers. This function also saves the current component parameter values, as defined in the *Configure* dialog or as modified by appropriate APIs. This function is called by the *CAN_Sleep()* function.
- *void CAN_RestoreConfig(void)* restores the component configuration and nonretention registers. This function also restores the component parameter values to what they were prior to calling the *CAN_Sleep()* function.

9.9 S/PDIF Transmitter (SPDIF_Tx)

SoC 3/5's *SPDIF_Tx* component[95] provides a simple way to add a digital audio output to any design.[96] It formats incoming audio- and meta-data to create a *S/PDIF* bit stream appropriate for optical, or coaxial, digital audio. This component, shown in Fig. 9.23, supports interleaved and separated audio. The *SPDIF_Tx* component receives audio data from DMA, as well as, channel status information. Although the channel status DMA will be managed by the component, alternatively, this data can be handled separately to better control a given system. *SPDIF_Tx* provides a fast solution whenever an *S/PDIF* transmitter is essential, including, e.g., digital audio players, computer audio interfaces and audio mastering equipment.

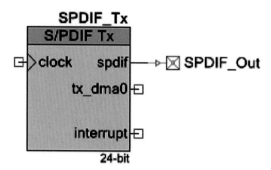

Fig. 9.23 PSoC 3/5's S/PDIF transmitter component.

The supported features of the SPDIF_Tx include:

---

[95]This component can be used in conjunction with an I2S component and an external ADC to convert from analog audio to digital audio.

[96]S/PDIF refers to the Sony Philips digital interface data link layer protocol and an associated physical layer specification. This protocol is often used to transfer compressed digital audio and has no defined data rate.

- conforming to IEC-60958, AES/EBU, AES3 standards for Linear PCM Audio Transmission,
- configurable audio sample lengths (8/16/24),
- or channel status bits generator for consumer applications,
- DMA support,
- sample rate support for clock/128 (up to 192 kHz),

and,

- independent left and right channel FIFOs, or interleaved stereo FIFOs.

9.9.1 SPDIF_Tx component I/O Connections[97]

The following are the available I/O connections for PSoC 3/5LP's SPDIF_Tx component:

- clock—The clock rate must be two times the desired data rate for the *spdif* output, e.g., production of 48-kHz audio, would require a clock frequency given by:

$$(2)(48\,\text{kHz})(64) = 6.144\,\text{MHz} \tag{9.1}$$

- *spdif*—Serial data output.
- *sck*—Serial clock output.
- *interrupt*—Interrupt output.
- *tx_DMA0* - DMA request output for audio FIFO 0 (Channel 0 or Interleaved).
- *tx_DMA1*—DMA request for audio FIFO 1 (Channel 1) output. Displays, if *Separated* under the *Audio Mode* parameter is selected.
- *cst_DMA0**—Request for channel status FIFO 0 (Channel 0) output. Displays, if the *checkbox* under the *Managed DMA* parameter is deselected.
- *cst_DMA1**—Request for channel status FIFO 1 (Channel 1) output. Displays, if the checkbox under the Managed DMA parameter is deselected.

9.9.2 SPDIF_Tx API

The *SPDIF_Tx API* supports the following functions:

- *void SPDIF_Start(void)* starts the *S/PDIF* interface, and the channel status DMA, if the component is configured to handle the channel status DMA. It also enables the Active mode power template bits, or clock gating as appropriate, starts the generation of the *S/PDIF* output with channel status, but the audio data is set to all 0s. It also allows the *S/PDIF* receiver to lock on to the component's clock.
- void *SPDIF_Stop(void)* disables the *S/PDIF* interface and the active mode power template bits or clock gating, as appropriate. The *S/PDIF* output is set to 0. The audio data and channel data FIFOs

---

[97] An asterisk (*) in the list of indicates that the I/O may be hidden on the symbol under the conditions listed in the description of that I/O.

are cleared. The *SPDIF_Stop()* function calls *SPDIF_DisableTx()* and stops the managed channel status DMA.

- *void SPDIF_Sleep(void)* is the preferred routine to prepare the component for sleep.[98] The *SPDIF_Sleep()* routine saves the current component state and then calls *SPDIF_Stop()* and *SPDIF_SaveConfig()* saves the hardware configuration, disables the active mode power template bits, or clock gating, as appropriate, sets the spdif output to 0. *SPDIF_Sleep()* should be called *CyPmSleep()* or *CyPmHibernate()* are called.

- *void SPDIF_Wakeup(void)* restores the *SPDIF* configuration and nonretention register values. The component is stopped, regardless of its state before sleep. The *SPDIF_Start()* function must be called explicitly to start the component again.[99]

- *void SPDIF_EnableTx(void)* enables the audio data output in the S/PDIF bit stream. Transmission will begin at the next X, or Z, frame.

- *void SPDIF_DisableTx(void)* disables the audio output in the S/PDIF bit stream. Transmission of data will stop at the next rising edge of the clock and a constant 0 value will be transmitted.

- *void SPDIF_WriteTxByte(uint8 wrData, uint8 channelSelect)* writes a single byte into the audio data FIFO. The component status should be checked before this call to confirm that the audio data FIFO is not full. *uint8 wrData* contains the audio data to transmit. *uint8 channelSelect* contains the constant for *Channel* to write. See channel status macros below. In the interleaved mode this parameter is ignored.

- *void SPDIF_WriteCstByte(uint8 wrData, uint8 channelSelect)* writes a single byte into the specified channel status FIFO. The component status should be checked before this call to confirm that the channel status FIFO is not full. *uint8 wrData* contains the status data to transmit and *uint8 channelSelect* the constant for the Channel to be written to.

- *void SPDIF_SetInterruptMode(uint8 interruptSource)* sets the interrupt source for the S/PDIF interrupt. Multiple sources may be ORed (Figs. 9.24, 9.25, and 9.26).

SPDIF Tx Interrupt Source	Value
AUDIO_FIFO_UNDERFLOW	0x01
AUDIO_0_FIFO_NOT_FULL	0x02
AUDIO_1_FIFO_NOT_FULL	0x04
CHST_FIFO_UNDERFLOW	0x08
CHST_0_FIFO_NOT_FULL	0x10
CHST_1_FIFO_NOT_FULL	0x20

Fig. 9.24 SPDIF—interrupt mode values.

---

[98] *SPDIF_Sleep()* should be called before calling *CyPmSleep()* or *CyPmHibernate()*.

[99] Calling *SPDIF_Wakeup()* without first calling *SPDIF_Sleep()* or *SPDIF_SaveConfig()* may produce unexpected behavior.

SPDIF Status Masks	Value	Type
AUDIO_FIFO_UNDERFLOW	0x01	Clear on read
AUDIO_0_FIFO_NOT_FULL	0x02	Transparent
AUDIO_1_FIFO_NOT_FULL	0x04	Transparent
CHST_FIFO_UNDERFLOW	0x08	Clear on read
CHST_0_FIFO_NOT_FULL	0x10	Transparent
CHST_1_FIFO_NOT_FULL	0x20	Transparent

Fig. 9.25 SPDIF—status mask values.

Name	Description
SPDIF_SPS_22KHZ	Clock rate is set for 22-kHz audio
SPDIF_SPS_44KHZ	Clock rate is set for 44-kHz audio
SPDIF_SPS_88KHZ	Clock rate is set for 88-kHz audio
SPDIF_SPS_24KHZ	Clock rate is set for 24-kHz audio
SPDIF_SPS_48KHZ	Clock rate is set for 48-kHz audio
SPDIF_SPS_96KhZ	Clock rate is set for 96-kHz audio
SPDIF_SPS_32KHZ	Clock rate is set for 32-kHz audio
SPDIF_SPS_64KHZ	Clock rate is set for 64-kHz audio
SPDIF_SPS_192KHZ	Clock rate is set for 192-kHz audio
SPDIF_SPS_UNKNOWN	Clock rate is not specified.

Fig. 9.26 SPDIF—frequency values.

The *SPDIF* component formats incoming audio data and metadata to create the *S/PDIF* bit stream. This component receives audio data from DMA, as well as, channel status information. Most of the time, the channel status DMA is managed by the component. However, there is an option that allows the data to be specified separately, to better control a system.

9.9.3 S/PDIF Data Stream Format

The audio and channel status data are independent byte streams, packed with the least significant byte and bit first. The number of bytes used for each sample is the minimum number of bytes to hold a sample. Any unused bits will be padded with zeros, starting at the left-most bit. The audio data stream can be a single byte stream, or it can be two byte streams. In the case of a single byte stream, the left and right channels are interleaved with a sample for the left channel first followed by the right channel. In the two stream case, the left and right channel byte streams use separate FIFOs. The status byte stream is always two byte streams.

9.9.4   S/PDIF and DMA Transfers

The S/PDIF interface is a continuous interface that requires an uninterrupted stream of data. For most applications, this requires the use of DMA transfers to prevent the underflow of the audio data or channel status FIFOs. Typically, the Channel Status DMA occurs entirely using two channel status arrays and can be modified using macros. However, data can be provided by an external DMA or CPU to allow flexibility. The S/PDIF can drive up to four DMA components, depending on the component configuration. DMA configuration, using PSoC Creator's DMA Wizard, should be based on Table 9.10.

Table 9.10 SPDIF DMA configuration parameters.

Name of the DMA Source Destination in the DMA Wizard	Direction	DMA Request Signal	DMA Request Type	Description
SPDIF_TX_FIFO_0_PTR	Destination	tx_dma0	Level	Transmit FIFO for Channel 0 or Interleaved Audio Data
SPDIF_TX_FIFO_1_PTR	Destination	tx_dma1	Level	Transmit FIFO for Channel 1 or Interleaved Audio Data
SPDIF_CST_FIFO_0_PTR	Destination	cst_dma0	Level	Transmit FIFO for Channel 0 or Interleaved Status Data
SPDIF_CST_FIFO_1_PTR	Destination	cst_dma1	Level	Transmit FIFO for Channel 1 or Interleaved Status Data

9.9.5   S/PDIF Channel Encoding

S/PDIF is a single-wire serial interface. The bit clock is embedded within the *S/PDIF* data stream. The digital signal is coded using the *Biphase Mark Code* (BMC), which is a kind of phase modulation. The frequency of the clock is twice the bit-rate. Every bit of the original data is represented as two logical states, which, together, form a cell. The logical level at the start of a bit is always inverted to the level at the end of the previous bit. To transmit a one in this format, there is a transition in the middle of the data bit boundary. If there is no transition in the middle, the data is considered a zero.

9.9.6 S/PDIF Protocol Hierarchy

The *S/PDIF* signal format is shown in Fig. 9.27. Audio data is transmitted in sequential blocks each of which contains 192 frames, each of which consists of two *subframes* that are the basic units into which digital audio data is organized.

A subframe, shown in Fig. 9.28, contains a preamble pattern, an audio sample that may be up to 24 bits wide, a validity bit that indicates whether the sample is valid, a bit containing user data, a bit containing the channel status, and an even parity bit for this subframe. There are three types of preambles: X, Y and Z. Preamble Z indicates the start of a block and the start of subframe channel 0. Preamble X indicates the start of a channel 0 subframe when not at the start of a block. Preamble Y always indicates the start of a channel 1 subframe.

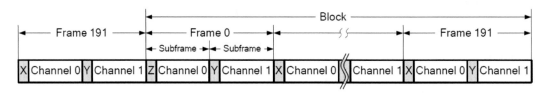

Fig. 9.27 S/PDIF block format.

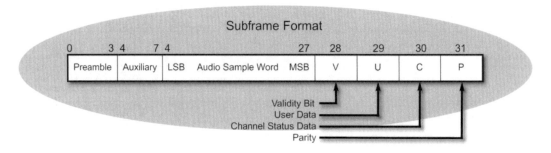

Fig. 9.28 *S/PDIF* subframe format.

9.9.7 *S/PDIF* Error Handling

There are two error conditions for the *S/PDIF* component that can occur, if the audio is emptied and a subsequent read occurs (transmit underflow) or the channel status FIFO is emptied and subsequent read occurs (status underflow). If transmit underflow occurs, the component forces the constant transmission of zeros for audio data and continue correct generation of all framing and status data. Before transmission begins again, the transmission must be disabled, the FIFOs should be cleared, data for transmit must be buffered, and then transmission re-enabled. This underflow condition can be monitored by the CPU using the component status bit *AUDIO_FIFO_UNDERFLOW*.[100] While the component is started, if the status underflow occurs, the component will send all 0s for channel status

---

[100] An interrupt can also be configured for this error condition.

with the correct generation of X, Y, Z framing and correct parity. The audio data is continuous, not impacted.

To correct channel status data transmission, the component must be stopped and restarted again. This underflow condition can be monitored by the CPU using the status bit *CHST_FIFO_UNDERFLOW*. An interrupt can also be configured for this error condition. If the component doesn't manage DMA, the status data must be buffered before restarting the component.

9.9.7.1 Enabling

Audio data transmission has dedicated enabling. When the component is started, but not enabled, the *S/PDIF* output with channel status is generated, but the audio data is set to all zeros. This allows the *S/PDIF* receiver to lock on the component clock and the transition into the enabled state occurs at the X or Z frame.

The *SPDIF_Tx* component is implemented as a set of configured UDBs as shown in Fig. 9.29.

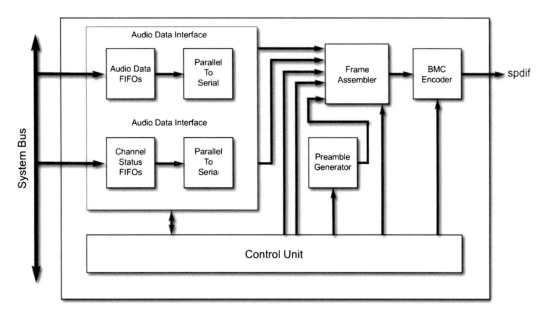

Fig. 9.29 A block diagram of the implementation of SPDIF_Tx.

The incoming audio data is received through the system bus interface and can be provided via the CPU, or DMA. The data is byte-wide, with the least significant byte first, and is stored in an audio buffer, i.e., one or two FIFOs, depending on the component configuration). The Channel Status stream has its own dedicated interface. As with the audio data, there are two Channel Status FIFOs and the channel status is byte-wide data, with the least significant byte occurring first. One byte is consumed from these FIFOs every eight samples. Both audio and status data are converted from parallel to serial form. The User Data are not defined in the S/PDIF standard and may be ignored by some receivers, so they are sent as constant zeros. The validity bit, when low, indicates the audio sample is fit for conversion to analog. This bit is sent as constant zeros. The preamble patterns are generated in the *Preamble Generator* block and are transmitted in serial form. This is all the data required to form the SPDIF subframe structure, except for the parity bit which is calculated in the *Frame Assembler* block during assembling all the inputs in the subframe structure. The output of the *Frame Assembler* block goes to *BMC Encoder* where the data is encoded in a spdif format. The *Control Unit* block gets the control data from the *System Bus* interface and returns the status of component operation to the bus. It controls all other blocks during data transmission.

9.9.8 S/PDIF Channel Encoding

S/PDIF is a single-wire serial interface and the bit clock is embedded within the S/PDIF data stream. The digital signal is coded using *Biphase Mark Code* (BMC), which is a kind of phase modulation. The frequency of the clock is twice the bit-rate. Every bit of the original data is represented as two logical states which together form a cell. The logical level at the start of a bit is always inverted to the level at the end of the previous bit. To transmit a '1' in this format, there is a transition in the middle of the data bit boundary. If there is no transition in the middle, the data is considered a '0' (Fig. 9.30).

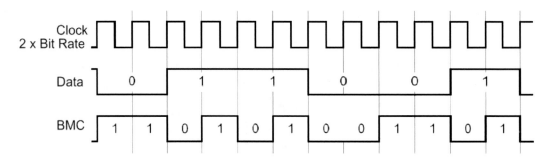

Fig. 9.30 S/PDIF channel encoding timing.

9.9.9 SPDIF Registers

The transmit control and status registers, shown in Figs. 9.31 and 9.32, for SPDIF are defined as follows:

- Enable/disable SPDIF_Tx component. When not enabled the component is in the reset state.
- txenable: Enable/disable audio data output in the S/PDIF bit stream.

SPDIF_Tx_CONTROL_REG

Bits	7	6	5	4	3	2	1	0
value				reserved			enable	txenable

Fig. 9.31 SPDIF control register.

- chst1_fifo_not_full: If set channel status FIFO 1 is not full.
- chst1_fifo_not_full: If set channel status FIFO 0 is not full.
- chst_fifo_underflow: If set channel status FIFOs underflow event has occurred.

SPDIF_Tx_STATUS_REG

Bits	7	6	5	4	3	2	1	0
value	reserved		chst1_fifo_not_full	chst1_fifo_not_full	chst1_fifo_underflow	audio1_fifo_not_full	audio0_fifo_not_full	audio_fifo_underflow

Fig. 9.32 SPDIF status register.

- audio1_fifo_not_full: If set audio data FIFO 1 is not full.
- audio0_fifo_not_full: If set audio data FIFO 0 is not full.
- audio_fifo_underflow: If set audio data FIFOs underflow event has occurred.

The register value may be read by the *SPDIF_Tx_ReadStatus()*. Bit 3 and bit 0 of the status register are configured in Sticky mode, which is a clear-on-read. In this mode, the input status is sampled each cycle of the status register clock. When the input goes high, the register bit is set and stays set regardless of the subsequent state of the input. The register bit is cleared on a subsequent read by the CPU.

9.10 Vector CAN (VCAN)

The Vector CANbedded environment[101] consists of a number of adaptive source code components that cover the basic communication and diagnostic requirements in automotive applications, e.g., ECUs.[102] The Vector CANbedded software suite is customer-specific and its operation varies according to the application and OEM.[103]

This PSoC Creator VCAN component, shown in Fig. 9.33, was designed for the Vector CANbedded suite to generically support the CANbedded structure, independent of the application. The PSoC 3 Vector CAN component was developed to allow easy integration of the Vector certified CAN driver.[104]

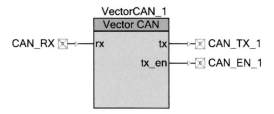

Fig. 9.33 PSoC3's Vector CAN component.

PSoC Creator's VCAN component features include:

---

[101] Vector Informatik GMBH provides a suite of software components for the automotive industry that serve as defacto standards in the automotive industry worldwide.

[102] Engine control units. (ECUs).

[103] Original equipment manufacturer (OEM).

[104] This component is used in conjunction with a CAN driver for PSoC 3 that is provided by Vector.

- CAN2.0 A/B protocol implementation,
- ISO 11898-1 compliant,
- Programmable bit rate up to 1 Mbps @ 8 MHz (BUS_CLK),
- Two- or three-wire interface to a external transceiver (Tx, Rx, and Tx Enable),

and,

- Driver provided and supported by Vector.

The Vector driver uses the CAN interrupt, allowing access. The *Vector_CAN_Init()* function sets up the CAN interrupt with the interrupt service routine *CanIsr_0()* generated by the Vector CAN configuration tool.

9.10.1 Vector CAN I/O Connections

This section describes the various input and output connections for the Vector CAN component. An asterisk (*) in the list of I/O's indicates that the I/O may be hidden on the symbol under the conditions listed in the description of that I/O.

- rx—CAN bus receive signal (connected to CAN RX bus of external transceiver).
- tx—CAN bus transmit signal, (connected to the CAN TX bus of the external transceiver).
- tx_en—External transceiver enable signal.[105]

The Vector CAN component is connected to the *BUS_CLK* clock signal. A minimum value of 8 MHz is required to support all standard CAN baud rates up to 1 Mbps.[106]

The Vector CAN Driver APIs use function pointers. The Keil compiler for PSoC 3 does function call analysis to determine how it can overlay function variables and arguments. When function pointers are present the compiler cannot adequately analyze the calling structure, so the *NOOVERLAY* option is selected to avoid problems that occur because of the use of function pointers. Further information on the handling of function pointers with the Keil compiler is available in the application note: Function Pointers in C51 (www.keil.com/appnotes/docs/apnt_129.asp).

In main the initialization process requires:

- including the v_inc.h file for the driver in main.c,
- enabling global interrupts, if required,
- calling the Vector_CAN_Start() function,
- calling the *CanInitPowerOn()* function (generated by the *Vector GENy* tool),

and,

- writing the necessary functionality using an API from Vector CAN and generated by the *Vector GENy* tool.

---

[105]This output displays when the Add Transceiver Enable Signal option is selected in the Configure dialog.

[106]The value of the *BUS_CLK* selected in the PSoC 3 project design-wide resources must be the same as the value selected in the Vector CAN driver configuration for bus timing.

9.10.2 Vector CAN API

Table 9.11 Vector CAN functions supported by PSoC Creator.

Function	Description
Vector_CAN_Start()	Initializes and enables the Vector CAN component using the Vector_CAN_Init() and Vector_CAN_Enable() functions.
Vector_CAN_Stop()	Disables the Vector CAN component.
Vector_CAN_GlobalIntEnable()	Enables Global Interrupts from CAN Core.
Vector_CAN_GlobalIntDisable()	Disables Global Interrupts from CAN Core.
Vector_CAN_Sleep()	Prepares the component for sleep.
Vector_CAN_Wakeup()	Restores the component to the state when Vector_CAN_Sleep() was called.
Vector_CAN_Init()	Initializes the Vector CAN component based on settings in the component customizer. Sets up the CAN interrupt with the interrupt service routine CanIsr_0() generated by the Vector CAN configuration tool.
Vector_CAN_Enable()	Enables the Vector CAN component.
Vector_CAN_SaveConfig()	Saves the component configuration
Vector_CAN_RestoreConfig()	Restores the component configuration.

PSoC Creator's Vector CAN component can be configured under software control as summarized in Fig. 9.11. By default, PSoC Creator assigns the instance name Vector_CAN_1 to the first instance of a component in a given design.[107] The instance name used becomes the prefix of every global function name, variable, and constant symbol. PSoC Creator provides the following application programming interface for the Vector CAN Component:

- *uint8 Vector_CAN_Start(void)* is the preferred method to begin component operation.
- *uint8 Vector_CAN_Start()* sets the *initVar* variable, calls the *Vector_CAN_Init()* function, and then calls the *Vector_CAN_Enable()* function.[108]
- *uint8 Vector_CAN_Stop(void)*disables the Vector CAN component. Return a value indicating whether the register is written and verified.
- *uint8 Vector_CAN_GlobalIntEnable(void)* enables global interrupts from the CAN Core.[109]
- *uint8 Vector_CAN_GlobalIntDisable(void)* disables global interrupts from the CAN Core. Return Value: Indication whether register is written and verified.
- *void Vector_CAN_Sleep(void)* is the preferred routine to prepare the component for sleep.
- *Vector_CAN_Sleep()* saves the current component state, then calls *Vector_CAN_SaveConfig()* and calls *Vector_CAN_Stop()* to save the hardware configuration.[110]
- *void Vector_CAN_Wakeup(void)* is the preferred routine for restoring the component to the state when*Vector_CAN_Sleep()* was called. *Vector_CAN_Wakeup()* calls

---

[107]The instance can be renamed to any unique value that follows PSoC Creator's syntactic rules for identifiers.

[108]Returns whether the register has been written and verified.

[109]The return value indicates whether the register has been written to and verified.

[110]*Vector_CAN_Sleep()* should be called before *CyPmSleep()* or *CyPmHibernate()*.

- *Vector_CAN_RestoreConfig()* to restore the configuration. If the component was enabled before *Vector_CAN_Sleep()* was called, *Vector_CAN_Wakeup()* will also re-enable the component. Calling *Vector_CAN_Wakeup()* without first calling *Vector_CAN_Sleep()* , or
- *Vector_CAN_SaveConfig()* may produce unexpected behavior.
- *void Vector_CAN_Init (void)* initializes, or restores, the component according to the customizer *Configure* dialog settings. It is not necessary to call *Vector_CAN_Init()* because
- *Vector_CAN_Start()* calls this function and is the preferred method to begin component operation. It sets up the CAN interrupt with the interrupt service routine *CanIsr_0()* generated by the *Vector CAN* configuration tool.
- *uint8 Vector_CAN_Enable(void)* activates the hardware and begins component operation. It is not necessary to call *Vector_CAN_Enable()* because the Vector_CAN_Start() it, which is the preferred method to begin component operation. The return value indicates whether the register is written and verified.
- *void Vector_CAN_SaveConfig(void)* saves the component configuration and non-retention registers, saves the current component parameter values, as defined in the *Configure* dialog or as modified by appropriate APIs. [111]
- *void Vector_CAN_RestoreConfig(void)* restores the component configuration and nonretention registers, restores the component parameters to calling *Vector_CAN_Sleep()*.[112] The global variable, *Vector_CAN_initvar,* is defined in Table 9.12.

Table 9.12 The global variable,*Vector_CAN_initVar.*

Variable	Description
Vector_CAN_initVar	Vector_CAN_initVar indicates whether the Vector CAN has been initialized. The variable is initialized to 0 and set to 1 the first time Vector_CAN_Start() is called. This allows the component to restart without reinitialization after the first call to the Vector_CAN_Start() routine. If reinitialization of the component is required, then the Vector_CAN_Init() function can be called before the Vector CAN Start() or Vector CAN Enable() function.

9.11   Inter-IC Sound Bus (I2S)

The Integrated Inter-IC Sound Bus (I2S) [8] is a serial bus interface standard used for connecting digital audio devices and based on a specification[113] developed by Philips Semiconductor. PSoC Creator's I2S component provides a serial bus interface for stereo audio data, is used primarily by audio ADC and DAC components and operates in master mode only. This component is bidirectional and therefore capable of functioning as a transmitter (Tx) and a receiver (Rx). The number of bytes used for each sample, whether for the right or left channel, is the minimum number of bytes to hold a sample (Fig. 9.34).
I2C features include:

---

[111] This function is called by the *Vector_CAN_Sleep()* function.

[112] Calling this function without first calling the *Vector_CAN_Sleep()* or *Vector_CAN_SaveConfig()* may produce unexpected behavior.

[113] I2S bus specification; February 1986, revised June 5, 1996.

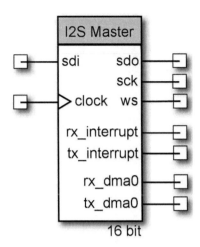

Fig. 9.34 The Inter-IC Sound Bus (I2S).

- 8-32 data bits per sample [3].
- 16, 32, 48, 64-bit word select period:6.144 MHz.
- Data rates up to 96 kHz.
- DMA support.
- Independent right and left channel FIFOs or interleaved stereo FIFOs.
- Independent enable of Tx and Rx.
- Tx and Rx FIFO interrupts.

9.11.1 Functional Description of the I2S Component

Left/Right and Rx/Tx Configuration—The configuration for the Left and Right channels, viz., the Rx and Tx direction, number of bits, and word-select period, are identical. If the application must have different configurations for Rx and Tx, then two unidirectional component instances should be used.

Data Stream Format
The data for Tx and Rx is independent byte streams that are packed with the most significant byte first and the most significant bit in bit 7 location of the first word. The number of bytes used for each sample, for the right or left channel, is the minimum number of bytes to hold a sample. Any unused bits will be ignored on Tx, and will be 0 on Rx.The data stream for one direction can be a single byte stream, or it can be two byte streams. In the case of a single byte stream, the left and right channels are interleaved with a sample for the left channel first followed by the right channel. In the two-stream case, the left and right channel byte streams use separate FIFOs.

DMA
The I2S has a *continuous interface* , i.e., it requires an uninterrupted stream of data. For most applications, this requires the use of DMA transfers to prevent the underflow of the Tx direction, or the overflow of the Rx direction. The I2S can drive up to two DMA components for each direction. PSoC Creator's DMA Wizard can be used to configure DMA operation as defined in Table 9.13.

Table 9.13 DMA and the I2S component.

Name of DMA Source/Destination in the DMA Wizard	Direction	DMA Request Signal	DMA Request Signal	Description
I2S_RX_FIFO_0_PTR	Source	rx_dma0	Level	Receive FIFO for Left or Interleaved Channel
I2S_RX_FIFO_1_PTR	Source	rx_dma1	Level	Receive FIFO for Right Channel
I2S_TX_FIFO_0_PTR	Destination	rx_dma0	Level	Transmit FIFO for Left of Interleaved Channel
I2S_TX_FIFO_1_PTR	Destination	tx_dma1	Level	Transmit FIFO for Right Channel

9.11.2 Tx and Rx Enabling

The Rx and Tx directions have separate enables. When not enabled, the Tx direction transmits all 0 values, and the Rx direction ignores all received data. The transition into, and out of, the enabled state occurs at a word select boundary such that a left/right sample pair is always transmitted, or received.

9.11.3 I2S Input/Output Connections

The I/O connections for the I2S component are:

- *sdi*—Serial data input.[114]
- *clock*—The clock rate must be two times the desired clock rate for the output serial clock (SCK). e.g., to produce 48-kHz audio with a 64-bit word select period, the clock frequency would be: $2 \times 48\,\text{kHz} \times 64 = 6.144\,\text{MHz}$.
- *sdo*—Serial data output. Displays if the Tx option is selected for the Direction parameter.
- *sck*—Output serial clock.
- *ws*—Word select output indicates the channel being transmitted.
- *rx_interrupt*—Rx direction interrupt.[115]
- *tx_interrupt*—Tx direction interrupt.[116]

---

[114]If this signal is connected to an input pin, the *Input Synchronized* selection for this pin should be disabled. This signal should already be synchronized to *SCK*, so delaying the signal with the input pin synchronizer could cause the signal to be shifted into the next clock cycle.

[115]Displays if an Rx option for the Direction parameter has been selected.

[116]Displays if a Tx option for the Direction parameter is selection.

- *rx_DMA0*—Rx direction DMA request for FIFO 0 (Left or Interleaved).[117]
- *rx_DMA1*—Rx direction DMA request for FIFO 1 (Right).[118] Displays if Rx DMA under the DMA Request parameter and Separated L/R under the Data Interleaving parameter for Rx are selected.
- *tx_DMA0*—Tx direction DMA request for FIFO 0 (Left or Interleaved).[119]
- *tx_DMA1*—Tx direction DMA request for FIFO 1 (Right) (Fig. 9.35).[120]

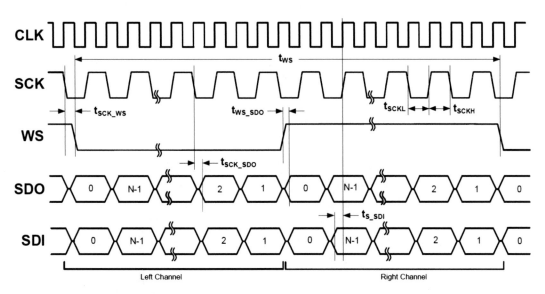

Fig. 9.35 I2S data transition timing diagram.

9.11.4 I2S Macros

By default, the *PSoC Creator Component* catalog contains three *Schematic Macro* implementations for the *I2S* component. These macros contain the *I2S* component already connected to digital pin components. The *Input Synchronized* option is unchecked on the SDI pin and the generation of APIs for all the pins is turned off. The Schematic Macros use the I2S component, configured for Rx only, Tx only, and both Rx and Tx directions, as shown in Figs. 9.36 and 9.37.

---

[117] Displays if Rx DMA under the DMA Request parameter is selected.

[118] Displays if Rx DMA under the DMA Request parameter and Separated L/R under the Data Interleaving parameter for Rx is selected.

[119] Displays if Tx DMA under the DMA Request parameter is selected.

[120] Displays if Tx DMA under the DMA Request parameter and Separated L/R under the Data Interleaving parameter for Tx are selected.

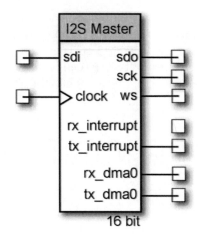

Fig. 9.36 I2S Tx and Rx.

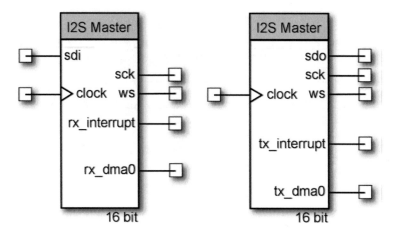

Fig. 9.37 I2S Rx only and I2S Tx only.

9.11.5 I2S APIs

- *void I2S_Start(void)* starts the I2S interface, enables the Active mode power template bits, or clock gating, as appropriate. Starts the generation of the *sck* and *ws* outputs. The Tx and Rx directions remain disabled.
- *void I2S_Stop(void)* disables the *I2S* interface and the Active mode power template bits or clock gating as appropriate. sets the *sck* and *ws* outputs to 0. disables the Tx and Rx directions and clears their FIFOs.
- *void I2S_EnableTx(void)* enables the Tx direction of the I2S interface.[121]
- *void I2S_DisableTx(void)* disables the Tx direction of the I2S interface.[122]
- void I2S_EnableRx(void) enables the Rx direction of the I2S interface.[123]

---

[121]Transmission begins at the next word select falling edge.

[122]Transmission of data stops and a constant 0 value is transmitted at the next word select falling edge.

[123]Data reception begins at the next word select falling edge.

- *void I2S_DisableRx(void)* disables the Rx direction of the I2S interface.[124]
- *void I2S_SetRxInterruptMode(uint8 interruptSource)* sets the interrupt source for the I2S Rx direction interrupt. Multiple sources may be ORed (Table 9.14).

Table 9.14 I2S Rx interrupt source.

I2S Rx Interrupt Source	Value
RX_FIFO_OVERFLOW	0x01
RX_FIFO_0_NOT_EMPTY	0x02
RX_FIFO_1_NOT_EMPTY	0x04

9.11.6 I2S Error Handling

Two error conditions can occur if the transmit FIFO is empty, and a subsequent read occurs, i.e., a transmit underflow, or the receive FIFO is full and a subsequent write occurs, i.e., a receive overflow. If the transmit FIFO becomes empty, and data is not available for transmission while transmission is enabled, i.e., a Transmit underflow, the component will force the constant transmission of 0s. Before transmission begins again, transmission must be disabled, the FIFOs should be cleared, data for transmit must be buffered, and then transmission re-enabled. The CPU can monitor this underflow condition using the transmit status bit *I2S_TX_FIFO_UNDERFLOW*. An interrupt can also be configured for this error condition. While reception is enabled, if the receive FIFO becomes full and additional data is received (Receive overflow), the component stops capturing data. Before reception begins again, reception must be disabled, the FIFOs should be cleared, and then reception re-enabled. The CPU can monitor this overflow condition using the receive status bit *I2S_RX_FIFO_OVERFLOW*. An interrupt can also be configured for this error condition.

9.12   Local Interconnect Network (LIN)

The LIN standard [9] was co-developed by a set of companies involved in the automotive industry.[125] It was intended from the outset to serve as a multiplexed communication system that was much simpler than the controller area network (CAN), or the serial peripheral interface(SPI) [17]. LIN functions as a subnetwork to CAN and is based on an architecture that supports only a single master, and multiple slaves. It is not as robust, has a smaller bandwidth/bit rate and offers less functionality than CAN, but it is much more economical. LIN targets low-cost automotive networks as a complement to the existing portfolio of automotive multiplex networks and is typically used for networking sunroof controls, rain detection systems, automatic headlight controls, door locks, interior lighting controls, etc.

---

[124]At the next word select falling edge, data reception is no longer sent to the receive FIFO.

[125]The original concept for the LIN protocol is attributed to Motorola but they were soon joined in supporting the new standard by Audi, BMW, Daimler Chrysler, Volkswagen and Volvo.The current version is LIN 2.0 and was issued in September 2003.

The LIN specification consists of an *API specification specification*, a *configuration/diagnostic specification*, a *physical layer specification*. a *node capability language specification* and *protocol specification*.

- The *API specification* describes the interface between the application program and the network.
- The *configuration/diagnostic specification* is a description of LIN services available above the data link layer associated with sending configuration and diagnostic messages. The physical layer specification defines clock tolerances, supported bit rates, etc.
- The *node capability language specification* defines the language format for certain types of LIN modules employed in plug and play applications.
- The *capability language specification* defines the format of the configuration file used to configure the LIN network.

The LIN system functions as an asynchronous communications system that operates without requiring a clock. Therefore, it functions as a single wire system[126] that does not require arbitration. Baud rates are limited to 20 kbits/s to avoid EMI issues. The master is responsible for determining the priority, and therefore the order of message transmission. The master employs a stable clock for reference and monitors data and check bytes, while controlling the error handler. The master controls the bus and transmits *Sync Break, Sync Byte*, and *ID* data fields. Two to sixteen slaves receive/transmit data when their respective IDs are transmitted by the master.[127] Slaves can transmit 1, 2, 4 or 8 data bytes at a time, together with a check-byte.

The main properties of the LIN bus are:

- Data format similar to the common serial UART format,
- Safe behavior with data checksums,
- Self-synchronization of slaves on master speed,
- Single-master, multiple-slaves (up to 16 slaves),
- Single-wire (max 40 m),

and,

- Speeds up to 19.2 kbps (choice is 2400, 9600, 19,200 bps) (Fig. 9.38).

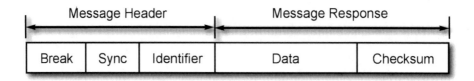

Fig. 9.38 The LIN message frame.

The message frame format employed by LIN consists of a *break* containing 13 bits followed by a delimiter of one bit which alerts all the nodes on the LIN bus and signals the start of a frame. This is

---

[126]Such systems are often referred to as *one-wire systems*, but in point of fact, an additional wire is required to provide a ground return for the system.

[127]It should be noted that a master can also serve as a slave.

immediately followed by a clock synchronization, or *sync* field (x55), that allows the slaves to adjust their respective internal baud rates to that of the bus. A message *identifier* (ID) follows the sync field that consists of a 6-bit message and a 2-bit parity field. IDs 0-59 are assigned to the signal-carrying data frames, 60–61 to the diagnostic data frames, 62 to user-defined extensions and 63 is reserved for future use.[128] The slaves *listen* for IDs and check the respective parities for which they are either a publisher or subscriber. The slave response consists of one-to-eight data bytes followed by an 8-bit checksum.[129]

9.12.1 LIN Slave

PSoC Creator's LIN Slave component implements a LIN 2.1 slave node on PSoC 3 and PSoC 5LP devices. Options for LIN 2.0, or SAE J2602-1, compliance are also available. This component consists of the hardware blocks necessary to communicate on the LIN bus, and an API to allow the application code to easily interact with the LIN bus communication. The component provides an API that conforms to the API specified by the LIN 2.1 Specification. This component provides a good combination of flexibility and ease of use. A customizer for the component is provided that allows all the LIN Slave parameters to be easily configured.

Supported features include:

- Automatic baud rate synchronization,
- Automatic configuration services handling,
- Automatic detection of bus inactivity,
- Customizer for fast and easy configuration,
- Editor for *.ncf/*.ldf files with syntax checking,
- Full LIN 2.1 or 2.0 Slave Node implementation,
- Fully implements a Diagnostic Class I Slave Node,
- Full transport layer support,
- Full error detection,
- Import of *.ncf/*.ldf files and *.ncf file export,

and,

- Supports compliance with SAE J2602-1 specification.

The LIN bus is based on a single wire, wired-AND, with a termination resistor placed at each node[130] The LIN slave component has the following I/O connections:

- RXD—a digital input terminal
- TXD—a digital output terminal that transmits the data sent via the LIN bus by the LIN node.

---

[128]ID 63 always employs the *classic* checksum algorithm.

[129]The classic checksum algorithm is used with LIN 1.3 nodes and the enhanced checksum algorithm is used with LIN 2.0. The enhanced checksum algorithm requires that the data values be summed and if the sum is greater than, or equal to 256, 255 is subtracted and the result is appended to the message response.

[130]Typical resistor values are 1 kΩ for each master, 30 kΩ for each slave, and the supply voltage ranges from 8 to 18 V.

9.12.2 PSoC and LIN Bus Hardware Interface

A LIN physical layer transceiver device is required when the PSoC LIN slave node is connected directly to a LIN bus. In such cases, the *txd* pin of the LIN component connects to the TXD pin of the transceiver, and the *rxd* pin connects to the RXD pin of the transceiver, as shown in Fig. 9.39. The LIN transceiver device is required because the PSoC's electrical signal levels are not compatible with the electrical signals on the LIN bus. Some LIN transceiver devices also have an *enable* or *sleep* input signal that is used to control the operational state of the device. The LIN component does not provide this control signal. Instead, a pin is used to output the desired signal to the LIN transceiver device, if this signal is needed.

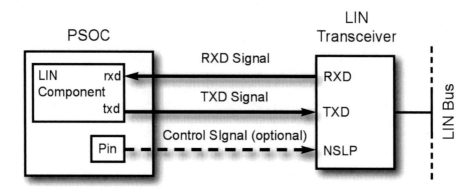

Fig. 9.39 The LIN bus physical layer.

9.13   LCD (Visual Communication)

Visual displays are often an important component of an embedded system for displaying important messages, certain parameter values and/or to facilitate debugging. PSoC 3 has as many as 64 built-in segment LCD drivers that be interfaced directly with a wide variety of segment, LCD, glass types. This gives it the capability to drive up to 768 pixels (16 commons × 48 segments).

The features supported by the PSoC 3 LCD drivers are:

- Adjustable refresh rate from 10 to 150 Hz. Configurable power modes, which allows power optimization,
- Direct drive with internal bias generation no other external hardware is required,
- Maximum 64 in-built LCD drivers (which includes both common and segment pin driver). No CPU intervention in LCD refresh,
- Static, 1/3, 1/4, 1/5 bias ratios. Supports 14-segment and 16-segment alphanumeric display, 7-segment numeric display, dot matrix, and special symbols,
- Support for both Type A and Type B waveforms,
- Support for LCD glass with up to 16 common lines,

and,

- Support for 14-segment and 16-segment alphanumeric display, 7-segment numeric display, dot matrix, and special symbols (Fig. 9.40).

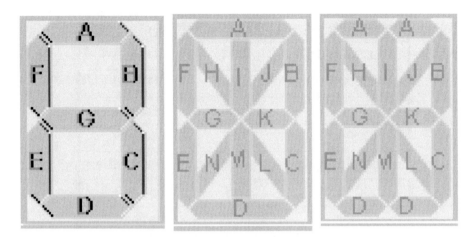

Fig. 9.40 Supported LCD segment types.

PSoC Creator's LCD component is based on a set of library routines that facilitate the use of one, two or four-line LCD modules that employ the *Hitachi HD44780* LCD display driver, 4-bit protocol. The Hitachi interface has proven to be a widely adopted standard for driving LCD displays of the type shown in Fig. 9.41.

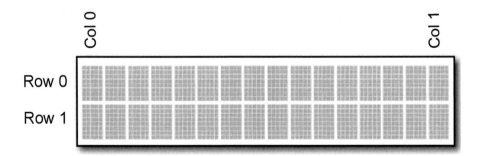

Fig. 9.41 Hitachi 2x16 LCD.

Each of the 32 segments shown in the figure consist of an array of 40 elements (8 × 5). This particular LCD is capable of displaying two rows of 16 characters[131] each and limited graphic displays. Seven logical port pins are used to transmit data bits 0–3, LCD enable,[132] register select[133] and read/not write[134] to the display's integral hardware controller as shown in Table 9.15. The

---

[131] Custom character sets are also supported.

[132] Strobed to confirm new data available.

[133] Select for either data or control input.

[134] Toggle for polling the LCD's ready bit.

LCD_Char_Position() function manages display addressing as follows: row zero, column zero is in the upper left corner with the column number increasing to the right, as shown in Fig. 9.41.[135]

Table 9.15 Logical to physical LCD port mapping.

Logical Port Pin	LCD Module Pin	Description
LCDPort_0	DB4	Data Bit 0
LCDPort_1	DB5	Data Bit 1
LCDPort_2	DB6	Data Bit 2
LCDPort_3	DB7	Data Bit 3
LCDPort_4	E	LCD Enable
LCDPort_5	RS	Register Select
LCDPort_6	R/!W	Read/not Write

9.13.1 Resistive Touch

PSoC Creator's resistive touchscreen component[136] is used to interface with a 4-wire resistive touch screen. The component provides a method to integrate and configure the resistive touch elements of a touchscreen with the emWin[137] Graphics library [23]. It integrates hardware-dependent functions that are called by the touchscreen driver supplied with emWin, when polling the touch panel (Fig. 9.42).

The supported I/O connections are xm, xp, ym, yp where

- xm is a digital I/O connection and designated as signal x− with low being active.
- xp is an analog/digital output connection designated as signal x+ from the x axis of the

The point of contact divides each layer in a series resistor network with two resistors, and a connecting resistor between the two layers. By measuring the voltage at this point, information about the position of the contact point orthogonal to the voltage gradient can be obtained. To get a complete set of coordinates, a voltage gradient must be applied once in the vertical and then in the horizontal direction. First, a supply voltage is applied to one layer and a measurement made of the voltage across the other layer; then the supply voltage is applied to the other layer and the opposite layer voltage is measured. When in touch mode, one of the lines is connected to detect touch activity.

---

[135] In a four-line display, writing beyond column 19 of row 0 can result in row 2 being corrupted because the addressing maps row 0, column 20 to row 2, column 0. This is not an issue in the standard 2x16 Hitachi module.

[136] This component provides a 4-wire resistive touch screen interface to read the touchscreen coordinates and measure the screen resistance It provides access to the functionality of the SEGGER emWin graphics library for translation of resistance to screen coordinates.

[137] emWin is a product of SEGGER Microcontroller that was designed to function as an efficient graphical user-interface that is processor- and graphical LCD controller-independent. (http://www.segger.com/embedded-software.html).

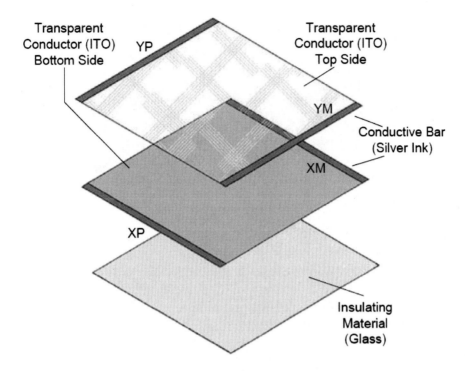

Fig. 9.42 Resistive touchscreen construction.

9.13.2 Measurement Methods

As shown in Fig. 9.43, a touch by a finger, or a stylus, can be uniquely defined by the measurement of three parameters, viz., the x-position, y-position and a third parameter related to the touch pressure. The latter measurement makes it possible to differentiate between finger and stylus contacts. The conductive bars are located on the opposite edges of the panel, as shown. Voltage applied to the layer produces a linear gradient across this layer. The conducting layers are oriented so that the conducting bars are orthogonal to each other, and voltage gradients in the respective layers are also orthogonal. An equivalent circuit for a resistive touchscreen, can be based on treating the conductive layers as resistors placed between the conductive bars in the corresponding layers. When the touchscreen is touched, a resistive connection is formed between the two layers, as shown in Fig. 9.43.

To measure a 4-wire touch sensor, a voltage (VCC) is applied to a conductive bar on one of the layers and the other conductive bar on the same layer is grounded, see Fig. 9.43. This creates a linear voltage gradient in this layer. One of the conductive bars in the other layer is connected to an ADC through a large impedance. The ADC reference is set to VCC, which makes the ADC range from 0 to the max ADC value. When the screen is touched, the ADC reading corresponds to the position on one of the axes. To obtain the second coordinate, the other layer must be powered and read by the ADC. VCC, GND, Analog hi-Z, and ADC input are switched between the two layers, as shown in the y-position measurement in Fig. 9.43. The second ADC reading corresponds to the position on the other axis. Finally, to obtain the touch pressure, two measurements of the cross-layer resistance are required. VCC is applied to a conductive bar on one of the layers while a conductive bar on the other layer is grounded. The voltages on the unconnected bars are then measured, as shown in Figs. 9.44c, d, respectively.

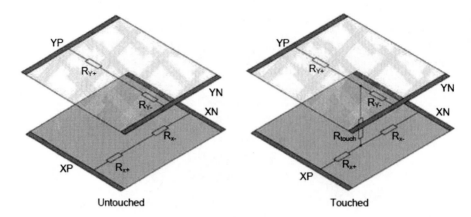

Fig. 9.43 Resistive touchscreen equivalent circuit.

Examination of Fig. 9.44a shows that an equivalent circuit for this case is given by

$$\frac{x}{AD_{max}} = \frac{v_{in}}{v_{ref}} = \frac{v_{in}}{v_{cc}} = \frac{i\,R_{x-}}{i(R_{-x} + R_{x+})} = \frac{R_{x-}}{R_{-plate}} \tag{9.2}$$

where $x = ADC$ value when the ADC input voltage is equal to v_{in}, $AD_{max} = 2^{ADC\_resolution}$, v_{ref} is the ADC reference voltage and R_{x_plate} is given by

$$R_{x\_plate} = R_{x-} + R_{x+} \tag{9.3}$$

A similar analysis of Fig. 9.44b–d gives

$$\frac{y}{AD_{max}} = \frac{R_{y-}}{R_{y-} + R_{y+}} = \frac{R_{y-}}{R_{y\_plate}} \tag{9.4}$$

$$\frac{z_1}{AD_{max}} = \frac{R_{x-}}{R_{x-} + R_{touch} + R_{y+}} \tag{9.5}$$

and,

$$\frac{z_2}{AD_{max}} = \frac{R_{x-} + R_{touch}}{R_{x-} + R_{touch} + R_{y+}} \tag{9.6}$$

Combining these equations yields

$$R_{touch} = R_{x\_plate}\left[\frac{x}{2^{ADC\_resolution}}\right]\left[\frac{z_2}{z_1} - 1\right] \tag{9.7}$$

and,

$$R_{touch} = R_{x\_plate}\left[\frac{x}{2^{ADC\_resolution}}\right]\left[\frac{2^{ADC\_resolution}}{z_1} - 1\right] - R_{y\,plate}\left[1 - \frac{y}{2^{ADC\_resolution}}\right] \tag{9.8}$$

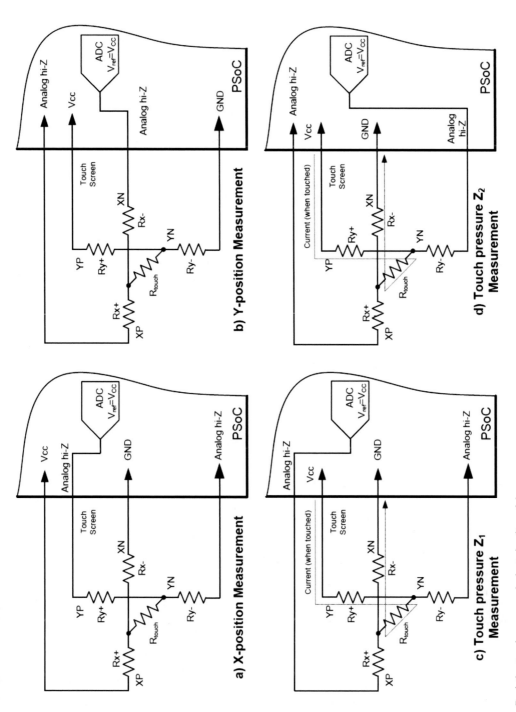

Fig. 9.44 Resistive touchscreen equivalent circuit models.

Equation 9.7 assumes that x_{plate}, x, z_1 and z_2 are known. R_{touch} can also be determined by evaluation of Eq. 9.8, assuming that the values of x_{plate} and y_{plate} are known. A flowchart is shown in Fig. 9.45 that represents the steps required to measure the touchscreen parameters.

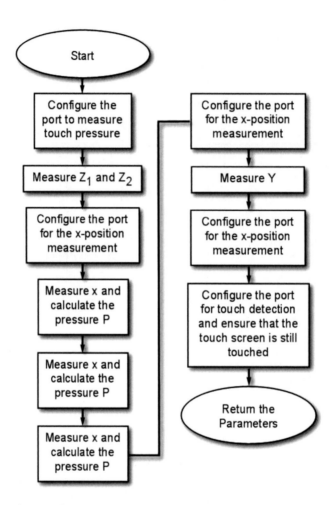

Fig. 9.45 Touchscreen flowchart for parameter measurement.

9.13.3 Application Programming Interface

PSoC Creator supports the following functions for resistive touchscreens:

- void ResistiveTouch_Start(void) calls the *ResistiveTouch_Init()* and *ResistiveTouch_Enable()* APIs.
- *void ResistiveTouch_Init(void)* calls the *Init* functions of the DelSig ADC or SAR ADC and AMux components.
- *void ResistiveTouch_Enable(void)* enables the DelSig ADC or SAR ADC and the AMux components.
- *void ResistiveTouch_Stop(void)* stops the DelSig ADC or SAR ADC and the AMux components.

- *void ResistiveTouch_ActivateX(void)* configures the pins for measurement of X-axis. void ResistiveTouch_ActivateY(void) configures the pins for measurement of Y-axis.
- *int16 ResistiveTouch_Measure(void)* returns the result of the A/D converter.
- *uint8 ResistiveTouch_TouchDetect(void)* detects a touch on the screen.
- void ResistiveTouch_SaveConfig(void) saves the configuration of the DelSig ADC or SAR ADC.
- *void ResistiveTouch_RestoreConfig(void)* restores the configuration of the DelSig ADC or SAR ADC.
- *void ResistiveTouch_Sleep(void)* prepares the DelSig ADC or SAR ADC for low-power modes by calling *SaveConfig* and *Stop* functions.
- void ResistiveTouch Wakeup(void) restores the DelSig ADC or SAR ADC after waking up from a low-power mode (Fig. 9.46).

	XP	XM	YP	YM
Touch	Res Pullup	Digital Hi-Z	Analog Hi-Z	Strong Drive
x- coordinate	Strong Drive	Strong Drive	Analog Hi-Z	Analog Hi-Z
y- coordinate	Analog Hi-Z	Analog Hi-Z	Strong Drive	Strong Drive

Fig. 9.46 Pin configurations for measurement of the touch coordinates.

- void LCD_Char_Start(void)—initializes the LCD hardware module as follows:
  - Enables 4-bit interface
  - Clears the display
  - Enables auto cursor increment
  - Resets the cursor to start position
  - If defined in PSoC Creator's Customizer GUI, a custom LCD character set is also loaded.
- *void LCD_Char_Stop(void)* turns off the LCD display screen.
- *void LCD_Char_PrintString(char8 * string)* writes a null-terminated string of characters to the screen beginning at the current cursor location.aaa
- *void LCD_Char_Position(uint8 row, uint8 column)* moves the cursor to the specified location.
- *void LCD_Char_WriteData(uint8 dByte)* writes data to the LCD RAM in the current position. The position is then incremented/decremented depending on the specified entry mode.
- *void LCD_Char_WriteControl(uint8 cByte)* writes a command byte to the LCD module.[138]
- *void LCD_Char_ClearDisplay(void)* clears the contents of the screen, resets the cursor location to be row and column zero and calls *LCD_Char_WriteControl()* with the appropriate argument to activate the display.

9.13.4 Capacitive Touchscreens

A capacitive touchscreen [13] can be used, as an alternative to resistive touchscreens , and consists of an insulator, e.g., glass, coated with a transparent conductor such as indium tin oxide (ITO). Because a human body is also an electrical conductor, touching the surface of the screen results in a distortion

---

[138]Different LCD models can have their own commands.

of the screen's electrostatic field that is measurable as a change in the screen's capacitance. The location that is touched can be determined by a variety of technologies, and subsequently, can then be sent to the controller for processing. Unlike its resistive counterpart, a capacitive touchscreen is not compatible with most types of electrically insulating materials, e.g., gloves. A special capacitive stylus, or glove with fingertips that generate static electricity is required. This disadvantage especially affects a capacitive touchscreen's usability in consumer electronics, such as touch tablet PCs and capacitive smart phones in cold weather.

Surface capacitance applications have only one side of the insulator that is coated with a conductive layer. A small voltage is applied to that layer, resulting in a uniform electrostatic field. When a conductor, such as a human finger, touches an uncoated surface, a capacitor is formed, dynamically. The sensor's controller can determine the location of a touch, indirectly, based on the change in the capacitance, as measured from the four corners of the surface. The controller has a limited resolution, is prone to false signals from parasitic capacitive couplings, and requires calibration during manufacturing. It is therefore most often used in simple applications such as industrial controls and kiosks.

Projected capacitive touch (*PCT*) is a capacitive technology, consisting of an insulator, such as glass or foil, coated with a transparent conductor, e.g., copper, antimony tin oxide (ATO), nanocarbon, or indium tin oxide (ITO), that permits more accurate and flexible operation by etching, rather than coating, a conductive layer. An X-Y grid is formed, either by etching a single layer to form a grid pattern of electrodes, or by etching two separate, perpendicular layers of a conductive material, with parallel lines or tracks to form a grid, comparable to that of the pixel grid found in many LCD displays. A higher resolution PCT allows operation without direct contact.

PCT is a more robust solution than resistive touch technology because the PCT layers are made from glass. Depending on the implementation, an active, or passive, the stylus can be used instead of, or in addition to, a finger. This is common with point-of-sale devices that require a signature capture. Gloved fingers may be sensed, depending on the implementation and gain settings. Conductive smudges and similar interference on the panel surface can interfere with the performance. Such conductive smudges come mostly from sticky, or perspiring fingertips, especially in high humidity environments. Collected dust, which adheres to the screen due to the moisture from fingertips, can also be a problem.

There are two types of PCT: *Self Capacitance* and *Mutual Capacitance*. If a finger. which is also a conductor, touches the surface of the screen, the local electrostatic field, created by the application of a voltage to each row and column, distorts the local electrostatic field and hence the effective capacitance, and this distortion can be measured to obtain the finger coordinates. Currently, mutual capacitive technology is more common than PCT technology. In mutual capacitive sensors, there is a capacitor at every intersection of each row and each column, e.g., a 16-by-14 array has 224 independent capacitors.

A voltage is applied to the rows or columns so that a finger, or conductive stylus, close to the surface of the sensor changes the local electrostatic field, thereby reducing the mutual capacitance. The capacitance change at each point on the grid can be measured to accurately determine the touch location by measuring the voltage on the other axis. Mutual capacitance allows multi-touch operation where multiple fingers, palms, or styli, can be accurately tracked at the same time. Self-capacitance sensors can have the same X-Y grid as mutual capacitance sensors, but the columns and rows operate independently. With self-capacitance, the capacitive load of a finger is measured as a current on each column, or row, electrode. This method produces a stronger signal than the mutual capacitive method, but it is unable to detect accurately more than one finger, which results in *ghosting*, or misplaced location sensing.

9.14 Recommended Exercises

9-1 Give examples of when each of the communications protocols discussed in this chapter might be used to provide the most efficient and cost-effective transmission channel.

9-2 Explain why a twisted pair of conductors is used when deploying communications protocols such as USB. What is the significance of the use of 90 Ω impedance wiring in such cases? Can 50 or 72 Ω, impedance cable be used instead? If not, why not? And if so, what are the constraints on their use, if any?

9-3 Calculate the CRC for a string of bytes consisting of 01010101, 00000000, 11111111, 00001111, 00000011, 01010101, 11110000 and 10101010.

9-4 Explain the advantages and disadvantages of using parity checks, versus cyclic redundancy, to ensure data integrity.

9-5 Prepare a table comparing each of the communication protocols discussed in this chapter with respect to parameters such as path differences, transmission speeds, handshaking techniques, multiple masters, multiple slave support, error detection methods, etc.

9-6 When transmitting multiple bits in the form of bytes, are parallel transmission paths always capable of transmitting data faster than serial paths? If not, give an example of a situation for which serial transmission can be faster than parallel transmission.

9-7 Explain how arbitration works for each of the protocols discussed in this chapter, if applicable. In particular, treat the case of multiple masters, and slaves, operating in the same network.

9-8 Estimate the propagation delay of individual bits when transmitted in serial fashion over a distance of 5, 30, and 1000 m. State all your assumptions.

9-9 What are the advantages of the USB protocol that have led to its largely replacing the once ubiquitous RS232 protocol?

9-10 Why do many automotive and other applications often employ multiple communications protocols in the same environment, e.g., why are CAN, LIN and FlexRay sometimes employed in the same vehicle?

References

1. M. Ainsworth, PSoC 3 to PSoC 5LP Migration Guide. AN77835. Document No. 001-77835 Rev.*D1. Cypress Semiconductor (2017)
2. R. Ball, R. Pratt, *Engineering Applications of Microcomputers Instrumentation and Control* (Prentice Hall, Englewood Cliffs, 1984)
3. Delta Sigma ADC and I2C Master testbench with PSoC 3/5LP. CE95301. Cypress Semiconductor (2012)
4. A.N. Doboli, E.H. Currie, *Introduction to Mix-Signal, Embedded Design* (Springer, Berlin, 2010)
5. J. Eyre, J. Bier, The evolution of DSP processors. IEEE Signal Proc. Mag. **17**(2), 44–51 (2000)
6. F^{20}MC/FR Family, All Series, Method of Confirming Data in Serial Communications. AN206373. Document No. 002-06373 Rev.*B1. Cypress Communications. Cypress Semiconductor (1917)
7. I2C-bus specification and user manual (PDF). Rev. 6. NXP. 2014-04-04. UM10204. Archived (PDF) from the original on 2013-05-11. UM10204I2C-bus specification and user manual Rev. 6 (April 4, 2014)

8. Inter-IC Sound Bus (I2S)2.40. Document Number: 001-85020 Rev. *A. Cypress Semiconductor (2013)
9. Local Interconnect Network. https://en.wikipedia.org/wiki/Local_Interconnect_Network
10. R. Murphy, USB 101: An Introduction to Universal Serial Bus 2.0. Cypress Application Note AN57294 (2011)
11. R. Murphy, PSoC 3 and PSoC 5LP–Introduction to Implementing USB Data Transfers. AN56377. Document No. 001-56377 Rev.*M1. Cypress Semiconductor (2017)
12. R. Murphy, PSoC 3 and PSoC 5LP USB General Data Transfer with Standard HID Drivers. AN82072. Document No. 001-82072 Rev. *F. Cypress Semiconductor (2017)
13. S. Paliy, A. Bilynskyy, PSoC 1—Interface to Four-Wire Resistive Touchscreen. Application Note AN2376. Cypress Semiconductor Corporation (2011)
14. P. Phalguna, PSoC 3 and PSoC 5LP SPI Bootloader. AN84401. Document No. 001-84401 Rev.*D. Cypress Semiconductor (2017)
15. PSoC 3 Technical Reference Manual (TRM). Document No. 001-50235 Rev. *M (2020)
16. M. Ranjith, PSoC 3and PSoC 5LP–Getting Started with Controller Area Network(CAN). AN52701. Document No.001-52701 Rev. *L1. Cypress Semiconductor (2017)
17. Serial Peripheral Interface (SPI). https://en.wikipedia.org/wiki/Serial_Peripheral_Interface
18. Serial Peripheral Interface (SPI) Master 2.50. Document Number: 001-96814 Rev. *D. Cypress Semiconductor (2017)
19. Serial Peripheral Interface (SPI) Slave 2.70. Document Number: 001-96790 Rev. *C. Cypress Semiconductor (2017)
20. V. Shankar Ka, PSoC 3 and PSoC 5 Interrupts. AN54460. Document No. 001-54460 Rev. *D 1. Cypress Semiconductor
21. UART Full Duplex and printf() Support with PSoC 3/4/5LP. CE210741. Document No. 002-10741 Rev.*C. Cypress Semiconductor (2017)
22. Universal Asynchronous Receiver Transmitter (UART 2.50). Document Number: 001-97157 Rev. *D. Cypress Semiconductor (2017)
23. User's Reference Manual for emWin V5.14. SEGGER Microcontroller GmbH & Co. KG (2012)

Phase-Locked Loops

10

Abstract

Phase-locked loops (Two seminal papers should be considered when studying the origin of phase-looked loops, viz. (E.V. Appleton, Proc Camb Philos Soc 21(Part III):231, 1922–1923) and (H. de Bellescize, L'Onde Electrique 11:230–240, 1932).) (PLLs) are electronic circuits that employ negative feedback to lock the output phase of a signal to the signal's input phase by detecting the phase error between input and output as shown in Fig. 10.1 and reducing the resulting error to zero (For the purposes of this discussion, and, in the present context, zero shall mean reducing the error to an acceptable level.). A voltage-controlled oscillator (Two types of oscillators are commonly used in PLLs, viz., harmonic (sinusoidal waveforms) and relaxation (sawtooth or triangular waveforms).) (VCL) is employed to reduce the phase error to zero (It should be noted that the input signal, and/or the reference signal generated by the VCL, may be sinusoidal.). The simplest phase detector consists of an XOR gate configured as shown in Fig. 10.3. Phase-locked loops are widely employed in RF applications, telecommunications, a wide variety of digital circuits, digital computers and in many other applications. They can be used to demodulate a signal, recover a signal from a noisy communication channel, generate a stable frequency at multiples of an input frequency (frequency synthesis (U.L. Rhode, Digital PLL frequency synthesizers, 1983), or distribute precisely timed clock pulses in digital logic circuits such as microprocessors.

10.1 PLL Use and Application

Many highly stable oscillators, e.g., quartz oscillators maintained in well-controlled temperature environments,[1] are available only for a limited range of frequencies (Fig. 10.1). However, the need for stable oscillators [11] outside such ranges can be addressed by employing the technique illustrated in Fig. 10.2. The VCO's output signal is divided by N before being asserted to the input of the phase detector.

---

[1] Temperature-controlled oscillators are often referred to as TCXOs.

© Springer Nature Switzerland AG 2021
E. H. Currie, *Mixed-Signal Embedded Systems Design*,
https://doi.org/10.1007/978-3-030-70312-7_10

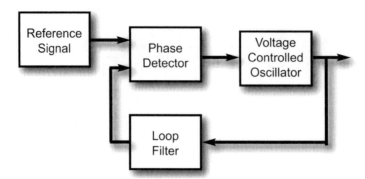

Fig. 10.1   Schematic diagram of generic PLL.

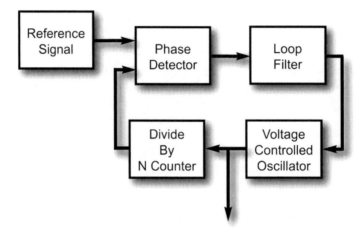

Fig. 10.2   PLL output greater than the reference signal.

PLLs, in general, can be classified into four distinct groups [13], viz.,

- Analog PLLs (LPLL)[2]
- Software PLLs (SPLL)[3]
- All Digital PLL (ADPLL)
- Digital PLL (DPLL)—This type of PLL is usually an

Although analog PLLs are in widespread use, digital phase-locked loops [5] are becoming increasingly important in a wide variety of applications because of their increased reliability, smaller size, lower cost, speed, and improved performance. Analog, or linear, PLLs (LPPLs) and digital PLLs (DPLLs) differ in a number of respects, e.g.,

- LPLLs can be slower to "lock-in" than digital PLLs.
- DPLLs can operate a much lower frequencies than analog PLLs due in part to the fact that LPLLs rely on an analog lowpass filter.
- LPLLs can have temperature and power supply dependencies that adversely and materially affect performance.

---

[2]Sometimes referred to as linear phase-locked loops (LPLLs).

[3]Typically implemented by using digital signal processors.

• Some LPLLs include analog multipliers that can be sensitive to DC drift.

10.2 Phase Detection (See Figs. 10.3 and 10.4)

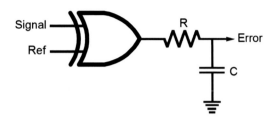

Fig. 10.3 A simple phase detector implementation.

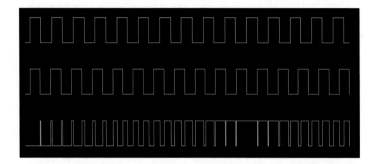

Fig. 10.4 XOR phase detector's input and output.

10.3 Voltage-Controlled Oscillators

A voltage-controlled oscillator's output frequency is typically determined by an external resistor/-capacitor combination[1] and a DC control voltage. PLLs are often used in tracking filters, AM detectors, FSK decoders, modems, telemetry transmitters/receivers, tone decoders, high frequency clocks for microprocessors, etc. [2]. The phrase "PLL Bandwidth" refers to the range over which the PLL does not lose "lock" with respect to the reference frequency and jitter.[4]

Prior to the assertion of an input signal, a PLL is said to be in the "free-running" mode. The assertion of a signal causes the VCO to enter a mode which results in a change in the VCO's frequency. This mode is known as the "capture mode". Once the VCO frequency and the input signal frequency are the same, the PLL is described as being in "phase-locked mode". From this point on, the VCO will "track" the input signal's frequency.

Regardless of the type of phase detector, there are two inputs,[5] one from the input signal and the other from the VCO. The output of the phase detector is a DC voltage that is proportional to the phase difference between the input signal and the VCO and is referred to as the "error voltage". The PLL

---

[4]A PLL with a high bandwidth yields a fast lock time but tracks any jitter of the reference clock source, which can then appear on the output of the PLL [3]. Conversely, if a low bandwidth PLL is employed the lock time is increased, but the reference clock jitter is removed.

[5]In some PLL applications [6], the input signal is asserted from a fixed frequency source, e.g., a quartz-based oscillator.

also employs a low pass filter to remove high-frequency noise [9], and determines parameters such as transient response, bandwidth and lock/capture ranges.

Assuming that the input signal is given by:

$$v_{in}(t) = A sin[\omega_c + \theta(t)] \tag{10.1}$$

and the output signal is given by:

$$v_o(t) = A cos[\omega_c + \theta_o(t)] \tag{10.2}$$

The output of the phase detector can be expressed as:

$$v_{pd}(t) = K_m A_i A_o sin[\omega_c t + \theta_o(t)] \tag{10.3}$$

$$= \frac{K_m A_i A_o}{2} \left(sin[\theta_i(t) - \theta(t)] + sin[2\omega_c t + \theta_i(t) + \theta_o(t)] \right) \tag{10.4}$$

$$\approx K_d[sin\theta_e(t)]f(t) \tag{10.5}$$

where the phase error, θ_e, is given by:

$$\theta_e(t) = \theta_i(t) - \theta_o(t) \text{ and } K_d = \frac{K_m A_i A_o}{2} \tag{10.6}$$

and the second term of Eq. 10.4 has been removed by a low pass filter (Fig. 10.5).

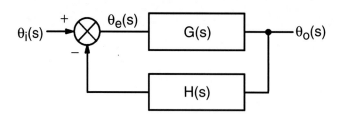

Fig. 10.5 Phase detector.

10.4   Modeling a Voltage-Controlled Oscillator

There are four basic types of voltage-controlled oscillators, viz.,

- Crystal Oscillators—there are a variety of crystal, typically quartz) oscillators as illustrated in part in Fig. 13.5.
- Relaxation Oscillators[6]—a FSM machine with no stable states (Fig. 10.6).
- Ring Counters employ an odd number of inverters connected as shown in Fig. 10.7).
- Resonant Oscillator—this type of oscillator involves a resonant circuit placed in the positive feedback loop of a voltage to current amplifier [8].

---

[6] Also known as astable multivibrators.

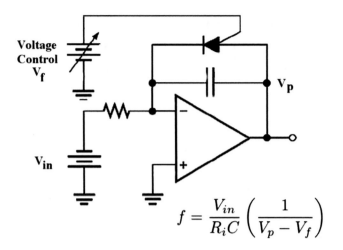

$$f = \frac{V_{in}}{R_i C}\left(\frac{1}{V_p - V_f}\right)$$

Fig. 10.6 Example of a voltage controlled relaxation oscillator.

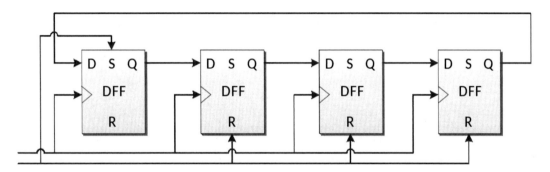

Fig. 10.7 Example of a ring counter.

- YIG Oscillators[7] employ "YIG spheres"[8]
- Negative Resistance Oscillators[9]—employ devices with negative volt-ampere characteristics.

Figure 10.8 is an example of a comparator-based, relaxation oscillator whose frequency of oscillation[10] is given by:

$$f = \frac{1}{2ln(3)RC} \tag{10.7}$$

---

[7]YIG-based (Yttrium Iron Garnet) oscillators are very high Q extremely stable frequency sources, exhibit minimal phase jitter, have very linear tuning characteristics and therefore, are of particular interest when making critical measurements. They are typically operating in frequency ranges from 2–18 GHz, e.g., 2–8 GHz, 8–12 GHz, etc. [4].

[8]YIG crystals are prepared as spheres that function as a ferromagnetic material in which precessing electrons resonate at a particular frequency, when the spheres are in the presence of both an externally-applied, a static magnetic field and an RF field. The latter is applied in a different direction from that of the externally applied, static magnetic field.

[9]Gunn Diodes, varactors, tunnel diodes, certain plasma devices, etc., exhibit negative resistance characteristics that can be employed to create oscillators. The phrase "Negative resistance" refers to regions of such devices I–V characteristics in which there is an inverse, and typically nonlinear, relationship between current (I) and voltage (V).

[10]Assuming that the supply voltages are symmetric, i.e. plus and minus V.

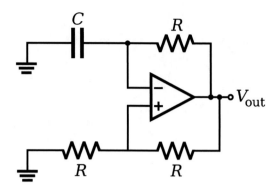

Fig. 10.8 A typical OpAmp-based relaxation oscillator. [8]

10.5   Phase and Frequency

Phase and frequency are related by the following equations:

$$\omega = \frac{d\theta}{dt} \tag{10.8}$$

and,

$$\omega = 2\pi f \tag{10.9}$$

Assuming a linear model for a VCO, and that the "free-running" frequency[11] of the VCO is given by ω_c:

$$\Delta\omega = K_o V_c(t) \tag{10.10}$$

or,

$$\omega_o = \omega_c + K_o V_c(t) \tag{10.11}$$

and therefore:

$$\theta(t) = \int_{-\infty}^{t} K_o V_c(t) dt \tag{10.12}$$

The output of the VCO is given by:

$$x_{vco}(t) = A sin[2\pi f_c t + \theta(t)] = A sin\left[2\pi f_c t + \int_{-\infty}^{t} K_o V_c(t) dt\right] \tag{10.13}$$

and therefore:

$$x_{vco} = A sin[2\pi f_c t + K_o V_v(t) t] \tag{10.14}$$

---

[11] The free-running frequency is defined as the frequency of the VCO when there is no input voltage.

10.6   PLL Lowpass Filter

If the input to the phase detector is given by:

$$x_{in} = A_c cos[2\pi f_c t + \phi(t)]]$$ (10.15)

and the output of the VCO is given by:

$$x_{vco} = -A_v sin[2\pi f_c t + \theta(t)]$$ (10.16)

The input to the LPF is given by:

$$x_{in}(t)x_{vco}(t) = -A_v sin[2\pi f_c t + \theta(t)][A_c cos[2\pi f_c t + \phi(t)]$$ (10.17)

Given that:

$$sin(a)cos(A) = \frac{1}{2}[sin(A - B) + sin(A + B)]$$ (10.18)

Equation 10.17 becomes:

$$x_{in}x_{vco} = \frac{1}{2}A_c A_v sin[\phi(t) - \theta(t)] - \frac{1}{2}A_c A_v sin[4\pi f_c t + \theta(t) + \phi(t)]$$ (10.19)

The low-pass filter removes the carrier's second harmonic so that the output of the phase detector is given by:

$$e_d = \frac{1}{2}A_c A_v sin[\phi(t) - \theta(t)]$$ (10.20)

10.6.1 Phase and Jitter

Jitter refers to the temporal variations in the occurrence of a signal that would otherwise be expected to show no temporal variation. Sources of jitter include:

- Connectors
- Phase-lock loop implementation circuitry
- Quartz crystal thermal noise
- Cabling
- PCB Traces
- Mechanical vibration/shock
- Power supply and/or ground noise

10.6.2 Phase Noise

Oscillators do not produce signals consisting of only a single frequency. This arises as a result of the presence of a variety of both internal and external noise sources and manifests itself as "phase noise" [7] which can be observed in both the temporal and frequency domains and modeled as:

$$v_{osc}(t) = [V_0 + v_{noise}(t)][cos(\omega_0 t + \theta_0 + \phi_{noise}(t))] \qquad (10.21)$$

or,

$$v_{osc}(t) = [V_0 + v_{noise}(t)][cos(\omega_0 t + \theta_1(t))] \qquad (10.22)$$

$$= [V_0 + v_{noise}(t)][cos(\omega_1 t)] \qquad (10.23)$$

where θ_1 represents the time-dependent phase component of the oscillation and ω_1 is the time dependent angular frequency.[12] Thus, random variations in phase noise produces additional spectral components that degrade the oscillators spectral purity. This effect can be minimized by requiring that the oscillator have a very high Q factor.[13]

10.7   Key PLL Parameters

There are a number of important parameters that characterize PLL's [10], viz.,

- Output amplitude.
- Power consumption.
- Required supply voltages.
- Loop bandwidth—this parameter defines the speed characteristics of the control loop.
- Transient behavior—settling time, overshoot, etc., are characterized by this parameter.
- Steady state errors—time and/or phase error.
- Phase noise—this parameter is dependent upon both the PLL bandwidth and the phase noise introduced by the VCO.
- Lock-in range[14]—the frequency range over which the PLL can track the input frequency and "lock-in" [6].
- Capture range—This parameter is defined as the range over which the PLL is able to "lock-in" assuming that it begins in the unlocked state. This range, though less than the lock-in range, is defined by the cutoff frequency of the LPF.
- Spectral purity of the output—relative amplitude of primary frequency versus the sidebands.

Careful consideration should be given to each of these factors before incorporating a PLL into any design.

---

[12]Mechanical shock and/or vibration can also cause phase noise [7].

[13]Phase noise can also be further reduced by employing low noise resistors, e.g., small resistance values, low-loss PCB dielectric, inductors, capacitors and in some applications discrete components as opposed to integrated circuits.

[14]This is determined primarily by the phase detector and VCO characteristics

10.8 Digital Phase-Locked Loops

PLLs in general can be classified into four distinct groups[13], viz.,

- Analog PLLs (LPLL)[15]
- Software PLLs (SPLL)[16]
- All Digital PLL (ADPLL)
- Digital PLL (DPLL)—This type of PLL is usually an employed when jitter, phase noise, power and die area are of concern.

Although analog PLLs are in widespread use, digital phase-locked loops [14] are becoming increasingly important in a wide variety of applications because of their increased reliability, smaller size, lower cost, speed, and improved performance. Analog, or linear, PLLs (LPPLs) and digital PLLs (DPLLs) differ in a number of respects, e.g.,

- LPLLs can be slower to "lock-in" than digital PLLs.
- DPLLs can operate at much lower frequencies than analog PLLs due in part to the fact that LPLLs rely on an analog, lowpass filter.
- LPLLs can have temperature and power supply dependencies that adversely and materially affect performance.
- Some LPLLs include analog multipliers that can be sensitive to DC drift.

10.9 PSoC 3 and PSoC 5LP Phase-Locked Loop

The PSoC Phase-Locked Loop (PLL) allows designers to generate a clock signal from the existing clocks in the PSoC system. It produces an output frequency equal to the input frequency multiplied by the ratio P/Q, where P can range from 4 to 256, and Q can range from 1 to 16. The gain range is therefore 0.25 to 256 times the input frequency. The PLL uses a voltage-controlled oscillator (VCO) [12] to generate the new output clock. This clock is divided by P, and compared to the input clock divided by Q by the phase frequency detector (PFD). The PFD output is filtered, and used to trim the VCO. This achieves an output 2 clock that is equal in frequency to the input clock multiplied by P and divided by Q. This topology is shown in Fig. 10.9.

The PLL input, output, and intermediate frequencies are limited to certain ranges. A PSoC PLL can generate frequencies between 24 and 80 MHz, given an input clock between 1 and 48 MHz. The intermediate signal, equal to the input frequency divided by Q, must be between 1 and 3 MHz.[17]

---

[15]Sometimes referred to as linear phase-locked loop s (LPLL).

[16]Typically implemented by using digital signal processors.

[17]The exact limitations on the input, output, and intermediate frequency of the PLL vary between parts, and may be found in the applicable device datasheet.

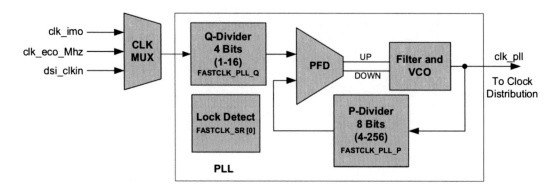

Fig. 10.9 Block diagram of the PSoC 3 and 5LP Phased-Lock Loop.

10.10 PSoC 3 and PSoC 5LP PLL Topology

The PLL introduces no frequency inaccuracy, and consumes less power at a given output frequency than the IMO. Thus, when a clock is desired within the operational range of the PLL, it is recommended that the IMO be run at the minimum speed of 3 MHz, and the PLL be used to generate the desired output frequency. The resulting PLL output clock will be within the accuracy specification percentages of the input clock. The PLL may be configured during design time using PSoC Creator's "Clocks" design wide resources tab. A PLL may be reconfigured at runtime using register writes or the provided API. PLL reconfiguration APIs are provided by PSoC Creator in all PSoC 3 and PSoC 5LP projects. These APIs are documented in the System Reference Guide.

PSoC 3's on-chip PLL can be used to boost the clock frequency of the selected clock input, e.g., IMO, MHz ECO, and DSI clock) to run at the maximum operating frequency. The Clock frequencies can be synthesized from 24–80 MHz. The PLL output is routed to the clock distribution network to serve as one of the input sources. A 4-bit input divider Q (FASTCLK_PLL_Q) is used with the reference clock and an 8-bit feedback divider P(FASTCLK_PLL_P). The outputs of these two dividers are then compared and locked, resulting in an output frequency that is P/Q times the input reference clock. The PLL achieves frequency lock in less than 250 μs, and lock status (FASTCLK_PLL_SR[0]) indicates locked status, allowing the output clock to be routed to the clock trees.

Note that when a PLL parameter is changed or the PLL is enabled or disabled, it takes four bus clock cycles (50 μs at 3 MHz) for the corresponding status to be reflected in the FAST-CLK_PLL_SR[0] status bit. Additionally, the FASTCLK_PLL_SR[0] bit is not updated while the PLL is disabled. This delay must be incorporated in the firmware before reading the status bit. The PLL's charge pump current (Icp) can be configured using bits 6:4 of register FASTCLK_PLL_CFG1. This bit-field must be set to $0x01$ (2 μA) when the output frequency 67 MHz. It must be set to $0x02$ (3% μA) when the output frequency is   67 MHz. The PLL takes inputs from the IMO, the crystal oscillator MHz ECO, or the DSI, which can be an external clock.

For low power operation, the PLL must be disabled before entering SLEEP/HIBERNATE mode to assure clean entry into SLEEP/HIBERNATE and wakeup. The PLL can be re-enabled after wakeup and when it is locked; then it can be used as a system clock. SLEEP/HIBERNATE mode is available only when the PLL is disabled.

10.11 Recommended Exercises

10-1 Determine the change in frequency for a voltage-controlled oscillator (VCO) with a transfer function of $K_O = 2.5\,\text{kHz/V}$ and a DC input voltage change of $\Delta V_O = 0.8\,\text{V}$.

10-2 Calculate the voltage at the output of a phase comparator with a transfer function of $K_D = 0.5\,\text{V/rad}$ and a phase error of $V_{theta} = 0.75\,\text{rads}$.

10-3 Determine the hold in range, (i.e. the maximum change in frequency) for a phase lock loop with an open-loop gain of $K_V = 20\,\text{kHz/rad}$.

10-4 Find the phase error necessary to produce a VCO frequency shift of $\Delta f = 10\,\text{kHz}$ for an open-loop gain of $K_V = 40\,\text{kHz/rad}$.

10-5 Given $f_{osc} = 1.2\,\text{MHz}$ at $VCO_{in} = 4.5\,\text{V}$ and $f_{osc} = 380\,\text{kHz}$ at $VCOin = 1.6\,\text{V}$. Find K_o.

10-6 A phase-locked loop has a center frequency of $\omega_\theta = 105\,\text{rad/s}$, $K_O = 103\,\text{rad/s}$ per V, and $K_D = 1\,\text{V/rad}$. There is no other gain in the loop. Determine the overall transfer function H(s) for (a) The loop filter $F(s) = 1$ (all-pass filter), (b) The loop filter F(s) is shown below, (c) Loop filter F(s) as in part (b), XOR for the phase detector and $V_{DD} = 5V$. (d) Natural frequency ω_n and damping factor ζ for part (c) (Fig. 10.10).

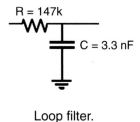

R = 147k

C = 3.3 nF

Loop filter.

Fig. 10.10 Loop filter.

10-7 Determine the change in frequency for a voltage-controlled oscillator (VCO) with a transfer function of $K_0 = 2.5\text{kHz/V}$ and a DC input voltage change of $\Delta V_O = 0.8\text{V}$.

10-8 Calculate the voltage at the output of a phase comparator with a transfer function of $K_D = 0.5\,\text{V/rad}$ and a phase error of $V_\Theta = 0.75\,\text{rads}$.

10-9 Determine the hold in range, (i.e. the maximum change in frequency) for a phase lock loop with an open-loop gain of $K_v = 20\,\text{kHz/rad}$

10-10 Find the phase error necessary to produce a VCO frequency shift of $\Delta f = 10\,\text{kHz}$ for an open-loop gain of $K_v = 40\,\text{kHz/rad}$

10-11 Given $f_{osc} = 1.2\,\text{MHz}$ at $\text{VCO}_{in} = 4.5\,\text{V}$ and $f_{osc} = 380\,\text{kHz}$ at $\text{VCO}_{in} = 1.6\,\text{V}$. Find K_Q

10-12 A PLL is locked onto a 2 MHz incoming signal with peak amplitude of 0.35V and a phase angle of 75°. The VCO's peak amplitude is 0.25 V and the phase angle is 180°. Determine the VCO frequency and the control voltage fed back to the VCO?

References

1. R. Adler, A study of locking phenomena in oscillators. Proc. IRE Waves Electrons **34**, 351–357 (1946)
2. D. Banerjee, PLL *Performance, Simulation and Design*, 3rd edn. (Dean Banerjee Publications, 2003)
3. R.E. Best, *Phase Locked Loops*, 5th edn. (McGraw-Hill, New York, 2003). ISBN: 0071412018. 3
4. M. Curtin, P. O'Brien, Phase-Locked Loops for High-Frequency Receivers and Transmitters Part 1, Analog Dialogue, 33-3, Analog Devices. (1999). Part 2, Analog Dialogue, 33-5, Analog Devices, (1999) Part 3, Analog Dialogue, 33-7, Analog Devices (1999)
5. Fundamentals of Phase-Locked Loops (PLLs). MT-086Tutorial. Analog Devices
6. F.M. Gardner, *Phaselock Techniques*, 2nd edn. (Wiley, London, 1979). ISBN: 0-47-104294-3
7. A. Hajimiri, T.H. Lee, A general theory of phase noise in electrical oscillators. IEEE J. Solid State Circuits **33**(2), 179–194 (1998)
8. R.G. Irvine, *Operational Amplifier Characteristic and Applications*, 3rd edn. (Prentice-Hall, Englewood, 1994)
9. D.B. Leeson, A simple model of feedback oscillator noise. Proc. IEEE **54**, 329–330 (1966)
10. C. Lindsey, C.M. Chie (eds.), *Phase-Locked Loops and Their Applications* (IEEE Press, New York, 1987)
11. U.L. Rhode, *Digital PLL Frequency Synthesizers* (Prentice-Hall, Englewood, 1983)
12. Voltage Controlled Oscillator in PSoC 3 / PSoC 5. http://www.cypress.com
13. M.A. Wickert, *Phase Locked Loops with Applications*. ECE 5675/4675 Lecture Notes (Springer, Berlin, 2011)
14. D.H. Wollaver, *Phase-Locked Circuit Design* (PTR Prentice Hall, New Jersey, 1991)

Analog Signal Processing

11

Abstract

In this chapter, The discussion focuses on mixed-signal processing (However, it has not been possible, in this textbook, to engage in a detailed discussion of signal processing. Instead, certain common, signal processing applications will be discussed, e.g., mixing and other examples that involve both analog and digital signal processing, commonly encountered in embedded systems.), and in particular the various components are often incorporated into an embedded system to provide the necessary functionality for a particular application. As in the previous chapters, PSoC 3/5LP serves the various needs for illustrative examples of key concepts throughout this chapter. It should be noted that many of the blocks, also referred to as "modules", found in devices such as PSoC 3 and PSoC 5LP are in reality repeated instantiations of some fundamental hardware components with variations whose characteristics are controlled and/or defined by registers. Therefore, the discussion makes occasional references to the controlling and other related registers merely to highlight the functionality at a lower level of abstraction and to emphasize the fact that the behavior of the various modules can be changed dynamically under program control, and in real-time.

11.1 Mixed-Signal Evolution

Prior to 1970, applications of digital technology was somewhat limited as a result of the fact that vacuum tubes and the associated analog technology had dominated the world of electronic applications for more than two-thirds of the twentieth century. Although Transistor-Transistor Logic (TTL) was developed in 1961, the first widely-used, commercial versions, known as the "Texas Instruments 5400 Series" did not appear until 1963. This was soon followed, circa 1966, by Texas Instruments 7400 series which was widely adopted as the defacto standard for hardware logic components.

The 7400 series low-cost, and the relative ease with which digital logic-based systems could be developed lead designers to make more and more use of microcontrollers and digital techniques. This was further motivated by the fact that analog components such as resistors, capacitors, inductors, as well as, vacuum tubes, tend to exhibit some degree of variation in component values as a function of aging, temperature, vibration, humidity, etc., which could substantially alter a system's performance.

© Springer Nature Switzerland AG 2021
E. H. Currie, *Mixed-Signal Embedded Systems Design*,
https://doi.org/10.1007/978-3-030-70312-7_11

However, given the fact that the real-world is predominantly analog, it was necessary to combine both analog and digital techniques in implementing an embedded system in order to meet the requirements of increasingly more complex, and sophisticated embedded systems.[1] Most embedded systems involve both analog and digital signal processing techniques for handling I/O requirements, data acquisition and storage, data/signal conditioning, etc., and are hence often referred to as "mixed-signal" systems. PSoC 3 and PSoC 5LP include both analog and digital modules that are interoperable, and highly configurable, as is shown in this chapter. Their mixed-signal architectures allow them to address a myriad of embedded system applications.

The reader would be well advised to bear in mind that in mixed-signal design it is often best to *"Redden Caesari quae sunt Caesaris"*[2] and use the digital/analog techniques and components that best meet the application's requirements and in the most beneficial combination. For example, in filter design, and deployment, there are frequency ranges for which digital techniques are quite unsuitable, in spite of their ability, in principle at least, to often provide far superior filtering than that of traditional analog techniques.

11.2   Analog Functions

Embedded systems are often called upon to handle a wide variety of inputs that include both digital and analog control/command signals. Analog functions employed in implementing embedded systems include:

- **Analog-to-Digital Conversion** Many transducers used in conjunction with embedded systems produce voltages or currents that are related to a parameter, or parameters, that the transducer is experiencing as inputs. Analog-to-digital conversion of the outputs of such transducers may be required to prepare these signals for acceptance and processing by the embedded system.
- **Current and Voltage Sensing** Transducers used in conjunction with embedded systems introduce a variety of voltage and current levels, some of which may be required to be in the form of analog signals that fall within certain ranges in terms of power, current and/or voltage.
- **Current and Voltage Output** Embedded systems are often required to provide specific current and voltage levels to external devices in ranges beyond the capacity of microcontrollers which are generally limited to output currents on the order of 25 milliamperes and voltages less than ± 12 vdc.
- **Analog Filters**[3]

Many embedded systems include the ability to recover signals, remove interference, etc. and both digital and analog filters each play an important role in such cases. The most common types of filters are defined as:

- highpass filters that pass frequencies above a certain frequency and block lower frequencies,
- lowpass filters that pass frequencies below a certain frequency and block higher frequencies,
- bandpass filters that pass all frequencies within a specified range and block all other frequencies,

---

[1] Software-Defined Radios (SDRs), cell phones, digital television, etc., are excellent examples of a merging of digital and analog, i.e., mixed-signal, techniques to provide increasingly more sophisticated receivers and transmitters.

[2] "Render unto Caesar those things which are Caesar's ...".

[3] Assumed for the purpose of this section to be ideal filters.

- notch filters, also referred to as bandstop or band-reject, filters that are extremely narrow-band filters with steep sides that remove a narrow band of frequencies while passing, those frequencies above and below that band with constant, e.g., unity, gain,
- allpass filters that pass frequencies within a specified range without altering their magnitude but varies the phase delay, and
- adaptive filters that can change their characteristics to meet changing conditions encountered by an embedded system.

- **Analog Mixing (Up-Conversion and Down Conversion)** Some transducers and other signal sources produce modulated carriers and the embedded system must be able to extract the data signal in such cases. Similarly, some embedded applications require outputs to external devices in the form of modulated carriers.
- **Current-to-Voltage and Voltage-to-Current conversion** In addition to the fact that input signals can be in the form of current or voltage signals, the ability to convert from one to the other may be required for output signals provided by an embedded system.
- **Analog Signal Pre- and/or Post-Conditioning** It is often necessary to subject input signals to some form of pre- or post-conditioning, e.g., filtering, voltage/current level shifts, up/down frequency conversion, etc., prior to/or after their processing by the embedded system
- **Amplification**[4] Amplification is an important consideration in many systems, whether as part of a signal conditioning requirement for input signals, or for driving external devices such as motors and other actuators.
- **Current/Voltage-to-Frequency Conversion** Depending on the type of I/O devices involved in an embedded system, it may be necessary to convert current and/or voltage to frequency.
- **Frequency-to-Voltage/Current Conversion** Depending on the type of I/O devices involved in an embedded system it may be necessary to convert frequencies to voltages/currents.
- **Pulse Width Modulation/Demodulation** PWMs are often used to provide proportional control of external devices, e.g., motors, illuminators, etc., and demodulation of pulse width modulated signals.
- **Integration** Integration of signals is sometimes employed as part of the input signal conditioning, or for other reasons, in an embedded system.
- **Pulse Shaping** Pulse re-shaping may be required to restore a signal that has been distorted, e.g., to improve pulse width, pulse height, overall shape and/or timing.
- **Differentiation** Some embedded systems require that analog signals be differentiated.
- **Analog Voltage-to-Reference Voltage Comparison** Many embedded systems employ comparators to compare an input signal to a reference value which serves as a threshold for action, or inaction, by the embedded system.
- **Track-and-Hold amplifier** Track-and-Hold amplifiers are used to maintain a signal input level, i.e., a sample, for a period of time to allow processing of the sample to be completed before the system accepts the next sample to be processed.
- **Unity Gain Buffer** Such buffers are used to avoid overloading a previous stage, or input source.
- **Voltage Summing** Allows an embedded system to sum multiple input/output signals.
- **Logarithmic Amplifier** "Log amps" are often used with transducers, or other devices, with a wide dynamic range in order to bring both high and low-level signals into an acceptable range for either input to, or output from, an embedded system.
- **Exponential Amplifier** These amplifiers can be used with sensors, or other sources, producing logarithmic signals.

---

[4] Amplification of this type is sometimes referred to as "multiplication".

- **Instrumentation Amplifier** These amplifiers are often used to make high accuracy, non-perturbing voltage/current measurements in an embedded system application.
- **Digital-to-Analog Conversion** OpAmps are sometimes used in conjunction with other components to provide an analog output of a digital input.

Operational amplifiers frequently play an important role with respect to providing these types of analog functions, as shall be shown in the remaining discussions in this chapter. Much of the discussion that follows focuses on idealized operational amplifiers.

11.2.1 Operational Amplifiers (OpAmps)

The "Operational Amplifier" or "OpAmp" [5] was developed in the 1930s, under a Federal grant, with the specific goal, among others, of finding a replacement for mechanical integrators used in various military applications,[6] e.g., the ball and disk integrator[7] that was subject to slippage and therefore error [11]. The result was a vacuum tube-based design, that was to serve as a key component in a wide variety of military and civilian applications, and ultimately become the basic building block for a significant number of the early analog computers. The following example clearly illustrates how one might use such devices as the basis for analog computers capable of solving, inter alia, differential equations[8] [8].

Consider the differential equation

$$\frac{d^2x}{dt^2} = -\omega^2 x \tag{11.1}$$

Given an electrical, or mechanical, integrator with the following property[9]

$$f(t)_{out} = -\int g(t)_{in} tothe \tag{11.2}$$

it follows that substituting the LHS of Eq. 11.1 into Eq. 11.2 yields $-dx/dt$ which can then be substituted into Eq. 11.2 for a second time to produce x. If x is then multiplied by $-\omega^2$, the result is equal to d^2x/dt^2. The OpAmp configuration capable of solving Eq. 11.1 is shown in Fig. 11.1. This configuration is sometimes used as a sine wave generator to produce tones.

It should be noted that this method of solving differential equations, and in particular complex systems of differential equations was to remain in popular use for almost 50 years. Much of the early verification of various aspects of chaos theory was carried out on analog computer systems.

The earliest commercially available operational amplifiers, circa 1941, were general-purpose, DC-coupled,[10] voltage amplifiers that had high gain and employed a feedback loop. The passive components employed in the feedback loop, and for input, usually a capacitor or resistor, allowed

---

[5] Sometimes referred to as an "OpAmp".

[6] Airborne Sextants, fire control systems, etc.

[7] Hence the name operational amplifier because it was to perform the "operation" of integration.

[8] Analog computers have also proven superior to digital computers in solving certain types of partial differential nonlinear equations.

[9] Unless otherwise noted, functions such as f(t) and g(t) are assumed to be "well-behaved".

[10] DC-coupled refers to the fact that the amplifiers could handle both DC and AC signals.

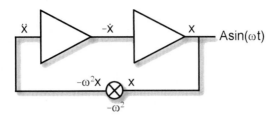

Fig. 11.1 An analog computer solution of a differential equation.

Table 11.1 Examples of OpAmp applications.

OpAmp Applications	
Differentiator	Integrator
Summer	Subtractor
Multiplier (Amplifier)	Differential Amplifier
Preamp	Buffer
Precision Rectifier	Voltage Clamp
Oscillator	Waveform Generator
Pulse Shaper	Comparator
Analog Filter	Current/Voltage Regulator
Voltage-to-Current Converter	Current-to-Voltage Converter
Voltage-to-Frequency Converter	Frequency-to-Voltage Converter
Constant Current Source	Constant Voltage Source
Transimpedance Amplifier	Voltage Follower
Reference Voltage Supply	Current Injector
Phase Lead/Lag	Time Delay
Absolute Value	Peak Follower
AC to DC Converter	Full Wave Rectifier
Rate Limiter	

operational amplifiers to serve as integrators, differentiators, summers, scalers, multipliers, followers, etc. A short list of possible operational amplifier configurations is shown in Table 11.1.

They were followed by a succession of solid-state devices, initially based on discrete transistors (1961) and ultimately in the form of integrated circuits, most notably the μA709 (1965) operating at significantly lower supply voltages, e.g., ±15 vdc.

However, these early solid-state devices were prone to oscillate, sometimes at such high frequencies, that the then commonly available oscilloscopes had a hard time "seeing" the oscillations. This type of oscillation was to plague some designers so much that they referred to operational amplifiers as "operational oscillators". Oscillation, and other problems associated with the μA709, were resolved in 1968 with the introduction of the μA741 which remains the low-cost OpAmp of choice for many applications to this day.

The development of the Field Effect Transistor led to the introduction of FET-based OpAmps as the next step in the evolution of operational amplifiers providing much higher input impedances[11] and therefore significantly lower input currents and the capability of operating at much higher frequencies. Additionally, the requirement for external dual power supplies for OpAmps was removed by the introduction of devices such as the $LM324$ (1972) which has multiple OpAmps in a single package and operates from a single external supply.[12]

Modern operational amplifiers are usually classified in terms of their input/output type as a

- voltage-controlled voltage source (VCVS) whose gain is represented by A_o and defined as the ratio of the output voltage to the input voltage (v_o/v_i),
- voltage-controlled current source (VCCS) whose gain is represented by the symbol g_m as the ratio of the output current to the input current,
- a current-controlled voltage source (CCVS) represented by the symbol r_m and defined as the ratio of the output voltage to the input current (v_o/i_i).

or,

- current-controlled current source (CCCS) represented by the symbol A_i and defined as the ratio of the output current to the input current (i_o/i_i).

Examples of just a few of the many applications of operational amplifiers are given in Table 11.1.

11.3   Fundamental Linear System Concepts

Before proceeding with a discussion of operational amplifiers, analog/digital filters and other topics discussed in this chapter, some fundamental concepts must be introduced. Important definitions and figures of merit related to operational amplifiers will be presented to help characterize the behavior of operational amplifiers in a variety of configurations commonly found in, or related to, embedded systems. It has of course not been possible to cover these topics in great detail, but a number of references are provided that should be of help for those interested in more detailed discussions [15,24].

11.3.1 Euler's Equation

Leonhard Euler (1707–1783) a Swiss physicist and mathematician made a number of important contributions to science including his discovery that:

$$e^{j\theta} = \cos(\theta) + \mathrm{jsin}(\theta) \tag{11.3}$$

and therefore:

$$e^{-j\theta} = \cos(\theta) - \mathrm{jsin}(\theta) \tag{11.4}$$

---

[11] The input impedance of a typical $\mu A741$ is of the order of $2\,\mathrm{M\Omega}$. OpAmps with input impedances that exceed $10^{12}\,\Omega$ are now available.

[12] It should be noted that the $LM324$ is inherently a dual supply system. However, by employing a "virtual" ground it is possible to operate its OpAmps using only a single supply.

which leads to the important results that:

$$\sin(\theta) = \frac{e^{j\theta} - e^{-j\theta}}{2j} \tag{11.5}$$

$$\cos(\theta) = \frac{e^{j\theta} + e^{-j\theta}}{2} \tag{11.6}$$

and because:

$$\theta = \omega t = 2\pi f t = \frac{2\pi t}{T} \tag{11.7}$$

it follows that:

$$\sin(\omega t) = \frac{e^{j\omega t} - e^{-j\omega t}}{2j} \tag{11.8}$$

$$\cos(\omega t) = \frac{e^{j\omega t} + e^{-j\omega t}}{2} \tag{11.9}$$

so that for well-behaved functions, i.e., functions that are continuous, periodic, etc., can be expressed as an infinite complex exponential series, viz.,

$$f(t) = \sum_{k=-\infty}^{\infty} g_k e^{-jk\omega_0 t} \tag{11.10}$$

which expresses a continuous, periodic function in the time-domain as an infinite sum of discrete values in the frequency domain and ω_0 the fundamental frequency and its harmonics represented by $k\omega_0$.

If the function f(t) is an aperiodic, continuous-time signal it can be expressed in terms of a complex integral, known as the Fourier Transform, as:

$$f(t) = \frac{1}{2\pi} \int_{-\infty}^{\infty} G(j\omega)e^{j\omega t} d\omega = \int_{-\infty}^{\infty} G(j2\pi f)e^{j2\pi f t} df \tag{11.11}$$

11.3.2 Impulse Characterization of a System

By determining the response of a LTI system to a very fast input pulse it becomes possible to ascertain the system's response to an arbitrary input. This type of analysis is facilitated by an important class of functions known as generalized functions which are particularly useful in understanding the behavior of embedded systems.

Two of these functions are the Kronecker and Dirac delta functions. These functions have some unique properties, e.g., the Dirac delta function, also known as the unit impulse function, is defined as:

$$\delta = \begin{cases} \infty & \text{if } x = 0 \\ 0 & \text{if } x \neq 0 \end{cases} \tag{11.12}$$

subject to the constraint that:

$$\int_{-\infty}^{\infty} \delta(x)dx = 1 \tag{11.13}$$

The Dirac delta function also has the property called *sampling* or *sifting* , viz.,

$$\int_{-\infty}^{\infty} f(x)\delta(x - x_0)dx = f(x_0) \tag{11.14}$$

The Kronecker delta function[13] is given by:

$$\delta_{ij} = \begin{cases} 1 & \text{if } i = j \\ 0 & \text{if } i \neq j \end{cases} \tag{11.15}$$

or, as an integer function as:

$$\delta[n] = \begin{cases} 1 & \text{if } n = 0 \\ 0 & \text{if } n \neq 0 \end{cases} \tag{11.16}$$

Thus:

$$\sum_{i=-\infty}^{\infty} a_i \delta_{ij} = a_j \tag{11.17}$$

In the case of continuous-time systems the Dirac delta function is used as the impulse function and for discrete-time systems the Kronecker delta function is used. A system's response to an impulse function is called the *impulse response function*. As shown in a later section of this chapter, the Laplace transform [19] of the impulse response function is the system's transfer function. Both the Kronecker and the Dirac delta functions are mathematical models of a real-world pulse that can be used to determine the behavior of discrete- and continuous-time systems, respectively.

11.3.3 Fourier, Laplace and Z Transforms

Engineers, scientists and a variety of technologists often rely on a family of mathematical tools known collectively as "transforms". These powerful tools, part of a field of mathematics referred to as "Operational Calculus", make it possible to analyze, in considerable detail, a wide variety of physical systems and phenomena. By transforming, i.e., mapping, a problem into a different function space in which the problem reduces to a set of algebraic equations[14] which are generally simpler to manipulate it is often possible to gain considerable insight into the characteristics and behavior of a system while avoiding what can be substantial mathematical analysis challenges in the original space.

---

[13]The Dirac Delta function is sometimes confused with the Kronecker Delta function. The former is defined as a "continuous" function which when integrated has an area of unity. Specifically, it has a height h and width epsilon ϵ.

[14]The author shall use the term equation or equations as opposed to formula or formulas to delineate between a mnemonic and an algorithm, e.g., as in the case of H_2O and a relationship between variables, such as $F = ma$, respectively.

Inverse transforms are also available which allow the completed analysis to then be returned to the spatial/temporal domain from which it originated.

Signals can be broadly classified as either periodic, aperiodic, or discrete, or, alternatively, as continuous or periodic. Powerful tools are often required to investigate such a diversity of signals and their processing.

In this and other chapters use will be made of the:

- Laplace transform[15] which originally developed as a technique for solving ordinary differential equations (linear). It provides a method of mapping continuous time-domain functions to the s-domain where $s = \sigma + j\omega$ that is defined in bilateral form as:

$$\mathcal{L}\{f(t)\} = \int_{-\infty}^{\infty} f(t)e^{-st}dt \qquad (11.18)$$

- Fourier transform—The Fourier transform is equivalent to the Laplace transform, when $s = j\omega$, and is a method of solution of differential equations that provides the steady-state response of a system [26]. It can also be used to map discrete-time signals, that are continuous,[16] periodic functions to the frequency domain.[17]

$$\mathcal{F}(\omega) = \int_{-\infty}^{\infty} f(t)e^{-j\omega t}d\omega \qquad (11.19)$$

- Fourier Series is a method of expressing well-behaved, continuous function in terms of an infinite series, or approximated by a partial sum thereof, consisting of sine and/or cosine terms. This series produces a frequency domain representation of a periodic, continuous-time signal.

$$f(x) = a_0 + \sum_{n=1}^{N} [a_n cos(nx) + bsin(nx) \qquad (11.20)$$

$$a_n = \int_{-\pi}^{\pi} f(x)cos(nx)dx \quad n \geqslant 0 \qquad (11.21)$$

$$b_n = \int_{-\pi}^{\pi} f(x)sin(nx)dx \quad n \geqslant 1 \qquad (11.22)$$

and the

- Z-transform which is the discrete-time equivalent of the Laplace transform and is a mapping from the time-domain to the z-domain and expressed as:

---

[15]MatLab provides, as part of its symbolic toolbox, laplace() and ilaplace() functions to computer Laplace transform and the inverse Laplace transform of a function, respectively.

[16]The reader is cautioned to delineate between continuous-time functions and the mathematical meaning of "continuous" when referring to aperiodic functions.

[17]The development of Digital Fourier Transforms, i.e., those based on discrete data, is widely attributed to Carl Friedrich Gauss as a method to interpolate the orbits of asteroids Pallas and Juno. These asteroids are found in an asteroid belt between Jupiter and Mars and were discovered in 1802 by Heinrich W. Olbersy based on observed data. The Fast Fourier Transform (FFT) [26] was developed to greatly reduce the time required to compute the DFT by Cooley and Tukey in 1965. It is generally regarded as one of the most important algorithms of the twentieth century.

$$\mathbf{Z}\{x[n]\} = \mathbf{X}(z) = \sum_{n=-\infty}^{\infty} x[n]z^{\{-n\}} \tag{11.23}$$

11.3.4 Linear Time Invariant Systems (LTIs)

A system which is definable in terms of a single input x(t) signal and a single output signal y(t) (SISO) such that there exists an F(x(t)) for which:

$$y(t) = F(x(t)) \tag{11.24}$$

is said to be "linear"[18] if:

$$F(x_1(t) + x_2(t)) = F(x_1(t)) + F(x_2(t)) \tag{11.25}$$

and,

$$F(ax(t)) = aF(x(t)) \qquad\qquad \forall a \in \Re \tag{11.26}$$

Furthermore, if

$$y(t - T) = F(x(t - T)) \qquad\qquad \forall T \in \Re \tag{11.27}$$

then, the system is said to be linear and time-invariant, or equivalently LTI. Linearity gives rise to a number of important benefits but perhaps the greatest of which, in the present context, is superposition which allows the response of a LTI system to be determined by inputting the individual components of a signal into a system, determining the output in each case and then summing the individual responses to obtain the overall response of the system to the composite input signal [23].

If there exists a function h(t) referred to as the impulse function such that:

$$y(t) = \int_{-\infty}^{\infty} h(v)x(t - v)dv \tag{11.28}$$

the system can also be said to be LTI. Conversely, if a system is LTI then there exists an impulse function, $h(v)$. Equation 11.28 is called the *convolution integral* and $h(v)$ is referred to as the "unit impulse response". A step function can also be used to characterize a system just as completely as the unit impulse. However, for the purposes of these discussions an impulse function will suffice.

The existence of an impulse response function for a system allows an arbitrary input to be represented as a set of impulse functions of the appropriate amplitude and the response of the system to each such impulse function determined so that the response to an arbitrary input can be viewed as the sum of the responses to the impulse functions making up the input signal. The process of decomposing the input signal into a series of impulses is referred to as *impulse decomposition*. The combining of the resulting impulse responses is referred to as *synthesis*. However, there is an even simpler technique

---

[18]It is often suggested that nonlinear system with nonlinear terms that are deemed "small" can be treated as linear. However, in some systems it is the existence of small terms and not their magnitude which determines whether the system will behave in a quasi-linear fashion or is capable of becoming significantly nonlinear. If the signal levels are sufficiently low it may be possible to constrain a system to operating in a linear region, e.g. as is often the case with transistors.

which relies on the impulse being known and the existence of an analytic expression for the input. This process is known as *convolution* and is discussed in a later section.

The characterization of systems and signals of the types under discussion in this and subsequent chapters typically depend less on the shape of the time-domain input waveform, and more on their respective amplitude (gain) [30], frequency and phase of its spectral components. Therefore, the ability to map time-domain functions that fully embody the system's characteristics into the frequency domain is an important part of predicting behavior. Any LTI[19] system can be characterized, in principle at least, by its transfer function which is simply a function that gives the output of the system as a function of the input,[20] or in more formal Laplace terms, a transform function, H(s), for a LTI system, is a linear mapping by the Laplace transform,[21] $\mathcal{L}\{f(t)\}$, of the input, referred to X(s), to the output, Y(s) where the Laplace Transform is defined as:

$$\mathcal{L}\{f(t)\} = \int_0^\infty f(t)e^{-st}dt \tag{11.29}$$

and its inverse,[22] as:

$$\mathcal{L}^{-1}\{f(s)\} = -\frac{1}{2\pi j} \int_{\alpha-j\infty}^{\alpha+j\infty} f(s)e^{st}ds \tag{11.30}$$

and for which $s = \sigma + j\omega$.

Example 11.1 MatLab provides a very convenient method for finding the inverse Laplace transform of a complex function in the form of the "ilaplace" operator which is available in MatLab's Symbolic Toolbox. Assuming that:

$$H(s) = \frac{a_m s^m + a_{m-1}s^{m-1} + \ldots + a_1 s + a_0}{b_n s^n + b_{n-1}s^{n-1} + \ldots + b_1 s + b_0} = \frac{s(s-7)(s+4)}{(s+2)(s^2+5s+6)}$$

[MatLab]

```
>> ilaplace((s * (s − 7))/((s + 2) * (s² + 5 * s + 6)))
ans = 30*exp(-3*t)+(-29+18*t)*exp(-2*t)
```

11.3.5 Impulse and Impulse Response Functions

A transfer function is defined in the Laplace domain as the ratio of the output function to the input function assuming that the initial conditions are zero.

---

[19]Linear Time Invariant systems, in addition to being linear, exhibit no explicit time dependence. Such systems are completely characterized by the system's impulse response, or equivalently its step response.

[20]Assuming zero initial conditions.

[21]The Laplace transform allows LTI systems to be mapped to the frequency domain and completely characterized by their respective frequency transfer function H(s).

[22]Using the integral form of the inverse Laplace transform requires integration in the complex plain and it is often preferable to instead rely on partial fraction expansion, i.e., a sum of simpler fractions, and tables of known transforms.

Therefore, if:

$$\mathcal{L}\{y(t)\} = Y(s) \tag{11.31}$$

and:

$$\mathcal{L}\{x(t)\} = X(s) \tag{11.32}$$

then the transfer function for a given system leads to the following:

$$H(s) = \frac{Y(s)}{X(s)} \tag{11.33}$$

$$Y(s) = H(s)X(s) \tag{11.34}$$

$$y(t) = \mathcal{L}^{-1}\{H(s)X(s)\} \tag{11.35}$$

Table 11.2 shows that the Laplace transfer of the Dirac delta function is:

$$\mathcal{L}\{\delta(t)\} = 1 \tag{11.36}$$

so that if $x(t) = \delta(t)$, $Y(s) = (1)H(s)$ and therefore the *impulse response* of a system occurs when an *impulse function* is applied to the input.

The computation of the Laplace transform of a function f(t) is relatively straightforward because it involves integration in the real domain as opposed to the computation of the inverse Laplace transfer

Table 11.2 Some common Laplace transforms.

Time domain	Description	Frequency domain
δ	Unit impulse	1
A	Step	$\dfrac{A}{s}$
t	Ramp	$\dfrac{1}{s^2}$
e^{-at}	Exponential decay	$\dfrac{1}{s+a}$
$sin(\omega t)$	Sine function	$\dfrac{\omega}{s^2+\omega^2}$
$cos(\omega t)$	Cosine function	$\dfrac{s}{s^2+\omega^2}$
te^{-at}		$\dfrac{1}{(s+a)^2}$
$t^2 e^{-at}$		$\dfrac{2}{(s+a)^3}$
$e^{-at}sin(\omega t)$	Decaying sine	$\dfrac{\omega}{(s+a)^2+\omega^2}$
$e^{-at}sin(\omega t)$	Decaying cosine	$\dfrac{s+a}{(s+a)^2+\omega^2}$

which is defined in terms of integration in the complex domain. In most cases the computation of the Laplace transform is straightforward and the explicit and sometimes tedious computation of the inverse based on integration in the complex plane can be avoided by employing tables of Laplace transform inverses. Table 11.2 shows some of the more common Laplace Transforms. A combination of partial fraction expansion and utilization of such tables is much easier than having to employ integration techniques in the complex plane. MATLAB [4] provides an even simpler approach as shown in Example 11.2.

Example 11.2 MATLAB can be used to find the impulse response of a system's transfer function, e.g.,

$$H(s) = \frac{s+2}{s^3 + 4s^2 + 5} = \frac{1s^1 + 2}{1s^3 + 4s^2 + 0s^1 + 5} \tag{11.37}$$

by the following:
[*MatLab*]

$$>> \text{num} = [1\ 2]$$
$$>> \text{den} = [1\ 4\ 0\ 5]$$
$$>> \text{impulse(num, den)}$$

The graphical result is shown in Figs. 11.2 and 11.3.

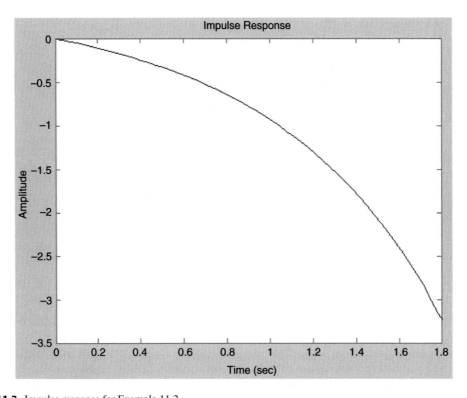

Fig. 11.2 Impulse response for Example 11.2.

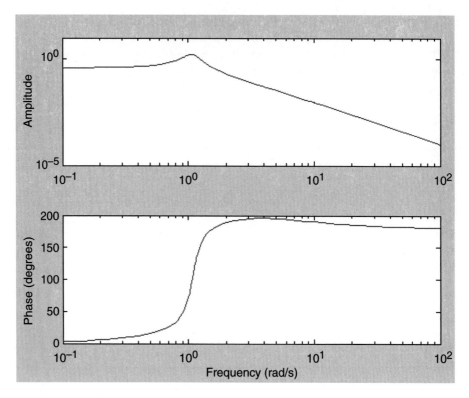

Fig. 11.3 Bode plot for Example 11.2.

11.3.6 Transfer, Driving and Response Functions

Consider a causal,[23] linear. time-invariant (LTI) system which has a single input and single output (SISO) that can be represented by an ordinary differential equation with constant coefficients, e.g.,

$$y^n + a_1 y^{(n-1)} + \cdots + a_{n-2}\ddot{y} + a_{n-1}\dot{y} + a_n y$$
$$= b_1 x^m + b_2 x^{(m-1)} + \ldots + b_{m-1}\ddot{x} + b_m \dot{x} + b_{m+1}$$

Taking the Laplace Transform of both sides yields:

$$Y(s) = H(s)X(s) \tag{11.38}$$

and therefore,

$$H(s) = \frac{Y(s)}{X(s)} \tag{11.39}$$

---

[23]Causal systems are defined as systems for which the output at a particular time, t_0, depends only on the input for $t \le t_0$ and not on any time in the future, i.e. for all $t > 0$.

where,

$$\mathcal{L}\{x(t)\} = \int_0^\infty x(t)e^{-st}dt = X(s) \tag{11.40}$$

and,

$$\mathcal{L}\{y(t)\} = \int_0^\infty y(t)e^{-st}dt = Y(s) \tag{11.41}$$

$Y(s)$ is referred to as the response function, $X(s)$ as the driving function and $H(s)$ as the transfer function. The most general form[24] of a transfer function for continuous-time systems of the type under discussion is represented by:

$$H(s) = \frac{a_m s^m + a_{m-1} s^{m-1} + \ldots + a_1 s + a_0}{b_n s^n + b_{n-1} s^{n-1} + \ldots + b_1 s + b_0} = \frac{M(s)}{N(s)} \tag{11.42}$$

and in an equivalent factored form as:

$$H(s) = \frac{(s - z_m)(s - z_{m-1})\ldots(s - z_2)(s - z_1)}{(s - p_n)(s - p_{n-1})\ldots(s - p_2)(s - p_1)} \tag{11.43}$$

As discussed in Sect. 11.10.3 Passive Filters of this chapter, the transfer function for a simple RC circuit is given by:

$$H(s) = \frac{sRC}{1 + sRC} \tag{11.44}$$

which has both a zero for $s = 0$ and a pole for $s = -1/RC$. Poles/Zeros refer to points in the complex plane for which the denominator/numerator of the transfer function becomes zero, respectively.

This can be formally expressed as

$$lim_{s \to z_i} H(s) = 0 \tag{11.45}$$

and

$$lim_{s \to p_i} H(s) = \infty \tag{11.46}$$

for the general form of a transfer function of the type shown in Eq. 11.43 which is expressed in terms of the roots of the denominator and numerator of a complex transfer function. If the system is to be stable, then the poles must lie in the left-hand side of the complex plane.

---

[24]It is assumed that, at least for the sake of this discussion, $H(s)$ is a rational function, i.e., it can be written as the ratio of two polynomials, which is usually the case.

11.3.7 Common Mode Voltages

OpAmps are inherently two input devices and therefore input signals[25] that are common to both must be taken into account when analyzing an OpAmp's characteristics. The *common-mode input* voltage is defined as:

$$v_{icm} = \frac{(v_{i1} + v_{i2})}{2} \tag{11.47}$$

Similarly the **common-mode output** voltage is defined as:

$$v_{ocm} = \frac{(v_{o1} + v_{o2})}{2} \tag{11.48}$$

11.3.8 Common Mode Rejection

Embedded systems employing OpAmps are often in environments containing a variety of sources of electronic noise,[26] as well as, signal. An important figure of merit for an OpAmp is the value of a parameter known as the Common Mode Rejection Ratio, or CMRR.

The output of an OpAmp can be expressed as

$$v_o = A_d(v_{i1} - v_{i2}) + A_{cm}\left[\frac{v_+ + v_-}{2}\right] \tag{11.49}$$

where A_d is the differential gain and A_{cm} is the common-mode gain. CMRR is a quantitative measure of a device's ability to reject common-mode signals, i.e., signals applied to both inputs and has been formally defined by the IEEE as:

$$CMRR = 10\,log_{10}\left[\frac{A_d^2}{A_{cm}^2}\right] = 20\,log_{10}\left[\frac{A_d}{|\,A_{cm}\,|}\right] \tag{11.50}$$

Obviously, it is desirable for the CMRR to be as low as possible, particularly when the signal of interest is small relative to the ambient common-mode signals such as signals originating from thermocouples, thermistors, etc.

Some versions of PSoC 3 and PSoC5 LP provide seven 8-pin ports which is equivalent to 56 GPIOs and therefore the latter can be employed for analog signal I/O. Because these ports are located in the analog part of the chip as are the analog globals, AGL[7:4] and AGR[7:4] connected to these ports, is some cases it is possibl to achieve some improvement in signal-to-noise ratio [9].

---

[25] In this case, inclusive of signals containing or representing noise. Note that in some environments, it is possible for the desired signal to appear on both inputs albeit, at different signal strengths.

[26] Noise is sometimes referred to as "the part you don't want", whereas the signal is defined as "the part you do want". OpAmps can also introduce noise [16].

11.3.9 Total Harmonic Distortion (THD)

An important parameter for many devices and applications with both inputs and outputs is known as Total Harmonic Distortion, or THD. Nonlinearities in a system can give rise to unwanted harmonics which are "injected" into a signal, and THD is an important measure of such effects. In the case of a pure sine wave, THD is defined as the ratio of the sum of the higher harmonics present to the first harmonic of the distorted signal, i.e.,

$$\text{THD} = \sum_{n=2}^{\infty} \frac{P_n}{P_1} = \sum_{n=2}^{\infty} \frac{V_n^2}{V_1^2} = \frac{P_{total} - P_1}{P_1} \tag{11.51}$$

where P_n is the power of the nth harmonic and P_{total} represents the total power of the distorted signal and V_n is the amplitude of the voltage of the nth harmonic. THD is sometimes also combined with noise and defined as:

$$\text{THD} + \text{N} = \frac{\sum \text{Harmonic Power} + \text{Noise Power}}{\text{Fundamental Power}} \tag{11.52}$$

If the output signal is weakly distorted, it is possible to use a Taylor series[27] expansion to model the output signal in terms of the input signal and thereby quantify the distortion [5]. That is,

$$v_o = a_0 + a_1 v_i^2 + a_3 v_i^3 + a_4 v_i^4 + \cdots = \sum_{n=0}^{\infty} a_n v_i^n \tag{11.53}$$

where,

$$a_n = \frac{1}{n!} \left[\frac{d^n v_o}{d v_i^2} \right]_{v_i=0} \tag{11.54}$$

which for an input of the form

$$v_i = V \cos(wt) \tag{11.55}$$

can be expressed as

$$v_o = \left[a_0 + \frac{1}{2} V^2 a_2 \right] + \left[a_1 + \frac{3}{4} V^2 a_3 \right] V^1 \cos(\omega t) + \left[\frac{a_2}{2} \right] V^2 \cos(2\omega t) + \left[\frac{a_3}{4} \right] V^3 \cos(3\omega t) + \cdots \tag{11.56}$$

where a_0 and a_1 represent the DC component and circuit gain, respectively. This result shows that second and third-order harmonics occur within the first four terms of this series which for many applications is sufficient to characterize the harmonics distortion. Second and third-order distortion is defined as

---

[27] The Taylor series is a series expansion of a function based on its value and that of its derivatives at a single point. In this particular case, the series is actually a Maclaurin Series since it is being evaluated at the origin, i.e., in the neighborhood of $v_i = 0$.

$$HD_2 = \frac{1}{2}\frac{a_2}{a_1}V \tag{11.57}$$

$$HD_3 = \frac{1}{4}\frac{a_3}{a_1}V^2 \tag{11.58}$$

and the total harmonic distortion is given by

$$THD = \sqrt{HD_2^2 + HD_3^2 + HD_4^2 + \cdots} \tag{11.59}$$

11.3.10 Noise

Noise[28] has been characterized as "... the part we don't want ..." and it is present in every real-world system to a greater or lesser degree. Before beginning a discussion of noise it will prove helpful to define some key concepts, e.g., methods for arriving at average values for a given parameter. Root-mean-square, or RMS as it is commonly referred to, is defined by the following:

$$\text{RMS value of } f(t) = \sqrt{\frac{1}{T}\int_0^T f^2(t)dt} \tag{11.60}$$

where T represents a characteristic time interval, e.g., the period of the function f(t). In the case of a truly random noise,[29] its average value will be zero, however, it does in power being dissipated, thus the RMS value of the noise is an important parameter when considering circuit/device noise.

Example 11.3 Assuming a frequency of 60 Hz and a wave form given by $f(t) = (1.697 \times 10^{-3})sin(t)$, Eq. 11.60 becomes

$$RMS\, f(t) = \sqrt{\frac{1}{T}\int_0^T [169.7\sin(t)]^2(t)dt} = \frac{1.697\text{x}10^{-3}}{\sqrt{2}} \approx 1.20\,\text{mV}$$

for $T = 1.66 \times 10^{-2}$, i.e. 60 Hz.

Noise is present in all circuits, and in multiple forms, including:

- *White Noise*[30] is a generic term that refers to any noise source for which noise as a function of frequency is constant, and usually within a specified range.

---

[28]"Like diseases, noise is never eliminated, just prevented, cured, or endured, depending on its nature, seriousness, and the cost/difficulty of treating it." from Analog-Digital Conversion Handbook, by D.H. Sheingold, Analog Devices.

[29]Noise that is in reality completely random probably doesn't exist, but for the sake of these discussions "relatively random" shall suffice.

[30]A true white noise source would be required to supply infinite energy across an infinite spectrum, therefore physical white noise sources are necessarily restricted to finite portions of the spectrum. Approximations to a white noise source are sometimes referred to as non-white, colored or pink noise sources.

- *Thermal*[31]—J. B. Johnson [13] was the first to report the existence of thermal noise by noting that the statistical fluctuation of electric charge in conductors results in a random variation in potential across a conductor. H. Nyquist [22] confirmed Johnson's observations by providing a theoretical basis for what Johnson had observed. The random motion of charge carriers gives rise to a what is approximately Gaussian noise, i.e., statistical noise with a probability density that is a *normal distribution*, i.e. Gaussian. In the case of resistors, the RMS value of the voltage associated with such noise is given by:

$$v_{rms} = \sqrt{4kT \Delta f R} \tag{11.61}$$

and therefore where k is Boltzmann's constant, T is the temperature of the resistor in Kelvin, Δf is the bandwidth of interest, and R is the value of the resistance. Note that if both sides of Eq. 11.61 are squared then

$$v_{rms}^2 = 4kT \Delta f R \Rightarrow \frac{v_{rms}^2}{R} = 4kT \Delta f = P \tag{11.62}$$

which is the noise power, P, dissipated in the resistor.[32] The noise power spectral density is a measure of the noise present in a 1 Hz bandwidth and is defined as:

$$P_{sd} = \frac{P}{\Delta f} = 4kT \tag{11.63}$$

and has units of V^2/Hz. Thermal noise, in the case of a MOS device, is modeled as a current source in parallel with drain and source. Noise in resistors can be modeled as a noise source in series with an ideal noise-free resistor with the noise power being expressed as the ratio of noise power to 1 mW and designated as dBm.

Example 11.4 Noise power relative to 1 mW can be expressed as

$$P_{rel} = 10 \log_{10} \left[\frac{P_{noise}}{1 \times 10^{-3}} \right] = 10 \log_{10} \left[P_{noise} \right] + 30 \quad \text{dBm}$$

so that in the case of thermal noise in a resistor, if $R = 50\,\Omega$, $\Delta f = 10\,\text{kHz}$ and $T = 300\,\text{K}$

$$P_{rel} = -134\,\text{dBm} \tag{11.64}$$

- *Flicker noise* is modeled as a voltage source in series with the gate for example of a CMOS, MOSFET or similar device and results from trapped charged carriers. It is inversely proportional to the frequency and is related to DC current flow. The average mean square value is given by:

---

[31] Thermal noise is also referred to as Johnson, Nyquist or Nyquist-Johnson noise.
[32] Some portion of the noise can also be distributed throughout any circuit that the resistor is connected to or coupled to electromagnetically. Because the noise power is directly proportional to Δf, one way to minimize circuit noise is to limit the bandwidth as much as possible.

$$\overline{e^2} = \int \left[\frac{K_e^2}{f} \right] df \qquad (11.65)$$

$$\overline{i^2} = \int \left[\frac{K_i^2}{f} \right] df \qquad (11.66)$$

where K_e and K_i are voltage and current constants, respectively, for the device under consideration and f is frequency.

- *Burst* noise is found in semiconductor devices and may be related to imperfections in semiconductor materials and heavy-ion implants. It occurs at rates less than 100 Hz.
- *Avalanche* noise is found in Zener diodes and occurs when a PN junction is in reverse breakdown mode. In such cases, the reversal of the electric field in the junction's depletion region allows electrons to develop sufficient kinetic energy to collide with the crystal lattice's atoms and thereby create additional electron-hole pairs. Avalanche noise sources are sometimes used as a "white noise" sources for testing filters, amplifiers, etc.[33]
- *1/f Noise* origin is unclear although it is known to be ubiquitous and that in many situations the transition between so-called "white noise"[34] and 1/f noise occurs in the region between 1 to 100 Hz.
- *Shot Noise* is created by current flow as a result of charges crossing a potential barrier such as that of a PN junction and is given by

$$\overline{i_n^2} = \overline{(i - i_D)^2} = 2 \int q \, i_D \, df \qquad (11.67)$$

where q is the charge on an electron[35] and df is the frequency differential. Note that shot noise is not a function of temperature and that its value is constant with respect to the frequency.

11.3.11 Multiple Noise Sources

Modern electronic devices contain multiple noise sources and therefore it is important to determine how such noise is to be combined in order to determine the overall noise signal. Noise sources can be internal, external or a combination of both. In addition to internal noise sources interacting with external sources to introduce noise in an embedded system, different parts of an embedded system can interact to produce noise. Although noise is a random process and therefore cannot be predicted, it is possible to predict the noise power, in some cases. Resistors, which are a common element in operational amplifier implementations, are sources of noise that can, in some cases, be a significant concern. An ideal resistor in series with a noise voltage source can be used to model actual resistors.

For example, two resistors independently giving rise to noise can be represented by

$$\overline{e_1^2} = \int 4kT R_1 df \qquad (11.68)$$

---

[33] A Zener diode operating in avalanche mode is capable of producing "white noise" up to frequencies as high as several hundred MHz.

[34] White noise is noise which contains equal amounts of noise at all frequencies.

[35] The charge on an electron is 1.62×10^{-19} C.

$$\overline{e_2^2} = \int 4kT R_2 df \qquad (11.69)$$

respectively.

If the average mean voltage, $\overline{E_{total}}^2$ is the voltage resulting from the two resistors being connected in series and E_{total} is given by:

$$E_{total} = e_1(t) + e_2(t) \qquad (11.70)$$

then

$$\overline{E_{total}(t)^2} = \overline{\left[e_1(t) + e_2(t)\right]^2} = \overline{e_1(t)^2} + \overline{e_2(t)^2} + \overline{2e_1(t)e_2(t)} \qquad (11.71)$$

However, in this case e_1 and e_2 are independent noise sources and therefore the average value of the product of $e_1(t)$ and $e_2(t)$ is zero and therefore:

$$\overline{E_{total}(t)^2} = \overline{e_1(t)^2} + \overline{e_2(t)^2} \qquad (11.72)$$

Thus the average mean square value of multiple noise sources is the sum of the average mean square value of the noise from each source whether the sources are current or voltage sources.

11.3.12 Signal-to-Noise Ratio

Because signals and noise coexist, it is important to have a measure of the relative strengths of each, in part, as a way of quantifying how significant noise is in a given system.

It is formally defined as the ratio of the signal power to the noise power and frequently, measured in dB.

Therefore:

$$SNR = \frac{P_{signal}}{P_{noise}} \qquad (11.73)$$

and, in terms of dB:

$$SNR_{dB} = 10 \log_{10}\left[\frac{P_{signal}}{P_{noise}}\right] = 20 \log_{10}\left[\frac{v_{signal}^2}{v_{noise}^2}\right] \qquad (11.74)$$

11.3.13 Impedance Matching

There are many situations for which it is necessary to consider how much power is being transferred from a source to a sink, i.e., to a load. In some cases it is desirable to deliver as much power from the source to the load as possible. However, in other situations, it is important to deliver as little power as possible to the load, or the next stage, if only to avoid loading the previous stage and degrading the signal.

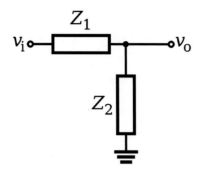

Fig. 11.4 Impedance matching example.

Assuming that Z_1 and Z_2 are the source and load impedances, respectively, as shown in Fig. 11.4 and that

$$Z_1 = R_1 \tag{11.75}$$

$$Z_2 = R_2 \tag{11.76}$$

so that

$$i_i = \frac{v_i}{R_1 + R_2} \tag{11.77}$$

and therefore

$$P = i_i^2 R_2 = \left[\frac{v_i}{(R_1 + R_2)} \right]^2 R_2 \tag{11.78}$$

and setting

$$\frac{dP}{dR_2} = \frac{d}{R_2} \left(\left[\frac{v_i}{(R_1 + R_2)} \right]^2 R_2 \right) = 0 \tag{11.79}$$

implies that

$$R_1 = R_2 \tag{11.80}$$

In which case the case second derivative can be shown to be negative and therefore Eq. 11.80 must hold, i.e., in order to deliver the maximum power to the load the resistance of the source must be equal to the resistance of the load.

In order to carry out a similar calculation for the case in which the source and the load have both resistive and reactive components, refer again to Fig. 11.4. The magnitude of the current passing through Z_1 and Z_2 is given by

$$| i_i | = \frac{| v_i |}{| Z_1 + Z_2 |} \tag{11.81}$$

and power dissipated in Z_2, i.e., the power dissipated in the resistive component of R_{Z2} is given by

$$P = i_{RMS}^2 R_2 = \left[\sqrt{\frac{1}{2\pi} \int_0^{2\pi} I^2 \sin^2(\omega t) dt} \, \right]^2 = \left[\frac{I}{\sqrt{2}} \right]^2 R_2 \qquad (11.82)$$

and therefore

$$P = \frac{1}{2} \left[\frac{|\, v_i \,|}{|\, Z_1 + Z_2 |} \right]^2 R_{Z2} = \frac{1}{2} \frac{|\, v_i \,|^2}{|\, Z_1 + Z_2 \,|^2} R_{Z2} = \frac{1}{2} \left[\frac{|\, v_i \,|^2}{(R_1 + R_2)^2 + (X_1 + X_2)^2} \right] R_{Z2} \qquad (11.83)$$

Again setting the derivative of power P with respect to Z_2 to zero yields the result that

$$R_1 + X_1 = R_2 - X_2 \qquad (11.84)$$

and therefore

$$R_1 = R_2 \qquad (11.85)$$

$$X_2 = -X_2 \qquad (11.86)$$

which is equivalent to requiring that Z_1 be the complex conjugate of Z_2.

11.4 OpAmps and Feedback

As discussed in Chap. 1, the generalized SISO system can be represented as shown in Fig. 11.5. Positive feedback is less frequently employed with operational amplifiers because feeding back a positive signal can cause the amplifier to saturate.[36] However, negative feedback is used with operational amplifiers in a variety of contexts and with a wide range of important and useful results. In what follows, it is assumed that the open-loop gain,[37] A_0, shall be constrained by $A_0 \in \mathbb{R}$ and $A_0 > 0$.

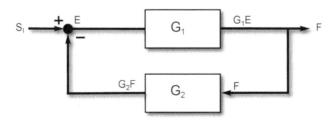

Fig. 11.5 A generalized SISO feedback system.

---

[36]Feedback amplifiers began to appear as early as the 1920s, as a result of the efforts of Harold S. Black, a Western Electric engineer interested in developing better repeater amplifiers [2, 3, 14].

[37]Open-loop gain is defined as the gain of the amplifier in the absence of feedback, either positive or negative.

In general,

$$\frac{f(t)}{s(t)} = \frac{G_1}{1 + G_1 G_2} \tag{11.87}$$

and in the present case

$$\frac{v_o}{v_i} = -\frac{A_0}{(1 - \beta A_0)} = \text{Closed Loop Gain} = A_f \tag{11.88}$$

where A_0 is the open-loop gain of the amplifier, β is the feedback coefficient and βA_0 is the loop gain. If $\beta A_0 >> 1$, then $A_f \approx \frac{1}{\beta}$ and if $\beta A_0 << 1$, then $A_f \approx A_0$. In the event that $\beta A_0 \approx 1$, the system can be expected to become unstable and oscillation may occur and relatively independent of the open-loop gain A_0.

11.4.1 The Ideal Operational Amplifier

The so-called "Ideal OpAmp"[38] is assumed to have the following characteristics:

- Infinite input impedance, regardless of the amplitude, or frequency, of the input signal, i.e. input current to both inputs is zero.
- Zero output impedance, regardless of the output frequency.
- Infinite "open-loop"[39] gain, where gain is defined as the ratio of the output voltage to the input voltage.
- Settling time is zero.[40]
- Zero input offset.[41]
- An infinite slew rate.[42]
- Introduces zero degrees of variation from a 180° of phase shift from input to output, as a function of frequency.
- No nonlinear effects at any frequency.
- No noise at any frequency.
- Output power is delivered to the load without internal loss within the OpAmp.

It is helpful to think of the ideal OpAmp in terms of the following five rules:

1. "For any output voltage in the linear operating region of an OpAmp with negative feedback, the inputs are at virtually the same potential." [14]

---

[38] While such an ideal device does not actually exist, OpAmps are available with input impedances as high as $10^6 \Omega$ for bipolar devices and $10^{12} \Omega$ for FET (Field Effect Transistor) devices, gains as high as 10^9, output impedances as low as 100Ω and a gain-bandwidth product of 20 MHz.

[39] Open-loop gain is the amplifier's gain without feedback.

[40] Settling time is the delay between an input and the associated response at the output an OpAmp.

[41] This implies that when the input is 0 V, the output is also 0 V.

[42] Slew rate is defined as the maximum rate of change with respect to time of the output voltage for all possible input voltages, typically in terms of volts/μsecond. This upper limit is caused, in the case of operational amplifiers, by limitations of charge and discharge rates of capacitors within the amplifier.

2. No current enters either of the OpAmp's input terminals.[43]
3. KCL[44] is to be applied liberally in analyzing various configurations of an OpAmp.
4. Input voltages times their respective closed-loop gains, add algebraically at the output.
5. Voltages applied to either input are multiplied by the non-inverting gain.

11.4.2 Non-ideal Operational Amplifiers

While the discussion in this chapter has thus far focused on ideal operational amplifiers, it is important to consider the characteristics of actual operational amplifiers [11] in order to appreciate in what manner, and to what extent, they deviate from their idealized counterpart. Although ideal operational amplifiers do not exist, in many cases they can be sufficiently approximated by real-world devices. The reality is that commercially available operational amplifiers often vary significantly from the ideal operational amplifier, e.g., the input impedance is not infinite but typically in the M Ω range, open-loop voltage gain ranges from 100K to 1M+, etc. A brief review of the comparisons between ideal OpAmps and real OpAmps follows.

- *Input impedance* of an OpAmp is characterized by two parameters: (1) common-mode impedance and (2) differential impedance. The former is the impedance of each of the inputs with respect to ground and the latter refers to the impedance between the two inputs. An ideal amplifier is assumed to have infinite input impedance. Real OpAmps have finite input impedances, although in some cases the impedance can be as high as 10^{14} Ω.
- *Output impedance* of a typical OpAmp is non-zero and nominally 100 Ω.
- *Input current* depends on the type of OpAmp input stage. For those with JFET or MOS, the input current can be in the range of 1–10 pA. While this represents a relatively small current, in the presence of large impedances, significant voltages can arise. In most cases the currents involved are different for the inputs which can give rise to an *offset voltage* as defined below.
- *Gain* —While the open-loop[45] gain is assumed to be infinite for the ideal OpAmp, in reality open-loop DC gains vary from 100,00 to 1,000,000+. For many applications employing negative feed, this range of gain can be quite acceptable. When real OpAmps are used with negative feedback[46] as is shown later in the chapter the closed-loop gain is a function of the amount of feedback employed.
- *Offset voltage*—Because the transistors in an operational amplifier are not actually identical, grounding the inputs does not assure that the output will be zero. The input bias currents associated with each of the inputs can be assumed to be different for each input. The offset voltage is by definition the input that is required to provide an output of 0 V.
- *Slewing* is the rate of change of the output is not infinite, as has been assumed for the ideal operational amplifier, due in part to capacitances within the OpAmp. Slew rates of 5 V/ms and higher are typical.
- *Saturation*—Dynamic range is often important when employing OpAmps, and therefore, the closer the output can be to the rails the better. However, it is possible to drive the output into "saturation"

---

[43]This is not to suggest that current applied to an input terminal is always zero, merely that the OpAmp itself is assumed to draw no input current. If there are other connections to the input terminals, the current that may appear to be being drawn by the OpAmp is in fact being routed to these additional input connections.

[44]KCL refers to Kirchhoff's Current law which states the sum of all currents into the node of a circuit must equal the sum of all currents out of the node.

[45]Open loop gain implies gain in the absence of any feedback.

[46]Positive feedback can cause the output to saturate, i.e., to be driven out of the linear range.

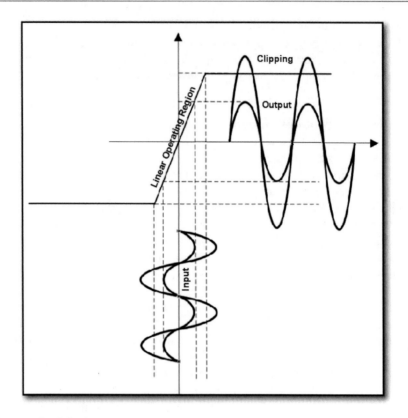

Fig. 11.6 An example of clipping.

if the gain is set sufficiently high to cause an output that attempts to exceed the supply voltage as shown in Fig. 11.6.

- *Power supply rejection ratio* is defined as:

$$PSRR = \frac{\Delta V_{ps}}{\Delta v_o} \qquad (11.89)$$

and is a measure of the effects of power supply voltage variations, including noise in the OpAmp's output. (The parameters ΔV_{ps} and Δv_o are expressed as RMS values.)

- *Power dissipation*—Although there are no intrinsic power limitations associated with ideal OpAmps solid-state devices are inherently power, current and voltage limited. A real OpAmp is generally limited to output currents that do not exceed 25 mA and maximum voltages of ± 15 V.

- *Settling time*, as shown in Fig. 11.7, consists of three components: (1) propagation delay through the OpAmp from input to output, (2) slewing time and (3) ring time. There is no response from the output to a given input during the propagation delay and the slewing rate determines how long before the output reaches its final value. Ringing is the damped oscillation centered on the final value.

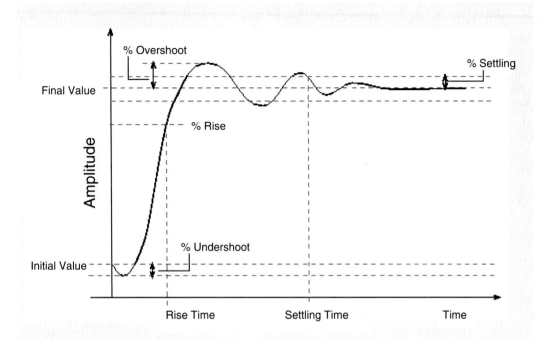

Fig. 11.7 Settling time for a real OpAmp.

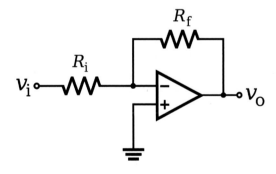

Fig. 11.8 Inverting amplifier configuration.

11.4.3 Inverting Amplifiers

An ideal inverting amplifier has the following transfer characteristic:

$$\frac{V_{out}}{V_{in}} = -A \tag{11.90}$$

Figure 11.8 shows the configuration of an inverting amplifier and since:

$$i_i = \frac{V_{in}}{R_i} = i_f = -\frac{V_o}{R_f} \tag{11.91}$$

$$V_o = -\frac{R_f}{R_i}V_i = AV_i \Rightarrow A = -\frac{V_o}{V_i} = \frac{R_f}{R_i} \tag{11.92}$$

for the ideal OpAmp.

11.4.4 Miller Effect

Operational amplifiers employing negative feedback are subject to a phenomenon first discovered with vacuum tubes known as the "Miller Effect" which arose because of unintended capacitive coupling between the input and the output. In the case of operational amplifiers, this effect can significantly reduce their performance at high frequencies [20].

As shown in Fig. 11.9, given an operational amplifier with a gain of A,[47] the input current is given by:

$$i = \frac{v_i - v_o}{Z_1} = \frac{v_i - Av_i}{Z_1} = v_i\left[\frac{1-A}{Z_1}\right] \tag{11.93}$$

and because the input impedance is given by:

$$Z_i = \frac{v_i}{i_i} \tag{11.94}$$

it follows that:

$$Z_i = \frac{Z_1}{1-A} \tag{11.95}$$

Therefore, if Z_1 is a capacitor the effective input capacitance is increased by a factor of $1 - A$, and if Z is an inductor, or a resistor, it is reduced by that same factor. If the feedback impedance Z_1 is replaced by Z_2 and Z_3 as shown in Fig. 11.9 , then

$$Z_2 = Z_3\left[1 - \frac{v_o}{v_i}\right] = Z_3(1-A) \tag{11.96}$$

$$Z_3 = \frac{AZ_1}{A-1} \approx Z_1 \tag{11.97}$$

Thus the Miller effect reflects the fact that parasitic capacitances can be viewed as equivalent to the presence of unintended input and output capacitances, i.e., Z_2 and Z_3, respectively, as shown in Fig. 11.9. A technique known as "compensation" is used in some cases to minimize the adverse effects of the Miller Effect. Replacing Z_1 with Z_2 and Z_3 means that the current through Z_1, Z_2 and Z_3 must be the same. If the gain A is sufficiently large, the input capacitance can function as a short and thus block the input signal.

Finally,

$$Z_3(v_i - v_o) = Z_2v_i \tag{11.98}$$

$$j\omega C_{out}(v_i - v_o) = j\omega C_{in}v_i \tag{11.99}$$

---

[47]Note that in most cases A < 0, i.e., it is negative.

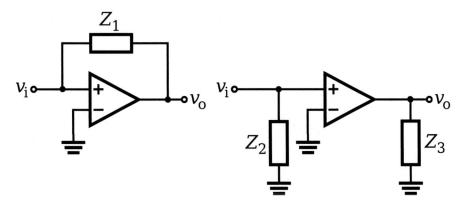

Fig. 11.9 Miller effect.

$$C_{out}\left[1 - \frac{v_o}{v_i}\right] = C_{in} \tag{11.100}$$

$$C_{in} = [1 - A]C_{out} \tag{11.101}$$

11.4.5 Noninverting Amplifier

The noninverting amplifier is configured as shown in Fig. 11.10 and

$$\frac{v_o}{v_i} = A \tag{11.102}$$

and,

$$v_0 = \left[\frac{R_1 + R_2}{R_1}\right]v_i \Rightarrow A = 1 + \frac{R_2}{R_1} \tag{11.103}$$

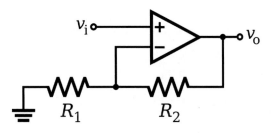

Fig. 11.10 A noninverting amplifier.

11.4.6 Summing Amplifier

Similarly, an ideal weighted[48] summing, or "adder", amplifier is based on an operational amplifier configured as shown in Fig. 11.11.

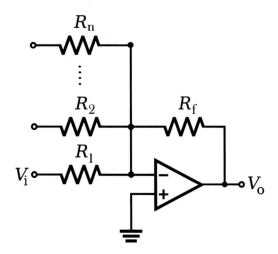

Fig. 11.11 Summing amplifier.

Because

$$v_o = \left[\frac{v_1}{R_1} + \frac{v_2}{R_2} + \cdots + \frac{v_n}{R_n} \right] R_f = \left[\frac{R_f v_1}{R_1} + \frac{R_f v_2}{R_2} + \cdots + \frac{R_f v_n}{R_n} \right] \tag{11.104}$$

and therefore,

$$V_o = \left[A_1 v_1 + A_2 v_2 + \cdots + A_n v_n \right] \tag{11.105}$$

and if,

$$R = R_1 = R_2 = \cdots = R_n \Rightarrow A = A_1 = A_2 = \cdots = A_n \tag{11.106}$$

Equation 11.105 becomes

$$v_o = A \left[v_1 + v_2 + \ldots + v_n \right] \tag{11.107}$$

---

[48]Resistor values can be selected to combine signals of different amplitudes.

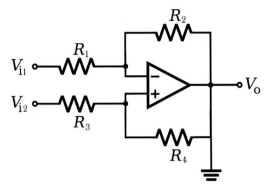

Fig. 11.12 Difference amplifier.

11.4.7 Difference Amplifier

In this example, as shown in Fig. 11.12, the fact that the difference amplifier circuit is linear allows superposition to be imposed so that by inspection, and referring to Eq. 11.103, it follows that:

$$v_o = A_2 v_{i2} - A_1 v_{i2} = \left[\frac{1 + \frac{R_1}{R_2}}{1} + \frac{R_3}{R_4}\right] v_{i2} - \left[\frac{R_2}{R_1}\right] v_{i1} \tag{11.108}$$

and if

$$R_1 = R_2 = R_3 = R_4 \tag{11.109}$$

then,

$$v_o = v_{i2} - v_{i1} \tag{11.110}$$

and the difference amplifier is referred to as a "differential amplifier".

11.4.8 Logarithmic Amplifier[49]

When dealing with a signal that has a large dynamic range, it can sometimes be difficult to keep high levels of the signal from masking the lower levels. One technique for addressing this problem is to use a logarithmic amplifier configuration, as shown in Fig. 11.13, to effectively expand the lower levels of a signal and compress the higher levels so that both fall into a detectable range that can best be handled by the embedded system. The current through a diode[50] is well known to be given by:

$$i_d = i_s[e^{\left(\frac{q v_d}{n V_T}\right)} - 1] \approx I_s e^{\left(\frac{q v_d}{n V_T}\right)} \tag{11.111}$$

---

[49]The phrases "Logarithmic Amplifier" and "Log Amp" though in common use for such circuits is of course a misnomer since the OpAmp in such applications is producing the logarithmic output of the input signal and not amplifying the input signal.

[50]Second order effects are not taken into account by this equation. The ideality factor reflects the importance of such second order considerations.

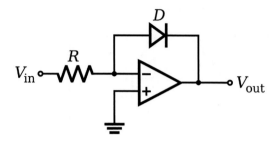

Fig. 11.13 An example of a logarithmic amplifier.

where v_d is the voltage across the diode, i_s is the reverse bias saturation current, V_T is the "thermal voltage"[51] and the so-called "ideality factor"[52] is assumed to be equal to one, in most cases.

Therefore,

$$i_d = \frac{v_i}{R} = \frac{i_s e^{\frac{qV_d}{V_T}}}{R} \tag{11.112}$$

can be expressed as

$$v_o = -V_T ln\left[\frac{v_i}{i_s R}\right] \Rightarrow v_d = -V_T ln[v_i] - constant \tag{11.113}$$

where $constant = i_s R$.

11.4.9 Exponential Amplifier

As shown in Fig. 11.14, an exponential amplifier can be configured by placing a diode at the input to the OpAmp with Eq. 11.111 representing the input current and therefore:

$$v_o = i_d R = i_s \left[e^{\frac{v_d}{(nV_T)}} - 1\right] R \approx I_s Re^{\frac{v_d}{nV_T}} = \alpha Re^{\beta v_i} \tag{11.114}$$

where $\alpha = I_s$ and $\beta = \frac{1}{nV_T}$.

11.4.10 OpAmp Integrator

One of the most common configurations of OpAmps, at least historically, has been as an integrator as illustrated in Fig. 11.15.

The current into the circuit is given by

---

[51] The thermal voltage, that at room temperature is ≈ 25 mV, is given by kT/q where T is the absolute temperature of the diode's PN junction, q is the charge on an electron and k is the Boltzmann constant.

[52] A diode's ideality factor reflects the extent to which the diode mimics the behavior predicted by the equation for an ideal diode.

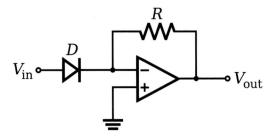

Fig. 11.14 An exponential amplifier (e^{v_i}).

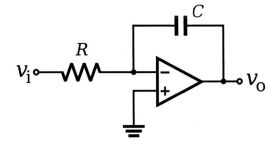

Fig. 11.15 An OpAmp configured as an integrator.

$$i = \frac{v_i}{R_i} = i_C = -C\frac{dv_0(t)}{dt} \Rightarrow \frac{dv_0(t)}{dt} = -\frac{v_i}{RC} \tag{11.115}$$

which is a first-order, linear differential equation whose general solution can be expressed as

$$v_0(t) = -\frac{1}{RC}\int_0^t v_i dt + constant \tag{11.116}$$

where the constant refers to the voltage on the capacitor at the start of the integration cycle, i.e., t=0. It should be noted that in some applications it is necessary to reset the integrator, typically by shorting the integrating capacitor, as for example in the case of a constant input voltage whose application is significantly greater than the RC time constant, because the time integral of a constant integrator is a linear function of time which could ultimately lead to saturation of the integrator. One application for this type of circuit has been in implementing the dual-slope, analog-to-digital converter.

11.4.11 Differentiator

An OpAmp can also be configured as a differentiator by using a resistor for feedback and a capacitor for input as shown in Fig. 11.16.

The amount of charge, q, stored in a capacitor is given by:

$$q = CV \tag{11.117}$$

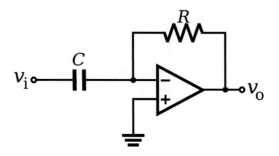

Fig. 11.16 An idealized differentiator.

Thus the output voltage[53] is given by:

$$i_i = \frac{dq}{dt} = C\frac{dv_i}{dt} = -\frac{v_o}{R}$$

(11.118)

and therefore,

$$v_o = -RC\frac{dv_i}{dt}$$

(11.119)

Unfortunately, differentiators tend to amplify high-frequency noise and therefore represent perhaps the least used configuration of OpAmps. In some applications, a resistor is placed in series with the input capacitor in order to limit the gain (R_f/R_i) of higher frequency components, while still allowing the low-frequency gain to be determined by the capacitor and feedback resistor. However, the cutoff frequency is subsequently determined by:

$$f_{cut\ off} = \frac{1}{2\pi R_i C}$$

(11.120)

11.4.12 Instrumentation Amplifiers

The so-called instrumentation amplifier, one configuration of which is shown in Fig. 11.17, is used in applications for which a small differential signal, often in the presence of a strong common-mode signal, must be measured. Instrumentation amplifiers are designed to ignore the common-mode signal while amplifying the differential input signal. Furthermore, such signals are often provided by sources of relatively low input impedance. This circuit provides very high input impedance that assures that the input signal is not subjected to an impedance that will degrade the input signal. As shown, this particular configuration consists of a differential input amplifier followed by a difference amplifier both of which are discussed in this chapter. The former providing very high input impedance and common-mode rejection while the latter provides single-ended output.

$$v_{o1} = v_{i2} + (v_2 - v_1)\left[1 + \frac{R_1}{2R_2}\right]$$

(11.121)

---

[53]Note that the positive input terminal is grounded and therefore the negative input terminal can be expected to be at ground also.

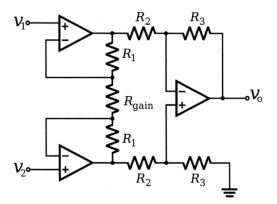

Fig. 11.17 A classical instrumentation amplifier configuration.

$$v_{on} = v_2 + (v_1 - v_2)\left[1 + \frac{R_1}{2R_2}\right] \tag{11.122}$$

$$v_{op} - v_{on} = (v_2 - v_1)\left[1 + \frac{R_1}{2R_2}\right] \tag{11.123}$$

$$v_{o2} = v_{cm} + \frac{v_d}{2}\left[1 + \frac{R_1}{2R_2}\right] \tag{11.124}$$

$$v_{o1} = v_{cm} - \frac{v_d}{2}\left[1 + \frac{R_1}{2R_2}\right] \tag{11.125}$$

and,

$$v_o = [v_{o1} - v_{o2}]\frac{R_4}{R_3} \tag{11.126}$$

$$v_o = v_d\left[1 + \frac{R_1}{2R_2}\right]\frac{R_4}{R_3} \tag{11.127}$$

In the sections that follow, the discussion will focus on various configurations of fundamental building blocks used in both PSoC 3 and PSoC 5LP, unless otherwise noted, that are referred to as the switched-capacitor and continuous-time (SC/CT) blocks.

11.4.13 Transimpedance Amplifier (TIA)

Embedded systems often make use of a variety of sensors some of which supply currents that are proportional to the parameter being sensed, e.g., photodetectors, photomultipliers, etc. Similarly, some external peripheral devices require current for input. As a result, there is a need for interfaces that convert current-to-voltage and/or voltage-to-current. OpAmps can be very useful when configured as shown in Fig. 11.18.

A specific example involving a photodetector is shown[54] in Fig. 11.19. The output of the transimpedance amplifier is a voltage proportional to the current flowing in the photodiode as the result of radiation detected from a laser.

The output voltage of this configuration is given by:

$$v_{out} = -i_{photo} R_f \tag{11.128}$$

A small capacitor, C_F, is sometimes used to assure that the transimpedance amplifier remains stable.

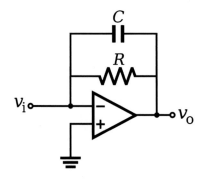

Fig. 11.18 A generic TIA.

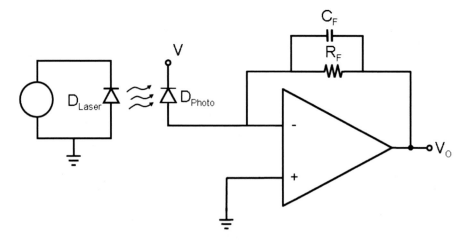

Fig. 11.19 A typical application of TIA and photodetector.

---

[54]It should be noted that the actual circuit would be subject to additional capacitances that are ignored in this discussion, viz, a capacitance introduced by the photodiode and the OpAmp's common-mode capacitance.

11.4.14 Gyrators

B.D.H. Telegen introduced a new circuit element, the Gyrator [31], which is a type of active impedance converter that can be used to simulate inductive loads, as shown in Fig. 11.20. Unlike its counterparts, viz., the capacitor, inductor, resistor and ideal transformer, the Gyrator does not exhibit reciprocity. Introduced initially as a passive device, the advent of the transistor, gyrator and subsequently the operational amplifier, made it possible to create active-gyrators.

Gyrators are referred to as negative impedance devices and thus can convert capacitive loads to inductive loads, and conversely. Although it is possible to implement gyrators using transistors, OpAmp-based gyrators are capable of providing significantly higher Q factors. For the circuit shown in Fig. 11.20 the inductance is given by Jayalalitha and Susan [12] (Fig. 11.21)

$$\frac{v_+}{v_{in}} = \frac{R}{R + j\omega C} \tag{11.129}$$

$$v_- = v_+ \tag{11.130}$$

$$i_1 = \frac{v_{in} - v_+}{R_L} \tag{11.131}$$

$$i_1 = \frac{1}{R_L}\left[1 - \frac{R}{R + \frac{1}{j\omega C}}R_L\right]v_{in} \tag{11.132}$$

$$i_2 = v_{in}\left[R + \frac{1}{j\omega C}\right]^{-1} \tag{11.133}$$

$$Z_{in} = \frac{v_{in}}{i_{in}} = \frac{v_{in}}{i_1 + i_2} \tag{11.134}$$

$$Z_{in} = \left[\frac{1 - \frac{R}{R+j\omega C}}{R}\right] \tag{11.135}$$

$$Z_{in} = R\left[R_L + \frac{1}{j\omega C}\right]\left[R + j\omega C - R\right]^{-1} \tag{11.136}$$

$$Z_{in} = R + j\omega R_L RC \tag{11.137}$$

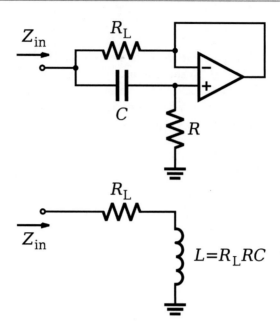

Fig. 11.20 Inductive load implemented as an OpAmp-based Gyrator.

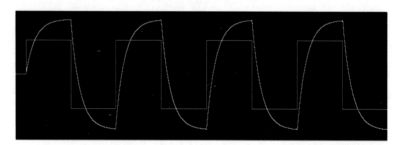

Fig. 11.21 Volt-ampere vs. Time characteristics for the gyrator shown in Fig. 11.20.

11.5   Capacitance Multiplier

An OpAmp and a small capacitor, C1, can be employed as a capacitance multiplier which can be very useful, for example, in the design of operational amplifier—based filters [18] as shown in Fig. 11.22. It simulates the simple RC circuit where the resistor has the same value as the resistor in the circuit being simulated (R3), but the capacitor C1 is N times smaller than C2.

Current flows from the input source through R to the capacitor (C_G). If R, for example, is 100 times larger than R_C, there is 1/100th the current through it into the capacitor. For a given input voltage, the rate of change in voltage in C_G is the same as in the equivalent C2 in Fig. 11.22b, but C2 appears to have 100 times the capacitance to make up for 1/100th the current.

The voltages across the two capacitors are the same, but the currents are not. The op-amp causes the negative input to be held at the same voltage as the voltage across C1. This means R2 has the same voltage across it as R3, and therefore the same current. Since the total current from V_{in} is the sum of the current in R1 and R2 and R2 is N times smaller than R1 the apparent charging current is $N + 1$ times larger than the current in C1 (Fig. 11.23).

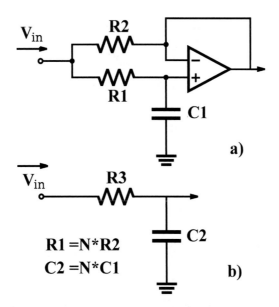

Fig. 11.22 An example of a capacitive multiplier.

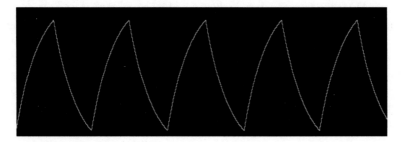

Fig. 11.23 Capacitive charge/discharge cycle.

11.6 Analog Comparators

It has been suggested that the comparator is the fundamental building block of mixed-signal design [32]. Comparators compare the differential voltage resulting from the voltages applied to both inputs and produce an output voltage that has the same sign and magnitude as that of one of the supply voltages. Analog comparators are basically differential amplifiers with extremely high open-loop gain and high slew rates.[55] They can be employed to compare one analog signal to another, e.g., a reference voltage, in terms of sign and magnitude. This is particularly useful for applications requiring the monitoring of various types of threshold which can be expressed as a voltage level, e.g., light levels, temperatures, fluid levels, etc., particularly in situations that require a rapid response to levels reaching some predefined threshold.

Generic OpAmps can be used for this purpose but have some potentially serious limitations in such applications, e.g., slower slew rates than obtainable with devices designed specifically to function solely as comparators. Also, comparators are designed to work with large differential inputs which

---

[55]Slew rates for comparators are usually expressed in terms of propagation delays.

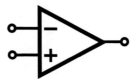

Fig. 11.24 An ideal comparator.

OpAmps are not. Although OpAmps have high input impedance and draw very little current they can be damaged by large differential voltages and their input characteristics in terms of impedance and input (bias) current, can depart significantly from their otherwise normal values when inputs exceed a few 100 mV. The idealized form of a comparator is shown in Fig. 11.24.

Because comparators are intended to function as nonlinear devices, they are not used with negative input feedback in order to avoid degrading their switching speeds. When used with positive feedback the comparator functions as a bistable[56] device. While similar to operational amplifiers in that idealized comparators are assumed to have infinite gain, require no input current, and have zero offset/bandwidth, unlike their OpAmp counterpart, comparators are designed to saturate and recover quickly, are not compensated, operate either in an open-loop mode or with positive feedback and typically have open collector, open-drain or open emitter outputs. Although an analog device, it functions as an analog input device with a "digital", i.e., binary, output.

If the voltage applied to the positive input terminal is greater than that applied to the negative input terminal, then the output voltage rapidly increases until it reaches the positive rail voltage.[57] Similarly, if the voltage applied to the positive terminal is less than that applied to the negative terminal, the output rapidly becomes equal to the negative supply voltage. But there is a potential problem, viz., what happens when the two inputs are such that the difference is "zero"[58] or perhaps more importantly when the difference between the two inputs is approximately zero.

If the inputs are such that

$$| V_+ - V_n | < \epsilon \tag{11.138}$$

for sufficiently small ϵ, noise can cause a transition, or multiple undesirable transitions to occur, and, in addition, the comparator may begin to function as a linear device with respect to output. This condition allows noise to be transmitted from the input to the output and therefore to devices external to the comparator. The problem also arises with OpAmps because of the inherent difficulty in establishing and/or maintaining either an absolute, or a differential value of 0 V.

Ideally the comparator should function in a manner that assures that when the differential voltage between the input terminals crosses zero the output state changes. Adding hysteresis establishes not one but two trigger points, $V_{+switch}$ and $V_{-switch}$, for a change of state.

In such cases, the hysteresis voltage is defined by:

---

[56]That is there are only two stable states, the output is either at the positive rail potential or the negative rail potential.

[57]The phrase "positive rail voltage is a term of art that refers to the positive supply voltage level. Most OpAmps operate between either positive and negative supply voltages or the equivalent by employing so-called "virtual grounds". In each case, the effective positive and negative supply voltages are referred to respectively as the positive and negative rails. If an OpAmp's output is "on one of the rails" meaning either at the positive or negative supply voltage, the output is said to be "saturated'."

[58]Zero volts in actual analog circuits is a topic in and of itself and shall not be treated in detail in this textbook except to note that in practice designs relying on a potential of precisely zero for functioning should be avoided.

$$V_{+switch} - V_{-switch} = V_{hysteresis} \tag{11.139}$$

If an offset voltage is present it becomes the mean value of $V_{+switch}$ and $V_{-switch}$ and not zero. Unfortunately, the offset voltage is a function of both the supply voltages and the temperature. However, using positive feedback can improve the situation substantially as is shown in the next section.

11.6.1 Schmitt Triggers

As discussed in the previous section, one of the challenges presented by comparators is their behavior near the threshold. As the voltage level approaches a threshold value, noise can cause a transition to occur prematurely. Perhaps worse is the possibility that noise could cause multiple premature transitions near threshold. The addition of hysteresis is one way to minimize this effect. The technique is to feedback some of the output signal to the positive input. Schmitt triggers are special cases of comparators that are typically used to improve pulse shape and as a way of generating very fast rise/fall time pulses. Pulses tend to degrade over time and the Schmitt trigger has been commonly employed to "sharpen-up"[59] such degraded pulses. The Schmitt trigger [29] was invented by Otto H. Schmitt,[60] circa 1937, in part to study squid nerves and has the interesting property of limited memory of prior events in the form of hysteresis which is a property exhibited by a variety of systems, e.g., those involving magnetic materials. Two bistable configurations of the Schmitt trigger are shown in Figs. 11.25 and 11.26. The positive input to the non-inverting Schmitt trigger can be derived by noting that:

$$v_+ = (v_o - v_i)\left[\frac{R_1}{R_1 + R_2}\right] + v_i = \left[\frac{R_1}{R_1 + R_2}\right]v_o + \left[\frac{R_2}{R_1 + R_2}\right]v_i \tag{11.140}$$

and therefore:

$$v_+ \approx \frac{R_1}{R_1 + R_2}v_o \tag{11.141}$$

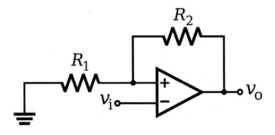

Fig. 11.25 Inverting Schmitt trigger.

---

[59]Perhaps an unfortunate use of the language, but the basic idea is to restore the pulse to fast rise and/or fall times, referred to by some as "squaring-up" the pulse.

[60]Schmitt described his invention as a simple hard valve circuit, i.e. vacuum tube, which provides positive off-on control with any differential from 0.1 to 20 V, while requiring less than a microampere to do so. Transition time was approximately 10 μs.

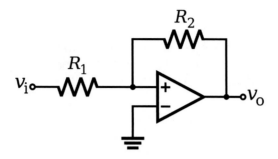

Fig. 11.26 Non-inverting Schmitt trigger.

which represents the amount of hysteresis. If the comparator is in a "saturated" state, for which the output voltage is equal to the "positive rail", e.g., $+15$ vdc and $R_2 = 14R_1$, then the hysteresis is ± 1 vdc.

11.6.2 Sample/Track-and-Hold Circuits

Sample/track and hold circuits[61] are used in a variety of contexts particularly when "sampling data" from sensors, and other devices. The basic concept is to sample a voltage by charging a capacitor at a particular point in time and to hold that voltage at a steady-state value until the next sampling period. In order for this technique to be effective it is important that the capacitor used have as small a "leakage value" as practicable, i.e., as close to zero as possible. The connection between the sensor, or other source, should be controlled by a switch with a very low "on" resistance and very low leakage currents, when switched to the "off" position.

PSoC Creator's *Sample/Track-and-Hold* component, shown in Fig. 11.27, provides a way to continuously sample a time-variant, analog signal and retain a sample value for a finite period of time. It supports both *Track-and-Hold* and *Sample-and-Hold* functions[62]

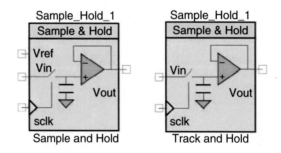

Fig. 11.27 PSoC 3/5LP's sample/hold and track/hold components.

---

[61] The *Sample-and-Hold* option samples the signal on the falling edge of the clock, or optionally on both the falling and rising edges of the clock. The *Track-and-Hold* mode samples the signal on the falling edge of the sample clock, but tracks the input signal while the sample clock remains low.

[62] These options are selectable in the customizer.

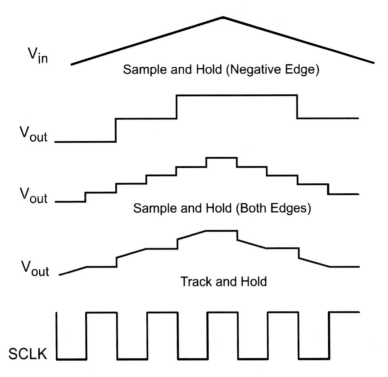

Fig. 11.28 Sample- and-Hold component's I/O signals.

11.6.2.1 Input/Output Connections

The following describes the various input and output signals, as shown in Fig. 11.28 for the Sample/Track-and-Hold component.[63]

- *Vin* is the *Sample/Track-and-Hold* component's analog input.
- *Vout* is the *Sample/Track-and-Hold*'s output. This signal can be routed to any pin or analog input, e.g., a comparator or ADC.
- *SCLK** is the sample clock input.
- *Vref** is an optional input[64] and is selected by the *Sample mode* parameter.[65]

11.6.2.2 Power

The *Power* parameter determines the initial drive power of the *Sample/Track-and-Hold* component and the speed with which this component reacts to changes in an input signal. There are four power settings available: *Minimum Power*, *Low Power*, *Medium Power* (default), and *High Power*.

---

[63] An asterisk (*) in the list of I/Os means that the I/O may be hidden on the symbol under the conditions listed in the description of that I/O.

[64] The *Vref mode* is used to select the reference voltage as *Internal* or *External*. If the *Vref mode* is set as Internal, the component takes the reference voltage from the internal source Vss, which is the ground signal internal to the component providing the amplifier reference.

[65] If the sample mode is *Sample-and-Hold* and *Vref* is *External*, then this pin is visible and is connected to a valid *Vref* source. If the sample mode is *Track-and-Hold*, this pin disappears from the symbol.

A *Minimum Power* setting results in the slowest response time and a *High Power* setting in the fastest response time.

- *void Sample_Hold_Start(void)* performs the required initialization and enables power to the block. The first time the routine is executed, the sample mode, clock edge, and power are set to their default values. When called to restart the mixer following a *Sample_Hold_Stop()* call, the current parameter settings are retained.
- *void Sample_Hold_Stop(void)* turns off the *Sample/Track-and-Hold* block, but does not affect *Sample-and-Hold* modes, or power settings.
- *void Sample_Hold_SetPower(uint8 power)* sets the drive power to one of four settings; minimum, low, medium, or high. Parameters: uint8 range: Sets full-scale range for Sample_Hold.
  Power Setting Notes Sample_Hold_MINPOWER Lowest active power and slowest reaction time Sample_Hold_LOWPOWER Low power and speed Sample_Hold_MEDPOWER Medium power and speed Sample_Hold_HIGHPOWER Highest active power and fastest reaction time
- void Sample_Hold_Sleep(void) is the preferred API to prepare the component for sleep. The *Sample_Hold_Sleep()* saves the current component state. Then it calls the *Sample_Hold_Stop()* and *Sample_Hold_SaveConfig()* to save the hardware configuration.[66]

11.6.2.3 Sample Clock Edge

This parameter provides the clock edge settings and only valid for the *Sample-and-Hold mode*. There are two types of edge settings: *Negative* and *Positive and Negative*.

void Sample_Hold_Start(void) performs the required initialization for the component and enables power to the block. The first time the routine is executed, the sample mode, clock edge, and power are set to their default values. When called to restart the mixer following a Sample_Hold_Stop() call, the current component parameter settings are retained.

void Sample_Hold_Stop(void) turns off the *Sample/Track-and-Hold* block.[67]

void Sample_Hold_SetPower(uint8 power) sets the drive power to one of four settings; minimum, low, medium, or high.

Allowable power settings for *Sample_Hold* are determined by the following:

- *Sample_Hold_MINPOWER* for the lowest active power and slowest reaction time,
- *Sample_Hold_LOWPOWER* for low power and speed,
- *Sample_Hold_MEDPOWER* for medium power and speed,

and,

- *Sample_Hold_HIGHPOWER* for the highest active power and fastest reaction time.

void Sample_Hold_Sleep(void) saves the current component state and is the preferred API to prepare the component for sleep. It then calls *Sample_Hold_Stop()* and *Sample_Hold_SaveConfig()* to save the hardware configuration.[68]

---

[66]The Sample_Hold_Sleep() function should be called before calling *CyPmSleep()* or *CyPmHibernate()*.

[67]It does not affect Sample-and-Hold modes or power settings.

[68]*Sample_Hold_Sleep()* should be called before calling *CyPmSleep()*, or *CyPmHibernate()*.

void Sample_Hold_Wakeup(void) is the preferred API to restore the component to the state when *Sample_Hold_Sleep()* was called. *Sample_Hold_Wakeup()* calls *Sample_Hold_RestoreConfig()* to restore the configuration. If the component was enabled before *Sample_Hold_Sleep()* was called, *Sample_Hold_Wakeup()* also re-enables the component.[69]

void Sample_Hold_Init(void) initializes, or restores, the component according to the customizer *Configure* dialog settings.[70]

void *Sample_Hold_Enable(void)* activates the hardware and begins component operation.[71]

void Sample_Hold_SaveConfig(void) is an empty function reserved for future use.

11.7 Switched-Capacitor Blocks

Switched-capacitor modules are based on a very simple and fundamental concept that allows resistors to be replaced with capacitor-switch combinations that function as resistors. This technique was developed in part as a result of the difficulty of fabricating precision resistor values at the chip level. Switches, capacitors and, operational amplifiers have proven relatively easy to manufacture at the chip level. Therefore, combinations of switches and capacitors are an attractive alternative to resistors, particularly in light of the fact that in such applications, capacitor switches exhibit acceptable temperature characteristics.[72]

The fundamental concept involved in switched-capacitors is illustrated in Fig. 11.29.

A capacitor is alternately connected to ground and/or an input/output voltage connection v_1 and v_2 by the clocks Θ_1 and Θ_2. With Switch 2 open and Switch 2 closed, the charge Cv_i is transferred to the capacitor. Switch 1 then opens and switch 2 closes allowing the charge, Cv_o to be transferred to the load.

Thus, a net charge,

$$\Delta q = C(v_o - v_i) \tag{11.142}$$

is transferred during each cycle of period T_s.

The operation of these switches is subject to the following requirements:

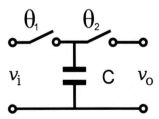

Fig. 11.29 The basic switched-capacitor configuration.

---

[69] Calling *Sample_Hold_Wakeup()* without first calling *Sample_Hold_Sleep()* or *Sample_Hold_SaveConfig()* may produce unexpected behavior.

[70] All registers are set to values determined by the customizer Configure dialog settings.

[71] It is not necessary to call *Sample_Hold_Enable()* because *Sample_Hold_Start()* calls it, which is the preferred method to begin component operation.

[72] However, some account should be made of capacitance temperature dependencies, particularly if there are wide swings in ambient temperature.

1. the Θ_1 and Θ_2 switches must not be closed at the same time,
2. the Θ_1 switch must be open before the Θ_2 switch closes,
3. the Θ_2 switch must be open before the Θ_1 switch closes,

and finally,

4. the frequency, f_s, used for switching must allow the capacitor to fully charge, and discharge during each cycle.[73]

By definition:

$$i = \frac{\Delta q}{\Delta t} \tag{11.143}$$

and,

$$R = \frac{v}{i} \Rightarrow R = \frac{v\Delta t}{\Delta q} = \frac{v\Delta t}{vC} = \frac{T_s}{C} = \frac{1}{f_s C} \tag{11.144}$$

it should be noted therefore, that the ratio of two resistances is simply the inverse ratio of their corresponding capacitive equivalents, i.e.,

$$\frac{R_1}{R_2} = \frac{\frac{1}{f_s C_1}}{\frac{1}{f_s C_2}} = \frac{C_2}{C_1} \tag{11.145}$$

and thus the ratio of switch capacitance is independent of the clocks, and consequently so are the equivalent resistances. Furthermore, while in effect the charge is being delivered in discrete packets by this method, just as charge in a resistor that has been subjected to an applied potential difference is also delivered in the form of quanta, each of which carries a fixed amount of charge, viz., the charge on an electron, 1.6×10^{-19} C.

Filters and other analog switched-circuits make frequent use of RC constants which can be implemented using only capacitances and switches. In addition to requiring less real estate than their resistor counterparts, switched-capacitors provide better linearity, closer tolerances ($\pm 1.0\%$), better matching ($\pm 0.1\%$) wider range, and allow the RC time constants to be varied by varying the switching frequency. Figure 11.30 shows a switched-capacitor integrator.

However, switched-capacitance inputs and outputs are subject to the same issue as that of any sampled system, viz., "what you find, depends on when you look".

11.7.1 Switched-Capacitor and Continuous-Time Devices

The PSoC 3/5LP switched-capacitor (SC) and continuous-time (CT) module, shown in Fig. 11.31, is a , highly optimized block that can be configured as a:

---

[73]This constraint is imposed to assure that the value of charge on the capacitor accurately represents the input voltage during that cycle.

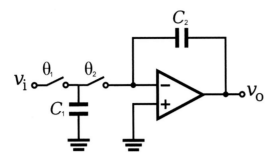

Fig. 11.30 A switched-capacitor integrator.

- CT, unity gain, amplifier,general-purpose
- CT, programmable gain amplifier,
- CT transimpedance amplifier,
- CT mixer,
- Delta-Sigma modulator [21],
- Operational amplifier,
- Sampled Mixer,

or,

- Track and hold amplifier,

with programmable power and bandwidth, routability to GPIO, routable reference selection, and sample-and-hold capability. The behavior of this block is controlled by settings as delineated in Tables 11.3, 11.4, and 11.5.

However, it should be noted that the SC/CT block in PSoC 3/5LP differs from that implemented in PSoC 1 in that the former has been optimized to carry out the specific functionality referenced above in terms of gain, bandwidth product, and slew rate. Therefore, the implementation of functionality such as integrators, differentiators and filters, using OpAmps and discrete external components, PSoC 3/5LP may be more appropriate choices, than the PSoC 1 [5].

The simplest configuration of the SW/CT block is the OpAmp. In this mode, the internal resistors and capacitors associated with the SW/CT block feed and input terminals are disconnected. This allows external components to be used for input and feedback and the techniques discussed in earlier sections to be employed. This mode can be selected by setting the MODE[2:0] bits in the SC[0...3]_CR0 to 000. The OpAmp is a two-stage, rail-to-rail amplifier with a folded cascode[74] first stage and Class A[75] second stage which is internally compensated. The value of the compensation capacitor and the drive strength of the output stage are both programmable to accommodate different

---

[74]Cascode refers to a two-stage amplifier which consists of a transconductance amplifier and a current buffer. It is capable of providing higher bandwidth, output impedance, input impedance and gain than a single-stage amplifier, while significantly improving input/output isolation because there is no direct coupling between input and output. This configuration is not subject to the Miller effect (cf. Sect. 11.4.4).

[75]Class A amplifiers produce outputs that are undistorted replications of the input and conduct during the entire input cycle. Alternatively, they are linear amplifiers that are always conducting.

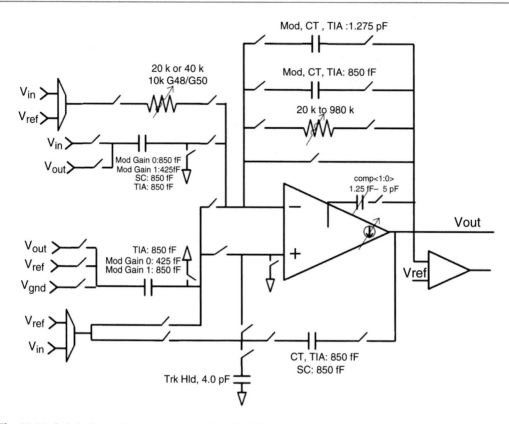

Fig. 11.31 Switched-capacitor and continuous-time block diagram.

loading conditions. The appropriate setting is a function of the minimum required slew rate[76] and load capacitance.

Table 11.3 Register- selectable operational modes.

SC_MODE[2:0]	Operational Mode
[000]	OpAmp
[001]	Transimpedance Amplifier
[010]	CT Mixer
[011]	DT Mixer NRZ S/H
[100]	Unity Gain Buffer
[101]	First-Order Modulator
[110]	Programmable Gain
[111]	Track & Hold Amplifier

The load current is given by:

$$i_{load} = C_{load} \left[\frac{\Delta v}{\Delta t} \right] \tag{11.146}$$

---

[76]Which is the desired rate of change of the signal with respect to time.

where C_{load} includes both the internal capacitance and the capacitance of the load.

Table 11.4 Miller capacitance between the amplifier output and output driver.

SC_COMP[1:0]	CMiller (pF)
00	1.30
01	2.60
10	3.90
11	5.20

Table 11.5 SC/CT block drive control settings.

SC_Drive[1:0]	$i_{load}(\mu A)$
2'b00	280
2'b01	420
2'b10	530
2'b11	650

Assuming a value of 10 pF for the internal capacitance, the drive controls SC_DRIVE[1:0] should be set according to the slew requirements at the output in the SC[0...3]_CR1[1:0]register bits.

This OpAmp configuration has three control options for modifying the closed-loop bandwidth and stability applicable to all configurations:

1. Current through the first stage of the amplifier (BIAS_CONTROL),
2. Miller capacitance between the amplifier input and the output stages (SC_COMP[1:0]),

and,

3. Feed back capacitance between the output stage and the negative input terminal (SC_REDC[1:0]).

The bias control doubles the current through the amplifier stage. The BIAS_CONTROL should be set to 1 to provide greater overall bandwidth once the circuit is stabilized rather than using the option of less current in the first stage. Bias current can be doubled by setting the SC[0...3]_CR2[0] register bit. The SC_COMP bits set the amount of compensation and directly affect the amplifier's gain-bandwidth. The Miller capacitance should be set to one of the four values for the SC[0...3]_CR[3:2] as shown in Table 11.6.

There is also an option related to the capacitance between the output driver and the negative input terminal that affects the stability-capacitance option. This option contributes to a higher frequency zero and a lower frequency pole which reduces the overall bandwidth and provides some additional phase margin at the unity gain frequency, depending on the CT configuration. Table 11.7 shows the available settings.

Table 11.6 Miller capacitance between the amplifier output and the output driver.

SC_COMP[1:0]	CMiller (pF)
00	1.30
01	2.60
10	3.90
11	5.20

Table 11.7 C_{FB} for CT mix, PGA, OpAmp, unity gain buffer, and T/H modes.

SC_REDC[1:0]	C_{FB}(pF)
00	0.00
01	1.30
10	0.85
11	2.15

PSoC 3 and PSoC 5LP each have 4 operational amplifiers configured as shown in Fig. 11.32 which have the following features:

- 25 ma drive capability,
- 3 MHz gain-bandwidth into a 200 pF load
- Low noise
- Less than 5 mV of offset
- Rail-to-rail capability to within:
    1. 50 mV of V_{ss} or V_{dd} for a 1 mA load.
    2. 500 mV of V_{ss} or V_{dda} for a 25 mA load.
- A slew rate of 3 Vμs for a 200 pF load.[77]

The OpAmps are configurable either as uncommitted OpAmps or as unity gain buffers. Access to the negative and positive inputs of the OpAmps is provided by muxes and analog switches. An analog global, local analog bus or reference voltage is connected to an input via a mux. A GPIO is connected to an input via an analog switch.

---

[77]Or the equivalent of 3 mV/s!

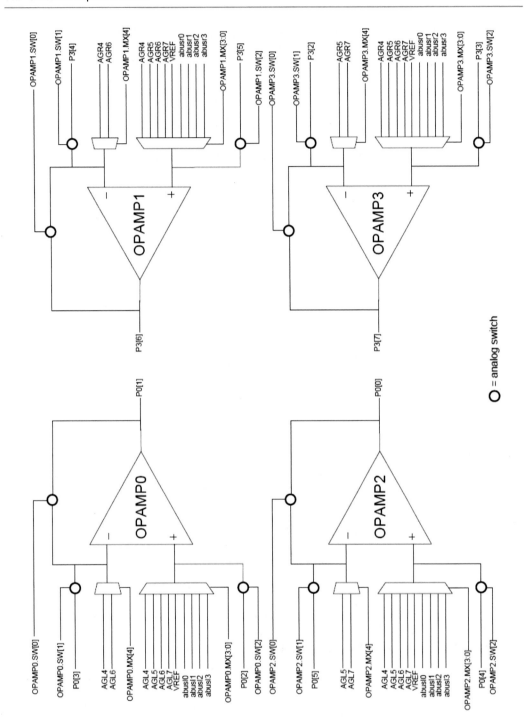

Fig. 11.32 PSoC 3/5LP operational amplifier connections.

11.7.1.1 PSoC 3/5LP OpAmps and PGAs

PSoC 3 provides Operational Amplifiers and Programmable Gain Amplifiers as shown in Fig. 11.33. The OpAmp is able to function either in a basic operational amplifier configuration or as a simple follower. PSoC 3/5LP OpAmps can also be used by employing available internal resistors, capacitors, and multiplexers, or in conjunction with external components, as shown in Fig. 11.34. Output current from the OpAmp should not exceed 25 mA and output loads should not exceed 10K Ω.

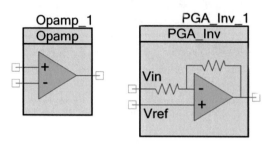

Fig. 11.33 PSoC Creator's graphical representation of an OpAmp and a PGA [25].

The programmable gain amplifier's gain can be set as one of the following values: 1 (default value), 2, 4, 8, 16, 24, 25, 48, or 50) shown in Table 11.8.

Table 11.8 PGA gain settings.

Gain Setting	Gain Value
PGA_GAIN_01	1
PGA_GAIN_02	2
PGA_GAIN_04	4
PGA_GAIN_08	8
PGA_GAIN_16	16
PGA_GAIN_24	24
PGA_GAIN_25	25
PGA_GAIN_32	24
PGA_GAIN_48	48
PGA_GAIN_50	50

There four power settings: minimum, low, medium (default value), or high. The power settings affect the PGA's response time with low power resulting in the slowest response time and high power the fastest. Vref_Input is a parameter that can be set either as *Internal Vss* which sets

$$V\mathrm{ref}_{\mathrm{input}} = Vss \tag{11.147}$$

i.e., the reference input is set to an internal ground, or as *External* which sets

$$V\mathrm{ref}_{\mathrm{input}} = External \tag{11.148}$$

for cases in which the reference input is to be connected to an arbitrary reference signal.

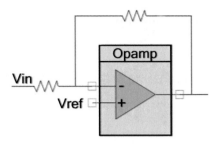

Fig. 11.34 PSoC 3/5LP operational amplifier configured as an inverting, variable gain OpAmp using external components [25].

```
Example 6.6: Sample C source program for initializing and
                    starting a PGA:

#include <device.h>
void main()
{
PGA_1_Start();
PGA_1_SetGain(PGA_1_GAIN_24);
PGA_1_SetPower(PGA_1_MEDPOWER);
}
```

The PGA is constructed from a generic SC/CT block. The gain is selected by adjusting two resistors, R_a and R_b that are shown in Fig. 11.35. R_a may be set to either 20k or 40kΩ. R_b may be set between 20k and 1000kΩ, to generate the possible gain values selectable in either a parameter dialog in PSoC Creator or via the *SetGain* function which selects the proper resistor values for the selected gain.

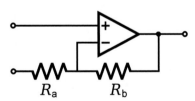

Fig. 11.35 PGA internal gain resistors.

The OpAmp component can be configured as either a follower or as an OpAmp that can be used in conjunction with external components. It can be used to drive loads that are less than 10K and provide a maximum driving current of 25 mA. When used as a follower, the negative input is inaccessible.

11.7.2 Continuous-Time Unity Gain Buffer

The CT, unity gain buffer is, as shown in Fig. 11.36, simply an OpAmp with the inverting input connected to the output and used when an internally generated signal is being used with high output impedance, e.g., a voltage DAC driving a load, or an external, high impedance source impedance driving a significant on-chip load, such as the continuous-time Mixer.

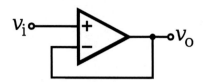

Fig. 11.36 An OpAmp configured as a unity gain buffer.

11.7.3 Continuous-Time, Programmable Gain Amplifier

The programmable gain amplifier (PGA) is a continuous-time OpAmp, configured as shown in Fig. 11.37, with selectable taps for the input and feedback resistors. It is selectable by setting MODE[2:0] bits in the SC[0,,3]_CR0 register to '110'. The PGA can be implemented as either a positive or negative gain topology, or as half of a differential amplifier. Gain is selected by setting bit [16] '1'(SC_GAIN) in the SCL[0...3]_CR1 register. If SC_GAIN is set to one, then the configuration is non-inverting with a gain of $(1 + \frac{R_{FB}}{R_{in}})$. If SC_GAIN is set to zero then the configuration is inverting with a gain of $-\frac{R_{FB}}{R_{in}}$. As shown in Fig. 11.38 it is possible to create a differential amplifier by connecting two PGAs as shown. A low impedance external resistor R_{LAD} is used to reduce gain error. The differential amplifier's gain is given by

$$v_{o+} - v_{o-} = A(vi+ - v_{i-}) \tag{11.149}$$

and the common-mode voltage of the output is also the common voltage input, viz.,

$$VCM = \frac{(v_{i+} - v_{i-})}{2} \tag{11.150}$$

11.7.4 Continuous-Time Transimpedance Amplifier

PSoC 3/5LP's transimpedance amplifier is a continuous-time OpAmp with a dedicated and selectable feedback resistor. The TIA configuration is selected by setting the MODE[2:0] bits in the SC[0...3]_CR0 register to '001'. The output of the transimpedance amplifier is a voltage that is proportional to the input current. The conversion gain is determined by the feedback resistor value, R_{fb}, so that:

$$v_o = v_{ref} - (i_i)R_{fb} \tag{11.151}$$

The output voltage, v_o, ie referenced to v_{ref} which can be a routed reference. A value for the feedback resistor, R_{fb}, can be selected programmatically as one of eight values over a range from $20k\,\Omega$ to $1.0\,M\Omega$, as shown in Table 11.9.

The inverting input shunt capacitance resulting from parasitic capacitances introduced by the analog global routing and at the input pin can adversely affect stability and therefore, an internal

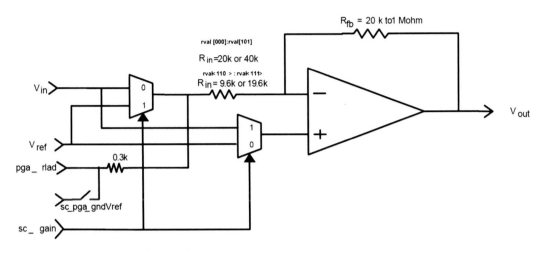

Fig. 11.37 CT PGA configuration [25].

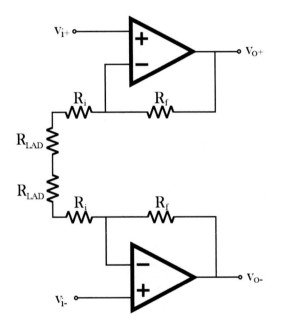

Fig. 11.38 Differential amplifier constructed from two PGAs.

shunt capacitance is employed to assure that the TIA remains stable. The feedback capacitance is set by the SC_REDC[1:0] bits in the SCL[0...3]_CR2 register bits [3:2] and the SCR[0...1]_CR2 register, bits [3:2], as shown in Table 11.10.

11.8   PSoC 3/5LP Comparators

Comparators make it possible to make fast comparisons between two voltages particularly with respect to other methods, such as using an ADC [1]. In some applications, a DAC is connected to the negative input to allow the reference voltage to be varied programmatically to allow the comparator

Table 11.9 Trans impedance amplifier feedback resistor values.

SC_RVAL[2:0]	Nominal RFB (kΩ)
000	20
001	30
010	40
011	80
100	120
101	250
110	500
111	1000

Table 11.10 TIA feedback capacitance settings.

SC_REDC[1:0]	C_{FB}(pF)
00	0.00
01	1.30
10	0.85
11	2.15

to be "adjustable". The positive input is connected to the voltage that is being compared to a reference value. In this case, the output goes high when the voltage that is being compared to the reference voltage is greater than the reference voltage. The output of the comparator can be sampled in software, or digitally routed to another component.

PSoC 3/PSoC 5 have four comparators configured as shown in Fig. 11.39. The configuration of the inputs to the comparators is controlled by the CMPx_SW0, CMPx_SW2, CMPx_SW3, CMPx_SW4, and CMPx_SW6 registers.

Inputs to the positive terminal can be analog globals, analog locals, the analog mux, and the comparator reference buffer. Inputs to the negative input can be from analog globals/locals/mux, and the voltage reference.

11.8.1 Power Settings

The PSoC 3/5LP comparators can operate in one of three power modes, viz., fast, slow and ultra-low power which are selected by the power mode select bits SEL[1:0] in CMPx_CR, the comparator control register. Power modes differ in response time and power consumption. Power consumption is maximum in the fast mode and minimum in the ultra-low power mode. The three-speed levels are provided to enable a comparator to be optimized for either speed or power consumption. Inputs to the comparators are via muxes whose inputs include analog globals (AGs), the local analog bus (ABUS), the Analog Mux Bus (AMUXBUS), and precision references. The output from each comparator is routed through a synchronization block to a two-input Lookup Table (LUT). The output of the LUT

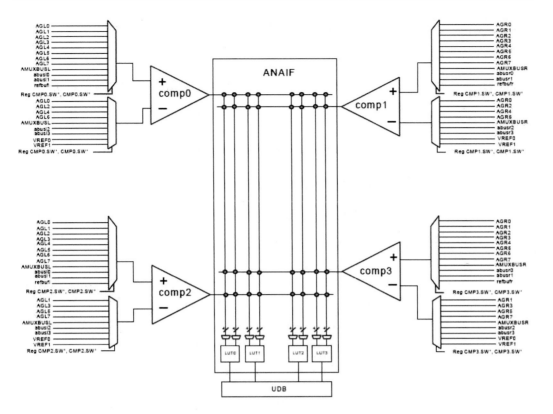

Fig. 11.39 Comparator block diagram.

is routed to the UDB Digital System Interface (DSI). The comparator can also be used to wake-up the device from sleep. An "x" used with a register name denotes the particular comparator number (x = 0–3). Connection to the positive input is from analog globals, analog locals, analog mux bus, and comparator reference buffer. Connection to the negative input is from analog globals, analog locals, analog mux bus, and voltage reference.

Comparator output can be passed through an optional glitch filter.[78] The glitch filter is enabled by setting the filter enable (FILT) bit in the control (CMPx_CR6) register. The output of the comparator is stored in the CMP_WRK register and can be read over the PHUB interface.

PSoC 3/5LP comparators have the following features:

- Low input offset
- Low power mode
- Multiple speed modes
- Output routable to digital logic blocks or pins
- Selectable output polarity
- User-controlled offset calibration
- Flexible input selection
- Speed power tradeoff
- Optional 10 mV input hysteresis

---

[78]Glitch filters are employed to remove transients, i.e., "glitches" in the output of a comparator.

- Low input offset voltage ($<1\,$mV)
- Glitch filter for comparator output
- Sleep wake-up

Four LUTs allow logic functions to be applied to comparator outputs. The LUT logic has two inputs:

- Input A is selected using MX_A[1:0] bits in LUT control (LUTx_CR1:0) register
- Input B is selected using MX_B[1:0] bits in LUT Control (LUTx_CR5:4) register

The logic function implemented in the LUT is selected using control bits (Q[3:0]) in the LUT Control register (LUTx_CR). The bit settings for various logic functions are given in Table 11.11. The output of the LUT is routed to the digital system interface of the UDB array. From the digital system interface of the UDB array, these signals can be connected to other blocks in the device, or an I/O pin.

The state of the LUT output is indicated in the LUT output (LUTx_OUT) bit in the LUT clear-on-read sticky[79] status (LUT_SR) register and can be read over PHUB interface. The LUT interrupt can be generated by all four LUTs and is enabled by setting the LUT mask (LUTx_MSK) bit in the LUT mask (LUT_MSK) register.

Table 11.11 Control words for LUT.

Control Word (Binary)	Output (A and B are LUT Inputs)
0000	False('0')
0001	A AND B
0010	A AND (NOT B)
0011	A
0100	(NOT A) AND B
0101	B
0110	A XOR B
0111	A OR B
1000	A NOR B
1001	A XNOR B
1010	NOT B
1011	A OR (NOT B)
1100	NOT A
1101	(NOT A) OR B
1110	A NAND B
1111	TRUE ('1')

---

[79] A sticky bit is a bit in a register that retains its value after the event that caused its value, has occurred.

11.8.1.1 Hysteresis

As discussed previously, hysteresis helps to avoid excessive toggling of the comparator output when the signals are noisy in applications that compare signals that are very close to each other in terms of sign and magnitude,. The 10 mV hysteresis level is enabled by setting the hysteresis enable (HYST) bit in the control (CMPx_CR5) register.

11.8.1.2 Wake-Up from Sleep

The comparator can run in sleep mode and the output can be used to wake-up the device from sleep. Comparator operation in sleep mode is enabled by setting the override (PD_OVERRIDE) bit in the control (CMPx_CR2) register.

11.8.1.3 Comparator Clock

The comparator output changes asynchronously but it can be synchronized with a clock. The clock source can be one of the four digitally-aligned analog clocks or any UDB clock. Clock selection is done by the mx_clk bits [2:0] of the CMP_CLK register. The selected clock can be enabled or disabled by setting or clearing the clk_en (CMP_CLK [3]) bit. Comparator output synchronization is optional and can be bypassed by setting the bypass_sync (CMP_CLK [4]) bit.

11.8.1.4 Offset Trim

Comparator offset is dependent on the common-mode input voltage to the comparator. The offset is factory trimmed for common-mode input voltages 0.1 V and Vdd −0.1 V to less than 1 mV. If the common-mode input range at which the comparator is to operate is known, a priori, a custom trim can be done to reduce the offset voltage further.

11.9 PSoC 3/5LP Mixers

PSoC 3 and PSoC 5LP provide two types of "mixers",[80] viz., continuous, and sampled. These components are single-ended, and not intended to function as "precision" mixers. The continuous-time configuration is suitable for multiplying and up-mixing. The discrete-time configuration (sampled) has a sample-and-hold capability and is appropriate for sampled- or down-mixing. A continuous-time mixer uses input switches to toggle between the inverting and non-inverting inputs of a programmable gain amplifier for which the gains are 1 and -1, respectively. If a fixed local oscillator is used as a sampling clock, the mixer can be used to perform the frequency conversion of a signal.

The PSoC 3/5LP mixer has the following features:

- Power settings are adjustable.
- Continuous-time up-mixing[81] with input frequencies up to 500 kHz and sample clock rates up to 1 MHz.
- Discrete-time, sample-and-hold mixing with input frequencies up to 1 MHz and sample clock rates to 4 MHz.
- Selectable reference voltages.

---

[80]Mixing in the present context is in actuality the multiplication of two signals resulting in the production of four output signals, viz., the sum, difference, and the two mixed signals.

[81]Up-mixing refers to mixing to signals and producing a signal at a frequency which is the sum of the two mixed-signal frequencies. Similarly, down-mixing produces a signal whose frequency is the difference in frequency of the two mixed signals.

11.9.1 Basic Mixing Theory

Before proceeding, some basic mixing concepts need to be introduced. The term "mixing" in the present context refers to the mixing, or more accurately stated, multiplying of two signals. Given two signals such as

$$y_1 = A_1 sin(\omega_1 t + \phi_1) \tag{11.152}$$

$$y_2 = A_2 sin(\omega_2 t + \phi_2) \tag{11.153}$$

The product, Y, is given by

$$Y = y_1 y_2 = A_1 A_2 [sin(\omega_1 t + \phi_1)][sin(\omega_2 t + \phi_2)] \tag{11.154}$$

but,

$$sin(u)sin(v) = \frac{1}{2}\left[cos(u - v) - cos(u + v)\right] \tag{11.155}$$

and therefore,

$$Y = \frac{A}{2}\left[cos[(\omega_1 - \omega_2)t + \Phi_1 - cos[(\omega_1 + \omega_2)t + \Phi_2]\right] \tag{11.156}$$

where $A = A_1 A_2$, $\Phi_1 = \phi_1 - \phi_2$ and $\Phi_2 = \phi_1 + \phi_2$. Thus, the result of mixing two sine wave signals is to produce the sum and difference of the two signals in terms of frequency which together with the two original signals, results in the presence of four signals. Note that the resulting sum and difference signals have also undergone a phase shift. Any of the resulting signals can, if necessary, be subsequently removed, e.g., by filtering.

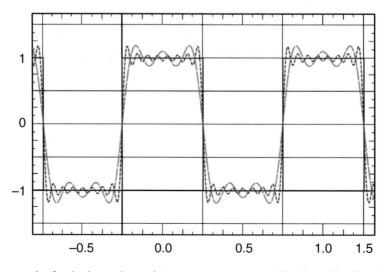

Fig. 11.40 An example of cosine harmonics used to represent a square wave (http://en.wikipedia.org).

If y_2 is a square wave, then it can be expressed as a trigonometric partial sum by the following:

$$y_2 = \frac{4}{\pi}\left[cos(\omega_2 t) - \frac{cos(3\omega_2 t)}{3} + \frac{cos(5\omega_2 t)}{5} - \cdots\right] \tag{11.157}$$

as illustrated in Fig. 11.40[82] [10].

If y_1 is defined as:

$$y_1 = cos(\omega_i t) \tag{11.158}$$

Then,

$$Y = y_1 y_2 = \left[A_1 cos(\omega_1 t)\right]\frac{4}{\pi}\left[cos(\omega_2 t) + \frac{cos(3\omega_2 t)}{3} + \frac{cos(5\omega_2 t)}{5} - \cdots\right] \tag{11.159}$$

so that,

$$Y = y_i y_{clk} = \frac{2A_1}{\pi}\left[cos(\omega_- t) - \frac{cos(3\omega_{3-} t)}{3} + \frac{cos(5\omega_{5-} t)}{5} - \cdots\right] \tag{11.160}$$

$$+ \frac{2A_1}{\pi}\left[cos(\omega_+ t) - \frac{cos(3\omega_{3+-} t)}{3} + \frac{cos(5\omega_{5+} t)}{5} - \cdots\right] \tag{11.161}$$

where,

$$\omega_+ = \omega_{clk} \tag{11.162}$$

$$\omega_{3+} = 3\omega_{clk} + \omega_i \tag{11.163}$$

$$\omega_{5+} = 5\omega_{clk} + \omega_i \tag{11.164}$$

$$\cdots \tag{11.165}$$

$$\omega_- = |\omega_{clk} - \omega_i| \tag{11.166}$$

$$\omega_{3-} = |3\omega_{clk} - \omega_i| \tag{11.167}$$

$$\omega_{5-} = |5\omega_{clk} - \omega_i| \tag{11.168}$$

$$\cdots \tag{11.169}$$

Thus, in this case, in addition to the sum and difference frequencies for v_i, and v_{clk} the third, fifth and all additional higher-order, odd harmonics are present in the output.[83] The unwanted harmonics can be removed by suitable filtering.

---

[82]The "ringing effects" seen at the corners of this square wave are known as the Gibbs', or Gibbs-Wilbraham', phenomenon [10] and was first observed by Wilbraham in 1848 [33] and subsequently "rediscovered" by Gibbs in 1898 who provided a much more rigorous mathematical foundation for it. This effect is commonly encountered in the course of processing digital signals, e.g., to the series.

[83]Note that the v_i and v_{clk} frequencies are not present in the output.

11.9.2 PSoC 3/5LP Mixer API

The PSoC 3/5LP mixer has an API consisting of three function calls:

- **void Mixer_Start(void)** powers up the Mixer. Performs all the required initialization for the mixer and enables power to the block. The first time the routine is executed, the input and feedback resistance values are configured for the operating mode selected in the design. When called to restart the mixer following a **Mixer_Stop()** call, the current component parameter settings are retained.
- **void Mixer_Stop(void)** powers down the Mixer. This does not affect mixer type or power settings.

and,

- **void Mixer_SetPower(uint8 power)**—Set drive power to one of the following four levels.
  - **Mixer_MINPOWER**—Lowest active power and slowest reaction time.
  - **Mixer_LOWPOWER**—Low power and speed.
  - **Mixer_MEDPOWER**—Medium power and speed.
  - **Mixer_HIGHPOWER** —Highest active power and fastest reaction time.

11.9.3 Continuous-Time Mixer

As shown in Fig. 11.41,

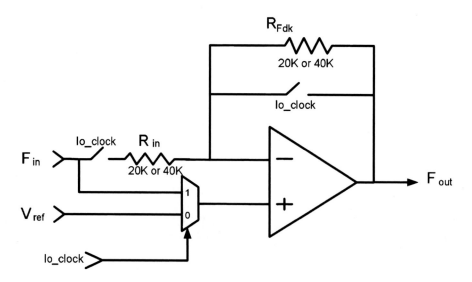

Fig. 11.41 PSoC 3/5LP configuration for a CT mixer [25].

The OpAmp is configured as a PGA that uses the lo_clock input signal to toggle between an inverting unity gain PGA and a non-inverting unity gain buffer. The output signal includes frequency components at $(F_{clk} \pm F_{in})$ plus terms at odd harmonics of the LO frequency $\pm$, the input signal frequency: $3 * F_{clk} \pm F_{in}, 5 * F_{clk} \pm F_{in}, 7 * F_{clk} \pm F_{in}$, etc. Continuous-time mode is preferable for "up-conversion" since it provides a much higher conversion gain than the sampled mixer. In order to assure optimal performance the value for F_{clk} should meet the Nyquist criteria,[84] viz.,

$$F_{clk} > 2F_{out} \tag{11.170}$$

An example of an implementation of the CT mixer input and output waveforms is shown in Fig. 11.42 and is based on a CT block.

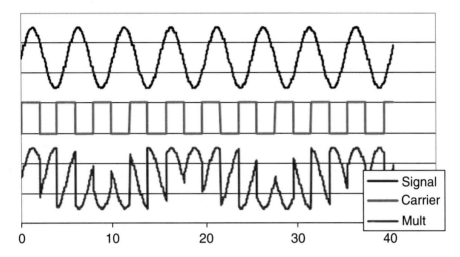

Fig. 11.42 An example of CT Mixer input and output waveforms [25].

11.9.4 Sampled-Mixer

Before beginning a discussion of the sampled-mixer it will be necessary to discuss two of the types of encoding used in digital systems, viz.,

- **Non-Return to Zero (NRZ-L)**—two distinct voltage levels are used to represent zeros and ones.x A voltage level of zero is not used to represent the binary value zero. Typically, a positive value is used for one and a negative value for zero with both being of the same absolute value.
- **Non-Return to Zero Inverted (NRZI)**—any transition to high or low represents a binary value of one, the absence of a transition represents a binary value of zero.

The sampled-mixer provided in PSoC 3/5LP is basically a NRZ Sample-and-Hold circuit[85] with a very fast response. Unlike the CT mixer which has an upper-frequency limit of 4 MHz, the sampled mixer can accept input frequencies as high as 14 MHz. The output of the sampled-mixer can be used

---

[84]Simply stated, the Nyquist criteria states that a sampled, band-limited, analog signal can be completely reconstructed if the sampling rate is twice the highest frequency component in the original analog signal.

[85]Sample-and-Hold circuits are used to sample a time-dependent signal and then hold that sample for a period of time to allow certain operations to be carried out with respect to the sample. A common method of "holding" is to store the sample via a capacitor that is capable of holding the sample for the time required without degrading it, i.e., with sufficiently low leakage current.

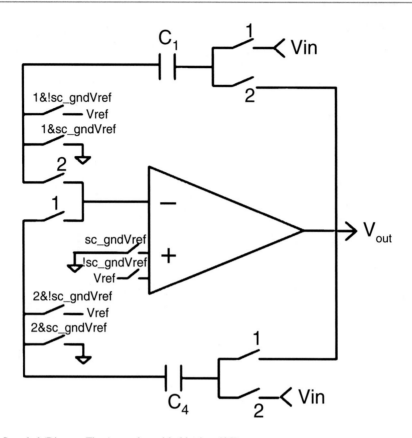

Fig. 11.43 Sampled (Discrete-Time) sample-and-hold mixer [25].

as input to an internal ADC[86] via analog routing, or in conjunction with an external device such as a ceramic filter.[87]

As mentioned previously, the sampled mixer shown in Fig. 11.43 is used primarily for downconversion which can be accomplished by removing the undesired products resulting from mixing the input frequency and the sample clock. The NRZ sample-and-hold functionality is based on alternately selecting one of two capacitors as the integrating capacitor. Thus, one capacitor, either C_1 or C_4, serves as the integrating capacitor, while the other is used to sample the input signal. This configuration is designed such that the input signal is sampled at a rate less than the input signal frequency and the integration of each new value occurs on the rising edge of f_{clk}.

If $f_{clk} > f_{in}/2$, then

$$f_{out} = | f_{in} - f_{clk} | + \text{aliasing components} \qquad (11.171)$$

If $f_{clk} < f_{in}/2$, then

$$f_{out} = | f_{in} - N f_{clk} | \qquad (11.172)$$

---

[86]If the output is routed to an internal ADC both the mixer and the ADC must employ the sample clock.

[87]For example the 455 kHz Murata Cerafil. 455 kHz is a standard frequency used in receivers as part of the intermediate frequency (IF) stage.

for the largest integer value of N such that $N f_{clk} < f_{in}$

For example, if the desired down-converted frequency is 500 kHz and the input frequency is 13.5 MHz, Eq. 11.172 is satisfied for values of $N = 7$ and $f_{clk} = 2$ MHz. Examples for $N = 1$ and $N = 3$ are shown in Figs. 11.44 and 11.45, respectively.

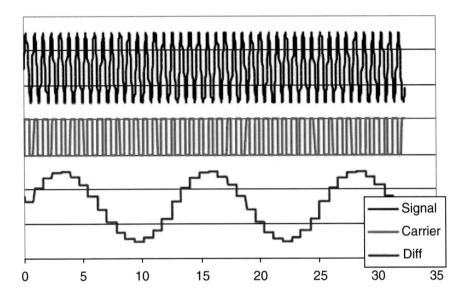

Fig. 11.44 Sampled mixer waveforms for $N = 1$.

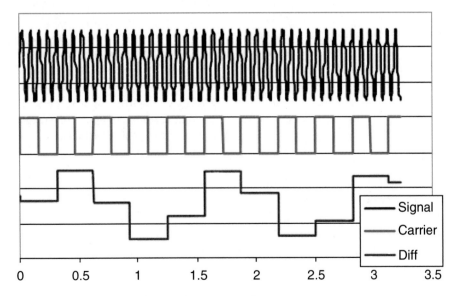

Fig. 11.45 Sampled mixer waveforms for $N = 3$.

```
Example 6.6: C source code for implementing a mixer that
employs an internal local oscillator

#include <device.h>
#include "Mixer_1.h"
#include "lo_clk.h"
void main()
{
/* Setup Local Oscillator Clock */
lo_clk_Enable();
lo_clk_SetMode(CYCLK_DUTY);
/* API Calls for Mixer Instance */
Mixer_1_Start();
Mixer_1_SetPower(Mixer_1_HIGHPOWER);
while (1)
{
}
}
```

11.10 Filters

Almost all embedded systems are confronted with the potential for noise altering the response and/or performance of the system. Various techniques exist for dealing with noise, depending on the source, type of noise, amplitude, spectral composition, etc. While the best cure for noise is to avoid it, the reality is that it is almost always present and must be dealt with directly. Analog filters are often used for this purpose and consist of two fundamental types, viz., passive and active. Digital filters are also used but typically require that the analog signal to be filtered first be converted to a digital format, processed, and then converted back to analog form. Although digital filters do have truly outstanding characteristics they also involve what can be significant and perhaps prohibitive overhead in some applications that may not be acceptable in some application. At very high frequencies, digital filters are less attractive than their analog counterparts for a variety of reasons.

11.10.1 Ideal Filters

Filters provide a method for separating signals, e.g., as in the case of an amplitude-modulated or frequency carrier,[88] allow signals to be restored by removing unwanted signals/noise, and restore a signal that may have been otherwise altered. Filters may be based on combinations of RLC components, rely on mechanical resonances, employ the piezoelectric effect, utilize acoustic wave techniques, etc., depending on the operating environment the application, frequency range and the desired filter characteristics.

Ideal filters can be characterized as:

---

[88]There are a variety of techniques for demodulating signals, and filters represent one such method, e.g. mixing techniques, a topic discussed in Sect. 11.8.1.4.

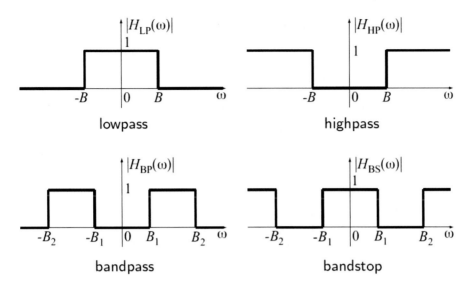

Fig. 11.46 "Brickwall"Transfer functions for ideal filters.

- **Lowpass (LPF)**—"passes" all frequencies below a certain frequency with no change in amplitude and specifiable phase shift.
- **Bandpass (BPF)**—"passes' all frequencies above a given frequency and below an upper frequency with no change in amplitude and specifiable phase shift.
- **Highpass (HPF)**—"passes" all frequencies above a certain frequency with the change in amplitude and specifiable phase shift.
- **Notch (NPF)**—"blocks" frequencies within a specified range.
- **Allpass** (APF)—"passes" all frequencies and alters only the phase, e.g., unity gain at all frequencies. The phase shift at the corner frequency for all frequencies is 90°. This type of filter is often used to match phase, introduce a delay and to create a 90° phase shift for certain types of circuits.

Ideal filters have a number of important characteristics that should be kept in mind when designing a filter for a particular application, viz., no attenuation in the passband with sharp cutoff and complete attenuation in the stopband. Such filters are referred to as "brickwall" filters and are represented graphically as shown in Fig. 11.46.

Thus, a filter is considered ideal if

$$| H(\omega) | = \begin{cases} 1, & \text{if } \omega \text{ is in the passband} \\ 0, & \text{if } \omega \text{ is in the stopband} \end{cases} \tag{11.173}$$

and,

$$\angle H(\omega) = \begin{cases} -\omega\tau, & \text{if } \omega \text{ is in the passband} \\ 0, & \text{if } \omega \text{ is in the stopband} \end{cases} \tag{11.174}$$

where τ is a positive constant. The phase shift for ideal filters is shown in Fig. 11.47.

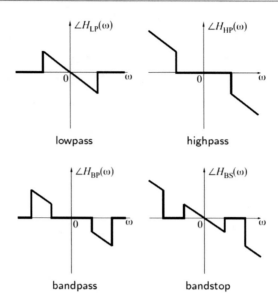

Fig. 11.47 Phase as a function of frequency for ideal filters.

11.10.2 Bode Plots

Filter design has been greatly facilitated by the availability of computer programs designed to handle the mathematical complexities. Their ability to display the results of the associated calculations significantly simplifies the designer's task. A common graphical characterization of a filter is the so-called Bode Plot.[89] A Bode plot is a graphical representation of a transfer function that represents a linear, time-invariant system in which the abscissa is usually the log of the frequency and the ordinate is the log of the system's gain. A typical Bode plot is shown in Fig. 11.48.

11.10.3 Passive Filters

Passive filters typically consist of combinations of resistors, capacitors and in some cases inductors[90] configured in various ways to provide the type of filter needed and the desired filter characteristics. Passive filters do not power supplies. However, passive filters are affected by changes in the components' capacitances, resistance, and inductance as a result of humidity, temperature, aging, vibration, etc., because any such variation can seriously, and adversely, alter a filter characteristics. Also, at low frequencies the physical size of components, given the inverse relationship between the physical size and the operating frequencies, can be a problem. However, even though passive filters are inherently stable, unlike their active filter counterparts which can oscillate, they tend to be more linear than active filters, and can be designed to handle arbitrarily large voltages, currents and frequencies. Active filters employing solid-state devices, such as operational amplifiers, tend to be frequency, current and voltage limited.

---

[89]This technique was introduced by Hendrik Bode in 1938, at Bell Laboratories, where he worked as an engineer.
[90]Inductors are often not used because of size, cost and other considerations.

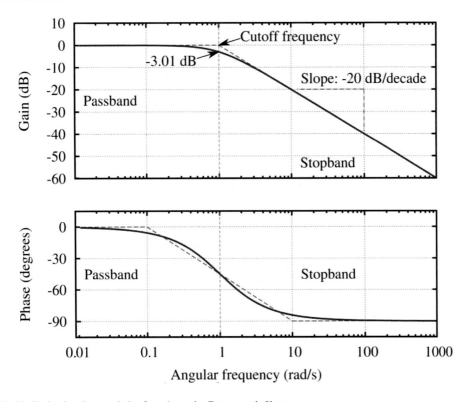

Fig. 11.48 Bode plot characteristics for a 1st-order Butterworth filter.

One of the simplest forms of passive filter consists of a combination of resistor and capacitor as shown in Fig. 11.49. Although this type of filter is simplicity itself, it is instructive to carry out a brief analysis in order to illustrate some important aspects of filters.

Treating this circuit as a voltage divider and defining $s = \sigma + j\omega$ leads to the result that:

$$v_r = \frac{sRC}{1 + sRC} v_i(s) \tag{11.175}$$

and,

$$v_c = \frac{1}{1 + sRC} v_i(s) \tag{11.176}$$

Therefore ,the transfer functions for the resistor and capacitor is given by:

$$H_r(s) = \frac{v_R}{v_i} = \frac{sRC}{1 + sRC} \tag{11.177}$$

$$H_c(s) = \frac{v_R}{v_i} = \frac{1}{1 + sRC} \tag{11.178}$$

Note that assuming that the excitation is steady-state, then $s = j\omega$, so that when:

$$s = -\frac{1}{RC} \tag{11.179}$$

Eqs. (11.177) and (11.178) become infinite and Eq. 11.179 is said to be a "pole"[91] for both transfer functions. In addition, Eq. 11.177 is zero when $s = 0$ which is referred to as a "zero" of the resistor's transfer function.

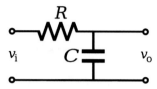

Fig. 11.49   A very simple, passive, lowpass filter.

Gain is defined as the ratio of the output voltage to the input voltage and therefore, for the capacitor:

$$A_C = \left|H_C\right| = \frac{1}{\sqrt{1 + (wRC)^2}} \tag{11.180}$$

and for the resistor:

$$A_R = \left|H_R\right| = \frac{wRC}{\sqrt{1 + (wRC)^2}} \tag{11.181}$$

Notice that as expected the capacitor gain approaches zero, but the resistor gain approaches unity, as the frequency increases. Thus, the RC combination shown in Fig. 11.49 functions as a lowpass filter if the output is taken across the capacitor and as a highpass filter if the output is taken across the resistor. Such filters are used, but because of the resistive component it does degrade the signal can making them unsuitable for some applications. One of the problems with this type of filter is that either resistive or reactive loads will change the characteristics of this filter and therefore, must be taken into account in designing a filter. An operational amplifier can be used to address both of these types of degradation but has the undesirable feature of increasing the amplitude of any noise present at the input terminals.[92]

The phase shift introduced by the resistor and capacitor is given by:

$$\Theta_r = tan^{-1}\left(\frac{1}{\omega RC}\right) \tag{11.182}$$

and,

$$\Theta_c = tan^{-1}(-\omega RC) \tag{11.183}$$

---

[91] Poles are defined as the "zeros" of the denominator and zeros are the "zeros" of the numerator of a rational transfer function.

[92] Filters employing operational amplifiers are referred to as "active" filters, because they require power supplies.

Active filters require external power, and may cost more, but may also be of much smaller size and offer better characteristics and performance than their passive counterparts. However, active filters can introduce noise into a system,[93] if not carefully designed and implemented. This can arise as a result of the noise introduced through the power supplies.

11.10.4 Analog Active Filters

Analog active[94] filters are characterized by a number of factors including the:

1. number of stages or sections employed in a cascaded fashion,[95]
2. cutoff frequency, i.e., the point at which the response of the filter falls below 3 dB,
3. response of the filter in the stopband,
4. gain as a function of frequency in the passband,
5. phase shift in the passband,
6. degree of ringing, if any,
7. rate of roll-off,

and,

8. transient response

Lowpass filters of the four types shown in Fig. 11.50 are commonly used, and of these, the Butterworth[96] filter is sufficient for most lowpass applications. Items 2–6 of this list can be determined by examining the Bode plots of the components shown in Fig. 11.48 for the phase and gain of a filter. A fifth type, the Bessel filter, is very flat in the passband but rolls off more slowly than Butterworth,Elliptic and Tchebychev filters.

11.10.4.1 Sallen–Key Filters (S-K)

PSoC 3 does not provide explicit internal support for analog LPF, BPF, HPF, and BSF because the switched-capacitance block designed for PSoC 3 has been optimized for other configurations. However, PSoC 3/5LP do have operational amplifiers that can be used in conjunction with external components to provide analog filtering. A popular design methodology in such cases was suggested by Sallen–Key [28]. This type of filter is referred to as a voltage-controlled, voltage source, or VCVS filter. It has gained widespread popularity, in part, because of its relative simplicity, ability to employ conventional operational amplifiers, excellent passband characteristics, relatively low cost, and required few components. Furthermore, multiple S–K filters can be cascaded without significant signal degradation.

---

[93] One potential noise source for active filters is the operational amplifier's power supply.

[94] An active filter is a combination of passive components and active components capable of adding gain. The latter therefore, requires input power.

[95] The advantage of cascading filters is that it increases the overall effective "order" of the filter which in turn increases the roll-off at a rate of 6 dB/octave times the equivalent order of the cascaded filters. One heuristic often employed to determine the order of each filter stage is to count the number of storage elements in each stage, e.g., the number of capacitors, which is usually the same as that stage's order.

[96] In 1930, S Butterworth published an important paper describing his design methodology for what became known as the Butterworth filter. He wound wire around cylinders that were 1.25 inches in diameter and 3 inches long to create resistors/inductors and placed capacitors inside the cylinders to complete the filter.

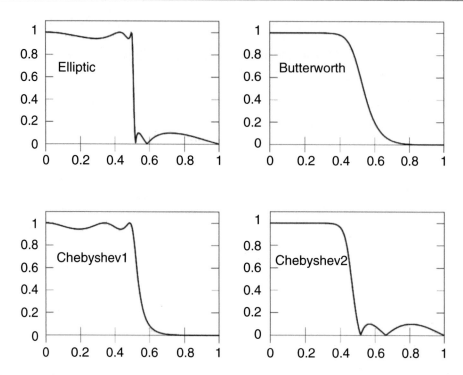

Fig. 11.50 Normalized graphs of common 5th-order lowpass filter configurations.

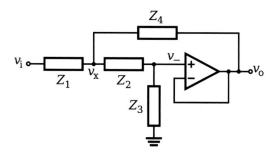

Fig. 11.51 The generic form of the Sallen–Key filter.

Assuming the generic configuration of the unity gain Sallen–Key filter shown in Fig. 11.51 the transfer function for this type of unity gain filter is given by:

$$\frac{v_o}{v_i} = \frac{Z_3 Z_4}{Z_1 Z_2 + Z_4(Z_1 + Z_2) + Z_3 Z_4} \tag{11.184}$$

By choosing different impedances for Z_1, Z_2, Z_3, Z_4 the Sallen–Key topology can be transformed into a filter exhibiting highpass, bandpass, and lowpass characteristics. The nominal frequency limit of a filter is referred to as the "corner" frequency which is the frequency at which the input signal is reduced to 50% of its maximum power[97] just prior to the output terminals.[98] Beyond that point, the attenuation is usually referred to in terms of dB/octave[99] or dB/decade.

A typical lowpass filter response curve is shown in Fig. 11.52.

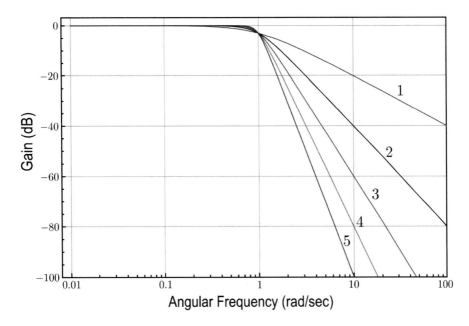

Fig. 11.52 Bode plot of an nth order Butterworth filters for $n = 1$–5.

As shown the "cutoff frequency" is defined as the point at which the response curve is -3 dB down and, in this particular case, the response curve falls off at the rate of -20 dB/decade. As can be seen from this figure, the order of the filter dramatically affects the rate of roll from the passband to the stopband.

11.10.4.2 Sallen–Key Unity-Gain Lowpass Filter
As shown in Fig. 11.53, a Sallen–Key lowpass filter can be configured by setting:

$$Z_1 = R_1 \tag{11.185}$$

$$Z_2 = R_2 \tag{11.186}$$

$$Z_3 = -\frac{j}{\omega C_1} = \frac{1}{sC_1} \tag{11.187}$$

$$Z_4 = -\frac{j}{\omega C_2} = \frac{1}{sC_2} \tag{11.188}$$

---

[97] The so-called "half power-point".

[98] This is also referred to as the "-3 dB down" point.

[99] An octave implies a doubling of frequency, e.g., -6 dB/octave implies that the signal is reduced by 50% if the frequency is doubled. Decade refers to a change in frequency by a factor of ten.

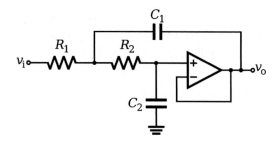

Fig. 11.53 A Sallen–Key unity-gain lowpass filter.

and after rearranging,

$$H(s) = \frac{\frac{1}{R_1 R_2 C_1 C_2}}{s^2 + \left[\frac{R_1 + R_2}{R_1 + R_2 C_1} \right] s + \frac{1}{R_1 R_2 C_1 C_2}} \tag{11.189}$$

which is in the form of the transfer function of a second-order, unity-gain, lowpass filter, i.e.,

$$H(s) = \frac{\omega_c^2}{s^2 + 2\zeta \omega_c s + \omega_c^2} \tag{11.190}$$

If ω_c^2 is defined as

$$\omega_c^2 = \frac{1}{R_1 R_2 C_1 C_2} = \omega_0 \omega_n \tag{11.191}$$

then

$$f_c = \frac{1}{2\pi \sqrt{R_1 R_2 C_1 C_2}} \tag{11.192}$$

and,

$$2\zeta = \frac{1}{Q} = \frac{\sqrt{R_1 R_2 C_1 C_2}}{R_1 C_1 + R_2 C_1} = \left[\frac{1}{R_1} + \frac{1}{R_2} \right] \frac{\sqrt{R_1 R_2 C_1 C_2}}{C_1} \tag{11.193}$$

After rearranging, Eq. 11.193 becomes

$$Q = \frac{\sqrt{R_1 R_2 C_1 C_2}}{C_2(R_1 + R_2)} \tag{11.194}$$

The Q or selectivity value determines the height and width of the frequency response and is defined for bandpass filter as

$$Q = \frac{f_c}{f_H - f_L} = \frac{f_c}{Bandwidth} \tag{11.195}$$

and

$$f_c = \sqrt{f_H f_L} \tag{11.196}$$

which is the geometric mean for the $-3\,\text{dB}$ points of f_H and f_L. The reader may be puzzled by the meaning of Q for a lowpass filter given how it is defined. In the case of a Butterworth lowpass filter[100] it is used as a measure of the filter's response, i.e. under-, critically- or highly-damped. Because R_1, R_2, C_1 and C_2 are independent variables, it is possible to simplify Eqs. 11.192 and 11.194 by choosing R_2 and C_2 to be integer multiples of R_1 and C_1, respectively.

11.10.4.3 Sallen–Key Highpass Filter

Similarly, the configuration for a Sallen–Key 2nd-order highpass filter is shown in Fig. 11.54. In this case, the transfer function for the S–K bandpass filter is given by

$$H(s) = \frac{s^2}{s^2 + \frac{R_1}{R_1 R_2 \left(\frac{C_1 C_2}{C_2 + C_2}\right)} s + \frac{1}{\left[R_1 R_2 \left(\frac{C_1 C_2}{C_1 + C_2}\right)\right](C_1 + C_2)}} \tag{11.197}$$

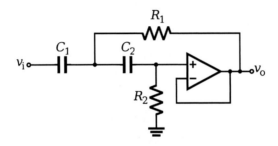

Fig. 11.54 A Sallen–Key highpass filter.

and the general for a 2nd-order high pass filter is given by

$$H(s) = \frac{s^2}{s^2 + 2\zeta \omega_c s + \omega_c^2} \tag{11.198}$$

and therefore,

$$Q = \frac{\sqrt{R_1 R_2 C_1 C_2}}{R_1 (C_1 + C_2)} \tag{11.199}$$

and

$$f_c = \frac{1}{2\pi \sqrt{R_1 R_2 C_1 C_2}} \tag{11.200}$$

---

[100] A 4th order Butterworth filter having a Q of .707 is maximally flat in the passband.

11.10.4.4 Sallen–Key Bandpass Filter

Similarly, the transfer function for the bandpass filter shown in Fig. 11.55 is given by

$$H(s) = \frac{\frac{R_a+R_b}{R_a}\frac{s}{R_1 C_1}}{s^2 + \left[\frac{1}{R_1 C_1} + \frac{1}{R_2 C_1} + \frac{1}{R_2 C_2} - \frac{R_b}{R_a R_f C_1}\right]s + \left[\frac{R_1+R_2}{R_1 R_f R_2 C_1 C_2}\right]} \tag{11.201}$$

The denominator is of the general form for bandpass filters and expressed as:

$$H(s) = \frac{G\omega_n^2 s}{s^2 + 2\xi\omega_0 s + \omega_0^2} \tag{11.202}$$

where G is the so-called inner gain[101] of the filter and given by

$$G = \frac{R_a + R_b}{R_a} \tag{11.203}$$

and the gain at the peak frequency[102] is given by

$$A = \frac{G}{G - 3} \tag{11.204}$$

and the center frequency by

$$f_0 = \frac{1}{2\pi}\sqrt{\frac{R_f + R_1}{R_1 R_2 R_f C_1 C_2}} \tag{11.205}$$

Setting

$$C_1 = C_2 \tag{11.206}$$

$$R_2 = \frac{R_1}{2} \tag{11.207}$$

$$R_a = R_b \tag{11.208}$$

yields

$$G = 2 \tag{11.209}$$

$$A = 2 \tag{11.210}$$

$$f_0 = \frac{1}{2\pi}\sqrt{\frac{R_f + R_1}{R_1^2 C_1^2 R_f}} = \frac{1}{2\pi R_1 C_1}\sqrt{1 + \frac{R_1}{R_f}} \tag{11.211}$$

---

[101] This is the gain determined by the negative feedback loop.
[102] Note that if the gain is $\leqslant 3$ the circuit will oscillate.

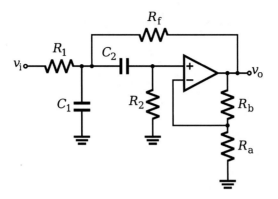

Fig. 11.55 A Sallen–Key bandpass filter.

11.10.4.5 An Allpass Filter

The phrase "allpass filter" is in some respects an oxymoron because such filters pass all frequencies at constant gain. The important characteristic of this particular type of so-called filter is that its phase response varies linearly with respect to the frequency which makes allpass filters useful. R and C can be used to form a lowpass filter as shown previously whose transfer function is given by

$$H(s) = \frac{1}{1 + sRC} \tag{11.212}$$

The current into the negative feedback loop is given by

$$\frac{v_i - v_{-input}}{R_f} = \frac{v_i - v_i H(s)}{R_f}$$

$$v_{+input} - i_f R_f = v_i H(s) - \left[\frac{v_i - vi H(s)}{R_f}\right] R_f$$

$$= [2H(s) - 1]v_i$$

$$= \left[\frac{2}{1 + sRC}\right] v_i = \left[\frac{1 - sRC}{1 + sRC}\right] v_i \tag{11.213}$$

and therefore,

$$|H| = 1 \tag{11.214}$$

and

$$\angle H = -2\tan^{-1}(\omega RC) \tag{11.215}$$

which for

$$\omega RC = 1 \Rightarrow \angle H = -90° \tag{11.216}$$

which means that gain is independent of frequency and phase is dependent on frequency (Fig. 11.56).

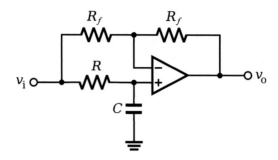

Fig. 11.56 A simple first-order allpass filter.

11.10.5 Pulse Width Modulator (PWM)

The pulse width modulator is a component that provides user-selectable pulse widths for use as single or continuous hardware timing control signals. The most common use of a PWM is to generate periodic waveforms with adjustable duty cycles. The PWM also provides optimized features for power control, motor control, switching regulators, and lighting control. It can also be used as a clock divider by supplying it with a clock input and using the terminal count, or a PWM output, as the divided clock output.

While PWMs, Timers, and Counters share many capabilities each provides very specific functionality. A Counter component is used in situations that require the counting of a number of events but also provides rising edge capture input as well as a compare output. A Timer component is used in situations focused on timing the length of events, measuring the interval of multiple rising and/or falling edges, or for multiple capture events. The PSoC 3/5LP PWM module is provided with an Application Programming Interface (API) that allows the designer to configure the PWM in software.

PSoC 3/5LP's PWM component provides compare outputs to generate single or continuous timing and control signals in hardware. The PWM is designed to provide an easy method of generating complex real-time events accurately with minimal CPU intervention. The PWM features include

- 8 or 16-bit resolution
- Configurable Capture
- Configurable Dead-band
- Configurable Hardware/Software Enable
- Configurable Trigger
- Multiple Configurable Kill modes.

and the PWM component may be combined with other analog and digital components to create custom peripherals. The PWM generates up to 2 left- or right-aligned PWM outputs, or 1 center-aligned or dual-edged PWM output. The PWM outputs are double $=$ buffered to avoid glitches due to duty-cycle changes while running. Left-aligned PWMs are used for most general-purpose PWM uses. Right-aligned PWMs are typically used only in special cases that require alignment opposite of left-aligned PWMs. Center-aligned PWMs are most often used in controlling an AC motor to maintain phase alignment. Dual-edge PWMs are optimized for power conversion where phase alignment must be adjusted.

The optional deadband provides complementary outputs with adjustable dead time where both outputs are low between each transition. The complementary outputs and dead time are most often used to drive power devices in half bridge configurations to avoid shoot-through currents and the resulting potential for damage. A kill input is also available that, when enabled, immediately disables the deadband outputs. Three kill modes are available to support multiple use scenarios. Two hardware dither[103] modes are provided to increase PWM flexibility. The first dither mode increases effective resolution by 2-bits when resources or clock frequency preclude a standard implementation in the PWM counter. The second dither mode uses a digital input to select one of the two PWM outputs on a cycle by cycle basis typically used to provide fast transient response in power converts.

The trigger and reset inputs allow the PWM to be synchronized with other internal, or external, hardware. The optional trigger input is configurable so that a rising edge starts the PWM. A rising edge on the reset input causes the PWM counter to reset its count, as if the terminal count was reached. The enable input provides hardware enable to gate PWM operation based on a hardware signal. An interrupt can be programmed to be generated under any combination of the following conditions; when the PWM reaches the terminal count or when a compare output goes high.

The clock input defines the signal to count and increments or decrements the counter on each rising or following edge of the clock. The reset input resets the counter to the period value and then normal operation continues. An enable input works in conjunction with the software enable and trigger input, if the latter is enabled.[104]

The kill input disables the PWM output(s). Several kill modes are supported all of which rely on this input to implement the final kill of the output signal(s). If deadband is implemented only the deadband outputs (ph1 and ph2) are disabled and the *pwm*, *pwm1*, and *pwm2* outputs are not disabled.[105] The *cmp_sel* input selects either pwm1 or pwm2 output as the final output to the PWM terminal. When the input is 0 (low) the PWM output is *pwm1* and when the input is 1(high) the PWM output is *pwm2* as shown in the configuration tool waveform viewer.[106]

The capture input forces the period counter value into the read FIFO. There are several modes defined for this input in the *Capture Mode* parameter.[107] When the Fixed Function PWM implementation is chosen, the capture input is always rising edge sensitive.

The trigger input enables the operation of the PWM. The functionality of this input is defined by the Trigger Mode and Run Mode parameters. After the Start API command, the PWM is enabled but the counter does not decrement until the trigger condition has occurred. The trigger condition is set with the Trigger Mode parameter.[108] The terminal count output is '1' when the period counter is equal to zero. In normal operation, this output will be '1' for a single cycle where the counter is reloaded with the period. If the PWM is stopped with the period counter equal to zero then this signal will remain high until the period counter is no longer zero. The interrupt output is the logical OR of the group of possible interrupt sources. This signal will go high while any of the enabled interrupt sources remain true (Fig. 11.57).

The pwm or pwm1 output is the first, or only, pulse width modulated output and is defined by the PWM Mode, compare modes(s), and compare value(s) as indicated in waveforms in the Configure

---

[103] *Dithering* is sometimes used as a method for reducing harmonic content and involves frequency modulation within a narrow band. In some applications dither is used when a PWM is being used to control a mechanical device, e.g., a valve or actuator, as a method of overcoming static friction by introducing some ripple into the actuating current.

[104] The enable input will not be visible in PSoC Creator if the *EnableMode* parameter is set to "Software Only." This input is not available when the *Fixed Function PWM* implementation is chosen.

[105] The kill input is not visible if the kill mode parameter in PSoC Creator is set to *Disabled*. When the *Fixed Function PWM* implementation is chosen kill will only kill the deadband outputs if deadband is enabled. It will not kill the comparator output when deadband is disabled.

[106] The *cmp_sel* input is visible when the PWM mode parameter is set to Hardware Select.

[107] The capture input is not visible if the *Capture Mode* parameter is set to *None*.

[108] The *trigger input* is not visible if the *trigger mode* parameter is set to *None*.

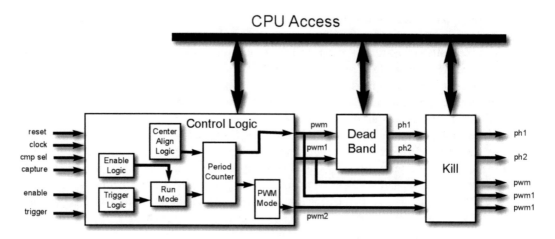

Fig. 11.57 A block diagram of PSoC 3/5LP's PWM architecture.

dialog in PSoC Creator. When the instance is configured in one output, Dual Edged, Hardware Select, Center Aligned, or Dither PWM Modes, then the output pwm is visible. Otherwise, the output pwm1 is visible with pwm2 the other pulse width signal. The pwm2 output is the second pulse width modulated output. The pwm2 output is only visible when the PWM Mode is set to Two Outputs.

The ph1 and ph2 outputs are the deadband phase outputs of the PWM. In all modes where only the PWM output is visible, these are the phased outputs of the PWM signal which is also visible. In the two output modes, these signals are the phased outputs of the pwm1 signal only.[109] The bit-width resolution of the period counter is 8–16 bits with 8-bits as the default value.

11.11 DC-DC Converters

Embedded systems frequently rely on multiple voltage and current sources for powering motors, sensors, displays, and various other types of peripherals. These power supplies typically employ a common input voltage, e.g., 12 V, for the supply and maintenance of the respective output voltages and currents that are critical for the proper operation of an embedded system. Thus, the embedded system may require only a single external power supply, and the system derives all the other voltages and currents required for its operation from this single source.

Traditionally, electronic systems have relied on so-called linear power supplies[110] for their operation, but the desire for more efficient power conversion, smaller physical size, etc., has given rise to switching power supplies. This type of power supply is capable of increasing the input voltage, lowering the input voltage and/or inverting the input voltage to provide the needed DC to DC conversion. The latter capability is important when employing devices that require either negative voltages, or plus and minus values of a given voltage, or voltages. As the name implies, switching power supplies rely on *switches* capable of handling the required power to facilitate the conversion from one DC value of voltage/current to another. In modern designs, *vertical metal oxide semiconductors* (VMOS) are often employed as such switches.

---

[109]Both of these outputs are visible if deadband is enabled in 2–4 or 2–256 modes and are not visible if deadband is disabled.

[110]Linear power supplies often employ a transformer, bridge rectifier, resistors, and a large capacitance.

Switching supplies have become popular, due in part to their ability to efficiently provide well-regulated, output voltages/currents, and in relatively small physical volumes.[111] These supplies, referred to generically as DC–DC converters,[112] rely on several basic concepts which depend, in part, on the relationship of the input voltage/current to the output voltage/current, e.g., if the output voltage is less than the input voltage, it is referred to as *step-down* converter[113] and, if it is higher, it is called a *step-up*[114] converter. A key element in DC–DC converters is an inductor which has the interesting characteristics of constraining the current slew rate through a power switch and the ability to store energy conservatively,[115] in its magnetic field, which is, in the case of the ideal inductor, expressible in terms of joules as,

$$E = \frac{1}{2}Li^2 \tag{11.217}$$

where L is the inductance in Henries and i is the current passing through the inductor.

An inductor, diode, and power switch can be configured to serve as a way in which to transfer power, or equivalently energy, from the input to the output of a DC–DC converter. Because a switching supply is inherently more efficient than a linear supply,[116] less energy is lost in the form of thermal heating, and smaller components can be used. The ability to store energy temporarily in an inductor makes it possible to output voltages that are lower or higher than the input voltage, and if necessary inverted. If an embedded design relies on multiple supply voltages for its proper operation, monitoring of these voltages can be very important to assure system reliability. Such monitoring can then be used to cause corrective action to be taken programmatically and/or alert users and/or other systems.

PSoC 3/5LP provide a number of components to facilitate the monitoring of power, voltage and/or current. PSoC 3/5LP's *voltage default detection* components allow voltages to be monitored and the existence of any voltage/current source that has a value, or values, outside a predefined range to be detected and, if required, acted upon. Voltage sequencing components can be used to control the *power-up,* and *power-down,* sequencing of power converters. Trim and margin components allow output voltages of converters to be adjusted and controlled to meet the various power supply requirements.

11.12 PSoC Creator's Power Supervision Components

Power supervision is an important component of any embedded system and PSoC Creator offers a number of components to facilitate voltage fault detection, interfacing to DC–DC power converters, monitoring of power rails, controlling DC–DC power converters, and so on.

---

[111] Switching supplies are inherently more noisy than linear supplies as a result of switching transients present in a switching supply.

[112] Or, "DC-to-DC converters".

[113] This type of converter is often referred to as a *buck* converter.

[114] Also known as a *boost* converter.

[115] Except for some typically minor I^2R losses, i.e. Ohmic heating.

[116] Linear supplies rely on resistive elements to provide the voltage drops required for voltage conversion and regulation

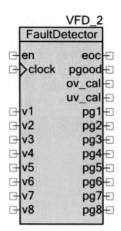

Fig. 11.58 Voltage fault component.

11.12.1 Voltage Fault Detector (VFD)

The Voltage Fault Detector component[117] provides a simple way to monitor up to 32 voltage inputs against user-defined over- and under- voltage limits without using the ADC component and without having to write any firmware. This component simply outputs a good/bad status result (*power good* or *pg[x]*) for each voltage being monitored. The component operates entirely in hardware without any intervention from PSoC's CPU core resulting in a known, fixed-fault, detection latency. The *Voltage Fault Detector* component, shown in Fig. 11.58 is capable of interfacing to up to 32 voltage inputs and is responsible for determining the status of those voltages by comparing them to a user-defined *under-voltage* (UV) threshold, an *over-voltage* (OV) threshold, or both.

Clock is used to set the time base for the component and should be set to 16× the desired multiplexing frequency. When internal OV, and UV, thresholds are generated by VDACs, the multiplexing frequency is largely determined by the VDAC update rate. When the VDACs are configured for 0–1 V range, the multiplexing frequency must not exceed 500 kHz (clock = 8 MHz) factoring in the VDAC update rate, plus DMA time, to adjust DACs and analog settling time. When the VDACs are configured for 0–4 V range, the multiplexing frequency cannot exceed 200 kHz (clock = 3.2 MHz).

When external references are selected, the user can set the timebase to a frequency that meets the system requirements. In that case, the VDAC settling time does not need to be factored in, because the VDACs are not present and the OV and/or UV thresholds will be common across the entire voltage set to be monitored. Therefore, the frequency is limited only by the analog voltage settling time and the maximum frequency of operation of the component's state machine.[118]

In either case, since DMA is involved and needs to run to completion within the time window dictated by the multiplexing frequency selected, this component inherently dictates a minimum *BUS_CLK* frequency. The component minimum *BUS_CLK:clock* ratio for this component is 2:1.

- *Enable* is a synchronous active high signal gates the clock input to the state machine controller.[119]

---

[117]This component is not supported in the PSoC 5 device.

[118]A practical limit might be 12 MHz.

[119]One purpose of this input is to support VDAC calibration.

- *Over Voltage Reference* is an analog input[120] In this case, the user provides an over voltage threshold that replaces the internal OV VDAC. This can come from a PSoC pin or through a separate instantiation of a VDAC, for example.
- *Under Voltage Reference* is an analog input. The user must provide an under-voltage threshold that replaces the internal UV VDAC, e.g., from a PSoC pin, or through a separate instantiation of a VDAC.
- *Voltages* are analog inputs that are the voltages that are to be monitored.
- *Power Good* is a global output that is an active high signal indicating all voltages are within range *The(Individual:Active* high signal indicating v[x] is within range.)
- *End of Cycle* is an output pulse that is active high after every voltage input has been compared to its reference threshold(s) and indicates the end of one complete comparison cycle. For example, this signal could be used to capture the reference voltage VDACs for calibration purposes.

The *Trim and Margin* component provide a simple way to adjust and control the output voltage of up to 24 DC-DC converters to meet system power supply requirements. Users of this component simply enter the power converter's nominal output voltages, voltage trimming range, margin high and margin low settings into PSoC Creator's intuitive, easy-to-use graphical configuration GUI and the component takes care of the rest. The component will also assist the user to select appropriate external passive component values based on performance requirements.

11.12.2 Trim and Margin

The component's APIs make it possible to manually trim the power converter output voltages to any desired level, within the operational limits of the power converter. Real-time active trimming, or margining, is supported via a continuous background task with an update frequency controlled by the application (Fig. 11.59).

The component supports:

- most adjustable DC–DC converters or regulators including LDOs, switchers,and modules,
- up to 24 DC–DC converters,
- 8–10-bit resolution PWM pseudo-DAC outputs,
- real-time, closed-loop active trimming when used in conjunction with the Power Monitor, component

and

- built-in margining.

The Trim and Margin component should be used in any application that requires PSoC to adjust, and control, the output voltage of multiple DC–DC power converters.

11.12.3 Trim and Margin I/O Connections

The trim and margin component has the following I/O connections:

---

[120]This is accessible only when the *ExternalRef* parameter is true.

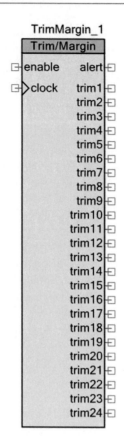

Fig. 11.59 Trim margin component.

- *clock* is used to drive the PWM pseudo-DAC outputs.
- *enable* enables the PWMs and is used as a clock enable to the PWMs.
- *alert* is asserted when closed-loop trimming/margining is not achievable because PWM is at min, or max, duty cycle, but the desired power converter output voltage has not been achieved. It remains asserted as long as the alert condition exists on any output.
- *trim[1..24]* terminals are the PWM outputs that pass through an external RC filter to produce an analog control voltage that adjusts the output voltage of the associated power converter. The number of terminals is determined by the *Number of Voltages* parameter.

11.13 Voltage Sequencer

PSoC Creator's *Voltage Sequencer* component provides a simple way to impose power-up and power-down, sequencing of up to 32 power converters, inclusive, to meet the user-defined system requirements (Fig. 11.60).

Once the sequencing requirements have been entered into PSoC Creator's easy-to-use graphical configuration GUI. The *Voltage Sequencer* will automatically take care of the sequencing implementation. This component should be used in applications requiring sequencing of multiple DC-DC power converters. The component can be directly connected to the *enable* (en) and *power good* (pg) pins of

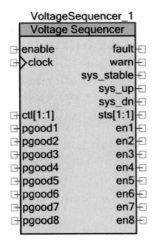

Fig. 11.60 Voltage sequencer component.

the DC-DC power converter circuits, for *sequencing-only* applications. The *Voltage Sequencer* can be connected to PSoC Creator's *Power Monitor* or *Voltage Fault Detector* components. The *Power Monitor* and *Voltage Fault Detector* components are available in the *Power Supervision* category of PSoC Creator's component catalog.

Supported features include:

- Autonomous (standalone) or host driven operation,
- Power converter circuits with logic-level enable inputs and logic-level power good (*pgood*) status outputs,
- Sequencing, and monitoring, of up to 32 power converter rails,

and,

- Sequence order, timing, and inter-rail dependencies can be configured via PSoC Creator.

11.13.1 Voltage Sequencer I/O

The following describes the various I/O connections of PSoC Creator's *Voltage Sequencer* component.[121]

- *Enable* is a global enable pin that can be used to initiate a power-up, or a power down sequence.
- *Clock* is an input timing source used by the component.
- *System Stable* is an active-high signal that is asserted when all power converters have powered up successfully, i.e., all sequencer state machines are in the ON state, and have been running normally for a user-defined amount of time.
- *System Up* is an active high signal is asserted when all power converters have powered up successfully (all sequencer state machines are in the ON state).

---

[121] An asterisk (*) in the list of I/Os indicates that the I/O may be hidden on the symbol under the conditions listed in the description of that I/O.

- *System Down* is an active high signal that is asserted when all power converters have powered down successfully, i.e., all sequencer state machines are in the OFF state.
- *Warning*[122] is an active high signal that is asserted when one, or more, power converters did not shut down within the user-specified time period.
- *Fault*[123] is an active high signal that is asserted when a fault condition has been detected on one, or more, power converters. This terminal should not be connected to an interrupt component because this component has a buried interrupt service routine that needs to respond to faults, as soon as possible.
- *Sequencer Controls* are general-purpose inputs with a user-defined polarity that may be used to gate power-up sequencing state changes, force partial or complete power-down sequencing, or both.[124]
- *Sequencer Status* outputs are general-purpose outputs with a user-defined polarity that can be asserted, and de-asserted, at any point throughout the sequencing process to indicate the sequencer's progress.[125]
- *Power Converter Enables*Output Power converter enable outputs. When asserted, these outputs enable the selected power converter so that it will begin regulating power to its output.
- *Power Converter Power Goods* are power converter, *power-good* status inputs. These signals may come directly from the power converter status output pins, or be derived from within PSoC by ADC monitoring of power converter voltage outputs, e.g., by using the *Power Monitor* component, or over-voltage/under-voltage window comparator threshold detection, using the *Voltage Fault Detector* component.

11.14   Power Monitor Component

Features:

- Interfaces to up to 32 DC–DC power converters.
- Measures power converter output voltages and load currents using a DelSig-ADC.
- Monitors the health of the power converters generating warnings and faults based on user-defined thresholds.
- Support for measuring other auxiliary voltages in the system (Fig. 11.61).

11.14.1   Power Converter Voltage Measurements

For power converter voltage measurements, the ADC can be configured into single-ended mode (0–4.096 V range). The ADC can also be configurable into differential mode (2.048 V range) to support remote sensing of voltages where the remote ground reference is returned to PSoC over a PCB trace. In cases where the analog voltage to be monitored equals or exceeds, Vdda or the ADC range, the use

---

[122]This terminal is visible when the checkbox labeled Disable TOFF_MAX warnings on the Power Down tab of the Configure dialog is deselected.

[123]The use of this terminal should be restricted to driving other logic, or pins.

[124]These terminals are visible when a non-zero value is entered into the number of control inputs parameter on the General tab of the Configure dialog.

[125]These terminals are visible when a non-zero value is entered into the Number of status outputs parameter on the General tab of the Configure dialog.

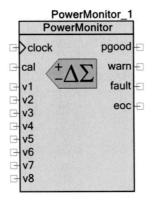

Fig. 11.61 Power monitor component.

of external resistor dividers is recommended to scale down the monitored voltages to an appropriate range.

11.14.2 Power Converter Current Measurements

For power converter load current measurements, the ADC can be configured into differential mode (± 64 mV or ± 128 mV range) to support voltage measurement across a high-side series shunt resistor on the outputs of the power converters. Firmware APIs convert the measured differential voltage into the equivalent current based on the external resistor component value used (Fig. 11.62).

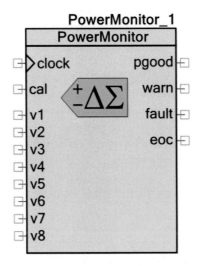

Fig. 11.62 Power monitor component.

The ADC can also be configured into single-ended mode (0–4.096 V range) to support connection to external current sense amplifiers (CSAs) that convert the differential voltage drop across the shunt resistor into a single-ended voltage or to support power converters or hot-swap controllers that integrate similar functionality.

11.14.3 Auxiliary Voltage Measurements

Up to 4 auxiliary input voltages can be connected to the ADC to measure other system inputs. The ADC can be configured into a single-ended (0–4.096 V) or a differential mode ($\pm$2.048 V or $\pm$64 mV ranges) to measure the auxiliary input voltages.

11.14.4 ADC Sequential Scanning

The ADC will sequence through all power converters and auxiliary inputs, if enabled, in a round-robin fashion, taking voltage and load current measurements. This component will measure the voltages of all the power converters in the system, but can be configured to measure currents from a subset of the power converters, including no current measurements at all. Doing so will minimize the number of IO's required, and will minimize the overall ADC scan time. This component needs some knowledge of components external to PSoC 3/5LP for 2 reasons: (1) scaling factors for input voltages that have been attenuated to meet IO input range limits or ADC dynamic range limits where applicable, and (2) scaling factors for current measurements, e.g., series resistor, series inductor or CSA gain, etc.

11.14.5 I/O Connections

PSoC Creator's *Power Monitor* component has the following I/O connections:

- *clock* is an input signal is used to drive all digital output signals. The maximum frequency used for this clock is 67 MHz
- *cal* is an analog input that is the calibration voltage used to calibrate the 64 mV or 128 mV differential voltage ADC range settings.[126] When the *cal* pin is employed, a POR calibration occurs automatically as part of the *PowerMonitor_Start()* API to calibrate 64 mV or 128 mV differential voltage ADC range. For subsequent calibrations to occur at run time, PowerMonitor_Calibrate() should be used.[127]
- *v[x]* are analog inputs that connect to the power converter output voltage as seen by their loads. These inputs can be either direct connections to the power converter outputs, or scaled versions using external scaling resistors. Every power converter will have the voltage measurement enabled. The component supports a maximum of 32 voltage input terminal pins and the unused terminals are hidden.
- *i[x]* are analog inputs that enable this component to measure power converter load currents. This could be a differential voltage measurement across a shunt resistor along with the corresponding v[x] input or could be a single-ended connection to an external CSA. Current monitoring is optional on a power converter by power converter basis. When the differential *v[x]* voltage measurement is selected for a power converter in the component customizer, current measurement is disabled

---

[126]This signal is an optional input connection.

[127]The maximum input voltage applied to this pin should not exceed 100% of the differential ADC range used, i.e., either the 64 or 128 mV range.

for that power converter, in order to limit the number of IOs. In that case, the i[x] terminal is replaced by the *rtn[x]* terminal representing the differential voltage measurement return path. This component supports a maximum of 24 current input terminals and the unused terminals are hidden. These terminals are mutually exclusive with respect to the associated rtn[x] input terminals.

- *rtn[x]* are analog inputs connected to a ground reference point that is physically close to the power converter.[128]
- *aux[x]* Since this component embeds the only available *DelSig ADC* converter, the aux[x] analog inputs enable users to connect other auxiliary voltage inputs for measurement by the ADC. Up to 4 auxiliary input terminals are available and these terminals will be hidden if the user does not enable auxiliary input voltage monitoring in the component customizer.
- *aux_rtn[x]* are analog inputs that can be connected to the auxiliary input voltage ground reference point. Up to 4 aux_rtn[x] terminals are available.
- *eoc* is a digital output signal that is an active high pulse indicating that the ADC conversion has been completed, for the current sample set. This terminal is pulsed high for one clock cycle when an ADC measurement has been taken from every analog input, i.e., voltages, currents, or auxiliary. It can also be used to generate an application-specific interrupt for the MCU core, or to drive other hardware, e.g., to connect it to a pin to measure the ADC update rate for all the inputs. or used to run custom firmware filtering algorithms, once all the samples are gathered.
- *pgood* is an output terminal that is driven as an active high, when all power converter voltages and currents, if measured, are within a user-specified operating range. Individual power converters can be masked from participating in the generation of the *pgood* output. An option exists in the customizer to make this terminal a bus to expose the individual pgood status outputs for each converter.
- *warn* is an output terminal that is driven active high when one or more power converter voltages or currents,if measured, are outside the user-specified nominal range, but not by enough to be considered a fault condition.
- *fault* is an output terminal that is driven active high when one, or more, power converter voltages, or, if measured, currents that are outside the user-specified nominal range to such a degree that it is to be considered a fault condition.

11.15 PSoC Creator's Fan Controller Component

System cooling is a critical component of any high-performance electronic system. As circuit miniaturization continues, increasing demands are placed on system designers to improve the efficiency of their thermal management designs. Several factors conspire to make this a difficult task. Fan manufacturers specify duty-cycle to RPM relationships in their datasheets with tolerances as high as $\pm 20\%$. To guarantee a fan will run at the desired speed, system designers would therefore, need to run the fans at speeds 20% higher than nominal to ensure that any fan from that manufacturer will provide sufficient cooling. This results in excessive acoustic noise and higher power consumption.

Real-time, closed-loop control of fan speeds is possible using any standard microcontroller-based device running custom firmware algorithms, but this approach requires frequent CPU interrupts and constantly consumes processing power. An open-loop control mode where the firmware is responsible for controlling fan speeds and the associated application programming interfaces (APIs) are employed make the task quick and easy. The *FanController* can also be configured in a closed-loop control mode

---

[128]These terminals are mutually exclusive with the associated *i[x]* input terminals.

where the programmable logic resources in PSoC take care of fan control autonomously, freeing the CPU completely to do other important system management tasks [17].

Fig. 11.63 A typical four-wire fan.

A typical 4-wire, brushless, DC fan is shown in Fig. 11.63. Two of the four wires are used to supply power to the fan, the other two wires are used for speed control and monitoring. Fans come in standard sizes, e.g., 40, 80, and 120 mm. The most important specification when selecting a fan for a cooling application is how much air the fan can move. This is specified either as cubic feet per minute (ft^3/min), or cubic meters per minute (m^3/min). The size, shape, and pitch of the fan blades all contribute to the fan's capability to move air. Obviously, smaller fans will need to run at a higher speed than larger fans to move the same volume of air in a given time frame. Applications that are space-constrained and need smaller fans due to physical dimension limitations generate significantly more acoustical noise.

This is an unavoidable tradeoff that needs to be made to meet system-level needs. To manage acoustic noise generation, the *FanController* component can be configured to drive the fans at the minimum possible speed to maintain safe operating temperature limits. This also extends the operating life of the fan compared to systems that run all fans at full speed all the time. The FanController component interacts with fans by driving via PWM speed control signals and monitoring fan actual speeds by measuring the tachometer (TACH) pulse trains. The component can be configured to automatically regulate fan speeds and detect fan failures based on the TACH inputs.

With 4-wire fans, speed control of 4-wire fans is often accomplished by using a pulse-width modulator (PWM).[129] Fan manufacturers specify the PWM duty cycle versus nominal fan speed by providing either a table or a graph, of the type shown in Fig. 11.64. The *FanController* component provides a graphical user interface where designers can enter this information which then automatically configures and optimizes the firmware and hardware inside PSoC to control fans with these parameters. It is important to note that at low duty cycles, not all fans behave the same way. Some fans stop rotating as the duty cycle approaches 0%, while others rotate at a nominal specified minimum RPM. In both cases, the duty cycle to RPM relationship can be non-linear, or unspecified. When

---

[129]Increasing the duty cycle of the PWM control signal will increase fan speed.

entering duty cycle to RPM information into the *FanController* component customizer interface, select two data points from the linear region where the behavior of the fan is well-defined.

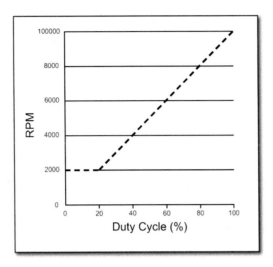

Fig. 11.64 Fan RPM versus duty cycle.

3-wire and *4-wire* DC fans include hall-effect sensors that sense the rotating magnetic fields generated by the rotor as it spins. The output of the hall-effect sensor is a pulse-train that has a period inversely proportional to the rotational speed of the fan. The number of pulses that are produced per revolution depends on how many poles are used in the electromechanical construction of the fan. For the most common 4-pole brushless DC fan, the tachometer output from the hall-effect sensor will generate two high and low pulses per fan revolution.

If the fan stops rotating due to mechanical failure or other fault, the tachometer output signal will remain static at either a logic low or logic high level. The *FanController* component measures the period of the tachometer pulse train for all fans in the system using a custom hardware implementation. The firmware APIs provided convert the measured tachometer periods into RPM to enable the development of fan control algorithms that are firmware based. That same hardware block can generate alerts when it detects that a fan has stopped rotating, referred to as a stall event.

At the cabling level, wire color coding is not consistent across manufacturers, but the connector pin assignment is standardized. Figure 11.65 shows the connector pin-outs, when viewed looking into the connector with the cable behind it.[130] The keying scheme that was chosen also enables 4-wire fans to connect to control boards that were designed to support 3-wire fans, i.e., without PWM speed control signal, without modification.

The *FanController* component interacts with fans by driving the PWM speed control signals and monitoring fan actual speeds by measuring the tachometer *(TACH)* pulse trains. The component can be configured to automatically regulate fan speeds and detect fan failures based on the *TACH* inputs.

PSoC Creator's *FanController* component supports the following features:

- 4-pole and 6-pole motors,
- 25, 50 kHz or user-specified PWM frequencies,

---

[130]The connectors are keyed to prevent incorrect insertion into the fan controller board.

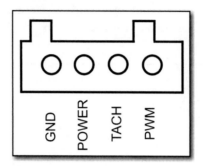

Fig. 11.65 4-wire fan connector pin-outs.

- customizable alert pin for fan fault reporting,
- fan speeds up to 25,000 RPM
- fan stall/rotor lock detection on all fans,
- firmware controlled or hardware controlled fan speed regulation individual or banked PWM outputs with tachometer inputs,

and,

- up to 16 PWM controlled, 4-wire, brushless, DC fans.

The *Fan Controller* component enables rapid development of fan controller solutions using PSoC 3/5LP. It is a system-level solution that encapsulates all the necessary hardware blocks including PWMs, tachometer input capture timer, control registers, status registers, and a DMA controller reducing development time and effort.

PSoC Creator's graphical user interface allows the fan's electromechanical parameters such as duty cycle-to-RPM mapping and physical fan bank organization to be taken into account. Performance parameters such as PWM frequency, resolution, and open/closed-loop control methodology can be configured through the same user interface. Once the system parameters are entered, the component delivers the most optimal implementation, saving resources within PSoC to enable the integration of other thermal management and system management functionality.[131]

The *Fan Controller* component can be used in thermal management applications to drive and monitor 4-wire, PWM-based, DC, cooling fans. If the application requires more than 16 fans, multiple instances of the *Fan Controller* component can be employed. Similarly, if fans are organized in banks, one *Fan Controller* component per bank can be employed or a single component that controls handles all the banks.

PSoC Creator supported fan parameters include:

- *Motor Support* specifies the number of high-low pulses that appear on the fan's tachometer output per revolution.
- *Number of banks* specifies the number of fan banks.
- *PWM Resolution* specifies the duty cycle resolution for the modulated PWM signal that drives the fans to control rotational speed.

---

[131]Designs using PSoC Creator's Fan Controller component should not use PSoC3's low power sleep or hibernate modes. Entering those modes prevents the Fan Controller component from controlling and monitoring the fans.

- *PWM Frequency*[132] specifies the frequency of the modulated PWM signal that drives the fans.
- *Duty A(%), RPM A* specifies one data point on the duty cycle-to-RPM transfer function for the selected fan or bank of fans. The RPM A parameter specifies the speed the fan nominally runs at when driven by a PWM with a duty cycle of Duty A.[133]
- *Duty B (%), RPM B* specifies a 2nd data point on the duty cycle-to-RPM transfer function for the selected fan or bank of fans. The *RPM B* parameter specifies the speed the fan nominally runs at when driven by a PWM with a duty cycle of *Duty B (%)*.[134]
- *Initial RPM* specifies the initial RPM of an individual fan. The value of *Initial RPM* is converted into a duty cycle and set as the initial duty cycle for an individual fan.[135]

11.15.1 *FanController* API Functions

- *clock* is an input for a user-defined clock source for the fan control PWMs. It is present only when the External Clock option is selected in the component customizer.
- *tach1..16* are tachometer signal inputs from each fan that enables the Fan Controller to measure fan rotational speeds. The component is designed to work with 4-pole DC fans that produce 2 high-low pulse trains per rotation on their tachometer output or 6-pole DC fans that produce 3 high-low pulse trains. tach2..16 inputs are optional.
- *fan1..16*[136] are PWM outputs with variable duty cycles to control the speed of the fans. These output terminals are replaced by the bank1..8 outputs if fan banking is enabled.[137]
- *bank1..8* are PWM outputs with variable duty cycles to control the speed of the fan banks. These outputs appear only when banking is enabled.
- *alert* is an active high output terminal asserted when fan faults are detected (if enabled).
- *eoc* is the End-of-Cycle output and pulsed high each time the tachometer block has measured the speed of all fans in the system. This can be used to synchronize firmware algorithms to the Fan Controller hardware by connecting the terminal to a *Status Register* component, or to an Interrupt component.

PSoC Creator's *FanController* component supports a number of unique features that are important in applications involving one of more fans, e.g., the *Tolerance* parameter determines the acceptable variance in a fan's speed with respect to the desired speed [17]. This parameter enables fine-tuning of the hardware control logic to match the selected fan's electromechanical characteristics.[138] The *Acoustic Noise Reduction* parameter restricts allowable audio fan noise by constraining the positive rate of change in speed, i.e., the fan's rate of acceleration. This is accomplished by allowing the

---

[132]The default setting is 25 kHz.

[133]The *Fan Controller* component can drive PWM duty cycles down to 0%, even if Duty A (%) is set to a non-zero value. The valid range for the Duty A (%) parameter is 0–99. The default setting is 25. A valid range for the RPM A parameter is 500–24,999, inclusive. The default setting is 1000.

[134]The Fan Controller component is capable of driving PWM duty cycles up to 100% even if Duty B (%) is set below 100%. The valid range for the Duty B (%) parameter is 1–100 and the default setting is 100. A valid range for the RPM B parameter is $501 \cdots 25,000$, inclusive and the default setting is 10,000.

[135]*Initial RPM* be set lower than the *RPM A* parameter.

[136]$fan2 \cdots 16$ are optional.

[137]The maximum number of fan outputs, and associated tach inputs, is limited to 12 to minimize digital resource utilization, for hardware UDB mode.

[138]The valid range for this parameter is $1 \cdots 10\%$. Default setting is 1%. If 8-bit PWM resolution is selected in the FanController Fans tab, a Tolerance parameter setting of 5% is recommended.

PWM's duty cycle to increase gradually. *Alerts* are triggered by fan stall or rotor lock of the fan and can be used to illuminate an LED to visually signal these conditions (Tables 11.12 and 11.13).

The following is a code fragment illustrating the basic code required to support a fan and two switches that vary the fan's duty cycle:

```
/* Duty cycles expressed in percent */
#define MIN_DUTY 0
#define MAX_DUTY 100
#define INIT_DUTY 50
#define DUTY_STEP 5
void main() { uint16 dutyCycle = INIT_DUTY;
/* Initialize the Fan Controller */
FanController_Start();
/* API uses Duty Cycles Expressed in Hundredths of a Percent */
FanController_SetDutyCycle(1, dutyCycle*100);

/* Check for Button Press to Change Duty Cycle */
if((!SW1_Read()) || (!SW2_Read()))
{
/* Increase Duty Cycle */
if(!SW1_Read())
{
if(dutyCycle > MIN_DUTY)
dutyCycle -= DUTY_STEP; }
/* SW2 = Increase Duty Cycle */
else {
if((dutyCycle += DUTY_STEP) > MAX_DUTY)
dutyCycle = MAX_DUTY;
}
/* Adjust Duty Cycle of the Fan Bank */
FanController_SetDutyCycle(1, dutyCycle*100);

/* Switch Debounce */
CyDelay(250);
}
}
```

11.16 Recommended Exercises

11-1 Design a Sallen–Key bandpass filter that will pass signals above 70 Hz and below 1500 Hz. Sketch the Bode plot for this filter and graph the filters phase-shift characteristics over the same domain.

11-2 A certain sensor has an output impedance of $Z_{out} = 1000 + j250\Omega$ and measures pressure. Assume that the sensor provides a current that is directly proportional to the pressure sensed that is expressed mathematically as:

Table 11.12 Table of *Fan Controller* functions. (Part 1).

Function	Description
FanController_Start()	Start the component
FanController_Sopt()	Stop the component and disable hardware blocks
FanController_Init()	Initializes the component
FanController_Enable()	Enables hardware blocks inside the component
FanController_EnableAlert()	Enables alerts from the component
FanController_DisableAlert()	Disables alerts from the component
FanController_SetAlertMode()	Configures alert sources
FanController_GetAlertMode()	Returns currently enabled alert sources
FanController_SetAlertMask()	Enables masking of alerts from each fan

Table 11.13 Table of *Fan Controller* functions. (Part 2).

Function	Description
FanController_GetAlertMask()	Returns alert masking status of each fan
FanController_GetAlertSource()	Returns pending alert source(s)
FanController_GetFanStallStatus()	Returns a bit mask representing the stall status of each fan
FanController_GetFanSpeedStatus()	Returns a bit mask representing the speed regulation status of each fan in hardware control mode
FanController_SetDutyCycle()	Sets the PWM duty cycle for the specified fan or fan bank
FanController_GetDutyCycle()	Returns the PWM duty cycle for the specified fan or fan bank
FanController_SetDesiredSpeed()	Sets the desired fans speed for the specified fan in hardware control mode
FanController_GetDesiredSpeed()	Returns the desired fans speed for the specified fan in hardware control mode
FanController_GetActualSpeed()	Returns the actual speed for the specified fan
FanController_OverrideHardwareControl()	Enables firmware to override hardware (UDB) fan control

$$P = K(P_{system} - P_{ambient})\mu amps \qquad (11.218)$$

Design a system that will interface with the sensor and produce a voltage that varies linearly over the range of 0–5 V DC and is proportional to the net measured system pressure.

11-3 Design an analog system that solves the following Lorenz equation system utilizing OpAmps and four-quadrant multipliers such as Analog Devices AD633ANZ.

$$\frac{dx}{dt} = \sigma(y - x)$$

$$\frac{dy}{dt} = x(\rho - z) - y \hspace{4cm} (11.219)$$

$$\frac{dz}{dt} = xy - \beta z$$

11-4 Describe the potential impact of using non-ideal OpAmps in a hardware system whose design is based on ideal OpAmps. What are the principle limitations of popular OpAmps?

11-5 Design a precision voltage supply that employs a 5 V Zener diode as a reference voltage and is capable of supplying up to 100 mA at 5 Vdc.

11-6 Using and PSoC Creator OpAmps design a function generator capable of producing a sine, square and ramp functions whose output can be varied from 100–1000 Hz. The output impedance must be 50 Ω.

11-7 Design a digital voltmeter using PSoC Creator that can measure voltages from 0–10 V and has an input impedance of 10 MΩ.

11-8 Using a switched capacitor approach for any resistors needed, an OpAmp and PSoC Creator design a "R-2R" digital-to-analog converter based on the resistor network shown in Fig. 11.66. Show how you would extend your design to create an A/D converter.

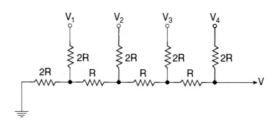

Fig. 11.66 R-R network.

11-9 Using the governing equations which are shown below for a PID (Proportional, Integral, and Derivative) controller), show how you would implement them in PSoC Creator.
The governing equations are:

$$u(t) = K_p \left(e(t) + K_i \int_0^t e(t') dt' + K_d \frac{de(t)}{dt} \right)$$

or, equivalently:

$$u(t) = K_p \left(e(t) + \frac{1}{T_i} \int_0^t e(t') dt' + T_d \frac{de(t)}{dt} \right)$$

where K_p, K_i, and K_d coefficients of the proportional, integral, and derivative terms respectively (aka denoted P, I, and D, respectively) and T_i and T_d are the integral time constant and the derivative time constant, respectively.

Compute the error e(t) that is the difference between the setpoint $SP = r(t)$ and the process variable, $PV = y(t)$. The error correction comprised of an expression that is a function of proportional, integral, and derivative terms and should be calculated in your design and displayable on a four digit display.

11-10 A given OpAmp is assumed to be ideal and has a DC gain of 120 dB, an AC gain of 40 dB at 100 kHz, and a slew rate of 1.79 V/µs. The rail-to-rail voltages are ±12 V DC. A 100 kΩ resistor is connected between the OpAmp's output and the negative input terminal. The negative input terminal is connected to ground by a 4.7 kΩ resistor. The input signal is connected to a 22 kΩ resistor which is connected to the positive input terminal. What are the input resistance, output resistance and gain of this circuit?

Lab measurements show that the OpAmp has an offset voltage of 0.89 mV and an input bias current of 9.4 nA. If the positive input is grounded, what is the output voltage? What is the unity-gain bandwidth of this amplifier?

11-11 Show how to design a translinear analog multiplier[139] by using OpAmps, diodes and resistors and explain the limitations of such a design. Compare your design with the results produced by a Gilbert Cell multiplier (mixer) [6,7].

11-12 Repeat Exercise 11-11 using only OpAmps and resistors [27].

11-13 Design a simple system consisting of an ADC and PWM to illuminate an LED. The illumination of the LED, in terms of brightness, is to be directly proportional to the analog input voltage. Employ a second PWM to flash the LED on and off at the rate of one flash per second, if the input voltage is negative.

11-14 Show the C language code required in main.c in order to implement your design for Exercise 6-1 in PSoC Creator. Explain what modifications to your code would be required in order to allow a user to change the flash rate to 0.5, 0.75, 1.1.25, 1.5, 1.75, and 2 s for a negative input voltage.

11-15 Design a Sallen–Key bandpass filter that will filter out all frequencies greater than 1 MHz and all below 1 kHz, assuming that the resistor values used must fall between 1 and 10 kΩ. State the capacitive values required and any assumptions that you are making about the characteristics of the operational amplifier.

11-16 Employing components provided by PSoC Creator, explain how to use a PWM to create an arbitrarily variable duty-cycle, analog voltage that can be applied to an external device.

11-17 A photomultiplier is employed to measure light intensity at very low levels. Its output current is directly proportional to the intensity of the light that is incident upon the photomultiplier. Describe a circuit, using PSoC Creator components, that would accept the photomultiplier's output current and display the measured intensity on an LCD. Assume that the photomultiplier's output current is a logarithmic function of the incident light level.

---

[139]Barrie Gilbert introduced the concept of translinear devices in a paper he gave in 1996.

11-18 A mirror is attached to the rotating shaft of a DC motor. This device is to be used to "freeze" the motion of a rotating device, e.g., a fan. Describe the drive circuit for the motor assuming that PWM, and tach controls are employed to allow the user to vary the rotational speed of the mirror and maintain that speed via a tach-based, feedback loop.

11-19 An oscilloscope probe is needed that has an extremely high input impedance. Describe a circuit using OpAmps that could be used for this purpose. This "probe" must not pass signals with components higher in frequency than 100 Hz.

11-20 Explain the advantages and disadvantages of digital filters versus analog filters and the limitations of each from a manufacturing, cost and performance perspective.

11-21 Design a fan controller system that utilizes four thermocouples, PWMs and tachometers together with a table that specifies how to respond to different, ambient, temperature conditions within an enclosure.

11-22 Select a digital filter type from this chapter and design a bandpass filter with the same bandpass characteristics as that of the analog filter in Example 6.2.

11-23 Calculate the maximum output current available from a 1-watt DC-DC converter, assuming that the input is 12 vdc and the output is 5 vdc. Repeat the calculation assuming that the input voltage varies from 7.5–24 vdc, and plot the output current as a function of the input voltage. Assuming that the converter has an analog input that allows the input to be adjusted from 4.5–5.5 vdc, extend your design to use PSoC Creators power management components to sense and correct any deviations from 5 vdc of the output. Add LEDs to your design to illuminate when the output falls with the range of 4.9–5.1 V.

11-24 Design an analog filter to remove 60 Hz signals from the input to an embedded system using capacitance values in the range from 1 to 10 μfd. If necessary, you may use combinations of such capacitors, in series or parallel, to obtain optimum performance of your filter design.

11-25 Using a PSoC Creator OpAmp component and a lookup table of stored sine values, design a sine wave generator capable of producing a sine wave output of amplitude ± 1 V (RMS) with frequency 100 Hz. The lookup table is to contain 256 values. Add external potentiometers to your design that will allow the amplitude and/or frequency to be varied over a range of $\pm 10\%$.

11-26 A temperature sensor, located 300 ft from the embedded system, amplitude modulates a 1000 Hz sine wave and transmits the signal to the embedded signal every 10 s. The signal lines are susceptible to 60 Hz noise which must be removed from the input signal prior to processing by the embedded system. Using PSoC Creator, design a system that (a) removes any unwanted noise from the input signal, (b) demodulates the signal to obtain the amplitude at an instance in time, (c) converts that value to a digital value, and d) displays the result on an LCD screen driven by the embedded system. Assume that the maximum amplitude of the input signal is 10 V(RMS) corresponding to 100 °C, 0 ° C corresponds to 0.1 V(RMS), and that the displayed temperature must be accurate to within a tenth of a degree C.

Find the impulse-response for each. Do these equations represent stable and/or causal systems?

References

1. M.D. Anu, A. Mohan, PSoC 3 and PSoC 5LP-ADC Data Buffering Using DMA. AN61102. Document No. 001-61102Rev. *L1. Cypress Semiconductor (2018)
2. H.S. Black, Stabilized feedback amplifiers. Bell Syst. Tech. J. **13**(1), 1 (1934)
3. H.S. Black, Inventing the negative feedback amplifier. IEEE Spectr. (1977)
4. G.E. Carlson, *Signal and Linear System Analysis (with MATLAB)*, 2nd edn. (Wiley, London, 1998)
5. A.N. Doboli, E.H. Currie, *Introduction to Mix-Signal, Embedded Design* (Springer, Berlin, 2010)
6. B. Gilbert, A high-performance monolithic multiplier using active feedback. IEEE J. Solid State Circuits **9**(6), 364–373 (1974)
7. B. Gilbert, Translinear circuits: an historical overview. Analog Integr. Circ. Sig. Process **9**, 95–118 (1996). https://doi.org/10.1007/BF00166408
8. *Handbook of Operational Amplifier Applications* (Burr-Brown Research Corporation, Tucson, 1963)
9. M. Hastings, PSoC 3 and PSoC 5LP–Pin Selection for Analog Designs. AN58304. Document No. 001-58304Rev. *H. Cypress Semiconductor (2017)
10. E. Hewitt, R.E. Hewi, *The Gibbs–Wilbraham Phenomenon: An Episode in Fourier Analysis. Archive for History of Exact Sciences*, vol. 21 (Springer, Berlin, 1979)
11. R.G. Irvine, *Operational Amplifier. Characteristics and Applications*, 3rd edn. (Prentice Hall, Englewood, 1994)
12. D.S. Jayalalitha, D. Susan, Grounded simulated inductor—a review. Middle-East J. Sci. Res. **15**(2), 278–286 (2013)
13. J.B. Johnson, Thermal agitation of electricity in conductors. Phys. Rev. **32**, 97 (1928)
14. W. Jung, *OpAmp Applications Handbook* (Newnes, 1994)
15. E.W. Kamen, B.S. Heck, *Fundamentals of Signals and Systems (Using the Web and MATLAB)*, 3rd edn. (Pearson Prentice Hall, Englewood, 2007)
16. A. Kay, *Operational Amplifier Noise: Techniques and Tips for Analyzing and Reducing Noise*, 1st edn. (Newnes, 2012)
17. J. Konstas, PSoC 3 and LP Intelligent Fan Controller. Cpress Semiconductor (2011)
18. D. Lancaster, *Active Filter Cookbook*, 2nd edn. (Newnes, 1996)
19. D. Meador, *Analog Signal Processing with Laplace transforms and Active Filter Design* (Delmar, 2002)
20. J.H. Miller, Dependence on the input impedance of a three-electrode vacuum tube upon the load in the plate circuit. Sci. Pap. Bur. Stand. **15**(351), 367–385 (1920)
21. M.S. Nidhin, Accurate Measurement Using PSoC 3 and PSoC 5LP Delta-Sigma ADCs. AN84783A. Document No. 001-84783Rev. *D1. Cypress Semiconductor (2017)
22. H. Nyquist , Certain topics in telegraph transmission theory. Trans. AIEE **47**(2), 617–644 (1928)
23. S.J. Orfanidis, *Introduction to Signal Processing* (Prentice Hall, Englewood, 2010)
24. A.D. Poularikas, S. Seely, *Elements of Signals and Systems* (Krieger Publishing, Malabar, 1994)
25. PSoC 3, PSoC 5 Architecture TRM (Technical Reference Manual). Document No. 001-50234 Rev D, Cypress Semiconductor (2009)
26. R.W. Ramirez, *The FFT Fundamentals and Concepts* (Prentice Hall, Englewood, 1985)
27. V. Riewruja, A. Rerkratn, Analog multiplier using operational amplifiers. Indian J. Pure Appl. Phys. **48**, 67–70 (2010)
28. R.P. Sallen, A practical method of designing RC filters. IRE Trans. Circuit Theory **2**,75–84 (1955)
29. O.H. Schmitt, A thermionic trigger. J. Sci. Instrum. **XV**, 24–26 (1938)
30. D. Sweet, Peak Detection with PSoC 3 and PSoC 5LP. AN60321. Document No. 001-60321Rev.*I. Cypress Semiconductor (2017)
31. B.D.H. Tellegen, The gyrator. A new electric circuit element. Philips Res. Rep. **3**, 81–101 (1948)
32. D. Van Ess, Application Note: Comparator with Independently Programmable Hysteresis Thresholds (AN2310). Cypress Semiconductor (2005), pp. 1–3, pp 1–2
33. H. Wilbraham, Camb. Dublin Math. J. **3**, 198 (1848)

Digital Signal Processing

12

Abstract

Prior to the advent of the microprocessor, much of what could be considered as signal processing relied heavily on analog techniques as opposed to digital signal processing. Analog components are subject to temperature dependencies, drift, aging, ambient humidity dependencies/variations induced variances, electromagnetic fields, and a host of other perturbations, not to mention, variances in component values supplied by manufacturers, etc. However, in today's world, in spite of the availability of a wide variety of digital components (Van Eßet al., Laboratory manual for introduction to mixed-signal embedded design. Cypress University Alliance. Cypress Semiconductor, 2008) that are relatively unaffected by such considerations, there is still a need for the processing of analog signals utilizing analog components and devices. For example, analog filtering can be done very effectively while often avoiding processing time and word-length issues. Digital filtering can provide much better filter characteristics in some cases while minimizing phase shift. This chapter reviewed some of the more classical, analog approaches to signal processing. The PSoC family of devices allows the designer to create arbitrarily complex mixed-signal systems (Ashby, My First Five PSoC 3 Designs. Spec. £001-58878 Rev. *C. Cypress Semiconductor, 2013; Narayanasamy, Designing an efficient PLC using a PSoC, 2011).

12.1 Digital Filters

The conventional wisdom that digital techniques are universally superior to that of their analog counterparts is not always correct. In principle, the concept of converting all incoming signals to digital form and then carrying out whatever programmatic operations may be required, on those signals, to determine what actions should, or should not be taken, if any, would seem to be the best approach.

Digital filters certainly have much to offer when compared to their analog counterpart by virtue of their superior performance in terms of passband ripple, greater stopband attenuation, greatly reduced design times, better signal-to-noise characteristics,[1] less nonlinearity but are not necessarily

---

[1] Digital filters do introduce some noise in terms of quantization noise, as a result of the conversion, from analog-to-digital and digital-to-analog, of filtered signals, etc.

© Springer Nature Switzerland AG 2021
E. H. Currie, *Mixed-Signal Embedded Systems Design*,
https://doi.org/10.1007/978-3-030-70312-7_12

the answer for all embedded systems. In cases for which the responsiveness of the system is a primary concern the processing overhead, i.e., latency, associated with digital filters can in some cases make them inapplicable. Digital filters are generally more complex than analog filters have good EMI and magnetic noise immunity, very stable with respect to temperature and time, provide excellent repeatability, but do not, generally speaking, offer the dynamic range of analog filters or have the capability operate over as wide a frequency range of a comparable analog filter.

Since digital filters are discrete-time devices, difference equations are often used to model their behavior, e.g,

$$y_n = -a_1 y_{n-1} - a_2 y_{n-2} - \cdots - a_N y_{n-N} + b_0 x_n + \cdots + b_{n-M} \tag{12.1}$$

$$= -\sum a_k y_{n-k} + \sum b_k x_{n-k} \tag{12.2}$$

where a_k and $b_k \in \mathbb{Z}$.

In the z-domain, the transfer function for a LTI IIR digital filter [12] is of the general form given by:

$$H(z) = \frac{b_0 + b_1 z^{-1} + b_2 z^{-1} + b_3 z + \ldots + b_n z^{-n}}{1 + a_1 z^{-1} + a_1 z^{-1} + a_1 z + \ldots + a_m z^{-m}} \tag{12.3}$$

If n > m then the filter is said to be an nth-order filter and conversely, if m > n, then the filter is said to be an mth order filter.

Digital filters are modeled in terms of adders, multipliers, and positive and negative delays.

There are two fundamental types of digital filters:

- Finite Impulse Response (FIR) filters that are nonrecursive, stable, linear with respect to phase, are relatively insensitive to coefficient quantization errors and depend on either the difference of contiguous samples or weighted averages. The latter serves as a lowpass filter and the former as a highpass filter. The impulse function for an FIR filter is of finite duration. Such filters can be expressed mathematically as:

$$y[n] = \sum_{k=0}^{M} b_k [n - k] \tag{12.4}$$

where M is the number of feed-forward taps.

MATLAB provides support, in the MATLAB Signal Processing Toolbox, for window-based FIR filters by providing
1. b = fir1(n,Wn)
2. b = fir1(n,Wn,'ftype')
3. b = fir1(n,Wn,window)
4. b = fir1(n,Wn,'ftype',window)
5. b = fir1(...,'normalization')
   for "windowed" linear-phase FIR digital filter design.
- Infinite Impulse Response (IIR) filters have impulse responses of infinite duration and can be expressed mathematically as:

$$y[n] = -\sum_{k=1}^{N} a_k y[n - k] + \sum_{k=1}^{M} b_k x[n - k] \tag{12.5}$$

where N, M are the number of feedforward taps and feedback taps, respectively, and a_{k} is the kth feedback tap.[2] Because the output of an IIR filter output is a function of both the previous M outputs and N inputs, it is a recursive filter whose impulse response is of infinite duration. IIR filters typically require fewer numbers of multiplications than their FIR counterparts, can be used to create filters with characteristics of analog filters but are sensitive to coefficient quantization errors.[3]

In addition to butter, cheby1, cheby2, elliptic, and Bessel which represent a complete filter design suite, MatLab support for IIR filters includes:

1. buttord
2. cheb1ord,
3. cheb2ord

and,

4. ellipord

12.2 Finite Impulse Response (FIR) Filters

Finite impulse filters, of the type shown in Fig. 12.1 are causal (non-recursive), inherently stable (BIBO[4]), do not need feedback, are relatively insensitive to coefficient quantization errors, and capable of providing the same delay for all components of the input signal.[5] In addition, FIR filters are further characterized by the fact that their impulse response is finite. FIR filters are quantifiable in terms of the following linear constant-coefficient difference (LCCD) equation:

$$y[n] = b_0 x[n] + b_1 x[n - N] + \ldots + b_N x[n - N] = \sum_{k=0}^{N} b_k x[n - k] \qquad (12.6)$$

where N is the number of "taps",[6] y[n] is the output at the discrete-time instance n and similarly x[n] is each of the input samples. Note that this type of filter does not depend on previous values of y. To determine the impulse response of this particular configuration,

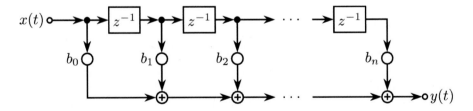

Fig. 12.1 A generic finite impulse response filter of order n.

---

[2]If $a_k = 0$, then this expression reverts to that of the FIR filter.

[3]Because digital filters introduce quantization errors, the positions of poles, and zeros, in the complex plane can shift which is referred to as coefficient quantization error.

[4]Bounded-Input-Bounded-Output.

[5]A very important consideration for audio and video applications.

[6]The number of taps is a measure of the number of terms in Eq. 12.6 which is $(N + 1)$ and N is the order of the filter. The filter coefficients, b_j, are referred to as the jth feedforward taps.

$$x[n] = \delta[n] \tag{12.7}$$

so that Eq. 12.6 becomes

$$y[n] = b_0\delta[n] + b_1\delta[n-N] + \ldots + b_N\delta[n-N] = \sum_{k=0}^{N} b_k\delta[n-k] = b_n \tag{12.8}$$

One of the most commonly encountered types of FIR filters is known as a "Moving Average Filter" which can be either LP, BP, or HP. It is based on a concept of averaging a number of samples to produce each of the output values, i.e.,

$$y[n] = \frac{1}{M}\sum_{k=0}^{M-1} x[n+k] \tag{12.9}$$

where M is the number of samples in the average. As simple as this technique is, it turns about to be an excellent method for removing random noise while retaining a sharp step response.

12.3   Infinite Impulse (IIR) Response Filters

The infinite Impulse Response or IIR filter has an impulse response that is infinite in duration.[7]
    The transfer function for an IIR filter is of the form:

$$H(z) = \frac{p_0 + p_1 z^{-1} + p_2 z^{-2} + \ldots + p_M z^{-M}}{d_0 + d_1 z_{-1} + d_2 z^{-2} + \ldots + d_N z^{-N}} \tag{12.10}$$

In terms of a difference equation, it is defined by a set of recursion coefficients and the following equation:

$$y[n] = -\sum_{k=1}^{M} a_k y[n-k] + \sum_{k=1}^{N} b_k x[n-k] \tag{12.11}$$

The transfer function for IIR filters for which m where a_k is the kth feedback tap which depends on previous outputs, M is the number of feedback taps and N is the number of feedforward taps. The a_k coefficients are referred to as the recursive or "reverse" coefficients, and the b_k coefficients are called the "forward" coefficients. Therefore, unlike its FIR counterpart, the IR filter output is a function of the previous outputs and inputs which is a characteristic common to all IIR structures and is responsible for the infinite duration of the impulse response. Note that if $a_k = 0$, then Eq. 12.11 becomes identical to Eq. 12.6.

---

[7] Since the IIR filter is a "recursive" filter and has the property that its impulse response is in terms of exponentially decaying sinusoids, and therefore, infinitely long. Of course, in real-world systems, at some point, the responses fall below the round-off noise level, and thereafter may safely be ignored.

12.4   Digital Filter Blocks (DFBs)

PSoC 3/5LP have support for filter components called Digital Filter Blocks (DFBs)[8] that have two separate filtering channels. The DFB has its own multiplier and accumulator which supports 24-bit x 24-bit multiplication and a 48-bit accumulator. This combination is used to provide a Finite Impulse Response (FIR) filter with a computation rate of approximately one FIR tap for each clock cycle.

The DFB features include:

- data alignment support options for I/O samples,
- one interrupt and two DMA request channels,
- three semaphore bits programmatically accessible,
- two usage models for block operation and streaming,
- cascading of 2–4 stages per channel with each stage having their own filter class, filter type, window type, # of filter taps,[9] center frequency, and bandwidth specifications.

and,

- two streaming data channels.

The DFB is implemented as a 24-bit, fixed point, programmable, limited scope, DSP engine as shown in Figure 12.2.

The DFB supports two streaming data channels, where programming instructions, historic data, and filter coefficients, and results are stored locally with new periodic data samples received from the other peripherals and blocks through the PHUB interface. In addition, the system software can load sample and coefficient data in/out of the DFB data RAMs, and/or reprogram them for different operations in block mode. This allows for multi-channel processing or deeper filters than supported in local memory. The block provides software configurable interrupt (DFB_INTR_CTRL) and two DMA channel requests (DFB_DMA_CTRL). Three semaphore bits are available for system software to interact with the DFB code (DFB_SEMA).

Data movement is typically controlled by the system DMA controller, but it can also be moved directly by the CPU. The typical usage model is for data to be supplied to the DFB over the system bus, from another on-chip system data source such as an ADC. The data typically passes through main memory or is directly transferred through DMA. The DFB processes this data and passes the result to another on-chip resource, such as a DAC, or main memory, via DMA, on the system bus. Data movement in, or out of, the DFB is typically controlled by the system DMA controller but can be moved directly by the CPU.

The DFB consists of subcomponents, viz., a

1. controller,
2. bus interface,
3. Datapath,

and,

4. Address Calculations Units (ACUs).

---

[8]Only one filter component can be incorporated into a design at a time.
[9]Filter taps are limited to a maximum of 128 taps for each channel.

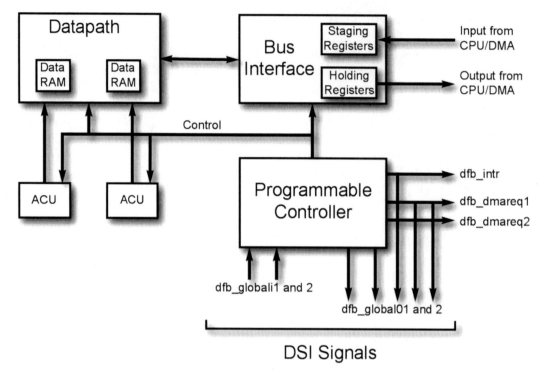

Fig. 12.2 The digital filter block diagram.

The DFB's programmable controller has three memories[10] and a relatively small amount of logic and consists of a RAM-based state machine, RAM-based control store, program counters, and "next state" control logic, as shown in Fig. 12.2. Its function is to control the address calculation units and the Datapath and to communicate with the bus interface to move data in, and out of, the Datapath.

The Datapath subblock is a 24-bit fixed point, numerical processor containing a Multiply and Accumulate (MAC) capability, a multi-function Arithmetic Logic Unit (ALU), sample/coefficient/Data RAM (Data RAM is shown in Fig. 12.2), and data routing, shifting, holding, and rounding functions. The Datapath block is the calculation unit inside the DFB.

The addressing of the two data RAMs in the Datapath block are controlled by the two (identical) Address Calculation Units (ACUs), one for each RAM. These three sub-functions make up the core of the DFB block and are wrapped with a 32-bit DMA-capable AHB-Lite Bus Interface with Control/Status registers.

These three sub-functions make up the core of the DFB block and are wrapped with a 32-bit DMA-capable AHB-Lite Bus Interface with Control/Status registers. The Controller consists of a RAM-based state machine, a RAM-based control store, program counters, and next state control logic. Its function is to control the address calculation units and the Datapath, and to communicate with the bus interface to move data in and out of the Datapath.

Proprietary assembly code and an assembler allow the user to write assembly code to implement the data transform the DFB should perform. Alternatively, a "wizard" is provided to facilitate digital filter design for both FIR and FII filters. The wizard allows the designer to set either one or two data

---

[10]The code that embodies the data transform function of the DFB resides in these memories.

stream channels, referred to as Channel A and Channel B, of a filter component that is passing data either in, or out, using DMA transfers or register writes, via firmware and an integral co-processor. The filter has 128 taps that determine the frequency responses of the filter. Either channel can be configured to produce an interrupt in response to receiving a data-ready event which in turn enables the interrupt output. Filter can also be implemented in Verilog [13].

12.5 PSoC 3/5LP Filter Wizard

PSoC Creator includes a powerful wizard for configuring digital IIR and FIR filters. The wizard is shown in Fig. 12.3, that provides a graphical representation of the filter by displaying the response for each of the following in a color-coded format:

1. **Amplitude**—gain is displayed graphically as a function of frequency.
2. **Phase**—phase is displayed graphically as a function of frequency.
3. **Group Delay**—occurs when the phase is a function of frequency is non-linear. If the frequency components of a signal are propagated through a device with no Group Delay, then the components experience the same time delay. If some frequencies are traveling faster than others then there is group delay and distortion results.
4. **Impulse Response**—the impulse response completely characterizes the filter.
5. **Tone Input Wave**—an input signal (sinusoid) is shown graphically for a bandpass filter
6. **Tone Response Wave**—the response to the tone input into a bandpass filter is shown graphically.

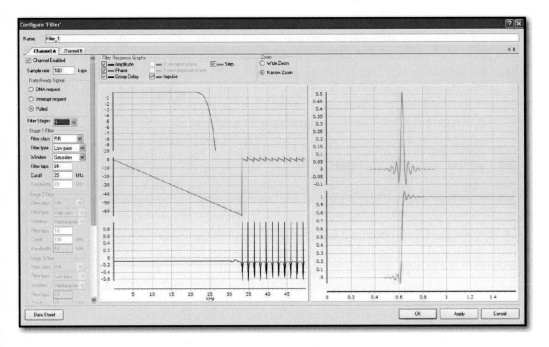

Fig. 12.3 PSoC 3/5LP's filter configuration wizard.

Several different types of "windows"[11] are supported that offer various combinations of band width transition, passband ripple, and stopband attenuation characteristics:

1. **Rectangular**—large passband ripple, sharp roll-off, and poor stopband attenuation. Rarely used because of the large ripple effect as a result of the Gibb's[12] phenomenon.

$$w(n) = 1 \tag{12.12}$$

2. **Hamming**[13]—smoothed passband, wider transition band and better stopband attenuation than

$$w(n) = 0.54 - 0.46 \cos\left[\frac{2\pi n}{(N-1)}\right] \tag{12.13}$$

3. **Gaussian**—Wider transmission band, but greater stopband attenuation and smaller stopband lobes than Hamming.

$$w(n) = exp\left[-\frac{1}{2}\left(\frac{\frac{2n}{N-1}-1}{\sigma}\right)^2\right] \qquad \sigma \leq 0.5 \tag{12.14}$$

4. **Blackman**—provides a steeper roll-off than its Gaussian counterpart, but similar stopband attenuation although larger lobes in the stopband.

$$w(n) = a_0 - a_1 \cos\left[\frac{2\pi n}{(N-1)}\right] + a_2 \cos\left[\frac{4\pi n}{(N-1)}\right] \tag{12.15}$$

Windows are often employed to deal with undesirable behavior taking places at the edges of a filter's characteristics. They are introduced either by multiplication in the time domain, or by convolution in the frequency domain.

The sample data rate is the design rate but the operational rate is determined by the data source driving the filter. There are no decimation or interpolation stages and therefore, the sample rate is the same throughout each channel. The maximum sample for a channel is:

$$f_{sMax} = \frac{Clk_{bus}}{Channel\,Depth + 9} \tag{12.16}$$

where Clk_{bus} is the bus clock speed and Channel Depth is the total number of taps used for a given channel.

If both channels are used then Eq. 12.16 becomes:

$$f_{sMax} = \frac{Clk_{bus}}{Channel\,Depth_A + Channel\,Depth_B + 19} \tag{12.17}$$

---

[11] Windows, or more properly, windows functions are functions that are defined as zero outside a specified interval. This type of function is also referred to as a "tapering" or "apodization" function. Because such functions have a value of zero outside of the interval over which they are nonzero, multiplying a window function by a signal is equivalent to viewing the signal through a window, and thereby constraining "spectral leakage".

[12] The best known Gibb's phenomenon is the so-called ringing effect observed and the leading and trailing edges of a square wave.

[13] This should not be confused with Hanning Window which is defined as $w(n) = 0.5(1 - \cos[(2\pi n)/M])$ for $0 \leq n \leq M$ and zero for all other values of n. It is sometimes referred to as the "raised cosine" window.

The filter component can alert the system of the availability of data either via a DMA request that is specific to each channel. or through an interrupt request shared between the two channels, or the status register can be polled to check for new data ready. Although the output holding register is doubled buffered, it is important to remove the data from the output before it's overwritten. Each channel can have as many as four cascaded[14] stages assuming that sufficient resources are available. The cutoff frequency parameter is used to set the 'edge" of the passband frequencies for Lowpass, Highpass, and Sinc4 filters.

The filters center frequency is defined as the arithmetic mean of the upper and lower cutoff frequencies for the bandpass stop and pass filters:

$$f_c = \frac{f_u + f_l}{2} \tag{12.18}$$

and the bandwidth (BW) is defined as:

$$BW = f_u - f_l \tag{12.19}$$

The wizard allows the designer to:

- view graphical representations of the frequency response, phase delay, group delay, and impulse and step responses,
- select from 1 to 4 filter stages,
- zoom in and out to provide a view of the filter's responses with either a linear or logarithmic frequency scale, respectively, over a frequency range from DC to the Nyquist frequency.
- enable or disable the filter cascade's response to a positive step function,
- enable or disable the filter cascade's response to a unit impulse,
- select a tone[15] input wave at the center frequency of the bandpass
- implement both FIR and FII filters

This function is capable of acting as a filter that

12.5.1 Sinc Filters

The normalized sinc function, shown in Fig. 12.4 is defined as:

$$sinc(x) = \frac{sinc(\pi x)}{\pi x} \tag{12.20}$$

and can be used as the basis for a sinc-based lowpass digital filter with a linear phase characteristic. A related function, shown in Fig. 12.5 called the rect function is defined as

$$rect(t) = \sqcap(t) = \begin{cases} 0 & \text{if } |t| > \frac{1}{2} \\ \frac{1}{2} & \text{if } |t| = \frac{1}{2} \\ 1 & \text{if } |t| < \frac{1}{2}. \end{cases} \tag{12.21}$$

---

[14]Cascading of stages refers to the use of multiple filters that are interconnected so that the output of one filter stage becomes the input of another filter stage.

[15]The tone is a sine wave whose frequency is the center frequency of the bandpass filter.

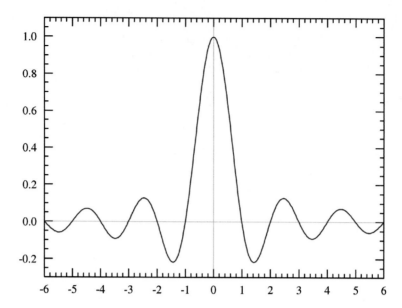

Fig. 12.4 The "normalized" sinc function.

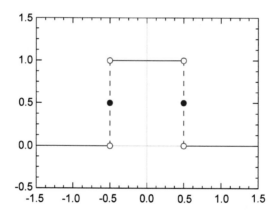

Fig. 12.5 The rectangular or Rect function.

These two functions are related by the fact that Fourier Transform of the rect function is the sinc function and the inverse Fourier Transform of rect is the sinc function, i.e., given that the Fourier Transform and its inverse are defined by

$$\mathcal{F}(\omega) = \int_{-\infty}^{\infty} f(t)e^{-i\omega t}\,dt \qquad (12.22)$$

and

$$\mathcal{F}^{-1}(t) = \int_{-\infty}^{\infty} F(f)e^{i\omega t}\,d\omega \qquad (12.23)$$

It follows that

$$F(\omega) = \int_{-\infty}^{\infty} \frac{sin(\omega t)}{\pi \omega} dx = \text{rect}(\omega) \qquad (12.24)$$

$$F^{-1}(t) = \int_{-\infty}^{\infty} rect(t)e^{i2\pi ft} df = \frac{1}{\sqrt{2\pi}} sin\left[\frac{\omega t}{2\pi}\right] \qquad (12.25)$$

Thus, the impulse function for a rect function in the frequency domain is the sinc function in the time domain and conversely, a rect function in the time domain is mapped to a sinc function in the frequency domain. If the sinc function is "convolved" with an input signal, theoretically an ideal lowpass filter could be realized. However, the fact that the sinc function extends to $\pm\infty$ presents a problem. One approach is to just cut off all the points on the sinc curve beyond a certain point and then examine the Bode plot to determine the resulting effect.

This technique can lead to an undesirable ripple in the passband and outside of the passband as a result of the steepness of the truncated ends of the sinc function and outside the passband. Another possibility is to employ so-called window functions, e.g., the Blackman or Hamming windows, that are multiplied by the truncated sinc function. This results in a steeper roll-off and less ripple in the passband and stopbands. If the stopband attenuation is a major concern, the Blackman window should be employed but it will result in some degradation of the roll-off. If roll-off is the primary concern, then the Hamming window is the better choice.

12.6 Data Conversion

Embedded systems by necessity are required to carry out various operations with digital and analog data [6]. Input of digital data can be provided by external communications channels [7], digital sensors, or other digital sources. Analog inputs often have to be converted to equivalent digital data to permit numerical and logic processing, storage in memory, etc. In addition, digital data may have to be converted to its analog equivalent to allow it to control external devices such as motors, other actuators, etc., or for other purposes. Thus, analog-to-digital, and digital-to-analog, conversion is an important capability for many embedded systems.

12.7 Analog to Digital Conversion

Since the world is essentially analog, it is not surprising to find that embedded systems make extensive use of analog-to-digital and digital-to-analog techniques in order to get the real-world into the computational domain. Although different architecturally, one is rarely to be found without the other nearby. Arguably, much of the world at increasingly finer grain levels appear not to be continuous but discrete, embedded systems are typically dealing with an environment full of continuous sources some of which of necessity must be monitored by the embedded system. Digital-to-analog converters, of necessity, must bridge the gap between the discrete-value environment of the digital domain and that of the continuous values,[16] to allow embedded systems to communicate and to some extent control external processes.

---

[16] Often after some lowpass filtering.

Analog-to-digital converters are presented with continuous-valued inputs, continuous-time signals, and expected to provide digital equivalents in the form of discrete values and times to what has become ever-increasing degrees of resolution.

12.8   Basic ADC Concepts

There are a number of important and very fundamental concepts involved in the use and deployment of analog-to-digital converters including:

- *Aliasing* is the introduction of spurious signals as a result of sampling at a rate below the Nyquist criteria, i.e., a rate less than the high-frequency component in the input signal.
- *Resolution* refers to the number of quantization levels of an ADC, e.g., an 8-bit ADC has a resolution of 256.
- *Dither* refers to the addition of a small amount of white noise to a low level, periodic signal prior to conversion by an ADC. The addition of the noise is shown in Figs. 12.6 and 12.7. Note that in Fig. 12.7 the noise added before the analog-to-digital conversion is subtracted at the output and is referred to as "subtractive dithering".

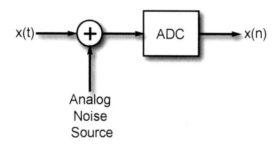

Fig. 12.6   Simple dither application using an analog noise source.

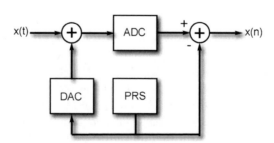

Fig. 12.7   Dither application using a digital noise source.

- *Sampling rate* refers to the number of samples per unit of time. It is usually chosen to be at least twice the highest frequency component in the signal being sampled.
- *Oversampling* is the process of collecting more samples than would otherwise be needed to accurately reproduce a signal that reduces the in-band quantization noise by a factor equal to the square root of the oversampling ratio, e.g., reducing the noise by a factor of two, increases the

effective processing gain of 3dB. Keep in mind that we are only talking about broadband noise here. Other sources of noise and other errors cannot simply be removed by oversampling.

- *Undersampling* is a technique used in conjunction with ADCs that allows it to function as a mixer. Thus, a high-frequency signal can be input and the output of the ADC is a lower frequency. However, this technique does require digital filtering in order to recover the signal of interest.
- *Decimation* refers to the use in oversampling applications and the subsequent discarding of samples after conversion in a way that does not significantly alter the accuracy of the measurement.
- *Quantization* is the process of breaking a continuous signal into discrete samples or quantum.
- *Quantization error* is defined as the arithmetic difference between an actual signal and its quantized, digital value.
- *Dynamic range* is the range between the noise floor and the maximum output level.

12.8.1 Delta-Sigma ADC

The delta-sigma[17] modulator emerged from the early development of pulse code modulation technology and was originally developed in 1946. However, it remained dormant until 1952 when it appeared again in various publications including a related patent application. Its appeal was the fact that it could offer increased data transmission since it could transmit the changes, i.e., "deltas", in value between consecutive samples, instead of transmitting the actual values of the sample. A comparator is used as a one-bit ADC and the output of the comparator was then converted to an analog signal, using a 1-bit DAC, the output of which was then subtracted from the input signal after it had passed through an integrator. Delta-Sigma modulators rely on techniques known as "over-sampling" and noise-shaping in order to provide the best performance.

A greatly simplified example of a Delta-Sigma modular is shown in Fig. 12.8

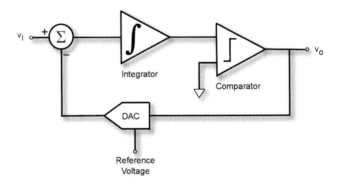

Fig. 12.8 An example of a first-order, $\Delta\Sigma$ modulator.

In this example, the input voltage, v_i, is added to the output of the single-bit, digital-to-analog converter, and the sum is then integrated and the output applied to the input of the comparator. The output of the comparator is either high or low, i.e., one or zero, depending on whether the output of the integrator output is $\geqslant$ zero or negative, respectively. The output of the comparator is then provided

---

[17]The literature refers to both "delta-sigma" and "sigma-delta" modulators. Purists argue that the proper name is Delta-Sigma since the signal passes through the delta phase prior to the sigma phase. However, this distinction is comparable to that between Tweedledee and Tweedledum.

to the input of the DAC and the process is repeated. The output of the DAC is ± the reference voltage (Table 12.1).

Example 6.7 - As a quantitative, illustrative example, consider the following:

Let the reference voltage be +/-2.5 V and assume that the input voltage is 1 V Then when the process begins the output of the DAC is zero so that 1+ 0 = 1 which when integrated becomes 1 and the comparator outputs a one which is then applied to the DAC with the result that the DAC output becomes 2.5 V. When this is summed with the input the total is -1.5 V which is integrated to produce an output from the integrator of -0.5 V. The comparator then outputs a zero and the process continues.

Table 12.2 shows the results of the process outlined in Example 6-7. Figure 12.9 shows the configuration of a 2nd-order, Delta-Sigma modulator (Fig. 12.10). A graphical representation of the various associated signals of a Delta-Sigma modulator with a sinusoidal input is shown in Fig. 12.11.

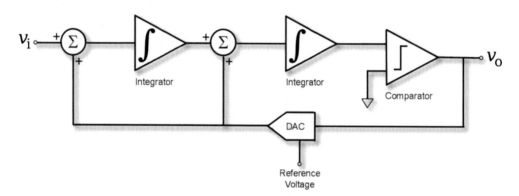

Fig. 12.9 An example of a second-order, Delta-Sigma modulator.

12.8.2 PSoC 3/5LP Delta-Sigma Converter

The architecture of PSoC 3/5LP includes a very high SNR/resolution Delta-Sigma ADC that employs oversampling, noise shaping, averaging and decimation. A Delta-Sigma Analog-to-Digital Converter (ADC) has two main components: a modulator and a decimator. The modulator converts the analog input signal to a high data rate (oversampling), low resolution (usually 1 bit) bitstream, the average value of which gives the average of the input signal level. This bitstream is passed through a decimation filter to obtain the digital output at high resolution and lower data rate. The decimation filter is a combination of down-counter and a digital low pass (averaging) filter that averages the bitstream to get the digital output.

Features of the PSOC 3/5LP Delta-Sigma converter include:

- 12- to 20-bit resolution
- An optional input buffer with RC lowpass filter

- Configurable gain rom 0.25–256
- Differential/single-ended inputs
- Gain ad offset correction
- Incremental continuous modes
- Internal and external reference options
- Reference filtering for low noise

The PSoC 3/5LP uses a 3rd-order modulator with a high impedance front end buffer followed by an RC filter. The modulator sends out a high data rate bitstream in thermometric format. The output of the modulator is passed on to the analog interface that converts the output to two's complement (4 bit) and passes it on to the decimation filter. The decimation filter converts the output of the modulator into a lower data rate, but sufficiently high-resolution output.

The input impedance of the modulator is too low for some applications and therefore, higher input impedance, low noise, independent buffers have been provided for each of the differential inputs. These buffers can be bypassed/powered down by setting DSM_BUF0[1], DSM_BUF1[1] and/or $DSM\_BUF0[0]$, $DSM\_BUF1[0]$, respectively. The buffers have adjustable gains $(1, 2 4 or 8)$ determined by $DSM\_BUF1[3:2]$. The buffers can operate either in a level-shifted mode to allow the input level to be shifted above zero and rail-to-rail when the input is rail-to-rail. input to the buffers can be from analog globals, analog locals, the analog mux bus, reference voltages, and V_{ssa}.

The PSoC 3/5LP Delta-Sigma modulator consists of three active, OpAmp-based integrators (INT1, INT2, and INT3), an active summer, a programmable quantize, and switched-capacitor feedback DAC as shown in Fig. 12.10.

The three active integrators function as a 3rd-order modulator whose transfer function together with the quantizer provides highpass noise shaping. Increasing the order of the modulator improves the highpass filter response and lowers noise present in the signal frequency band. The three integrators and quantize stages are followed by an active summer. The analog input and the output of all three OpAmp stages are then summed. The summer output is quantized by a quantizer that is programmable to output 2, 3, or 9 levels. The DAC connects the quantized output back to the first stage OpAmp input. It is this feedback DAC that ensures that the average of the quantized output is equal to the average input signal level.

The quantization level can be set as 2, 3, or 9. The lowest level provides the best linearity and the highest provides the best SNR. The number of quantization levels is configured in DSM_CR0[1:0] register bits. The quantized output is stored in the register DSM_OUT1. The quantized outputs data in a format referred to as the thermometric[18] and is illustrated by the pattern of the output levels shown in Table 12.1.

12.8.3 Successive Approximation Register ADC

As shown in Fig. 12.12, there are four components that make up a Successive Approximation Register ADC (SAR):

1. A voltage DAC that converts the SAR output to an analog voltage which can then be used to compare with the input voltage.
2. A comparator that compares the analog input to the DAC output.

---

[18]In the thermometric format, the number of ones increases from LSB to MSB as the quantization level increases.

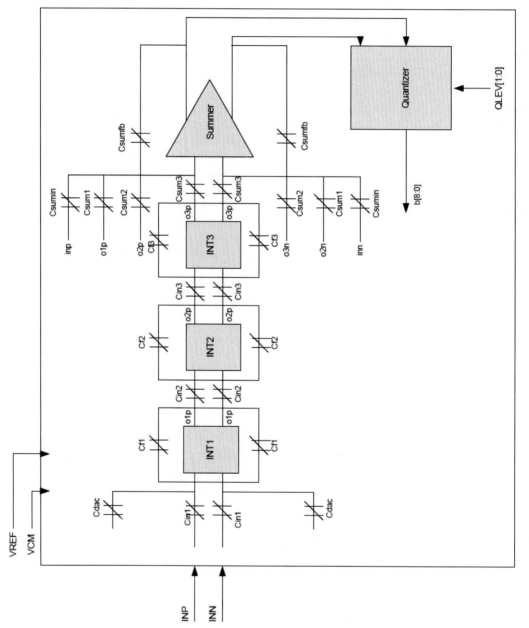

Fig. 12.10 Delta-Sigma modulator block diagram.

Table 12.1 Quantize output data.

Level	Quantizer Output Data
2 Level Quantizer	
Level 1	00000000
Level 2	11111111
3 Level Quantizer	
Level 1	00000000
Level 2	00001111
Level 3	11111111
9 Level Quantizer	
Level 1	00000000
Level 2	00000001
Level 3	00000011
Level 4	00000111
Level 5	00001111
Level 6	00011111
Level 7	00111111
Level 8	01111111
Level 9	11111111

3. A successive approximation register which, based on the output of the comparator, provides the appropriate input to the DAC.
4. A Track and hold circuit that holds an input value constant during the conversion at which point it loads another sample of the input.

PSoC LP has an 8-bit voltage DACs and therefore, the SAR is limited to 8-bit resolution. Although PSoC 3 does not currently have explicit support for a SAR, it is possible to construct a SAR based on the resources that are available in PSoC 3 [8]. SAR logic must set or reset a given bit based on the output of the comparator.

In a typical implementation, this operation is repeated 8 times until the SAR generates an "end of conversion" signal and latches the data. The VDAC can accept data from the DAC bus and therefore, data can be transferred directly from the SAR to VDAC without incurring CPU overhead. However, it is necessary to generate a strobe using SAR logic when data is available on the DAC bus so that the VDAC will produce a corresponding output voltage.

The limitation on the speed of conversion is set primarily by:

1. The speed with which the comparator is able to resolve differences between v_{i} and the output of the DAC.

Table 12.2 Delta-sigma example output.

Iteration	Input	DAC Output	Sum	Integrator Output	Comparator Output	Mean Voltage Output
0	1	0	0	0	0	0
1	1	0	1	1	1	2.5
2	1	2.5	-1.5	-0.5	0	0
3	1	-2.5	3.5	3	1	0.83
4	1	2.5	-1.5	1.5	1	1.25
5	1	2.5	-1.5	0	1	1.5
6	1	2.5	-1.5	-1.5	0	0.83
7	1	-2.5	3.5	2	1	1.07
8	1	2.5	-1.5	0.5	1	1.25
9	1	2.5	-1.5	-1	0	0.83
10	1	-2.5	3.5	2.5	1	1
11	1	2.5	-1.5	1	1	1.14
12	1	2.5	-1.5	-0.5	0	0.83
13	1	-2.5	3.5	3	1	0.96
14	1	2.5	-1.5	1.5	1	1.07
15	1	2.5	-1.5	0	1	0.94
16	1	2.5	-1.5	-1.5	0	1.03
17	1	-2.5	3.5	2	1	1.11
18	1	2.5	-1.5	0.5	1	0.92
19	1	2.5	-1.5	-1	0	1
20	1	-2.5	3.5	2.5	1	1.07
21	1	2.5	-1.5	1	1	0.91
22	1	2.5	-1.5	-0.5	0	0.98
23	1	-2.5	3.5	3	1	1.04
24	1	2.5	-1.5	1.5	1	1

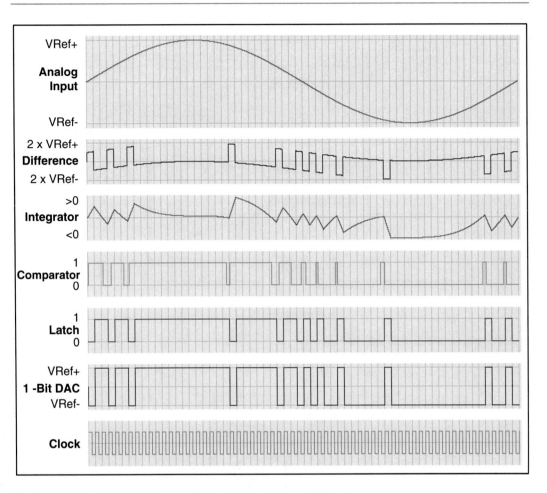

Fig. 12.11 First order, Delta-Sigma modulator signals with sinusoidal input.

2. The DAC's settling time is a function of the settling time for the MSB.
3. Overhead introduced by the latency of the various components of the SAR ADC. Such factors include the sample-and-hold acquisition time, sample-and-hold settling time, the EOC recognition time by the CPU, etc.

The change of state of the comparator signals that the binary representation of the input signal has been found and that the data can then be accessed programmatically.

12.8.4 Analog MUX

A multiplexer, or *Mux*, is a device that allows one, or more, inputs to be switched, and/or combined, programmatically to one, or more, outputs. These inputs/outputs may be either digital or analog, respectively. Typically, muxes are controlled by digital signals consisting of one or more binary inputs. Depending on the "address" represented by the binary inputs one of the input sources is connected to the output of the mux, as shown in Fig. 12.13. Analog muxes are often used when sampling multiple analog inputs with an analog-to-digital converter.

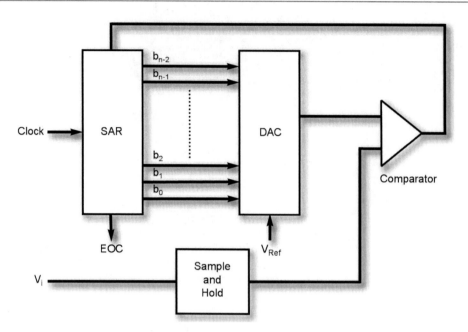

Fig. 12.12 A schematic diagram of a SAR ADC.

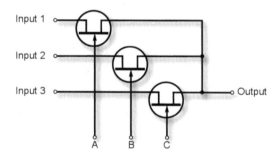

Fig. 12.13 A simple analog multiplexer.

The analog multiplexer, or AMux, is a passive device that can combine multiple analog signals, or multiple pairs thereof, in a single signal, or pair, to allow the output signal to be routed to a single input of some other component. The AMux also allows multiple input simultaneous input connections to be routed to a single connection. The AMux employs individual switches that connect blocks to analog busses and analog busses to pins.

Unlike most hardware multiplexers, the AMux is a collection of independent switches that are controlled by firmware, and not by hardware. This makes AMux much more flexible than other types of multiplexers because it allows more than one signal at a time to be connected to the common output signal. Note that in the "Differential Mode" the firmware will not allow the differential signals to be connected to each other and instead treats such cases as two parallel multiplexers controlled by the same signal.

12.8.4.1 Allowable Input/Output Connections

There are various types of input/output connections that are supported by the AMux:

- aN (Analog)—the AMux supports 2–32 analog inputs, inclusive.
- bN (Analog)[19]—the paired inputs (aN, bn) are only used when the Mux Type parameter is set to "Differential".b
- y (Analog)—This is a required connection and is the output of the AMux.
- x (Analog)—the "x" signal is the output connection when using the AMux in a differential mode. Its output is determined by the "void AMux_Select(void)" function.

In setting up an AMux, certain parameters must be set in order to achieve the desired configuration, i.e.:

- **Channels**—this parameter specifies the number of single or paired inputs and may have a value of 2–32, inclusive.
- **MuxType**—this parameter determines whether a single input per connection (Single)[20] or dual input per connection (Differential) is to be used. If two, or more, input signals have different signal references, the "Differential" mode must be used. This mode is often employed when the output of the mux is connected to an ADC with a differential input.

12.8.4.2 The AMux API

The application programming interface or API for the AMux provides programmatic access to various routines that allow the designer to configure the AMux. By default, PSoC Creator assigns the instance name "AMux_1" to the first instance of AMux. The API function calls for AMux are:

- **void AMux_n_Start**—disconnects all channels.[21]
- **void AMux_n_Stop**—disconnects all channels.[22]
- **void AMux_n_Select(uint8 chan)**—disconnects all other channels and then connects the selected channel (chan) signal.
- **void AMux_n_FastSelect(uint8 chan)**—disconnects the last connection made with either FastSelect or Select function calls and then connects the "init8 chan".[23]
- **void AMux_n_Connect(uint8 chan)**—connects the given channel to the common signal without affecting any previous channel connection.
- **void AMux_n_Disconnect(chan)**—disconnect only the specified channel from the output.
- **void AMux_n_DisconnectAll(void)**—disconnect all channels.

---

[19]This type of I/O may be hidden on the symbol under the conditions listed in the description of that I/O.

[20]Single refers to cases in which each input signal is referenced with respect to a common signal, e.g., V_ssa.

[21]With respect to AMux function calls, there are no return values, side effects, or parameters to specify, unless otherwise noted.

[22]The Stop API call is not required, but is provided for "compatibility' reasons.

[23]If the Connect function was used to select a channel prior to calling *FastSelect*, the channel selected will not be disconnected, which is useful when parallel signals need to be connected.

12.8.5 Analog/Digital Virtual MUX

PSoC 3/5LP have support for analog/digital "virtual" muxes which are analogous to hardware muxes in that they connect a selected input to an output. However, unlike their hardware counterpart, virtual muxes are can not be dynamically controlled. They can be used at the schematic level to chose from a variety of different sources, e.g., to select from a number of different clock sources. The actual connection to be made is selected at build time. The default number of inputs[24] is two with a maximum of sixteen and the selected input Virtual muxes do not consume any resources but merely connect a predefined input to the output.

12.8.6 PSoC 3/5LP Delta-Sigma ADC (ADC_DelSig)

The Delta-Sigma ADC provided in PSoC 3/5LP supports resolutions from 8–20 bits, continuous mode operation, an adjustable sample rate (10–375,000 sps), a high input impedance input buffer, and selectable input buffer gain all of which makes it ideal for sampling signals over a wide range of frequencies. Whether used to sample input from strain gauges, thermocouples, or other forms of high precision but low amplitude sensors, the ADC_DelSig is designed to spread the quantization noise across a sufficiently wide spectrum to allow it to be moved out of the input signal's bandwidth and then filtered by a lowpass filter.

The Delta-Sigma ADC is an inherently three-terminal device with an optional fourth and fifth pin for a start of conversion (SOC)which occurs as a result of the presence of a rising edge and an external clock source, respectively. The other three pins are positive input, negative input, and end of conversion (EOC). This positive input is used for a positive analog signal input to the ADC_DelSig. The conversion result is a function of the positive minus the voltage reference, which is either negative or V_{ssa}.[25]

The ADC_DelSig's negative input functions as the reference input and the result of a conversion is a function of the positive input minus the negative input. If the ADC_INPUT_Range option is selected for this device, then the following modes are available:

$$0.0 \pm 1.024\,\text{V (Differential)} - \text{Input} \pm \text{Vref}$$

$$0.0 \pm 2.048\,\text{V (Differential)} - \text{Input} \pm 2\text{Vref}$$

$$0.0 \pm 0.512\,\text{V (Differential)} - \text{Input} \pm \text{Vref}/2$$

$$0.0 \pm 0.256\,\text{V (Differential)} - \text{Input} \pm \text{Vref}/4$$

User-definable parameters for the ADC_DelSig include the following:

- **Variable power settings**: Low, Medium or High.
- **Conversion modes**: Continuous, Fast Filter or FIR.
- **Resolution**: 8, 9, 10, 11, 12, 13, 14, 15, 16, 17, 18, 19 or 20.
- **Input buffer gain**: 1, 2, 4 or 8[26] Start of conversion:
- **Start of Conversion**: Can be initiated at the hardware, or the software level.

---

[24]The number of inputs is specified by *NumInputTerminals*.

[25]V_{ssa} is an analog ground.

[26]The input buffer gain can also be disabled.

- **Conversion Rate**: 10 to 375,000 samples second
- **Clock Source**: External or internal.
- **Input Range**:

    0.0–1.024 V(Single-Ended)
    0.0–1.024 V(Single-Ended)
    V_{ssa} to V_{dda}, (Single-Ended)
    0.0 ± 1.024 V (Differential) Negative Input $\pm V_{ref}$
    0.0 ± 2.048 V (Differential) Negative Input $\pm 2(V_{ref})$
    0.0 ± 0.512 V(Differential) Negative Input $\pm V_{ref}/2$
    0.0 ± 0.256 V (Differential) Negative Input $\pm V_{ref}/4$

The ADC_DelSig consists of three blocks: an input amplifier, a 3rd-order Delta-Sigma modulator, and a decimator as shown in Fig. 12.14.

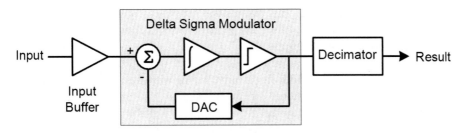

Fig. 12.14 The PSoC 3 ADC_DelSig block diagram.

The input amplifier provides a high impedance input and a user-selectable input gain. The decimator block contains a 4 stage CIC decimation filter[27] and a post-processing unit. The CIC filter operates on the data sample directly from the modulator. The post-processing unit optionally performs gain, offset, and simple filter functions on the output of the CIC decimator filter. Decimation is a combination of downsampling and filtering where downsampling refers to the discarding of samples particularly in cases when oversampling[28] is employed. In some cases, lowpass filtering is employed prior to downsampling in order to remain consistent with the Nyquist criteria.

12.8.7 I/O Pins

In order for an embedded system to interact with the real-world, there must, obviously, be provision for hardware connections for both input and output. Microcontrollers have I/O pins for this purpose and for PSoC 3/5LP the I/O pins are as important, as is their respective potential configurability. In addition to the pins being configurable at the schematic level, they can also be configured dynamically through program control. At the schematic level, the pins component is definable as analog, digital input, digital output, or bidirectional with an initial state of high or low.

Implementing embedded systems with PSoC 3/PSoC 5LP typically requires extensive use of various types of I/O pins including:

---

[27] Cascaded integrator comb filters (CIC) are linear-phase FIR filters and are more efficient than conventional FIR filters.
[28] Oversampling is sampling at rates greater than the Nyquist criteria.

- Analog Pins
- Digital Input Pins
- Digital Output Pins

and,

- Digital Bidirectional Pins

PSoC 3/5LP's Pins components can be configured into complex combinations of input, output, bidirectional, and analog I/O connections to provide both on- and off-device signals via physical I/O pins. A pin component can have 1–64 pins inclusive with a default value of 1 pin. It provides access to external data via an appropriately configured physical I/O pin and allows electrical characteristics to be associated with one or more pins. These characteristics are then used by PSoC Creator to automatically place and route the signals constrained within the component. Pins can be deployed from schematics and/or software. To access a Pins component from component APIs, the component must be contiguous and non-spanning. This ensures that the pins are guaranteed to be mapped into a single physical port. Pins components that span ports, or are not contiguous, can only be accessed from a schematic, or with the global per-pin APIs.[29]

An analog Pins component may also support digital input, or output, connections, or both, and bidirectional connections, e.g., analog with digital input, analog with digital output, analog with digital input/output, and analog with bidirectional digital I/O. Digital input pins can also support digital output and analog connections. Digital output pins can support digital input and analog connections. Bidirectional pins can support analog connections. When the Pins component is used in conjunction with an internal reference voltage (Vref) an SIO pin must be used, however, Vref can only be used with another digital connection, i.e. analog pins cannot be used. Digital pins can be used with an IRQ but not an analog pin. There are eight available drive modes for a pin as shown in Fig. 12.15 .

The drive modes for pins includes:

- Strong drive
- High impedance analog
- High impedance digital
- Open drain drives high
- Open drain drives low
- Resistive pull-up
- Resistive pull-down
- Resistive pull-up and pull-down
- Resistive pull-up/pull-down

The defaults for drive modes are high impedance for analog, digital and digital I/O and open drain (drives low) for bidirectional. All other

12.8.8 Digital to Analog Converters (DACs)

Digital to analog converters are an important component in embedded systems that allows the system to convert digital data into its analog equivalent for driving actuators, motors, switches, etc. The

---

[29]#defines are created for each pin in the Pins component to be used with global APIs.

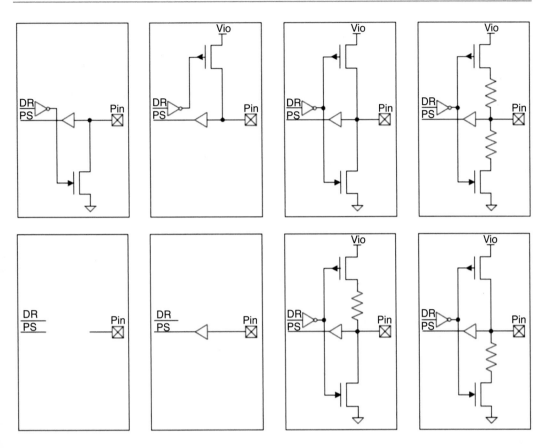

Fig. 12.15 PSoC 3/5LP pin drive modes.

selection of a DAC is based on a number of factors, one of which is the desired resolution which determines the number of analog levels that the DAC is capable of producing and the N-bit resolution, where N is the power of 2 representing the number of possible output levels. Dividing the number of levels into the DAC's maximum output voltage determines the voltage step size, i.e.,

$$\text{Output Voltage Step Size} = \frac{\text{Maximum Output Voltage}}{2^N} \tag{12.26}$$

Another factor is the sampling frequency, which refers to the maximum rate of output that the DAC is capable of producing. This is an important consideration when the accuracy, or fidelity, of the output analog signal is of concern. If the Nyquist-Shannon condition is to be met then the DAC must be capable of producing analog values at a rate of at least twice the highest frequency component to be included in the output. A third concern is the so-called monotonicity of the output from a DAC. In particular, if the output voltage is assumed to be increasing, or decreasing, each actual output step must represent a monotonic increase, or decrease, in the output. Dynamic range is also a consideration and is a function of the resolution of the DAC and the noise floor. Total harmonic distortion is a figure of merit for DACs and may need to be taken into account when selecting a DAC, depending on the application.

PSoC 3/5LP DACs generate either a voltage or a current output and employ a current mirror[30] architecture in which the current is mirrored from a reference source to a mirror DAC. Calibration and value current mirrors are responsible for the 8-bit calibration [DACx.TR] and the 8-bit DAC value. The current is then diverted into the scaler to generate the current corresponding to the DAC value. The DAC value can either be given from the register DACx.D, or 8 lines from the UDB [5]. This selection is made using the DACx.CR1[6] bit. The DAC is strobed to get its output to change for the input code. The strobe control is enabled by the DACx.STROBE[2] bit. The strobe sources for the DAC can be selected from the bus write strobe, analog clock strobe to any UDB signal strobe. This selection is based on the setting in DACx.STROBE[2:0].

- **Voltage (VDAC) Mode**—The current is routed through resistors according to the range, and voltage across it provided as output. The outputs from the PSoC 3/5LP DACs are single-ended in both IDAC and VDAC modes.
- **Current (IDAC) Mode**—The two mirrors for the current source and sink provide output as a current source or current sink, respectively. These mirrors also provide range options in the current mode.

12.8.9 PSoC 3/5LP Voltage DAC (VDAC8)

The VDAC8 is an 8-bit voltage digital-to-analog converter that can be configured in various ways depending on the application. It is controllable via hardware, software, or a combination of both hardware and software. It can be employed as a fixed or programmable voltage source with:

- a CPU, DMA or UDB data source text,
- Software, or clock-driven, output strobe,
- Two ranges: 1.020V and 4.096V full scale
- Voltage output

12.8.9.1 Input/Output Connections
When used as a VDAC, the output is an 8-bit digital-to-analog conversion voltage to support applications where reference voltages are needed. The reference source is a voltage reference from the Analog reference block called VREF(DAC). The DAC can be configured to work in voltage mode by setting the DACx.CR0 [5] register.

In this mode, there are two output ranges selected by register DACx.CR0 [3:2].

- 0–1.024 V
- 0–4.096 V

Both output ranges have 255 equal steps.

The VDAC is implemented by driving the output of the current DAC through resistors and obtaining a voltage output. Because no buffer is used, any DC current drawn from the DAC affects the output level. Therefore, in this mode any load connected to the output should be capacitive. The VDAC is capable of converting up to 1 Msps. However, the DAC is slower in 4V mode than 1V mode,

---

[30]Current mirrors are circuits designed to accurately replicate a reference current referred to as the "golden current source", and sometimes includes scaling of the replicated current. Current mirrors can be thought of as ideal current amplifiers. Golden current sources are expected to be relatively temperature, and voltage-independent.

because the resistive load to Vssa is 4 times larger. In 4V mode, the VDAC is capable of converting up to 250 ksps. The VDAC8's output can be routed to any analog compatible pin on PSoC 3/5LP.

An 8-bit wide data signal, i.e., data[0:7], connects the VDAC8 directly to the DAC Bus. The DAC Bus may be driven by UDB based components, control registers, or routed directly from GPIO pins. Input is enabled by setting the *Data_Source* parameter to "DAC Bus". data[7:0] input should be used when the hardware is capable of setting the proper value without CPU intervention and the strobe option should be set as External. For many applications, this input is not required but instead, the CPU or DMA will write a value directly to the data register. In firmware, the *SetRange()* function or directly writing a value to the *VDAC8_n_Data* register (assuming an nth instance name) should be used.

In strobe input mode, the data is transferred from the VDAC8 register to the DAC on the next positive edge of the strobe signal. If this parameter is set to "Register Write"the pin will disappear from the symbol and any write to the data registers will be immediately transferred to the DAC. For audio or periodic sampling applications, the same clock used to clock the data into the DAC could also be used to generate an interrupt. Each rising edge of the clock would transfer data to the DAC and cause an interrupt to get the next value loaded into the DAC register.

The output voltage is determined by:

$$v_o = 1.020 \left[\frac{value}{256} \right] \text{volts} \tag{12.27}$$

or,

$$v_o = 4.096 \left[\frac{value}{256} \right] \text{volts} \tag{12.28}$$

depending upon the selected output range and $0 \leq (value) \leq 255$.

12.8.10 PSoC 3/5LP Current DAC (IDAC8)

When used as an IDAC, the output is an 8-bit digital-to-analog conversion current. This is done by setting the*DACx.CR0 [5]* register. The reference source is a current reference from the analog reference called *IREF(DAC)*. In this mode, there are three output ranges selected by register *DACx.CR0 [3:2]*.

- 0–2.048 mA, 8 μA/bit
- 0–256 μA, 1 μA/bit
- 0–32 μA, 0.125 μA/bit

For each level, there are 255 equal steps of M/256 where $M = 2.048$ mA, 256 μA, or 32 μA. In the 2.048 mA configuration, the block is intended to output a current into an external 600 Ω load. The IDAC is capable of converting up to 8 Msps. The user also has the option of selecting the output as either a current source, or a current sink. This is controlled by the *DACx.CR1[14]* register. This selection can also be made by using a UDB input. UDB control for the source/sink selection is enabled using the *DACx.CR1[2]* bit. Separate muxes are used for current and voltage modes.

It is possible to achieve a higher resolution current output DAC by summing the outputs of two 8-bit current DACs, each one having a different segment of the input bus for input, as shown in Fig. 12.16.

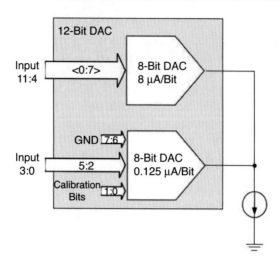

Fig. 12.16 Higher resolution current DAC configuration.

The range of the two DACs used partially overlap.

For example, the implementation of a 12-bit DAC using two 8-bit DACs require: One DAC scaled to the range 0–2.048 mA and the second one scaled to the range 0–32 μA. The middle 4 bits of the lowest range DAC are used as inputs to the lower 4 bits. This architecture may have problems if there is a mismatch between the two DACs, and therefore, adjustment and scaling may be required. The last two bits of the LSB DAC are used for minor calibration requirements.

12.9   PSoC 3/5LP Gates

PSoC 3/5LP provide a powerful suite of digital functions that are interoperable with their analog counterparts. In addition to digital functionality such as counters, timers, cyclic redundancy check modules, pulse width modulators, quadrature decoders, shift registers, pseudo-random sequence (PRS) generators, and precision illumination signal modulators (PrISM) a full set of logic functions such as AND, OR, (NOT), NOR, NAND, XOR, XNOR, and inverters is also provided to provide all the basic operations. Arbitrarily complex combinations of these logic components allow the designer to create logic configurations [10] for a wide variety of situations involving PSoC 3/5's functions allow analog and digital blocks. Except for the inverter, which functions as a NOT gate, all the included logic gates [4] have two digital inputs as a default. The inverter has a single input and a single output but the other gates can have as many as eight digital inputs (*NumTerminals*), inclusive. A second parameter *TerminalWidth* defines the number of bus connections that can be attached to the same number of discrete logic gates in parallel.

All the digital logic gates used are converted to their VHDL[31] equivalents and reduced to a sum of products and then placed into the Universal Digital Block's (UDB) Programmable Logic Devices (PLD). This process results in digital logic gates being automatically optimized and placed into the PSoC device. Resource usage is dependent upon the specific logic created and can not be determined prior to project compilation in PSoC Creator.

---

[31] The Very High-level Design Language (VHDL) was created as a hardware description language for the development of high-speed integrated circuits and has evolved into an industry-standard language for describing digital systems.

12.9.1 Gate Details

Logic levels for the PSoC 3/5LP gates are defined as:

- True $= 1 =$ high logic-level
- False $=== $ low logic-level

The AND gate, shown symbolically in Fig. 12.17 functions in the same manner as a logical AND operator, viz., the output is true when all inputs are true and is otherwise false (Tables 12.3 and 12.4).

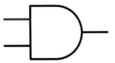

Fig. 12.17 AND gates perform logical multiplication.

Table 12.3 AND gate truth table.

Input 1	Input 2	Output
0	0	0
0	1	0
1	0	0
1	1	1

Table 12.4 OR gate truth table.

Input 1	Input 2	Output
0	0	0
0	1	1
1	0	1
1	1	1

The OR gate, shown symbolically in Fig. 12.18, functions in the same manner as a logical OR gate, viz., the output is true if any input is true and false if all inputs are false.

The inverter, shown symbolically in Fig. 12.19, which is also referred to as a NOT gate, performs a logical inversion function, viz., the output state of the inverter is the inverse state of the input (Table 12.5).

Fig. 12.18 OR gates perform logical addition.

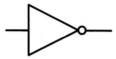

Fig. 12.19 An inverter functions as a NOT gate.

Table 12.5 NOT gate truth table.

Input	Output
1	0
0	1

The NAND gate, shown symbolically in Fig. 12.20, is the equivalent to a logical AND gate followed by a logical inverter. If all the inputs to the NAND gate are true, the output is false, otherwise, the output is true (Table 12.6).

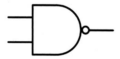

Fig. 12.20 The NAND gate functions as the combination of a logical NAND and NOT gate.

Table 12.6 NAND gate truth table.

Input 1	Input 2	Output
0	0	1
0	1	1
1	0	1
1	1	0

The NOR gate, shown symbolically shown in Fig. 12.21, functions as a logical OR gate followed by a logical NOT gate, viz., the output is true if all the inputs are false, otherwise the output is false (Table 12.7).

The XOR (exclusive-OR) gate, shown symbolically in Fig. 12.22, is useful as a parity generator. It has two or more inputs and one output. As shown in the Table 12.8, the XOR's output is true

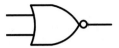

Fig. 12.21 The NOR gate functions as a combination of a logical OR and NOT gates.

Table 12.7 NOR Gate truth table.

Input 1	Input 2	Output
0	0	1
0	1	0
1	0	0
1	1	0

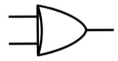

Fig. 12.22 An exclusive-OR gate.

Table 12.8 Exclusive-OR (XOR) truth table.

Input 1	Input 2	Input 3	Output
0	0	0	0
0	0	1	1
0	1	0	1
0	1	1	0
1	0	0	1
1	0	1	0
1	1	0	0
1	1	1	1

when there is an odd number of true inputs. Otherwise, the output is false. The XNOR gate, shown symbolically in Fig. 12.23, is an exclusive-NOR gate that functions as a logical XOR gate followed by a logical NOT gate, viz., the output is true when there is an even number of true inputs and otherwise the output is false (Table 12.9).

Table 12.9 Exclusive-NOR (XNOR) truth table.

Input 1	Input 2	Input 3	Output
0	0	0	1
0	0	1	0
0	1	0	0
0	1	1	1
1	0	0	1
1	0	1	0
1	1	0	0
1	1	1	1

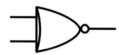

Fig. 12.23 An exclusive-NOR (XNOR) gate.

12.9.2 Tri-State Buffer (Bufoe 1.10)

The PSoC 3/5LP Tri-State Buffer (Bufoe) component is a four-terminal, non-inverting buffer with an active high output enable signal shown symbolically in Fig. 12.24. When the output enable signal is true, the buffer functions as a standard buffer. When the output enable signal is false, the buffer turns off. It is used to interface to a shared bus, e.g., I^2C. Bufoe's should be used with an I/O pin and not be used in conjunction with internal logic.

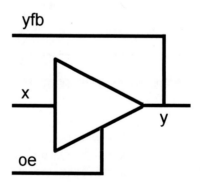

Fig. 12.24 A Bufoe is a buffer with an output enable signal (oe).

The four connections are:

- **x**—Input to the Bufoe.
- **oe** (output enable)—The Bufoe is enabled when oe is '1' and otherwise the output is in a high impedance state (referred to as "tri-stated").
- **y**—This connection is connected to the output of the buffer. When oe is true ('1'), this connection is an output, and y has the same value as x. When oe is false ('0'), this connection may be used as an input.
- **yfb** (output)—This is the feedback signal from the y connection. When oe is true ('1') the yfb and y have the same value as x. When oe is false ('0'), yfb has the same value seen at y irrespective of x.

12.9.3 D Flip-flop

A "D flip-flop", shown in Fig. 12.25 is a bistable device that can be used to store a digital value that can be preset or reset asynchronously.

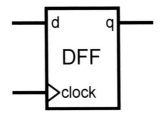

Fig. 12.25 PSoC 3/5LP's D flip-flop.

It functions nominally as a three-terminal device with signal input (d), clock input (clock), and output (q) and is frequently used to implement sequential logic. The D flip-flop output (q) tracks with the D flip-flop output (q) so that it can serve as a storage device.

A fourth input terminal called the *asynchronous preset* (ap) is accessible when the *PresetOrReset* parameter is set to Preset. The*ArrayWidth* parameter, whose default setting is '1', allows an array of D flip-flops to be created when the input or output is a bus. The *PresetOrReset* parameter controls whether the *asynchronous preset* input or *asynchronous reset* (ar) is visible with a default of "None". All D flip-flop components in the same UDB must have the same ar or ap input. In addition D Flip-Flop components in the same PLD must have the same clock signal. Resources The D Flip-Flop uses one macrocell. If the *ArrayWidth* parameter is greater than 1, the D flip-flop uses a number of macrocells equal to *ArrayWidth*.

12.9.4 Digital Multiplexer and Demultiplexer

The PSoC 3/5LP *digital multiplexer* is used to select 1 of n inputs and the digital demultiplexer is used to dynamically route a signal to one of n outputs, under firmware or hardware control. The most common control method is to connect the mux select signals to a control register using a bus. The control register is then used to select the input or output for the mux/demux. Another option is to drive the select signals from the hardware control logic to provide dynamic hardware routing. Tables 12.10 and 12.11 show the truth tables for a 4-input multiplexer and a 4-output demultiplexer, respectively.

There are three parameters that control multiplexers and demuxiplexers:

- *NumInputTerminals* determines the number of inputs of a multiplexer. The default is 4. The acceptable values are 2, 4, 8, and 16 and the corresponding select input widths are 1, 2, 3 and *NumOutputTerminals* determines the number of outputs of a de-multiplexer. The default is 4. The acceptable values are 2, 4, 8, and 16 and the corresponding select input widths are 1, 2, 3 and 4.
- *TerminalWidth* is used to create an array of parallel multiplexers or demultiplexers when the inputs and outputs are buses. It defines the bus width of the inputs and outputs and has a default value of 1. The width of the Select input is not affected by this parameter.

Table 12.10 4-input multiplexer truth table.

Select[1]	Select[2]	Input 3	Input 2	Input 1	Input 0	Output
0	0	X	X	X	0	0
0	0	X	X	X	1	1
0	1	X	X	0	X	0
0	1	X	X	1	X	1
1	0	X	0	X	X	0
1	0	X	1	X	X	1
1	1	0	X	X	X	0
1	1	1	X	X	X	1

Table 12.11 4-output demultiplexer truth table.

Select[1]	Select[2]	Input	Output 3	Output 2	Output 1	Output 0
0	0	0	0	0	0	0
0	0	1	0	0	0	1
0	1	0	0	0	0	0
0	1	1	0	0	1	0
1	0	0	0	0	0	0
1	0	1	0	1	0	0
1	1	0	0	0	0	0
1	1	1	1	0	0	0

12.9.5 Lookup Tables (LUTs)

PSoC 3/5LP have a lookup table component[1] that can be used to provide any logic function with as many as five inputs and eight outputs, inclusive. Such functions are implemented by creating logic equations that are implemented in the UDB PLDs. The LUT should be used at any time that a particular input combination should generate a specific set of outputs. The LUT allows an easy method of specifying the input to output relationship without having to generate specific gate level combinatorial logic. The use of the optional registered output mode allows the generation of sequential logic. State machines may also be created by registering the outputs and routing some of the outputs back to the LUT inputs. The LUT can configure all its outputs for all the possible input combinations. Additionally, it can be configured to register the output data on the rising edge of an input clock. Because the LUT is a hardware-only block, it does not have any software configuration options. The default LUT is configured with two inputs and two outputs, and the "Register Outputs" option is not selected.

The *Clock* input of the LUT is only available if the "Register Outputs" option is selected. All outputs will be registered on the rising-edge of this clock. Any clock in the system can be selected however it should be noted that if any of the outputs go to an I/O they will not work correctly if the LUT is operating faster than the fastest I/O operating speed of the type of PSoC used, e.g., 33 MHz in the case of the PSoC 3.

12.9.6 Logic High/Low

Logic High/Low components are provided as part of the PSoC 3/5LP architectures to provide constant digital values used to hard code digital inputs to in part optimize resource usage. The logic high and logic low functions are used for inputs that remain constant, e.g., for enabling timers, '1' counters, etc. Logic low is defined as '0' and logic High as '1'.

12.9.7 Registers

PSoC 3/5LP provide two types of very special registers, i.e., control registers and status register. The former is used to control/interact with a module and the latter is used when the firmware needs status information about a module. Thus, the status register allows the firmware to read digital signals and the control register can be used as a configuration register to allow the firmware[32] to specify the desired behavior of the digital system. The status register has a clock input and eight connections for status input, $status_0 - status_7$. The number of inputs depends on the *NumInputs* parameter and the firmware queries the input signals by reading the status register. The firmware sets the values of the output terminals for the control register by writing to it. The number of outputs depends on the *NumOutputs* parameter represents the number of the output terminals (specified as 1–8) with a default value of 8.

The *Bit0Mode -Bit7Mode* parameters are definable in PSoC Creator and used to set specific bits of the Status Register to be held high after being registered until a read is executed which also clears all the registered values. The settings are: *Transparent* and *Sticky* (Clear on Read). By default, a CPU read of this register will transparently read the state of the associated routing net. This mode can be used for a transient state that is computed and registered internally in the UDB.

---

[32]Note: The terms firmware and software are used throughout this text interchangeably and it is left as an exercise for the reader to determine which, if either, is more appropriate, within a given context.

In the *Sticky Status, with Clear on Read* mode, the associated routing net is sampled on each cycle of the status and control clock and if the signal is high in a given sample, it is captured in the status bit and remains high, regardless of the subsequent state of the associated route. When the CPU firmware reads the status register, the bit is cleared. Clearing of the status register is independent of the mode and will occur even if the block clock is disabled, it is based on the bus clock and occurs as part of the read operation.

```
Example 6.5: Sample C source code for reading/writing from/to
the status/control registers.

include <device.h>
void main()
{
uint8 value;
value = Status_Reg_1_Read();
}


#include <device.h>
void main()
{
uint8 value;
Control_Reg_1_Write(0x3E);
value = Control_Reg_1_Read();
}
```

12.9.8 PSoC 3/5LP Counters

PSoC 3/5LP architecture includes counters and timers which are important in most embedded systems. These counters are capable of counting up, down, or up-and-down and are configurable to allow them to operate as 8, 16, 24, or 32-bit counters. Options include compare out and capture input. Additionally, enable and reset inputs can be synchronized with other PSoC components and the period of a count is programmable.

Counters are particularly useful in situations that require the "counting" of events and if required capturing the current count value for programmatic use or to compare an output for hardware synchronization and.or signaling. In the simplest configuration Counters count either up or down and utilize a single input from either other components internal to PSoC, or an I/O pin. A "count event" occurs with each rising edge of the input and continues until the terminal count is reached at which point the Counter is "reloaded". In the case of a "down" counter the terminal count when the Counter reaches zero and subsequently the Counter is reloaded with the Period value. Counters also have optional "capture" functions that allow the current count to be captured for comparison or software processing.

Up/down-counters are similar to up- and down-counters, but there are some important differences. One configuration provides a count input and a direction input. When active a '1' on the up and down input forces the counter to increment by one on a rising edge of the count input, a '0' on the up and down input causes the counter to decrement by one on a rising edge of the count input. The other configuration provides an up-count input and a down count input. The counter will increment or decrement based on which respective count input had a rising edge. This version of the counter requires an additional oversample clock input while all other versions do not. On counter underflow and overflow, flags are set and the period reloaded allowing glitch-proof counter expansion in firmware. During each clock cycle, the optional compare output compares the current count to the compare value. The compare mode is configurable to all the standard comparison modes providing several waveform options. The compare output provides a logic-level that may be routed to I/O pins and other component inputs.

An optional capture input copies the current count value into a storage location on a rising edge. Firmware can be used to read the capture value at any time without timing restrictions as long as the capture FIFO[33] has room. The Capture FIFO allows storage for a maximum of 4 capture values. The enable and reset inputs allow the Counter to be synchronized with other internal or external hardware. The Counter enable signal may be generated by a software API, the hardware compare input, or the AND of both. For the hardware Enable input, the counter only counts while the Enable input is high. A rising edge on the reset input causes the counter to reset its count as if the terminal count was reached. If the reset input remains high, the counter will remain in reset. An interrupt can be programmed to be generated under any combination of the following conditions: when the counter reaches the terminal count, the comparator output is asserted, or a capture event has occurred.

In the default mode the counter counts the number of rising edge events on the count input. The counter can also be used as a clock divider by supplying a clock signal to the count input and using the compare, or terminal count, outputs as the divided clock output. Furthermore, the counter can be employed as a frequency counter by using a known period on the enable input of the counter while counting the signal's input. After the enable period, the counter will contain the number of rising edges measured during that period allowing calculation of the input frequency. The up- and down-counter may be used to measure complementary events such as the output of a quadrature decoder to measure a sensor's position data. A timer component is a better choice for timing the length of events, measuring the interval of multiple rising and/or falling edges, or for multiple capture events. Another option is to use a PWM when multiple compare outputs are involved because of the support a PWM provides for center alignment, output kill, and deadband[34] outputs.

The input and output connections for PSoC 3/5LP's counter component includes a clock input which defines the oversample clock rate required to increment on upCnt, or decrement on dwnCnt, or alternatively cause neither an upCnt or dwnCnt. The count input is the input connection for the signal to be counted. A counter value is either incremented or decremented depending on the assigned direction or pin usage selected for the *Clock Mode* parameter. The reset input resets the counter to the starting value. For the "Up Counter" configuration, the starting value is zero and For "Down Counter", "Count Input and Direction" and "Clock With UpCnt & DwnCnt" configurations, the starting value is set to the current period register value.

12.9.8.1 UDB Implementation of a Counter

When the UDB mode is selected for a Counter component:

- The *Resolution* parameter defines the bit-width resolution of the counter. This value may be set to 8, 16, 24, or 32 for maximum count values of 255, 65535, 16777215, and 4294967295 respectively,
- The *Compare Mode* (software option) parameter configures the operation of the Compare output signal which is the status of a compare between the compare value parameter and current counter

---

[33]FIFO refers to a First-In-First-Out device in which the sequential order of the input data is the same order in which the output data occurs.

[34]Deadband refers to a signal range for which nothing happens. It is often employed to prevent oscillation of a device or "hunting". A deadband is analogous to the mechanical backlash in a gear system.

value. It defines the initial setting loaded into the control register which can be updated at any time to re-configure the compare operation of the counter.

1. *Less Than*—The Counter value is less than the compare value
2. *Less Than Or Equal To*—The Counter value is less than or equal to the compare value
3. *Equal To*—The Counter value is equal to the compare value
4. *Greater Than*—The Counter value is greater than the compare value
5. *Greater Than Or Equal To*—The Counter value is greater than or equal to the compare value
6. *Software Controlled*—The compare mode can be set during runtime with the SetCompare-Mode() API call to any one of the 5 compare modes listed above.

- The *Clock Mode* can be up-counter, down-counter, count input and direction, and count with upCnt and dwnCnt. This parameter configures the desired clocking and direction control method. The value is an enumerated type and can be set to any of the following options:
  1. *Count Input + Direction*—The counter is a bi-directional counter counting up while the up_ndown input is high on each rising edge of the input clock and counting down while up_ndown is low on each rising edge of the input clock.
  2. *Clock with UpCnt DwnCnt*—The counter is a bi-directional counter incrementing the counter by 1 for each rising edge on the upCnt input, and decrementing the counter by 1 for each rising edge of the dwnCnt input.
  3. *Up Counter*—The counter is an up-counter only configured to increment on any rising edge of the input clock signal while the counter is enabled.
  4. *Down Counter*—The counter is a down-counter only configured to decrement on any rising edge of the input clock signal while the counter is enabled.
- The *Period* parameter defines the max count's value (or rollover point) for the counter. This parameter defines the initial value loaded into the period register which can be changed at any time by the software with the Counter_WritePeriod() API. The limits of this value are defined by the Resolution parameter. For 8, 16, 24 and 32-bit Resolution parameters the maximum value of the Period value is defined as $(2^8) - 1$, $(2^{16}) - 1$, $(2^{24}) - 1$, and $(2^{32}) - 1$ or $255, 65535, 16777215, and 4294967295$ respectively. When Clock Mode is configured as "Clock with UpCnt & DwnCnt" or "Count Input and Direction" the counter is set to the period at start and any time the counter overflows at all 0xFF or underflows at all 0x00.
- The *Capture Mode* parameter configures the implementation of the capture input. This value is an enumerated type and can be set to any of the following values:
  1. *None*: No capture implemented and the capture input pin is hidden
  2. *Rising Edge*: Capture the counter value on any rising edge of the capture input Falling *Edge*: Capture the counter value on any falling edge of the capture input
  3. *Either Edge*: Capture the counter value on any edge of the capture input
  4. For the *Software Controlled* mode, the mode is set at runtime by setting the *Compare Mode* bits in the control register Counter_CTRL_CAPMODE_MASK with the enumerated capture mode types defined in the Counter.h header file.
- The *Enable Mode* parameter configures the enable implementation of the counter. This value is an enumerated type and can be set to any of the following options:
  1. *Software*: The Counter is enabled based on the enable bit of the control register only.
  2. *Hardware*: The Counter is enabled based on the enable input only.
  3. *Software And Hardware*: The Counter is enabled if, and only if, both the input and the control register bits are active.
- The *Reload Counter* parameters allow the counter value to be reloaded when one or more of the following selected events occur. The counter is reloaded with its start value (for an up-counter this is reloaded to a value of Zero, for a down-counter this is reloaded to the max counts or period

value). This configuration is OR'd with all the other Reload Counter parameters to provide the final reload trigger to the counter.

1. On *Capture* The counter value will be reloaded when a capture event has occurred. By default, this parameter is set to false. This parameter is only shown when UDB is selected for Implementation.
2. On *Compare*—The counter value will be reloaded when a compare true event has occurred. By default, this parameter is set to false. This parameter is only shown when the UDB is selected for Implementation.
3. On *Reset*—The counter value will be reloaded when a reset event has occurred. By default, this parameter is set to true. This parameter is always shown, but it is only active when UDB is selected for Implementation.
4. On *TC*—The counter value will be reloaded when the counter has overflowed (in count up mode) or underflowed (in count down mode). By default, this parameter is set to true. This parameter is always shown, but it is only active when UDB is selected for Implementation. When the clock mode is set to "Clock with UpCnt & DwnCnt" this option reloads to the period value when the count is 0x00 or all 0xFF. This configuration is OR'd with all the other reload parameters to provide the final reload trigger to the counter.

- The *Interrupt* parameters allow the initial interrupt sources to be configured. These values are OR'd with any of the other Interrupt parameters to give a final group of events that can trigger an interrupt. The software can re-configure this mode at any time; this parameter simply defines an initial configuration.

1. On *TC*—This option is always available; it is set to false by default.
2. *On Capture*—This option is set to false by default. It is always shown, but it is only active when UDB is selected for Implementation.
3. On *Compare*—This option is set to false by default. It is always shown, but it is only active when UDB is selected for Implementation.

and

- The *Compare Value* (Software Option) parameter defines the initial value loaded into the compare register of the counter. This value is used in conjunction with the Compare Mode parameter selected to define the operation of the compare output. This value can be any unsigned integer value from 0 to $(2^R esolution - 1)$, but it must be less than the max_counts or Period value. If the value is allowed to be larger than max_counts, the compare output would be a constant '0' or '1' value and is therefore, not allowed.

12.9.8.2 Clock Selection

The *Counter* component's clock/count input can be any signal whose rising edges are to be counted. When configured to utilize the fixed-function timer block in the device, the clock input to the *Counter* component has the following restrictions:

1. The clock input must be from a user-defined clock that is synchronized to the bus clock or directly from the bus clock via a clock defined using the existing clock feature, and with a source of the bus clock.
2. If the frequency of the clock matches the bus clock, then the clock must be a direct connection to the bus clock using the existing clock scheme listed earlier. A user-defined clock with a frequency that matches the bus clock will generate an error during the build process.

The Timer, Counter and PWM components share a common set of internal requirements and are therefore, implemented in PSoC 3/5LP as fixed-function blocks. When the Fixed Function implementation of a Counter, Timer or PWM is to be employed, certain limitations are imposed, viz., operation is restricted to

- 8 or 16-bits only
- Down count only
- Reload on Reset and
- Terminal Count only
- Interrupt on Terminal Count only

When configured to utilize the fixed-function timer block, the clock input to the Counter component will have the following restrictions:

1. The clock input must be from a user-defined clock that is synchronized to the bus clock or directly from the bus clock (via a clock defined using the existing clock feature, and with a source of the bus clock).
2. If the frequency of the clock matches the bus clock, then the clock must be a direct connection to the bus clock (again using the existing clock scheme listed earlier). A user-defined clock with a frequency that matches the bus clock will generate an error during the build process.

Thus, the default configuration of the Counter component provides a very simple counter that increments a count value on every rising edge of the clock input. The count input is the signal whose rising edge is counted and the reset input provides a hardware mechanism for resetting the count value. Since this is configured as an up-counter by default, when a reset event occurs on the reset input, the counter value is reset to zero. Terminal count indicates in real-time whether the counter value is at the terminal count (Maximum value or Period). The period is programmable to be any value from 1 to $(2^{Resolution}) - 1$.

The compare output is a real-time indicator that the count value compares to the compare value as defined in the compare configuration. The compare configuration is set in the control register for the component and can be set by software at any time. The default Maximum Count (Period) is set to $2^{Resolution} - 1$ and the compare value is set to $1/2$ of that number. The counter increments on any rising edge clock until it rolls over at the terminal count.

A simple extension of the default configuration provides a clock divider with a programmable duty cycle. If a clock input is applied to the counter clock input with the default period and compare parameter settings, the compare output will be a 50% duty cycle clock with 1/256th the frequency of the input clock. This is because the default compare configuration is less than or equal to which would have a high state on the compare output from 0 to 127 and a low signal from 128 to 255. Any even number period setting can have a 50% duty cycle if the compare value or compare configuration is changed. Adding hardware enable functionality to the basic counter allows a frequency counter function to be implemented. If the *Enable* input is driven by a known period signal, such as a 1kHz Clock starting with a counter value of 0x00 and an up-counter implementation, the frequency of an input signal is easily determined.

12.9.9 Timers

Timers are a form of counter designed to time the interval between hardware events. They are ubiquitous in embedded systems being used to determine elapsed times between events, periods of recurrent events, triggers for various types of events, elapsed time since an event or function last occurred, etc. Some timers have the same features found in *Counters* and *PWMs*. Typical uses of PSoC 3/5LP timers include recording the number of clock cycles between events, measuring the number of clock cycles between two rising edges generated for example by a tachometer sensor, or the measurement of the period and duty cycle of a PWM input.

For PWM measurement, a PSoC 3/5LP timer is configured to start on a rising edge, capture the next falling edge and then capture and stop on the next rising edge. An interrupt on the final capture signals the CPU that all the captured values are available in the FIFO. The PSoC 3/5 timer can be used as a clock divider by driving a clock into the clock input and using the terminal count output as the divided clock output. In general, timers share many features with counters, and PWMs. A counter is better used in situations that require the counting of a number of events but also provides rising edge capture input and compare the output. A PWM is more appropriate for situations requiring multiple compare outputs with control features like center alignment, output kill, and deadband outputs.

PSoC 3/5LP timers are designed to provide an easy method for timing complex, real-time events with great accuracy and minimal CPU overhead. Timers of this type only count down from a predefined state defined by the "period value" which is inversely proportional to the timer's clock frequency. Therefore, the minimal time interval that can be measured is determined by the timer's clock.

The maximum timer interval that can be measured is given by:

$$T_{max} = (Timer\ Clock\ Frequency)(Timer\ Resolution) \tag{12.29}$$

The Timer component provided with PSoC includes a function known as "capture". This function is an extremely useful feature in that it makes it possible to "capture" the Timer's "count" at any particular moment and save that value in a FIFO[35] storage location. The FIFO is capable of storing four such values after which the data in the FIFO will be overwritten by new "captured" data.[36] This data can be accessed programmatically, that is by the firmware, without destroying the data read from the FIFO.

PSoC 3/5LP timers are specifically designed to provide an easy method of timing complex real-time events accurately with minimal CPU intervention and may be combined with other analog and digital components to create complex peripherals [9, 11]. PSoC 3/5LP timers count only in the down direction starting from the period value and require a single clock input. The input clock period is the minimum time interval able to be measured. The maximum timer measurement interval is the input clock period multiplied by the resolution of the timer. The signal interval to be captured may be routed from an I/O pin, or other internal component outputs. Once started, the timer operates continuously and reloads the timer period value on reaching the terminal count. The timer capture input is the most useful feature of the timer because on a capture event the current timer count is copied into a storage location. Firmware can then read the capture value at any time without timing restrictions as long as the capacity of the capture FIFO is not exceeded.

However, it is important to not write to the FIFO when it is full to avoid overwriting the oldest value. If the oldest value is overwritten, the newly captured value will be returned in its place the next time the FIFO is read. It is up to the software to keep track of the amount of data that is written to the FIFO if unwanted overwriting of its data is to be avoided.

The Capture FIFO allows storage of up to 4 capture values. The capture event may be generated by software, rising or falling edges, or all edges allowing great measurement flexibility. To further assist in measurement accuracy of fast signals an optional 7-bit counter may be used to capture every

---

[35]FIFO = First-In-First-Out

[36]The oldest data is overwritten first, and therefore, the newest data is returned the next time the FIFO is "read".

n[2..127] of the configured edge type. The trigger and reset inputs allow the timer to be synchronized with other internal or external hardware. The optional trigger input is configurable so that a rising edge, falling edge, or all edges to start the timer counting. A rising edge on the reset input causes the counter to reset its count as if the terminal count was reached.

PSoC 3/5LP timers support:

- 8-, 16-, 24- or 32-bit resolution,
- implementation as a Fixed Function or UDB device,
- a 4-deep capture FIFO, an optional capture edge counter,
- configurable hardware/software enable
- continuous or single-shot running modes.

12.9.9.1 PSoC 3/5LP Timer I/O Connections

Timer I/O connections for a PSoC 3/5LP timer include:

- a clock input that determines the operating frequency of the timer,
- a capture input that copies the period counter value to a 4-sample FIFO in the UDB, or alternatively to a single sample register in the Fixed Function block,
- a capture_out output that is an indicator of when a hardware capture has been triggered
- an interrupt output that is a copy of the interrupt source a terminal count (tc) output that goes high if the current count value is equal to the terminal count (zero)[37]
- a reset input that resets the period counter, to the period value, and the capture counter. This reset function is synchronous and requires at least one rising edge of the clock.
- an enable input that enables the period counter to decrement on each rising edge of the clock. If the enable value is low, then the outputs remain active but the timer does not change states,

12.9.10 Shift Registers

Shift registers are sequential logic circuits that typically consist of cascaded flip-flops sharing a common clock with the output of one flip-flop serving as the input for the next flip-flop in the chain. They are available in a number of configurations as discrete devices, e.g.,

- serial-in, serial-out shift registers[38]
- Serial-In-Parallel-Out shift (SIPO) registers
- parallel in serial out shift registers
- bidirectional (reversible) shift registers[39]

PSoC 3/5LP's Shift Register component provides synchronous shifting of data into and out of a parallel register. The parallel register can be read or written to by the CPU or DMA. The Shift Register component provides universal functionality similar to standard 74xxx series logic shift registers including: 74164, 74165, 74166, 74194, 74299, 74595 and 74597.

---

[37] The terminal count output is a "zero-compare" of the period counter value, i.e., if the period counter is zero the output will be high.

[38] Serial shift registers, in their simplest form, allow shifting in only one direction, i.e., from the input towards the output, often referred to as "right-shift" or "left-shift".

[39] Bidirectional shift registers allow shifts in either direction, e.g., left-to-right or right-to-left.

In most applications, the Shift Register is used in conjunction with other components and logic to create higher-level application-specific functionality, such as a counter to count the number of bits shifted. In general usage, the PSoC 3/5LP shift register functions as a 1–32 bit shift register that shifts data on the rising edge of the clock input.

The shift direction is configurable and allows a right shift where the MSB shifts in the input and the LSB shift out the output or a left shift where the LSB shift in the input and the MSB shifts out the output. The reset input (active high) causes the entire shift register contents to be set to all zeros. The reset input is synchronous to the clock input. The shift register value may be read by the CPU or DMA at any time.

A rising edge on the optional store input transfers the current shift register value to the FIFO from where it can later be read by the CPU. The store input is asynchronous to the clock input. The shift register value may be written by the CPU or DMA at any time. A rising edge on the optional load input transfers pending FIFO data (already written by CPU or DMA) to the shift register. The load input is asynchronous to the clock input. The Shift Register component may generate an interrupt signal on any combination of the following signals; Load, Store or Reset.

12.9.11 Pseudo-Random Sequence Generator (PRS)

The PSoC 3/5LP pseudo-random sequence generator[40] supported in the PSoC 3/5LP architecture can be used to provide a pseudo-random bitstream or random bits as required. It utilizes a Galois[41] linear feedback shift register[42] (LFSR) to produce the bitstream based on maximal code length, or period. Setting the *Enable Input* on the PRS allows the PRS to run continuously and it can be started with a nonzero seed value. By implementing the LFSR in hardware it is possible to generate very fast pseudo-random sequences. GPS, spread spectrum, video games, cryptography, noise generators, and many other applications make use of such sequences. A simple example of a Galois PRNG employing D-type flip-flops and exclusive OR (XOR) gates is shown in Fig. 12.26. For the implementation shown the initial state can be arbitrarily chosen except for all zeros because in that case, the system would remain in the "zero" state.

The PRS has the following features:

- continuous, or single-step, run modes,
- an enable input for synchronized operation with other components,
- a computed pseudo-random number can be read directly from the LFSR,

---

[40] A pseudo-random number generator does not generate a truly random sequence of values because ultimately it will repeat the sequence.

[41] Galois was a nineteenth-century French mathematician who made some significant contributions to Group Theory and the algebra of polynomials.

[42] A LFSR is a finite state machine which consists of a combination shift register and XOR function in which a seed value is placed in a shift register and is shifted one bit to the right and if the bit value shifted from the rightmost bit position is a 1, then the register is XORed with a mask, otherwise, the bit register is shifted one bit position to the right again and then the process of examining the bit and determining if the register should be subjected to an XOR with the mask is repeated. This process continues for a long as required to produce the required pseudo-random bitstream. In some applications, single-stepping is employed to produce individual random bit values as required. The number of bits that are generated before the sequence repeats are referred to as its "period". Maximal period LFSRs generate 2^{n-1} bits before repeating where n is the bit length of the register. A 32-bit LFSR will produce in excess of 4 billion bits before repeating. Each of the bit positions in the shift register that has an effect on the next state is referred to as a "tap". The speed of the LFSR in generating pseudo-random bitstreams is largely a result of the minimal use of combinational logic.

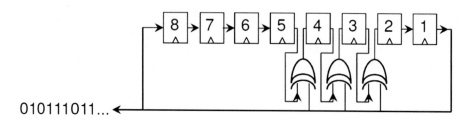

Fig. 12.26 A simple example of a Galois PRNG.

- either a standard of custom polynomial/seed value can be employed,
- a serial output bitstream
- a PRS sequence lengths 2–64 bits in length

The repeating sequence of states of an LFSR allows it to be used as a divider, or as a counter when a non-binary sequence is acceptable. LFSR counters have simpler feedback logic than natural binary counters or Gray code counters and therefore, can operate at higher clock rates. However, it is necessary to ensure that the LFSR never enters an all-zeros state, for example by presetting it at start-up to any other state in the sequence. The PRS has an enable input, a clock input, and a serial bitstream output. The clock input is used in continuous mode only and the output is synchronized and when operated in continuous modes The PRS runs as long as the *Enable* input is held high.

12.9.12 Precision Illumination Signal Modulation (PrISM)

PSoC 3/5LP have precision illumination signal modulation (PrISM) components that use linear feedback shift registers (LFSRs) of the type discussed in Sect. 12.9.11 to generate a pseudo-random bitstream sequence and up to two user-adjustable, pseudo-random, pulse densities. ranging from 0 to 100. The PrISM runs continuously after started and as long as the Enable input is held high. Its pseudo-random number generator may be started with any valid seed value excluding 0.

The result is modulation technology that significantly reduces low-frequency flicker and radiated electromagnetic interference (EMI) which are common problems with high brightness LED designs. The PrISM is also useful in other applications requiring this capability, such as motor controls and power supplies.

12.9.13 Quadrature Decoder

A quadrature decoder [14] is used to decode the output of a quadrature encoder. A quadrature encoder senses the current position, velocity, and direction of an object (for example, mouse, trackball, robotic axles, and others). It can also be used for precision measurement of speed, acceleration, and position of a motor's rotor and with rotary knobs to determine user input.

The PSoC Quadrature Decoder (QuadDec) Component transitions of a pair of digital signals to be counted. The signals are often provided by a speed/position feedback system mounted on a motor, or trackball. The signals, typically called A and B, are positioned 90° out of phase with each other, which results in a Gray code output[43] which is necessary in order to avoid glitches. It also allows the

---

[43] A Gray code is a sequence of bits in which only one bit changes on each count.

detection of direction and relative position. A third optional signal, named *Index*, is used as a reference to establish an absolute position once per rotation.

The index input detects a reference position for the quadrature encoder. If an index input is provided, when inputs A, B, and index are zero, the counter is reset to zero. Additional logic is typically added to gate the index pulse. Index gating allows the counter to be reset only during one of many possible rotations, e.g., as in the case of a linear actuator that only resets the counter when the far limit of travel has been reached. This limit is signaled by a mechanical limit switch whose output is AND-ed with the Index pulse.

The clock input is required for sampling and glitch filtering of the inputs. If glitch filtering is used then the filtered outputs will not change until three successive samples of the input are the same value. For effective glitch filtering, the sample clock period should be greater than the maximum time during which glitching is expected to take place. A counter can be incremented/decremented at a resolution of $1\times, 2\times$, or $4\times$ the frequency of the A and B inputs as shown in Fig. 12.27. The clock input frequency should be greater than, or equal to, $10\times$ the maximum A or B input frequency (Fig. 12.28).

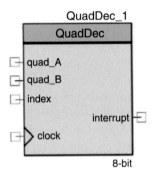

Fig. 12.27 QuadDec component.

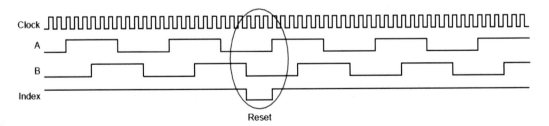

Fig. 12.28 QuadDec timing diagram.

An interrupt output is provided following the occurrence of one or more counter overflow/under-flow, counter rest due to an index input or invalid state transition from the A and B inputs (Fig. 12.29).

The counter size is defined in terms of the number of bits. The counter holds the current position encoded by the quadrature encoder. A counter size should be selected that is large enough to encode the maximum position in both the positive and negative directions. The 32-bit counter implements the lower 16 bits in the hardware counter and the upper 16 bits in software to reduce hardware resource usage. Available settings include: 8, 16, or 32 bits.

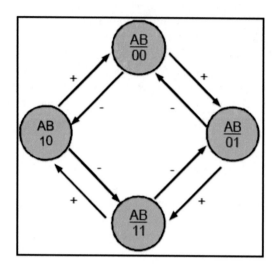

Fig. 12.29 QuadDec state diagram.

A field is provided in PSoC Creator that determines whether to apply digital glitch filtering to all inputs. Filtering can be applied to reduce the probability of having miscounts due to glitches on the inputs. Some filtering is already done using hysteresis on the GPIOs, but additional filtering may be required. If selected, filtering is applied to all inputs. The filtered outputs do not change until three successive samples of the input are the same value. For effective filtering, the period of the sample clock should be greater than the maximum time during which glitching is expected to take place.

```
Example 6.7 Sample source code demonstrating the use of the quadrature
decoder and writing the position information to an LCD.

#include <device.h>
void main()
{
uint8 stat;
uint16 count16;
uint8 i;
CYGlobalIntEnable;
LCD_1_Start();
QuadDec_1_Start();
QuadDec_1_SetInterruptMask(QuadDec_1_COUNTER_RESET |
QuadDec_1_INVALID_IN);
stat = QuadDec_1_GetEvents();
LCD_1_Position(1, 0);
LCD_1_PrintInt16(stat);
while(1)
{
CyDelay();
count16 = QuadDec_GetCounter();
LCD_1_Position(2, 0);
LCD_1_PrintInt16(count16);
}
}
```

12.9.14 The QuadDec Application Programming Interface

Application Programming Interface (API) routines allow the component to be configured using the software. The following table lists and describes the interface to each function. Each function is described in more detail in Appendix A.

By default, PSoC Creator assigns the instance name "QuadDec_1" to the first instance of a component in a given design. However, it can be renamed to any unique value that follows the syntactic rules for identifiers. The instance name becomes the prefix of every global function name, variable, and constant symbol. For readability, the instance name used in the following table is "QuadDec." (Figs. 12.30 and 12.31)

Global Variables

Function	Description
QuadDec_initVar	QuadDec_initVar indicates whether the Quadrature Decoder has been initialized. The variable is initialized to 0 and set to 1 the first time QuadDec_Start() is called. This allows the component to restart without reinitialization after the first call to the QuadDec_Start() routine. If reinitialization of the component is required, then the QuadDec_Init() function can be called before the QuadDec_Start() or QuadDec_Enable() function.
QuadDec_count32SoftPart	High 16 bits of 32-bit counter value is stored in this variable.
QuadDec_swStatus	Status register value is stored in this variable.

Fig. 12.30 QuadDec global variables.

Function	Description
QuadDec_Start()	Initializes UDBs and other relevant hardware
QuadDec_Stop()	Turns off UDBs and other relevant hardware
QuadDec_GetCounter()	Reports the current value of the counter
QuadDec_SetCounter()	Sets the current value of the counter
QuadDec_GetEvents()	Reports the current status of events
QuadDec_SetInterruptMask()	Enables or disables interrupts due to the events
QuadDec_GetInterruptMask()	Reports the current interrupt mask settings
QuadDec_Sleep()	Prepares the component to go to sleep
QuadDec_Wakeup()	Prepares the component to wake up
QuadDec_Init()	Initializes or restores default configuration provided with the customizer
QuadDec_Enable()	Enables the Quadrature Decoder
QuadDec_SaveConfig()	Saves the current user configuration
QuadDec_RestoreConfig()	Restores the user configuration

Fig. 12.31 The QuadDec functions.

12.9.15 Cyclic Redundancy Check (CRC)

A cyclic redundancy check (CRC) [13] provides the basis upon which the integrity, or lack thereof, of binary data has been compromised and is very useful when transferring such data from location to the other. The receiver of the data simply recomputes the check compares to the CRC value transferred with the dais a method of determining who the integrity, or lack thereof, of digital data that is often used when transferring data from one spatial domain to another. It is equivalent to conducting polynomial long division and retaining only the remainder. The remainder is then appended to the data and transmitted to another location. Upon arrival, the division process is repeated and the transmitted remainder is compared to the locally calculated one. If they do not agree some systems return a negative acknowledgment to the transmitter, causing the data to be retransmitted. This type of check is supported by PSoC 3/5LP and in the default configuration is used to compute the value for a serial bit-stream of arbitrary length that is sampled on the rising edge of the data clock. The value is either reset to 0 before starting or can optionally be seeded with an initial value. Following the computation of the for a particular bitstream, the computed value is available. s are often used for checking the integrity of stored data as well as transmitted data.

The default use of PSoC's Cyclic Redundancy Check (CRC) component is to compute the CRC of a serial bitstream of any arbitrary length. The input data is sampled on the rising edge of the data clock. A CRC value is reset to 0 before starting, or alternatively, can optionally be seeded with an initial value. On completion of processing the bitstream, the computed CRC value may be read out. This component can be used as a checksum to detect alteration of data during transmission or storage. CRCs are popular because they are simple to implement in binary hardware, are easy to analyze mathematically, and are excellent for detecting common errors, particularly single bit errors, caused by noise in transmission channels (Fig. 12.32).

Polynomial Name	Polynomial	Use
Custom	User defined	General
CRC-1	$x + 1$	Parity
CRC-4-ITU	$x^4 + x + 1$	ITU G.704
CRC-5-ITU	$x^5 + x^4 + x^2 + 1$	ITU G.704
CRC-5-USB	$x^5 + x^2 + 1$	USB
CRC-6-ITU	$x^6 + x + 1$	ITU G.704
CRC-7	$x^7 + x^3 + 1$	telecom systems, MMC
CRC-8-ATM	$x^8 + x^2 + x + 1$	ATM HEC
CRC-8-CCITT	$x^8 + x^7 + x^3 + x^2 + 1$	1-Wire bus

Fig. 12.32 Polynomial names.

PSoC Creator allows the developer to select from a variety of CRC computational schemes including custom configurations (Fig. 12.33).

PSoC 3/5LP's component has provisions for inputting data and a clock signal with the result of the check being accessible programmatically.

```
Example 6.4 The following C source code illustrates how to computer
a   using PSoC 3/5LP's   component.
#include <device.h>
void main()
uint32 CRC_val = 0;
uint16 CRC_part1 = 0;
uint16 CRC_part2 = 0;
uint8 i = 0;
uint8 j = 0;

clock_Enable();
id_Enable();
LCD_Start();
_Start();
for(i=0;i<4;i++)
{
for(j=0;j<=13;j+=5)
{
CRC_val = _Read();
CRC_part2 = HI16(CRC_val);
CRC_part1 = LO16(CRC_val);

LCD_Position(i, j);
LCD_PrintInt16(CRC_part2);
j+=4;
LCD_Position(i, j);
LCD_PrintInt16(CRC_part1);
CyDelay(500);
}
}
for(;;){}
}
```

12.10 Recommended Exercises

12-1 N inverters are connected as shown. Assuming that the delay introduced by each inverter is $10\,\mu s$, derive an expression in terms of the number of inverters and the the delays for each inverter that defines the frequency of oscillation of the inverter network. Under what conditions will this circuit not oscillate.

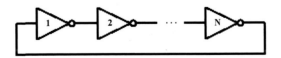

12-2 Explain the basic differences between the Delta-Sigma and SAR A/D converters. What are the advantages/disadvantages of each? In some applications, the sampling takes place at frequencies much higher than the Nyquist sampling rate. Explain the reason(s) for employing such a high sampling rate. Explain the role of decimation in such cases.

Polynomial Name	Polynomial	Use
CRC-8-Maxim	$x^8 + x^5 + x^4 + 1$	1-Wire bus
CRC-8	$x^8 + x^7 + x^6 + x^4 + x^2 + 1$	General
CRC-8-SAE	$x^8 + x^4 + x^3 + x^2 + 1$	SAE J1850
CRC-10	$x^{10} + x^9 + x^5 + x^4 + x + 1$	General
CRC-12	$x^{12} + x^{11} + x^3 + x^2 + x + 1$	telecom systems
CRC-15-CAN	$x^{15} + x^{14} + x^{10} + x^8 + x^7 + x^4 + x^3 + 1$	CAN
CRC-16-CCITT	$x^{16} + x^{12} + x^5 + 1$	XMODEM,X.25, V.41, Bluetooth, PPP, IrDA, CRC-CCITT
CRC-16	$x^{16} + x^{15} + x^2 + 1$	USB
CRC-24-Radix64	$x^{24} + x^{23} + x^{18} + x^{17} + x^{14} + x^{11} + x^{10} + x^7 + x^6 + x^5 + x^4 + x^3 + x + 1$	General
CRC-32-IEEE802.3	$x^{32} + x^{26} + x^{23} + x^{22} + x^{16} + x^{12} + x^{11} + x^{10} + x^8 + x^7 + x^5 + x^4 + x^2 + x + 1$	Ethernet, MPEG2
CRC-32C	$x^{32} + x^{28} + x^{27} + x^{26} + x^{25} + x^{23} + x^{22} + x^{20} + x^{19} + x^{18} + x^{14} + x^{13} + x^{11} + x^{10} + x^9 + x^8 + x^6 + 1$	General
CRC-32K	$x^{32} + x^{30} + x^{29} + x^{28} + x^{26} + x^{20} + x^{19} + x^{17} + x^{16} + x^{15} + x^{11} + x^{10} + x^7 + x^6 + x^4 + x^2 + x + 1$	General
CRC-64-ISO	$x^{64} + x^4 + x^3 + x + 1$	ISO 3309
CRC-64-ECMA	$x^{64} + x^{62} + x^{57} + x^{55} + x^{54} + x^{53} + x^{52} + x^{47} + x^{46} + x^{45} + x^{40} + x^{39} + x^{38} + x^{37} + x^{35} + x^{33} + x^{32} + x^{31} + x^{29} + x^{27} + x^{24} + x^{23} + x^{22} + x^{21} + x^{19} + x^{17} + x^{13} + x^{12} + x^{10} + x^9 + x^7 + x^4 + x + 1$	ECMA-182

Fig. 12.33 CRC polynomials.

s

12-3 Explain the advantages and disadvantages of digital filters versus analog filters and the limitations of each from a manufacturing, ost and performance perspective.

12-4 Design a fan controller system that utilizes four thermocouples, PWMs and tachometers together with a table that specifies how to respond to different, ambient, temperature conditions within an enclosure.

12-5 Select a digital filter type from this chapter and design a bandpass filter with the same bandpass characteristics as that of the analog filter in Example 2.

12-6 Produce signal-flow-graphs for each of the following difference equations:

$$y[n] = x[n] - x[n-2]$$

$$y[n] = x[n] - x[n-1] - y[n-1]$$

Find the impulse-response for each. Do these equations represent stable and/or causal systems?

12-7 What happens in the logic circuit shown below when S=R=1 and the clock transitions from 1 to 0?

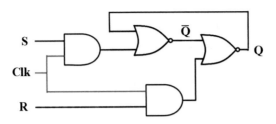

12-8 Create a design for the FSM shown below using a LUT component. Implement your design in PSoC Creator.

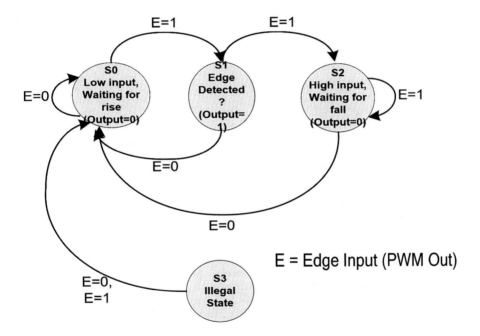

12-9 Given the following equations for computing quantization errors, prove that

$$\bar{e} = \frac{1}{Q} \int_{-Q/2}^{Q/2} e\, de = 0, \quad \text{and} \quad \overline{e^2} = \frac{1}{Q} \int_{-Q/2}^{Q/2} e^2\, de = \frac{Q^2}{12}$$

where Q is defined as

$$Q = \frac{R}{2^B}$$

(B is the number of bits in each sample and R the full scale range.)
Show that the quantization error can be expressed as

$$e_{rms} = \sqrt{e^2} = \frac{Q}{\sqrt{12}}$$

12-10 Based on Example 12-9, determine how many bits are required, in selecting an A/D converter, to provide a full scale range of 10 V and and quantization less than 1 mV. What is the quantization error for the number of bits you have selected. If the dynamic range is defined as

$$SNR = 20log_{10}\frac{R}{Q} = B$$

show that SNR = 6B dB and calculate the dynamic range for the A/D you have selected.

12-11 Produce signal-flow-graphs for each of the following difference equations:

$$y[n] = x[n] - x[n-2] \tag{12.30}$$

$$y[n] = x[n] - x[n-1] - y[n-1] \tag{12.31}$$

References

1. M. Ainsworth, PSoC 3 to PSoC 5LP Migration Guide. AN77835. Document No. 001-77835 Rev.*D. Cypress Semiconductor (2017)
2. R. Ashby, My First Five PSoC 3 Designs. Spec. £001-58878 Rev. *C. Cypress Semiconductor (2013)
3. Cyclic Redundancy Check (CRC). Document Number: 002-20387 Rev. *A. Cypress Semiconductor (2017)
4. Digital Logic Gates 1.0. Document Number: 001-50454 Rev. *F. Cypress Semiconductor (2017)
5. T. Dust, G. Reynolds, Designing PSoC Creator Components with UDB Datapaths. AN82156. Document No. 001-82156 Rev. *I1. Cypress Semiconductor (2018)
6. M. Hastings, PSoC 3 and PSoC 5LP: Getting More Resolution from 8-Bit DACs. AN64275. Document No. 001-64275 Rev.*G. Cypress Semiconductor (2017)
7. B.P. Lathi, *Modern Digital and Analog Communication Systems* (Oxford University Press, Oxford, 1995)
8. A. Mohan, SAR ADC in PSoC 3. AN60832. Cypress Semiconductor April (2010)
9. R. Narayanasamy, Designing an efficient PLC using a PSoC (2011). www.eetimes.com/designing-an-efficient-plc-using-a-psoc/#
10. D. Van Ess, *Learn Digital Design with PSoC, a Bit at a Time* (CreateSpace Independent Publishing Platform, 2014)
11. D. Van Eß, E. Currie, A.N. Doboli, *Laboratory Manual for Introduction to Mixed-Signal Embedded Design*. Cypress University Alliance. Cypress Semiconductor (2008)
12. D. Van Ess, P. Sekar, PSoC 1, PSoC 3, PSoC 4, and PSoC 5LP -Single-Pole Infinite Impulse Response (IIR) Filters. AN2099. Document No. 001-38007 Rev. *J. Cypress Semiconductor (2017)
13. M. Vijay Kumar, F. Antonio Rohit De Lima, PSoC Creator -Implementing Programmable Logic Designs with Verilog. AN82250. Document NUMBER:001-82250 Rev.*J. Cypress Semiconductor (2018)
14. Quadrature Decoder (QuadDec) 3.0. Document Number: 001-96233 Rev. *B. Cypress Semiconductor (2017)

The Pierce Oscillator

13

Abstract

The Pierce Oscillator (Pierce, Electrical System, US patent 2,133,642, filed Feb. 25, 1924, issued Oct. 18, 1938) has a number of desirable characteristics. It will work at any frequency in the range from 1 kHz to 200 MHz. It has very good short-term stability because the crystal's source and load impedances are primarily capacitive as opposed to resistive, thus providing high in-circuit Q. The circuit provides a large output signal and simultaneously drives the crystal at a low power level. Large shunt capacitances to ground on both sides of the crystal make the oscillation frequency relatively insensitive to stray capacitance and giving the circuit a high immunity to noise. The Pierce configuration does have one disadvantage, viz., it needs a relatively high gain amplifier to compensate for relatively high gain losses in the circuitry surrounding the crystal. Quartz crystals are quite capable of oscillation well into the 300+MHz range[1] and and as low as .5 HZ depending how the crystal was "cut" (Lee et al., A 10-MHz micromechanical resonator Pierce reference oscillator for communications, in *Digest of Technical Papers, the 11th International Conference on Solid-State Sensors & Actuators (Transducers'01)*, Munich, June 10–14 (2001), pp. 1094-1097). Quartz crystals can vibrate in several different modes which is determined by the "cut". As shown in Fig. 13.6 there are several ways to "slice" the crystal and produce a thin planar quartz crystal. AT is the most commonly used cut is and it operates in a thickness shear vibration mode.

13.1 Historical Background

In 1880, Pierre and Jacque Curie reported observing[2] the presence of surface charges on certain types of crystalline materials.[3]

> "Those crystals having one or more axes whose ends are unlike, that is to say, hemihedral crystals with oblique faces, have the special physical property of giving rise to two electrical poles of opposite signs at the extremities of these axes when they are subjected to a change in temperature. This is the phenomenon known under the

---

[1] For AT-cut crystals.

[2] This 1880 discovery was made using little more than tinfoil, magnets, a jewelers saw, and wire.

[3] Examples include tourmaline, Rochelle salt, topaz, cane sugar, cadmium sulfide, lithium niobate, zinc oxide, lead zirconium titanate, etc.

© Springer Nature Switzerland AG 2021
E. H. Currie, *Mixed-Signal Embedded Systems Design*,
https://doi.org/10.1007/978-3-030-70312-7_13

name of pyro-electricity...". "We have found a new method for the development of polar electricity in these same crystals, consisting in subjecting them to variations in pressure along hemihedral axes...".

Paradoxically, they failed to note that applying a potential across a crystal[9] with appropriate orientation caused the crystal to deform. Thus, the principle of reciprocity applies to piezoelectric materials. It was Walter Cady (1920) who sought to use the latter effect to create the first crystal resonators [1] based on his observation that piezoelectric crystals would exhibit resonance effects by oscillating mechanically at a particular frequency[4] when a time-varying electric field was applied and thus the crystal resonator, often referred to as the crystal oscillator was born.
Cady defined piezoelectricity [1] as

"...electric polarization produced by mechanical strain in crystals belonging to certain classes, the polarization being proportional to the strain and changing sign with it."

G.W. Pierce(1923) demonstrated a vacuum tube oscillator using a crystal [5] and arguably remains the basis for the most often used quartz-based oscillator since its inception. In 1925, K.S. van Dyke proved that a two-electrode, piezoelectric, resonator has the equivalent circuit shown in Fig. 13.1. Such polarization can be introduced by compression, tension, bending, shear, and torsion of quartz and other crystals. This polarization results in a potential appearing across the crystal. This circuit can be described as a series resonant circuit shunted by a capacitor [3].[5]

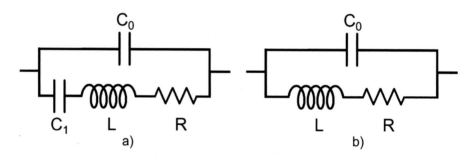

Fig. 13.1 (a) Series and (b) parallel resonance in a crystal.

Quartz consists of an arrangement of silicon dioxide ($Si O_2$ atoms which has a particular shape based on the allowed bonding geometry for silicon and oxygen atoms. It has a number of direction-dependent properties, i.e., anisotropy.

13.2   Q Factor

An important consideration when dealing with resonant circuits is that of Q factor which is defined as

$$Q \overset{\text{def}}{=} 2\pi \left[\frac{\text{Energy Stored/Cycle}}{\text{Energy Dissipated/Cycle}} \right] \tag{13.1}$$

---

[4]As shall be shown, there are two frequencies at which such as crystal can be shown to oscillate, viz., if the crystal is exhibiting series resonance or the crystal is exhibiting parallel resonance. Typically, and these two frequencies are within 1% of each other.
[5]Which is often referred to as a "tuned circuit".

$$Q = 2\pi f_r \left[\frac{\text{Energy Stored}}{\text{Power Loss}} \right] \tag{13.2}$$

$$= \frac{\text{Reactance}}{\text{Resistance}} \tag{13.3}$$

when a system is at resonance.[6] Note that the energy dissipated per cycle is precisely the amount of energy required to maintain the resonance condition at constant amplitude and frequency, f_r. The tacit assumption is that the only energy loss is resistive and that the energy stored is contained in the inductive and capacitive reactances present in the system. Typical crystal resonators can have Q factors of 100,000 and temp stabilities of less than 1 ppm/°C.

13.2.1 Barkhausen Criteria

The Barkhausen Criteria[7] defines the necessary, though not sufficient conditions, for a linear feedback circuit to oscillate.[8] As shown in Fig. 13.2, A represents the amplifier gain, $\beta(j\omega)$ the transfer function for the feedback loop.

If the absolute value of the loop gain, $|\beta(j\omega)|$, equals one and the phase shift through the feedback path is zero, or an integer multiple of 2π. The gain can be derived by breaking the feedback path and computing

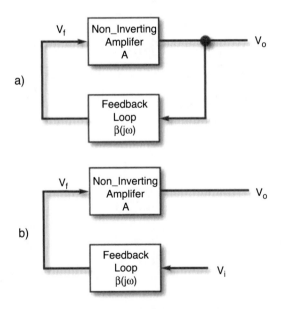

Fig. 13.2 Barkhausen criteria schematic (**a**) Closed-loop circuit, (**b**) Open-loop circuit.

---

[6]Note that the factor, 2π, was introduced to simplify some types of computation.

[7]Heinrich Georg Barkhausen (1881–1956) [3] was a physicist who defined the necessary conditions for an oscillation to occur in linear, positive feedback circuits, in 1921. His criterion is also used to avoid undesired oscillation in circuits with negative feedback. The restriction to linear circuits makes it inapplicable to circuits with non-linear feedback, e.g. tunnel diodes, varactors, negative resistance devices, etc.

[8]It is assumed that there is sufficient noise, in such a circuit, to initiate oscillation.

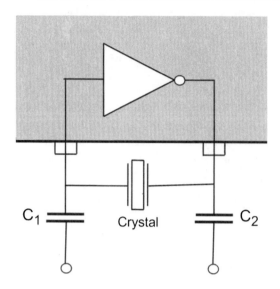

Fig. 13.3 PSoC-based Pierce oscillator.

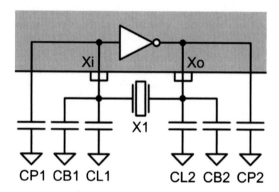

Fig. 13.4 PSoC-based Pierce oscillator (stray capacitance case).

$$G = \frac{V_o}{V_f} = \left[\frac{V_f}{V_i}\right]\left[\frac{V_o}{V_f}\right] = \beta A(j\omega) \tag{13.4}$$

External oscillator support is provided for in both PSoC 3 and PSoC 5 as shown in Fig. 13.3. The inverting amplifier shown in the figure is internal to the PSoC 3 and PSoC 5. Capacitors C1, C2, and the crystal form a pi-network which is loaded by the two capacitors to ensure the required frequency of oscillation. The pi-network create a 180° phase shift at the resonant frequency. The inverting amplifier provides an additional 180° phase shift for a total shift of 360°, 360 degrees of phase shift exist in the circuit causing this configuration to oscillate at the appropriate frequency.

Not shown are the parasitic capacitances[9] that can affect the applied load capacitance and should be considered when a high degree of frequency accuracy is required. The total combination of these capacitances should equal the rated load capacitance of the crystal X1, e.g., 12.5 pF. Figure 13.4.

---

[9] Such capacitances arise as a result of stray PCB and pin capacitances.

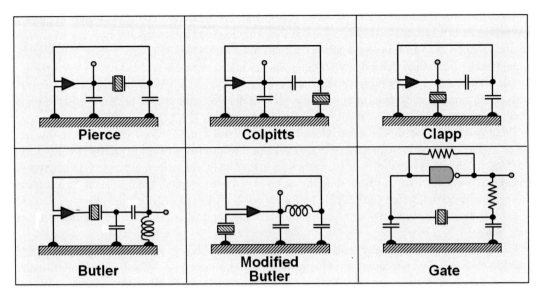

Fig. 13.5 Some basic oscillator types [11].

The Pierce oscillator[10]

$$Z(j\omega) = \frac{j\omega}{\omega^2}\left[\frac{\omega^2 L_1 C_1 - 1}{C_0 + C_1 - \omega^2 L_1 C_1 C_0}\right] \tag{13.5}$$

"Series-resonance" occurs when the following condition is met:

$$f_{serial} = \left[\frac{1}{2\pi\sqrt{LC_1}}\right] \tag{13.6}$$

and, "parallel resonance" occurs when:

$$f_{parallel} = \frac{1}{2\pi\sqrt{LC_1}}\sqrt{1 + \frac{C_1}{C_0}} = f_{serial}\sqrt{1 + \frac{C_1}{C_0}} \tag{13.7}$$

The choice of oscillator circuit type depends on factors such as the desired frequency stability, input voltage and power, output power and waveform, tunability, design complexity, cost, and the crystal unit's characteristics.

Although there are a great many extant oscillator circuits, three are the most commonly employed, viz., the Pierce, Colpitts, and Clapp oscillators. They are basically the same circuit except that the RF ground points are not the same, as shown in Fig. 13.5. The Butler and modified Butler are also similar to each other; in each, the emitter current is the crystal current. The gate oscillator is a Pierce-type that uses a logic gate plus a resistor in place of the transistor in the Pierce oscillator. (Some gate oscillators use more than one gate) [6].

---

[10]George Washington Pierce (1872–1956), a professor of physics at Harvard University, did extensive research on rectifiers, mercury vapor discharge tubes, magnetostriction and crystal oscillators [8], which resulted in him being issued fifty-three patents and the introduction of the Pierce oscillator.

All of the incarnations of quartz-based oscillators have the property that the resonance frequency is temperature dependent. In many applications that variation in resonant frequency is not a factor in that the variation as a function of temperature is acceptable within the oscillators ambient temperature range, In cases where such dependence is of concern, the quartz crystal is maintained in a temperature-stabilized environment frequently referred to as a crystal oven [3]. Mechanical vibration, eddy currents produced by ambient magnetic fields, humidity, crystal defects, atmospheric pressure variations and other effects can adversely affect the resonant frequency stability [10, 11].

The ground point location of a Pierce oscillator, which has a profound effect on its performance, makes it generally superior to the others, e.g., with respect to the effects of stray reactances and biasing resistors, which appear mostly across the capacitors in the circuit, rather than the crystal. In the case of the Colpitts oscillator, a larger part of the strays appears across the crystal, and the biasing resistors are also across the crystal, which can degrade performance. The Clapp oscillator is less popular because connecting the collector directly to the crystal can result in spurious oscillation and additional losses (Fig. 13.6).

The Pierce family usually operates at "parallel resonance", although it can be designed to operate at series resonance by connecting an inductor in series with the crystal. The Butler family usually operates at (or near) series resonance. The Pierce oscillator can be designed to operate with the crystal current above or below the emitter current [7]. Gate oscillators are common in digital systems when high stability is not a major consideration (Figs. 13.7 and 13.8).

13.3   External Crystal Oscillators

PSoC 3 and PSoC 5LP kHz and MHz external oscillators are Pierce oscillators, that utilizes an inverting amplifier to assert feedback across a crystal connected across the input and output of the amplifier as shown in Fig. 13.3.[11] The crystal input and output (Xi and Xo) pins are also labeled in the diagram. In this scheme, there are three distinct parts. The crystal or resonator (X1) in the pi-network is physically designed to oscillate at the desired frequency. The load capacitors (CL1, CL2) in the pi-network load the crystal, or resonator, to ensure proper operation.

The inverting amplifier amplifies the output of the pi-network and drives the input terminal with the inverse signal. The pi-network made up of a crystal or resonator and capacitors exists to provide 180 degrees of phase shift at the resonant frequency. When combined with the inverting amplifier, 360 degrees of phase shift exist in the circuit, resulting in resonant oscillation at the desired frequency.

Two types of resonators can be used with the MHz ECO, viz., crystal resonators and ceramic resonators. Crystals have tighter initial frequency tolerances and a higher cost [4].

- Ceramic resonators are usually found in smaller packages, often with integrated pi-network capacitors. Crystal resonators offer initial frequency tolerances as low as tens of parts per million (PPM), or thousandths of a percent. Ceramic resonators offer initial frequency tolerances as low as thousands of PPM, or tenths of a percent. PSoC 3 and PSoC 5LP's Internal Main Oscillator (IMO) is rated between 1% and 7% accurate depending on the clock frequency generated. Lower frequency IMO outputs are more accurate.
- Crystal Resonator—Resonator selection should begin with the choice of desired frequency accuracy. Frequency accuracy requirements are determined based upon the end application. If the frequency accuracy requirement is 1000 PPM or better, a crystal resonator should be selected. If

---

[11]The gray region of this figure indicates that the inverting amplifier is internal to the PSoC 3 and PSoC 5LP devices, and all other components are external.

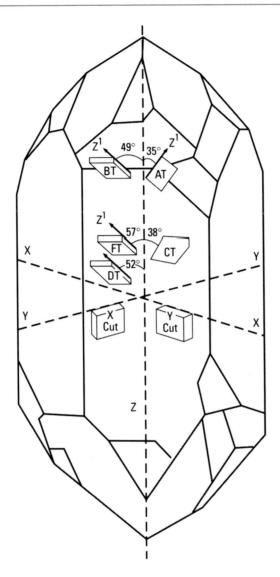

Fig. 13.6 Examples of various quartz plane "cuts".

only 50,000 PPM or less frequency accuracy is required, a ceramic resonator may be used. Thus, frequency selection is another important resonator selection factor.[12] Lower frequency ECOs have more design tolerance, consume less current, and start up faster. PSoC 3 and PSoC 5LP's MHz ECO can operate with resonators in the range of 4 to 25 MHz.

Other important selection factors include resonator packaging, load capacitance, and drive level. A lower load capacitance makes ECO circuit design easier. Resonators in larger packages typically have lower load capacitances, so larger packages are beneficial. Finally, crystals with a high drive level rating are more likely to be compatible with PSoC 3 and PSoC 5LP's MHz ECOs.

---

[12]The MHz resonator does not need to resonate at the same frequency as that required for clocking a particular design.

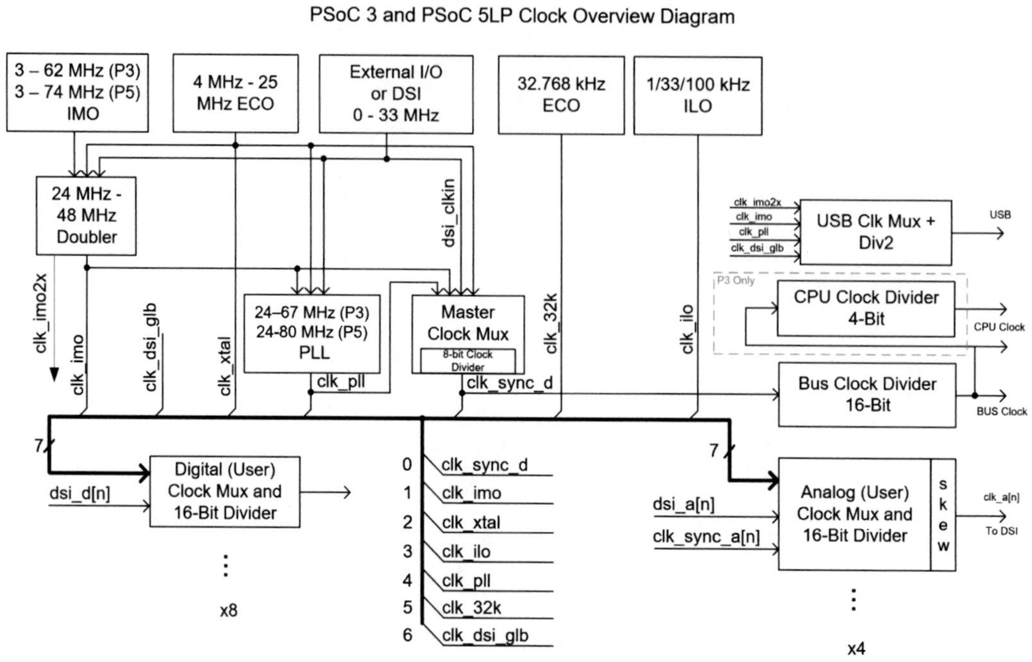

PSoC 3 and PSoC 5LP Clock Overview Diagram

Fig. 13.7 PSoC 3 and PSoC 5LP clock block diagram.

Ceramic resonators are different from crystal resonators in that they have extremely high drive level specifications, and sometimes have integrated load capacitors. When ordering ceramic resonators with integrated load capacitors, the built-in capacitance values may be specified.

- kHz Crystal—PSoC 3 and PSoC 5LP's kHz ECO circuitry works with 32.768 kHz parallel resonant crystals. This frequency is used because it is 2^{15} Hz, and a 15-bit counter can generate a 1 Hz signal that is useful for Real-Time Clocks (RTCs).[13] The kHz crystal oscillators typically have initial frequency tolerances of tens of PPM or thousandths of a percent. In real-time clocks, 11.5 PPM of error will lead to a clock drift of one second per day.
- kHz Crystal Selection—PSoC 3 and P Soc 5LP's kHz ECO can operate with parallel resonant 32.768 kHz crystals with a load capacitance of 6 pF or 12.5 pF. Acceptable crystals must meet these guidelines, and also have frequency accuracy and packaging acceptable in the application. Frequency accuracy is specified as initial, across temperature, and across time.
- Pi-network Capacitors—Both crystal and ceramic resonators must be loaded with the proper capacitance in order to oscillate at the correct frequency. The pi-network capacitors should be chosen based on the resonator's load capacitance specification and parasitic capacitances.

Parasitic capacitances include PCB trace capacitance and microcontroller pin capacitance. These parasitic capacitances are shown in Fig. 13.4 as CP1, CP2 for pin capacitance, and CB1,CB2 for PCB trace capacitance. PCB trace capacitance can be measured after manufacturing with an LCR meter, or calculated before manufacturing using the physical properties of the traces and PCB. Pin capacitance can be determined through measurement, or, by checking the GPIO DC Specifications table in the device datasheet. Both traces should have the same capacitance if proper PCB layout practices are followed. Pin and trace capacitances can range from 0.1 to 10 pF.

---

[13] The kHz ECO is sometimes described as a 32 kHz ECO to save space, but it operates at 32.768 kHz.

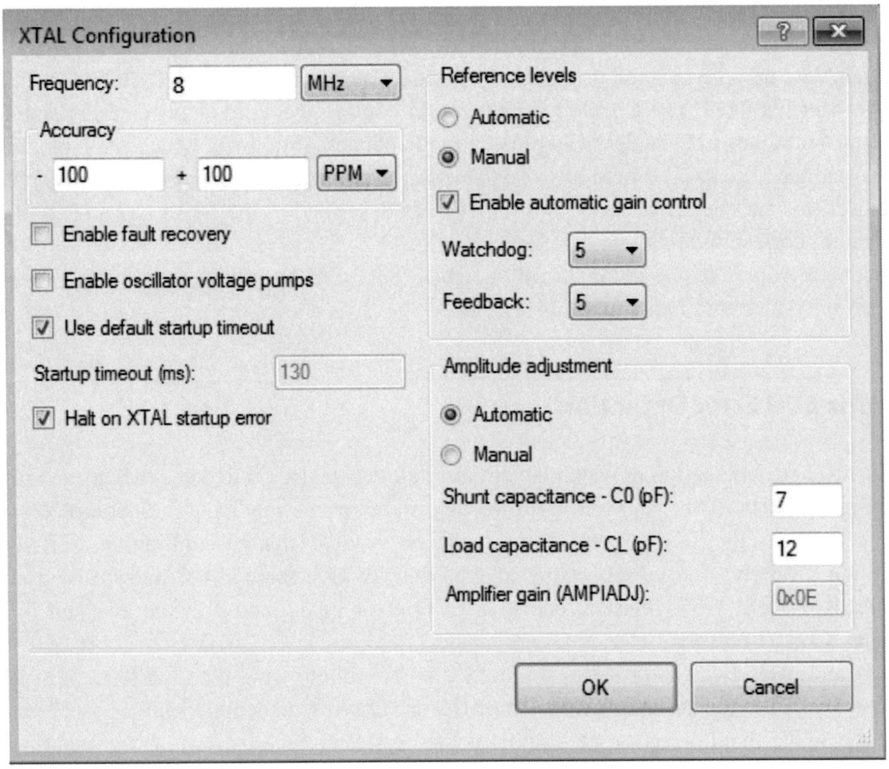

XTAL Configuration

Frequency: 8 MHz ▼

Accuracy

- 100   + 100   PPM ▼

☐ Enable fault recovery

☐ Enable oscillator voltage pumps

☑ Use default startup timeout

Startup timeout (ms): 130

☑ Halt on XTAL startup error

Reference levels

○ Automatic

◉ Manual

☑ Enable automatic gain control

Watchdog: 5 ▼

Feedback: 5 ▼

Amplitude adjustment

◉ Automatic

○ Manual

Shunt capacitance - C0 (pF): 7

Load capacitance - CL (pF): 12

Amplifier gain (AMPIADJ): 0x0E

OK   Cancel

Fig. 13.8 PSoC 3 and PSoC 5LPClock overview diagram.

13.4 ECO Automatic Gain Control

The Automatic Gain Control(AGC), allows the MHz ECO to increase or reduce the inverting amplifier gain to increase or decrease oscillation amplitude. This can increase or decrease the drive level. The AGC monitors the amplitude of the crystal input waveform and compares it to a reference value.

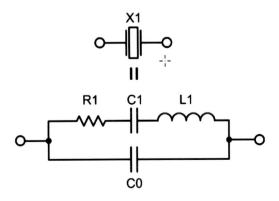

Fig. 13.9 PSoC Creator MHz ECO configuration dialog.

If the amplitude is higher or lower than desired, the inverting amplifier gain is modified in hardware. The reference value is generated based on the vref_sel_fb bits of the FASTCLK_XMHZ_CFG1 register. The AGC may be enabled or disabled in PSoC Creator using the MHz ECO Configuration dialog, shown in Fig. 13.9. The Enable automatic gain control checkbox controls this feature. AGC feedback register values may be deselected using the Feedback dropdown, which only appears when the AGC is enabled. The AGC should be left disabled in most designs. It should only be enabled if the ECO circuit is not meeting drive level requirements, as described in the Drive Level section. Lower feedback values correspond to lower drive level results.

The feedback value can be automatically set by PSoC Creator based on the ECO's operating frequency, if the Automatic radio button is selected.

13.5   MHz ECO Error Detection

Watchdog—The error detection or watchdog circuitry allows the ECOs to detect when each resonator is oscillating properly. This result is used to determine when the ECO has completed startup and can be used to clock the system. It may also be used to change configurations if the ECO stops working properly. The current value of the MHz ECO error bit is stored in the xerr bit of the FASTCLK_XMHZ_CSR register. The kHz ECO error bit is stored in the ana_stat bit of the SLOWCLK_X32_CRregister.

The kHz and MHz ECO error status bits may also be polled using the CyXTAL_32KHZ_Read Status() and CyXTAL_ReadStatus()APIs. The MHz ECO error detection circuitry works similarly to that of the AGC, monitoring the amplitude of the crystal input waveform and comparing it to a reference value. If the amplitude is lower than the desired value, the ECO error bit is asserted. The reference value is generated based on the vref_sel_wd bits of the FASTCLK_XMHZ_CFG1 register.

This value may be configured in PSoC Creator using the Design Wide Resources interface, in the MHz ECO Configuration dialog. The ECO error detection threshold is set using the Watchdog dropdown. This value should only be changed if the ECO is found to have greater than expected noise susceptibility, or if crystal failures are not being detected. It can be automatically set by PSoC Creator based on the ECOs operating frequency if the Automatic radio button is selected by default.

The kHz ECO error detection does not require configuration, unlike the MHz ECO. The fault recovery setting allows the designer to choose what behavior to carry out when the ECO fails to oscillate. If fault recovery is enabled, then the device automatically enables and switches to the IMO clock in case of ECO failure. Fault recovery may be enabled using the Enable fault recovery checkbox in the MHz ECO Configuration dialog.

The Halt on XTAL startup error checkbox controls the behavior of the PSoC firmware during MHz ECO startup. If this box is checked, the firmware will automatically halt device startup if the MHz ECO does not stabilize within the specified startup timeout. The device will instead go into a special error function, whose contents can be edited. This function is called CyClockStartupError(), and is located in the cyfitter_cfg.csource file.5.2.3.

13.6   Amplitude Adjustment

Amplitude adjustment allows modification of the gain of the inverting amplifier in the Pierce oscillator circuit shown in Fig. 13.3. Modifying the gain of the inverting amplifier has multiple effects on the performance of the MHz ECO, including the metrics of negative resistance, drive level, and startup time that are explained in the ECO Performance Testing and Improvement section below. Amplitude adjustment configuration can be performed manually or automatically in PSoC Creator. The automatic selection will choose an inverting amplifier gain based on the frequency, shunt capacitance, and load capacitances entered into the tool. Manual selection allows the user to choose an amplifier gain value (AMPIADJ).

13.7 Real-Time Clock

The kHz ECO can be used to source a real-time clock for applications that must accurately keep track of time. PSoC Creator provides the RTC component, which allows the designer to quickly implement a real-time clock in PSoC 3 and PSoC 5LP using a kHz ECO. For more information about how to implement a Real-Time Clock, refer to the RTC Component datasheet in PSoC Creator under the Component Catalog.

13.8 Resonator Equivalent Circuit

Crystal and ceramic resonators can be modeled as a combination of basic passive components. The equivalent model is shown in Fig. 13.10. The series combination of R1, C1, and L1 is known as the motional arm. R1 is known as the motional resistance or effective series resistance (ESR). C1 is known as the motional capacitance. L1 is known as the motional inductance. C0 is known as the shunt capacitance. Some or all of these characteristics may be specified by the crystal manufacturer.

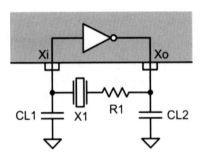

Fig. 13.10 Resonator equivalent circuit.

13.8.1 ECO Frequency Accuracy

The frequency of the clock generated by the ECO will always deviate from the desired frequency by some amount. It is important to understand how much it will deviate, and how to improve it. This section describes the various factors that impact frequency performance. Frequency accuracy is a measurement of the maximum expected deviation from the expected frequency. This maximum will be a sum of the individual causes of frequency inaccuracy. Each of the individual causes is listed below. In designs, all of these factors should be calculated and summed to determine overall system maximum frequency deviation, and thus frequency accuracy.

13.8.2 Initial Frequency Tolerance

Initial frequency tolerance, sometimes called simply frequency tolerance, describes the maximum expected deviation in resonant frequency of a resonator at room temperature when the proper load capacitance is in place. This contributor of frequency error is the base to which all other contributors should be added. It is usually the largest contributor of frequency error.

13.8.3 Frequency Temperature Variation

Frequency temperature variation, sometimes called frequency stability or temperature stability describes the increase in maximum expected deviation of the resonant frequency of a resonator across its operating temperature range.

13.8.4 Resonator Aging

Resonator aging describes an increase in frequency deviation that occurs over the operating life of the resonator. This spec is usually given in units of PPM/year. The resonator frequency deviation at a given age can be easily calculated by multiplying the aging value times the age in years. The term aging is also sometimes used to describe the damage that can occur to a crystal if its drive level specification is exceeded. This type of aging is usually not specified by crystal vendors, as it implies that the ECO circuit is not operating within specifications.

13.8.5 Load Capacitance Trim Sensitivity

A variation of the actual load capacitance has a tendency to "pull" the resonant frequency of the resonator away from its design frequency. This effect is known as "trim sensitivity" or sometimes pullability. This effect can be both good and bad. However, the designer can vary the resonant frequency altering the values of the capacitors used. However, capacitor values can vary as a result of manufacturing variances and/or temperature effects that in turn can cause the design frequency to vary from application-to-application and/or as a result of ambient temperature variations. Trim sensitivity is measured in PPM of added clock error per pF of variation in total actual load capacitance. Trim sensitivity is a function of the resonator's rated load capacitance(CL),shunt capacitance(C0), and motional capacitance(C1).

13.9 ECO Trim Sensitivity

$$\text{Trim Sensitivity} = S = \frac{C_1 * 1,000,000}{2 * (C_O + C_L)^2} \frac{(PPM)}{(pF)} \tag{13.8}$$

Equation (13.8) is a linearized, second order function relating trim sensitivity to the values of C_0 and C_L. As an illustrative example, consider a 32.768 kHz crystal quartz crystal with the following characteristics: $C_1 = 0.0035\,\text{pF}$, $C_0 = 1.6\,\text{pF}$, and $C_L = 12.5\,\text{pF}$

$$S = \frac{C_1(1 \times 10^6)}{2(C_0 + C_L)} = \frac{0.0035(1 \times 10^6)}{2(1.6 + 12.5)^2} = 8.8 \frac{PPM}{pF} \tag{13.9}$$

It should be noted that with 1 pF in error in the applied load capacitance, the frequency will shift by 8.8 PPM or 0.29 Hz.

Trim sensitivity implies requirements for load capacitor values and temperature coefficients. It should be considered along with resonator frequency accuracy and system performance requirements to determine what load capacitors to use.

13.9.1 Negative Resistance Characteristics

Negative resistance is a phenomena that certain devices exhibit in which the slope of their V-I characteristic becomes negative, i.e., increasing the applied voltage results in a reduction in current, as for example in the case of plasmas, tunnel diodes, certain combinations of bipolar transistors, and resonators. In the present context, it is defined as the amount of series resistance that can be added to the effective series resistance of the resonator without stopping the ECO from starting up properly and therefore, it is a measurement of reliability of the ECO, higher being better.[14] Alternatively, the ratio of -R/ESR is cam be compared to some arbitrary value, usually 3 or 5, to determine if -R is large enough.

Negative resistance is measured, as shown in Fig. 13.11 by connecting a resistor in series with the resonator, and noting the ECO clock output. A fixed value resistor should be used[15] The resistance should be increased until the ECO no longer starts up.[16] The highest value that allows startup should be added to the ESR specification of the resonator, and this should be considered the negative resistance of the ECO. The relationship between negative resistance, the maximum value of R1 and ESR is given by

$$- R = R_{IMAX} + ESR \tag{13.10}$$

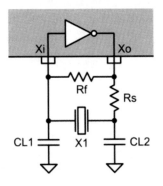

Fig. 13.11 Configuration required for negative resistance test.

---

[14]Negative resistance is also known as "-R" or "oscillation allowance".

[15]Potentiometers should be not used due to their non-ideal characteristics that can lead to an incorrect measurement.

[16]Negative resistance is lowest at high temperature and low VCCA and VCCD, so this corner should be tested.

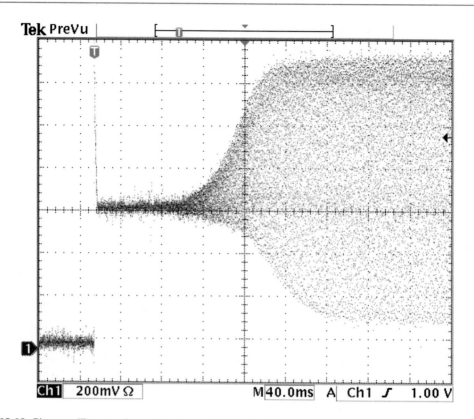

Fig. 13.12 Pierce oscillator topology with series and feedback resistors.

13.9.2 Series and Feedback Resistors

The resistors R_s and R_f shown in Fig. 13.12 can be used to improve MHz oscillator behavior under certain circumstances. The series resistor R_s reduces the power dissipated by the resonator to meet drive level specifications. The feedback resistor R_f allows noise feedback across the resonator at startup, decreasing startup time.

13.10 Drive Level

Drive level refers to the resonator power dissipation. If the ECO circuit drive level exceeds the resonator's specifications, resonator aging may occur and adversely affect the accuracy of the design frequency. The introduction of Rs creates a power divider with the ESR[17] of the crystal, where ESR is defined as:

$$ESR = R_s \cdot \left(1 + \frac{C_0}{C_L}\right)^2 \tag{13.11}$$

and Drive Level (DL) is defined as:

$$DL = (I_{RMS}^2)ESR \tag{13.12}$$

---

[17]Equivalent series resistance (ESR) is the real value of the crystal impedance, when the oscillator matches the load capacitance (C_L), impedance-wise.

Typical R_s values near ESR specifications are on the order of tens or hundreds, of ohms. Ceramic resonators, unlike crystal resonators, do not typically have drive level specifications. Thus, series resistors are not needed in their ECO circuits.[18]

Initial startup can be hampered in low noise environments, requiring the addition of a feedback resistor R_f. If added, this feedback resistor should be on the order of $5 - 15M\Omega$. In a balanced load capacitor configuration, both capacitor values are equal, otherwise, the load is said to be unbalanced.

Whether the configuration is balanced or unbalanced,[19] the combination of capacitances should still load the resonator according to its specifications. In the unbalanced configuration, the ratio of capacitances should be approximately three to one. This criterion should be used in conjunction with the basic criterion, as expressed in Eq. (13.1), to determine the capacitor values.[20]

13.10.1 Reducing Clock Power Consumption with ECOs

The use of an ECO eliminates the need for internal clock sources. Disabling these clock sources offers power savings. The IMO may be disabled using the CyIMO_Stop() API. Disabling the IMO saves hundreds or thousands of μA depending on the IMO's configuration. The power consumption of the ECOs themselves can also be optimized. Decreasing the drive level of the MHz ECO using the AGC reduces its current consumption. The kHz ECO has multiple power modes that may be selected using the SLOWCLK_X32_CR register, or using the CyXTAL_32KHZ_SetPowerMode()API provided by PSoC Creator.[21]

13.10.2 ECO Startup Behavior

An example of a 32 kHz ECO startup output waveform is shown in Fig. 13.13. Initially, when the part is held in reset, the crystal output is pulled low. During startup, the pin is pulled to logic high, and then when crystal startup commences, the output goes to the bias voltage. Crystal startup can range in length depending upon system configuration. Throughout the course of crystal startup, the amplitude of oscillations increases until it reaches its steady-state amplitude. The MHz ECO configuration firmware can cause startup code execution to halt depending on the project settings. See the MHz ECO Error Detection Watchdog section for more details. When observing oscillator input or output waveforms, a high impedance probe or a buffer such as an op-amp follower should be used. Otherwise, the probe will load the oscillator, and change ECO behavior cause it to stop oscillating.

13.10.3 32-kHz ECO Startup Behavior

Using an External Clock Signal on the ECO Pins. If the MHz or kHz ECO is not being used, an external clock signal may be routed onto the ECO clock nets using the kHz or MHz crystal input

---

[18]ECO circuits start-up occurs due to the presence of ambient thermal noise and electromagnetic interference. The pi-network acts as a filter for this noise, resonating at the proper frequency, with the inverting amplifier increasing the amplitude of the signal.

[19]If the output capacitor is larger than the input capacitor, the oscillator will be more stable but consume more current. If the input capacitor is larger than the output capacitor, the oscillator will consume less current but be less stable.

[20]A typical pair of unbalanced values for a 12.5 pF, 32.768 kHz crystal are $C_{in} = CL1 = 15$ pF and $Cout = CL2 = 47$ pF.

[21]ECO power consumption can be further reduced using unbalanced pi-network capacitors.

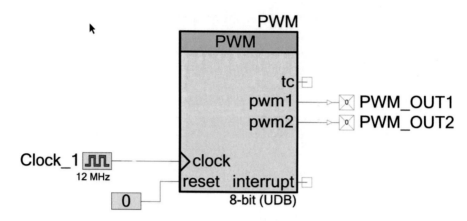

Fig. 13.13 $32k$ Hz ECO startup behavior.

pins. This allows the use of these clock nets in place of the usual DSI clock net in the clock tree. External clocks can be routed into the part through the kHz and MHz XtalIn pins. The XtalOut pins should be allowed to float. Ideally, the external signals should be rail-to-rail sine or square waves. If the amplitude of the signals is too low, they may not be properly translated into digital signals. These signals must be within the frequency ratings of the ECOs, either at 32.768 kHz or 4–25 MHz. No additional firmware is required. Just set up the oscillator, as if using a crystal, in the Clocks section of the Design Wide Resources in PSoC Creator. External clocks can also be routed into the part using the GPIOs.

13.11 Using Clock Resources in PSoC Creator

After an external oscillator has been configured, it is abstracted to a simple Clock Component in PSoC Creator. Multiple Clock Components can be incorporated into the schematic and connected to various components, or logic. Figure 13.14 shows a clock connected to a PWM.

13.12 Recommended Exercises

13-1 A negative feedback oscillator has the following loop gain transfer function:

$$H(s) = \frac{A}{s^3 + as^2 + bs + 1} \tag{13.13}$$

where a and b are defined by the value of certain circuit components and A is the dc gain. Assume that a,b, and A are known. Derive expressions for f in rad/sec and the minimum value of A required to sustain oscillation.

13-2 Five digital inverters are connected in series and the output of the last inverter in the chain is connected to the input of the first inverter. Does this circuit oscillate? What happens if an additional inverter is added? Derive an expression for the frequency of oscillation and state a rule regarding which combinations of inverters oscillate and which do not as a function of the number of inverters used.

13-3 Explain the behavior shown on the figure shown below of an oscillator, after oscillation begins as a function of A. What would you expect the behavior to be as power to the circuit goes to zero?

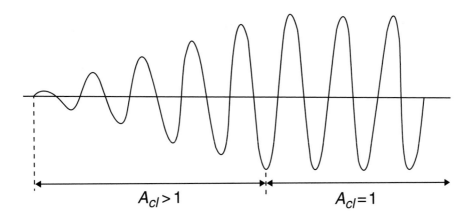

$A_{cl} > 1$ $A_{cl} = 1$

13-4 A Pierce oscillator can be created from a FET-based Colpitts oscillator simply by replacing the Colpitt's inductor with a crystal. Compare the performance of each.

13-5 Consider a Pierce oscillator operating in an ambient temperature environment of $30\,°C$ whose only temperature dependence is attributable to the series equivalent capacitance of the crystal's equivalent circuit. Assume that the temperature stability of the crystal is 3.750 ppm/°C, and calculate the percent variation in the frequency of the oscillator if the ambient temperate varies by $+/- 13.6\,°C$. Determine the percent variation the in oscillator's frequency if the ambient temperature changes from $25\,°C$ to $15\,°C$.

13-6 Design a FET-based CMOS Pierce oscillator assuming that $R_s = 83\,Ω$, $C_0 = 2.74\,pF$, $C_1 = 12.7\,pF$ and $Q = 96,000$ for the crystal and the MOS values $K_n = 100\,μA/V^2$; $K_n = 100\,μA/V^2$; $|V_{Tn}| = 1\,V$; $C_{iss} = 50\,pF$ at $V_{GS} = 0\,V$; and $C_{rss} = 5\,pF$ at $V_{GS} = 0\,V$.

13-7 The ideal amplifier shown below has infinite gain and a gain of A_v. Find this linear oscillator's frequency.

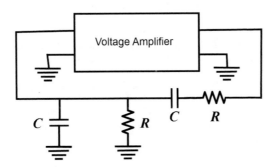

Voltage Amplifier

C R C R

13-8 A particular quartz crystal can be modeled as a series $R_s L_s C_s$ circuit that is connected in parallel with a second capacitor C_p. Assuming that $R_s = 4.7\,Ω$, $L_s = 2.768\,mH$, $C_s = 0.09965\,pF$, and $C_p = 29.38\,pF$, find the series resonant frequency, the parallel resonant frequency, and Q.

13-9   A capacitive load C_L is added to a crystal which changes the resonant frequency by the amount δf. The values for C_0, C_1 and C_L are 5 pF, 14 fF, and 20 pF respectively. If C_L increases or decreases by 10 fF, what is the change in frequency? If the aging of C_L is $2 \times 10 - 9$ per day, what is δf per day?
Hint:

$$\frac{\Delta f}{f_{\text{oscillator}}} \approx \frac{\Delta f}{f_{\text{resonator}}} + \frac{1}{2Q_L}\left[1 + \left(\frac{2f_f Q_L}{f}\right)^2\right]^{-1/2} d\varphi\left(f_f\right) \qquad (13.14)$$

where Q_L is the loaded Q of the resonator, and $d\phi(f_f)$ is a small change in loop phase at an offset frequency f_f away from carrier frequency f. Increasing Q_L can can result in noise reduction.[22]

13-10   Derive the following expression:

$$\delta f \equiv \frac{\Delta f}{f} \cong \frac{C_1}{2\left(C_0 + C_L\right)} \qquad (13.15)$$

Explain the relative impact and therefore,

$$\frac{\Delta(\delta f)}{\Delta C_L} \cong -\frac{C_1}{2\left(C_0 + C_L\right)^2} \qquad (13.16)$$

where

13-10   Derive the following expression for an oscillator that has a tuned circuit that includes filters and matching circuits to minimize unwanted modes of oscillation where BW is the bandwidth of the filter, f_f is the frequency offset of the center frequency of the filter, Q_L is the loaded Q of the resonator and Q_C and L_C are the tuned circuit's Q, inductance and capacitance, respectively. [10]

$$\frac{\Delta f}{f_{\text{oscillator}}} \approx \frac{d\varphi\left(f_f\right)}{2Q_L} \approx \left(\frac{1}{1 + \frac{2f_f}{BW}}\right)\left(\frac{Q_C}{Q}\right)\left(\frac{dC_C}{C_C} + \frac{dL_C}{L_C}\right) \qquad (13.17)$$

References

1. W.G. Cady, *Piezoelectricity*. McGraw-Hill; reprint (1964) (Dover, New York, 1946)
2. M. Frerking, *Crystal Oscillator Design and Temperature Compensation* (Springer, Berlin, 1978)
3. D.B. Leeson, A simple model of feedback oscillator noise spectrum. Proc. IEEE **54**(2), 329–330 (1966)
4. S. Lee, M.U. Demirci, C.T.-C. Nguyen, A 10-MHz micromechanical resonator Pierce reference oscillator for communications, in *Digest of Technical Papers, the 11th International Conference on Solid-State Sensors & Actuators (Transducers'01)*, Munich, June 10–14 (2001), pp. 1094–1097
5. R.J. Matthys, *Crystal Oscillator Circuits* (revised ed.) (Krieger Publishing, Malabar, 1992)
6. Oscillator Design Considerations. AN0016.0, Rev 1.30. Silicon Labs. http://www.silabs.com
7. G.W. Pierce, Electrical System, US patent 2,133,642, filed Feb. 25, 1924, issued Oct. 18 (1938)
8. W. Sansen, Design of crystal oscillators, in *Analog Design Essentials. The International Series in Engineering and Computer Science*, vol. 859 (Springer, Boston, 2006)
9. J.R. Vig, Quartz crystal resonators and oscillators. J. Vig. IEEE.org. Rev. 8.5.3.9. (2008)
10. J.R. Vig, et al., Acceleration, vibration, and shock effects-guidelines for the measurement of environmental sensitivities of precision oscillators, IEEE Standards Project P1193, in *Proceedings of the 1992 IEEE Frequency Control Symposium* (IEEE, Piscataway, 1992), pp. 763–791
11. E.A. Vittoz, M.G.R. Degrauwe, S. Bitz, High-performance crystal oscillator circuits: theory and application. IEEE J. Solid State Circuits **23**(3) (1988)

---

[22] Short-term resonator instabilities may occur at offset frequencies smaller than the resonator's half-bandwidth.

PSoC 3/5LP Design Examples

14

Abstract

This chapter includes several illustrative examples of designs commonly encountered in embedded systems that employ PSoC:

- Peak detection based on a sample-and-hold technique
- Full wave rectification
- Analog to digital conversion
- Waveform generation and
- Signal modulation/demodulation

14.1 Peak Detection

A peak detector detects the peaks of an input waveform and produces an output based on the detected peaks. The output of the peak detector depends on the type of peak detector used. Some peak detectors produce a digital output consisting of information about when positive and negative peaks of a waveform occur. In this case, the digital information can also be used to determine the direction of the slope of the input waveform. Other peak detectors produce an analog output with a magnitude equal to the last detected peak, or the magnitude of the maximum peak encountered. Accurately detecting the peaks in an input waveform can be useful in a variety of different applications.

One method for constructing a peak detector uses a comparator and a down-mixer acting as a sample-and-hold circuit [5]. When the slope of the input is positive, the output of the peak detector will be high and when the slope of the input is negative, the output of the peak detector will be low. Positive peaks are represented by a falling edge on the output, and negative peaks are represented by a rising edge.

A down-mixer is used as a sample-and-hold that delays the input signal sampled at time t_1 and then compares it to the input signal at time t_2. The output of the sample-and-hold is held on the falling edge of the sample (LO) clock. The comparator is clocked on the rising edge of the sample clock to ensure the sampled signal is stable and appropriately delayed from the input signal. There is approximately 10 mV of hysteresis built into the comparator. This helps ensure that slowly moving voltages or slightly noisy voltages will not cause the output of the comparator to oscillate. Enabling

© Springer Nature Switzerland AG 2021
E. H. Currie, *Mixed-Signal Embedded Systems Design*,
https://doi.org/10.1007/978-3-030-70312-7_14

hysteresis in the comparator is recommended for most input signals to reduce false peak detections. Figure 14.1 shows a PSoC Creator schematic of such a circuit.

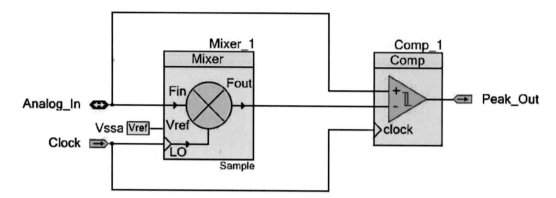

Fig. 14.1 PSoC Creator schematic for a peak detector, with sample-and-hold.

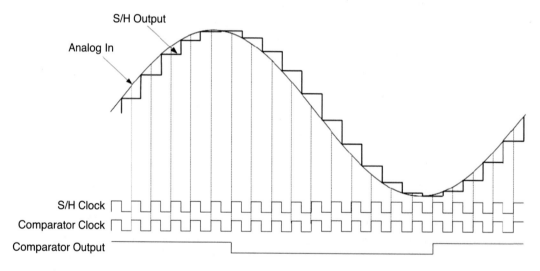

Fig. 14.2 Sample-and-hold peak detection waveform.

When the slope of the input waveform is positive, the sample-and-hold output is less than the input waveform at each rising edge of the comparator clock, so the output of the comparator is high. When the slope of the input waveform is negative, the sample-and-hold output is greater than the input waveform at each rising edge of the comparator clock, so the comparator output is low (Figs. 14.2, 14.3, 14.4, 14.5, and 14.6).

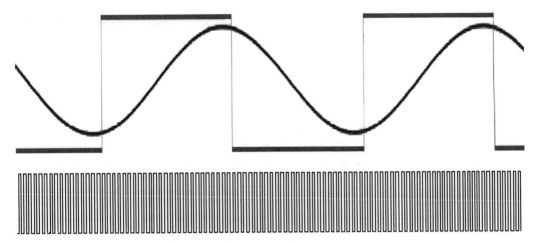

Fig. 14.3 Correct clock selection.

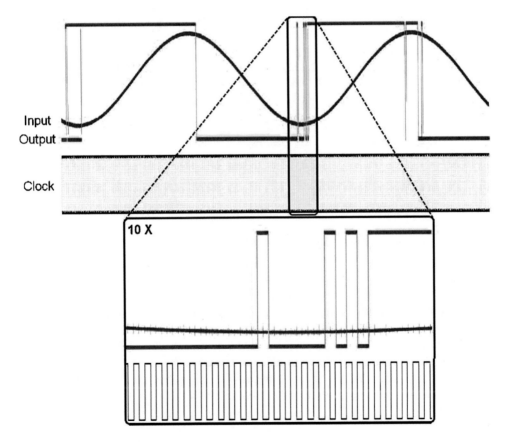

Fig. 14.4 Results of a clock speed which is too fast.

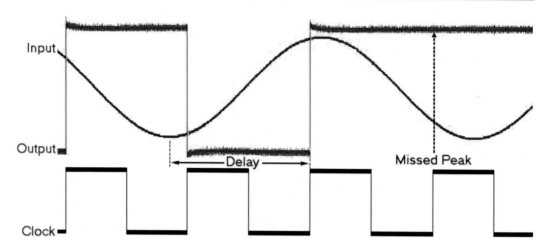

Fig. 14.5 Results of a clock speed which is too slow.

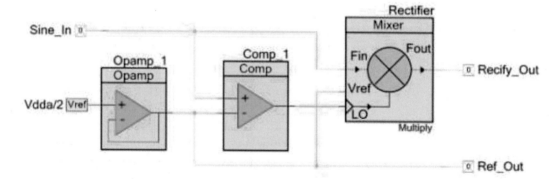

Fig. 14.6 Full-wave rectifier.

14.2   Debouncing Techniques

Embedded Systems are often employed in conjunction with various switching techniques for both analog and digital circuits. Such circuits that involve mechanical switching often involve, perhaps a purely unintended consequence, transients that arise as a result of mechanical switch contacts engaging in a phenomenon known as "bouncing", an example of which is shown in Fig. 14.7. In the case of analog circuits, these transients may be of relatively little concern due to circuit capacitance, short duration of the transients, etc. However, these same transients in digital circuits can prove quite troublesome and may cause serious losses of integrity. It is possible to use a simple RC filter, in some cases, to filter out the damped oscillation of bouncing switch contacts. In digital circuits, this method may not be effective in which case the transients may result in loss of data integrity or other dysfunction.

Fortunately, PSoC Creator provides a very useful component, as shown in Fig. 14.8, known as the "Debouncer" [1]. This component can be used as a hardware solution in that it is not necessary to write code to carry out the debouncing function.[1]

---

[1]It is also possible to employ a software method of debouncing as is show in Sect. 14.3.1.

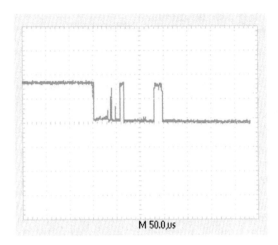

Fig. 14.7 Switch bounce transition from high to low.

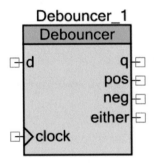

Fig. 14.8 PSoC Creator's debouncing module.

14.3 Sampling and Switch Debouncing

PSoC 3/5LP make it possible to handle bouncing switch contacts in software or hardware as indicated by the following examples. While there are a number of techniques to debounce switches, each of them requires some form of sampling of the state of an input pin, at a periodic rate. The sample period is chosen to be greater than the anticipated transition time of the signal, i.e., the time it takes for the signal to stabilize at the new state when the switch is opened or closed. During the transition time, the state of the signal is essentially unknown, i.e., at any time it could be either 0 or 1. Sampling should not occur more than once during that period to avoid extra transitions being detected and the maximum sample rate is the inverse of the maximum transition time. It may be necessary to check both the low-to-high and high-to-low transition times.

In practice, the sample rate should be set much lower than the expected transition time, but fast enough so that the system is responsive when the switch is opened or closed. A rate of 10 to 200 samples per second is usually appropriate.

14.3.1 Switch Debouncing Using Software

The simplest way to sample a switch is to poll the input pin, that is, program the CPU to read the pin's input value at regular time intervals. As shown below, the pin is sampled at an interval that is controlled by the PSoC Creator's *CyDelay()* function. The easiest way to detect a transition on the input is to use two variables, for current and previous values of the pin, and compare them for transition events. This example monitors one pin, in bit 0 of each element of the "switches" array. The array elements are monitored in their entirety, with the assumption that bits 1 to 7 are always zero. If multiple pins are to be monitored, then either use a separate pair of variables should be used for each pin, or a pair of variables should be used, with a bit defined for each pin. In addition, there are trade-offs in code size, execution speed, and RAM memory usage, for each method. With PSoC 3/5LP, it is possible to monitor a pin in software and hardware at the same time.

```
uint8 count; /* # of transitions of input pin 'SW' */

CY_ISR(SwInt_ISR)
{
SwReset_Write(1); /* clear interrupt source */
count++;
} /* end of SwInt_ISR() */

void main()
{
uint8 temp; /* local copy of count variable */

/* Initialization code */
. . .

for(;;) /* do forever */
{
/* Select a copy of the shared count variable, and display it.
 * This will ensure that the interrupt handler will not change the
 * the count variable while it is being displayed.
 */
CYGlobalIntDisable; /* macro */
temp = count;
CYGlobalIntEnable; /* macro */
LCD_Position(0, 14); /* row, column */
LCD_PrintHexUint8(temp);
}
} /* end of main() */

void main()
{
/* Initialization code */

. . .

/* Init switch variables */
uint8 switches[2] = {0, 0}; /* [0] = current, [1] = previous */
switches[0] = switches[1] = SW_Read(); /* 0 = pressed, 1 = not pressed

/* Init display */
LCD_Start();
LCD_Position(0, 0); /* row, column */
LCD_PrintString("Raw Count = ");
LCD_Position(1, 0); /* row, column */
```

```
LCD_PrintString("Filt. Count = ");

for(;;) /* do forever */
{
/* Place your application code here. */
/* Grab a copy of the shared count variable, and display the copy.
* This is so that the interrupt handler won't change the count
* variable while it's being displayed.
*/
CyGlobalIntDisable; /* macro */
temp = count;
CyGlobalIntEnable; /* macro */
LCD_Position(0, 14); /* row, column */
LCD_PrintHexUint8(temp);

/* Periodically sample the input pin, and display the filtered count */
CyDelay(50); /* msec */

/* Update the current and previous switch read values */
switches[1] = switches[0];
switches[0] = SW_Read();

/* Increment counter if a switch transitions either way */
if (switches[0] != switches[1])
{
filtered_count++;
}

/* Display the current value in filtered count variable */
LCD_Position(1, 14); /* row, column */
LCD_PrintHexUint8(filtered_count);
}
} /* end of main() */
```

14.3.2 Hardware Switch Debouncing

Although debouncing can effectively be performed in software, hardware debouncing can be just as effectively handled in most applications and uses fewer CPU cycles. An easy way employ hardware debouncing using PSoC is to poll using a status register and a clock components.as shown in Fig. 14.9.

The code required source code is content...

```
uint8 filtered_count; /* # of
    filtered transitions of input
    pin 'SW' */
CY_ISR(FiltInt_ISR){
        /* No need to clear any
            interrupt source;
            interrupt component
            should be
        * configured for
            RISING_EDGE mode.*/
        /* Read the debouncer
            status reg just to
            clear it, no need to
```

```
                    check  its * contents  in
                        this  application . */
                    FiltReg_Read ();
                        filtered_count ++;}
            /* end  of  FiltInt_ISR ()  */
```

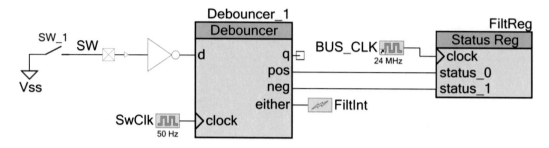

Fig. 14.9 An example of a hardware debouncing circuit.

14.4   PSoC 3/5LP Amplitude Modulation/Demodulation

There are several methods of modulating a carrier, e.g., amplitude, frequency [2] and phase shift.[2] [3] In all three cases, a carrier is modulated by a second signal that carries the information to be transmitted and both of these signals can be created and mixed within PSoC.

PSoC 3 and PSoC 5LP can be used to produce an AM modulated signal [6] by using the mixer component in what is referred to as the "Up-Mixer" mode. In this mode, a square wave serves as the carrier and it is mixed with a second signal which represents the information to be transmitted. signal. After the two signals have been mixed, i.e.multiplied. The modulated carrier is then filtered to remove harmonics.

The term modulation index in terms of the modulated carrier is defined as the ratio of the maximum peak of the modulated signal to the carrier amplitude peak. Over-modulation occurs when the modulated signal is greater than the peak amplitude of the carrier resulting in loss of information. The transmitted information can be extracted by using a process known as "coherent demodulation" in which the modulated carrier is multiplied, i.e., mixed, with an unmodulated carrier of the same frequency, e.g., a square wave of the same frequency as that of carrier and then by passing the input AM signal through a zero crossing detector (ZCD). The square wave and the AM signal are given to the Mixer component in "Down-Mixer" mode. The output of the mixer is filtered by a low-pass filter (LPF) to obtain the transmitted information.

Amplitude modulation is a well-known phenomenon that has been used in a wide variety of applications and consists of a carrier signal upon which is imposed the information to be transmitted. In many applications, both the carrier and the modulation signal are occurring at low power levels. This is satisfactory except in cases in which the modulated carrier is to be transmitted over long

---

[2]Both phase and frequency [2] shift modulation/demodulation can be implemented using PSOC 3 or PSoC 5LP components which does not require any CPU cycles by using PSoC 3 and PSoC 5LP analog and digital blocks.

distances, in which case a linear amplifier is required.[3] In what follows, it shall be assumed that the modulation/demodulation takes place within a PSoC device and therefore the power levels will be low (Figs. 14.10, 14.11, and 14.12).[4]

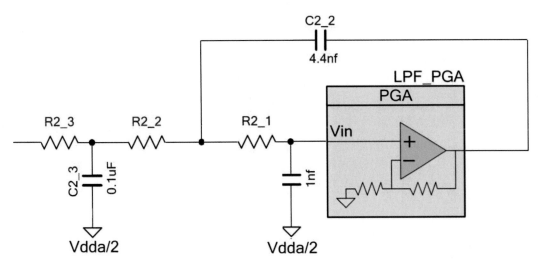

Fig. 14.10 Example of a lowpass filter (LP).

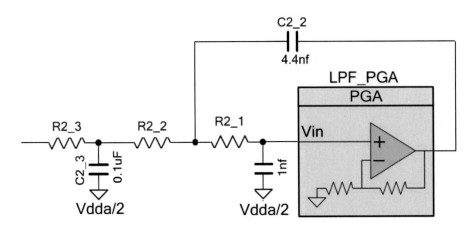

Fig. 14.11 Example of a bandpass filter (BP).

If the modulating signal is given by

$$m(t) = A_m \cos(2\pi f_m t) \tag{14.1}$$

---

[3]A linear amplifier is used instead of a nonlinear amplifier to avoid undesirable distortion of the modulated carrier resulting in lost information. Linear in the present context simply means that the output is the product o the magnitude of the input signal at every point of time and A which is the amplification or gain factor.

[4]Obviously, should a higher level modulated signal be involved for transmitting, receiving or both, external hardware can be used to keep the signals processed by the PSoC device within its acceptable operating ranges.

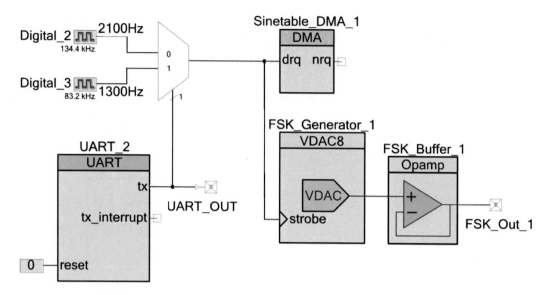

Fig. 14.12 Example of frequency shift keying (FSK) circuit.

and the carrier signal is given by

$$c(t) = \cos{(2\pi f_c t)} \tag{14.2}$$

then, if a constant value K is added to m

$$AM = (K + m(t)) \times c(t) = K\cos{(2\pi f_c t)} + A_m \cos{(2\pi f_m t)} \times \cos{(2\pi f_c t)} \tag{14.3}$$

which shows that the original carrier and two additional signals each of different frequencies will be present if K is not equal to zero. K is sometimes referred to as the offset coefficient. However, if $K = 0$ then the carrier is not present and the carrier is said to be suppressed. This mode is known as double sideband (Fig. 14.13).[5]

$$AM = m(t) \times c(t) = A_m \cos{(2\pi f_m t)} \times \cos{(2\pi f_c t)} \tag{14.4}$$

14.5   PSoC Creator's WaveDAC8 Component

Function Generation using the WaveDAC8 [4] is easily accomplished by using the WaveDAC8 component and provides the following important features (Fig. 14.14):
    The WaveDAX8 provides has the following features:

- Standard and arbitrary waveform generation
- Output may be voltage, current sink, or current source
- Hardware selection between two waveforms

---

[5]Removing the carrier is often done when transmitting RF signals because the carrier does not carry any information but can represent a significant amount of transmission power. One of the two sidebands can also be removed since either sideband, upper or lower carries the same message.

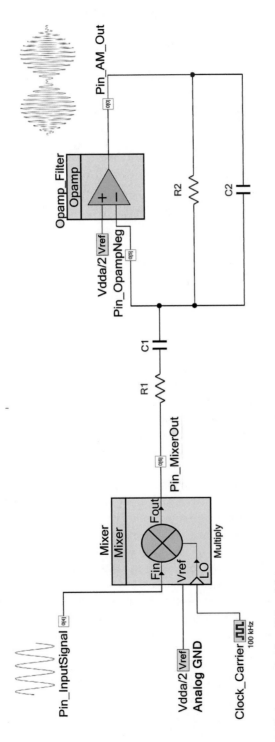

Fig. 14.13 An example of an AM modulator.

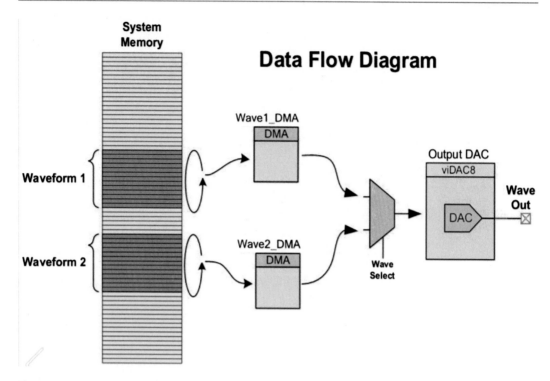

Fig. 14.14 WaveDac8 dataflow diagram utilizing a LUT in system memory.

- External clock input may be used to change the output waveform frequency
- Waveform tables can be up to 4000 points
- Predefined sine, triangle, square, and sawtooth waveforms are included
- Allows waveform arrays to be changed during run time
- Single line of C code required to initiate waveform output

The key element of PSoC Create's WaveDac8 component is either a current or voltage DAC, viz., the IDAC8 or VDAC8. Two DMA channels are provided, Wave1_DMA and Wave2_DMA, allowing data to be transferred from its respective locations in system memory as shown in Fig. 14.15. Internally, the WaveDAC8 contains a voltage or current DAC, two direct memory access (DMA) channels, an optional opamp follower. and a clock when the internal clock option is selected. The user interface and the API handle configuration of the DAC, DMA, and wave table generation. No knowledge of the DAC or DMA is required to take full advantage of the WaveDAC8 component.

The waveform(s) produced are determined by the user and PSoC Creator. Wave type, amplitude, offset, phase and number of samples are user-defined and the sample rate is user-defined as either an external or internal clock can be used.

Supported waveforms include:

- sine wave
- square wave
- saw tooth wave
- arbitrary (drawn)

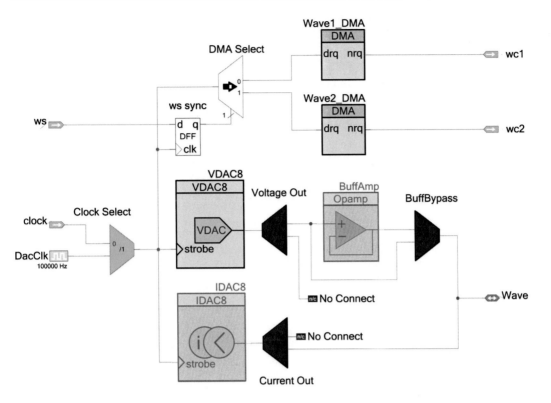

Fig. 14.15 WaveDac8 dataflow diagram for two signals stored in system memory.

- arbitrary (table-driven)[6]

This component can also be used to provide modulation and signal mixing by using multiple WaveDAC8's as illustrated in Fig. 14.15.

Arbitrary waveforms, either drawn or table-driven, are supported. Either of two waveforms can be selected by enabling external pins on the component making it possible to create modulated output signals.

The sampling rate and waveform frequency/period are related by

$$waveform_period = \frac{Samples}{SampleRate} \tag{14.5}$$

and,

$$waveform_period = \frac{SampleRate}{Samples} \tag{14.6}$$

If the waveform is produced from a lookup table (LUT), the data will be transferred via DMA from Flash memory to the DAC. The WaveDac8's DMA channel is shared by the CPU, as well as, other DMA channels. A minimum of 10 clock cycles is required to transfer each sample to the DAC. Therefore, the clock should be a minimum of ten times faster than the sample rate. If the sample rate is greater than 4 Msps, data should be transferred from Flash to SRAM and then to the DAC to avoid

---

[6]The data file format must have integer values, ranging from 0 to 255, that are comma-separated values (CSV).

Flash wait states. Multiple WaveDac8s require that the sample rate be at least 10–15 times the total sample rates of all the WaveDac8s.

The source code required to employ a WaveDAC8 component can be as simple as:

```
#include <project.h>

        int main() {
        WaveDAC8_1_Start();   Start WaveDAC8
        for (;;)              ; Loop forever
```

The sample rate is given by:

$$Sample\ Rate = (Output\ Freq)(Number\ Of\ Samples)$$
$$= (60)(64) = 3.84x10^3 ksps$$

(14.7)

The nth table entry is given by:

$$Sine\ Value_n = \left(\sin\left[n\left(\frac{360}{64}\right)\right]\left[Scale\ Factor\right]\right)Zero$$

(14.8)

where the sample number ranges from 0 to 63, the Scale factor is 127, and Zero is the value for which the DAC produces zero output, i.e.,127.

The maximum output frequency that can be achieved using the 8 bit DAC can be calculated as:

$$Maximum\ Frequency = \frac{Maximum\ Sample\ Rate}{Number\ of\ Samples}$$

(14.9)

Adding an RC filter to the output will provide a smoother sine wave. The filter should have a corner frequency (fc) above the fundamental frequency and below the sampling frequency, e.g., for a fundamental frequency of 60 Hz, the Sampling Freq is 120 Hz. If the corner frequency is 80 Hz, then a suitable RC low pass filter could consist of a 2 kΩ resistor and a 1 μf capacitor (Fig. 14.16).

As an example, consider a 64-entry look-up table (LUT) designed to drive a WaveDac8 to produce a 60 Hz sine wave. The nth table entry is given by:

Listing 14.1 The values in this case for the sine wave are are expressed in CSV format as:

```
        const char SineTable64[] = {
        127, 139, 152, 164, 176, 187, 198, 208,
217, 225,
        233, 239, 244, 249, 252, 253, 254, 253,
252, 249,
        244, 239, 233, 225, 217, 208, 198, 187,
176, 164,
        152, 139, 127, 115, 102, 90, 78, 67, 56,
46, 37,
        29, 21, 15, 10, 5, 2, 1, 0, 1, 2, 5, 10,
15, 21,
        29, 37, 46, 56, 67, 78, 90, 102, 115};
//64 samples stored in ROM
```

Function	Description
void WaveDAC8_Start(void)	Starts the DAC and DMA channels.
void WaveDAC8_Stop(void)	Disables DAC and DMA channels.
void WaveDAC8_Init(void)	Initializes or restores the component according to the customizer Configure dialog settings.
void WaveDAC8_Enable(void)	Activates the hardware and begins component operation.
void WaveDAC8_Wave1Setup(uint8 * wavePtr, uint16 sampleSize)	Sets the array and size of array used for waveform generation for waveform 1.
void WaveDAC8_Wave2Setup(uint8 * wavePtr, uint16 sampleSize)	Sets the array and size of array used for waveform generation for waveform 2.
void WaveDAC8_StartEx(uint8 * wavePtr1, uint16 sampleSize1, uint8 * wavePtr2, uint16 sampleSize2)	Sets the arrays and sizes of arrays used for waveform generation for both waveforms and starts the DAC and DMA channels.
void WaveDAC8_SetSpeed(uint8 speed)	Set drive mode / speed of the DAC.
void WaveDAC8_SetRange(uint8 range)	Set current or voltage range.
void WaveDAC8_SetValue(uint8 value)	Set 8-bit DAC value.
void WaveDAC8_DacTrim(void)	Set the trim value for the given range.
void WaveDAC8_Sleep(void)	Stops and saves the user configuration.
void WaveDAC8_Wakeup(void)	Restores and enables the user configuration.
void WaveDAC8_SaveConfig(void)	This function saves the component configuration. This function will also save the current component parameter values, as defined in the Configure dialog or as modified by appropriate APIs. This function is called by the WaveDAC8_Sleep() function.
void WaveDAC8_RestoreConfig(void)	This function restores the component configuration. This function will also restore the component parameter values to what they were before calling the WaveDAC8_Sleep() function.

Fig. 14.16 Function calls supported by WaveDAC8.

References

1. M. Ainsworth, Switch Debouncer and Glitch Filter with PSoC 3, PSoC 4, and PSoC 5LP. AN60024. Document No. 001-60024Rev.*P1. Cypress Semiconductor (2017)
2. T. Dust, PSoC®3 and PSoC 5LP: Low-Frequency FSK Modulation and Demodulation. Document No. 001-60594Rev.*J1. AN60594. Cypress Semiconductor (2017)
3. R. Fosler, Srinivas NVNS. PSoC 3and PSoC 5LP Phase-Shift Full-Bridge Modulation and Control. Document No. 001-76439Rev. *A1. AN76439. Cypress Semiconductor (2017)
4. M. Hastings, PSoC®3/PSoC 5LP Easy Waveform Generation with the WaveDAC8 Component. AN69133. Cypress Semiconductor (2017)
5. D. Sweet, Peak Detection with PSoC®3 and PSoC 5LP. Document No. 001-60321Rev.*I1. AN60321. Cypress Semiconductor (2017)
6. P. Vibhute, AM Modulation and Demodulation. Document No. 001-62582Rev.*F2. Cypress Semiconductor (2017)

PSoC Creator Function Calls

15

Abstract

PSoC Creator supports a wide range of components that a designer can use in developing mixed-signal embedded systems. Each component is supported by function calls that allow the designer to control the underlying block's functionality. In this chapter, the function calls are detailed in terms of the specific functions supported by each PSoC Creator component. Readers are encouraged to use the Cypress Semiconductor datasheets included, as PDF files, in PSoC Creator to learn the specific details of how to implement each component. PSoC 3 and PSoC 5 LP have a number of components that are common to both. However, there are some differences in the suite of components provided for each. For example, the Delta Sigma analog to digital converter is found in PSoC 3 and not in PSoC which instead provides a successive approximation register (SAR) and sequencing SAR. The latter can be used when sampling multiple sources.

This chapter begins with PSoC Creator components which are unique to PSoC 3 and then treats those which are common to both PSoC 3 and PSoC 5LP. Sections 15.1 to 15.16.2, inclusive, describe the components supported by PSoC Creator for PSoC 3 and their respective functions/function calls. Similarly, Sects. 15.17 to 15.127.1 describe components that are common to both PSoC 3 and PSoC 5LP and their respective functions/function calls.

15.1 Delta-Sigma Analog to Digital Converter 3.30 (ADC_DelSig)

PSoC 3 Family of Devices—This component provides a low-power, low-noise front end for precision measurements that produces 16-bit audio at high speed and low resolution for communications processing; high-precision 20-bit low-speed conversions for sensors such as strain gauges, thermocouples, other high-precision sen The ADC_DelSig is used in a continuous operation mode sors when used for processing audio information When scanning multiple sensors, the ADC_DelSig is used in one of the multi-sample modes. Single-point high-resolution measurements use the ADC_DelSig is used in single-sample mode. Delta-sigma converters use oversampling to spread the quantization noise across a wider frequency spectrum. This noise is shaped to move most of it outside the input signal's bandwidth. An internal low-pass filter is used to filter out the noise outside the desired input Delta Sigma Analog to Digital Converter (ADC_DelSig) Delta Sigma Analog to Digital Converter (ADC_DelSig) signal bandwidth. This makes delta-sigma converters good for both high-

© Springer Nature Switzerland AG 2021
E. H. Currie, *Mixed-Signal Embedded Systems Design*,
https://doi.org/10.1007/978-3-030-70312-7_15

speed medium resolution (8 to 16 bits) applications and low-speed high-resolution (16 to 20 bits) applications. The sample rate can be adjusted between 10 and 384,000 samples per second, depending on mode and resolution. Choices of conversion modes simplify interfacing to single streaming signals such as audio, or multiplexing between multiple signal sources. The ADC_DelSig is composed of three blocks: an input amplifier, a third-order delta-sigma modulator and a decimator. The input amplifier provides a high-impedance input and a user-selectable input gain. A decimator block contains a four-stage CIC decimation filter and a post-processing unit. A CIC filter operates on the data sample directly from the modulator. The post-processing unit optionally performs gain, offset and simple filter functions on the output of the CIC decimator filter. Also supported are selectable resolutions, 8 to 20 bits, eleven input ranges for each resolution and sample rates from 8 sps to 384 ksps. Operational modes include: single sample, multi-sample, continuous mode, multi-sample (Turbo) high input impedance input buffer, selectable input buffer gain (1, 2, 4, 8) or input buffer bypass, multiple internal or external reference options, automatic power configuration and up to four run-time ADC configurations. Features include: selectable resolutions, 8–20 bits, sample rates from 8 sps to 384 ksps, single and multi-sampling, continuous mode, high input impedance input buffer, selectable input buffer gain (1, 2, 4,8) or input buffer bypass , multiple internal or external reference options, automatic power configuration and up to four run-time ADC configurations [27].

15.1.1 ADC_DelSig Functions 3.30

- **ADC_Start( )**—Sets the initVar variable, calls the ADC_Init( ) function and then calls the ADC_Enable( ) function.
- **ADC_Stop( )**—Stops ADC conversions and powers down.
- **ADC_SetBufferGain( )**—Selects input buffer gain (1,2,4,8).
- **ADC_StartConvert( )**—Starts conversion.
- **ADC_StopConvert( )**—Stops conversions.
- **ADC_IRQ_Enable( )**—Enables interrupts at the end of conversion.
- **ADC_I_ Disable( )**—Disables interrupts.
- **ADC_IsEndConversion( )**—Returns a nonzero value if conversion is complete.
- **ADC_GetResult8( )**—Returns an 8-bit conversion result.
- **ADC_GetResult16( )**—Returns a 16-bit conversion result.
- **ADC_GetResult32( )**—Returns a 32-bit conversion result.
- **ADC_Read8( )**—Starts ADC conversions, waits for the conversion to be complete, stops ADC conversion and returns the signed 8-bit value of result.
- **ADC_Read32( )**—Starts ADC conversions, waits for the conversion to be complete, stops ADC conversion and returns the signed 32-bit value of result.
- **ADC_SetOffset( )**—Sets the offset used by the ADC_CountsTo_mVolts( ), ADC_CountsTo_uVolts( ) and ADC_CountsTo_Volts( ) functions.
- **ADC_SelectConfiguration( )**—Sets one of up to four ADC configurations.
- **ADC_SetGain( )**—Sets the gain used by the ADC_CountsTo_mVolts( ), ADC_CountsTo_uVolts( ) and ADC_CountsTo_Volts( ) functions.
- **ADC_CountsTo_mVolts( )**—Converts ADC counts to millivolts.
- **ADC_CountsTo_uVolts( )**—Converts ADC counts to microvolts.
- **ADC_CountsTo_Volts( )**—Converts ADC counts to floating-point volts.
- **ADC_Sleep( )**—Stops ADC operation and saves the user configuration.
- **ADC_Wakeup( )**—Restores and enables the user configuration.
- **ADC_Init( )**—Initializes or restores the ADC using the Configure dialog settings.

- **ADC_Enable( )**—Enables the ADC.
- **ADC_SaveConfig( )**—Saves the current configuration.
- **ADC_RestoreConfig( )**—Restores the configuration.
- **ADC_SetCoherency( )**—Sets the coherency register.
- **ADC_SetGCOR( )**—Calculates a new GCOR value and sets the GCOR registers.

15.1.2 ADC DelSig Function Calls 3.30

- **void ADC_Start(void)**—Sets the initVar variable, calls the ADC_Init( ) function and then calls the ADC_Enable( ) function. This function configures and powers up the ADC, but does not start conversions. By default, the ADC is configured for Config1. Use the ADC_SelectConfiguration( ) function to select an alternate configuration afterward.
- **void ADC_Stop(void)**—Disables and powers down the ADC.
- **void ADC_SetBufferGain(uint8 gain)**—Sets the input buffer gain.
- **void ADC_StartConvert(void)**—Forces the ADC to initiate a conversion.
- **void ADC_StartConvert(void)**—Forces the ADC to initiate a conversion. In Single Sample mode, call this API to start a single conversion. When the conversion completes, use ADC_IsEndConversion( ) API to check or wait on this event, the ADC will halt. If the ADC_StartConvert( ) function is called while the conversion is in progress, the next conversion start is queued and a new conversion will start after finishing the current conversion. If you want to start a new conversion without waiting for the current conversion to finish, then stop the current conversion by calling ADC_StopConvert( ) After stopping the conversion, restart the conversion by calling ADC_StartConvert( ). In Multi Sample, Continuous or Multi Sample (Turbo) modes, call this API to start continuous ADC conversions until either the ADC_StopConvert( ) or ADC_Stop( ) functions are executed.
- **void ADC_StopConvert(void)**—Forces the ADC to stop all conversions. If the ADC is in the middle of a conversion, the ADC will be reset and not provide a result for that partial conversion.
- **void ADC_IRQ_Enable(void)**—Enables interrupts at the end of a conversion. Global interrupts must also be enabled for the ADC interrupts to occur. To enable global interrupts, use the enable global interrupt macro "CYGlobalIntEnable;" in main.c, before interrupts occur.
- **void ADC_IRQ_Disable(void)**—Disables interrupts at the end of a conversion.
- **uint8 ADC_IsEndConversion(uint8 retMode)**—Checks for ADC end of conversion. This function provides the programmer with two options. In one mode this function immediately returns with the conversion status. In the other mode, the function does not return (blocking)
- **int8 ADC_GetResult8(void)**—Returns a signed 8-bit value. The largest positive signed 8-bit value that can be represented is 127, but in single-ended 8-bit mode, the maximum positive value is 255. Hence, for an 8-bit single-ended mode, use the ADC_GetResult16( ) function instead. Note that if the ADC resolution is set greater than 8 bits, the LSB of the result is returned.
- **int16 ADC_GetResult16(void)**—Returns a 16-bit result for a conversion that has a resolution of 8 to 16 bits. If the resolution is set greater than 16 bits, it will return the 16 least significant bits of the result. When the ADC is configured for 16-bit single-ended mode, use the ADC_GetResult32( ) function instead. This function returns only signed 16-bit results, which allows a maximum positive value of 32767, not 65535.
- **int32 ADC_GetResult32(void)**—Returns a 32-bit result for a conversion that has a resolution of 8 to 20 bits.
- **int8 ADC_Read8(void)**—This function simplifies getting results from the ADC when only a single reading is required. When called, it will start ADC conversions, wait for the conversion

to be complete, stop ADC conversion and return the result. This is a blocking function and will not return until the result is ready. When the ADC is configured for 8-bit single ended mode, the ADC_Read16( ) function should be used instead. This function returns only signed 8-bit values. The maximum positive signed 8-bit value is 127, but in singled-ended 8-bit mode, the maximum positive value is 255.

- **int16 ADC_Read16(void)**—This function simplifies getting results from the ADC when only a single reading is required. When called, it will start ADC conversions, wait for the conversion to be complete, stop ADC conversion and return the result. This is a blocking function and will not return until the result is ready. When the ADC is configured for 16-bit single ended mode, the ADC_Read32( ) function should be used instead. This function returns only signed 16-bit values, which allows a maximum positive value of 32767, not 65535.
- **int32 ADC_Read32(void)**—This function simplifies getting results from the ADC when only a single reading is required. When called, it will start ADC conversions, wait for the conversion to be complete, stop ADC conversion and return the result. This is a blocking function and will not return until the result is ready. Returns a 32-bit result for a conversion with a result that has a resolution of 8 to 20 bits.
- **void ADC_SetOffset(int32 offset)**—Sets the ADC offset which is used by the functions ADC_CountsTo_uVolts( ), ADC_CountsTo_mVolts( ) and ADC_CountsTo_Volts( ) to subtract the offset from the given reading before calculating the voltage conversion.
- **void ADC_SetGain(int32 adcGain)**—Sets the ADC gain in counts per volt for the voltage conversion functions below. This value is set by default by the reference and input range settings. It should only be used to further calibrate the ADC with a known input or if an external reference is used.
- **void ADC_SelectConfiguration(uint8 config, uint8 restart)**—Sets one of up to four ADC configurations. This API first stops the ADC and then initializes the registers with the default values for the new configuration. The custom GGOR register value, set by ADC_SetGCOR( ) API for particular configuration, is not overwritten to default value. If the value of the second parameter restart is 1, then ADC will be revalue is zero, then you must call ADC_Start( ) and ADC_StartConvertIf this value is zero, then you must call ADC_Start( ) and ADC_StartConvert( ) to restart the conversion. ADC_Start( ) API should be called before first ADC_SelectConfiguration API usage for initialization and correct operation. as a 16-bit integer. For example, if the ADC measured 0.534 V, the return value would be 534 mV. The calculation of voltage depends on the value of the voltage reference. When the Vref is based on Vdda, the value used for Vdda is set for the project in the System tab of the Design Wide Resources (DWR).
- **int32 ADC_CountsTo_uVolts(int32 adcCounts)**—Converts the ADC output to microvolts as a 32-bit integer. The calculation of voltage depends on the value of the voltage reference. When the Vref is based on Vdda, the value used for Vdda is set for the project in the System tab of the Design Wide Resources (DWR).
- **float32 ADC_CountsTo_Volts(int32 adcCounts)**—Converts the ADC output to volts as a floating-point number. For example, if the ADC measures a voltage of 1.2345 V, the returned result would be +1.2345 V. The calculation of voltage depends on the value of the voltage reference. When the Vref is based on Vdda, the value used for Vdda is set for the project in the System tab of the Design Wide Resources (DWR).
- **void ADC_Sleep(void)**—The ADC_Sleep( ) function checks to see if the Component is enabled and saves that state. Then it calls the ADC_Stop( ) function and calls ADC_SaveConfig( ) to save the user configuration. Call the ADC_Sleep( ) function before calling the CyPmSleep( ) or the CyPmHibernate( ) function.

- **void ADC_Wakeup(void)**—The ADC_Wakeup( ) function calls the ADC_RestoreConfig( ) function to restore the user configuration. If the Component was enabled before the ADC_Sleep( ) function was called, the DC_Wakeup( ) function will re-enable the Component.
- **void ADC_Init(void)**—Initializes or restores the Component parameters per the Configure dialog settings. You are not required to call this function if ADC_Start( ) is called.
- **void ADC_Enable(void)**—Enables the clock and power for ADC.
- **void ADC_SaveConfig(void)**—This function saves the Component configuration. This will save non-retention registers. This function will also save the current Component parameter values, as defined in the Configure dialog or as modified by appropriate APIs. This function is called by the ADC_Sleep( ) function.
- **void ADC_RestoreConfig(void)**—This function restores the Component configuration. This will restore non-retention registers. This function will also restore the Component parameter values to what they were prior to calling the ADC_Sleep( ) function.
- **void ADC_SetCoherency(uint8 coherency)**—This function allows you to change which of the ADC's three-word results will trigger a coherency unlock. The ADC's result will not be updated until the set byte is read by either the ADC or DMA. By default, the LSB is the coherency byte. If DMA or if a custom API is written where the LSB is not the last byte read, this use this API to set the last byte of the ADC result that is read. If a multi-byte read is performed is performed either by DMA or the ARM processor, the coherency can be set to any byte in the last word read.
- **uint8 ADC_SetGCOR(float gainAdjust)**—This function calculates a new GCOR (ADC Gain) value and writes it into the GCOR registers. The GCOR value is a 16-bit value that represents a gain of 0 to 2. The ADC result is multiplied by this value before it is placed in the ADC output registers. When executing the function, the old GCOR value is multiplied by gainAdjust input and reloaded into the GCOR register. The GCOR value is normalized based on the GVAL register. The value, calculated by this API, is also stored into the RAM for each active configuration and used by SelectConfiguration( ) API to initialize the GCOR register.
- **int16 ADC_iReadGCOR(void)**—This function returns the current GCOR register value, normalized based on the GVAL setting. For example, if the GCOR value is $0x0812$ and the GVAL register is set to 11 (0x0B), the returned value is shifted by four bits to the left. (Actual GCOR value = 0x0812, returned value = 0x8120)

15.2   Inverting Programmable Gain Amplifier (PGA_Inv) 2.0

The component utilizes implements a programmable gain (-1.0 (0 dB) and -49.0 (+33.8 dB)) OpAmp in inverting mode and based on a SC/CT block. The gain can be selected PSoC Creator or changed at run time.[1] The input of the PGA_Inv operates from rail to rail, but the maximum input swing (difference between Vin and Vref) is limited to VDDA/Gain. Output of the PGA_Inv is class A and is rail-to-rail for sufficiently high load resistance. The PGA_Inv is used when an input signal has insufficient amplitude and the preferred output polarity is the inverse of the input. A PGA_Inv can be placed in front of a comparator, ADC, or mixer to increase the signal amplitude. A unity gain PGA_Inv following another gain stage or buffer can be used to generate differential outputs. Features include: Gain steps from -1 to -49, High input impedance and adjustable power settings [56].

---

[1] The maximum bandwidth is limited by the gain-bandwidth of the OpAmp and is reduced as the gain is increased.

15.2.1 PGA_Inv) Functions 2.0

- **PGA_Inv_Start( )**—Starts the PGA_Inv.
- **PGA_Inv_Stop( )**—Powers down the PGA_Inv.
- **PGA_Inv_SetGain( )**—Sets gain to predefined constants.
- **PGA_Inv_SetPower( )**—Sets drive power to one of four settings.
- **PGA_Inv_Sleep( )**—Stops and saves the user configurations.
- **PGA_Inv_Wakeup( )**—Restores and enables the user configurations.

15.2.2 PGA_Inv) Function Calls 2.0

- **void PGA_Inv_Inv_Start(void)**—Turns on the PGA_Inv and sets the power level.
- **void PGA_Inv_Stop(void)**—Turns off PGA_Inv and enables its lowest power state.
- **void PGA_Inv_SetGain(uint8 gain)**—Sets gain of amplifier between -1 and -49. uint8 gain sets the gain to a specific value.
- **void PGA_Inv_SetPower(uint8 power)**—Sets the drive power to one of four settings: minimum, low, medium, or high. uint8 power sets the power level to one of four settings: minimum, low, medium, or high.
- **void PGA_Inv_Sleep(void)**—This is the preferred routine to prepare the component for sleep. The PGA_Inv_Sleep( ) function saves the current component state. Then it calls the PGA_Inv_ Stop( ) function and calls PGA_Inv_SaveConfig( ) to save the hardware configuration. Call the PGA_Inv_Sleep( ) function before calling the CyPmSleep( ) or the CyPmHibernate( ) function.
- **void PGA_Inv_Wakeup(void)**—This is the preferred routine to restore the component to the state when PGA_Inv_Sleep( ) was called. The PGA_Inv_Wakeup( ) function calls the PGA_Inv_RestoreConfig( ) function to restore the configuration. If the component was enabled before the PGA_Inv_Sleep( ) function was called, the PGA_Inv_Wakeup( ) function will also re-enable the component.[2]
- **void PGA_Inv_Init(void)**—Initializes or restores the component according to the customizer Configure dialog settings. It is not necessary to call PGA_Inv_Init( ) because the PGA_Inv_Start( ) routine calls this function and is the preferred method to begin component operation. All registers will be set to values according to the customizer Configure dialog.
- **void PGA_Inv_Enable(void)**—Activates the hardware and begins component op PGA_Inv_ Enable( ) because the PGA_Inv_Start( ) routine calls this function, which is the preferred method to begin component operation.
- **void PGA_Inv_SaveConfig(void)**—Empty function. Provided for future use.
- **void PGA_Inv_RestoreConfig(void)**—Empty function. Provided for future use.

15.3 Programmable Gain Amplifier (PGA) 2.0

This component is a noninverting high input impedance SC/CT-based OpAmp with programmable gain 1 (0 dB) and 50 (+34 dB). In addition to wide bandwidth it has selectable input voltage reference. Gain is selected using the configuration window or changed at run time using the provided API. The maximum bandwidth is limited by the gain-bandwidth product of the OpAmp. The input of the

---

[2]Calling the PGA_Inv_Wakeup( ) function without first calling the PGA_Inv_Sleep( ), or the PGA_Inv_SaveConfig( ) function may produce unexpected behavior.

PGA operates from rail to rail, but the maximum input swing (difference between Vin and Vref) is limited to VDDA/Gain. Output of the PGA is class A and is rail-to-rail for sufficiently high load resistance. A PGA is used when an input signal has insufficient amplitude. A PGA can be put in front of a comparator, ADC, or mixer to increase the amplitude of the signal to these components. The PGA can be used as a unity gain amplifier to buffer the inputs of lower impedance blocks, including mixers or inverting PGAs. A unity gain PGA can also be used to buffer the output of a VDAC or reference. Features include: gain steps from 1 to 50, high input impedance, selectable input reference and adjustable power settings [66].

15.3.1 PGA Functions 2.0

- **PGA_Start( )**—Starts the PGA.
- **PGA_Stop( )**—Powers down the PGA.
- **PGA_SetGain( )**—Sets gain to predefined constants.
- **PGA_SetPower( )**—Sets drive power to one of four settings.
- **PGA_Sleep( )**—Stops and saves the user configurations.
- **PGA_Wakeup( )**—Restores and enables the user configurations.
- **PGA_Init( )**—Initializes or restores default PGA configuration.
- **PGA_Enable( )**—Enables the PGA.
- **PGA_SaveConfig( )**—Empty function. Provided for future use.
- **PGA_RestoreConfig( )**—Empty function. Provided for future use.

15.3.2 PGA Function Calls 2.0

- **void PGA_Start(void)**—This is the preferred method to begin component operation. This function turns on the amplifier with the power and gain based on the settings provided during the configuration or the current values after PGA_Stop( ) has been called.
- **void PGA_SetGain(uint8 gain)**—This function sets the amplifier gain to a value between uint8 gain:
- **void PGA_SetPower(uint8 power)**—This function sets the drive power to one of four settings; minimum, low, medium, or high.
- **void PGA_SetGain(uint8 gain)**—This function sets the amplifier gain to a value between 1 and 50.
- **void PGA_Sleep(void)**—This is the preferred API to prepare the component for sleep. The PGA_Sleep( ) API save the current component state. Then it calls the PGA_Stop( ) function and calls PGA_SaveConfig( ) to save the hardware configuration.[3]
- **void PGA_Wakeup(void)**—This is the preferred API to restore the component to the state when PGA_Sleep( ) was called. The PGA_Wakeup( ) function calls the PGA_RestoreConfig( ) function to restore the configuration. If the component was enabled before the PGA_Sleep( ) function was called, the PGA_Wakeup( ) function will also re-enable the component.[4]

---

[3]Call the PGA_Sleep( ) function before calling the CyPmSleep( ) or the CyPmHibernate( ) function.
[4]Calling the PGA_Wakeup( ) function without first calling the PGA_Sleep( ) or PGA_SaveConfig( ) function may produce unexpected behavior.

- **void PGA_Init(void)**—This function initializes or restores the component according to the customizer Configure dialog settings.[5]
- **void PGA_Enable(void)**—This function activates the hardware and begins component operation.[6]
- **void PGA_SaveConfig(void)**—Empty function. Provided for future use.
- **void PGA_RestoreConfig(void)**—Empty function. Provided for future use.

15.4   Trans-Impedance Amplifier (TIA) 2.0

The Trans-Impedance Amplifier (TIA) component is an OpAmp-based, current-to-voltage amplifier with resistive gain and user-selected bandwidth that converts an external current to a voltage for use with sensors. e.g., photodiodes and other current sources. The TIA's gain is expressed in Ohms, ranging from 20 k to 1.0 MΩ. Some sensors have significant output capacitance and therefore, feedback capacitance is needed to assure stability. Feedback capacitance in the TIA is used to guarantee stability. The TIA's programmable feedback also limits the bandwidth of broadband noise. Features include: selectable gain and corner frequency, capacitive compensation, variable power settings and selectable input reference voltages [98].

15.4.1 TIA Functions 2.0

- **TIA_Start( )**—Powers up the TIA. Powers down the TIA.
- **TIA_Stop( )**—Powers up the TIA. Powers down the TIA.
- **TIA_SetPower( )**—Sets drive power to one of four levels.
- **TIA_SetResFB( )**—Sets the resistive feedback to one of eight values.
- **TIA_SetCapFB( )**—Stops and saves the user configurations.
- **TIA_Sleep( )**—Sets the capacitive feedback to one of four values.
- **TIA_Wakeup( )**—Restores and enables the user configurations.
- **TIA_Init( )**—Initializes or restores default TIA configuration.
- **TIA_Enable( )**—Enables the TIA
- **TIA_SaveConfig( )**—Empty function. Provided for future use.
- **TIA_RestoreConfig( )**—Empty function. Provided for future use.

15.4.2 TIA Function Calls 2.0

- **void TIA_Start(void)**—Performs all the required initialization for the component and enables power to the amplifier.
- **void TIA_Stop(void)**—Powers down TIA to its lowest power state and disables output.
- **void TIA_SetCapFB(uint8 cap_feedback)**—Set the amplifier capacitive feedback value.
- **void TIA_SetCapFB(uint8 cap_feedback)**—Set the amplifier capacitive feedback value.
- **void TIA_Sleep(void)**—This is the preferred API to prepare the component for sleep. The TIA_Sleep( ) function saves the current component state. Then it calls the TIA_Stop( ) function

---

[5]It is not necessary to call PGA_Init( ) because the PGA_Start( ) API calls this function and is the preferred method to begin component operation.

[6]It is not necessary to call PGA_Enable( ) because the PGA_Start( ) API calls this function, which is the preferred method to begin component operation.

and calls TIA_SaveConfig( ) to save the hardware configuration. Call the TIA_Sleep( ) function before calling the CyPmSleep( ) or the CyPmHibernate( ) function. Refer to the PSoC Creator System Reference Guide for more information about power management functions.

- **void TIA_Wakeup(void)**—This is the preferred routine to restore the component to the state when TIA_Sleep( ) was called. The item TIA_Wakeup( ) function calls the TIA_RestoreConfig( ) function to restore the configuration. If the component was enabled before the TIA_Sleep( ) function was called, the TIA_Wakeup( ) function will also re-enable the component.
- **void TIA_Init(void)**—Initializes or restores the component according to the customizer Configure dialog settings. It is not necessary to call TIA_Init( ) because the TIA_Start( ) routine calls this function and is the preferred method to begin component operation.
- **void TIA_Enable(void)**—Activates the hardware and begins component operation. It is not necessary to call TIA_Enable( ) because he TIA_Start( ) routine calls this function, which is the preferred method to begin component operation.
- **void TIA_RestoreConfig(void)**—Empty function. Provided for future use.

15.5   SC/CT Comparator (SCCT_Comp) 1.0

The SC/CT Comparator (SC/CT_Comp) component[7] provides a hardware solution to compare two analog input voltages. The implementation uses a mode of the Switched Capacitor/Continuous Time (SC/CT) analog block to implement the comparator. The output can be digitally routed to another component. A reference or external voltage can be connected to either input. You can also invert the output of the comparator using the Polarity parameter. Features include: output routable to digital logic blocks or pins and selectable output polarity [77].

15.5.1 SC/CT Comparator (SC/CT_Comp) Functions 1.0

- **Comp_Start( )**—Initializes the component with default customizer values and enables operation.
- **Comp_Stop( )**—Turns off the component.
- **Comp_Sleep( )**—Stops component operation and saves the user configuration.
- **Comp_Wakeup( )**—Restores and enables the user configuration.
- **Comp_Init( )**—Initializes or restores default component configuration.
- **Comp_Enable( )**—Enables the component.

15.5.2 SC/CT Comparator (SC/CT_Comp) Function Calls 1.0

- **void Comp_Start(void)**—Performs all of the required initialization for the component and enables power to the block. The first time the routine is executed, the component is initialized to the configuration from the customizer. When called to restart the component following a Comp_Stop( ) call, the current component parameter settings are retained.
- **void Comp_Stop(void)**—Turns off the component. It will disable the related SC/CT block.

---

[7]The SC/CT_Comp is not as high performance component as the dedicated comparator. Hysteresis support is not available. However, it is still useful in applications that don't have strict requirements for offset voltage and response time parameters and the number of required comparators exceeds the available dedicated comparators.

- **void Comp_Sleep(void)**—This is the preferred API to prepare the component for low power mode operation (disable for this case). If the component is enabled it configures the comparator for low power operation.[8]
- **void Comp_Wakeup(void)**—This is the preferred API to restore the component to the state before Comp_Sleep( ) was called.
- **void Comp_Init(void)**—Initializes or restores the component according to the customizer settings. It is not necessary to call Comp_Init( ) because the Comp_Start( ) API calls this function and is the preferred method to begin component operation.
- **void Comp_Enable(void)**—Activates the hardware and begins component operation. It is not necessary to call Comp_Enable( ) because the Comp_Start( ) API calls this function, which is the preferred method to begin component operation.

15.6 Mixer 2.0

This component is a single-ended modulator that can be used for frequency conversion of an input signal using a fixed Local Oscillator (LO) signal as the sampling clock.Signals can be moved between frequency bands or to encode and decode signals or to convert signal power at one frequency into power at another frequency to make signal processing easier, e.g., shifting higher frequencies to baseband. The mixer typically used in conjunction with an off-chip filter. Alternatively, the output can be used to drive an on-chip ADC through internal routing. The component offers two configurations; As an up-mixer, continuous-time balance mixer, operates as a switching multiplier or a down-mixer, discrete-time, sample-and-hold mixer. The component accepts two signals at different frequencies as inputs and outputs a mixture of signals at multiple frequencies, including the sum and difference of the input signal and the local oscillator signal. Typically, the unwanted frequency components in the output signal are removed by filtering. Features include: single-ended mixer, Continuous-time up mixing (input frequencies up to 500 kHz and sample clock up to 1 MHz), discrete-time, sample-and-hold down mixing (input frequencies up to 14 MHz, sample clock up to 4 MHz), adjustable power settings and selectable reference voltage [61].

15.7 Mixer Functions 2.0

- **Mixer_Start( )**—Powers up the Mixer.
- **Mixer_Stop( )**—Powers down the Mixer.
- **Mixer_SetPower( )**—Sets drive power to one of four levels.
- **Mixer_Wakeup( )**—Restores and enables the user configuration.
- **Mixer_Init( )**—Initializes or restores default Mixer configuration.
- **Mixer_Enable( )**—Enables the Mixer.
- **Mixer_SaveConfig( )**—Empty function. Provided for future use.
- **Mixer_RestoreConfig( )**—Empty function. Provided for future use.

---

[8]Call the Comp_Sleep( ) function before calling the CyPmSleep( ) or the CyPmHibernate( ) function.

15.8   Mixer Function Calls 2.0

- **void Mixer_Start(void)**—Performs all the required initialization for the component and enables power to the block. The first time the routine is executed, the input and feedback resistance values are configured for the operating mode selected in the design. When called to restart the mixer following a Mixer_Stop( ) call, the current component parameter settings are retained.
- **void Mixer_Stop(void)**—Turns off the Mixer block.
- **void Mixer_SetPower(uint8 power)**—Sets the drive power to one of four settings; minimum, low, medium, or high.
- **void Mixer_Sleep(void)**—This is the preferred API to prepare the component for sleep. The Mixer_Sleep( ) API saves the current component state. and then calls the Mixer_Stop( ) function and Mixer_SaveConfig( ) to save the hardware configuration. C all the Mixer_Sleep( ) function before calling the CyPmSleep( ) or the CyPmHibernate( ) unction.
- **void Mixer_Wakeup(void)**—This is the preferred API to restore the component to the state when Mixer_Sleep( ) was called. The Mixer_Wakeup( ) function calls the Mixer_RestoreConfig( ) function to restore the configuration. If the component was enabled before the Mixer_Sleep( ) function was called, the Mixer_Wakeup( ) function will also re-enable the component.
- **void Mixer_Init(void)**—Initializes or restores the component according to the customizer Configure dialog settings. It is not necessary to call Mixer_Init( ) because the Mixer_Start( ) API calls this function and is the preferred method to begin component operation.
- **void Mixer_Enable(void)**—Activates the hardware and begins component operation. It is not necessary to call Mixer_Enable( ) because the Mixer_Start( ) API calls this function, which is the preferred method to begin component operation.
- **void Mixer_SaveConfig(void)**—Empty function. Provided for future use.
- **void Mixer_RestoreConfig(void)**—Empty function. Provided for future use.

15.9   Sample/Track and Hold Component 1.40

This component provides a way to sample a continuously varying analog signal and to hold or freeze its value for a finite period of time. It supports both Track and Hold and Sample and Hold functions, which can be selected in the customizer. Features include: two operating modes: Sample and Hold, Track and Hold and four power mode settings [75].

15.9.1 Sample/Track and Hold Component 1.40 Functions 1.40

- **Sample_Hold_Start( )**—Configures and enables power of Sample/Track and Hold.
- **Sample_Hold_Stop( )**—Turns off the Sample/Track and Hold block.
- **ample_Hold_SetPower( )**—Sets the drive power of Sample/Track and Hold.
- **Sample_Hold_Sleep( )**—Puts the Sample/Track and Hold into sleep mode.
- **Sample_Hold_Wakeup( )**—Wakes up Sample/Track and Hold.
- **Sample_Hold_Init( )**—Initializes the Sample/Track and Hold component.
- **Sample_Hold_Enable( )**—Activates the hardware and begins component operation.
- **Sample_Hold_SaveConfig( )**—Empty function. Provided for future use.
- **Sample_Hold_RestoreConfig( )**—Empty function. Provided for future use.

15.9.2 Sample/Track and Hold Component Function Calls 1.40

- **void Sample_Hold_Start(void)**—Performs all the required initialization for the component and enables power to the block. The first time the routine is executed, the sample mode, clock edge and power are set to their default values. When called to restart following a Sample_Hold_Stop( ) call, the current component parameter settings are retained.
- **void Sample_Hold_Stop(void)**—Turns off the Sample/Track and Hold block.
- **void Sample_Hold_SetPower(uint8 power)**—Sets the drive power to one of four settings; minimum, low, medium, or high.
- **void Sample_Hold_Sleep(void)**—This is the preferred API to prepare the component for sleep. The SampleHold_Sleep( ) API saves the current component state. Then it calls the Sample_Hold_Stop( ) function and calls Sample_Hold_SaveConfig( ) to save the hardware configuration. Call the Sample_Hold_Sleep( ) function before calling the CyPmSleep( ) or the CyPmHibernate( ) function.
- **void Sample_Hold_Wakeup(void)**—This is the preferred API to restore the component to the state when Sample_Hold_Sleep( ) was called. The Sample_Hold_Wakeup( ) function calls the Sample_Hold_RestoreConfig( ) function to restore the configuration. If the component was enabled before the Sample_Hold_Sleep( ) function was called, the Sample_Hold_Wakeup( ) function also re-enables the component.
- **void Sample_Hold_Wakeup(void)**—This is the preferred API to restore the component to the state when Sample_Hold_Sleep( ) was called. The Sample_Hold_Wakeup( ) function calls the Sample _Hold_RestoreConfig( ) function to restore the configuration. If the component was enabled before the Sample_Hold_Sleep( ) function was called, the Sample_Hold_Wakeup( ) function also re-enables the component.
- **void Sample_Hold_Enable(void)**—Activates the hardware and begins component operation. It is not necessary to call Sample_Hold_Enable( ) because the Sample_Hold_Start( ) API calls this function, which is the preferred method to begin component operation.
- **void Sample_Hold_SaveConfig(void)**—Empty function. Provided for future use.
- **void SampleHold_RestoreConfig(void)**—Empty function. Provided for future use.

15.10 Controller Area Network (CAN) 3.0

The Controller Area Network (CAN) controller implements the CAN2.0A and CAN2.0B specifications as defined in the Bosch specification and conforms to the ISO-11898-1 standard. The CAN Component is certified by the C&S group GmbH based on the standard protocol an data link layer conformance tests. Features include: The Controller Area Network (CAN) controller implements the CAN2.0A and CAN2.0B specifications as defined in the Bosch specification and conforms to the ISO-11898-1 standard. The CAN Component is certified by the C&S group GmbH based on the standard protocol an data link layer conformance tests. Features include: CAN2.0A and CAN2.0B protocol implementation, ISO 11898-1 compliant, supports standard 11-bit and extended 29-bit identifiers, programmable bit rate up to 1 Mbps, up to 16 receive mailboxes with hardware message filtering, up to 8 transmit message mailboxes with programmable transmit priority: Round-Robin and Fixed, two-wire or three-wire interface to external transceiver (tx, rx and tx enable), supports listen-only mode of operation and supports single-shot transmission, as well as internal and external loopback modes [21].

15.10.1 Controller Area Network (CAN) Functions 3.0

- **CAN_Start( )**—Sets the initVar variable, calls the CAN_Init( ) function and then calls the CAN_Enable( ) function.
- **CAN_Stop( )**—Disables the CAN.
- **CAN_GlobalIntEnable( )**—Enables global interrupts from CAN Component. Disables global interrupts from CAN Component.
- **CAN_GlobalIntDisable( )**—Disables global interrupts from CAN Component
- **CAN_SetPreScaler( )**—Sets prescaler for generation of the time quanta from the BUS_CLK/SYSCLK
- **CAN_SetArbiter( )**—Sets arbitration type for transmit buffers.
- **CAN_SetTsegSample( )**—Configures: Time segment 1, Time segment 2, Synchronization Jump Width and Sampling Mode.
- **CAN_SetRestartType( )**—Sets reset type.
- **CAN_SetSwapDataEndianness( )**—Enables or disables the endian swapping of CAN data bytes. (This function is not available for PSoC 3/PSoC 5LP part families.)
- **CAN_SetEdgeMode( )**—Sets Edge mode.
- **CAN_RXRegisterInit( )**—Writes only receive CAN registers.
- **CAN_SetOpMode( )**—Sets Operation mode.
- **CAN_SetErrorCaptureRegisterMode( )**—Sets the error capture register mode to free running or error capture mode. (This function is not available for PSoC 3/PSoC 5LP part families.)
- **CAN_ReadErrorCaptureRegister( )**—This function returns the value of error capture register.
- **CAN_ArmErrorCaptureRegister( )**—This function arms the error capture register when the ECR is in error capture mode. (This function is not available for PSoC 3/PSoC 5LP part families.)
- **CAN_GetTXErrorFlag( )**—Returns the flag that indicates if the number of transmit errors equals or exceeds 0x60.
- **CAN_GetRXErrorFlag( )**—Returns the flag that indicates if the number of receive errors equals or exceeds 0x60.
- **CAN_GetTXErrorCount( )**—Returns the number of transmit errors.
- **CAN_GetRXErrorCount( )**—Returns the number of receive errors.
- **CAN_GetErrorState( )**—Returns error status of the CAN Component.
- **CAN_SetIrqMask( )**—Sets to enable or disable particular interrupt sources.
- **CAN_ArbLostIsr( )**—Clears Arbitration Lost interrupt flag.
- **CAN_OvrLdErrorIsr( )**—Clears Overload Error interrupt flag.
- **CAN_BitErrorIsr( )**—Clears Bit Error interrupt flag.
- **CAN_BitStuffErrorIsr( )**—Clears Bit Stuff Error interrupt flag.
- **CAN_AckErrorIsr( )**—Clears Acknowledge Error interrupt flag.
- **CAN_MsgErrorIsr( )**—Clears Form Error interrupt flag.
- **CAN_CrcErrorIsr( )**—Clears CRC Error interrupt flag.
- **CAN_BusOffIsr( )**—Clears Bus Off interrupt flag. Places CAN Component to Stop mode.
- **CAN_SSTErrorIsr( )**—Clears SST error flag and removes the failed message from the transmit mailbox. (This function is not available for PSoC 3/PSoC 5LP part families.)
- **CAN_RtrAutoMsgSentIsr( )**—Clears RTR Auto Message sent flag. (This function is not available for PSoC 3/PSoC 5LP part families.)
- **CAN_StuckAtZeroIsr( )**—Clears StuckAtZeroFlag. Places CAN Component to Stop mode. (This function is not available for PSoC 3/PSoC 5LP part families.)
- **CAN_MsgLostIsr( )**—Clears Message Lost interrupt flag.
- **CAN_MsgTXIsr( )**—Clears Transmit Message interrupt flag.

- **CAN_MsgRXIsr( )**—Clears Receive Message interrupt flag and call appropriate handlers for Basic and Full interrupt based mailboxes.
- **CAN_RxBufConfig( )**—Configures all receive registers for particular mailbox.
- **CAN_TxBufConfig( )**—Configures all transmit registers for particular mailbox.
- **CAN_SendMsg( )**—Sends an message from one of the Basic mailboxes.
- **CAN_SendMsg0-7( )**—Checks if mailbox 0-7 has untransmitted messages waiting for arbitration.
- **CAN_TxCancel( )**—Cancels transmission of a message that has been queued for transmission.
- **CAN_ReceiveMsg0-15( )**—Acknowledges receipt of new message.
- **CAN_ReceiveMsg( )**—Clears Receive particular Message interrupt flag.
- **CAN_Sleep( )**—Prepares CAN Component to go to sleep
- **CAN_Wakeup( )**—Prepares CAN Component to wake up
- **CAN_Init( )**—Initializes or restores the CAN per the Configure dialog settings.
- **CAN_Enable( )**—Enables the CAN.
- **CAN_SaveConfig( )**—Saves the current configuration.
- **CAN_RestoreConfig( )**—Restores the configuration.

15.10.2 Controller Area Network (CAN) Function Calls 3.0

- **uint8 CAN_Start(void)**—Sets the initVar variable, calls the CAN_Init( ) function and then calls the CAN_Enable( ) function. This function sets the CAN Component into run mode and starts the counter if polling mailboxes available.
- **uint8 CAN_GlobalIntEnable(void)**—This function enables global interrupts from the CAN Component.
- **uint8 CAN_GlobalIntDisable(void)**—This function disables global interrupts from the CAN Component.
- **uint8 CANSetPreScaler(uint16 bitrate)**—This function sets the prescaler for generation of the time quanta from the BUSCLK/SYSCLK. Values between 0x0 and 0x7FFF are valid.
- **uint8 CAN_SetArbiter(uint8 arbiter)**—This function sets the arbitration type for transmit buffers. Types of arbiters are Round Robin and Fixed priority.
- **uint8 CAN_SetTsegSample(uint8 cfgTseg1, uint8 cfgTseg2, uint8 sjw, uint8 sm)**—This function configures: Time segment 1, Time segment 2, Synchronization Jump Width and Sampling Mode.
- **uint8 CAN_SetRestartType(uint8 reset)**—This function sets reset type. Types of reset are Automatic and Manual. Manual reset is the recommended setting.
- **uint8 CAN_SetSwapDataEndianness(uint8 swap)**—This function selects whether the data byte endianness of the CAN receive and transmit data fields has to be swapped or not swapped. This is useful to match the data byte endianness to the endian setting of the processor or the used CAN protocol. This function is not applicable to PSoC 3/PSoC 5LP part families.
- **uint8 CAN_SetEdgeMode(uint8 edge)**—This function sets Edge Mode. Modes are 'R' to 'D' (Recessive to Dominant) and Both edges are used.
- **uint8 CAN_RXRegisterInit(uint32 *regAddr, uint32 config)**—This function writes CAN receive registers only.
- **uint8 CAN_SetOpMode(uint8 opMode)**—This function sets Operation Mode.
- **uint8 CAN_SetErrorCaptureRegisterMode(uint8 ecrMode)**—This function sets the error capture register mode. The 2 modes are possible: free running and error capture mode. item **uint32 CAN_ReadErrorCaptureRegister(void)**—This function returns the value of error capture register.

- **uint8 CAN_ArmErrorCaptureRegister(void)**—This function arms the error capture register when the ECR is in error capture mode, by setting the ECR_STATUS bit in the ECR register
- **uint8 CAN_GetTXErrorFlag(void)**—This function returns the flag that indicates if the number of transmit errors equals or exceeds 0x60.
- **uint8 CAN_GetRXErrorFlag(void)**—This function returns the flag that indicates if the number of receive errors equals or exceeds 0x60.
- **uint8 CAN_GetTXErrorCount(void)**—This function returns the number of transmit errors.
- **uint8 CAN_GetRXErrorCount(void)**—This function returns the number of receive errors.
- **uint8 CAN_GetErrorState(void)**—This function returns the error status of the CAN Component.
- **uint8 CAN_GetErrorState(void)**—This function returns the error status of the CAN Component.
- **void CAN_ArbLostIsr(void)**—This function is the entry point to the Arbitration Lost Interrupt. It clears the Arbitration Lost interrupt flag. It is only generated if the Arbitration Lost Interrupt parameter is enabled.
- **CAN_OvrLdErrorIsr(void)**—This function is the entry point to the Overload Error Interrupt. It clears the Overload Error interrupt flag. It is only generated if the Overload Error Interrupt parameter is enabled.
- **CAN_BitErrorIsr(void)**—This function is the entry point to the Bit Error Interrupt. It clears Bit Error interrupt flag. It is only generated if the Bit Error Interrupt parameter is enabled.
- **void CAN_BitStuffErrorIsr(void)**—This function is the entry point to the Bit Stuff Error Interrupt. It clears the Bit Stuff Error interrupt flag. It is only generated if the Bit Stuff Error Interrupt parameter is enabled.
- **void CAN_AckErrorIsr(void)**—This function is the entry point to the Acknowledge Error Interrupt. It clears the Acknowledge Error interrupt flag. It is only generated if the Acknowledge Error Interrupt parameter is enabled.
- **void CAN_MsgErrorIsr(void)**—This function is the entry point to the Form Error Interrupt. It clears the Form Error interrupt flag. It is only generated if the Form Error Interrupt parameter is enabled.
- **void CAN_CrcErrorIsr(void)**—This function is the entry point to the CRC Error Interrupt. It clears the CRC Error interrupt flag. It is only generated if the CRC Error Interrupt parameter is enabled.
- **void CAN_BusOffIsr(void)**—This function is the entry point to the Bus Off Interrupt. It puts the CAN Component in Stop mode. It is only generated if the Bus Off Interrupt parameter is enabled. Enabling this interrupt is recommended.
- **void CAN_SSTErrorIsr(void)**—This function is the entry point to the single shot transmission error Interrupt. It is only generated if the single shot transmission is enabled. Generated when the mailbox set for single shot transmission experienced an arbitration loss or a bus error during transmission.
- **void CAN_RtrAutoMsgSentIsr(void)**—This function is the entry point to the RTR automatic message sent Interrupt. It is only generated if RTR message sent interrupt parameter is enabled.
- **void CAN_StuckAtZeroIsr(void)**—This function is the entry point to the stuck at dominant bit Interrupt. It is only generated if Stuck at zero interrupt parameter is enabled. Enabling this interrupt is recommended.
- **void CAN_MsgLostIsr(void)**—This function is the entry point to the Message Lost Interrupt. It clears the Message Lost interrupt flag. It is only generated if the Message Lost Interrupt parameter is enabled.
- **void CAN_MsgTXIsr(void)**—This function is the entry point to the Transmit Message Interrupt. It clears the Transmit Message interrupt flag. It is only generated if the Transmit Message Interrupt parameter is enabled.

- **void CAN_MsgRXIsr(void)**—This function is the entry point to the Receive Message Interrupt. It clears the Receive Message interrupt flag and calls the appropriate handlers for Basic and Full interrupt based mailboxes. It is only generated if the Receive Message Interrupt parameter is enabled. Enabling this interrupt is recommended.
- **uint8 CAN_RxBufConfig(const CAN_RX_CFG *rxConfig)**—This function configures all receive registers for a particular mailbox. The mailbox number contains CAN_RX_CFG structure.
- **uint8 CAN_TxBufConfig(const CAN_TX_CFG *txConfig)**—This function configures all transmit registers for a particular mailbox. The mailbox number contains CAN_TX_CFG structure.
- **uint8 CAN_SendMsg(const CANTXMsg *message)**—This function sends a message from one of the Basic mailboxes. The function loops through the transmit message buffer designed as Basic CAN mailboxes. It looks for the first free available mailbox and sends from it. There can only be three retries.
- **uint8 CAN_SendMsg0-7(void)**—These functions are the entry point to Transmit Message 0-7. This function checks if mailbox 0-7 already has untransmitted messages waiting for arbitration. If so, it initiates transmission of the message. Only generated for Transmit mailboxes designed as Full.
- **void CAN_TxCancel(uint8 bufferId)**—This function cancels transmission of a message that has been queued for transmission. Values between 0 and 7 are valid.
- **void CAN_ReceiveMsg0-15(void)**—These functions are the entry point to the Receive Message 0-15 Interrupt. They clear Receive Message 0-15 interrupt flags. They are only generated for Receive mailboxes designed as Full interrupt based.
- **void CAN_ReceiveMsg(uint8 rxMailbox)**—This function is the entry point to the Receive Message Interrupt for Basic mailboxes. It clears the Receive particular Message interrupt flag. It is only generated if one of the Receive mailboxes is designed as Basic.
- **void CAN_Sleep(void)**—This is the preferred routine to prepare the Component for sleep. The CAN_Sleep( ) routine saves the current Component state. Then it calls the CAN_Stop( ) function and calls CAN_SaveConfig( ) to save the hardware configuration. Call the CAN_Sleep( ) function before calling the CyPmSleep( ) or the CyPmHibernate( ) function.
- **void CAN_Wakeup(void)**—This is the preferred routine to restore the Component to the state when CAN_Sleep( ) was called. The CAN_Wakeup( ) function calls the CAN_RestoreConfig( ) function to restore the configuration. The function restores CAN Rx and Tx buffer control register configurations provided by the customizer. If the Component was enabled before the CAN_Sleep( ) function was called, the CAN_Wakeup( ) function will also re-enable the Component.
- **uint8 CAN_Init(void)** Initializes or restores the Component according to the customizer Configure dialog settings. It is not necessary to call CAN_Init( ) because the CAN_Start( ) routine calls this function and is the preferred method to begin Component operation.
- **uint8 CAN_Enable(void)**—Activates the hardware and begins Component operation. It is not necessary to call CAN_Enable( ) because the CAN_Start( ) routine calls this function, which is the preferred method to begin Component operation.
- **void CAN_SaveConfig(void)**—This function saves the Component configuration and non-retention registers. This function also saves the current Component parameter values, as defined in the Configure dialog or as modified by appropriate APIs. This function is called by the CAN_Sleep( ) function. This function is applicable only for PSoC 3/PSoC 5LP part families.
- **void CAN_RestoreConfig(void)**—This function restores the Component configuration and non-retention registers. This function also restores the Component parameter values to what they were prior to calling the CAN_Sleep( ) function. This function is applicable only for PSoC 3/PSoC 5LP part families.

15.11 Vector CAN 1.10

The Vector CANbedded environment consists of a number of adaptive source code components that cover the basic communication and diagnostic requirements in automotive applications. The Vector CANbedded software suite is customer specific and its operation will vary according to application and OEM. This component for the Vector CANbedded suite is written to generically support the CANbedded structure regardless of the flavor of the particular OEM application. The Vector CAN component developed for PSoC 3 allows easy integration of the Vector certified CAN driver. Features include: CAN2.0 A/B protocol implementation, ISO 11898-1 compliant, CAN2.0 A/B protocol implementation, ISO 11898-1 compliant, programmable bit rate up to 1 Mbps, two or three wire interface to external transceiver (Tx, Rx and Tx Enable) and driver is provided and supported by Vector [103].

15.11.1 Vector CAN Functions 1.10

- **Vector_CAN_Start( )**—Initializes and enables the Vector CAN component using the Vector_CAN_Init( ) and Vector_CAN_Enable( ) functions.
- **Vector_CAN_Stop( )**—Disables the Vector CAN component.
- **Vector_CAN_GlobalIntEnable( )**—Enables Global Interrupts from CAN Core.
- **Vector_CAN_GlobalIntDisable( )**—Disables Global Interrupts from CAN Core.
- **Vector_CAN_Sleep( )**—Prepares the component for sleep.
- **Vector_CAN_Wakeup( )**—Restores the component to the state when Vector_CAN_Sleep( ) was called.
- **Vector_CAN_Init( )**—Initializes the Vector CAN component based on settings in the component customizer.
- **Vector_CAN_Enable( )**—Enables the Vector CAN component.
- **Vector_CAN_SaveConfig( )**—Saves the component configuration.
- **Vector_CAN_RestoreConfig( )**—Restores the component configuration.

15.11.2 Vector CAN Function Calls 1.10

- **uint8 Vector_CAN_Start(void)**—This is the preferred method to begin component operation. Vector_CAN_Start( ) sets the initVar variable, calls the Vector_CAN_Init( ) function and then calls the Vector_CAN_Enable( ) function.
- **uint8 Vector_CAN_Stop(void)**—Disables the Vector CAN component.
- **uint8 Vector_CAN_GlobalIntEnable(void)**—This function enables global interrupts from the CAN Core.
- **uint8 Vector_CAN_GlobalIntDisable(void)**—This function disables global interrupts from the CAN Core
- **void Vector_CAN_Sleep(void)**—This is the preferred routine to prepare the component for sleep. The Vector_CAN_Sleep( ) routine saves the current component state. Then it calls the Vector_CAN_SaveConfig( ) function and calls Vector_CAN_Stop( ) to save the hardware configuration. Call the Vector_CAN_Sleep( ) function before calling the CyPmSleep( ) or CyPmHibernate( ) functions.

- **void Vector_CAN_Wakeup(void)**—This is the preferred routine to restore the component to the state when Vector_CAN_Sleep( ) was called. The Vector_CAN_Wakeup( ) function calls the Vector_CAN_RestoreConfig( ) function to restore the configuration. If the component was enabled before the Vector_CAN_Sleep( ) function was called, the Vector_CAN_Wakeup( ) function will also re-enable the component.
- **void Vector_CAN_Sleep(void)**—This is the preferred routine to prepare the component for sleep. The Vector_CAN_Sleep( ) routine saves the current component state. Then it calls the Vector_CAN_SaveConfig( ) function and calls Vector_CAN_Stop( ) to save the hardware configuration. Call the Vector_CAN_Sleep( ) function before calling the CyPmSleep( ) or CyPmHibernate( ) functions.
- **void Vector_CAN_Wakeup(void)**—This is the preferred routine to restore the component to the state when Vector_CAN_Sleep( ) was called. The Vector_CAN_Wakeup( ) function calls the Vector_CAN_RestoreConfig( ) function to restore the configuration. If the component was enabled before the Vector_CAN_Sleep( ) function was called, the Vector_CAN_Wakeup( ) function will also re-enable the component.
- **uint8 Vector_CAN_Enable(void)**—Activates the hardware and begins component operation. It is not necessary to call Vector_CAN_Enable( ) because the Vector_CAN_Start( ) routine calls this function, which is the preferred method to begin component operation.
- **void Vector_CAN_SaveConfig(void)**—This function saves the component configuration. This will save nonretention registers. This function will also save the current component parameter values, as defined in the Configure dialog or as modified by appropriate APIs. This function is called by the Vector_CAN_Sleep( ) function.
- **void Vector_CAN_RestoreConfig(void)**—This function restores the component configuration. This will restore nonretention registers. This function will also restore the component parameter values to what they were prior to calling the Vector_CAN_Sleep( ) function.

15.12 Filter 2.30

This component includes a filter design feature, which greatly simplifies the design and implementation processes and supports two streaming channels that can be streamed directly to ROM or other hardware blocks (such as the ADC) using DMA. The filtered results can likewise be transferred using DMA, interrupts, or polling methods. The DFB's 128 data and coefficient locations are shared as needed between the two filter channels and this information is used to guide the choice of filter implementation. It reports (but does not set) the minimum bus clock frequency required to execute the filtering within the declared sample interval. This clock can hen be set in the design-wide resource manager [44].

15.12.1 Filter Functions 2.30

- **Filter_Start( )**—Configures and enables the Filter component's hardware for interrupt, DMA and filter settings.
- **Filter_Stop( )**—Stops the filters from running and powers down the hardware.
- **Filter_Read8( )**—Reads the current value on the Filter's output holding register. Byte read of the most significant byte.
- **Filter_Read16( )**—Reads the current value on the Filter's output holding register. Two-byte read of the most significant bytes.

- **Filter_Read24( )**—Reads the current value on the Filter's output holding register. Three-byte read of the data output holding register.
- **Filter_Write8( )**—Writes a new 8-bit sample to the Filter's input staging register.
- **Filter_Write16( )**—Writes a new 16-bit sample to the Filter's input staging register.
- **Filter_Write24( )**—Writes a new 24-bit sample to the Filter's input staging register.
- **Filter_ClearInterruptSource( )**—Writes the Filter_ALL_INTR mask to the status register to clear any active interrupts.
- **Filter_IsInterruptChannelA( )**—Identifies whether Channel A has triggered a data-ready interrupt.
- **Filter_IsInterruptChannelB( )**—Identifies whether Channel B has triggered a data-ready interrupt.
- **Filter_Sleep( )**—Stops and saves the configuration.
- **Filter_Wakeup( )**—Restores and enables the configuration.
- **Filter_Init( )**—Initializes or restores default Filter configuration.
- **Filter_Enable( )**—Enables the Filter.
- **Filter_SaveConfig( )**—Saves the configuration of Filter nonretention registers.
- **Filter_RestoreConfig( )**—Restores the configuration of Filter nonretention registers.
- **Filter_SetCoherency( )**—Sets the key coherency byte in the coherency register.

15.12.2 Filter Function Calls 2.30

- **void Filter_Start(void)**—This is the preferred method to begin component operation. Configures and enables the Filter component's hardware for interrupt, DMA and filter settings.
- **void Filter_Stop(void)**—Stops the Filter hardware from running and powers it down.
- **uint8 Filter_Read8(uint8 channel)**—Reads the highest order byte of Channel A's or Channel B's output holding register.
- **uint16 Filter_Read16(uint8 channel)**—Reads the two highest-order bytes of Channel A's or Channel B's output holding register.
- **uint32 Filter_Read24(uint8 channel)**—Reads all three bytes of Channel A's or Channel B's output holding register. uint8 channel: Which filter channel should be read. Options are Filter_CHANNEL_A and Filter_CHANNEL_B.
- **void Filter_Write8(uint8 channel, uint8 sample)**—Writes to the highest-order byte of Channel A's or Channel B's input staging register.
- **void Filter_Write16(uint8 channel, uint16 sample)**—Writes to the two highest-order bytes of Channel A's or Channel B's input staging register.
- **void Filter_Write24(uint8 channel, uint32 sample)**—Writes to all three bytes of Channel A's or Channel B's input staging register.
- **void Filter_ClearInterruptSource(void)**—Writes the Filter_ALL_INTR mask to the status register to clear any active interrupt.
- **uint8 Filter_IsInterruptChannelA(void)**—Identifies whether Channel A has triggered a data-ready interrupt.
- **uint8 Filter_IsInterruptChannelB(void)**—Identifies whether Channel B has triggered a data-ready interrupt.
- **void Filter_SetCoherency(uint8 channel, unit8 byte_select)**—Sets the value in the DFB coherency register. This value determines the key coherency byte. The key coherency byte is the software's way of telling the hardware which byte of the field will be written or read last when an update to the field is desired.

- **void Filter_SetCoherencyEx(uint8 regSelect, uint8 key)**—Configures the DFB coherency register for each of the staging and holding registers. Allows multiple registers with the same configuration to be set at the same time. This API should be used when the coherency of the staging and holding register of a channel is different.
- **void Filter_SetDalign(uint8 regSelect, uint8 state)**—Configures the DFB dalign register for each of the staging and holding registers. Allows multiple registers with the same configuration to be set at the same time.
- **void Filter_Sleep(void)**—Stops the DFB operation. Saves the configuration registers and the component enable state. Should be called just before entering sleep.
- **void Filter_Wakeup(void)**—This is the preferred API to restore the component to the state when Filter_Sleep( ) was called. The Filter_Wakeup( ) function calls the Filter_RestoreConfig( ) function to restore the configuration. If the component was enabled before the Filter_Sleep( ) function was called, the Filter_Wakeup( ) function will also re-enable the component.
- **void Filter_Init(void)**—Initializes or restores the component according to the customizer Configure dialog settings. It is not necessary to call Filter_Init( ) because the Filter_Start( ) API calls this function and is the preferred method to begin component operation.
- **void Filter_Enable(void)**—Activates the hardware and begins component operation. It is not necessary to call Filter_Enable( ) because the Filter_Start( ) API calls this function, which is the preferred method to begin component operation.
- **void Filter_SaveConfig(void)**—This function saves the component configuration and nonretention registers. It also saves the current component parameter values, as defined in the Configure dialog or as modified by appropriate APIs. This function is called by the Filter_Sleep( ) function.
- **void Filter_RestoreConfig(void)**—This function restores the component configuration and nonretention registers. It also restores the component parameter values to what they were before calling the Filter_Sleep( ) function.

15.13 Digital Filter Block Assembler 1.40

The digital filter block (DFB) in PSoC 3 and PSoC 5LP is a 24-bit fixed point, programmable, limited-scope DSP engine that can be used as a mini-DSP processor in application. The DFB component allows you to directly configure the DFB using its assembly instructions. The component assembles the instructions entered in the code editor and generates the corresponding hex code words that are then loaded into the DFB. It also includes a simulator, which can aid in simulating and debugging the assembly instructions. Features include: an editor to enter the assembler instructions to configure the DFB block, an assembler that converts the assembly instructions to instruction words, supports simulation of the assembly instructions, supports a code optimization option that provides a mechanism to incorporate up to 128 very large instruction words inside the DFB Code RAM, hardware signals such as DMA requests, DSI inputs and outputs and interrupt lines and supports semaphores to interact with the system software and the option to tie the semaphores to hardware signals [31].

15.13.1 Digital Filter Block Assembler Functions 1.40

- **DFB_Stop( )**—Turns off the run bit. If DMA control is used to feed the channels, allows arguments to turn off one of the TD channels.
- **DFB_Pause( )**—Pauses DFB and enables writing to the DFB RAM.

- **DFB_Resume( )**—Disables writing to the DFB RAM, clears any pending interrupts, disconnects the DFB RAM from the data bus andruns the DFB.
- **DFB_SetCoherency( )**—Sets the coherency key to low/mid/high byte based on the coherencyKey parameter that is passed to the DFB.
- **DFB_SetDalign( )**—Allows 9- to 16-bit input and output samples to travel as 16-bit values on the AHB bus.
- **DFB_LoadDataRAMA( )**—Loads data to RAMA DFB memory.
- **DFB_LoadDataRAMB( )**—Loads data to RAMB DFB memory.
- **DFB_LoadInputValue( )**—Loads the input value into the selected channel.
- **DFB_GetOutputValue( )**—Gets the value from one of the DFB output holding registers.
- **DFB_SetInterruptMode( )**—Assigns the events that will trigger a DFB interrupt.
- **DFB_GetInterruptSource( )**—Looks at the DFB_SR register to see which interrupt sources have been triggered.
- **DFB_ClearInterrupt( )**—Clears the interrupt request.
- **DFB_SetDMAMode( )**—Assigns the events that will trigger a DMA request for the DFB.
- **DFB_SetSemaphores( )**—Sets semaphores specified with a 1.
- **DFB_ClearSemaphores( )**—Clears semaphores specified with a 1.
- **DFB_GetSemaphores( )**—Checks the current status of the DFB semaphores and returns that value.
- **DFB_SetOutput1Source( )**—Chooses which internal signals will be mapped to output 1.
- **DFB_SetOutput2Source( )**—Chooses which internal signals will be mapped to output 2.
- **DFB_Sleep( )**—Prepares the DFB component to go to sleep.
- **DFB_Wakeup( )**—Prepares DFB Component to wake up.
- **DFB_Enable( )**—Enables the DFB hardware block. Sets the DFB run bit. Powers on the DFB block.
- **DFB_SaveConfig(void)**—Saves the user configuration of the DFB nonretention registers. This routine is called by DFB_Sleep( ) to save the component configuration before entering sleep.
- **DFB_RestoreConfig( )**—Restores the user configuration of the DFB nonretention registers. This routine is called by DFB_Wakeup( ) to restore the component configuration when exiting sleep.

15.13.2 Digital Filter Block Assembler Function Calls 1.40

- **void DFB_Start(void)**—This function initializes and enables the DFB component using the DFB_Init( ) and DFB_Enable( ) functions.
- **void DFB_Stop(void)**—This function turns off the run bit. If DMA control is used to feed the channels, DFB_Stop( ) allows arguments to turn off one of the TD channels.
- **void DFB_Pause(void)**—This function pauses the DFB and enables writing to the DFB RAM. Turns off the run bit, connects the DFB RAM to the data bus and clears the DFB run bit and passes the control of all DFB RAMs onto the bus.
- **void DFB_Resume(void)**—This function disables writing to the DFB RAM, clears any pending interrupts, disconnects the DFB RAM from the data bus and runs the DFB. It passes the control of all DFB RAM to the DFB and then sets the run bit.
- **void DFB_SetCoherency(uint8 coherencyKeyByte)**—This function sets the coherency key to low, med, or high byte based on the coherencyKeyByte parameter that is passed to the DFB. Note that the function directly writes to the DFB Coherency register. Therefore the coherency for all the registers must be specified when passing the coherencyKeyByte parameter.DFB_SetCoherency( ) allows you to select which of the three bytes of each of STAGEA, STAGEB, HOLDA, and HOLDB

will be used as the key coherency byte. Coherency refers to the HW added to this block to protect against block malfunctions. This is needed in cases where register fields are wider than the bus access, which leaves intervals when fields are partially written or read (incoherent). The key coherency byte is the way the SW tells the HW which byte of the field will be written or read last when you want to update the field. When the key byte is written or read, the field is flagged coherent. If any other byte is written or read, the field is flagged incoherent.

- **void DFB_SetDalign(uint8 dalignKeyByte)**—This feature allows 9- to 16-bit input and output samples to travel as 16-bit values on the AHB bus. These bits, when set high, cause an 8-bit shift in the data to all access of the corresponding staging and holding registers. Note that this function directly writes to the DFB Data Alignment register. Therefore the alignment for all registers must be specified when passing the dalignKeyByte parameter. Since the DFB datapath is MSB aligned, it is convenient for the system SW to align values on bits 23:8 of the Staging and Holding register to bits 15:0 of the bus. This is because a transfer from a PHUB spoke that is 16 or even 8 bits wide goes to the DFB spoke which is 32 bits wide. The Dalign allows the DFB to justify data so that transfers to and from these different size spokes can happen more efficiently.
- **void DFB_LoadDataRAMA(int32 * ptr, uint32 * addr, uint8 size)**—This function loads data to the DFB RAM A memory.
- **void DFB_LoadDataRAMB(uint32 * ptr, uint32 * addr, uint8 size)**—This function loads data to DFB RAM B memory.
- **void DFB_LoadInputValue(uint8 channel, uint32 sample)**—This function loads the input value into the selected channel.[9]
- **int32 DFB_GetOutputValue(uint8 channel)**—This function gets the value from one of the DFB Output Holding registers.
- **void DFB_SetInterruptMode(uint8 events)**—This function assigns the events that trigger a DFB interrupt.
- **uint8 DFB_GetInterruptSource(void)**—This function looks at the DFB_SR register to see which interrupt sources have been triggered.
- **uint8 DFB_GetInterruptSource(void)**—This function looks at the DFB_SR register to see which interrupt sources have been triggered.
- **void DFB_SetDMAMode(uint8 events)**—This function assigns the events that trigger a DMA request for the DFB. Two different DMA requests can be triggered.
- **void DFB_SetSemaphores(uint8 mask)**—This function sets semaphores specified with a 1.
- **void DFB_ClearSemaphores(uint8 mask)**—This function clears semaphores specified with a 1.
- **uint8 DFB_GetSemaphores(void)**—This function checks the current status of the DFB semaphores and returns that value.
- **void DFB_SetOutput1Source(uint8 source)**—This function allows you to choose which internal signals are mapped to output 1.
- **void DFB_SetOutput2Source(uint8 source)**—This function allows you to choose which internal signals are mapped to output 2.
- **void DFB_Sleep(void)**—This is the preferred routine to prepare the component for sleep. The DFB_Sleep( ) routine saves the current component state. Then it calls the DFB_Stop( ) function and calls DFB_SaveConfig( ) to save the hardware configuration. Call the DFB_Sleep( ) function before calling the CyPmSleep( ) or the CyPmHibernate( ) function.
- **void DFB_Wakeup(void)**—This is the preferred routine to restore the component to the state when DFB_Sleep( ) was called. The DFB_Wakeup( ) function calls the DFB_RestoreConfig( ) function

---

[9]The write order is important. When the high byte is loaded, the DFB sets the input ready bit. Careful attention must be paid to the byte order, if coherency or data alignment is changed.

to restore the configuration. If the component was enabled before the DFB_Sleep( ) function was called, the DFB_Wakeup( ) function will also re-enable the component.

- **void DFB_Init(void)**—This function initializes or restores the default DFB component configuration provided with the customizer: powers on the DFB (PM_ACT_CFG) and the RAM (DFB_RAM_EN), moves CSA/CSB/FSM/DataA/DataB/Address calculation unit (ACU) data to the DFB RAM using an 8051/ARM core, changes RAM DIR to DFB, sets the interrupt mode, sets the DMA mode, sets the DSI outputs and clears all semaphore bits and pending interrupts.
- **void DFB_Enable(void)**—This function enables the DFB hardware block, sets the DFB run bit and powers on the DFB block.
- **void DFB_SaveConfig(void)**—This function saves the component configuration and nonretention registers. It also saves the current component parameter values, as defined in the Configure dialog or as modified by appropriate APIs. This function is called by the DFB_Sleep( ) function.
- **void DFB_RestoreConfig(void)**—This function restores the component configuration and nonretention registers. It also restores the component parameter values to what they were before calling the DFB_Sleep( ) function.

15.14   Power Monitor 8, 16 and 32 Rails 1.60

Power Converter Voltage Measurements For power converter voltage measurements, the ADC can be configured into single-ended mode (0-4.096 V range or 0-2.048 V range). The ADC can also be configurable into differential mode ($\pm$2.048 V range) to support remote sensing of voltages where the remote ground reference is returned to PSoC over a PCB trace. In cases where the analog voltage to be monitored equals or exceeds Vdda or the ADC range, external resistor dividers are recommended to scale the monitored voltages down to an appropriate range [65].

Power Converter Current Measurements For power converter load current measurements, the ADC can be configured into differential mode ($\pm$64 mV or $\pm$128 mV range) to support voltage measurement across a high-side series shunt resistor on the outputs of the power converters. Firmware APIs convert the measured differential voltage into the equivalent current based on the external resistor component value used. The ADC can also be configured into single-ended mode (matching the selected voltage measurement range) to support connection to external current sense amplifiers (CSAs) that convert the differential voltage drop across the shunt resistor into a single ended voltage or to support power converters or hot-swap controllers that integrate similar functionality. Features include: interfaces to up to 32 DC-DC power converters, measures power converter output voltages and load currents using a DelSig-ADC, monitors the health of the power converters generating warnings and faults based on user-defined thresholds, support for measuring other auxiliary voltages in the system and support for 3.3 V and 5 V chip power supplies.

15.14.1   Power Monitor Functions for 8, 16 and 32 Rails 1.60

- **PowerMonitor_Start( )**—Initializes the Power Monitor with default customizer values.
- **PowerMonitor_Stop( )**—Disables the component. ADC sampling stops.
- **PowerMonitor_Init( )**—Initializes the component. Includes running self-calibration.
- **PowerMonitor_Enable( )**—Enables hardware blocks within the component and starts scanning.
- **PowerMonitor_EnableFault( )**—Enables generation of the fault signal.
- **PowerMonitor_DisableFault( )**—Disables generation of the fault signal.

- **PowerMonitor_SetFaultMode( )**—Configures fault sources from the component.
- **PowerMonitor_GetFaultMode( )**—Returns enabled fault sources from the component.
- **PowerMonitor_SetFaultMask( )**—Enables or disables faults from each power converter through a mask.
- **PowerMonitor_GetFaultMask( )**—Returns fault mask status of each power converter.
- **PowerMonitor_GetFaultSource( )**—Returns pending fault sources from the component.
- **PowerMonitor_GetOVFaultStatus( )**—Returns over voltage fault status of each power converter. The status is reported regardless of the Fault Mask.
- **PowerMonitor_GetUVFaultStatus( )**—Returns under voltage fault status of each power converter. The status is reported regardless of the Fault Mask.
- **PowerMonitor_GetOCFaultStatus( )**—Returns over current fault status of each power converter. The status is reported regardless of the Fault Mask.
- **PowerMonitor_EnableWarn( )**—Enables generation of the warning signal.
- **PowerMonitor_DisableWarn( )**—Disables generation of the warning signal.
- **PowerMonitor_SetWarnMode( )**—Configures warning sources from the component.
- **PowerMonitor_GetWarnMode( )**—Returns enabled warning sources from the component.
- **PowerMonitor_SetWarnMask( )**—Enables or disables warnings from each power converter through a mask.
- **PowerMonitor_GetWarnMask( )**—Returns warning mask status of each power converter.
- **PowerMonitor_GetWarnSource( )**—Returns pending warning sources from the component.
- **PowerMonitor_GetOVWarnStatus( )**—Returns over voltage warning status of each power converter. The status is reported regardless of the Warning Mask.
- **PowerMonitor_GetUVWarnStatus( )**—Returns under voltage warning status of each power converter. The status is reported regardless of the Warning Mask.
- **PowerMonitor_GetOCWarnStatus( )**—Returns over current warning status of each power converter. The status is reported regardless of the Warning Mask.
- **PowerMonitor_SetUVWarnThreshold( )**—Sets the power converter under voltage warning threshold for the specified power converter.
- **PowerMonitor_GetUVWarnThreshold( )**—Returns the power converter under voltage warning threshold for the specified power converter.
- **PowerMonitor_SetOVWarnThreshold( )**—Sets the power converter over voltage warning threshold for the specified power converter.
- **PowerMonitor_GetOVWarnThreshold( )**—Returns the power converter over voltage warning threshold for the specified power converter.
- **PowerMonitor_SetUVFaultThreshold( )**—Sets the power converter under voltage fault threshold for the specified power converter.
- **PowerMonitor_GetUVFaultThreshold( )**—Returns the power converter under voltage fault threshold for the specified power converter.
- **PowerMonitor_SetOVFaultThreshold( )**—Sets the power converter over voltage fault threshold for the specified power converter.
- **PowerMonitor_GetOVFaultThreshold( )**—Returns the power converter over voltage fault threshold for the specified power converter.
- **PowerMonitor_SetOCWarnThreshold( )**—Sets the power converter over current warning threshold for the specified power converter.
- **PowerMonitor_GetOCWarnThreshold( )**—Returns the power converter over current warning threshold for the specified power converter.
- **PowerMonitor_SetOCFaultThreshold( )**—Sets the power converter over current fault threshold for the specified power converter.

- **PowerMonitor_GetOCFaultThreshold( )**—Returns the power converter over current fault threshold for the specified power converter.
- **PowerMonitor_GetConverterVoltage( )**—Returns the power converter output voltage for the specified power converter.
- **PowerMonitor_GetConverterCurrent( )**—Returns the power converter load current for the specified power converter.
- **PowerMonitor_GetAuxiliaryVoltage( )**—Returns the voltage for the auxiliary input.
- **PowerMonitor_Calibrate( )**—Calibrates the ADC across the various range settings.
- **PowerMonitor_SetAuxiliarySampleMode( )**—Sets the ADC sample mode for the selected auxiliary input.
- **PowerMonitor_GetAuxiliarySampleMode( )**—Returns the ADC sample mode for the selected auxiliary input.
- **PowerMonitor_RequestAuxiliarySample( )**—Requests and returns a single unfiltered on-demand sample result of the specified auxiliary input.

15.14.2 Power Monitor Functions for 8, 16 and 32 Rails 1.60

- **void PowerMonitor_Start(void)**—Enables the component. Calls the Init( ) API if the component has not been initialized before. Calls Enable( ) API. This API requires global interrupts enabled in the CPU core. To enable global interrupts, call the enable global interrupt macro "CyGlobalIntEnable" in your main.c file before PowerMonitor_Start( ) API is called.
- **PowerMonitor_Stop (void)**—Disables the component. ADC sampling stops.
- **PowerMonitor_Init(void)**—Initializes the component. Includes running self-calibration.
- **void PowerMonitor_Enable(void)**—Enables hardware blocks within the component and tarts scanning.
- **void PowerMonitor_EnableFault(void)**—Enables generation of the fault signal. Specifically which fault sources are enabled is configured using the PowerMonitor_SetFaultMode( ) and the PowerMonitor_SetFaultMask( ) APIs. Fault signal generation is automatically enabled by Init( ).
- **PowerMonitor_DisableFault(void)**—Disables generation of the fault signal.
- **void PowerMonitor_SetFaultMode(uint8 faultMode)**—Configures fault sources from the component. Three fault sources are available: OV, UV and OC. This is set to the customizer setting by Init( ).
- **int8 PowerMonitor_GetFaultMode(void)**—Returns enabled fault sources from the component.
- **void PowerMonitor_SetFaultMask(uint32 faultMask)**—Enables or disables faults from each power converter through a mask. Masking applies to all fault sources. Masking applies for Fault generation and Power Good generation. By default all power converters have their fault masks enabled.
- **uint32 PowerMonitor_GetFaultMask(void)**—Returns fault mask status of each power converter. Masking applies to all fault sources.
- **uint8 PowerMonitor_GetFaultSource(void)**—Returns pending fault sources from the component. This API can be used to poll the fault status of the component. Alternatively, if the fault pin is used to generate interrupts to PSoC's CPU core, the interrupt service routine can use this API to determine the source of the fault. In either case, when this API returns a non-zero value, the GetOVFaultStatus( ), GetUVFaultStatus( ) and GetOCFaultStatus( ) APIs can provide further information on which power converter(s) caused the fault. The fault source bits are sticky and are only cleared by calling the relevant Get Status APIs.

- **uint32 PowerMonitor_GetOVFaultStatus(void)**—Returns over voltage fault status of each power converter. The status is reported regardless of the Fault Mask.
- **uint32 PowerMonitor_GetUVFaultStatus(void)**—Returns under voltage fault status of each power converter. The status is reported.
- **uint32 PowerMonitor_GetOCFaultStatus(void)**—Returns over current fault status of each power converter. The status is reported regardless of the Fault Mask.
- **void PowerMonitor_EnableWarn(void)**—Enables generation of the warning signal. Specifically which warning sources are enabled is configured using the PowerMonitor_SetWarnMode( ) and the PowerMonitor_SetWarnMask( ) APIs. Warning signal generation is automatically enabled by Init( ).
- **void PowerMonitor_DisableWarn(void)**—Disables generation of the warning signal.
- **void PowerMonitor_SetWarnMode(uint8 warnMode)**—Configures warning sources from the component. Three warning sources are available: OV, UV and OC. This is set to the customizer setting by Init( ).
- **PowerMonitor_GetWarnMode(void)**—Returns enabled warning sources from the component.
- **void PowerMonitor_SetWarnMask(uint32 warnMask)**—Enables or disables warnings from each power converter through a mask. Masking applies to all warning sources. By default all power converters have their warning masks enabled.
- **uint32 PowerMonitor_GetWarnMask(void)**—Returns warning mask status of each power converter. Masking applies to all warning sources.The warning pin is used to generate interrupts to PSoC's CPU core, the interrupt service routine can use this API to determine the source of the warning. In either case, when this API returns a non-zero value, the GetOVWarnStatus( ), GetUVWarnStatus( ) and GetOCWarnStatus( ) APIs can provide further information on which power converter(s) caused the warning.
- **uint8 PowerMonitor_GetWarnSource(void)**—Returns pending warning sources from the component. This API can be used to poll the warning status of the component. Alternatively, if the warning pin is used to generate interrupts to PSoC's CPU core, the interrupt service routine can use this API to determine the source of the warning. In either case, when this API returns
- **uint32 PowerMonitor_GetOVWarnStatus(void)**—Returns over voltage warning status of each power converter. The status is reported regardless of the Warning Mask.
- **uint32 PowerMonitor_GetUVWarnStatus(void)**—Returns under voltage warning status of each power converter. The status is reported regardless of the Warning Mask.
- **uint32 PowerMonitor_GetUVWarnStatus(void)**—Returns under voltage warning status of each power converter. The status is reported regardless of the Warning Mask.
- **PowerMonitor_SetUVWarnThreshold(uint8 converterNum, uint16 uvWarnThreshold)**— Sets the power converter under voltage warning threshold for the specified power converter.
- **uint16 PowerMonitor_GetUVWarnThreshold(uint8 converterNum)**—Returns the power converter under voltage warning threshold for the specified power converter
- **void PowerMonitor_SetOVWarnThreshold(uint8 converterNum, uint16 ovWarnThreshold)**—Sets the power converter over voltage warning threshold for the specified power converter
- **uint16 PowerMonitor_GetOVWarnThreshold(uint8 converterNum)**—Returns the power converter under voltage warning threshold for the specified power converter.
- **PowerMonitor_SetUVFaultThreshold(uint8 converterNum, uint16 uvFaultThreshol)**—Sets the power converter under voltage fault threshold for the specified power converter.
- **uint16 uvFaultThreshold**—Specifies the under voltage fault threshold in mV The range of this value is runtime checked if this value exceeds maximum range API does nothing. Use API PowerMonitor_GetUVFaultThreshold for checking valid range.

- **void PowerMonitor_GetUVFaultThreshold(uint8 converterNum)**—Returns the power converter under voltage fault threshold for the specified power converter.
- **void PowerMonitor_SetOVFaultThreshold(uint8 converterNum, uint16 ovFaultThreshold)**—Sets the power converter over voltage fault threshold for the specified power converter.
- **uint16 ovFaultThreshold**—Specifies the over voltage fault threshold in mV The range of this value is runtime checked if this value exceeds maximum range API does nothing. Use API PowerMonitor_GetOVFaultThreshold for checking valid range.
- **uint16 PowerMonitor_GetOVFaultThreshold(uint8 converterNum)**—Returns the power converter under voltage fault threshold for the specified power converter.
- **PowerMonitor_SetOCWarnThreshold(uint8 converterNum, float ocWarnThreshold)**—Sets the power converter over current warning threshold for the specified power converter.
- **float PowerMonitor_GetOCWarnThreshold(uint8 converterNum)**—Returns the power converter over current warning threshold for the specified power converter.
- **void PowerMonitor_SetOCFaultThreshold(uint8 converterNum, float ocFaultThreshold)**—Sets the power converter over current fault threshold for the specified power converter.
- **float ocFaultThreshold**—Specifies the over current fault threshold in Amperes. The range of this value is runtime checked if this value exceeds maximum range API does nothing. Use API PowerMonitor_GetOCFaultThreshold for checking valid range.
- **float PowerMonitor_GetOCFaultThreshold(uint8 converterNum)**—Returns the power converter over current fault threshold for the specified power converter.
- **uint16 PowerMonitor_GetConverterVoltage(uint8 converterNum)**—Returns the power converter output voltage for the specified power converter. If averaging is enabled the value returned is the average value.
- **float PowerMonitor_GetConverterCurrent(uint8 converterNum)**—Returns the power converter load current for the specified power converter. f averaging is enabled the value returned is the average value.
- **float PowerMonitor_GetAuxiliaryVoltage(uint8 auxNum)**—Returns the voltage for auxiliary input in units of Volts (V) independent of the ADC range setting for aux inputs.
- **PowerMonitor_Calibrate(void)**—Calibrates the ADC across the various range settings. If "cal" input pin is exposed, then a valid voltage should be provided to this input pin. The cal voltage should not exceed 100% of ADC range (±64 mV or ±128 mV) as specified in the General tab window. This voltage will be used to calibrate the low range (either ±64 mV or ±128 mV) ADC configurations.
- **void PowerMonitor SetAuxiliarySampleMode(uint8 auxNum, uint8 sampleMode)**—Sets the ADC sample mode for the selected auxiliary input. Note: all auxiliary inputs are set to continuous sampling mode by default.
- **uint8 PowerMonitor GetAuxiliarySampleMode(uint8 auxNum)**—Returns the ADC sample mode for the selected auxiliary input.
- **float PowerMonitor RequestAuxiliarySample(uint8 auxNum)**—Requests and returns a single unfiltered on-demand sample result of the specified auxiliary input. Calling this API will cause the normal ADC conversion sequence to be interrupted in order to obtain the requested sample as soon as possible. The API may also be called when the auxiliary input sample mode is set to continuous. It doesn't affect on continuous auxiliary measurements.

15.15   ADC Successive Approximation Register 3.10 (ADC_SAR)

PSoC 5LP family of devices—The ADC Successive Approximation Register (ADC_SAR) Component provides medium-speed (maximum 1-msps sampling), medium-resolution (12 bits maximum), analog-to-digital conversion. Features include: 12-bit resolution at up to 1 msps maximum, four power modes, selectable resolution and sample rate, single-ended or differential input. This component provides medium speed A/D conversion (1-msps sampling rate) at 12-bit resolution [4].

15.15.1   ADC_SAR Functions 3.10

- **ADC_Start( )**—Powers up the ADC and resets all states
- **ADC_Stop( )**—Stops ADC conversions and reduces the power to the minimum
- **ADC_SetPower( )**—Sets the power mode
- **ADC_SetResolution( )**—Sets the resolution of the ADC
- **ADC_StartConvert( )**—Starts conversions
- **ADC_StopConvert( )**—Stops conversions
- **ADC_IRQ_Enable( )**—An internal IRQ is connected to the eoc. This API enables the internal ISR.
- **ADC_IRQ_Disable( )**—An internal IRQ is connected to the eoc. This API disables the internal ISR.
- **ADC_IsEndConversion( )**—Returns a nonzero value if conversion is complete
- **ADC_GetResult8( )**—Returns a signed 8-bit conversion result
- **ADC_GetResult16( )**—Returns a signed 16-bit conversion result
- **ADC_SetOffset( )**—Sets the offset of the ADC
- **ADC_SetScaledGain( )**—Sets the ADC gain in counts per 10 volts
- **ADC_CountsTo_Volts( )**—Converts ADC counts to floating-point volts
- **ADC_CountsTo_mVolts( )**—Converts ADC counts to millivolts
- **ADC_CountsTo_uVolts( )**—Converts ADC counts to microvolts
- **ADC_Sleep( )**—Stops ADC operation and saves the user configuration
- **ADC_Wakeup( )**—Restores and enables the user configuration
- **ADC_Init( )**—Initializes the default configuration provided with the customizer
- **ADC_Enable( )**—Enables the clock and power for the ADC
- **ADC_SaveConfig( )**—Saves the current user configuration
- **ADC_RestoreConfig( )**—Restores the user configuration

15.15.2   ADC_SAR Function Calls 3.10

- **void ADC_Start(void)**—This is the preferred method to begin Component operation. ADC_Start( ) sets the initVar variable, calls the ADC_Init( ) function and then calls the ADC_Enable( ) function.
- **void ADC_Stop(void)**—Stops ADC conversions and reduces the power to the minimum.
- **void ADC_SetPower(uint8 power)**—Sets the operational power of the ADC. You should use the higher power settings with faster clock speeds.
- **void ADC_SetResolution(uint8 resolution)**—Sets the resolution for the GetResult16( ) and GetResult8( ) APIs.

- **void ADC_StartConvert(void)**—Forces the ADC to initiate a conversion. In free-running mode, the ADC runs continuously. In software trigger mode, the function also acts as a software version of the SOC and every conversion must be triggered by ADC_StartConvert( ). This function is not available when the Hardware Trigger sample mode is selected.
- **void ADC_StopConvert(void)**—Forces the ADC to stop conversions. If a conversion is currently executing, that conversion will complete, but no further conversions will occur. This function is not available when the Hardware Trigger sample mode is selected.
- **void ADC_IRQ_Enable(void)**—Enables interrupts to occur at the end of a conversion. Global interrupts must also be enabled for the ADC interrupts to occur. To enable global interrupts, call the enable global interrupt macro "CYGlobalIntEnable;" in your main.c file before enabling any interrupts.
- **void ADC_IRQ_Disable(void)**—Disables interrupts at the end of a conversion.
- **uint8 ADC_IsEndConversion(uint8 retMode)**—Immediately returns the status of the conversion or does not return (blocking) until the conversion completes, depending on the retMode parameter.
- **int8 ADC_GetResult8(void)**—Returns the result of an 8-bit conversion. If the resolution is set greater than 8 bits, the function returns the LSB of the result. ADC_IsEndConversion( ) should be called to verify that the data sample is ready.
- **int16 ADC_GetResult16(void)**—Returns a 16-bit result for a conversion with a result that has a resolution of 8 to 12 bits.
- **ADC_IsEndConversion( )**—Should be called to verify that the data sample is ready.
- **void ADC_SetOffset(int16 offset)**—Sets the ADC offset, which is used by ADC_CountsTo_Volts( ), ADC_CountsTo_mVolts( ) and ADC_CountsTo_uVolts( ), to subtract the offset from the given reading before calculating the voltage conversion.
- **void ADC_SetScaledGain(int16 adcGain)**—Sets the ADC gain in counts per 10 volts for the voltage conversion functions that follow. This value is set by default by the reference and input range settings. It should only be used to further calibrate the ADC with a known input or if the ADC is using an external reference.
- **float ADC_CountsTo_Volts(int16 adcCounts)**—Converts the ADC output to volts as a floating-point number. For example, if the ADC measured 0.534 volts, the return value would be 0.534. The calculation of voltage depends on the value of the voltage reference. When the Vref is based on Vdda, the value used for Vdda is set for the project in the System tab of the Design Wide Resources (DWR).
- **int16 ADC_CountsTo_mVolts(int16 adcCounts)**—Converts the ADC output to millivolts as a 16-bit integer. For example, if the ADC measured 0.534 volts, the return value would be 534. The calculation of voltage depends on the value of the voltage reference. When the Vref is based on Vdda, the value used for Vdda is set for the project in the System tab of the Design Wide Resources (DWR).
- **int32 ADC_CountsTo_uVolts(int16 adcCounts)**—Converts the ADC output to microvolts as a 32-bit integer. For example, if the ADC measured 0.534 volts, the return value would be 534000. The calculation of voltage depends on the value of the voltage reference. When the Vref is based on Vdda, the value used for Vdda is set for the project in the System tab of the Design Wide Resources (DWR).
- **void ADC_Sleep(void)**—This is the preferred routine to prepare the Component for sleep. The ADC_Sleep( ) routine saves the current Component state, then it calls the ADC_Stop( ) function. Call the ADC_Sleep( ) function before calling the CyPmSleep( ) or the CyPmHibernate( ) function. See the PSoC Creator System Reference Guide for more information about power-management functions.

- **void ADC_Wakeup(void)**—This is the preferred routine to restore the Component to the state when ADC_Sleep( ) was called. If the Component was enabled before the ADC_Sleep( ) function was called, the ADC_Wakeup( ) function also re-enables the Component.
- **void ADC_Init(void)**—Initializes or restores the Component according to the customizer Configure dialog settings. It is not necessary to call ADC_Init( ) because the ADC_Start( ) routine calls this function and is the preferred method to begin Component operation.
- **void ADC_Enable(void)**—Activates the hardware and begins Component operation. The higher power is set automatically depending on clock speed. The ADC_SetPower( ) API description contains the relation of the power from the clock rate. It is not necessary to call ADC_Enable( ) because the ADC_Start( ) routine calls this function, which is the preferred method to begin Component operation.
- **void ADC_SaveConfig(void)**—This function saves the Component configuration and non-retention registers. It also saves the current Component parameter values, as defined in the Configure dialog or as modified by the appropriate APIs. This function is called by the ADC_Sleep( ) function.
- **void ADC_RestoreConfig(void)**—This function restores the Component configuration and non-retention registers. It also restores the Component parameter values to what they were before calling the ADC_Sleep( ) function.

15.16 Sequencing Successive Approximation ADC 2.10 (ADC_SAR_Seq)

PSoC 5LP family of devices—The Sequencing SAR ADC Component enables makes it possible for you to configure and then use the different operational modes of the SAR ADC on PSoC 5LP. You also have schematic level and firmware level support for seamless use of the Sequencing SAR ADC in PSoC Creator designs and projects. You are able to configure multiple analog channels that are automatically scanned with the results placed in individual SRAM locations. Features include: Selectable resolution (8, 10 or 12 bit) and sample rate (up to 1 Msps) and can scan. up to 64 single ended or 32 differential channels automatically, or just a single input[10] [80].

15.16.1 ADC_SAR_Seq Functions 2.10

- **ADC_SAR_Seq_Start( )**—Powers up the ADC_SAR_Seq and resets all states.
- **ADC_SAR_Seq_Stop( )**—Stops ADC_SAR_Seq conversions and reduces the power to the minimum.
- **ADC_SAR_Seq_SetResolution( )**—Sets the resolution of the ADC_SAR_Seq.
- **ADC_SAR_Seq_StartConvert( )**—Starts conversions.
- **ADC_SAR_Seq_StopConvert( )**—Stops conversions.
- **ADC_SAR_Seq_IRQ_Enable( )**—An internal IRQ is connected to the eoc. This API enables the internal ISR.
- **ADC_SAR_Seq_IRQ_Disable( )**—An internal IRQ is connected to the eoc. This API disables the internal ISR.
- **ADC_SAR_Seq_IsEndConversion( )**—Returns a nonzero value if conversion is complete.

---

[10]Only GPIOs can be connected to the channel inputs. The actual maximum number of input channels depends on the number of routable analog GPIOs available on a specific PSoC part and package.

- **ADC_SAR_Seq_GetAdcResult( )**—Returns a signed 16-bit conversion result available in the ADC SAR Data Register not the result buffer.
- **ADC_SAR_Seq_GetResult16( )**—Returns a signed16-bit conversion result for specified channel.
- **ADC_SAR_Seq_SetOffset( )**—Sets the offset of the ADC_SAR_Seq.
- **ADC_SAR_Seq_SetScaledGain( )**—Sets the ADC_SAR_Seq gain in counts per 10 volts.
- **ADC_SAR_Seq_CountsTo_Volts( )**—Converts ADC_SAR_Seq counts to floating-point volts.
- **ADC_SAR_Seq_CountsTo_mVolts( )**—Converts ADC_SAR_Seq counts to millivolts.
- **ADC_SAR_Seq_CountsTo_uVolts( )**—Converts ADC_SAR_Seq counts to microvolts.
- **ADC_SAR_Seq_Sleep( )**—Stops ADC_SAR_Seq operation and saves the user configuration.
- **ADC_SAR_Seq_Wakeup( )**—Restores and enables the user configuration.
- **ADC_SAR_Seq_Init( )**—Initializes the default configuration provided with the customizer.
- **ADC_SAR_Seq_Enable( )**—Enables the clock and power for the ADC_SAR_Seq.
- **ADC_SAR_Seq_SaveConfig( )**—Saves the current user configuration.
- **ADC_SAR_Seq_RestoreConfig( )**—Restores the user configuration.

15.16.2 ADC_SAR_Seq Function Calls 2.10

- **void ADC_SAR_Seq_Start(void)**—This is the preferred method to begin Component operation. ADC_SAR_Seq_Start( ) sets the initVar variable, calls the ADC_SAR_Seq_Init( ) function and then calls the ADC_SAR_Seq_Enable( ) function. It is not recommended to call this API a second time without stopping the Component first. If the initVar variable is already set, this function only calls the ADC_SAR_Seq_Enable( ) function.
- **void ADC_SAR_Seq_Stop(void)**—Stops ADC_SAR_Seq conversions and reduces the power to the minimum.[11]
- **void ADC_SAR_Seq_StartConvert(void)**—This forces the ADC to initiate a conversion. In free-running mode, the ADC_SAR_Seq runs continuously. In software trigger mode, the function also acts as a software version of the SOC and every conversion must be triggered by ADC_SAR_Seq_StartConvert( ). In Hardware trigger mode this function is unavailable.
- **void ADC_SAR_Seq_StopConvert(void)**—This forces the ADC_SAR_Seq to stop conversions. If a conversion is currently executing, that conversion completes, but no further conversions happen. This only applies to free-running mode.
- **void ADC_SAR_Seq_IRQ_Enable(void)**—This enables interrupts to occur at the end of a conversion. Global interrupts must also be enabled for the ADC interrupts to occur. To enable global interrupts, call the enable global interrupt macro "CyGlobalIntEnable;" in the main.c file before enabling any interrupts.
- **void ADC_SAR_Seq_IRQ_Disable(void)**—Disables interrupts at the end of a conversion.
- **uint32 ADC_SAR_Seq_IsEndConversion(uint8 retMode)**—Immediately returns the status of the conversion or does not return (blocking) until the conversion completes, depending on the retMode parameter.
- **int16 ADC_SAR_Seq_GetAdcResult(void)**—Gets the data available in the SAR DATA register, not the results buffer.

---

[11]This API does not power down the ADC, but reduces power to a minimum level. This device has a defect that causes connections to several analog resources to be unreliable when the device is not powered. The unreliability manifests itself in silent failures (for example, unpredictable bad results from analog Components) when the Component using that resource is stopped.

- **int16 ADC_SAR_Seq_GetResult16(uint16 chan)**—Returns the conversion result for channel "chan".
- **void ADC_SAR_Seq_SetOffset(int32 offset)**—Sets the ADC_SAR_Seq offset, which is used by ADC_SAR_Seq_CountsTo_Volts( ), ADC_SAR_Seq_CountsTo_mVolts( ) and ADC_SAR_Seq_CountsTo_uVolts( ), to subtract the offset from the given reading before calculating the voltage conversion.
- **void ADC_SAR_Seq_SetScaledGain(int32 adcGain)**—Sets the ADC_SAR_Seq gain in counts per 10 volts for the voltage conversion functions that follow. This value is set by default by the reference and input range settings. It should only be used to further calibrate the ADC_SAR_Seq with a known input or if the ADC_SAR_Seq is using an external reference. To calibrate the gain, supply close to reference voltage to ADC inputs and measure it by multimeter. Calculate the gain coefficient using following formula.

$$adcGain = \frac{counts \, x \, 10}{V_{measured}} \tag{15.1}$$

Where the counts are returned from ADC_SAR_Seq_GetResult16( ) value, measured by multimeter in volts.

- **float32 ADC_SAR_Seq_CountsTo_Volts(int16 adcCounts)**—Converts the ADC_SAR_Seq output to volts as a floating-point number. For example, if the ADC_SAR_Seq measured 0.534 volts, the return value would be 0.534. The calculation of voltage depends on the value of the voltage reference. When the V_ref is based on V_dda, the value used for Vdda is set for the project in the System tab of the Design Wide Resources (DWR).
- **int32 ADC_SAR_Seq_CountsTo_mVolts(int16 adcCounts)**—Converts the ADC_SAR_Seq output to millivolts as a 32-bit integer. For example, if the ADC_SAR_Seq measured 0.534 volts, the return value would be 534. The calculation of voltage depends on the value of the voltage reference. When the Vref is based on Vdda, the value used for Vdda is set for the project in the System tab of the Design Wide Resources (DWR).
- **int32 ADC_SAR_Seq_CountsTo_uVolts(int16 adcCounts)**—Converts the ADC_SAR_Seq output to microvolts as a 32-bit integer. For example, if the ADC_SAR_Seq measured 0.534 volts, the return value would be 534000. The calculation of voltage depends on the value of the voltage reference. When the Vref is based on Vdda, the value used for Vdda is set for the project in the System tab of the Design Wide Resources (DWR).
- **void ADC_SAR_Seq_Sleep(void)**—This is the preferred routine to prepare the Component for sleep. The ADC_SAR_Seq_Sleep( ) routine saves the current Component state. Then it calls the ADC_SAR_Seq_Stop( ) function and calls ADC_SAR_Seq_SaveConfig( ) to save the hardware configuration. Call the ADC_SAR_Seq_Sleep( ) function before calling the CyPmSleep( ) or the CyPmHibernate( ) function. See the PSoC Creator System Reference Guide for more information about power-management functions.
- **void ADC_SAR_Seq_Wakeup(void)**—This is the preferred routine to restore the Component to the state when ADC_SAR_Seq_Sleep( ) was called. The ADC_SAR_Seq_Wakeup( ) function calls the ADC_SAR_Seq_RestoreConfig( ) function to restore the configuration. If the Component was enabled before the ADC_SAR_Seq_Sleep( ) function was called, the ADC_SAR_Seq_Wakeup( ) function also re-enables the Component.
- **void ADC_SAR_Seq_Init(void)**—Initializes or restores the Component according to the customizer Configure dialog settings. It is not necessary to call ADC_SAR_Seq_Init( ) because the ADC_SAR_Seq_Start( ) routine calls this function and is the preferred method to begin Component operation.

- **void ADC_SAR_Seq_Enable(void)**—Activates the hardware and begins Component operation. It is not necessary to call ADC_SAR_Seq_Enable( ) because the ADC_SAR_Seq_Start( ) routine calls this function, which is the preferred method to begin Component operation.
- **void ADC_SAR_Seq_SaveConfig(void)**—This function saves the Component configuration and non-retention registers. It also saves the current Component parameter values, as defined in the Configure dialog or as modified by the appropriate APIs. This function is called by the ADC_SAR_Seq_Sleep( ) function.
- **void ADC_SAR_Seq_RestoreConfig(void)**—This function restores the Component configuration and non-retention registers. It also restores the Component parameter values to what they were before calling the ADC_SAR_Seq_Sleep( ) function.

15.17 Operational Amplifier (OpAmp) 1.90

PSoC 3 and 5LP family of devices—The OpAmp component provides a low-voltage, low-power operational amplifier and may be internally connected as a voltage follower. The inputs and output may be connected to internal routing nodes, directly to pins, or a combination of internal and external signals. The OpAmp is suitable for interfacing with high-impedance sensors, buffering the output of voltage DACs, driving up to 25 mA; and building active filters in any standard topology. Supported features include: Unity gain bandwidth >3.0 MHz, input offset voltage 2.0 mV max, rail-to-rail inputs and output, output direct low resistance connection to pin, 25-mA output current, programmable power and bandwidth and internal connection for follower saves a pin [63].

15.17.1 OpAmp Functions 1.90

- **Opamp_Start( )**—Turns on the OpAmp and sets the power level to the value chosen during the parameter selection.
- **Opamp_Stop( )**—Disables OpAmp (power down).
- **Opamp_SetPower+Sets**—Sets the power level.
- **Opamp_Sleep( )**—Stops and saves the user configuration.
- **Opamp_Wakeup( )**—Restores and enables the user configuration.
- **Opamp_Init( )**—Initializes or restores default OpAmp configuration.
- **Opamp_Enable( )**—Enables the OpAmp.
- **Opamp_RestoreConfig( )**—Empty function. Provided for future.

15.17.2 OpAmp Function Calls 1.90

- **void Opamp_Start(void)**—Turns on the OpAmp and sets the power level to the value chosen during the parameter selection.
- **void Opamp_Stop(void)**—Turns off the OpAmp and enable its lowest power state.
- **void Opamp_SetPower(uint8 power)**—Sets the power level.
- **void Opamp_Sleep(void)**—This is the preferred routine to prepare the component for sleep. The OpAmp_Sleep( ) routine saves the current component state. Then it calls the OpAmp_Stop( ) function and calls OpAmp_SaveConfig( ) to save the hardware configuration. Call the OpAmp_Sleep( ) function before calling the CyPmSleep( ) or the CyPmHibernate( ) function.

- **void Opamp_Wakeup(void)**—This is the preferred routine to restore the component to the state when OpAmp_Sleep( ) was called. The Opamp_Wakeup( ) function calls the OpAmp_RestoreConfig( ) function to restore the configuration. If the component was enabled before the OpAmp_Sleep( ) function was called, the OpAmp_Wakeup( ) function will also re-enable the component.
- **void Opamp_Init(void)**—Initializes or restores the component according to the customizer Configure dialog settings. It is not necessary to call OpAmp_Init( ) because the OpAmp_ Start( ) routine calls this function and is the preferred method to begin component operation.
- **void Opamp_Enable(void)**—Activates the hardware and begins component operation. It is not necessary to call OpAmp_Enable( ) because the OpAmp_Start( ) routine calls this function, which is the preferred method to begin component operation.
- **void Opamp_SaveConfig(void)**—Empty function. Provided for future use.
- **void Opamp_RestoreConfig(void)**—Empty function. Provided for future use.

15.18   Analog Hardware Multiplexer (AMUX) 1.50

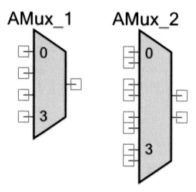

The Analog Hardware Multiplexer (AMuxHw) Component is used to provide hardware-switchable connections from GPIOs to analog resource blocks (ARBs). Features include: single-ended or differential inputs, mux or switch mode, from 1 to 256 inputs, hardware controlled and bidirectional (passive) [6].

15.18.1   Analog Hardware Multiplexer (AMUX 1.50) Functions 1.50

- **AMux_Init( )**—Disconnects all channels.
- **AMux_Start( )**—Disconnects all channels.
- **AMux_Stop( )**—Disconnects all channels.
- **AMux_Select( )**—Disconnects all channels, then connects "chan". When *AtMostOneActive is* true, this is implemented as AMux_FastSelect( ).
- **AMux_Connect( )**—Connects "chan" signal, but does not disconnect other channels. When *AtMostOneActive* is true.
- **AMux_FastSelect( )**—Disconnects the last channel that was selected by the AMux_Select( ) or AMux_FastSelect( ) function, then connects the new signal "chan".
- **AMux_Disconnect( )**—Disconnects only "chan" signal.

- **AMux_DisconnectAll( )**—Disconnects all channels.

15.18.2 Analog Hardware Multiplexer (AMUX) Function Calls 1.50

- **void AMux_Init(void)**—Disconnects all channels.
- **void AMux_Start(void)**—Disconnects all channels.
- **AMux_Stop(void)**—Disconnects all channels.
- **void AMux_Select(uint8 chan)**—The AMux_Select( ) function first disconnects all other channels, then connects the given channel. When *AtMostOneActive* is true, this is implemented as AMux_FastSelect( ).
- **void AMux_FastSelect(uint8 chan)**—This function first disconnects the last connection made with the AMux_FastSelect( ) or AMux_Select( ) functions, then connects the given channel. The AMux_FastSelect( ) function is similar to the AMux_Select( ) function, except that it is faster because it only disconnects the last channel selected rather than all possible channels.
- **void AMux_Connect(uint8 chan)**—This function connects the given channel to the common signal without affecting other connections. When *AtMostOneActive* is true, this function is not available.
- **void AMux_Disconnect(uint8 chan)**—Disconnects only the specified channel from the common terminal.
- **void AMux_DisconnectAll(void)**—Disconnects all channels.

15.19 Analog Hardware Multiplexer Sequencer (AMUXSeq) 1.80

The analog multiplexer sequencer (AMuxSeq) Component [7] is used to connect one analog signal at a time to a different common analog signal, by breaking and making connections in hookup-order sequence. The AMuxSeq is primarily used for time division multiplexing when it is necessay to multiplex multiple analog signals into a single source or destination. Because the AMuxSeq Component is passive, it can be used to multiplex input or output signals. It has a simpler and faster API than the AMux and should be used when multiple simultaneous connections are not required and the signals will always be accessed in the same order [8].

15.19.1 Analog Multiplexer Sequencer Functions (AMUXSeq) 1.80

- **AMuxSeq_Init( )**—Disconnects all channels
- **AMuxSeq_Start( )**—Disconnects all channels
- **AMuxSeq_Stop( )**—Disconnects all channels
- **AMuxSeq_Next( )**—Disconnects the previous channel and connects the next one in the sequence.
- **AMuxSeq_DisconnectAll( )**—Disconnects all channels
- **AMuxSeq_GetChannel( )**—The currently connected channel is returned. If no channel is connected, -1 is returned.

15.19.2 Analog Multiplexer Sequencer Function Calls (AMUXSeq) 1.80

- **AMuxSeq_GetChannel( )**—The currently connected channel is returned. If no channel is connected, −1 is returns.
- **void AMuxSeq_Init(void)**—Disconnects all channels. The next time AMuxSeq_Next( ) is called, the first channel is selected.
- **void AMuxSeq_Start(void)**—Disconnects all channels. The next time AMuxSeq_Next( ) is called, the first channel is selected.
- **void AMuxSeq_Stop(void)**—Disconnects all channels. The next time AMuxSeq_Next( ) is called, the first channel is selected.
- **void AMuxSeq_Next(void)**—Disconnects the previous channel and connects the next one in the sequence. When AMuxSeq_Next( ) is called for the first time or after AMuxSeq_Init( ), AMuxSeq_Start( ), AMuxSeq_Enable( ), AMuxSeq_Stop( ), or AMuxSeq_DisconnectAll( ), it connects channel 0.
- **void AMuxSeq_DisconnectAll(void)**—This function disconnects all channels. The next time AMuxSeq_Next( ) is called, the first channel will be selected. i
- **nt8 AMuxSeq_GetChannel(void)**—The currently connected channel is returned. If no channel is connected, −1 is returned.

15.20 Analog Virtual Mux 1.0

Virtual mux components are similar to conventional muxes in that they connect a selected input to an output. For a conventional mux, the input selection can be dynamically controlled by a control signal. For a virtual mux, the input selection is determined by an expression that evaluates to a constant when used within a design. The purpose of the virtual mux is to pick one input at build time. There are two separate virtual mux components: one analog and one digital. Features include: selects 1 of up to 16 inputs, selection is static and configurable number of inputs [104].

15.20.1 Analog Virtual Mux Functions 1.0

NONE.

15.21 Comparator (Comp) 2.00

The Comparator (Comp) component provides a hardware solution to compare two analog input voltages. The output can be sampled in software or digitally routed to another component. Three speed levels are provided to allow you to optimize for speed or power consumption. A reference or external voltage can be connected to either input. The output of the comparator can be inverted by using the Polarity parameter. Additional features include: low input offset, user controlled offset calibration, multiple speed modes, low-power mode, output routable to digital logic blocks or pins, selectable output polarity and a configurable operation mode during Sleep [19].

15.21.1 Comparator Functions (Comp) 2.00

- **Comp_Start( )**—Initializes the Comparator with default customizer values.
- **Comp_Stop( )**—Turns off the Comparator.
- **Comp_SetSpeed( )**—Sets speed of the Comparator.
- **Comp_ZeroCal( )**—Zeros the input offset of the Comparator.
- **Comp_GetCompare( )**—Returns compare result.
- **Comp_LoadTrim( )**—Writes a value to the Comparator trim register.
- **Comp_Sleep( )**—Stops Comparator operation and saves the user configuration.
- **Comp_Wakeup( )**—Restores and enables the user configuration.
- **Comp_Init( )**—Initializes or restores default Comparator configuration.
- **Comp_Enable( )**—Enables the Comparator.
- **Comp_SaveConfig( )**—Empty function. Provided for future use.
- **Comp_RestoreConfig( )**—Empty function. Provided for future use.
- **Comp_PwrDwnOverrideEnable( )**—Enables Comparator operation in sleep mode. Only valid for PSoC 3 silicon.
- **Comp_PwrDwnOverrideDisable( )**—Disables Comparator operation in sleep mode. Only valid for PSoC 3 silicon.

15.21.2 Comparator Function Calls (Comp) 2.00

- **void Comp_Start(void)**—This is the preferred method to begin component operation. Comp_Start( ) sets the initVar variable, calls the Comp_Init( ) function and then calls the Comp_Enable( ) function.
- **void Comp_Stop(void)**—Disables and powers down the comparator.
- **void Comp_SetSpeed(uint8 speed)**—This function selects one of three speed modes for the comparator. The comparator power consumption increases for the faster speed modes.
- **uint16 Comp_ZeroCal(void)**—Performs custom calibration of the input offset to minimize error for a specific set of conditions: comparator reference voltage, supply voltage and operating temperature. A reference voltage in the range at which the comparator will be used must be applied to the negative input of the comparator while the offset calibration is performed. The comparator component must be configured for Fast or Slow operation when calibration is performed. The calibration process will not work correctly if the comparator is configured in Low Power mode.
- **Comp_GetCompare(void)**—This function returns a nonzero value when the voltage connected to the positive input is greater than the negative input voltage. This value is not affected by the Polarity parameter. This value always reflects a noninverted state. uint8: Comparator output state. Nonzero value when the positive input voltage is greater than the negative input voltage; otherwise, the return value is zero.
- **void Comp_LoadTrim(uint16 trimVal)**—This function writes a value into the comparator trim register.
- **void Comp_SaveConfig(void)**—This function saves the component configuration and non-retention registers. It also saves the current component parameter values, as defined in the Configure dialog or Empty function. Implemented for future usage. No effect by calling this function.
- **void Comp_RestoreConfig(void)**—This function restores the component configuration and non-retention registers. It also restores the component parameter values to what they were prior to calling the Comp_Sleep( ) function. Implemented for future use. No effect by calling this function.

- **void Comp_Sleep(void)**—This is the preferred routine to prepare the component for sleep. The Comp_Sleep( ) routine saves the current component state. Then it calls the Comp_Stop( ) function and calls Comp_SaveConfig( ) to save the hardware configuration. Call the Comp_Sleep( ) function before calling the CyPmSleep( ) or the CyPmHibernate( ) function.
- **vvoid Comp_Wakeup(void)**—This is the preferred routine to restore the component to the state when Comp_Sleep( ) was called. The Comp_Wakeup( ) function calls the Comp_RestoreConfig( ) function to restore the configuration. If the component was enabled before the Comp_Sleep( ) function was called, the Comp_Wakeup( ) function will also re-enable the component.ut first calling the Comp_Sleep( ) or Comp_SaveConfig( ) function may produce unexpected behavior.
- **void Comp_PwrDwnOverrideEnable(void)**—This is the power-down override feature. It allows the component to stay active during sleep mode. Before calling this API, the Comp_SetPower( ) API should be called with the Comp_LOWPOWER parameter to set the comparator power mode to Ultra Low Power. This is because Ultra Low Power is the only valid power mode for the comparator in Sleep mode.
- **void Comp_Enable(void)**—Activates the hardware and begins component operation. It is not necessary to call Comp_Enable( ) because the Comp_Start( ) routine calls this function, which is the preferred method to begin component operation.
- **void Comp_PwrDwnOverrideDisable(void)**—This is the power-down override feature. This function allows the comparator to stay inactive during sleep mode.
- **void Comp_Init(void)**—Initializes or restores the component according to the customizer Configure dialog settings. It is not necessary to call Comp_Init( ) because the Comp_Start( ) routine calls this function and is the preferred method to begin component operation.

15.22 Scanning Comparator 1.10

The Scanning Comparator (ScanComp) component provides a hardware solution to compare up to 64 pairs of analog input voltages signals using just one hardware comparator. The sampled comparator outputs can be enabled for connection in digital hardware. A reference or external voltage can be connected to each input. Features include: Scan up to 64 single ended or differential channels automatically[12] [76]

15.22.1 Scanning Comparator Functions 1.10

- **ScanComp_Start( )**—Performs all the required initialization for the component and enables power to the block.
- **ScanComp_Init( )**—Initializes or restores the component according to the customizer settings.
- **ScanComp_Enable( )**—Activates the hardware and begins component operation.
- **ScanComp_Stop( )**—Turns off the Scanning Comparator.
- **ScanComp_SetSpeed( )**—Sets the drive power and speed.
- **ScanComp_SetDACRange( )**—Sets the DAC to a new range.
- **ScanComp_GetDACRange( )**—Gets the DAC range setting
- **ScanComp_SetDACVoltage( )**—Sets the DAC output to a new voltage.
- **ScanComp_GetDACVoltage( )**—Gets the current DAC output voltage setting

---

[12]The number of input and output channels will be limited by the hardware available in the device being used, up to 64 outputs routable to digital logic blocks or pins and multiple comparison modes.

- **ScanComp_SetChannelDACVoltage( )**—Sets the DAC output for a specific channel to a new voltage.
- **ScanComp_GetChannelDACVoltage( )**—Gets the DAC output voltage for a specific channel.
- **ScanComp_GetCompare( )**—Gets the current comparison result for the selected channel.
- **ScanComp_GetInterruptSource( )**—Gets the pending interrupt requests from the selected block. Even masked interrupts are returned
- **ScanComp_GetInterruptSourceMasked( )**—Gets the pending interrupt requests from the selected block. Masked interrupts are not returned.
- **ScanComp_GetInterruptMask( )**—Gets the current interrupt mask from the selected block.
- **ScanComp_SetInterruptMask( )**—sets the interrupt masks for the selected block.
- **ScanComp_Sleep( )**—This is the preferred API to prepare the component for low power mode operation.
- **ScanComp_Wakeup( )**—This is the preferred API to restore

15.22.2 Scanning Comparator Function Calls 1.10

- **void ScanComp_Start(void)**—Performs all the required initialization for the component and enables power to the block. The first time the routine is executed, the component is initialized to the configuration from the customizer. Power/speed is set based on the configured sample rate and the comparator response time specs, or if an external clock is used, it is set to the maximum. When called to restart the comparator following a ScanComp_Stop( ) call, the current component parameter settings are retained.
- **void ScanComp_Init(void)**—Initializes or restores the component according to the customizer settings. It is not necessary to call ScanComp_Init( ) because the ScanComp_Start( ) API calls this function and is the preferred method to begin component operation.
- **void ScanComp_Enable(void)**—Activates the hardware and begins component operation. It is not necessary to call ScanComp_Enable( ) because the ScanComp_Start( ) API calls this function, which is the preferred method to begin component operation.
- **void ScanComp_Stop(void)**—Turns off the Scanning Comparator by turning off the comparator itself and halting the muxing of inputs and turning off the DAC if it is used.
- **void ScanComp_SetSpeed(uint8 speed)**—Sets the drive power and speed to one of three settings. Power/speed is set by ScanComp_Start( ) based on the configured sample rate and the comparator response time specs, or if an external clock is used, it is set to the maximum.
- **void ScanComp_SetDACRange(uint8 DACRange)**—Sets the DAC to a new range. Used only when the Internal DAC is selected.
- **uint8 ScanComp_GetDACRange(void)**—Gets the DAC range setting. Used only when the Internal DAC is selected.
- **void ScanComp_SetDACVoltage(uint8 DACVoltage)**—Sets the DAC output to a new voltage. Used only when the Internal DAC is selected.
- **uint8 ScanComp_GetDACVoltage(void)**—Gets the current DAC output voltage setting. Used only when the Internal DAC is selected.
- **void ScanComp_SetChannelDACVoltage(uint8 channel, uint8 DACVoltage)**—Sets the DAC output for a specific channel to a new voltage. Used only when the Internal DAC is selected and voltage is "Per channel".
- **uint8 ScanComp_GetChannelDACVoltage(uint8 channel)**—Gets the DAC output voltage for a specific channel. Used only when the Internal DAC is selected and voltage is "Per channel".

- **uint8 ScanComp_GetCompare(uint8 channel)**—Gets the current comparison result for the selected channel.
- **uint8 ScanComp_GetInterruptSource(uint8 inputBlock)**—Gets the pending interrupt requests from the selected block. This function can determine which of the channels generated an interrupt. Even masked interrupts are returned. This function clears the interrupt status for that input block.
- **uint8 ScanComp_GetInterruptSourceMasked(uint8 inputBlock)**—Gets the pending interrupt requests from the selected block. This function can determine which of the channels generated an interrupt. Masked interrupts are not returned. This function clears the interrupt status.
- **ScanComp_GetInterruptMask(uint8 inputBlock)**—Gets the current interrupt mask from the selected block. This function can determine which of the channels' interrupts are currently masked.
- **void ScanComp_SetInterruptMask(uint8 inputBlock, uint8 mask)**—Sets the interrupt masks for the set block of 8 or less channels.
- **void ScanComp_Sleep(void)**—This is the preferred API to prepare the component for low power mode operation. The scanning comparator cannot operate in sleep mode in
- **void ScanComp_Wakeup(void)**—This is the preferred API to restore the component to the state when ScanComp_Sleep( ) was called.

15.23   8-Bit Current Digital to Analog Converter (iDAC8) 2.00

The IDAC8 component is an 8-bit current output DAC (Digital to Analog Converter). The output can source or sink current in three ranges. The IDAC8 can be controlled by hardware, software, or by a combination of both hardware and software. This component supports three ranges: 2040 µA, 255 µA and 31.875 µA, current sink or source selectable, software- or clock-driven output strobe and data source may be CPU, DMA, or Digital components [1].

15.23.1   iDAC8 Functions 2.00

- **IDAC8_Stop( )**—Disables the IDAC8 and sets it to the lowest power state.
- **IDAC8_SetSpeed( )**—Sets DAC speed.
- **IDAC8_SetPolarity( )**—Sets the output mode to current sink or source.
- **IDAC8_SetRange( )**—Sets full-scale range for IDAC8.
- **IDAC8_SetValue( )**—Sets value between 0 and 255 with the given range.
- **IDAC8_Sleep( )**—Stops and saves the user configuration.
- **IDAC8_Wakeup( )**—Restores and enables the user configuration.
- **IDAC8_Init( )**—Initializes or restores default IDAC8 configuration.
- **IDAC8_Enable( )**—Enables the IDAC8.
- **IDAC8_SaveConfig( )**—Saves the current configuration.

15.23.2   iDAC8 Function Calls 2.00

- **void IDAC8_Start(void)**—This is the preferred method to begin component operation. IDAC8_Start( ) sets the initVar variable, calls the IDAC8_Init( ) function and then calls the IDAC8_Enable( ) function.
- **void IDAC8_Stop(void)**—Powers down IDAC8 to lowest power state and disables output.
- **void IDAC8_SetSpeed(uint8 speed)**—Sets DAC speed.

- **void IDAC8_SetPolarity(uint8 polarity)**—Sets output polarity to sink or source. This function is valid only if the Polarity parameters is set to either source or sink.
- **void IDAC8_SetRange(uint8 range)**—Sets full-scale range for IDAC8.
- **void IDAC8_SetValue(uint8 value)**—Sets value to output on IDAC8. Valid values are between 0 and 255.
- **void IDAC8_Sleep(void)**—This is the preferred API to prepare the component for sleep. The IDAC8_Sleep( ) API saves the current component state. Then it calls the IDAC8_Stop( ) function and calls IDAC8_SaveConfig( ) to save the hardware configuration. Call the IDAC8_Sleep( ) function before calling the CyPmSleep( ) or the CyPmHibernate( ) function.
- **void IDAC8_Wakeup(void)**—This is the preferred API to restore the component to the state when IDAC8_Sleep( ) was called. The IDAC8_Wakeup( ) function calls the IDAC8_RestoreConfig( ) function to restore the configuration. If the component was enabled before the IDAC_Sleep( ) function was called, the IDAC8_Wakeup( ) function will also re-enable the component.
- **void IDAC_Init(void)**—Initializes or restores the component according to the customizer Configure dialog settings. It is not necessary to call IDAC8_Init( ) because the IDAC8_Start( ) API calls this function and is the preferred method to begin component operation.
- **void IDAC8_Init(void)**—Initializes or restores the component according to the customizer Configure dialog settings. It is not necessary to call IDAC8_Init( ) because the IDAC8_Start( ) API calls this function and is the preferred method to begin component operation.
- **void IDAC8_Enable(void)**—Activates the hardware and begins component operation. It is not necessary to call IDAC8_Enable( ) because the IDAC8_Start( ) API calls this function, which is the preferred method to begin component operation.
- **void IDAC8_SaveConfig(void)**—This function saves the component configuration and non-retention registers. It also saves the current component parameter values, as defined in the Configure dialog or as modified by appropriate APIs. This function is called by the IDAC8_Sleep( ) function. Note In the DAC Bus mode, the values are not saved.
- **void IDAC8_RestoreConfig(void)**—This function restores the component configuration and non-retention registers. This function also restores the component parameter values to what they were before calling the IDAC8_Sleep( ) function. Note In the DAC Bus mode, the values are not restored.

15.24 Dithered Voltage Digital/Analog Converter (DVDAC) 2.10

The Dithered Voltage Digital to Analog Converter (DVDAC) component has a selectable resolution between 9 and 12 bits. Dithering is used to increase the resolution of its underlying 8-bit VDAC8. Only a small output capacitor is required to suppress the noise generated by dithering. Supported features include:two voltage ranges, 1 and 4 volts, two voltage ranges (1 and 4 volts), adjustable 9, 10, 11, or 12-bit resolution, dithered using DMA for zero CPU overhead and a single DAC block [35].

15.24.1 DVDAC Functions 2.10

- **DVDAC_Start( )**—Initializes the DVDAC with default customizer values.
- **DVDAC_Stop( )**—Disables the DVDAC and sets it to the lowest power state.
- **DVDAC_SetValue( )**—Sets the DVDACs output.
- **DVDAC_Sleep( )**—Stops and saves the user configuration.
- **DVDAC_WakeUp( )**—Restores and enables the user configuration.
- **DVDAC_Init( )**—Initializes or restores the default DVDAC configuration

- **DVDAC_Enable( )**—Enables the DVDAC.
- **DVDAC_SaveConfig( )**—Saves the non-retention DAC data register value.
- **DVDAC_RestoreConfig( )**—Restores the non-retention DAC data register value.

15.24.2 DVDAC Function Calls 2.10

- **DVDAC_Start(void)**—Performs all the required initialization for the component and enables power to the block. The first time the routine is executed, the component is initialized to the configured settings. When called to restart the DVDAC following a DVDAC_Stop( ) call, the current component parameter settings are retained.
- **void DVDAC_Stop(void)**—Stops the component and turns off the analog blocks in the DVDAC.
- **void DVDAC_SetValue(uint16 value)**—Sets the DVDACs output. The function populates the SRAM array based on the value and the resolution setting. That array is then transferred to the internal VDAC by DMA.
- **void DVDAC_Sleep(void)**—This is the preferred API to prepare the component for sleep. The DVDAC_Sleep( ) API saves the current component state. Then it calls the DVDAC_Stop( ) function and calls DVDAC_SaveConfig( ) to save the hardware configuration. Call the DVDAC_Sleep( ) function before calling the CyPmSleep( ) or the CyPmHibernate( ) function.
- **void DVDAC_Wakeup(void)**—This is the preferred API to restore the component to the state when DVDAC_Sleep( ) was called. The DVDAC_Wakeup( ) function calls the DVDAC_RestoreConfig( ) function to restore the configuration. If the component was enabled before the DVDAC_Sleep( ) function was called, the DVDAC_Wakeup( ) function will also re-enable the component.
- **void DVDAC_Init(void)**—Initializes or restores the component according to the customizer Configure dialog settings. It is not necessary to call DVDAC_Init( ) because the
- **DVDAC_Start( )**—API calls this function and is the preferred method to begin component operation.
- **void DVDAC_Enable(void)**—Activates the hardware and begins component operation. It is not necessary to call DVDAC_Enable( ) because the DVDAC_Start( ) API calls this function, which is the preferred method to begin component operation.
- **void DVDAC_SaveConfig(void)**—This function saves the component configuration and non-retention registers. This function is called by the DVDAC_Sleep( ) function.
- **void DVDAC_RestoreConfig(void)**—This function restores the component configuration and non-retention registers. This function is called by the DVDAC_Wakeup( ) function.

15.25 8-Bit Voltage Digital to Analog Converter (VDA8) 1.90

The VDAC8 component is an 8-bit voltage output Digital to Analog Converter (DAC). The output range can be from 0 to 1.020 V (4 mV/bit) or from 0 to 4.08 V (16 mV/bit). The VDAC8 can be controlled by hardware, software, or a combination of both hardware and software. Voltage output ranges: 1.020-V and 4.080-V full scale. Features include: software- or clock-driven output strobe and the data source can be CPU, DMA, or Digital components [2].

15.25.1 VDA8 Functions 1.90

- **VDAC8_Start( )**—Initializes the VDAC8 with default customizer values.
- **VDAC8_Stop( )**—Disables the VDAC8 and sets it to the lowest power state.
- **VDAC8_SetSpeed( )**—Sets DAC speed.
- **VDAC8_SetValue( )**—Sets value between 0 and 255 with the given range.
- **VDAC8_SetRange( )**—Sets range to 1 or 4 volts.
- **VDAC8_Sleep( )**—Stops and saves the user configuration.
- **VDAC8_WakeUp( )**—Restores and enables the user configuration.
- **VDAC8_Init( )**—Initializes or restores default VDAC8 configuration.
- **VDAC8_Enable( )**—Enables the VDAC8.
- **VDAC8_SaveConfig( )**—Saves non-retention DAC data register value.
- **VDAC8_RestoreConfig( )**—Restores non-retention DAC data register value.

15.25.2 VDA8 Function Calls (VDA8) 1.90

- **void VDAC8_Start(void)**—This is the preferred method to begin component operation.
- **void VDAC8_Start( )**—sets the initVar variable, calls the VDAC8_Init( ) function, calls the VDAC8_Enable( ) function and powers up the VDAC8 to the given power level. A power level of 0 is the same as executing the VDAC_Stop( ) function. PatVar variable is already set, this function only calls the VDAC8_Enable( ) function.
- **void VDAC8_Stop(void)**—Powers down VDAC8 to lowest power state and disables output.
- **void VDAC8_SetSpeed(uint8 speed)**—Set DAC speed.
- **void VDAC8_SetRange(uint8 range)**—Sets range to 1 or 4 volts.
- **void VDAC8_SetValue(uint8 value)**—Sets value to output on VDAC8. valid values are between 0 and 255.
- **void VDAC8_Sleep(void)**—This is the preferred API to prepare the component for sleep. The VDAC8_Sleep( ) API saves the current component state. Then it calls the VDAC8_Stop( ) function and calls VDAC8_SaveConfig( ) to save the hardware configuration. Call the VDAC8_Sleep( ) function before calling the CyPmSleep( ) or the CyPmHibernate( ) function.
- **void VDAC8_Wakeup(void)**—This is the preferred API to restore the component to the state when VDAC8_Sleep( ) was called. The VDAC8_Wakeup( ) function calls the VDAC8_RestoreConfig( ) function to restore the configuration. If the component was enabled before the VDAC8_Sleep( ) function was called, the VDAC8_Wakeup( ) function will also re-enable the component.
- **void VDAC8_Init(void)**—Initializes or restores the component according to the customizer Configure dialog settings. It is not necessary to call VDAC8_Init( ) because the VDAC8_Start( ) API calls this function and is the preferred method to begin component operation.
- **void VDAC8_Enable(void)**—Activates the hardware and begins component operation. It is not necessary to call VDAC8_Enable( ) because the VDAC8_Start( ) API calls this function, which is the preferred method to begin component operation.
- **void VDAC8_SaveConfig(void)**—This function saves the component configuration and non-retention registers. This function will also save the current component parameter values, as defined in the Configure dialog or as modified by appropriate APIs. This function is called by the VDAC8_Sleep( ) function.[13]

---

[13]Note: In the DAC Bus mode, the values are not saved.

- **void VDAC8_RestoreConfig(void)**—This function restores the component configuration and non-retention registers. This function will also restore the component parameter values to what they were before calling the VDAC8_Sleep( ) function.[14]

15.26 8-Bit Waveform Generator (WaveDAC8) 2.10

The WaveDAC8 component provides a simple and fast solution for automatic periodic waveform generation. A high-level interface allows you to select a predefined waveform or a custom arbitrary waveform. Two separate waveforms can be defined then selected with an external pin to create a modulated output. The input clock can also be used to change the sample rate or modulate the output. This component supports standard and arbitrary waveform generation, arbitrary waveform may be drawn manually or imported from file, output that may be voltage or current, sink or source, voltage output can be buffered or direct from DAC, hardware selection between two waveforms, waveforms may be up to 4000 points and predefined sine, triangle/square/sawtooth waveforms [3].

15.26.1 WaveDAC8 Functions 2.10

- **void WaveDAC8_Start(void)**—Starts the DAC and DMA channels.
- **void WaveDAC8_Stop(void)**—Disables DAC and DMA channels.
- **void WaveDAC8_Init(void)**—Initializes or restores the component according to the customizer Configure dialog settings.
- **void WaveDAC8_Enable(void)**—Activates the hardware and begins component operation.
- **void WaveDAC8_Wave1Setup(uint8 * wavePtr, uint16 sampleSize)**—Sets the array and size of array used for waveform generation for waveform 1.
- **void WaveDAC8_Wave2Setup(uint8 * wavePtr, uint16 sampleSize)**—Sets the array and size of array used for waveform generation for waveform 2.
- **void WaveDAC8_StartEx(uint8 * wavePtr1, uint16 sampleSize1, uint8 * wavePtr2, uint16 sampleSize2)**—Sets the arrays and sizes of arrays used for waveform generation for both waveforms and starts the DAC and DMA channels.
- **void WaveDAC8_SetSpeed(uint8 speed)**—Set drive mode/speed of the DAC.
- **void WaveDAC8_SetRange(uint8 range)**—Set current or voltage range.
- **void WaveDAC8_SetValue(uint8 value)**—Set 8-bit DAC value.
- **void WaveDAC8_DacTrim(void)**—Set the trim value for the given range.
- **void WaveDAC8_Sleep(void)**—Stops and saves the user configuration.
- **void WaveDAC8_Wakeup(void)**—Restores and enables the user configuration.
- **void WaveDAC8_SaveConfig(void)**—This function saves the component configuration. This function will also save the current component parameter values, as defined in the Configure dialog or as modified by appropriate APIs. This function is called by the WaveDAC8_Sleep( ) function.
- **void WaveDAC8_RestoreConfig(void)**—This function restores the component configuration. This function will also restore the component parameter values to what they were before calling the WaveDAC8_Sleep( ) function.

---

[14]In the DAC Bus mode, the values are not restored.

15.26.2 WaveDAC8 Function Calls 2.10

- **void WaveDAC8_Start(void)**—Performs all the required initialization for the component and enables power to the block. The first time the routine is executed, the range, polarity if any and power, i.e., speed settings are configured for the operating mode selected in the design. When called to restart the WaveDAC8 following a WaveDAC8_Stop( ) call, the current component parameter settings are retained. When the external clock is used this function should be called before the clock is started to cause correct waveform generation. Otherwise the first sample may be undefined.
- **void WaveDAC8_Stop(void)**—Turn off the WaveDAC8 block. Does not affect WaveDAC8 type or power settings.
- **void WaveDAC8_Wave1Setup(uint8 *WavePtr, uint16 SampleSize)**—Selects a new waveform array for the waveform 1 output. The WaveDAC8_Stop function should be called prior to calling this function and WaveDAC8_Start should be called to restart waveform.
- **void WaveDAC8_Wave2Setup(uint8 *WavePtr, uint16 SampleSize)**—Select a new waveform array for the waveform 2 output. The WaveDAC8_Stop function should be called prior to calling this function and WaveDAC8_Start should be called to restart waveform.
- **void WaveDAC8_StartEx(uint8 *WavePtr1, uint16 SampleSize1, uint8 *WavePtr2, uint16 SampleSize2)**—Select new waveform arrays for both waveform outputs and starts the WaveDAC8. The WaveDAC8_Stop function should be called prior to calling this function.
- **void WaveDAC8_Init(void)**—Initializes or restores the component according to the customizer Configure dialog settings. It is not necessary to call WaveDAC8_Init( ) because the WaveDAC8_Start( ) API calls this function and is the preferred method to begin component operation.
- **void WaveDAC8_Enable(void)**—Activates the hardware and begins component operation. It is not necessary to call WaveDAC8_Enable( ) because the WaveDAC8_Start( ) API calls this function, which is the preferred method to begin component operation.
- **void WaveDAC8_SetSpeed(uint8 speed)**—Sets the drive mode/speed to one of the settings.
- **void WaveDAC8_Enable(void)**—Activates the hardware and begins component operation. It is not necessary to call WaveDAC8_Enable( ) because the WaveDAC8_Start( ) API calls this function, which is the preferred method to begin component operation.
- **void Waved ac8_SetSpeed(uint8 speed)**—Sets the drive mode speed to one of the settings.
- **void WaveDAC8_SetValue(uint8 value)**—Sets the output of the DAC to the desired value. It is preferable to use this function when the clock is stopped. If this function is used during normal operation (clock is running), the predefined waveform may be interrupted.
- **void WaveDAC8_SetRange (uint8 range)**—Sets the DAC range to one of the settings.
- **void WaveDAC8_DacTrim(void)**—Sets the proper predefined trim calibration value for the present DAC mode and range.
- **void WaveDAC8_Sleep(void)**—This is the preferred API to prepare the component for sleep. The WaveDAC8_Sleep( ) API saves the current component state. Then it calls the WaveDAC8_Stop( ) function and calls WaveDAC8_SaveConfig( ) to save the hardware configuration. Call the WaveDAC8_Sleep( ) function before calling the CyPmSleep( ) or the CyPmHibernate( ) function.
- **void WaveDAC8_Wakeup(void)**—This is the preferred API to restore the component to the state when WaveDAC8_Sleep( ) was called. The WaveDAC8_Wakeup( ) function calls the WaveDAC8_RestoreConfig( ) function to restore the configuration. If the component was enabled before the WaveDAC8_Sleep( ) function was called, the WaveDAC8_Wakeup( ) function will also re-enable the component.

- **void WaveDAC8_SaveConfig(void)**—saves the component configuration. This function will also save the current component parameter values, as defined in the Configure dialog or as modified by appropriate APIs. This function is called by the WaveDAC8_Sleep( ) function.
- **void WaveDAC8_RestoreConfig(void)**—restores the component configuration. This function will also restore the component parameter values to what they were before calling the WaveDAC8_Sleep( ) function.

15.27   Analog Mux Constraint 1.50

Routing is strict. All of the devices connected to the net with the resource constraint must have a direct hardware connection to the resource. If the resources do not have a hardware connection to the specified constraint, an error will occur [9].

15.27.1 Analog Mux Constraint Functions 1.50

NONE.

15.28   Net Tie 1.50

The Net Tie Component connects two analog routes to each other. Each of the routes may have a different analog resource constraint. It is used to split an analog route for fine-grained control of analog, e.g., when one or both of the signals connected to the Net Tie Component have an Analog Constraint. Features include: connects two analog routes, a constrained analog route with an unconstrained analog route or two analog routes with different routing resource constraints [62].

15.28.1 Net Tie 1.50

NONE.

15.29   Analog Net Constraint 1.50

The Analog Net Constraint Component allows you to define the route of the analog signal to which it is connected. This is an advanced feature that is not needed for most designs, and should be used with caution. Features include: limits analog routing of a signal to a specific routing resource and all terminals on the signal must connect directly to the routing resource[15] [10].

15.29.1 Analog Net Constraint Functions 1.50

NONE.

---

[15]Routing is strict. All of the devices connected to the net with the resource constraint must have a direct hardware connection to the resource. If the resources do not have a hardware connection to the specified constraint, an error will occur.

15.30 Analog Resource Reserve 1.50

The Analog Resource Reserve component reserves a global analog routing resource so that the resource can be safely used by firmware-based manual analog routing. This is an advanced feature that is not needed for most designs, and should be used with caution. The Analog Resource Reserve component is used when the firmware intends to modify analog routing registers. The Analog Resource Reserve component protects against conflicting use of analog resources by firmware and automatic analog routing. Features include: prevents an analog router from using a global analog routing resource, and allows safe firmware access to a global analog routing resource [11].

15.30.1 Analog Resource Reserve Functions 1.50

NONE.

15.31 Stay Awake 1.50

On certain devices, to avoid signal integrity issues, specific analog blocks disconnect their terminals when the device goes to sleep. This also disconnects any routes (static or dynamic) that use the block terminal as a via. Routes that must stay awake during device sleep are identified by using the Stay Awake component, which has a single connection and no parameters. The net to which the Stay Awake component is attached is routed without using the affected analog block terminals. The Stay Awake component is used to establish a route where a net needs to remain connected while the device is asleep or hibernating [90].

15.31.1 Stay Awake Functions 1.50

NONE.

15.32 Terminal Reserve 1.50

The Terminal Reserve component reserves the analog routing resource connected to a component, such as the analog wire connected to a comparator or pin. This is an advanced feature that is not needed for most designs, and should be used with caution. When to Use a Terminal Reserve The Terminal Reserve component is used when user firmware modifies the analog routing registers that connect to the specified terminal. The Terminal Reserve component protects against conflicting use of analog resources by user firmware and automatic analog routing. Features include: prevents an analog router from using an analog block terminal routing resource and allows safe firmware access to an analog block terminal routing resource [92].

15.33 Terminal Reserve Functions 1.50

NONE.

15.34 Voltage Reference (Vref) 1.70

This component is a provides a voltage reference for analog blocks and provides Routable bandgap stable precision voltage references, viz., 0.256 and 1.024 volts for Vdda, Vssa, Vccd, Vddd and Vbat. Multiple instances of this component can be used in a given PSoC application. However, it should be noted that this voltage reference is not intended to source or sink current. If the intended usage is to drive a signal, a buffer must be used, e.g., an OpAmp, to provide the necessary source/sink current [106].

15.35 Capacitive Sensing (CapSense CSD) 3.5

Capacitive Sensing, using a Delta-Sigma Modulator (CapSense CSD) component, is a versatile and efficient way to measure capacitance in applications such as touch sense buttons, sliders, touchpad and proximity detection. The following references are available on the Cypress website: Started with CapSense and PSoC 3 and PSoC 5LP CapSense® Design Guide. Features include: Support for user-defined combinations of button, slider, touchpad and proximity capacitive sensors, automatic SmartSense™ tuning or manual tuning with integrated PC GUI, high immunity to AC power line noise, EMC noise and power supply voltage changes, optional two scan channels (parallel synchronized), which increases sensor scan rate and shield electrode support for reliable operation in the presence of water film or droplets [15].

15.35.1 Capacitive Sensing Functions 3.5

- **CapSense_Start( )**—Preferred method to start the component. Initializes registers and enables active mode power template bits of the subcomponents used within
- **CapSense_Stop( )**—Disables component interrupts and calls CapSense_ClearSensors( ) to reset all sensors to an inactive state.
- **CapSense_Sleep( )**—Prepares the component for the device entering a low-power mode. Disables Active mode power template bits of the sub-components used within CapSense, saves non-retention registers and resets all sensors to an inactive state.
- **CapSense_Wakeup( )**—Restores CapSense configuration and non-retention register values after the device wake from a low power mode sleep mode.
- **CapSense_Init( )**—Initializes the default CapSense configuration provided with the customizer.
- **CapSense_Enable( )**—Enables the Active mode power template bits of the subcomponents used within CapSense.
- **CapSense_SaveConfig( )**—Saves the configuration of CapSense non-retention registers. Resets all sensors to an inactive state.
- **CapSense_RestoreConfig( )**—Restores CapSense configuration and non-retention register values.

15.35.2 Capacitive Sensing Function Calls (CapSense CSD) 3.5

- **void CapSense_Start(void)**—This is the preferred method to begin component operation. CapSense_Start( ) calls the CapSense_Init( ) function and then calls the CapSense_Enable( ) function. Initializes registers and starts the CSD method of the CapSense component. Resets all sensors to an inactive state. Enables interrupts for sensors scanning. When SmartSense tuning mode is selected, the tuning procedure is applied for all sensors. The CapSense_Start( ) routine must be called before any other API routines.
- **void CapSense_Stop(void)**—Stops the sensor scanning, disables component interrupts and resets all sensors to an inactive state. Disables Active mode power template bits for the subcomponents used within CapSense.
- **void CapSense_Stop(void)**—Stops the sensor scanning, disables component interrupts and resets all sensors to an inactive state. Disables Active mode power template bits for the subcomponents used within CapSense.
- **void CapSense_Sleep(void)**—This is the preferred method to prepare the component for device low-power modes. Disables Active mode power template bits for the subcomponents used within CapSense. Calls CapSense_SaveConfig( ) function to save customer configuration of CapSense non-retention registers and resets all sensors to an inactive state.
- **void CapSense_Wakeup(void)**—Restores the CapSense configuration and non-retention register values. Restores the enabled state of the component by setting Active mode power template bits for the subcomponents used within CapSense.
- **void CapSense_Init(void)**—Initializes the default CapSense configuration provided by the customizer that defines component operation. Resets all sensors to an inactive state.
- **void CapSense_Enable(void)**—Enables Active mode power template bits for the subcomponents used within CapSense.
- **void CapSense_SaveConfig(void)**—Saves the configuration of CapSense non-retention registers. Resets all sensors to an inactive state.
- **void CapSense_RestoreConfig(void)**—Restores CapSense configuration and non-retention registers.

15.35.3 Capacitive Sensing Scanning Specific APIs 3.50

These API functions are used to implement CapSense sensor scanning.

- **CapSense_ScanSensor( )**—Sets scan settings and starts scanning a sensor or group of combined sensors on each channel.
- **CapSense_ScanEnabledWidgets( )**—The preferred scanning method. Scans all the enabled widgets.
- **CapSense_IsBusy( )**—Returns the status of sensor scanning.
- **CapSense_SetScanSlotSettings( )**—Sets the scan settings of the selected scan slot (sensor or pair of sensors).
- **CapSense_ClearSensors( )**—Resets all sensors to the non-sampling state.
- **CapSense_EnableSensor( )**—Configures the selected sensor to be scanned during the next scanning cycle.
- **CapSense_DisableSensor( )**—Disables the selected sensor so it is not scanned in the next scanning cycle.
- **CapSense_ReadSensorRaw( )**—Returns sensor raw data from the CapSense_sensorRaw[ ] array.

- **CapSense_SetRBleed( )**—Sets the pin to use for the bleed resistor (Rb) connection if multiple bleed resistors are used.

15.35.4 Capacitive Sensing API Function Calls 3.50

- **void CapSense_ScanSensor(uint8 sensor)**—Sets scan settings and starts scanning a sensor or pair of sensors on each channel. If two channels are configured, two sensors can be scanned at the same time. After scanning is complete, the ISR copies the measured sensor raw data to the global raw sensor array. Use of the ISR ensures this function is non-blocking. Each sensor has a unique number within the sensor array. This number is assigned by the CapSense customizer in sequence.
- **void CapSense_ScanEnabledWidgets(void)**—This is the preferred method to scan all the enabled widgets. Starts scanning a sensor or pair of sensors within the enabled widgets. The ISR continues scanning sensors until all enabled widgets are scanned. Use of the ISR ensures this function is non-blocking.All widgets are enabled by default except proximity widgets. Proximity widgets must be manually enabled as their long scan time is incompatible with the fast response required of other widget types.
- **uint8 CapSense_IsBusy (void)**—Returns the status of sensor scanning.
- **void CapSense_SetScanSlotSettings(uint8 slot)**—Sets the scan settings provided in the customizer or wizard of the selected scan slot (sensor or pair of sensors for a two-channel design). The scan settings provide an IDAC value (for IDAC configurations) for every sensor, as well as resolution. The resolution is the same for all sensors within a widget.
- **void CapSense_ClearSensors(void)**—Resets all sensors to the nonsampling state by sequentially disconnecting all sensors from the Analog MUX Bus and connecting them to the inactive state.
- **void CapSense_EnableSensor(uint8 sensor)**—Configures the selected sensor to be scanned during the next measurement cycle. The corresponding pins are set to Analog HI-Z mode and connected to the Analog Mux Bus. This also affects the comparator output.
- **void CapSense_DisableSensor(uint8 sensor)**—Disables the selected sensor. The corresponding pins are disconnected from the Analog Mux Bus and put into the inactive state.
- **uint16 CapSense_ReadSensorRaw(uint8 sensor)**—Returns sensor raw data from the global CapSense_sensorRaw[ ] array. Each scan sensor has a unique number within the sensor array. This number is assigned by the CapSense customizer in sequence. Raw data can be used to perform calculations outside the CapSense provided framework.
- **void CapSense_SetRBleed(uint8 rbleed)**—Sets the pin to use for the bleed resistor (Rb) connection. This function can be called at run time to select the current Rb pin setting from those defined in the customizer. The function overwrites the component parameter setting. This function is available only if Current Source is set to External Resistor. This function is effective when some sensors need to be scanned with different bleed resistor values. For example, regular buttons can be scanned with a lower value of bleed resistor. The proximity detector can be scanned less often with a larger bleed resistor to maximize proximity detection distance. This function can be used in conjunction with the CapSense_ScanSensor( ) function.

15.35.5 Capacitive Sensing High-Level APIs 3.50

- **CapSense_InitializeSensorBaseline(   )**—Loads   the   CapSense_sensorBaseline [sensor] array element with an initial value by scanning the selected sensor.

- **CapSense_InitializeEnabledBaselines( )**—Loads the CapSense_sensorBaseline[] array with initial values by scanning enabled sensors only. This function is available only for two-channel designs.
- **CapSense_InitializeAllBaselines( )**—Loads the CapSense_sensorBaseline[] array with initial values by scanning all sensors.
- **CapSense_UpdateSensorBaseline( )**—The historical count value, calculated independently for each sensor, is called the sensor's baseline. This baseline updated uses a low-pass filter with k = 256.
- **CapSense_UpdateEnabledBaselines( )**—Checks the CapSense_sensorEnableMask[]array and calls the CapSense_UpdateSensorBaseline( ) function to update the baselines for enabled sensors.
- **CapSense_EnableWidget( )**—Enables all sensor elements in a widget for the scanning process.
- **CapSense_DisableWidget( )**—Disables all sensor elements in a widget from the scanning process. **CapSense_CheckIsWidgetActive( )**—Compares the selected widget to the CapSense_signal[] array to determine if it has a finger-press.
- **CapSense_CheckIsAnyWidgetActive( )**—Uses the CapSense_CheckIsWidgetActive( ) function to find if any widget of the CapSense CSD component is in active state.
- **CapSense_GetCentroidPos( )**—Checks the CapSense_signal[] array for a finger press in a linear slider and returns the position.
- **CapSense_GetRadialCentroidPos( )**—Checks the CapSense_signal[] array for a finger press in a radial slider widget and returns the position.
- **CapSense_GetTouchCentroidPos( )**—If a finger is present, this function calculates the X and Y position of the finger by calculating the centroids within the touchpad.
- **CapSense_GetMatrixButtonPos( )**—If a finger is present, this function calculates the row and column position of the finger on the matrix buttons.

15.35.6 Capacitive Sensing Hi-Level Function Calls 3.50

- **void CapSense_InitializeSensorBaseline(uint8 sensor)**—Loads the CapSense_sensorBaseline [sensor] array element with an initial value by scanning the selected sensor (one-channel design) or a pair of sensors (two-channel design). The raw count value is copied into the baseline array for each sensor. The raw data filters are initialized if enabled.
- **void CapSense_InitializeEnabledBaselines(void)**—Scans all enabled widgets. The raw count values are copied into the CapSense_sensorBaseline[ ] array for all sensors enabled in scanning process. Initializes CapSense_sensorBaseline[ ] with zero values for sensors disabled from the scanning process. The raw data filters are initialized if enabled. This function is available only for two-channel designs.
- **void CapSense_InitializeAllBaselines(void)**—Uses the CapSense_InitializeSensorBaseline( ) function to load the CapSense_sensorBaseline[ ] array with initial values by scanning all sensors. The raw count values are copied into the baseline array for all sensors. The raw data filters are initialized if enabled.
- **void CapSense_UpdateSensorBaseline(uint8 sensor)**—The sensor's baseline is a historical count value, calculated independently for each sensor. Updates the CapSense_sensorBaseline [sensor] array element using a low-pass filter with k = 256. The function calculates the difference count by subtracting the previous baseline from the current raw count value and stores it in CapSense_signal[sensor]. If the auto reset option is enabled, the baseline updates independent of the noise threshold. If the auto reset option is disabled, the baseline stops updating if the signal is

greater than the noise threshold and resets the baseline when the signal is less than the minus noise threshold. Raw data filters are applied to the values if enabled before baseline calculation.

- **void CapSense_UpdateEnabledBaselines(void)**—Checks the CapSense_sensorEnableMask[] array and calls the CapSense_UpdateSensorBaseline( ) function to update the baselines for all enabled sensors.
- **void CapSense_EnableWidget(uint8 widget)**—Enables the selected widget sensors to be part of the scanning process.
- **void CapSense_DisableWidget(uint8 widget)**—Disables the selected widget sensors from the scanning process.
- **uint8 CapSense_CheckIsWidgetActive(uint8 widget)**—Compares the selected sensor CapSense_signal[ ] array value to its finger threshold. Hysteresis and debounce are considered. Hysteresis and debounce are considered. If the sensor is active, the threshold is lowered by the hysteresis amount. If it is inactive, the threshold is increased by the hysteresis amount. If the active threshold is met, the debounce counter increments by one until reaching the sensor active transition, at which point this API sets the widget as active. This function also updates the sensor's bit in the CapSense_sensorOnMask[ ] array. The touchpad and matrix buttons widgets need to have active sensor within col and row to return widget active status.
- **uint8 CapSense_CheckIsAnyWidgetActive(void)**—Compares all sensors of the CapSense _signal[ ] array to their finger threshold. Calls Capsense_CheckIsWidgetActive( ) for each widget so that the CapSense_sensorOnMask[ ] array is up to date after calling this function.
- **uint16 CapSense_GetCentroidPos(uint8 widget)**—Checks the CapSense_signal[ ] array for a finger press within a linear slider. The finger position is calculated to the API resolution specified in the CapSense customizer. A position filter is applied to the result if enabled. This function is available only if a linear slider widget is defined by the CapSense customizer.
- **uint16 CapSense_GetRadialCentroidPos(uint8 widget)**—A position filter is applied to the result if enabled. This function is available only if a radial slider widget is defined by the CapSense customizer.Checks the CapSense_signal[ ] array for a finger press within a radial slider. The finger position is calculated to the API resolution specified in the CapSense customizer.
- **uint8 CapSense_GetTouchCentroidPos(uint8 widget, uint16* pos)**—If a finger is present on touchpad, this function calculates the X and Y position of the finger by calculating the centroids within the touchpad sensors. The X and Y positions are calculated to the API resolutions set in the CapSense customizer. Returns a "1" if a finger is on the touchpad. A position filter is applied to the result if enabled. This function is available only if a touchpad is defined by the CapSense customizer.
- **uint8 CapSense_GetMatrixButtonPos(uint8 widget, uint8* pos)**—If a finger is present on matrix buttons, this function calculates the row and column position of the finger. Returns a "1" if a finger is on the matrix buttons. This function is available only if a matrix buttons are defined by the CapSense customizer.
- **void CapSense_TunerStart(void)**—Initializes the CapSense CSD component and EZI2C component. Also initializes baselines and starts the sensor scanning loop with the currently enabled sensors. All widgets are enabled by default except proximity widgets. Proximity widgets must be manually enabled as their long scan time is incompatible with the fast response required of other widget types.
- **void CapSense_TunerComm(void)**—Executes communication functions with Tuner GUI. Manual mode: Transfers sensor scanning and widget processing results to the Tuner GUI from the CapSense CSD component. Reads new parameters from Tuner GUI and apply them to the CapSense CSD component. Auto (SmartSense): Executes communication functions with Tuner GUI. Transfer sensor scanning and widget processing results to Tuner GUI. The auto tuning

parameters also transfer to Tuner GUI. Tuner GUI parameters are not transferred back to the CapSense CSD component. This function is blocking and waits while the Tuner GUI modifies CapSense CSD component buffers to allow new data.

- **void CapSense_SetAllRbsDriveMode(uint8 mode)**—Sets the drive mode for all pins used by bleed resistors (Rb) within the CapSense component. Only available when Current Source is set to External Resistor.
- **void CapSense_SetAllSensorsDriveMode(uint8 mode)**—Sets the drive mode for all pins used by capacitive sensors within the CapSense component.
- **void CapSense_SetAllCmodsDriveMode(uint8 mode)**—Sets the drive mode for all pins used by CMOD capacitors within the CapSense component.
- **void CapSense_SetAllRbsDriveMode(uint8 mode)**—Sets the drive mode for all pins used by bleed resistors (Rb) within the CapSense component. Only available when Current Source is set to External Resistor.

15.36 File System Library (emFile) 1.20

The emFile component provides an interface to SD cards formatted with a FAT file system. The SD card specification includes multiple hardware interface options for communication with an SD card. This component uses the SPI interface method for communication. Up to four independent SPI interfaces can be used for communication with one SD card each. Both FAT12/16 and FAT32 file system formats are supported. This component provides the physical interface to the SD card and works with the emFile library licensed from SEGGER Microcontroller to provide a library of functions to manipulate a FAT file system. Features include: Up to four Secure Digital (SD) cards in SPI mode, FAT12/16 or FAT32 format, optional integration with an Operating System (OS) and Long File Name (LFN) handling [43].

15.36.1 File System Library Functions 1.20

- **emFile_Sleep( )**—Prepares emFile to enter sleep mode.
- **emFile_Wakeup( )**—Restores emFile after coming out of sleep mode.
- **emFile_SaveConfig( )**—Saves the SPI Master configuration used by the HW driver.
- **emFile_RestoreConfig( )**—Restores the SPI Master configuration used by the HW Driver.

15.36.2 File System Library Function Calls 1.20

- **void emFile_Sleep(void)**—Prepares emFile to go to sleep.
- **void emFile_Wakeup(void)**—Restores emFile after coming out of sleep mode.[16]
- **void emFile_SaveConfig(void)**—Saves the SPI Master configuration used by the HW driver. This function is called by emFile_Sleep( ).
- **void emFile_RestoreConfig(void)**—Restores the SPI Master, used by the HW Driver.[17]

---

[16]Calling the emFile_Wakeup( ) function without first calling the emFile_Sleep( ) or emFile_SaveConfig( ) function may produce unexpected behavior.

[17]Calling this function without first calling the emFile_Sleep( ) or emFile_SaveConfig( ) function may produce unexpected behavior.

15.37 EZI2C Slave 2.00

The EZI2C Slave component implements an I2C register-based slave device. It is compatible with I2C Standard-mode, Fast-mode and Fast-mode Plus devices as defined in the NXP I2C-bus specification. The master initiates all communication on the I2C bus and supplies the clock for all slave devices. The EZI2C Slave supports standard data rates up to 1000 kbps and is compatible[18] with multiple devices on the same bus. The EZI2C Slave is a unique implementation of an I2C slave in that all communication between the master and slave is handled in the ISR (Interrupt Service Routine) and requires no interaction with the main program flow. The interface appears as shared memory between the master and slave. Once the EZI2C_Start( ) function is executed, there is little need to interact with the API. Features include: Industry standard NXP® I2C bus interface, emulates common I2C EEPROM interface, only two pins (SDA and SCL) required to interface to I2C bus, standard data rates of 50/100/400/1000 kbps, high level APIs require minimal user programming, supports one or two address decoding with independent memory buffers and memory buffers provide configurable Read/Write and Read-Only regions [41].

15.37.1 EZI2C Slave Functions 2.00

- **EZI2C_Stop( )**—Stops responding to I2C traffic. Disables interrupt.
- **EZI2C_EnableInt( )**—Enables component interrupt, which is required for most component operations.
- **EZI2C_DisableInt( )**—Disables component interrupt. The EZI2C_Stop( ) API does this automatically.
- **EZI2C_SetAddress1( )**—Sets the primary I2C slave address.
- **EZI2C_GetAddress1( )**—Returns the primary I2C slave address.
- **EZI2C_SetBuffer1( )**—Sets up the data buffer to be exposed to the master on a primary slave address request.
- **EZI2C_GetActivity( )**—Checks component activity status.
- **EZI2C_Sleep( )**—Stops I2C operation and saves I2C configuration. Disables component interrupt.
- **EZI2C_Wakeup( )**—Restores I2C configuration and starts I2C operation. Enables component interrupt.
- **EZI2C_Init( )**—Initializes I2C registers with initial values provided from customizer.
- **EZI2C_Enable( )**—Activates the hardware and begins component operation.
- **EZI2C_SaveConfig( )**—Saves the current user configuration of the EZI2C component.
- **EZI2C_RestoreConfig( )**—Restores non-retention I2C registers.

15.37.2 EZI2C Slave Function Calls 2.00

- **void EZI2C_Start(void)**—This is the preferred method to begin component operation. EZI2C_Start( ), calls the EZI2C_Init( ) function and then calls the EZI2C_Enable( ) function.

---

[18]The I2C peripheral is non-compliant with the NXP I2C specification in the following areas: analog glitch filter, I/O VOL/IOL, I/O hysteresis. The I2C Block has a digital glitch filter (not available in sleep mode). The Fast-mode minimum fall-time specification can be met by setting the I/Os to slow speed mode. See the I/O Electrical Specifications in "Inputs and Outputs" section of device, cf. the datasheet for details.

It must be executed before I2C bus operation. This function enables the component interrupt because interrupt is required for most component operations.

- **void EZI2C_Stop(void)**—Disables I2C hardware and component interrupt. The I2C bus is released, if it was locked up by the component.
- **void EZI2C_EnableInt(void)**—Enables component interrupt. Interrupts are required for most operations. Called inside EZI2C_Start( ) function.
- **void EZI2C_DisableInt(void)**—Disables component interrupt. This function is not normally required because the EZI2C_Stop( ) function calls this function.
- **void EZI2C_SetAddress1(uint8 address)**—Sets the primary I2C slave address. This address is used by the master to access the primary data buffer.
- **uint8 EZI2C_GetAddress1(void)**—Returns the primary I2C slave address. This address is the 7-bit right-justified slave address and does not include the R/W bit.
- **void EZI2C_SetBuffer1(uint16 bufSize, uint16 rwBoundary, volatile uint8* dataPtr)**—Sets up the data buffer to be exposed to the master on a primary slave address request.
- **uint8 EZI2C_GetActivity(void)**—Returns a non-zero value if an I2C read or write cycle has occurred since the last time this function was called. The activity flag resets to zero at the end of this function call. The Read and Write busy flags are cleared when read, but the BUSY flag is only cleared when slave is free that is, the master finishes communication with the slave generating Stop or repeated Start condition.
- **void EZI2C_Sleep(void)**—This is the preferred method to prepare the component before device enters sleep mode. The Enable wakeup from Sleep Mode selection influences this function implementation:
  1. **Unchecked:** Checks current EZI2C component state, saves it and disables the component by calling EZI2C_Stop( ) if it is currently enabled. EZI2C_SaveConfig( ) is then called to save the component non-retention configuration registers.
  2. **Checked:** If a transaction intended for component is in progress during this function call, it waits until the current transaction is completed. All subsequent I2C traffic intended for component is NAKed until the device is put to sleep mode. The address match event wakes up the device. Call the EZI2C_Sleep( ) function before calling the CyPmSleep( ) or the CyPmHibernate( ) function.
- **void EZI2C_Wakeup(void)**—This is the preferred method to prepare the component for active mode operation (when device exits sleep mode). The Error! Reference source not found. selection influences this function implementation:
  1. **Unchecked:** Restores the component non-retention configuration registers by calling EZI2C_RestoreConfig( ). If the component was enabled before the EZI2C_Sleep( ) function was called, EZI2C_Wakeup( ) re-enables it.
  2. **Checked:** Disables the backup regulator of I2C hardware. The incoming transaction continues as soon as the regular EZI2C interrupt handler is set up (global interrupts has to be enabled to service EZI2C component interrupt).
- **void EZI2C_Init(void)**—Initializes or restores the component according to the Configure dialog settings. It is not necessary to call EZI2C_Init( ) because the EZI2C_Start( ) API calls this function, which is the preferred method to begin component operation.
- **void EZI2C_Enable(void)**—Activates the hardware and begins component operation. Calls EZI2C_EnableInt( ) to enable the component interrupt. It is not necessary to call EZI2C_Enable( ) because the EZI2C_Start( ) API calls this function, which is the preferred method to begin component operation.
- **void EZI2C_SaveConfig(void)**—The Error! Reference source not found. Selection influences this function implementation:

1. **Unchecked:** Stores the component non-retention configuration registers.
2. **Checked:** Enables backup regulator of the I2C hardware. If a transaction intended for component executes during this function call, it waits until the current transaction is completed and I2C hardware is ready to enter sleep mode. All subsequent I2C traffic is NAKed until the device is put into sleep mode.

- **void EZI2C_RestoreConfig(void)**—The Error! Reference source not found. Selection influences this function implementation:
  1. **Unchecked:** Restores the component non-retention configuration registers to the state they were in before I2C_Sleep( ) or I2C_SaveConfig( ) was called.
  2. **Checked:** Disables the backup regulator of the I2C hardware. Sets up the regular component interrupt handler and generates the component interrupt if it was wake up source to release the bus and continue in-coming I2C transaction.
- **uint8 EZI2C_GetAddress2(void)**—Returns the secondary I2C slave address. This address is the 7-bit right-justified slave address and does not include the R/W bit.
- **void EZI2C_SetAddress2(uint8 address)**—Sets the secondary I2C slave address. This address is used by the master to access the secondary data buffer.
- **void EZI2_SetBuffer2(uint16 bufSize, uint8*dataPtr)**—Sets up the data buffer to be exposed to the master on a secondary slave address request.

15.38   I2C Master/Multi-Master/Slave 3.5

The I2C component supports I2C slave, master and multi-master configurations. The I2C bus is an industry-standard, two-wire hardware interface developed by Philips. The master initiates all communication on the I2C bus and supplies the clock for all slave devices. The I2C component supports standard clock speeds up to 1000 kbps. It is compatible[19] with I2C Standard-mode, Fast-mode and Fast-mode Plus devices as defined in the NXP I2C-bus specification. The I2C component is compatible with other third-party slave and master devices.[20] Features include: Industry-standard NXP® I2C bus interface, supports slave, master, multi-master and multi-master-slave operation, requires only two pins (SDA and SCL) to interface to I2C bus, supports standard data rates of 100/400/1000 kbps and high-level APIs require minimal user programming [52].

15.38.1   I2C Master/Multi-Master/Slave Functions 3.5

Generic Functions
- **I2C_Start( )**—Initializes and enables the I2C component. The I2C interrupt is enabled and the component can respond to I2C traffic.
- **I2C_Stop( )**—Stops responding to I2C traffic (disables the I2C interrupt).
- **I2C_EnableInt( )**—Enables interrupt, which is required for most I2C operations.
- **I2C_DisableInt( )**—Disables interrupt. The I2C_Stop( ) API does this automatically.

---

[19]The I2C peripheral is non-compliant with the NXP I2C specification in the following areas: analog glitch filter, I/O VOL/IOL, I/O hysteresis. The I2C Block has a digital glitch filter (not available in sleep mode). The Fast-mode minimum fall-time specification can be met by setting the I/Os to slow speed mode. See the I/O Electrical Specifications in "Inputs and Outputs" section of device datasheet for details.

[20]The component datasheet covers both the fixed hardware I2C block and the UDB version.

- **I2C_Sleep( )**—Stops I2C operation and saves I2C non-retention configuration registers (disables the interrupt). Prepares wake on address match operation if Wakeup from Sleep Mode is enabled (disables the I2C interrupt).
- **I2C_Wakeup( )**—Restores I2C non-retention configuration registers and enables I2C operation (enables the I2C interrupt).
- **I2C_Init( )**—Initializes I2C registers with initial values provided from the customizer.
- **I2C_Enable( )**—Activates I2C hardware and begins component operation.
- **I2C_SaveConfig( )**—Saves I2C non-retention configuration registers (disables the I2C interrupt).
- **I2C_RestoreConfig( )**—Restores I2C non-retention configuration registers.

15.38.2 I2C Master/Multi-Master/Slave Function Calls 3.5

- **void I2C_Start(void)**—This is the preferred method to begin component operation. I2C_Start( ) calls the I2C_Init( ) function and then calls the I2C_Enable( ) function. I2C_Start( ) must be called before I2C bus operation. This API enables the I2C interrupt. Interrupts are required for most I2C operations. You must set up the I2C Slave buffers before this function call to avoid reading or writing partial data while the buffers are setting up. I2C slave behavior is as follows when enabled and buffers are not set up: I2C Read transfer Returns 0xFF until the read buffer is set up. Use the I2C_SlaveInitReadBuf( ) function to set up the read buffer. I2C Write transfer Send NAK because there is no place to store received data. Use the I2C_SlaveInitWriteBuf( ) function to set up the read buffer.
- **void I2C_Stop(void)**—This function disables I2C hardware and component interrupt. Releases the I2C bus if it was locked up by the device and sets it to the idle state.
- **void I2C_EnableInt(void)**—This function enables the I2C interrupt. Interrupts are required for most operations.
- **void I2C_DisableInt(void)**—This function disables the I2C interrupt. This function is not normally required because the I2C_Stop( ) function disables the interrupt.
- **void I2C_Sleep(void)**—This is the preferred method to prepare the component before device enters sleep mode. The Enable wakeup from Sleep Mode selection influences this function implementation:
  1. **Unchecked:** Checks current I2C component state, saves it and disables the component by calling I2C_Stop( ) if it is currently enabled. I2C_SaveConfig( ) is then called to save the component non-retention configuration registers.
  2. **Checked:** If a transaction intended for component executes during this function call, it waits until the current transaction is completed. All subsequent I2C traffic intended for component is NAKed until the device is put to sleep mode. The address match event wakes up the device. Call the I2C_Sleep( ) function before calling the CyPmSleep( ) or the CyPmHibernate( ) function.
- **void I2C_Init(void)**—This function initializes or restores the component according to the customizer Configure dialog settings. It is not necessary to call I2C_Init( ) because the I2C_Start( ) API calls this function, which is the preferred method to begin component operation.
- **void I2C_Enable(void)**—This function activates the hardware and begins component operation. It is not necessary to call I2C_Enable( ) because the I2C_Start( ) API calls this function, which is the preferred method to begin component operation. If this API is called, I2C_Start( ) or I2C_Init( ) must be called first.
- **void I2C_SaveConfig(void)**—The Enable wakeup from Sleep Mode selection influences this function implementation:
  1. **Unchecked:** Stores the component non-retention configuration registers.

2. **Checked:** Disables the master, if it was enabled before and enables backup regulator of the I2C hardware. If a transaction intended for component executes during this function call, it waits until the current transaction is completed and I2C hardware is ready to enter sleep mode. All subsequent I2C traffic is NAKed until the device is put into sleep mode.

- **void I2C_RestoreConfig(void)**—The Enable wakeup from Sleep Mode selection influences this function implementation:
  1. **Unchecked:** Restores the component non-retention configuration registers to the state they were in before $I2C_Sleep( )$ or $I2C_SaveConfig( )$ was called.
  2. **Checked:** Enables master functionality, if it was enabled before and disables the backup regulator of the I2C hardware. Sets up the regular component interrupt handler and generates the component interrupt if it was wake up source to release the bus and continue in-coming I2C transaction.

15.38.3 Slave Functions 3.50

- **I2C_SlaveStatus( )**—Returns the slave status flags.
- **I2C_SlaveClearReadStatus( )**—Returns the read status flags and clears the slave read status flags.
- **I2C_SlaveClearWriteStatus( )**—Returns the write status and clears the slave write status flags.
- **I2C_SlaveSetAddress( )**—Sets the slave address, a value between 0 and 127 (0x00 to 0x7F).
- **I2C_SlaveInitReadBuf( )**—Sets up the slave receive data buffer. (master <- slave)
- **I2C_SlaveInitWriteBuf( )** Sets up the slave write buffer. (master -> slave)
- **I2C_SlaveGetReadBufSize( )**—Returns the number of bytes read by the master since the buffer was reset.
- **I2C_SlaveGetWriteBufSize( )**—Returns the number of bytes written by the master since the buffer was reset.
- **I2C_SlaveClearReadBuf( )**—Resets the read buffer counter to zero.
- **I2C_SlaveClearWriteBuf( )**—Resets the write buffer counter to zero.

15.38.4 I2C Master/Multi-Master/Slave Function Calls 3.50

- **uint8 I2C_SlaveStatus(void)**—This function returns the slave's communication status.
- **uint8 I2C_SlaveClearReadStatus(void)**—This function clears the read status flags and returns their values. The I2C_SSTAT_RD_BUSY flag is not affected by this function call.
- **uint8 I2C_SlaveClearWriteStatus(void)**—This function clears the write status flags and returns their values. The I2C_SSTAT_WR_BUSY flag is not affected by this function call.
- **void I2C_SlaveSetAddress(uint8 address)**—This function sets the I2C slave address
- **void I2C_SlaveInitReadBuf(uint8 * rdBuf, uint8 bufSize)**—This function sets the buffer pointer and size of the read buffer. This function also resets the transfer count returned with the I2C_SlaveGetReadBufSize( ) function.
- **void I2C_SlaveInitWriteBuf(uint8 * wrBuf, uint8 bufSize)**—This function sets the buffer pointer and size of the write buffer. This function also resets the transfer count returned with the I2C_SlaveGetWriteBufSize( ) function.
- **uint8 I2C_SlaveGetReadBufSize(void)**—This function returns the number of bytes read by the I2C master since an I2C_SlaveInitReadBuf( ) or I2C_SlaveClearReadBuf( ) function was executed. The maximum return value is the size of the read buffer.

- **uint8 I2C_SlaveGetWriteBufSize(void)**—This function returns the number of bytes written by the I2C master since an I2C_SlaveInitWriteBuf( ) or I2C_SlaveClearWriteBuf( ) function was executed. The maximum return value is the size of the write buffer.
- **void I2C_SlaveClearReadBuf(void)**—This function resets the read pointer to the first byte in the read buffer. The next byte the master reads will be the first byte in the read buffer.
- **void I2C_SlaveClearWriteBuf(void)**—This function resets the write pointer to the first byte in the write buffer. The next byte the master writes will be the first byte in the write buffer.

15.38.5 I2C Master and Multi-Master Functions 3.50

- **I2C_MasterStatus( )**—Returns the master status.
- **I2C_MasterClearStatus( )**—Returns the master status and clears the status flags.
- **I2C_MasterWriteBuf( )**—Writes the referenced data buffer to a specified slave address. item 2C_MasterReadBuf( )—Reads data from the specified slave address and places the data in the referenced buffer.
- **I2C_MasterSendStart( )**—Sends only a Start to the specific address.
- **I2C_MasterSendRestart( )**—Sends only a Restart to the specified address.
- **I2C_MasterSendStop( )**—Generates a Stop condition.
- **I2C_MasterWriteByte( )**—Writes a single byte. This is a manual command that should only be used with the I2C_MasterSendStart( ) or I2C_MasterSendRestart( ) functions.
- **I2C_MasterReadByte( )**—Reads a single byte. This is a manual command that should only be used with the I2C_MasterSendStart( ) or I2C_MasterSendRestart( ) functions.
- **I2C_MasterGetReadBufSize( )**—Returns the byte count of data read since the I2C_MasterClear ReadBuf( ) function was called.
- **I2C_MasterGetWriteBufSize( )**—Returns the byte count of the data written since the I2C_MasterClearWriteBuf( ) function was called.
- **I2C_MasterClearReadBuf( )**—Resets the read buffer pointer back to the beginning of the buffer.
- **I2C_MasterClearWriteBuf( )**—Resets the write buffer pointer back to the beginning of the buffer.

15.38.6 I2C Slave Function Calls 3.5

- **uint8 I2C_MasterStatus(void)**—This function returns the master's communication status.
- **uint8 I2C_MasterClearStatus(void)**—This function clears all status flags and returns the master status.
- **uint8 I2C_MasterWriteBuf(uint8 slaveAddress, uint8 * wrData, uint8 cnt, uint8 mode)**—This function automatically writes an entire buffer of data to a slave device. After the data transfer is initiated by this function, the included ISR manages further data transfer in byte-by-byte mode. Enables the I2C interrupt.
- **uint8 I2C_MasterReadBuf(uint8 slaveAddress, uint8 * rdData, uint8 cnt, uint8 mode)**—This function automatically reads an entire buffer of data from a slave device. Once this function initiates the data transfer, the included ISR manages further data transfer in byte by byte mode. Enables the I2C interrupt
- **uint8 I2C_MasterSendStart(uint8 slaveAddress, uint8 R_nW)**—This function generates a Start condition and sends the slave address with the read/write bit. Disables the I2C interrupt.
- **uint8 I2C_MasterSendRestart(uint8 slaveAddress, uint8 R_nW)**—This function generates a restart condition and sends the slave address with the read/write bit.

- **uint8 I2C_MasterSendStop(void)**—Generates Stop condition on the bus. The NAK is generated before Stop in case of a read transaction. At least one byte has to be read if a Start or ReStart condition with read direction was generated before. This function does nothing if Start or Restart conditions failed before this function was called.
- **uint8 I2C_MasterWriteByte(uint8 theByte)**—This function sends one byte to a slave. A valid Start or Restart condition must be generated before calling this function. This function does nothing if the Start or Restart conditions failed before this function was called.
- **uint8 I2C_MasterReadByte(uint8 acknNak)**—Reads one byte from a slave and generates ACK or prepares to generate NAK. The NAK will be generated before Stop or ReStart condition by SCB_MasterSendStop( ) or SCB_MasterSendRestart( ) function appropriately. This function is blocking. It does not return until a byte is received or an error occurs. A valid Start or Restart condition must be generated before calling this function. This function does nothing and returns a zero value if the Start or Restart conditions failed before this function was called.
- **uint8 I2C_MasterGetReadBufSize(void)**—This function returns the number of bytes that have been transferred with an I2C_MasterReadBuf( ) function.
- **uint8 I2C_MasterGetWriteBufSize(void)**—This function returns the number of bytes that have been transferred with an I2C_MasterWriteBuf( ) function.
- **void I2C_MasterClearReadBuf (void)**—This function resets the read buffer pointer back to the first byte in the buffer.
- **void I2C_MasterClearWriteBuf (void)**—This function resets the write buffer pointer back to the first byte in the buffer.

15.39 Inter-IC Sound Bus (I2S) 2.70

The Integrated Inter-IC Sound Bus (I2S) is a serial bus interface standard used for connecting digital audio devices together.[21] This component operates in master mode only. It operates as a transmitter (Tx) and receiver (Rx). The data for Tx and Rx function as independent byte streams. The byte streams are packed with the most significant byte first and the most significant bit in bit 7 of the first word. The number of bytes used for each sample (a sample for the left or right channel) is the minimum number of bytes to hold a sample. Features include Master only single- and multi-channel (up to 10 channels) I2S support, 8 to 32 data bits per sample, 16-, 32-, 48-, or 64-bit word select period, data rates up to 96 kHz with 64-bit word select period, audio clip detection in I2S Rx mode and byte swap audio samples to match USB audio class endianness requirements [54].

15.39.1 Inter-IC Sound Bus (I2S) Functions 2.70

- **I2S_Start( )**—Starts the I2S interface.
- **I2S_Stop( )**—Disables the I2S interface.
- **I2S_Init( )**—Initializes or restores default I2S configuration.
- **I2S_Enable( )**—Enables the I2S interface.
- **I2S_SetDataBits( )**—Sets the number of data bits for each sample.
- **I2S_EnableTx( )**—Enables the Tx direction of the I2S interface.
- **I2S_DisableTx( )**—Disables the Tx direction of the I2S interface.
- **I2S_SetTxInterruptMode( )**—Sets the interrupt source for the I2S Tx direction interrupt.

---

[21]Philips® Semiconductor (I2S bus specification; February 1986, revised June 5, 1996).

- **I2S_ReadTxStatus( )**—Returns state in the I2S Tx status register.
- **I2S_WriteByte( )**—Writes a single byte into the Tx FIFO.
- **I2S_ClearTxFIFO( )**—Clears out the Tx FIFO.
- **I2S_EnableRx( )**—Enables the Rx direction of the I2S interface.
- **I2S_DisableRx( )**—Disables the Rx direction of the I2S interface.
- **I2S_SetRxInterruptMode( )**—Sets the interrupt source for the I2S Rx direction interrupt.
- **I2S_ReadRxStatus( )**—Returns state in the I2S Rx status register.
- **I2S_ReadByte( )**—Returns a single byte from the Rx FIFO.
- **I2S_ClearRxFIFO( )**—Clears out the Rx FIFO.
- **I2S_SetPositiveClipThreshold( )**—Sets the positive clip detection threshold.
- **I2S_SetNegativeClipThreshold( )**—Sets the negative clip detection threshold.
- **I2S_Sleep( )**—Saves configuration and disables the I2S interface.
- **I2S_Wakeup( )**—Restores configuration and enables the I2S interface.

15.39.2 Inter-IC Sound Bus (I2S) Function Calls 2.70

- **void I2S_Start(void)**—This function starts the I2S interface. It starts the generation of the sck and ws outputs. The Tx and Rx directions remain disabled.
- **void I2S_Stop(void)**—This function disables the I2S interface. The sck and ws outputs are set to 0. The Tx and Rx directions are disabled and their FIFOs are cleared.
- **void I2S_Init(void)**—This function initializes or restores default I2S configuration provided with customizer. It does not clear data from the FIFOs and does not reset Component hardware state machines. It is not necessary to call I2S_Init( ) because the I2S_Start( ) routine calls this function and I2S_Start( ) is the preferred method to begin Component operation.
- **void I2S_Enable(void)**—This function enables the I2S interface. It starts the generation of the sck and ws outputs. The Tx and Rx directions remain disabled. It is not necessary to call I2S_Enable( ) because the I2S_Start( ) routine calls this function and I2S_Start( ) is the preferred method to begin Component operation.
- **cystatus I2S_SetDataBits(uint8 dataBits)**—This function sets the number of data bits for each sample. The Component must be stopped before calling this API. The API is only available when the Bit resolution parameter is set to Dynamic.
- **void I2S_EnableTx(void)**—This function enables the Tx direction of the I2S interface. Transmission begins at the next word select falling edge.
- **void I2S_DisableTx(void)**—This function disables the Tx direction of the I2S interface. Transmission of data stops and a constant 0 value is transmitted at the next word select falling edge.
- **void I2S_SetTxInterruptMode(interruptSource)/void I2S_SetTxInterruptMode(channel, interruptSource)**—This macro sets the interrupt source for the specified Tx stereo channel. Multiple sources may be ORed.[22]
- **void I2S_EnableTx(void)**—This function enables the Tx direction of the I2S interface. Transmission begins at the next word select falling edge.
- **void I2S_DisableTx(void)**—This function disables the Tx direction of the I2S interface. Transmission of data stops and a constant 0 value is transmitted at the next word select falling edge.

---

[22]The macro expects channel parameter only when more than one stereo channel is selected for Tx direction.

- **void I2S_SetTxInterruptMode(interruptSource)/void I2S_SetTxInterruptMode(channel, interruptSource)**—This macro sets the interrupt source for the specified Tx stereo channel. Multiple sources may be ORed.[23]
- **void I2S_DisableRx(void)**—This function disables the Rx direction of the I2S interface. At the next word select falling edge, data received is no longer sent to the receive FIFO.
- **void I2S_SetRxInterruptMode(interruptSource)/void I2S_SetRxInterruptMode(channel, interruptSource)**—This macro sets the interrupt source for the specified Rx stereo channel. Multiple sources may be ORed.[24]
- **uint8 I2S_ReadRxStatus(void)/uint8 I2S_ReadRxStatus(channel)**—This macro returns the status of the specified stereo channel(s). In a multi-channel configuration, the status bits of stereo channel 0 are combined with stereo channel 1 and the bits of channel 2 are combined with channel 3. Therefore the API will return the combined status of stereo channel 0 and stereo channel 1 if the status for channel 0 or channel 1 is requested.[25]
- **uint8 I2S_ReadByte(wordSelect)/uint8 I2S_ReadByte(channel, wordSelect)**—This macro returns a single byte from the Rx FIFO. You have to check the Rx status before this call to confirm that the Rx FIFO is not empty.[26]
- **void I2S_ClearRxFIFO(void)**—This function clears out the FIFOs for all Rx channels. Any data present in the FIFOs is lost. Call this function only when the Rx direction is disabled.
- **void I2S_SetPositiveClipThreshold(posThreshold)/void I2S_SetPositiveClipThreshold (channel, posThreshold)**—This macro sets the 8-bit positive clip detection threshold for the specified channel. This API is available if the Rx clip detection parameter is selected in the Configure dialog.The macro expects channel parameter only when more than one stereo channel is selected for Rx direction.
- **void I2S_SetNegativeClipThreshold(negThreshold)/void I2S_SetNegativeClipThreshold (channel, negThreshold)**—This API sets the 8-bit negative clip detection threshold for the specified channel. This API is available if the Rx clip detection parameter is selected in the Configure dialog.[27]
- **void I2S_Sleep(void)**—This is the preferred routine to prepare the Component for sleep. The I2S_Sleep( ) routine saves the current Component state and calls the I2S_Stop( ) function. The sck and ws outputs are set to 0. The Tx and Rx directions are disabled. Call the I2S_Sleep( ) function before calling the CyPmSleep( ) or the CyPmHibernate( ) function. Refer to the PSoC Creator System Reference Guide for more information about power management functions.
- **void I2S_Wakeup(void)**—This is the preferred routine to prepare the Component operation after exit from sleep. Starts the generation of the sck and ws outputs if the Component operated before sleep. Enables Rx and/or Tx direction according to their states before sleep.

15.39.3 Macro Callback Functions 2.70

- **I2C_ISR_EntryCallback I2C_ISR_ENTRY_CALLBACK**—Used at the beginning of the I2C_ISR( ) interrupt handler to perform additional application-specific actions.

---

[23]The macro expects channel parameter only when more than one stereo channel is selected for Tx direction.

[24]The macro expects channel parameter only when more than one stereo channel is selected for Rx direction.

[25]The macro expects channel parameter only when more than one stereo channel is selected for Rx direction.

[26]The macro expects channel parameter only when more than one stereo channel is selected for Rx direction.

[27]The macro expects channel parameter only when more than one stereo channel is selected for Rx direction.

- **I2C_ISR_ExitCallback I2C_ISR_EXIT_CALLBACK**—Used at the end of the I2C_ISR( ) interrupt handler to perform additional application-specific actions.
- **I2C_WAKEUP_ISR_EntryCall back_I2C_WAKEUP ISR_ENTRY_CALLBACK**—Used at the beginning of the I2C_WAKEUP_ISR( ) interrupt handler to perform additional application specific actions.
- **I2C_TMOUT_ISR_EntryCallback I2C_TMOUT_ISR_ENTRY_CALLBACK**—Used at the beginning of the I2C_ISR4( ) interrupt handler to perform additional application-specific actions.
- **I2C_TMOUT_ISR_ExitCallback_I2C_TMOUT_ISR_EXIT_ CALLBACK**—Used at the end of the I2C_ISR4( ) interrupt handler to perform additional application-specific actions.
- **I2C_SwPrepareReadBuf_Callback_I2C_SW_PREPARE_READ   _BUF_CALLBACK**— Used in the I2C_ISR( ) interrupt handler to perform additional application-specific actions.
- **I2C_SwAddrCompare_EntryCallback_I2C_SW_ADDR_COMPARE_ENTRY_ CALLBACK**—Used in the I2C_ISR( ) interrupt handler to perform additional application-specific actions.
- **I2C_SwAddrCompare_Exit_Callback_I2C_SW_ADDR_COMPARE_EXIT_CALLBACK**— Used in the I2C_ISR( ) interrupt handler to perform additional application-specific actions.
- **I2C_HwPrepareReadBuf_Callback_I2C_HW_PREPARE_READ_BUF_CALLBACK**— Used in the I2C_ISR( ) interrupt handler to perform additional application-specific actions.

15.40 MDIO Interface Advanced 1.20

The MDIO Interface component supports the Management Data Input/Output, which is a serial bus defined for the Ethernet family of IEEE 802.3 standards for the Media Independent Interface (MII). The MII connects Media Access Control (MAC) devices with Ethernet physical layer (PHY) circuits. The component is compliant with IEEE 802.3 Clause 45. Features include: Used in conjunction with Ethernet products, configurable physical address, supports up to 4.4 MHz in the clock bus (dc), compliant with IEEE 802.3 Clause 45 and automatically allocates memory for the register spaces that can be configured through an intuitive, easy-to-use graphical configuration GUI [60].

15.40.1 MDIO Interface Functions 1.20

- **MDIO_Interface_Start( )**—Initializes and enables the MDIO Interface.
- **MDIO_Interface_Stop( )**—Disables the MDIO Interface.
- **MDIO_Interface_Init( )**—Initializes default configuration provided with customizer.
- **MDIO_Interface_Enable( )**—Enables the MDIO Interface.
- **MDIO_Interface_EnableInt( )**—Enables the interrupt output terminal.
- **MDIO_Interface_DisableInt( )**—Disables the interrupt output terminal.
- **MDIO_Interface_SetPhyAddress( )**—Sets the physical address for the MDIO Interface.
- **MDIO_Interface_UpdatePhyAddress( )**—Updates the physical address of the MDIO Interface.
- **MDIO_Interface_SetDevAddress( )**—Sets the device address for the MDIO Interface.
- **MDIO_Interface_GetData( )**—Returns the value stored in the given address.
- **MDIO_Interface_SetData( )**—Sets the argument value in the given address.
- **MDIO_Interface_SetBits( )**—Sets the specified bits at the specified address.
- **MDIO_Interface_GetAddress( )**—Returns the last address written by the MDIO Host.
- **MDIO_Interface_SetData( )**—Sets the argument value in the given address.

- **MDIO_Interface_SetBits( )**—Sets the specified bits at the specified address.
- **MDI Interface_GetConfiguration( )**—Returns a pointer to the configuration array of the given register space.
- **MDIO_Interface_SetData( )**—Sets the argument value in the given address.
- **MDIO_Interface_SetBits( )**—Sets the specified bits at the specified address.
- **MDIO_Interface_PutData( )**—Sets the data to be transmitted to the MDIO Host.
- **MDIO_Interface_ProcessFrame( )**—Processes the last frame received from the MDIO Host.
- **MDIO_Interface_Sleep( )**—Stops the MDIO Interface and saves the user configuration.
- **MDIO_Interface_SetData( )**—Sets the argument value in the given address.
- **MDIO_Interface_SetBits( )**—Sets the specified bits at the specified address.
- **MDIO_Interface_Wakeup( )**—Restores the user configuration and enables the MDIO Interface.
- **MDIO_Interface_SetData( )**—Sets the argument value in the given address.
- **MDIO_Interface_SetBits( )**—Sets the specified bits at the specified address.
- **MDIO_Interface_SaveConfig( )**—Saves the current user configuration.
- **MDIO_Interface_RestoreConfig( )**—Restores the user configuration.

15.40.2 MDIO Interface Function Calls 1.20

- **void MDIO_Interface_Start(void)**—This is the preferred method to begin component operation. This function sets the initVar variable, calls the MDIO_Interface_Init( ) function and then calls the MDIO_Interface_Enable( ) function.
- **void MDIO_Interface_Stop(void)**—Disables the MDIO Interface. If the component is configured to operate in Advanced mode, this function disables all internal DMA channels.
- **void MDIO_Interface_Init(void)**—Initializes or restores default MDIO Interface configuration provided with customizer. Initializes internal DMA channels if the component is configured to operate in Advanced mode. It is not necessary to call MDIO_Interface_Init( ) because the MDIO_Interface_Start( ) routine calls this function, which is the preferred method to begin component operation.
- **void MDIO_Interface_Enable(void)**—Activates the hardware and begins component operation. It is not necessary to call MDIO_Interface_Enable( ) because the MDIO_Interface_Start( ) routine calls this function, which is the preferred method to begin component operation.
- **void MDIO_Interface_EnableInt(void)**—Enables the terminal output interrupt.
- **void MDIO_Interface_DisableInt(void)**—Disables the terminal output interrupt.
- **void MDIO_Interface_SetPhyAddress(uint8 phyAddr)**—Sets the 5-bit or 3-bit physical address for the MDIO slave device. For 3-bit address, the two most significant address bits from an MDIO frame will be ignored. For example, if 3-bit physical address is set to 0x4, the component will respond to the following physical addresses from an MDIO frame: 0x04, 0x0C, 0x14 and 0x1C.
- **void MD_Interface_UpdatePhyAddress(void)**—Updates the physical address based on the current value of phy_addr input signal. If a Firmware option is set for the Physical address parameter, the address is set equal to the default value from the customizer.
- **void MDIO_Interface_SetDevAddress(uint8 devAddr)**—Sets the 5-bit device address for the MDIO Interface.
- **cystatus MDIO_Interface_GetData(uint16 address, const uint16 *regData,uint16 numWords)**—Returns N values starting from the given address. If any address does not belong to the allocated register space, it returns an error. This API is only available in Advanced mode.
- **cystatus MDIO_Interface_SetData(uint16 address, const uint16 *regData,uint16 numWords)**—Writes N values starting from the given address. If any address does not belong to

the allocated register space or the register space is located in Flash, it returns an error. This API is only available in Advanced mode and if at least one register space is located in SRAM.

- **cystatus MDIO_Interface_SetBits(uint16 address, uint16 regBits)**—Sets the bits at the given address. If the address does not belong to the allocated register space or the register space is located in Flash, it returns an error. This API is only available in Advanced mode and if at least one register space is located in SRAM.
- **uint16 MDIO_Interface_GetAddress(void)**—Returns the last address written by the MDIO Host.
- **uint8 MDIO_Interface_GetConfiguration(uint8 regSpace)**—Returns a pointer to the configuration array of the given register space.
- **uint8 MDIO_Interface_GetConfiguration(uint8 regSpace)**—Returns a pointer to the configuration array of the given register space.
- **void MDIO_Interface_ProcessFrame(uint8* opCode, uint16* regData)**—Processes and parses the last frame received from the host. Only available in Basic mode.
- **void MDIO_Interface_Sleep(void)**—This is the preferred routine to prepare the component for sleep. The MDIO_Interface_Sleep( ) routine saves the current component state. Then it calls the MDIO_Interface_Stop( ) function and calls MDIO_Interface_SaveConfig( ) to save the hardware configuration. Call the MDIO_Interface_Sleep( ) function before calling the CyPmSleep( ) or the CyPmHibernate( ) function.
- **void MDIO_Interface_Wakeup(void)**—This is the preferred routine to restore the component to the state when MDIO_Interface_Sleep( ) was called. The MDIO_Interface_Wakeup( ) function calls the MDIO_Interface_RestoreConfig( ) function to restore the configuration. If the component was enabled before the MDIO_Interface_Sleep( ) function was called, the MDIO_Interface_Wakeup( ) function also re-enables the component.
- **void MDIO_Interface_SaveConfig(void)**—This function saves the component configuration and non-retention registers. It also saves the current component parameter values, as defined in the Configure dialog or as modified by appropriate APIs. This function is called by the MDIO_Interface_Sleep( ) function
- **void MDIO_Interface_RestoreConfig(void)**—This function restores the component configuration and non-retention registers. It also restores the component parameter values to what they were before calling the MDIO_Interface_Sleep( ) function.

15.41 SMBus and PMBus Slave 5.20

The System Management Bus (SMBus) and Power Management Bus (PMBus) Slave component provides a simple way to add a well-known communications method to a PSoC 3-, PSoC 4-, or PSoC 5LP-based design. The SMBus is a two-wire interface that is often used to interconnect a variety of system management chips to one, or more host, systems. It uses I2C with some extensions as the physical layer. There is also a protocol layer, which defines classes of data and how that data is structured. Both the physical layer and protocol layer add a level of robustness not originally embodied in the I2C specification. The SMBus Slave component supports most of the SMBus version 2.0 Slave device specifications with numerous configurable options. PMBus is an extension to the more generic SMBus protocol with specific focus on power conversion and power management systems. With some slight modifications to the SMBus protocol, the PMBus specifies application layer commands, which are not defined in the SMBus. The PMBus component presents all possible PMBus revision 1.2 commands and allows the selection of those commands relevant to an application. Features include

SMBus/PMBus Slave mode, SMBALERT# pin support, 25 ms Timeout, Configurable SMBus/PMBus commands and Packet Error Checking (PEC) support [85].

15.41.1 SMBus and PMBus Slave Functions 5.20

- **SMBusSlave_Start( )**—Initializes and enables the SMBus component. The I2C interrupt is enabled and the component can respond to the SMBus traffic.
- **SMBusSlave_Stop( )**—Stops responding to the SMBus traffic. Also disables the interrupt.
- **SMBusSlave_Init( )**—This function initializes or restores the component according to the customizer Configure dialog settings.
- **SMBusSlave_Enable( )**—Activates the hardware and begins component operation.
- **SMBusSlave_EnableInt( )**—Enables the component interrupt.
- **SMBusSlave_DisableInt( )**—Disables the interrupt.
- **SMBusSlave_SetAddress( )**—Sets the primary address.
- **SMBusSlave_SetAlertResponseAddress( )**—Sets the Alert Response Address.
- **SMBusSlave_SetSmbAlert( )**—Sets the value passed to the SMBALERT# pin.
- **SMBusSlave_SetSmbAlertMode( )**—Determines how the component responds to an SMBus master read at the Alert Response Address.
- **SMBusSlave_HandleSmbAlertResponse( )**—Called by the component when it responds to the Alert Response Address issued by the host and the SMBALERT Mode is set to MANUAL_MODE.
- **SMBusSlave_GetNextTransaction( )**—Returns a pointer to the next transaction record in the transaction queue. If the queue is empty, the function returns NULL.
- **SMBusSlave_GetTransactionCount( )**—Returns the number of transaction records in the transaction queue.
- **SMBusSlave_CompleteTransaction( )**—Causes the component to complete the currently pending transaction at the head of the queue.
- **SMBusSlave_GetReceiveByteResponse( )**—Returns the byte to respond to a "Receive Byte" protocol request.
- **SMBusSlave_HandleBusError( )**—Called by the component whenever a bus protocol error occurs.
- **SMBusSlave_StoreUserAll( )**—Saves the Operating register store in RAM to the User register store in Flash.
- **SMBusSlave_RestoreUserAll( )**—Verifies the CRC field of the User register store and then copies the contents of the User register store to the Operating register store.
- **SMBusSlave_EraseUserAll( )**—Erase the User register store in Flash.
- **SMBusSlave_RestoreDefaultAll( )**—Verifies the signature field of the Default register store and then copies the contents of the Default register store to the Operating register store.
- **SMBusSlave_StoreComponentAll( )**—Update the parameters of other components in the system with the current PMBus settings.
- **SMBusSlave_RestoreComponentAll( )**—Updates the PMBus Operating register store with the current configuration parameters of other components in the system.
- **SMBusSlave_Lin11ToFloat( )**—Converts the argument "linear11" to floating point and returns it.
- **SMBusSlave_FloatToLin11( )**—Takes the argument "floatvar" (a floating point number) and converts it to a 16-bit LINEAR11 value (11-bit mantissa + 5-bit exponent), which it returns.
- **SMBusSlave_Lin16ToFloat( )**—Converts the argument "linear16" to floating point and returns it.
- **SMBusSlave_FloatToLin16( )**—Takes the argument "floatvar" (a floating point number) and converts it to a 16-bit LINEAR16 value (16-bit mantissa), which it returns.

15.41.2 SMBus and PMBus Slave Function Calls 5.20

- **void SMBusSlave_Start(void)**—This is the preferred method to begin component operation. SM-BusSlave_Start( ) calls the SMBusSlave_Init( ) function and then calls the SMBusSlave_Enable( ) function.
- **void SMBusSlave_Stop(void)**—This function stops the component and disables the interrupt. It releases the bus if it was locked up by the device and sets it to the idle state.
- **void SMBusSlave_Init(void)**—This function initializes or restores the component according to the customizer Configure dialog settings. It is not necessary to call SMBusSlave_Init( ) because the SMBusSlave_Start( ) API calls this function, which is the preferred method to begin component operation.
- **void SMBusSlave_Enable(void)**—This function activates the hardware and calls EnableInt( ) to begin component operation. It is not necessary to call SMBusSlave_Enable( ) because the SMBusSlave_Start( ) API calls this function, which is the preferred method to begin component operation. If this API is called, SMBusSlave_Start( ) or SMBusSlave_Init( ) must be called first.
  void SMBusSlave_EnableInt(void)—This function enables the component interrupt. It is not required to call this API to begin the component operation since it is called in SMBusSlave_Enable( ).
- **void SMBusSlave_DisableInt(void)**—This function disables the component interrupt. This function is not normally required because the I2C_Stop( ) function disables the interrupt. The component does not operate when the interrupt is disabled.
- **void SMBusSlave_SetAddress(uint8 address)**—This function sets the primary slave address of the device.
- **void SMBusSlave_SetAlertResponseAddress(uint8 address)**—This function sets the Alert Response Address.
- **void SMBusSlave_SetSmbAlert(uint8 assert)**—This function sets the value to the SMBALERT# pin. As long as SMBALERT# is asserted, the component will respond to master read's to the Alert Response Address. The response will be the device's primary slave address. Depending on the mode setting, the component will automatically de-assert SMBALERT#, call the SMBus-Slave_HandleSmbAlertResponse( ) API, or do nothing.
- **void SMBusSlave_SetSmbAlertMode(uint8 alertMode)**—This function determines how the component responds to an SMBus master read at the Alert Response Address. When SMBALERT# is asserted, the SMBus master may broadcast a read to the global Alert Response Address to determine which SMBus device on the shared bus has asserted SMBALERT#. In Auto mode, SM-BALERT# is automatically de-asserted once the component acknowledges the Alert Response Address. In Manual mode, the component will call the API SMBusSlave_HandleSmbAlertResponse( ) where user code (in a callback function) is responsible for de-asserting SMBALERT#. In DO_NOTHING mode, the component will take no action.
- **void SMBusSlave_HandleSmbAlertResponse(void)**—This function is called by the component when it responds to the Alert Response Address and the SMBALERT Mode is set to MAN-UAL_MODE. This function defines a callback function where the user inserts code to run after the component has responded. For example, the user might update a status register and de-assert the SMBALERT# pin.
- **SMBusSlave_GetNextTransaction(void)**—This function returns a pointer to the next transaction record in the transaction queue. If the queue is empty, the function returns NULL. Only Manual Reads and Writes will be returned by this function, as the component will handle any Auto transactions on the queue. In the case of Writes, it is the responsibility of the user firmware servicing the Transaction Queue to copy the "payload" to the register store. In the case of Reads, it

is the responsibility of user firmware to update the contents of the variable for this command in the register store. For both, call SMBusSlave_CompleteTransaction( ) to free the transaction record. Note that for Read transactions, the length and payload fields are not used for most transaction types. The exception to this is Process Call and Block Process Call, where the block of data from the write phase will be stored in the payload field.

- **uint8 SMBusSlave_GetTransactionCount(void)**—This function returns the number of transaction records in the transaction queue.
- **void SMBusSlave_CompleteTransaction(void)**—This function causes the component to complete the currently pending transaction at the head of the queue. The user firmware transaction handler calls this function after processing a transaction. This alerts the component code to copy the register variable associated with the pending read transaction from the register store to the data transfer buffer so that the transfer may complete. It also advances the queue. Must be called for reads and writes.
- **uint8 SMBusSlave_GetReceiveByteResponse(void)**—This function is called by the component ISR to determine the response byte when it detects a "Receive Byte" protocol request. This function invokes a callback function where the user may insert their code to override the default return value of this function—which is 0xFF. This function will be called in ISR context. Therefore, user code must be fast, non-blocking and may only call re-entrant functions.
- **void SMBusSlave_HandleBusError(uint8 errorCode)**—This function is called by the component whenever a bus protocol error occurs. Examples of bus errors would be: invalid command, data underflow and clock stretch violation. This function is only responsible for the aftermath of an error since the component will already handle errors in a deterministic manner. This function is primarily for the purpose of notifying user firmware that an error has occurred. For example, in a PMBus device this would give user firmware an opportunity to set the appropriate error bit in the STATUS_CML register.
- **uint8 SMBusSlave_StoreUserAll(const uint8 * flashRegs)**—This function saves the Operating register store to the User register store in Flash. The CRC field in the register store data structure is recalculated and updated prior to the save. This function does not perform storing anything to Flash by default. Instead, it executes a user callback function where the user can implement an algorithm for storing Operating register store to Flash.
- **uint8 SMBusSlave_RestoreUserAll(const uint8 * flashRegs)**—This function verifies the CRC field of the User register store and then copies the contents of this register store to the Operating register store in RAM.
- **uint8 SMBusSlave_EraseUserAll(void)**—This function erases the User register store in Flash. The API does not erase the Flash by default. Instead, it contains a call to the callback routinewhere the user can implement an algorithm to erase the contents of the User register store in Flash.
- **uint8 SMBusSlave_RestoreDefaultAll(void)**—This function verifies the signature field of the Default register store and then copies the contents of the Default register store to the Operating register store in RAM.
- **uint8 SMBusSlave_StoreComponentAll(void)**—This function updates the parameters of other components in the system with the current PMBus settings. Because this action is very application specific, this function merely calls a user provided callback function. The only component provided firmware is a return value variable (retval) which is initialized to CYRET_SUCCESS and returned at the end of the function. The rest of the function must be user provided.
- **uint8 SMBusSlave_RestoreComponentAll(void)**—This function updates the PMBus Operating register store with the current configuration parameters of other components in the system. Because this action is very application specific, this function merely calls a user provided callback function. The only component provided firmware is a return value variable (retval) which is initialized to

CYRET_SUCCESS and returned at the end of the function. The rest of the function must be user provided.

- **float SMBusSlave_Lin11ToFloat (uint16 linear11)**—This function converts the argument "linear11" to floating point and returns it.
- **uint16 SMBusSlave_FloatToLin11 (float floatvar)**—This function takes the argument "floatvar" (a floating point number) and converts it to a 16-bit LINEAR11 value (11-bit mantissa + 5-bit exponent), which it returns.
- **float SMBusSlave_Lin16ToFloat(uint16 linear16, int8 inExponent)**—This function converts the argument "linear16" to floating-point and returns it. The argument Linear16 contains the mantissa. The argument inExponent is the 5-bit 2's complement exponent to use in the conversion.
- **uint16 SMBus_FloatToLin16(float floatvar, int8 outExponent)**—This function takes the argument "floatvar" (a floating -oint number) and converts it to a 16-bit LINEAR16 value (16-bit mantissa), which it returns. The argument outExponent is the 5-bit 2's complement exponent to use in the conversion.

15.42 Software Transmit UART 1.50

The Software Transmit UART (SW_Tx_UART) component is an 8-bit RS-232 data-format compliant serial transmitter, that is used to transmit serial data. It consists of firmware and a pin and therefore it is useful on devices without digital resources, or in projects where all digital resources are consumed. The SW_Tx_UART is supported in PSoC 3, PSoC 4 and PSoC 5LP with high accuracy Baud rates from 9,600 up to 115,200 bps and low Flash.ROM resource utilization [86].

15.42.1 Software Transmit UART Functions 1.50

- **SW_Tx_UART_Start( )**—Empty function, included for consistency with other components.
- **SW_Tx_UART_StartEx( )**—Configures the SW_Tx_UART to use the pin specified by the parameters.
- **SW_Tx_UART_Stop( )**—Empty function, included for consistency with other components.
- **SW_Tx_UART_PutChar( )**—Sends one byte via the Tx pin.
- **SW_Tx_UART_PutString( )**—Sends a NULL terminated string via the Tx pin.
- **SW_Tx_UART_PutArray( )**—Sends byteCount bytes from a memory array via the Tx pin.
- **SW_Tx_UART_PutHexByte( )**—Sends a byte in Hex representation (two characters, uppercase for A-F) via the Tx pin.
- **SW_Tx_UART_PutHexInt( )**—Sends a 16-bit unsigned integer in Hex representation (four characters, uppercase for A-F) via the Tx pin.
- **SW_Tx_UART_PutCRLF( )**—Sends a carriage return (0x0D) and a line feed (0x0A) via the Tx pin.

15.42.2 Software Transmit UART Function Calls 1.50

- **void SW_Tx_UART_Start(void)**—Empty function. Included for consistency with other components. This API is not available when PinAssignmentMethod is set to Dynamic.

- **void SW_Tx_UART_StartEx(uint8 port, uint8 pin)**—Configures the SW_Tx_UART to use the pin specified by the parameters. This API is only available when PinAssignmentMethod is set to Dynamic.
- **void SW_Tx_UART_Stop(void)**—Empty function. Included for consistency with other components.
- **void SW_Tx_UART_PutChar(uint8 txDataByte)**—Sends one byte via the Tx pin.
- **void SW_Tx_UART_PutString(const char8 string[])**—Sends a NULL terminated \string via the Tx pin.
- **void SW_Tx_UART_PutArray(const uint8 data[], uint16/uint32 byteCount)**—Sends byte-Count bytes from a memory array via the Tx pin.
- **void SW_Tx_UART_PutHexByte(uint8 txHexByte)**—Sends a byte in Hex representation (two characters, uppercase for A-F) via the Tx pin.
- **void SW_Tx_UART_PutHexInt(uint16 txHexInt)**—Sends a 16-bit unsigned integer in Hex representation (four characters, uppercase for A-F) via the Tx pin.
- **void SW_Tx_UART_PutCRLF( )**—Sends a carriage return (0x0D) and a line feed (0x0A) via the Tx pin.

15.43 S/PDIF Transmitter (SPDIF_Tx) 1.20

The SPDIF_Tx component provides a simple way to add digital audio output to any design. It formats incoming audio data and metadata to create the S/PDIF bit stream appropriate for optical or coaxial digital audio supporting both interleaved and separated audio. Audio data is received from DMA and channel status information. Typically, the channel status DMA will be managed by the component; however this data can be specifiedseparately to better control a system. This component conforms to IEC-60958, AES/EBU, AES3 standards for Linear PCM Audio Transmission, supports sample rates clock/128 (up to 192 kHz), has configurable audio sample lengths (8/16/24), includes channel status bits for consumer applications and has independent left and right channel FIFOs or interleaved stereo FIFOs [78].

15.43.1 S/PDIF Transmitter Functions 1.20

- **SPDIF_Start( )**—Starts the S/PDIF interface.
- **SPDIF_Stop( )**—Disables the S/PDIF interface.
- **SPDIF_Sleep( )**—Saves configuration and disables the SPDIF interface.
- **SPDIF_Wakeup( )**—Restores configuration of the S/PDIF interface.
- **SPDIF_EnableTx( )**—Enables the audio data output in the S/PDIF bit stream.
- **SPDIF_DisableTx( )**—Disables the audio output in the S/PDIF bit stream.
- **SPDIF_WriteTxByte( )**—Writes a single byte into the audio FIFO.
- **SPDIF_WriteCstByte( )**—Writes a single byte into the channel status FIFO.
- **SPDIF_SetInterruptMode( )**—Sets the interrupt source for the S/PDIF interrupt.
- **SPDIF_ReadStatus( )**—Returns state in the S/PDIF status register.
- **SPDIF_ClearTxFIFO( )**—Clears out the audio FIFO.
- **SPDIF_ClearCstFIFO( )**—Clears out the channel status FIFOs.
- **SPDIF_SetChannelStatus( )**—Sets the values of the channel status at run time.
- **SPDIF_SetFrequency( )**—Sets the values of the channel status for a specified frequency.
- **SPDIF_Init( )**—Initializes or restores default S/PDIF configuration.

- **SPDIF_Enable( )**—Enables the S/PDIF interface.
- **SPDIF_SaveConfig( )**—Saves configuration of S/PDIF interface.
- **SPDIF_RestoreConfig( )**—Restores configuration of S/PDIF interface.

15.43.2 S/PDIF Transmitter Function Calls 1.20

- **void SPDIF_Start(void)**—Starts the S/PDIF interface. Starts the channel status DMA if the component is configured to handle the channel status DMA. Enables Active mode power template bits or clock gating as appropriate. Starts the generation of the S/PDIF output with channel status, but the audio data is set to all 0s. This allows the S/PDIF receiver to lock on to the component's clock.
- **void SPDIF_Start(void)**—Starts the S/PDIF interface. Starts the channel status DMA if the component is configured to handle the channel status DMA. Enables Active mode power template bits or clock gating as appropriate. Starts the generation of the S/PDIF output with channel status, but the audio data is set to all 0s. This allows the S/PDIF receiver to lock on to the component's clock.
- **void SPDIF_Stop(void)**—Disables the S/PDIF interface. Disables Active mode power template bits or clock gating as appropriate. The S/PDIF output is set to 0. The audio data and channel data FIFOs are cleared. The SPDIF_Stop( ) function calls SPDIF_DisableTx( ) and stops the managed channel status DMA.
- **void SPDIF_Sleep(void)**—This is the preferred routine to prepare the component for sleep. The SPDIF_Sleep( ) routine saves the current component state. Then it calls the SPDIF_Stop( ) function and calls SPDIF_SaveConfig( ) to save the hardware configuration. Disables Active mode power template bits or clock gating as appropriate. The spdif output is set to 0. Call the SPDIF_Sleep( ) function before calling the CyPmSleep( ) or the CyPmHibernate( ) function.
- **void SPDIF_Wakeup(void)**—Restores SPDIF configuration and non-retention register values. The component is stopped regardless of its state before sleep. The SPDIF_Start( ) function must be called explicitly to start the component again.
- **void SPDIF_EnableTx(void)**—Enables the audio data output in the S/PDIF bit stream. Transmission will begin at the next X or Z frame.
- **void SPDIF_DisableTx(void)**—Disables the audio output in the S/PDIF bit stream. Transmission of data will stop at the next rising edge of clock and constant 0 value will be transmitted.
- **void SPDIF_WriteTxByte(uint8 wrData, uint8 channelSelect)**—Writes a single byte into the audio data FIFO. The component status should be checked before this call to confirm that the audio data FIFO is not full.
- **void SPDIF_WriteCstByte(uint8 wrData, uint8 channelSelect)** - Writes a single byte into the specified channel status FIFO. The component status should be checked before this call to confirm that the channel status FIFO is not full.
- **void SPDIF_SetInterruptMode(uint8 interruptSource)**—Sets the interrupt source for the S/PDIF interrupt. Multiple sources may be ORed.
- **uint8 SPDIF_ReadStatus(void)**—Returns state of the SPDIF status register.
- **void SPDIF_ClearTxFIFO(void)**—Clears out the audio data FIFO. Any data present in the FIFO will be lost. In the case of separated audio mode, both audio FIFOs will be cleared. Call this function only when transmit is disabled.
- **void SPDIF_ClearCstFIFO(void)**—Clears out the channel status FIFOs. Any data present in either FIFO will be lost. Call this function only when the component is stopped.

- **void SPDIF_SetChannelStatus(uint8 channel, uint8 byte, uint8 mask, uint8 value)**—Sets the values of the channel status at run time. This API is only valid when the component is managing the DMA.
- **uint8 SPDIF_SetFrequency(uint8 frequency)**—Sets the values of the channel status for a specified frequency and returns 1. This function only works if the component is stopped. If this is called while the component is started, a zero will be returned and the values will not be modified. This API is only valid when the component is managing the DMA.
- **void SPDIF_Init(void)**—Initializes or restores default S/PDIF configuration provided with customizer that defines interrupt sources for the component and channel status if the component is configured to handle the channel status DMA.
- **void SPDIF_Enable(void)**—Activates the hardware and begins component operation. It is not necessary to call SPDIF_Enable( ) because the SPDIF_Start( ) routine calls this function, which is the preferred method to begin component operation.
- **void SPDIF_SaveConfig(void)**—This function saves the component configuration. This will save non-retention registers. This function will also save the current component parameter values, as defined in the Configure dialog or as modified by appropriate APIs. This function is called by the SPDIF_Sleep( ) function.
- **void SPDIF_RestoreConfig(void)**—This function restores the component configuration. This will restore non-retention registers. This function will also restore the component parameter values to what they were prior to calling the SPDIF_Sleep( ) function.This routines is called by SPDIF_Wakeup( ) to restore component when it exits sleep.

15.44   Serial Peripheral Interface (SPI) Master 2.50

The SPI Master component provides an industry-standard, 4-wire master SPI interface. It can also provide a 3-wire (bidirectional) SPI interface. Both interfaces support all four SPI operating modes, allowing communication with any SPI slave device. In addition to the standard 8-bit word length, the SPI Master supports a configurable 3- to 16-bit word length for communicating with nonstandard SPI word lengths. SPI signals include: serial Clock (SCLK), Master-In, Slave-Out (MISO), Master-Out, Slave-In (MOSI), bidirectional Serial Data (SDAT) and slave Select (SS). Features include: 3- to 16-bit data width, four SPI operating modes, and a bit rate up to 18 Mbps [81].

15.44.1   Serial Peripheral Interface (SPI) Master Functions 2.50

- **SPIM_Start( )**—Calls both SPIM_Init( ) and SPIM_Enable( ). Should be called the first time the component is started.
- **SPIM_Stop( )**—Disables SPI Master operation.
- **SPIM_EnableTxInt( )**—Enables the internal Tx interrupt irq.
- **SPIM_EnableRxInt( )**—Enables the internal Rx interrupt irq.
- **SPIM_DisableTxInt( )**—Disables the internal Tx interrupt irq.
- **SPIM_DisableRxInt( )**—Disables the internal Rx interrupt irq.
- **SPIM_SetTxInterruptMode( )**—Configures the Tx interrupt sources enabled.
- **SPIM_SetRxInterruptMode( )**—Configures the Rx interrupt sources enabled.
- **SPIM_ReadTxStatus( )**—Returns the current state of the Tx status register.
- **SPIM_ReadRxStatus( )**—Returns the current state of the Rx status register.

- **SPIM_WriteTxData( )**—Places a byte/word in the transmit buffer which will be sent at the next available bus time.
- **SPIM_ReadRxData( )**—Returns the next byte/word of received data available in the receive buffer.
- **SPIM_GetRxBufferSize( )**—Returns the size (in bytes/words) of received data in the Rx memory buffer.
- **SPIM_GetTxBufferSize( )**—Returns the size (in bytes/words) of data waiting to transmit in the Tx memory buffer.
- **SPIM_ClearRxBuffer( )**—Clears the Rx buffer memory array and Rx FIFO of all received data.
- **SPIM_ClearTxBuffer( )**—Clears the Tx buffer memory array or Tx FIFO of all transmit data.[28]
- **SPIM_TxEnable( )**—If configured for bidirectional mode, sets the SDAT inout to transmit.
- **SPIM_TxDisable( )**—If configured for bidirectional mode, sets the SDAT inout to receive.
- **SPIM_PutArray( )**—Places an array of data into the transmit buffer.
- **SPIM_ClearFIFO( )**—Clears any received data from the Rx hardware FIFO.
- **SPIM_Sleep( )**—Prepares SPI Master component for low-power modes by calling SPIM_SaveConfig( ) and SPIM_Stop( ) functions.
- **SPIM_Wakeup( )**—Restores and re-enables the SPI Master component after waking from low-power mode.
- **SPIM_Init( )**—Initializes and restores the default SPI Master configuration.
- **SPIM_Enable( )**—Enables the SPI Master to start operation.
- **SPIM_SaveConfig( )**—Empty function. Included for consistency with other components.
- **SPIM_RestoreConfig( )**—Empty function. Included for consistency with other components.

15.44.2 Serial Peripheral Interface (SPI) Master Function Calls 2.50

- **void SPIM_Start(void)**—This function calls both SPIM_Init( ) and SPIM_Enable( ). This should be called the first time the component is started.
- **void SPIM_Stop(void)**—Disables SPI Master operation by disabling the internal clock and internal interrupts, if the SPI Master is configured that way.
- **void SPIM_EnableTxInt(void)**—Enables the internal Tx interrupt irq.
- **void SPIM_EnableRxInt(void)**—Enables the internal Rx interrupt irq.
- **void SPIM_DisableTxInt(void)**—Disables the internal Tx interrupt irq.
- **void SPIM_DisableRxInt(void)**—Disables the internal Rx interrupt irq.
- **void SPIM_SetTxInterruptMode(uint8 intSrc)**—Configures which status bits trigger an interrupt event.
- **void SPIM_SetRxInterruptMode(uint8 intSrc)**—uint8 intSrc: Bit field containing the interrupts to enable.
- **uint8 SPIM_ReadTxStatus(void)**—Returns the current state of the Tx status register. For more information, see Status Register.
- **uint8 SPIM_ReadRxStatus(void)**—Returns the current state of the Rx status register.
- **void SPIM_WriteTxData(uint8/uint16 txData)**—Places a byte/word in the transmit buffer to be sent at the next available SPI bus time.
- **uint8/uint16 SPIM_ReadRxData(void)**—Returns the next byte/word of received data available in the receive buffer.

---

[28] The Tx FIFO will be cleared only if software buffer is not used.

- **uint8 SPIM_GetRxBufferSize(void)**—Returns the number of bytes/words of received data currently held in the Rx buffer. If the Rx software buffer is disabled, this function returns 0 FIFO empty or 1 = FIFO not empty. If the Rx software buffer is enabled, this function returns the size of data in the Rx software buffer. FIFO data not included in this count.
- **uint8 SPIM_GetTxBufferSize(void)**—Returns the number of bytes/words of data ready to transmit currently held in the Tx buffer.
- **void SPIM_ClearRxBuffer(void)**—Clears the Rx buffer memory array and Rx hardware FIFO of all received data. Clears the Rx RAM buffer by setting both the read and write pointers to zero. Setting the pointers to zero indicates that there is no data to read. Thus, writing resumes at address 0, overwriting any data that may have remained in the RAM.
- **void SPIM_ClearTxBuffer(void)**—Clears the Tx buffer memory array of data waiting to transmit. Clears the Tx RAM buffer by setting both the read and write pointers to zero. Setting the pointers to zero indicates that there is no data to transmit. Thus, writing resumes at address 0, overwriting any data that may have remained in the RAM.[29]
- **void SPIM_TxEnable(void)**—If the SPI Master is configured to use a single bidirectional pin, this sets the bidirectional pin to transmit.
- **void SPIM_TxDisable(void)**—If the SPI master is configured to use a single bidirectional pin, this sets the bidirectional pin to receive.
- **void SPIM_PutArray(const uint8/uint16 buffer[], uint8 byteCount)**—Places an array of data into the transmit buffer.
- **void SPIM_ClearFIFO(void)**—Clears any data from the Tx and Rx FIFOs.[30]
- **void SPIM_Sleep(void)**—Prepares SPI Master to enter low-power mode . Calls SPIM_SaveConfig( ) and SPIM_Stop( ) functions.
- **void SPIM_Wakeup (void)**—Restores SPI Master configuration after exit low-power mode. Calls SPIM_RestoreConfig( ) and SPIM_Enable( ) functions. Clears all data from Rx buffer, Tx buffer and hardware FIFOs.
- **void SPIM_Init(void)**—Initializes or restores the component according to the customizer Configure dialog settings. It is not necessary to call SPIM_Init( ) because the SPIM _Start( ) routine calls this function and is the preferred method to begin component operation
- **void SPIM_Enable(void)**—Enables SPI Master for operation. Starts the internal clock if the SPI Master is configured that way. If it is configure for an external clock it must be started separately before calling this function. The SPIM_Enable( ) function should be called before SPI Master interrupts are enabled. This is because this function configures the interrupt sources and clears any pending interrupts from device configuration, and then enables the internal interrupts if there are any. A SPIM_Init( ) function must have been previously called.
- **void SPIM_SaveConfig(void)**—Empty function. Included for consistency with other components.
- **void SPIM_RestoreConfig(void)**—Empty function. Included for consistency with other components.

15.45   Serial Peripheral Interface (SPI) Slave 2.70

The SPI Slave provides an industry-standard, 4-wire slave SPI interface. It can also provide a 3-wire (bidirectional) SPI interface. Both interfaces support all four SPI operating modes, allowing

---

[29]If the software buffer is used, it does not clear data already placed in the Tx FIFO. Any data not yet transmitted from the RAM buffer is lost when overwritten by new data.

[30]Clears status register of the component.

communication with any SPI master device. In addition to the standard 8-bit word length, the SPI Slave supports a configurable 3- to 16-bit word length for communicating with nonstandard SPI word lengths.SPI signals include the standard Serial Clock (SCLK), Master In Slave Out (MISO), Master Out Slave In (MOSI), bidirectional Serial Data (SDAT), and Slave Select (SS). Features include: 3- to 16-bit data width, four SPI modes and bit rates up to 5 Mbps [82].

15.45.1 Serial Peripheral Interface (SPI) Slave Functions 2.70

- **SPIS_Start( )**—Calls both SPIS_Init( ) and SPIS_Enable( ). Should be called the first time the component is started.
- **SPIS_Stop( )**—Disables SPIS operation.
- **SPIS_EnableTxInt( )**—Enables the internal Tx interrupt irq.
- **SPIS_EnableRxInt( )**—Enables the internal Rx interrupt irq.
- **SPIS_DisableTxInt( )**—Disables the internal Tx interrupt irq.
- **SPIS_DisableRxInt( )**—Disables the internal Rx interrupt irq.
- **SPIS_SetTxInterruptMode( )**—Configures the Tx interrupt sources enabled.
- **SPIS_SetRxInterruptMode( )**—Configures the Rx interrupt sources enabled.
- **SPIS_ReadTxStatus( )**—Returns the current state of the Tx status register.
- **SPIS_ReadRxStatus( )**—Returns the current state of the Rx status register.
- **SPIS_WriteTxData( )**—Places a byte/word in the transmit buffer that will be sent at the next available bus time.
- **SPIS_WriteTxDataZero( )**—Places a byte/word in the shift register directly. This is required for SPI Modes where CPHA = 0.
- **SPIS_ReadRxData( )**—Returns the next byte/word of received data available in the receive buffer. Returns the size (in bytes/words) of received data in the Rx memory buffer.
- **SPIS_GetRxBufferSize( )**—Returns the size (in bytes/words) of data waiting to transmit in the Tx memory buffer.
- **SPIS_GetTxBufferSize( )**—Clears the Rx buffer memory array and Rx FIFO of all received data.
- **SPIS_ClearRxBuffer( )**—Clears the Rx buffer memory array and Rx FIFO of all received data.
- **SPIS_ClearTxBuffer( )**—Clears the Tx buffer memory array or Tx FIFO of all transmit data.
- **SPIS_TxEnable( )**—If configured for bidirectional mode, sets the SDAT inout to transmit.
- **SPIS_TxDisable( )**—If configured for bidirectional mode, sets the SDAT inout to receive.
- **SPIS_PutArray( )**—Places an array of data into the transmit buffer.
- **SPIS_ClearFIFO( )**—Clears the RX and TX FIFO's of all data for a fresh start.
- **SPIS_Sleep( )**—Prepares SPIS component for low-power modes by calling SPIS_SaveConfig( ) and SPIS_Stop( ) functions.
- **SPIS_Wakeup( )**—Restores and re-enables the SPIS component after waking from low-power mode.
- **SPIS_Init( )**—Initializes and restores the default SPIS configuration.
- **SPIS_Enable( )**—Enables the SPIS to start operation.
- **SPIS_SaveConfig( )**—Empty function. Included for consistency with other components. item **SPIS_RestoreConfig( )**—Empty function. Included for consistency with other components.

15.46 Serial Peripheral Interface (SPI) Slave Function Calls 2.70

- **void SPIS_Start(void)**—This is the preferred method to begin component operation. SPIS_Start( ) sets the initVar variable, calls the SPIS_Init( ) function, and then calls the SPIS_Enable( ) function.
- **void SPIS_Stop(void)**—Disables the SPI Slave component interrupts. Has no affect on the SPIS operation.
- **void SPIS_EnableTxInt(void)**—Enables the internal Tx interrupt irq.
- **void SPIS_EnableRxInt(void)**—Enables the internal Rx interrupt irq.
- **void SPIS_DisableTxInt(void)**—Disables the internal Tx interrupt irq
- **void SPIS_DisableRxInt(void)**—Disables the internal Rx interrupt irq
- **void SPIS_SetTxInterruptMode(uint8 intSrc)**—Configures the Tx interrupt sources that are enabled.
- **void SPIS_SetRxInterruptMode(uint8 intSrc)**—Configures the Rx interrupt sources that are enabled.
- **uint8 SPIS_ReadTxStatus(void)**—Returns the current state of the Tx status register. For more information, see the Status Register Bits section of this datasheet.
- **uint8 SPIS_ReadRxStatus(void)**—Returns the current state of the Rx status register. For more information see the Status Register Bits section of this datasheet.
- **void SPIS_WriteTxData(uint8/uint16 txData)**—Places a byte in the transmit buffer which will be sent at the next available bus time.
- **void SPIS_WriteTxDataZero(uint8/uint16 txData)**—Places a byte/word directly into the shift register for transmission. This byte/word will be sent to the master device during the next clock phase.
- **uint8/uint16 SPIS_ReadRxData(void)**—Reads the next byte of data received across the SPI.
- **uint8 SPIS_GetRxBufferSize(void)**—Returns the number of bytes/words of received data currently held in the Rx buffer.
- **uint8 SPIS_GetTxBufferSize(void)**—Returns the number of bytes/words of data ready to transmit currently held in the Tx buffer.
- **void SPIS_ClearRxBuffer(void)**—Clears the Rx buffer memory array and Rx hardware FIFO of all received data. Clears the Rx RAM buffer by setting both the read and write pointers to zero. Setting the pointers to zero indicates that there is no data to read. Thus, writing resumes at address 0, overwriting any data that may have remained in the RAM.
- **void SPIS_ClearTxBuffer(void)**—Clears the Tx buffer memory array of data waiting to transmit. Clears the Tx RAM buffer by setting both the read and write pointers to zero. Setting the pointers to zero indicates that there is no data to transmit. Thus, writing resumes at address 0, overwriting any data that may have remained in the RAM.
- **void SPIS_TxEnable(void)**—If the SPI Slave is configured to use a single bidirectional pin, this will set the bidirectional pin to transmit.
- **void SPIS_TxDisable(void)**—If the SPI Slave is configured to use a single bidirectional pin, this will set the bidirectional pin to receive.
- **void SPIS_PutArray(uint8/uint16 *buffer, uint8 byteCount)**—Writes available data from RAM/ROM to the Tx buffer while space is available. Keep trying until all data is passed to the Tx buffer. If using modes where CPHA = 0, call the SPIS_WriteTxDataZero( ) function before calling the SPIS_PutArray( ) function. .
- **void SPIS_ClearFIFO(void)**—Clears the RX and TX FIFO's of all data for a fresh start.[31]

---

[31] Clears the status register of the component.

- **void SPIS_Sleep(void)**—This is the preferred routine to prepare the component for low-power modes. The SPIS_Sleep( ) routine saves the current component state. Then it calls the SPIS_Stop( ) function to save the hardware configuration. Call the SPIS_Sleep( ) function before calling the CyPmSleep( ) or the CyPmHibernate( ) function.
- **void SPIS_Wakeup(void)**—This is the preferred routine to restore the component to the state when SPIS_Sleep( ) was called. The SPIS_Wakeup( ) function calls the SPIS_RestoreConfig( ) function to restore the configuration. If the component was enabled before the SPIS_Sleep( ) function was called, the SPIS_Wakeup( ) function will also re-enable the component. Clears all data from Rx buffer, Tx buffer, and hardware FIFOs.
- **void SPIS_Init(void)**—Initializes or restores the component according to the customizer Configure dialog settings. It is not necessary to call SPIS_Init( ) because the SPIS_Start( ) routine calls this function and is the preferred method to begin component operation.
- **void SPIS_Enable(void)**—Enables SPIS to start operation. Starts the internal clock if so configured. If an external clock is configured it must be started separately before calling this API. The SPIS_Enable( ) function should be called before SPIS interrupts are enabled. This is because this function configures the interrupt sources and then enables the internal interrupts if so configured. A SPIS_Init( ) function must have been previously called.
- **void SPIS_SaveConfig(void)**—Empty function. Included for consistency with other components.
- **void SPIS_RestoreConfig(void)**—Empty function. Included for consistency with other components.

15.47 Universal Asynchronous Receiver Transmitter (UART) 2.50

The UART Component provides asynchronous communications, commonly referred to as RS232 or RS485 and can be configured for Full/Half Duplex, RX only, or TX only versions. The different implementations vary only with respect to the amount of resources used. To process UART receive and transmit data, Independent circular receive and transit buffers, size configurable buffers are provided in SRAM and hardware FIFOs that ensure that data will not be missed and reduces time required for servicing the UART. Typically, the UART's baud rate, parity, number of data bits and number of start bits can reconfigured as required. "8N1," a common configuration for RS232 for eight data bits, no parity and one stop bit is the default configuration. UARTs can also be used in multidrop RS485 networks. A 9-bit addressing mode with hardware address detect and a TX output enable signal to enable the TX transceiver during transmissions is also supported. There have been many physical-layer and protocol-layer variations over time. These include, but are not limited to, RS423, DMX512, MIDI, LIN bus, legacy terminal protocols and IrDa. To support the commonly used UART variations, the Universal Asynchronous Receiver Transmitter (UART) support is provided, selection of the number of data bits, stop bits, parity, hardware flow control and parity generation and detection. A hardware-compiled option, allows the UART to output a clock and serial data stream consisting of data bits on the clock's rising edge. An independent clock and data output is provided for both the TX and RX to allow automatic calculation of the data CRC by connecting a CRC Component to the UART. 9-bit address mode with hardware address detection. Baud rates from 110 to 921,600 bps or arbitrary up to 4 Mbps, RX and TX buffers = 4 to 65,535, detection of Framing/Parity/Overrun errors, TX only/RX only optimized hardware, two out of three voting per bit, break signal generation and detection and 8x or 16x oversampling are a;so supported [102].

15.47.1 UART Functions 2.50

- **UART_GetChar( )**—Returns the next byte of received data.
- **UART_GetByte( )**—Reads the UART RX buffer immediately and returns the received character and error condition.
- **UART_GetRxBufferSize( )**—Returns the number of received bytes available in the RX buffer.
- **UART_ClearRxBuffer( )**—Clears the memory array of all received data.
- **UART_SetRxAddressMode( )**—Sets the software-controlled Addressing mode used by the RX portion of the UART.
- **UART_SetRxAddress1( )**—Sets the first of two hardware-detectable addresses.
- **UART_SetRxAddress2( )**—Sets the second of two hardware-detectable addresses.
- **UART_EnableTxInt( )**—Enables the internal interrupt irq.
- **UART_DisableTxInt( )**—Disables the internal interrupt irq.
- **UART_SetTxInterruptMode( )**—Configures the TX interrupt sources enabled.
- **UART_WriteTxData( )**—Sends a byte without checking for buffer room or status.
- **UART_ReadTxStatus( )**—Reads the status register for the TX portion of the UART.
- **UART_PutChar( )**—Puts a byte of data into the transmit buffer to be sent when the bus is available.
- **UART_PutString( )**—Places data from a string into the memory buffer for transmitting.
- **UART_PutArray( )**—Places data from a memory array into the memory buffer for transmitting.
- **UART_PutCRLF( )**—Writes a byte of data followed by a Carriage Return and Line Feed to the transmit buffer.
- **UART_GetTxBufferSize( )**—Returns the number of bytes in the TX buffer which are waiting to be transmitted.
- **UART_ClearTxBuffer( )**—Clears all data from the TX buffer Transmits a break signal on the bus.
- **UART_SetTxAddressMode( )**—Configures the transmitter to signal the next bytes as address or data.
- **UART_LoadRxConfig( )**—Loads the receiver configuration. Half Duplex UART is ready for receive byte.
- **UART_LoadTxConfig( )**—Loads the transmitter configuration. Half Duplex UART is ready for transmit byte.
- **UART_Sleep( )**—Stops the UART operation and saves the user configuration.
- **UART_Wakeup( )**—Restores and enables the user configuration UART_Init( ) Initializes default configuration provided with customizer.
- **UART_Enable( )**—Enables the UART block operation.
- **UART_SaveConfig( )**—Save the current user configuration.
- **UART_RestoreConfig( )**—Restores the user configuration.

15.47.2 UART Function Calls 2.50

- **void UART_Start(void)**—This is the preferred method to begin Component operation. UART_Start( ) sets the initVar variable, calls the UART_Init( ) function and then calls the UART_Enable( ) function.
- **uint8 UART_ReadControlRegister(void)**—Returns the current value of the control register.
- **void UART_WriteControlRegister(uint8 control)**—Writes an 8-bit value into the control register. Note that to change control register it must be read first using UART_ReadControlRegister function, modified and then written.

- **void UART_EnableRxInt(void)**—Enables the internal receiver interrupt.
- **void UART_DisableRxInt(void)**—Disables the internal receiver interrupt.
- **uint8 UART_ReadRxData(void)**—Returns the next byte of received data. This function returns data without checking the status. You must check the status separately.
- **uint8 UART_ReadRxStatus(void)**—Returns the current state of the receiver status register and the software buffer overflow status.
- **uint8 UART_GetChar(void)**—Returns the last received byte of data. UART_GetChar( ) is designed for ASCII characters and returns a uint8 where 1 to 255 are values for valid characters and 0 indicates an error occurred or no data is present.
- **uint16 UART_GetByte(void)**—Reads UART RX buffer immediately, returns received character and error condition.
- **uint8/uint16 UART_GetRxBufferSize(void)**—Returns the number of received bytes available in the RX buffer. RX software buffer is disabled (RX Buffer Size parameter is equal to 4): returns 0 for empty RX FIFO or 1 for not empty RX FIFO. RX software buffer is enabled: returns the number of bytes available in the RX software buffer. Bytes available in the RX FIFO do not take to account.
- **void UART_ClearRxBuffer(void)**—Clears the receiver memory buffer and hardware RX FIFO of all received data.
- **void UART_SetRxAddressMode(uint8 addressMode)**—Sets the software controlled Addressing mode used by the RX portion of the UART.
- **void UART_SetRxAddress1(uint8 address)**—Sets the first of two hardware-detectable receiver addresses.
- **void UART_SetRxAddress2(uint8 address)**—Sets the second of two hardware-detectable receiver addresses.
- **void UART_EnableTxInt(void)**—Enables the internal transmitter interrupt.
- **void UART_DisableTxInt(void)**—Disables the internal transmitter interrupt.
- **void UART_WriteTxData(uint8 txDataByte)**—Places a byte of data into the transmit buffer to be sent when the bus is available without checking the TX status register. You must check status separately.
- **uint8 UART_ReadTxStatus(void)**—Reads the status register for the TX portion of the UART.
- **void UART_PutChar(uint8 txDataByte)**—Puts a byte of data into the transmit buffer to be sent when the bus is be sent when the bus is available. This is a blocking API that waits until the TX buffer has room to hold the data.
- **void UART_PutString(const char8 string[])**—Sends a NULL terminated string to the TX buffer for transmission.
- **void UART_PutArray(const uint8 string[], uint8/uint16 byteCount)**—Places N bytes of data from a memory array into the TX buffer for transmission.
- **void UART_PutCRLF(uint8 txDataByte)**—Writes a byte of data followed by a carriage return (0x0D) and line feed (0x0A) to the transmit buffer.
- **uint8/uint16 UART_GetTxBufferSize(void)**—Returns the number of bytes in the TX buffer which are waiting to be transmitted. TX software buffer is disabled (TX Buffer Size parameter is equal to 4): returns 0 for empty TX FIFO, 1 for not full TX FIFO or 4 for full TX FIFO. TX software buffer is enabled: returns the number of bytes in the TX software buffer which are waiting to be transmitted. Bytes available in the TX FIFO do not take to account.
- **void UART_ClearTxBuffer(void)**—Clears all data from the TX buffer and hardware TX FIFO.

- **void UART_SendBreak(uint8 retMode)**—Transmits a break signal on the bus.[32] This can limit the break length for some UART variants. In these cases, the GPIO functionality can be used for longer break length generation.
- **void UART_SetTxAddressMode(uint8 addressMode)**—Configures the transmitter to signal the next bytes is address or data.
- **void UART_LoadRxConfig(void)**—Loads the receiver configuration in half duplex mode. After calling this function, the UART is ready to receive data.
- **void UART_LoadTxConfig(void)**—Loads the transmitter configuration in half duplex mode. After calling this function, the UART is ready to transmit data.
- **void UART_Wakeup(void)**—This is the preferred API to restore the Component to the state when UART_Sleep( ) was called. The UART_Wakeup( ) function calls the UART_RestoreConfig( ) function to restore the configuration. If the Component was enabled before the UART_Sleep( ) function was called, the UART_Wakeup( ) function will also re-enable the Component.
- **void UART_Init(void)**—Initializes or restores the Component according to the customizer Configure dialog settings. It is not necessary to call UART_Init( ) because the UART_Start( ) API calls this function and is the preferred method to begin Component operation.
- **void UART_Enable(void)**—Activates the hardware and begins Component operation. It is not necessary to call UART_Enable( ) because the UART_Start( ) API calls this function, which is the preferred method to begin Component operation.
- **void UART_SaveConfig(void)**—This function saves the Component configuration and non-retention registers. It also saves the current Component parameter values, as defined in the Configure dialog or as modified by appropriate APIs. This function is called by the UART_Sleep( ) function.
- **void UART_RestoreConfig(void)**—Restores the user configuration of non-retention registers.

15.47.3 UART Bootload Support Functions 2.50

- **UART_CyBtldrCommStart( )**—Starts the UART Component and enables its interrupt.
- **UART_CyBtldrCommStop( )**—Disables the UART Component and disables its interrupt.
- **UART_CyBtldrCommReset( )**—Resets the receive and transmit communication buffers.
- **UART_CyBtldrCommRead( )**—Allows the caller to read data from the bootloader host. This function manages polling to allow a block of data to be completely received from the host device.
- **UART_CyBtldrCommWrite( )**—Allows the caller to write data to the boot loader host. This function uses a blocking write function for writing data using UART communication Component.

15.47.4 UART Bootloader Support Function Calls 2.50

- **void UART_CyBtldrCommStart(void)**—Starts the UART communication Component.
- **void UART_CyBtldrCommStop(void)**—This function disables the UART Component and disables its interrupt.
- **void UART_CyBtldrCommReset(void)**—Resets the receive and transmit communication Buffers.
- **cystatus UART_CyBtldrCommWrite(const uint8 pData[], uint16 size, uint16 * count, uint8 timeOut)**—Allows the caller to write data to the boot loader host. This function

[32]The break signal length is defined by the UART bit time; the maximum value is 14 bits.

- **cystatus UART_CyBtldrCommWrite(const uint8 pData[], uint16 size, uint16 * count, uint8 timeOut)**—Allows the caller to write data to the boot loader host. This function uses a blocking write function for writing data using UART communication Component.

15.48 Full Speed USB (USBFS) 3.20

The USBFS Component provides a USB full-speed, Chap. 9 compliant device framework for constructing HID-based and generic USB devices and a low-level driver for the control endpoint that decodes and dispatches requests from the USB host. A GUI-based configuration dialog to aid in constructing descriptors, allows full device definition that can be imported and exported. USB Full speed device interface driver support for interrupt, control, bulk and isochronous transfer types, run-time support for descriptor set selection, USB string descriptors, USB HID class, Bootloader support, audio class (See the USBFS Audio section), MIDI devices (See the USBFS MIDI section), communications device class (CDC) (See the USBUART (CDC) section), mass storage device class (MSC) (See the USBFS MSC section), commonly used descriptor templates are provided with the Component and can be imported as needed in your design.A set of USB development tools, called SuiteUSB, is available free of charge when used with Cypress silicon [46].

15.48.1 USBFS Functions 3.20

- **USBFS_Start( )**—Activates the Component for use with the device and specific voltage mode.
- **USBFS_Init( )**—Initializes the Component's hardware.
- **USBFS_InitComponent( )**—Initializes the Component's global variables and initiates communication with host by pull-up D+ line.
- **USBFS_Stop( )**—Disables the Component.
- **SBFS_GetConfiguration( )**—Returns the currently assigned configuration. Returns 0 if the device is not configured.
- **USBFS_IsConfigurationChanged( )**—Returns the clear-on-read configuration state.
- **USBFS_GetInterfaceSetting( )**—Returns the current alternate setting for the specified interface.
- **USBFS_GetEPState( )**—Returns the current state of the specified USBFS endpoint.
- **USBFS_GetEPAckState( )**—Determines whether an ACK transaction occurred on this endpoint.
- **USBFS_GetEPCount( )**—Returns the current byte count from the specified USBFS endpoint.
- **USBFS_InitEP_DMA( )**—Initializes DMA for EP data transfers.
- **USBFS_Stop_DMA( )**—Stops DMA channel associated with endpoint.
- **USBFS_LoadInEP( )**—Loads and enables the specified USBFS endpoint for an IN transfer.
- **USBFS_LoadInEP16( )**—Loads and enables the specified USBFS endpoint for an IN transfer. This API uses the 16-bit Endpoint registers to load the data.
- **USBFS_ReadOutEP( )**—Reads the specified number of bytes from the Endpoint RAM and places it in the RAM array pointed to by pSrc. Returns the number of bytes sent by the host.
- **USBFS_ReadOutEP16( )**—Reads the specified number of bytes from the Endpoint buffer and places it in the system SRAM. Returns the number of bytes sent by the host. This API uses the 16-bit Endpoint registers to read the data.
- **USBFS_EnableOutEP( )**—Enables the specified USB endpoint to accept OUT transfers.
- **USBFS_DisableOutEP( )**—Disables the specified USB endpoint to NAK OUT transfers.
- **USBFS_SetPowerStatus( )**—Sets the device to self- powered or bus-powered.

- **USBFS_Force( )**—Forces a J, K, or SE0 State on the USB Dp/Dm pins. Normally used for remote wakeup.
- **USBFS_SerialNumString( )**—Provides the source of the USB device serial number string descriptor during run time.
- **USBFS_TerminateEP( )**—Terminates endpoint transfers.
- **BusPresent( )**—Determines VBUS presence for self-powered devices.
- **USBFS_Bcd_DetectPortType( )**—Determines if the host is capable of charging a downstream port.
- **USBFS_GetDeviceAddress( )**—Returns the currently assigned address for the USB device.
- **USBFS_EnableSofInt( )**—Enables interrupt generation when a Start-of-Frame (SOF) packet is received from the host.
- **USBFS_DisableSofInt**—Disables interrupt generation when a Start-of-Frame (SOF) packet is received from the host.

15.48.2 USBFS Function Calls 3.20

- **void USBFS_Start(uint8 device, uint8 mode)**—This function performs all required initialization for the USBFS Component. After this function call, the USB device initiates communication with the host the by pull-up D+ line. This is the preferred method to begin Component operation. Note that global interrupts have to be enabled because interrupts are required for USBFSComponent operation. PSoC 4200L devices: when USBFS Component configured to DMA with Automatic Buffer Management, the DMA interrupt priority is changed to the highest (priority 0) inside this function. PSoC 3/PSoC 5LP devices: when USBFS Component configured to DMA with Automatic Buffer Management, the Arbiter interrupt priority is changed to the highest (priority 0) inside this function.
- **void USBFS_Init(void)**—This function initializes or restores the Component according to the customizer Configure dialog settings. It is not necessary to call USBFS_Init( ) because the USBFS_Start( ) routine calls this function and is the preferred method to begin Component operation.
- **void USBFS_InitComponent(uint8 device, uint8 mode)**—This function initializes the Component's global variables and initiates communication with the host by pull-up D+ line.
- **void USBFS_InitComponent(uint8 device, uint8 mode)**—This function initializes the Component's global variables and initiates communication with the host by pull-up D+ line.
- **void USBFS_Stop(void)**—This function performs all necessary shutdowwn tasks required for the USBFS Component.
- **uint8 USBFS_GetConfiguration(void)**—This function gets the current configuration of the USB device.
- **uint8 USBFS_IsConfigurationChanged(void)**—This function returns the clear-on-read configuration state. It is useful when the host sends double SET_CONFIGURATION requests with the same configuration number or changes alternate settings of the interface. After configuration has been changed the OUT endpoints must be enabled and IN endpoint must be loaded with data to start communication with the host.
- **uint8 USBFS_GetInterfaceSetting(uint8 interfaceNumber)**—This function gets the current alternate setting for the specified interface. It is useful to identify which alternate settings are active in the specified interface.
- **uint8 USBFS_GetEPState(uint8 epNumber)**—This function returns the state of the requested endpoint.

- **uint8 USBFS_GetEPAckState(uint8 epNumber)**—This function determines whether an ACK transaction occurred on this endpoint by reading the ACK bit in the control register of the endpoint. It does not clear the ACK bit.
- **uint16 USBFS_GetEPCount(uint8 epNumber)**—This function returns the transfer count for the requested endpoint. The value from the count registers includes two counts for the two-byte checksum of the packet. This function subtracts the two counts.
- **void USBFS_InitEP_DMA(uint8 epNumber, const uint8 *pData)**—This function allocates and initializes a DMA channel to be used by the USBFS_LoadInEP( ) or USBFS_ReadOutEP( ) APIs for data transfer. It is available when the Endpoint Memory Management parameter is set to DMA. This function is automatically called from the USBFS_LoadInEP( ) and USBFS_ReadOutEP( ) APIs.
- **void USBFS_Stop_DMA(uint8 epNumber)**—This function stops DMA channel associated with endpoint. It is available when the Endpoint Buffer Management parameter is set to DMA. Call this function when endpoint direction is changed from IN to OUT or vice versa to trigger DMA re-configuration when USBFS_LoadInEP( ) or USBFS_ReadOutEP( ) functions are called the first time.
- **void USBFS_LoadInEP(uint8 epNumber, const uint8 pData[], uint16 length)**—This function performs different functionality depending on the Component's configured Endpoint Buffer Management. This parameter is defined in the Descriptor Root section in Component Configure window. Manual (Static/Dynamic Allocation): This function loads and enables the specified USB data endpoint for an IN data transfer. DMA with Manual Buffer Management: Configures DMA for a data transfer from system SRAM to endpoint buffer. Generates request for a transfer.
  DMA with Automatic Buffer Management:
  1. Configure DMA. This is required only once, with parameter pData is not NULL.
  2. Initiate DMA transaction on demand, with the pData pointer is NULL. Sets Data ready status: This generates the first DMA transfer and prepares data in endpoint buffer.
- **void USBFS_LoadInEP16(uint8 epNumber, const uint8 pData[], uint16 length)**—This function performs different functionality depending on the Component's configured Endpoint Buffer Management. This parameter is defined in the Descriptor Root section in Component Configure window. Manual (Static/Dynamic Allocation): This function loads and enables the specified USB data endpoint for an IN data transfer. DMA with Manual Buffer Management: Configures DMA for a data transfer from system SRAM to endpoint buffer. Generates request for a transfer. DMA with Automatic Buffer Management: 1. Configure DMA. This is required only once, with parameter pData is not NULL. 2. Initiate DMA transaction on demand, with the pData pointer is NULL. Sets Data ready status: This generates the first DMA transfer and prepares data in endpoint buffer.
- **uint16 USBFS_ReadOutEP(uint8 epNumber, uint8 pData[], uint16 length)**—This function performs different functionality depending on the Component's configured Endpoint Buffer Management. This parameter is defined in the Descriptor Root section in Component Configure window. Manual (Static/Dynamic Allocation): This function copies the specified number of bytes from endpoint buffer to system SRAM buffer. After data has been copied the endpoint is released to allow the host to write next data. The function does not support partial data reads therefore all received bytes has to be read at once. The length argument must be equal to the number of actually received bytes from the host. Call function USBFS_GetEPCount( ) to get actual number of received bytes. DMA with Manual Buffer Management: Configures DMA to transfer data from endpoint buffer to system SRAM buffer and generates a DMA request. The firmware must wait until the DMA completes the data transfer after calling the USB_ReadOutEP( ) API. For example, by checking endpoint state:

```
while  (USB\_OUT\_BUFFER\_FULL  ==  USBFS\_GetEPState
  (OUT_\EP))
{
}
```

- **void USBFS_EnableOutEP(uint8 epNumber)**—This function enables the specified endpoint for OUT transfers. Do not call this function for IN endpoints. The USBFS_EnableOutEP( ) has to be called to allow the host to write data into the endpoint buffer after DMA has completed transfer data from OUT endpoint buffer to system SRAM buffer. The function does not support partial data reads therefore all received bytes has to be read at once. The length argument must be equal to the number of actually received bytes from the host. Call function USBFS_GetEPCount( ) to get actual number of received bytes. DMA with Automatic Buffer Management: Configures DMA to transfer data from endpoint buffer to system SRAM buffer. Generally, this function should be called once to configure DMA for operation. Then use USBFS_EnableOutEP( ) to release endpoint to allow the host to write next data. The allocated buffer size and length parameter must be equal to endpoint maximum packet size. Note We recommend calling it with length equal to endpoint maximum packet size to read all received data bytes. Use return value to get actual number of received bytes.
- **void USBFS_DisableOutEP(uint8 epNumber)**—This function disables the specified USBFS OUT endpoint. Do not call this function for IN endpoints.
- **void USBFS_SetPowerStatus(uint8 powerStatus)**—This function sets the current power status. The device replies to USB GET_STATUS requests based on this value. This allows the device to properly report its status for USB Chap. 9 compliance. Devices can change their power source from self-powered to bus-powered at any time and report their current power source as part of the device status. You should call this function any time your device changes from self-powered to bus-powered or vice versa and set the status appropriately.
- **void USBFS_Force(uint8 state)**—This function forces a USB J, K, or SE0 state on the D+/D− lines. It provides the necessary mechanism for a USB device application to perform a USB Remote Wakeup.
- **void USBFS_SerialNumString(uint8 snString[])**—This function is available only when the User Call Back option in the Serial Number String descriptor properties is selected. Application firmware can provide the source of the USB device serial number string descriptor during run time. The default string is used if the application firmware does not use this function or sets the wrong string descriptor.
- **void USBFS_TerminateEP(uint8 epNumber)**—This function terminates the specified USBFS endpoint. This function should be used before endpoint reconfiguration.
- **uint8 USBFS_VBusPresent(void)**—Determines VBUS presence for self-powered devices.
- **uint8 USBFS_Bcd_DetectPortType (void)**—This function implements the USB Battery Charger Detection (BCD) algorithm to determine the type of USB host downstream port. This API is available only for PSoC 4 devices and should be called when the VBUS voltage transition (OFF to ON) is detected on the bus. If the USB device functionality is enabled, this API first calls USBFS_Stop( ) API internally to disable the USB device functionality and then proceeds to implement the BCD algorithm to detect the USB host port type. The USBFS_Start( ) API should be called after this API if the USB communication needs to be initiated with the host. This API is generated only if the 'Enable Battery Charging Detection' option is enabled in the 'Advanced' tab of the Component GUI. API implements the steps two to four of the BCD algorithm which are Data Contact Detect, Primary Detection, an Secondary Detection. The first step of BCD algorithm, namely VBUS detection, shall be handled at the application firmware level.

- **uint8 USBFS_GetDeviceAddress(void)**—This function returns the currently assigned address for the USB device.
- **void USBFS_EnableSofInt(void)**—This function enables interrupt generation when a Start-of-Frame (SOF) packet is received from the host.
- **void USBFS_DisableSofInt(void)**—This function disables interrupt generation when a Start-of-Frame (SOF) packet is received from the host.
- **uint8 USBFS_UpdateHIDTimer(uint8 interface)**—This function updates the HID Report idle timer and returns the status and reloads the timer if it expires.
- **uint8 USBFS_GetProtocol(uint8 interface)**—This function returns the HID protocol value for the selected interface.

15.48.3 USBFS Bootloader Support 3.20

The USBFS Component provides a USB full-speed, Chap. 9 compliant device framework for constructing HID-based and generic USB devices. It provides a low-level driver for the control endpoint that decodes and dispatches requests from the USB host. Additionally, the Component provides a GUI-based configuration dialog to aid in constructing your descriptors, allowing full device definition that can be imported and exported. Commonly used descriptor templates are provided with the Component and can be imported as needed in your design. Cypress offers a set of USB development tools, called SuiteUSB, available free of charge when used with Cypress silicon.[33] Features include: USB Full Speed device interface driver, support for interrupt, control, bulk and isochronous transfer types, run-time support for descriptor set selection, USB string descriptors, USB HID class support and Bootloader support, audio class support (See the USBFS Audio section), MIDI devices support (See the USBFS MIDI section), communications device class (CDC) support (See the USBUART (CDC) section) and a mass storage device class (MSC) support (See the USBFS MSC section).

15.48.4 USBFS Bootloader Support Functions 3.20

- **USBFS_CyBtldrCommStart( )**—Performs all required initialization for the USBFS Component, waits on enumeration and enables communication.
- **USBFS_CyBtldrCommStop( )**—Calls the USBFS_Stop( ) function.
- **USBFS_CyBtldrCommReset( )**—Resets the receive and transmit communication buffers.
- **USBFS_CyBtldrCommWrite( )**—Allows the caller to write data to the bootloader host. The function handles polling to allow a block of data to be completely sent to the host device.
- **USBFS_CyBtldrCommRead( )**—Allows the caller to read data from the bootloader host. The function handles polling to allow a block of data to be completely received from the host device.

15.48.5 USBFS Bootloader Support Function Calls 3.20

- **void USBFS_CyBtldrCommStart(void)**—This function performs all required initialization for the USBFS Component, waits on enumeration and enables communication.

---

[33] SuiteUSB is available from the Cypress website: http://www.cypress.com.

- **void USBFS_CyBtldrCommStop(void)**—This function performs all necessary shutdowwn tasks required for the USBFS Component.
- **void USBFS_CyBtldrCommReset(void)**—This function resets receive and transmit communication buffers.
- **cystatus USBFS_CyBtldrCommWrite(const uint8 pData[], uint16 size, uint16 *count, uint8 timeOut)**—Sends data to the host controller. A timeout is enabled. Reports the number of bytes successfully sent.
- **cystatus USBFS_CyBtldrCommRead(uint8 pData[], uint16 size, uint16 *count, uint8 time-Out)**—Receives data from the host controller. A timeout is enabled. Reports the number of bytes successfully read.

15.48.6 USB Suspend, Resume and Remote Wakeup 3.20

15.48.6.1 USB Suspend, Resume and Remote Wakeup Functions
- **SBFS_CheckActivity( )**—Returns the activity status of the bus since the last call the function.
- **USBFS_Suspend( )**—Prepares the USBFS Component to enter low power mode.
- **USBFS_Resume( )**—Prepares the USBFS Component for active mode operation after exit low power mode.
- **USBFS_RWUEnabled( )**—Returns current remote wakeup status.

15.48.6.2 USB Suspend, Resume and Remote Wakeup Function Calls 3.20
- **uint8 USBFS_CheckActivity(void)**—This function returns the activity status of the bus. It clears the hardware status to provide updated status on the next call of this function. It provides a way to determine whether any USB bus activity occurred. The application should use this function to determine if the USB suspend conditions are met.
- **void USBFS_Suspend(void)**—This function prepares the USBFS Component to enter low power mode. The interrupt on falling edge on Dp pin is configured to wakeup device when the host drives resume condition. The pull-up is enabled on the Dp line while device is in low power mode. The supported low power modes are Deep Sleep (PSoC 4200L) and Sleep (PSoC 3/ PSoC 5LP). Note For PSoC 4200L devices, this function should not be called before entering device Sleep mode. PSoC 4200L Sleep mode is only for CPU suspending. Note After enter low power mode, the data which is left in the IN or OUT endpoint buffers is not restored after wakeup and lost. Therefore it should be stored in the SRAM for OUT endpoint or read by the host for IN endpoint before enter low power mode.
- **void USBFS_Resume(void)**—This function prepares the USBFS Component for active mode operation after exit low power mode. It restores the Component active mode configuration such as device address assigned previously by the host, endpoints buffer and disables interrupt on Dp pin. The supported low power modes are Deep Sleep (PSoC 4200L) and Sleep (PSoC 3/PSoC 5LP). Note For PSoC 4200L devices, this function should not be called after exiting Sleep. Note To resume communication with the host, the data endpoints must be managed: the OUT endpoints must be enabled and IN endpoints must be loaded with data. For DMA with Automatic Buffer Management, all endpoints buffers must be initialized again before making them available to the host.
- **uint8 USBFS_RWUEnabled(void)**—This function returns the current remote wakeup status. If the device supports remote wakeup, the application should use this function to determine if remote wakeup was enabled by the host. When the device is suspended and it determines the conditions to

initiate a remote wakeup are met, the application should use the USBFS_Force( ) function to force the appropriate J and K states onto the USB bus, signaling a remote wakeup.

15.48.7 Link Power Management (LPM) Support

The ADC can be configured into single-ended mode (0-4.096 V range or 0-2.048 V range) For power converter voltage measurements. The ADC can also be configurable into differential mode ($\pm$2.048 V range) to support remote sensing of voltages where the remote ground reference is returned to PSoC over a PCB trace. In cases where the analog voltage to be monitored equals or exceeds Vdda or the ADC range, external resistor dividers are recommended to scale the monitored voltages down to an appropriate range. Interfaces to up to 32 DC-DC power converters, measures power converter output voltages and load currents using a DelSig-ADC, monitors the health of the power converters generating warnings and faults based on user-defined thresholds, support for measuring other auxiliary voltages in the system and 3.3 V and 5 V chip power supplies.

15.48.7.1 Link Power Management (LPM) Support Functions

- **USBFS_Lpm_GetBeslValue( )**—Returns the BESL value sent by the host.
- **USBFS_Lpm_RemoteWakeUpAllowed( )**—Return the remote wake up permission set for the device by host.
- **USBFS_Lpm_SetResponse( )**—Set the response for the received LPM token from host.
- **USBFS_Lpm_GetResponse( )**—Get the current response for the received LPM token from host.

15.48.7.2 Link Power Management (LPM) Support Function Calls

- **uint32 USBFS_Lpm_GetBeslValue (void)**—This function returns the Best Effort Service Latency (BESL) value sent by the host as part of the LPM token transaction.
- **uint32 USBFS_Lpm_RemoteWakeUpAllowed (void)**—This function returns the remote wakeup permission set for the device by the host as part of the LPM token transaction.
- **void USBFS_Lpm_SetResponse(uint32 response)**—This function configures the response in the handshake packet the device has to send when an LPM token packet is received.
- **uint32 USBFS_Lpm_Get response(void)**—This function returns the currently configured response value that the device will send as part of the handshake packet when an LPM token packet is received.

15.49 Status Register 1.90

The Status Register allows the firmware to read digital signals and supports up to an 8-bit status register and interrupts [89].

15.49.1 Status Counter Functions 1.90

- **StatusReg_Read( )**—Reads the current value of the status register
- **StatusReg_InterruptEnable( )**—Enables the status register interrupt
- **StatusReg_InterruptDisable( )**—Disables the status register interrupt
- **StatusReg_WriteMask( )**—Writes the value assigned to the mask register
- **StatusReg_ReadMask( )**—Returns the current interrupt mask value from the mask register

15.49.2 Status Register Function Calls 1.90

- **uint8 StatusReg_Read (void)**—Reads the value of a status register.
- **void StatusReg_InterruptEnable (void)**—Enables the status register interrupt. The default behavior is disabled. This is only valid if the status register generates an interrupt.
- **void StatusReg_InterruptDisable (void)**—Disables the status register interrupt. This is only valid, if the status register generates an interrupt.
- **void StatusReg_WriteMask (uint8 mask)**—Writes the current mask value assigned to the status register. This is only valid if the status register generates an interrupt.
- **uint8 StatusReg_ReadMask (void)**—Reads the current interrupt mask value assigned for the status register. This is only valid if the status register generates an interrupt.

15.50 Counter 3.0

This component provides a method to count events. It can implement a basic counter function and offers advanced features such as capture, compare output and count direction control. For PSoC 3 and PSoC 5LP devices component can be implemented using FF blocks or UDB. PSoC 4 devices support only UDB implementation. A UDB implementation typically has more features than an FF implementation. If the design is simple enough, a FF cab be used to conserve UDB resources for other purposes [22].

15.50.1 Counter Functions 3.0

- **Counter_Start( )**—Sets the initVar variable, calls the Counter_Init( ) function and then calls the Enable function
- **Counter_Stop( )**—Disables the Counter
- **Counter_SetInterruptMode( )**—Enables or disables the sources of the interrupt output
- **Counter_ReadStatusRegister( )**—Returns the current state of the status register
- **Counter_ReadControlRegister( )**—Returns the current state of the control register
- **Counter_WriteControlRegister( )**—Sets the bit-field of the control register
- **Counter_WriteCounter( )**—Writes a new value directly into the counter register
- **Counter_ReadCounter( )**—Forces a capture and then returns the capture value
- **Counter_ReadCapture( )**—Returns the contents of the capture register or the output of the FIFO
- **Counter_WritePeriod( )**—Writes the period register
- **Counter_ReadPeriod( )**—Reads the period register
- **Counter_WriteCompare( )**—Writes the compare register
- **Counter_ReadCompare( )**—Reads the compare register
- **Counter_SetCompareMode( )**—Sets the compare mode
- **Counter_SetCaptureMode( )**—Sets the capture mode
- **Counter_ClearFIFO( )**—Clears the capture FIFO
- **Counter_Sleep( )**—Stops the Counter and saves the user configuration
- **Counter_Wakeup( )**—Restores and enables the user configuration
- **Counter_Init( )**—Initializes or restores the Counter according to the Configure dialog settings
- **Counter_Enable( )**—Enables the Counter
- **Counter_SaveConfig( )**—Saves the Counter configuration
- **Counter_RestoreConfig( )**—Restores the Counter configuration

15.50.2 Counter Function Calls 3.0

- **void Counter_Start(void)**—This is the preferred method to begin component operation. Counter_Start( ) sets the initVar variable, calls the Counter_Init( ) function and then calls the Counter_Enable( ) function.
- **void Counter_Stop(void)**—This function disables the Counter only in software enable modes.
- **void Counter_SetInterruptMode(uint8 interruptSource)**—This function enables or disables the sources of the interrupt output.
- **uint8 Counter_ReadStatusRegister(void)**—This function returns the current state of the status register.
- **uint8 Counter_ReadControlRegister(void)**—This function returns the current state of the control register.
- **void Counter_WriteControlRegister(uint8 control)**—This function sets the bit field of the control register. It is available only if one of the modes defined in the control register is actually used.
- **void Counter_WriteCounter(uint8/16/32 count)**—This function writes a new value directly into the counter register.
- **uint8/16/32 Counter_ReadCounter(void)**—This function forces a capture then returns the capture value. The capture that occurs is not considered a capture event and does not cause the counter to be reset or trigger an interrupt.
- **uint8/16/32 Counter_ReadCapture(void)**—This function returns the contents of the capture register or the output of the FIFO (UDB only).
- **void Counter_WritePeriod(uint8/16/32 period)**—This function writes the period register.
- **void Counter_WriteCompare(uint8/16/32 compare)**—This function writes the compare register. It is available only for the UDB implementation.
- **uint8/16/32 Counter_ReadCompare(void)**—This function reads the compare register. It is available only for UDB implementation.
- **void Counter_SetCaptureMode(uint8 captureMode)**—This function sets the capture mode. It is available only for UDB implementation and when the Capture Mode parameter is set to Software Controlled.
- **void Counter_ClearFIFO(void)**—This function clears the capture FIFO. It is available only for UDB implementation.
- **void Counter_Sleep(void)**—This is the preferred routine to prepare the component for sleep. The Counter_Sleep( ) routine saves the current component state. Then it calls the Counter_Stop( ) function and calls Counter_SaveConfig( ) to save the hardware configuration. Call the Counter_Sleep( ) function before calling the CyPmSleep( ) or the CyPmHibernate( ) function.
- **void Counter_Wakeup(void)**—This is the preferred routine to restore the component to the state when Counter_Sleep( ) was called. The Counter_Wakeup( ) function calls the Counter_RestoreConfig( ) function to restore the configuration. If the component was enabled before the Counter_Sleep( ) function was called, the Counter_Wakeup( ) function also re-enables the component.
- **void Counter_Init(void)**—Initializes or restores the component according to the customizer Configure dialog settings. It is not necessary to call Counter_Init( ) because the Counter_Start( ) routine calls this function and is the preferred method to begin component operation.
- **void Counter_Enable(void)**—Activates the hardware and begins component operation. It is not necessary to call Counter_Enable( ) because the Counter_Start( ) routine calls this function, which is the preferred method to begin component operation. This function enables the Counter for either of the software controlled enable modes.

- **void Counter_SaveConfig(void)**—This function saves the component configuration and non-retention registers. It also saves the current component parameter values, as defined in the Configure dialog or as modified by appropriate APIs. This function is called by the Counter_Sleep( ) function.
- **void Counter_RestoreConfig(void)**—This function restores the component configuration and non-retention registers. It also restores the component parameter values to what they were before calling the Counter_Sleep( ) function.

15.51 Basic Counter 1.0

The Basic Counter component provides a selectable-width up-counter, implemented in PLD macrocells. This counter should be used when the bussed counter value needs to be routed, or when small, basic counter functionality is all that is necessary: Mux Sequencer: Connect the cnt output to the input of a mux to easily sequence signals, small Counter: Count level events on the en input without consuming any datapath resources or sas a mall timer to measure the number of clocks between events without consuming any datapath resources [12].

15.52 Basic Counter Functions 1.0

NONE.

15.53 Cyclic Redundancy Check (CRC) 2.50

The default use of the Cyclic Redundancy Check (CRC) Component is to compute the CRC from a serial bit stream of any length. The input data is sampled on the rising edge of the data clock. The CRC value is reset to 0 before starting or can optionally be seeded with an initial value. On completion of the bitstream, the computed CRC value may be read out. Features include: 1 to 64 bits, time Division Multiplexing mode, requires clock and data for serial bit stream input, serial data in, parallel result, standard [CRC-1 (parity bit), CRC-4 (ITU-T G.704), CRC-5-USB, etc.] or custom polynomial, standard or custom seed value and enable input provides synchronized operation with other Components [23].

15.53.1 CRC Functions 2.50

- **CRC_Start( )**—Initializes seed and polynomial registers with initial values. Computation of CRC starts on rising edge of input clock.
- **CRC_Stop( )**—Stops CRC computation.
- **CRC_Sleep( )**—Stops CRC computation and saves the CRC configuration.
- **CRC_Wakeup( )**—Restores the CRC configuration and starts CRC computation on rising edge of input clock.
- **CRC_Init( )**—Initializes the seed and polynomial registers with initial values.
- **CRC_Enable( )**—Starts CRC computation on rising edge of input clock.
- **CRC_SaveConfig( )**—Saves the seed and polynomial registers.
- **CRC_RestoreConfig( )**—Restores the seed and polynomial registers.

- **CRC_WriteSeed( )**—Writes the seed value.
- **CRC_WriteSeedUpper( )**—Writes the upper half of the seed value. Only generated for 33- to 64-bit CRC.
- **CRC_WriteSeedLower( )**—Writes the lower half of the seed value. Only generated for 33- to 64-bit CRC.
- **CRC_ReadCRC( )**—Reads the CRC value.
- **CRC_ReadCRCUpper( )**—Reads the upper half of the CRC value. Only generated for 33- to 64-bit CRC.
- **CRC_ReadCRCLower( )**—Reads the lower half of the CRC value. Only generated for 33- to 64-bit CRC.
- **CRC_WritePolynomial( )**—Writes the CRC polynomial value.
- **CRC_WritePolynomialUpper( )**—Writes the upper half of the CRC polynomial value. Only generated for 33- to 64-bit CRC.
- **CRC_WritePolynomialLower( )**—Writes the lower half of the CRC polynomial value. Only generated for 33- to 64-bit CRC.
- **CRC_ReadPolynomial( )**—Reads the CRC polynomial value.
- **CRC_ReadPolynomialUpper( )**—Reads the upper half of the CRC polynomial value. Only generated for 33- to 64-bit CRC.
- **CRC_ReadPolynomialLower( )**—Reads the lower half of the CRC polynomial value. Only generated for 33- to 64-bit CRC.

15.53.2 CRC Function Calls 2.50

- **void CRC_Start(void)**—Initializes seed and polynomial registers with initial values. Computation of CRC starts on rising edge of input clock.
- **void CRC_Stop(void)**—Stops CRC computation.
- **void CRC_Sleep(void)**—Stops CRC computation and saves the CRC configuration.
- **void CRC_Wakeup(void)**—Restores the CRC configuration and starts CRC computation on the rising edge of the input clock.
- **void CRC_Init(void)**—Initializes the seed and polynomial registers with initial values.
- **void CRC_Enable(void)**—Starts CRC computation on the rising edge of the input clock.
- **void CRC_SaveConfig(void)**—Saves the initial seed and polynomial registers.
- **void CRC_RestoreConfig(void)**—Restores the initial seed and polynomial registers.
- **void CRC_WriteSeed(uint8/16/32 seed)**—Writes the seed value.
- **void CRC_WriteSeedUpper(uint32 seed)**—Writes the upper half of the seed value. Only generated for 33- to 64-bit CRC.
- **void CRC_WriteSeedLower(uint32 seed)**—Writes the lower half of the seed value. Only generated for 33- to 64-bit CRC.
- **uint8/16/32 CRC_ReadCRC(void)**—Reads the CRC value.
- **uint32 CRC_ReadCRCUpper(void)**—Reads the upper half of the CRC value. Only generated for 33- to 64-bit CRC.
- **uint32 CRC_ReadCRCLower(void)**—Reads the lower half of the CRC value. Only generated for 33- to 64-bit CRC.
- **void CRC_WritePolynomial(uint8/16/32 polynomial)**—Writes the CRC polynomial value.
- **void CRC_WritePolynomialUpper(uint32 polynomial)**—Writes the upper half of the CRC polynomial value. Only generated for 33- to 64-bit CRC.

- **void CRC_WritePolynomialLower(uint32 polynomial)**—Writes the lower half of the CRC polynomial value. Only generated for 33- to 64-bit
- **uint8/16/32 CRC_ReadPolynomial(void)**—Reads the CRC polynomial value.
- **uint32 CRC_ReadPolynomialUpper(void)**—Reads the upper half of the CRC polynomial value. Only generated for 33- to 64-bit CRC.
- **uint32 CRC_ReadPolynomialLower(void)**—Reads the lower half of the CRC polynomial value. Only generated for 33- to 64-bit CRC.

15.54 Precision Illumination Signal Modulation (PRISM) 2.20

The Precision Illumination Signal Modulation (PrISM) component uses a linear feedback shift register (LFSR) to generate a pseudo random sequence. The sequence outputs a pseudo random bit stream, as well as up to two user-adjustable pseudo random pulse densities. The pulse densities may range from 0 to 100%. The LFSR is of the Galois form (sometimes known as the modular form) and uses the provided maximal length codes. The PrISM component runs continuously after it starts and as long as the enable input is held high. The PrISM pseudo random number generator can be started with any valid seed value, excluding 0. Features include: Programmable flicker-free dimming resolution from 2 to 32 bit, two pulse density outputs, programmable output signal density, serial output bit stream, continuous run mode, user-configurable sequence start value, standard or custom polynomials provided for all sequence lengths, kill input disables pulse density outputs and forces them low, enable input provides synchronized operation with other components, reset input allows restart at sequence start value for synchronization with other components and terminal Count Output for 8-, 16-, 24- and 32-bit sequence lengths [67].

15.54.1 PRISM Functions 2.20

- **PrISM_Start( )**—The start function sets polynomial, seed and pulse density registers provided by the customizer.
- **PrISM_Stop( )**—Stops PrISM computation.
- **PrISM_SetPulse0Mode( )**—Sets the pulse density type for Density0.
- **PrISM_SetPulse1Mode( )**—Sets the pulse density type for Density1.
- **PrISM_ReadSeed( )**—Reads the PrISM Seed register.
- **PrISM_WriteSeed( )**—Writes the PrISM Seed register with the start value.
- **PrISM_ReadPolynomial( )**—Reads the PrISM Polynomial register.
- **PrISM_WritePolynomial( )**—Writes the PrISM Polynomial register with the start value.
- **PrISM_ReadPulse0( )**—Reads the PrISM Pulse Density0 value register.
- **PrISM_WritePulse0( )**—Writes the PrISM Pulse Density0 value register with the new Pulse Density value.
- **PrISM_ReadPulse1( )**—Reads the PrISM Pulse Density1 value register.
- **PrISM_WritePulse1( )**—Writes the PrISM Pulse Density1 value register with the new Pulse Density value.
- **PrISM_Sleep( )**—Stops and saves the user configuration.
- **PrISM_Wakeup( )**—Restores and enables the user configuration
- **PrISM_Init( )**—Initializes the default configuration provided with the customizer.
- **PrISM_Enable( )**—Enables the PrISM block operation.
- **PrISM_SaveConfig( )**—Saves the current user configuration.

- **PrISM_RestoreConfig( )**—Restores the current user configuration.

15.54.2 PRISM Function Calls 2.20

- **void PrISM_Start(void)**—This is the preferred method to begin component operation. PrISM_Start( ) sets the initVar variable, calls the PrISM_Init( ) function and then calls the PrISM_Enable( ) function. The start function sets polynomial, seed and pulse density registers provided by the customizer. PrISM computation starts on the rising edge of the input clock.
- **void PrISM_Stop(void)**—Stops PrISM computation. Outputs remain constant.
- **void PrISM_SetPulse0Mode(uint8 pulse0Type)**—Sets the pulse density type for Density0. Less Than or Equal(<=) or Greater Than or Equal(>=).
- **uint8/16/32 PrISM_ReadSeed(void)**—Reads the PrISM seed register.
- **void PrISM_WriteSeed(uint8/16/32 seed)**—Writes the PrISM seed register with the start value.
- **uint8/16/32 PrISM_ReadPolynomial(void)**—Reads the PrISM polynomial.
- **void PrISM_WritePolynomial(uint8/16/32 polynomial)**—Writes the PrISM polynomial.
- **uint8/16/32 PrISM_ReadPulse0(void)**—Reads the PrISM PulseDensity0 value register.
- **void PrISM_WritePulse0(uint8/16/32 pulseDensity0)**—Writes the PrISM Pulse Density0 value register with the new Pulse Density value.
- **uint8/16/32 PrISM_ReadPulse1(void)**—Reads the PrISM Pulse Density1 value register.
- **void PrISM_WritePulse1(uint8/16/32 pulseDensity1)**—Writes the PrISM Pulse Density1 value register with the new Pulse Density value.
- **void PrISM_Sleep(void)**—This is the preferred API to prepare the component for sleep. The PrISM_Sleep( ) API saves the current component state. Then it calls the PrISM_Stop( ) function and calls PrISM_SaveConfig( ) to save the hardware configuration. Call the PrISM_Sleep( ) function before calling the CyPmSleep( ) or the CyPmHibernate( ) function.
- **void PrISM_Wakeup(void)**—this is the preferred API to restore the component to the state when PrISM_Sleep( ) was called. The PrISM_Wakeup( ) function calls the PrISM_RestoreConfig( ) function to restore the configuration. If the component was enabled before the PrISM_Sleep( ) function was called, the PrISM_Wakeup( ) function also re-enables the component.
- **void PrISM_Init(void)**—Initializes or restores the component according to the customizer Configure dialog settings. It is not necessary to call PrISM_Init( ) because the PrISM_Start( ) API calls this function and is the preferred method to begin the component operation.
- **void PrISM_Enable(void)**—Activates the hardware and begins component operation. It is not necessary to call PrISM_Enable( ) because the PrISM_Start( ) API calls this function, which is the preferred method to begin the component operation.
- **void PrISM_SaveConfig(void)**—This function saves the component configuration and non-retention registers. It also saves the current component parameter values, as defined in the Configure dialog or as modified by appropriate APIs. This function is called by the PrISM_Sleep( ) function.
- **void PrISM_RestoreConfig(void)**—This function restores the component configuration and non-retention registers. It also restores the component parameter values to what they were before calling the PrISM_Sleep( ) function.

15.55 Pseudo Random Sequence (PRS) 2.40

The Pseudo Random Sequence (PRS) component uses an LFSR to generate a pseudo random sequence, which outputs a pseudo random bit stream. The LFSR is of the Galois form (sometimes known as the modular form) and uses the provided maximal code length, or period. The PRS component runs continuously after starting as long as the Enable Input is held high. The PRS number generator can be started with any valid seed value other than 0. Features include: Time Division Multiplexing mode, serial output bit stream, continuous or single-step run modes, standard or custom polynomial, standard or custom seed value, enable input provides synchronized operation with other components, 2 to 64 bits PRS sequence length and computed pseudo random number can be read directly from the linear feedback shift register (LFSR) [68].

15.55.1 PRS Functions 2.40

- **PRS_Start( )**—Initializes seed and polynomial registers provided from customizer. PRS computation starts on rising edge of input clock.
- **PRS_Stop( )**—Stops PRS computation.
- **PRS_Sleep( )**—Stops PRS computation and saves PRS configuration.
- **PRS_Wakeup( )**—Restores PRS configuration and starts PRS computation on rising edge of input clock.
- **PRS_Init( )**—Initializes seed and polynomial registers with initial values.
- **PRS_Enable( )**—Starts PRS computation on rising edge of input clock.
- **PRS_SaveConfig( )**—Saves seed and polynomial registers.
- **PRS_RestoreConfig( )**—Restores seed and polynomial registers.
- **PRS_Step( )**—Increments the PRS by one when using API single-step mode.
- **PRS_WriteSeed( )**—Writes seed value.
- **PRS_WriteSeedUpper( )**—Writes upper half of seed value. Only generated for 33 to 64 bits PRS.
- **PRS_WriteSeedLower( )**—Writes lower half of seed value. Only generated for 33 to 64 bits PRS.
- **PRS_Read( )**—Reads PRS value.
- **PRS_ReadUpper( )**—Reads upper half of PRS value. Only generated for 33 to 64 bits PRS.
- **PRS_ReadLower( )**—Reads lower half of PRS value. Only generated for 33 to 64 bits PRS.
- **PRS_WritePolynomial( )**—Writes PRS polynomial value.
- **PRS_WritePolynomialUpper( )**—Writes upper half of PRS polynomial value. Only generated for 33 to 64 bits PRS.
- **PRS_WritePolynomialLower( )**—Writes lower half of PRS polynomial value. Only generated for 33 to 64 bits PRS. PRS_ReadPolynomial( ) Reads PRS polynomial value.
- **PRS_ReadPolynomialUpper( )**—Reads upper half of PRS polynomial value. Only generated for 33 to 64 bits PRS.
- **PRS_ReadPolynomialLower( )**—Reads lower half of PRS polynomial value. Only generated for 33 to 64 bits PRS.

15.55.2 PRS Function Calls 2.40

- **void PRS_Start(void)**—Initializes the seed and polynomial registers. PRS computation starts on the rising edge of the input clock.

- **void PRS_Stop(void)**—Stops PRS computation.
- **void PRS_Sleep(void)**—Stops PRS computation and saves the PRS configuration.
- **void PRS_Wakeup(void)**—Restores the PRS configuration and starts PRS computation on the rising edge of the input clock.
- **void PRS_Init(void)**—Initializes the seed and polynomial registers with initial values.
- **void PRS_Enable(void)**—Starts PRS computation on the rising edge of the input clock.
- **void PRS_SaveConfig(void)**—Saves the seed and polynomial registers.
- **void PRS_RestoreConfig(void)**—Restores the seed and polynomial registers.
- **void PRS_Step(void)**—Increments the PRS by one when API single-step mode is used.
- **void PRS_WriteSeed(uint8/16/32 seed)**—Writes the seed value.
- **void PRS_WriteSeedUpper(uint32 seed)**—Writes the upper half of the seed value. Only generated for 33 to 64 bits PRS.
- **void PRS_WriteSeedLower(uint32 seed)**—Writes the lower half of the seed value. Only generated for 33 to 64 bits PRS.
- **uint8/16/32 PRS_Read(void)**—Reads the PRS value.
- **uint32 PRS_ReadUpper(void)**—Reads the upper half of the PRS value.
- **void PRS_WritePolynomialUpper(uint32 polynomial)**—Writes the upper half of the PRS polynomial value. Only generated for 33 to 64 bits PRS.
- **void PRS_WritePolynomialLower(uint32 polynomial)**—Writes the lower half of the PRS polynomial value. Only generated for 33 to 64 bits PRS.
- **int8/16/32 PRS_ReadPolynomial(void)**—Reads the PRS polynomial value.
- **uint32 PRS_ReadPolynomialUpper(void)**—Reads the upper half of the PRS polynomial value. Only generated for 33 to 64 bits PRS.
- **uint32 PRS_ReadPolynomialLower(void)**—Reads the lower half of the PRS polynomial value. Only generated for 33 to 64 bits PRS.

15.56 Pulse Width Modulator (PWM) 3.30

The PWM component provides compare outputs to generate single or continuous timing and control signals in hardware. The PWM provides an easy method of generating complex real-time events accurately with minimal CPU intervention. PWM features may be combined with other analog and digital components to create custom peripherals. For PSoC 3 and PSoC 5LP devices, the component can be implemented using FF blocks or universal digital blocks (UDBs). A UDB implementation typically has more features than an FF implementation. If the design is simple enough, consider using FF and save UDB resources for other purposes. The PWM generates up to two left- or right-aligned PWM outputs or one center-aligned or dual-edged PWM output. The PWM outputs are double buffered to avoid glitches caused by duty cycle changes while running. Left-aligned PWMs are used for most general-purpose PWM uses. Right-aligned PWMs are typically only used in special cases that require alignment opposite of left-aligned PWMs. Center-aligned PWMs are most often used in AC motor control to maintain phase alignment. Dual-edged PWMs are optimized for power conversion where phase alignment must be adjusted. The optional dead band provides complementary outputs with adjustable dead time where both outputs are low between each transition. The complementary outputs and dead time are most often used to drive power devices in half-bridge configurations to avoid shoot-through currents and resulting damage. A kill input is also available that immediately disables the dead band outputs when enabled. Four kill modes are available to support multiple use scenarios. Two hardware dither modes are provided to increase PWM flexibility. The first dither mode increases effective resolution by two bits when resources or clock frequency preclude a standard implementation in the PWM counter. The second dither mode uses a digital input to select one of the two PWM outputs on a cycle-by-cycle basis; this mode is typically used to provide fast transient response in power converts. Trigger and reset inputs allow the PWM to be synchronized with other internal or external hardware. An optional trigger input is configurable with the Trigger Mode parameter. Only hardware trigger is available in the component for starts the PWM. The PWM cannot be triggered with an API call. A rising edge on the reset input causes the PWM counter to reset its count as if the terminal count was reached. The enable input provides hardware enable to gate PWM operation based on a hardware signal. An interrupt can be programmed to be generated under any combination of the following conditions: when the PWM reaches the terminal count or when a compare output goes high. Features include: 8- or 16-bit resolution, multiple pulse width output modes, configurable trigger, configurable capture, configurable hardware/software enable, configurable dead band, multiple configurable kill modes, customized configuration tool and a Fixed-function (FF) implementation for PSoC 3 and PSoC 5LP devices [70].

15.56.1 PWM Functions 3.30

- **PWM_Start( )**—Initializes the PWM with default customizer values.
- **PWM_Stop( )**—Disables the PWM operation. Clears the enable bit of the control register for either of the software controlled enable modes.
- **PWM_SetInterruptMode( )**—Configures the interrupts mask control of the interrupt source status register.
- **PWM_ReadStatusRegister( )**—Returns the current state of the status register.
- **PWM_ReadControlRegister( )**—Returns the current state of the control register.
- **PWM_WriteControlRegister( )**—Sets the bit field of the control register.
- **PWM_SetCompareMode( )**—Writes the compare mode for compare output when PWM Mode is set to Dither mode, Center Align mode or One Output mode.
- **PWM_SetCompareMode1( )**—Writes the compare mode for compare1 output into the control register.
- **PWM_SetCompareMode2( )**—Writes the compare mode for compare2 output into the control register.
- **PWM_ReadCounter( )**—Reads the current counter value (software capture).
- **PWM_ReadCapture( )**—Reads the capture value from the capture **FIFO**.
- **PWM_WriteCounter( )**—Writes a new counter value directly to the counter register. This will be implemented only for that currently running period.
- **PWM_WritePeriod( )**—Writes the period value used by the PWM hardware.
- **PWM_ReadPeriod( )**—Reads the period value used by the PWM hardware.
- **PWM_WriteCompare( )**—Writes the compare value when the instance is defined as Dither mode, Center Align mode or One Output mode.
- **PWM_ReadCompare( )**—Reads the compare value when the instance is defined as Dither mode, Center Align mode or One Output mode.
- **PWM_WriteCompare1( )**—Writes the compare value for the compare1 output.
- **PWM_ReadCompare1( )**—Reads the compare value for the compare1 output.
- **PWM_WriteCompare2( )**—Writes the compare value for the compare2 output
- **PWM_ReadCompare2( )**—Reads the compare value for the compare2 output.
- **PWM_WriteDeadTime( )**—Writes the dead time value used by the hardware in dead band implementation.

- **PWM_ReadDeadTime( )**—Reads the dead time value used by the hardware in dead band implementation.
- **PWM_WriteKillTime( )**—Writes the kill time value used by the hardware when the kill mode is set as Minimum Time.
- **PWM_ReadKillTime( )**—Reads the kill time value used by the hardware when the kill mode is set as Minimum Time.
- **PWM_ClearFIFO( )**—Clears all capture data from the capture FIFO.
- **PWM_Sleep( )**—Stops and saves the user configuration.
- **PWM_Wakeup( )**—Restores and enables the user configuration.
- **PWM_Init( )**—Initializes component's parameters to those set in the customizer placed on the schematic.
- **PWM_Enable( )**—Enables the PWM block operation.
- **PWM_SaveConfig( )**—Saves the current user configuration of the component.
- **PWM_RestoreConfig( )**—Restores the current user configuration of the component.

15.56.2 PWM Function Calls 3.30

- **void PWM_Start(void)**—This function intended to start component operation. PWM_Start( ) sets the initVar variable, calls the PWM_Init function and then calls the PWM_Enable function.
- **void PWM_Stop(void)**—Disables the PWM operation by resetting the seventh bit of the control register for either of the software-controlled enable modes. Disables the fixed-function block that has been chosen.
- **void PWM_SetInterruptMode(uint8 interruptMode)**—Configures the interrupts mask control of the interrupt source status register.
- **uint8 PWM_ReadControlRegister(void)**—Returns the current state of the control register. This API is available only if the enable mode is not "Hardware Only" or compare mode is software controlled at least for one channel.
- **void PWM_WriteControlRegister(uint8 control)**—Sets the bit field of the control register. This API is available only if the enable mode is not "Hardware Only" or compare mode is software controlled at least for one channel. See Control (FF) section for fixed function implementation.
- **void PWM_SetCompareMode(enum comparemode)**—Writes the compare mode for compare output when PWM Mode is set to Dither mode, Center Align mode or One Output mode.
- **void PWM_SetCompareMode1(enum comparemode)**—Writes the compare mode for compare1 output into the control register.
- **void PWM_SetCompareMode2(enum comparemode)**—Writes the compare mode for compare2 output into the control register. This API is valid only for UDB implementation and not available for fixed function PWM implementation.
- **uint8/16 PWM_ReadCounter(void)**—Reads the current counter value (software capture). This API is valid only for UDB implementation and not available for fixed function PWM implementation.
- **uint8/16 PWM_ReadCapture(void)**—Reads the capture value from the capture FIFO. This API is valid only for UDB implementation and not available for fixed function PWM implementation.
- **void PWM_WriteCounter(uint8/16 counter)**—Writes a new counter value directly to the counter register. This will be implemented for that currently running period and only that period. This API is valid only for UDB implementation and not available for fixed function PWM implementation.
- **void PWM_WritePeriod(uint8/16 period)**—Writes the period value used by the PWM hardware.
- **uint8/16 PWM_ReadPeriod(void)**—Reads the period value used by the PWM hardware.

- **void PWM_WriteCompare(uint8/16 compare)**—Writes the compare values for the compare output when the PWM Mode parameter is set to Dither mode, Center Aligned mode, or One Output mode.
- **uint8/16 PWM_ReadCompare(void)**—Reads the compare value for the compare output when the PWM Mode parameter is set to Dither mode, Center Aligned mode, or One Output mode.
- **void PWM_WriteCompare1(uint8/16 compare)**—Writes the compare value for the compare1 output.
- **uint8/16 PWM_ReadCompare1(void)**—Reads the compare value for the compare1 output.
- **void PWM_WriteCompare2(uint8/16 compare)**—Writes the compare value for the compare2 output. This API is valid only for UDB implementation and not available for fixed function PWM implementation.
- **uint8/16 PWM_ReadCompare2(void)**—Reads the compare value for the compare2 output. This API is valid only for UDB implementation and not available for fixed function PWM implementation.
- **void PWM_WriteDeadTime(uint8 deadband)**—Writes the dead time value used by the hardware in dead band implementation.
- **uint8 PWM_ReadDeadTime(void)**—Reads the dead time value used by the hardware in dead band implementation.
- **void PWM_WriteKillTime(uint8 killtime)**—Writes the kill time value used by the hardware when the Kill Mode is set to Minimum Time. This API is valid only for UDB implementation and not available for fixed function PWM implementation.
- **uint8 PWM_ReadKillTime(void)**—Reads the kill time value used by the hardware when the Kill Mode is set to Minimum Time. This API is valid only for UDB implementation and not available for fixed function PWM implementation.
- **void PWM_ClearFIFO(void)**—Clears the capture FIFO of any previously captured data. Here PWM_ReadCapture( ) is called until the FIFO is empty. This API is valid only for UDB implementation and not available for fixed function PWM implementation.
- **void PWM_Sleep(void)**—Stops and saves the user configuration.
- **void PWM_Wakeup(void)**—Restores and enables the user configuration.
- **void PWM_Init(void)**—Initializes component's parameters to those set in the customizer placed on the schematic. The compare modes are set by setting the respective bits of the control register. The interrupts are chosen as the output from the status register. If you are using fixed-function mode, the chosen fixed-function block is enabled. FIFO is cleared to enable FIFO full bit to be set in the status register. Usually called in PWM_Start( ).
- **void PWM_Enable(void)**—Enables the PWM block operation by setting the seventh bit of the control register. The outputs and component behavior will reflect component enable state after two clock cycles.
- **void PWM_SaveConfig(void)**—Saves the current user configuration of the component. The period, dead band, counter and control register values are saved.
- **void PWM_RestoreConfig(void)**—Restores the current user configuration of the component.

15.57 Quadrature Decoder (QuadDec) 3.0

The Quadrature Decoder (QuadDec) Component gives you the ability to count transitions on a pair of digital signals. The signals are typically provided by a speed/position feedback system mounted on a motor or trackball. The signals, typically called A and B, are positioned 90 degrees out of phase, which results in a Gray code output. A Gray code is a sequence where only one bit changes on each count.

This is essential to avoid glitches. It also allows detection of direction and relative position (Fig. 15.1). A third optional signal, named Index, is used as a reference to establish an absolute position once per rotation. Features include: Adjustable counter size: 8, 16, or 32 bits, counter resolution of 1x, 2x, or 4x the frequency of the A and B, Adjustable counter size: 8, 16, or 32 bits, counter resolution of 1x, 2x, or 4x the frequency of the A and B inputs for more accurate determination of position or speed, optional index input to determine absolute position, optional glitch filtering to reduce the impact of system-generated noise on the inputs for more accurate determination of position or speed, optional index input to determine absolute position, optional glitch filtering to reduce the impact of system-generated noise on the inputs, counter resolution of 1x, 2x, or 4x the frequency of the A and B, inputs, for more accurate determination of position or speed, optional index input to determine absolute position and optional glitch filtering to reduce the impact of system-generated noise on the inputs [71].

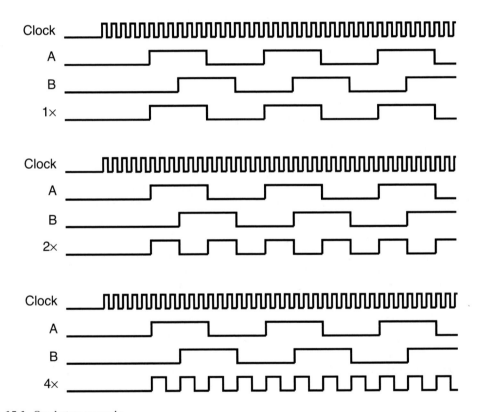

Fig. 15.1 Quadrature example.

15.57.1 QuadDec Functions 3.0

- **QuadDec_Start( )**—Initializes UDBs and other relevant hardware.
- **QuadDec_Stop( )**—Turns off UDBs and other relevant hardware.
- **QuadDec_GetCounter( )**—Reports the current value of the counter.
- **QuadDec_SetCounter( )**—Sets the current value of the counter.
- **QuadDec_GetEvents( )**—Reports the current status of events.

- **QuadDec_SetInterruptMask( )**—Enables or disables interrupts due to the events.
- **QuadDec_GetInterruptMask( )**—Reports the current interrupt mask settings.
- **QuadDec_Sleep( )**—Prepares the component to go to sleep.
- **QuadDec_Wakeup( )**—Prepares the component to wake up.
- **QuadDec_Init( )**—Initializes or restores default configuration provided with the customizer.
- **QuadDec_Enable( )**—Enables the Quadrature Decoder.
- **QuadDec_SaveConfig( )**—Saves the current user configuration.

15.57.2 QuadDec Function Calls 3.0

- **void QuadDec_Start(void)**—Initializes UDBs and other relevant hardware. Resets counter to 0 and enables or disables all relevant interrupts. Starts monitoring the inputs and counting.
- **void QuadDec_Stop(void)**—Turns off UDBs and other relevant hardware.
- **int8/16/32 QuadDec_GetCounter(void)**—Reports the current value of the counter.
- **void QuadDec_SetCounter(int8/16/32 value)**—Sets the current value of the counter.
- **uint8 QuadDec_GetEvents(void)**—Reports the current status of events. This function clears the bits of the status register.
- **void QuadDec_SetInterruptMask(uint8 mask)**—Enables or disables interrupts caused by the events. For the 32-bit counter, the overflow, underflow and reset interrupts cannot be disabled; these bits are ignored.
- **uint8 QuadDec_GetInterruptMask(void)**—Reports the current interrupt mask settings.
- **void QuadDec_Sleep(void)**—This is the preferred routine to prepare the component for sleep. The QuadDec_Sleep( ) routine saves the current component state. Then it calls the QuadDec_Stop( ) function and calls QuadDec_SaveConfig( ) to save the hardware configuration. Call the Quad-Dec_Sleep( ) function before calling the CyPmSleep( ) or the CyPmHibernate( ) function.
- **void QuadDec_Wakeup(void)**—This is the preferred routine to restore the component to the state when QuadDec_Sleep( ) was called. The QuadDec_Wakeup( ) function calls the Quad-Dec_RestoreConfig( ) function to restore the configuration. If the component was enabled before the QuadDec_Sleep( ) function was called, the QuadDec_Wakeup( ) function will also re-enable the component.
- **void QuadDec_Init(void)**—Initializes or restores the component according to the customizer Configure dialog settings. It is not necessary to call QuadDec_Init( ) because the QuadDec_Start( ) routine calls this function and is the preferred method to begin component operation.
- **void QuadDec_Enable(void)**—Activates the hardware and begins component operation. It is not necessary to call QuadDec_Enable( ) because the QuadDec_Start( ) routine calls this function, which is the preferred method to begin component operation.
- **void QuadDec_SaveConfig(void)**—This function saves the component configuration and non-retention registers. This function also saves the current component parameter values, as defined in the Configure dialog or as modified by appropriate APIs. This function is called by the QuadDec_Sleep( ) function.
- **void QuadDec_RestoreConfig(void)**—This function restores the component configuration and non-retention registers. This function also restores the component parameter values to what they were before calling the QuadDec_Sleep( ) function.

15.58 Shift Register (ShiftReg) 2.30

The Shift Register (ShiftReg) component provides synchronous shifting of data into and out of a parallel register. The parallel register can be read or written to by the CPU or DMA. The Shift Register component provides universal functionality similar to standard 74xxx series logic shift registers including: 74164, 74165, 74166, 74194, 74299, 74595 and 74597. In most applications the Shift Register component will be used in conjunction with other components and logic to create higher-level application-specific functionality, such as a counter to count the number of bits shifted. In general usage, the Shift Register component functions as a 2- to 32-bit shift register that shifts data on the rising edge of the clock input. The shift direction is configurable. It can be a right shift, where the MSB shifts in the input and the LSB shifts out the output, or a left shift, where the LSB shifts in the input and the MSB shifts out the output. The Shift Register value can be written by the CPU or DMA at any time. The rising edge of the component clock transfers pending FIFO data (previously written by the CPU or DMA) to the Shift Register when the load signal is set. A rising edge of the component clock transfers the current Shift Register value to the FIFO when a rising edge of the optional store input has been detected, where it can later be read by the CPU. Features include: Adjustable shift register size: 2 to 32 bits, simultaneous shift in and shift out, right shift or left shift, reset input forces shift register to all 0s, shift register value readable by CPU or DMA and shift register value writable by CPU or DMA [83].

15.58.1 ShiftReg Functions 2.30

- **ShiftReg_Start( )**—Starts the Shift Register and enables all selected interrupts
- **ShiftReg_Stop( )**—Disables the Shift Register
- **ShiftReg_EnableInt( )**—Enables the Shift Register interrupt
- **ShiftReg_DisableInt( )**—Disables the Shift Register interrupt
- **ShiftReg_SetIntMode( )**—Sets the interrupt source for the interrupt
- **ShiftReg_GetIntStatus( )**—Gets the Shift Register interrupt status
- **ShiftReg_WriteRegValue( )**—Writes a value directly to the shift register
- **ShiftReg_ReadRegValue( )**—Reads the current value from the shift register
- **ShiftReg_WriteData( )**—Writes data to the shift register input FIFO
- **ShiftReg_ReadData( )**—Reads data from the shift register output FIFO
- **ShiftReg_GetFIFOStatus( )**—Returns current status of input or output FIFO
- **ShiftReg_Sleep( )**—Stops the component and saves all non-retention registers
- **ShiftReg_Wakeup( )**—Restores all non-retention registers and starts component
- **ShiftReg_Init( )**—Initializes or restores default Shift Register configuration
- **ShiftReg_Enable( )**—Enables the Shift Register
- **ShiftReg_SaveConfig( )**—Saves configuration of Shift Register
- **ShiftReg_RestoreConfig( )**—Restores configuration of Shift Register

15.58.2 ShiftReg Function Calls 2.30

- **void ShiftReg_Start(void)**—This is the preferred method to begin component operation. ShiftReg_Start( ) sets the initVar variable, calls the ShiftReg_Init( ) function and then calls the ShiftReg_Enable( ) function. Note that one component clock pulse is required to start the component logic after this function is called.

- **void ShiftReg_Stop(void)**—Disables the Shift Register.
- **ShiftReg_EnableInt(void)**—Enables the Shift Register interrupts.
- **void ShiftReg_DisableInt(void)**—Disables the Shift Register interrupts.
- **void ShiftReg_SetIntMode(uint8 interruptSource)**—Sets the interrupt source for the interrupt. Multiple sources may be ORed together.
- **uint8 ShiftReg_GetIntStatus(void)**—Gets the interrupt status for the Shift Register interrupts.
- **void ShiftReg_WriteRegValue(uint8/16/32 shiftData)**—Writes a value directly to the Shift Register.
- **uint8/16/32 ShiftReg_ReadRegValue(void)**—Returns the current value from the shift register.
- **cystatus ShiftRe_WriteData(uint8/16/32 shiftData)**—Writes data to the shift register input FIFO. A data word is transferred to the shift register on a rising edge of the load input.
- **uint8/16/32 ShiftReg_ReadData(void)**—Reads data from the shift register output FIFO. A data word is transferred to the output FIFO on a rising edge of the store input.
- **uint8 ShiftReg_GetFIFOStatus(uint8 fifoId)**—Returns the current status of the input or output FIFO.
- **void ShiftReg_Sleep(void)**—This is the preferred routine to prepare the component for sleep. The ShiftReg_Sleep( ) routine saves the current component state. Then it calls the ShiftReg_Stop( ) function and calls ShiftReg_SaveConfig( ) to save the hardware configuration.Call the ShiftReg_Sleep( ) function before calling the CyPmSleep( ) or the CyPmHibernate( ) function.
- **void ShiftReg_Wakeup(void)**—This is the preferred routine to restore the component to the state when ShiftReg_Sleep( ) was called. The ShiftReg_Wakeup( ) function calls the ShiftReg_RestoreConfig( ) function to restore the configuration. If the component was enabled before the ShiftReg_Sleep( ) function was called, the ShiftReg_Wakeup( ) function will also re-enable the component. Note that one component clock pulse is required to return to normal operation after this function is called.
- **void ShiftReg_Init(void)**—Initializes or restores the component according to the customizer Configure dialog settings. It is not necessary to call ShiftReg_Init( ) because the ShiftReg_Start( ) routine calls this function and is the preferred method to begin component operation.
- **void ShiftReg_Enable(void)**—Activates the hardware and begins component operation. It is not necessary to call ShiftReg_Enable( ) because the ShiftReg_Start( ) routine calls this function, which is the preferred method to begin component operation.
- **void ShiftReg_SaveConfig(void)**—This function saves the component configuration and non-retention registers. This function also saves the current component parameter values, as defined in the Configure dialog or as modified by appropriate APIs. This function is called by the ShiftReg_Sleep( ) function.
- **void ShiftReg_RestoreConfig(void)**—This function restores the component configuration and non-retention registers. This function also restores the component parameter values to what they were prior to calling the ShiftReg_Sleep( ) function.

15.59 Timer 2.80

The Timer Component provides a method to measure intervals. It can implement a basic timer function and offers advanced features such as capture with capture counter and interrupt/DMA generation. For PSoC 3 and PSoC 5LP devices, the Component can be implemented using FF blocks or UDB. PSoC 4 devices support only the UDB implementation. A UDB implementation typically has more features than a FF implementation. If the design is simple enough, consider using FF and save UDB resources for other purposes. Note For PSoC 4 devices, there is also a Timer/Counter/Pulse Width Modulator (TCPWM) Component available for use. The following table shows the major feature differences between FF and UDB. There are also many specific functional differences between the FF and UDB implementations and differences between the FF implementation in different devices. Features include: Universal Digital Block (UDB) implementation for all devices. Features include: Fixed-function (FF) implementation for PSoC 3 and PSoC 5LP devices, 8-, 16-, 24-, or 32-bit timer, optional capture input, enable, trigger and reset inputs, for synchronizing with other Components and continuous or one shot run modes [95].

15.59.1 Timer Functions 2.80

- **Timer_Start( )**—Sets the initVar variable, calls the Timer_Init( ) function and then calls the Enable function.
- **Timer_Stop( )**—Disables the Timer.
- **Time_SetInterruptMode( )**—Enables or disables the sources of the interrupt output.
- **Timer_ReadStatusRegister( )**—Returns the current state of the status register.
- **Timer_ReadControlRegister( )**—Returns the current state of the control register.
- **Timer_WriteControlRegister( )**—Sets the bit-field of the control register.
- **Timer_WriteCounter( )**—Writes a new value directly into the counter register. (UDB only)
- **Timer_ReadCounter( )**—Forces a capture and then returns the capture value.
- **Timer_WritePeriod( )**—Writes the period register.
- **Timer_ReadPeriod( )**—Reads the period register.
- **Timer_ReadCapture( )**—Returns the contents of the capture register or the output of the FIFO.
- **Timer_SetCaptureMode( )**—Sets the hardware or software conditions under which a capture will occur.
- **Timer_SetCaptureCount( )**—Sets the number of capture events to count before capturing the counter register to the FIFO.
- **Timer_ReadCaptureCount( )**—Reports the current setting of the number of capture events.
- **Timer_SoftwareCapture( )**—Forces a capture of the counter to the capture FIFO.
- **Timer_SetTriggerMode( )**—Sets the hardware or software conditions under which a trigger will occur.
- **Timer_EnableTrigger( )**—Enables the trigger mode of the timer.
- **Timer_SetInterruptCount( )**—Sets the number of captures to count before an interrupt is triggered.
- **Timer_ClearFIFO( )**—Clears the capture FIFO.
- **Timer_Sleep( )**—Stops the Timer and saves its current configuration.
- **Timer_Wakeup( )**—Restores the Timer configuration and re-enables the Timer.
- **Timer_Init( )**—Initializes or restores the Timer per the Configure dialog settings.
- **Timer_Enable( )**—Enables the Timer.
- **Timer_SaveConfig( )**—Saves the current configuration of the Timer.
- **Timer_RestoreConfig( )**—Restores the configuration of the Timer.

15.59.2 Timer Function Calls 2.80

- **void Timer_Start(void)**—This is the preferred method to begin Component operation. Timer_Start( ) sets the initVar variable, calls the Timer_Init( ) function and then calls the Timer_Enable( ) function.
- **void Timer_Stop(void)**—For fixed-function implementations this disables the Timer and powers it down. For UDB implementations the Timer is disabled only in software enable modes.
- **void Timer_SetInterruptMode(uint8 interruptMode)**—Enables or disables the sources of the interrupt output.
- **uint8 Timer_ReadStatusRegister(void)**—Returns the current state of the status register.
- **uint8 Timer_ReadControlRegister(void)**—Returns the current state of the control register. This API is not available in the special case when the control register is not required (UDB implementation, enable mode is hardware only, capture mode not software controlled and trigger mode not software controlled).
- **void Timer_WriteControlRegister(uint8 control)**—Sets the bit field of the control register. This API is not available in the special case when the control register is not required (UDB implementation, enable mode is hardware only, capture mode not software controlled and trigger mode not software controlled).
- **void Timer_WriteCounter(uint8/16/32 counter)**—rites a new value directly into the counter register. This function is available only for the UDB implementation.
- **uint8/16/32 Timer_ReadCounter(void)**—Forces a capture and then returns the capture value.— Writes the period register.
- **void Timer_WritePeriod(uint8/16/32 period)**—Writes the period register.
- **uint8/16/32 Timer_ReadPeriod(void)**—Reads the period register.
- **void Timer_SetCaptureMode(uint8 captureMode)**—Sets the capture mode. This function is available only for the UDB implementation and when the Capture Mode parameter is set to Software Controlled.
- **void Timer_SetCaptureCount(uint8 captureCount)**—Sets the number of capture events to count before a capture is performed. This function is available only for the UDB implementation and when the Enable Capture Counter parameter is selected in the Configure dialog.
- **uint8 Timer_ReadCaptureCount(void)**—Reads the current value setting for the captureCount parameter as set in the Timer_SetCaptureCount( ) function. This function is only available for the UDB implementation and when the Enable Capture Counter parameter is selected in the Configure dialog.
- **void Timer_SoftwareCapture(void)**—Forces a software capture of the current counter value to the FIFO. This function is available only for UDB implementation.
- **void Timer_SetTriggerMode(uint8 triggerMode)**—Sets the trigger mode. This function is available only for UDB implementation and when Trigger Mode parameter is set to Software Controlled.
- **void Timer_EnableTrigger(void)**—Enables the trigger. This function is available only when Trigger Mode is set to Software Controlled.
- **void Timer_DisableTrigger(void)**—Disables the trigger. This function is available only when Trigger Mode is set to Software Controlled.
- **void Timer_SetInterruptCount(uint8 interruptCount)**—Sets the number of captures to count before an interrupt is generated for the InterruptOnCapture source. This function is available only when InterruptOnCaptureCount is enabled.
- **void Timer_ClearFIFO(void)**—Clears the capture FIFO. This function is available only for the UDB implementation.

- **void Timer_Sleep(void)**—This is the preferred routine to prepare the Component for sleep. Timer_Sleep( ) saves the current Component state. Then it calls the Timer_Stop( ) function and calls Timer_SaveConfig( ) to save the hardware configuration. Call the Timer_Sleep( ) function before calling the CyPmSleep( ) or the CyPmHibernate( ) function.
- **void imer_Sleep( )**—Without calling Timer_Stop( ).
- **void Timer_Wakeup(void)**—This is the preferred routine to restore the Component to the state when Timer_Sleep( ) was called. The Timer_Wakeup( ) function calls the Timer_RestoreConfig( ) function to restore the configuration. If the Component was enabled before the Timer_Sleep( ) function was called, the Timer_Wakeup( ) function also re-enables the Component.
- **void Timer_Init(void)**—Initializes or restores the Component according to the customizer Configure dialog settings. It is not necessary to call Timer_Init( ) because the Timer_Start( ) routine calls this function and is the preferred method to begin Component operation.
- **void Timer_Enable(void)**—Activates the hardware and begins Component operation. It is not necessary to call Timer_Enable( ) because the Timer_Start( ) routine calls this function, which is the preferred method to begin Component operation. This function enables the Timer for either of the software controlled enable modes.
- **void Timer_SaveConfig(void)**—This function saves the Component configuration and non-retention registers. It also saves the current Component parameter values, as defined in the Configure dialog or as modified by appropriate APIs. This function is called by the Timer_Sleep( ) function.
- **void Timer_RestoreConfig(void)**—This function restores the Component configuration and non-retention registers. It also restores the Component parameter values to what they were before calling the Timer_Sleep( ) function.

15.60 AND 1.0

Logic gates provide basic boolean operations. The output of a logic gate is a Boolean combinatorial function of the inputs. There are seven basic logic gates: AND, OR, Inverter (NOT), NAND, NOR, XOR, and XNOR. Features include: industry-standard logic gates, configurable number of inputs up to 8, and optional array of gates [32].

15.60.1 AND Functions 1.0

NONE.

15.61 Tri-State Buffer (Bufoe) 1.10

The Tri-State Buffer (Bufoe) component is a non-inverting buffer with an active high output enable signal. When the output enable signal is true, the buffer functions as a standard buffer. When the output enable signal is false, the buffer turns off. Features include: a Buffer with Output Enable signal and used to interface to a shared bus such as I2C and a Feedback signal. Tri-State Buffers should not be used for internal logic and can only be used with an I/O pin [99].

15.61.1 Tri-State Buffer (Bufoe) Functions 1.10

NONE.

15.62 D Flip-Flop 1.30

The D Flip Flop stores a digital value. Features include: asynchronous reset or preset, synchronous reset, preset, or both and configurable width for array of D Flip-Flops [25].

15.62.1 D Flip-Flop Functions 1.30

NONE.

15.63 D Flip-Flop w/ Enable 1.0

The D Flip Flop w/ Enable selectively captures a digital value. Features include: enable input allows d input to be selectively captured and configurable width for array of D Flip Flops with a single enable [24].

15.64 D Flip-Flop w/ Enable Functions 1.00

NONE.

15.65 Digital Constant 1.0

The Digital Constant provides a convenient way to represent digital values in designs. Features include: represents a digital value clearly on a schematic, display in hexadecimal or decimal and configurable width up to 32 bits [30].

15.65.1 Digital Constant Functions 1.00

NONE.

15.66 Lookup Table (LUT) 1.60

This component allows a Lookup Table (LUT) to be created that performs any logic function with up to five inputs and eight outputs. This is done by generating logic equations that are realized in the UDB PLDs. Optionally, the outputs can be registered. These registers are implemented in PLD macrocells. All macrocell flip-flops are initialized to a 0 value at power up and after any reset of the device. Features include: 1 to 5 inputs, 1 to 8 outputs, configuration tool and optionally registered outputs [59].

15.66.1 Lookup Table (LUT) Functions 1.60

NONE.

15.67 Digital Multiplexer and Demultiplexer 1.10

The Multiplexer component is used to select 1 of n inputs while the Demultiplexer component is used to route 1 signal to n outputs. The Multiplexer component implements a 2 to 16 input mux providing a single output, based on hardware control signals. The Demultiplexer component implements a 2 to 16 output demux from a single input, based on hardware control signals. Only one input or output connection may be made at a time. Features include: digital multiplexer, digital demultiplexer and up to 16 channels [33].

15.67.1 Digital Multiplexer and Demultiplexer Functions 1.10

NONE.

15.68 SR Flip-Flop 1.0

The SR Flip-Flop stores a digital value that can be set or reset. Features include: clocked for safe use in synchronous circuits, configurable width for array of SR Flip-Flops [87].

15.69 SR Flip-Flop Functions 1.0

NONE.

15.70 Toggle Flip-Flop 1.0

The Toggle Flip Flop captures a digital value that can be toggled. Features include: Features include: T input toggles Q value, configurable width for array of Toggle Flip Flops [97].

15.70.1 Toggle Flip-Flop 1.0 Functions

NONE.

15.71 Control Register 1.8

The Control Register allows the firmware to output digital signals and supports up to Up to 8-bits [20].

15.71.1 Control Register Functions 1.8

- **Control_Reg_Write( )**—Writes a byte to a control register
- **Control_Reg_Read( )**—Reads the current value assigned to a control register
- **Control_Reg_SaveConfig( )**—Saves the control register value
- **Control_Reg_RestoreConfig( )**—Restores the control register value
- **Control_Reg_Sleep( )**—Prepares the component for entering the low power mode
- **Control_Reg_Wakeup( )**—Restores the component after waking up from the low power mode

15.71.2 Control Register Function Calls 1.8

- **void Control_Reg_Write (uint8 control)**—Writes a byte to the control register
- **uint8 Control_Reg_Read (void)**—Reads the current value assigned to the control register
- **void Control_Reg_SaveConfig (void)**—Saves the control register value
- **void Control_Reg_RestoreConfig (void)**—Restores the control register value
- **void Control_Reg_Sleep (void)**—Prepares the component for entering the low power mode
- **void Control_Reg_Wakeup (void)**—Restores the component after waking up from the low power mode

15.72 Status Register 1.90

The Status Register allows the firmware to read digital signals and includes up to 8-bits and interrupt support [89].

15.72.1 Status Register Functions 1.90

- **StatusReg_Read( )**—Reads the current value of the status register
- **StatusReg_InterruptEnable( )**—Enables the status register interrupt
- **StatusReg_InterruptDisable( )**—Disables the status register interrupt
- **StatusReg_WriteMask( )**—Writes the value assigned to the mask register
- **StatusReg_ReadMask( )**—Returns the current interrupt mask value from the mask register

15.72.2 Status Register Function Calls 1.90

- **uint8 StatusReg_Read (void)**—Reads the value of a status register.
- **void StatusReg_InterruptEnable (void)**—Enables the status register interrupt. The default behavior is disabled. This is only valid if the status register generates an interrupt.
- **void StatusReg_InterruptDisable (void)**—Disables the status register interrupt. This is only valid if the status register generates an interrupt.
- **void StatusReg_WriteMask (uint8 mask)**—Writes the current mask value assigned to the status register. This is only valid if the status register generates an interrupt.
- **uint8 StatusReg_ReadMask (void)**—Reads the current interrupt mask value assigned for the status register. This is only valid if the status register generates an interrupt.

15.73 Debouncer 1.00

Mechanical switches and relays tend to make and break connections for a finite time before settling down to a stable state. Within this settling time, the digital circuit can see multiple transitions as the switch contacts bounce between make or break conditions. The Debouncer component takes an input signal from a bouncing contact and generates a clean output for digital circuits. The component will not pass the signal to the output until the predetermined period of time when the switch bouncing settles down. In this way, the circuit will respond to only one pulse generation performed by the pressing or releasing of the switch and not several state transitions caused by contact bouncing. Eliminates unwanted oscillations on digital input lines [5, 26].

15.73.1 Debouncer Functions 1.00

NONE

15.74 Digital Comparator 1.00

The Digital Comparator component provides a selectable-width, selectable-type comparator, implemented in PLD macrocells. Features include: 1 to 32 bit Configurable Digital Comparator and six selectable comparison operators [29].

15.74.1 Digital Comparator 1.00

NONE.

15.75 Down Counter 7-bit (Count7) 1.00

The Count7 Component is a 7-bit down counter with the count value available as hardware signals. This counter is implemented using a specific configuration of a universal digital block (UDB). To implement the counter, pieces of the control and status registers are used along with counter logic that is present in the UDB specifically for this function. Features include: 7-bit read/write period register, 7-bit count register that is read/write, automatic reload of the period to the count register on terminal count and routed load and enable signals [36].

15.75.1 Down Counter 7-bit (Count7) Functions 1.00

- **Count7_Start( )**—Performs all the required initialization for the Component and enables the counter.
- **Count7_Init( )**—Initializes or restores the Component according to the customizer settings.
- **Count7_Enable( )**—Enables the software enable of the counter.
- **Count7_Stop( )**—Disables the software enable of the counter.
- **Count7_WriteCounter( )**—This function writes the counter directly. The counter should be disabled before calling this function.

- **Count7_ReadCounter( )**—Reads the counter value.
- **Count7_WritePeriod( )**—Writes the period register.
- **Count7_ReadPeriod( )**—Reads the period register.
- **Count7_Sleep( )**—Is the preferred API to prepare the Component for low power mode operation.
- **Count7_Wakeup( )**—Is the preferred API to restore the Component to the state when Count7_Sleep( ) was called.
- **Count7_SaveConfig( )**—Saves the value of the Component's count register prior to entering the low power mode.
- **Count7_RestoreConfig( )**—Restores the value of Component's count register which was previously stored.

15.75.2 Down Counter 7-bit (Count7) Function Calls 1.00

- **void Count7_Start(void)**—Performs all the required initialization for the Component and enables the counter. The first time the routine is executed, the period is set as configured in the customizer. When called to restart the counter following a Count_Stop( ) call, the current period value is retained.
- **void Count7_Init(void)**—Initializes or restores the Component according to the customizer settings. It is not necessary to call Count7_Init( ) because the Count7_Start( ) API calls this function and is the preferred method to begin Component operation.
- **void Count7_Enable(void)**—Enables the software enable of the counter. The counter is controlled by a software enable and an optional hardware enable. It is not necessary to call Count7_Enable( ) because the Count7_Start( ) API calls this function, which is the preferred method to begin Component operation.
- **void Count7_Stop(void)**—Disables the software enable of the counter. This API halts the counter. Therefore, when you start it again (Count7_Start( ) call), it will continue counting from the last counter value.
- **void Count7_WriteCounter(uint8 count)**—This function writes the counter directly. The counter should be disabled before calling this function.
- **uint8 Count7_ReadCounter(void)**—This function reads the counter value.
- **void Count7_WritePeriod(uint8 period)**—This function writes the period register. The actual period is one greater than the value in the period register since the counting sequence starts with the period register value and counts down to 0 inclusive. The period of the counter output does not change until the counter is reloaded following the terminal count value of 0 or due to a hardware load signal.
- **uint8 Count7_ReadPeriod(void)**—This function reads the period register.
- **Count7_Sleep( )**—This is the preferred API to prepare the Component for low power mode operation. The Count7_Sleep( ) API saves the current Component state using Count7_SaveConfig( ) and disables the counter.
- **void Count7_Wakeup(void)**—This is the preferred API to restore the Component to the state when Count7_Sleep( ) was called. The Count7_Wakeup( ) function calls the Count7_RestoreConfig( ) function to restore the configuration.
- **void Count7_SaveConfig(void)**—This function saves the current counter value prior to entering low power mode. This function is called by the Count7_Sleep( ) function.
- **void Count7_RestoreConfig(void)**—This function restores the counter value, which was previously stored. This function is called by the Count7_Wakeup( ) function.

15.76 Edge Detector 1.00

The Edge Detector component samples the connected signal and produces a pulse when the selected edge occurs. Features include: detect rising edge, falling edge, or either edge.

15.76.1 Edge Detector Functions 1.00

NONE.

15.76.2 Digital Comparator 1.0

The Edge Detector component samples the connected signal and produces a pulse when the selected edge occurs. Detects Rising Edge, Falling Edge, or Either Edge [37].

15.76.3 Digital Comparator Functions 1.00

NONE.

15.77 Frequency Divider 1.0

The Frequency Divider component produces an output that is the clock input divided by the specified value. Features include: clock or arbitrary signal division by a specified value and enable and Reset inputs to control and align divided output [45].

15.77.1 Frequency Divider Functions 1.00

NONE.

15.78 Glitch Filter 2.00

Glitch filtering is the process of removing unwanted pulses from a digital input signal that is usually high or low. Glitches frequently occur on lines carrying signals from sources such as RF receivers. Electrical or in some cases even mechanical interference can trigger an unwanted glitch pulse from the receiver. This design outputs a '1' only when the current and previous N samples are '1' and a '0' only when the current and previous N samples are '0'. Otherwise, the output is unchanged from its current value [5, 47].

15.78.1 Glitch Filter Functions 2.00

NONE

15.79 Pulse Converter 1.00

The Pulse Converter component produces a pulse of known width when a pulse of any width is sampled on p_in. Terminals are provided for out_clk and sample_clk for configurability of sample rate and output pulse width [69].

15.79.1 Pulse Converter Functions 1.00

NONE.

15.80 Sync 1.00

The Sync component re-synchronizes a set of input signals to the rising edge of the clock signal. This component can be used when it is necessary to use a signal from one clock domain in another clock domain, the Sync component can be used to line up that signal's transitions to the clock domain of the destination. In this case the Sync component is clocked using the same clock as the destination and it can Synchronize 1 to 32 input signals [91].

15.81 Sync Functions 1.00

NONE.

15.82 UDB Clock Enable (UDBClkEn) 1.00

The universal digital block (UDB) Clock Enable (UDBClkEn) component supports precise control over clocking behavior. Features included: clock enable support and addition of synchronization on a clock when needed [101].

15.82.1 UDB Clock Enable (UDBClkEn) Functions 1.00

NONE.

15.83 LED Segment and Matrix Driver (LED_Driver) 1.10

The LED Segment and Matrix Driver (LED_Driver) component is a multiplexed LED driver that can handle up to 24 segment signals and 8 common signals. It can be used to drive 24 7-segment LEDs, eight 14/16-segment LEDs, eight RGB 7-segment LEDs, or a tri-color matrix of up to 192 LEDs in an 8x8 pattern. APIs are provided to convert alpha-numeric values to their segment codes and the brightness of each of the commons can be independently controlled. This component is supported for PSoC 3 and PSoC 5LP. Multiplexing the LEDs is an efficient way to save GPIO pins, however the commons must be multiplexed at a steady rate. To address this latter issue, the component uses PSoC's DMA and UDBs to multiplex the LEDs without CPU overhead. This eliminates cases of non-periodic updating as the multiplexing is handled solely using hardware. The CPU is thus used only when updating the display information and to change the brightness settings. When displaying the 7/14/16 segment digits, these digits do not have to be grouped as a single numerical display. An 8 digit display could be divided up into one 2-digit and two 3-digit displays for example. When operating in the LED matrix mode, the individual displays do not have to be arranged in a matrix, but instead can be various single or grouped LEDs. The component also supports displaying combined digits with annunciators. Features include: Features include: up to 8 RGB 7-segment digits, or 24 monochrome 7-segment digits, up to 8 14-segment or 16-segment digits, up to 192 LEDs in an 8x8 tri-color matrix,active high or active low commons, active high or active low segments, driver is multiplexed requiring no CPU overhead or interrupts, functions for numeric and string display using 7-, 14- and 16-segments, independent brightness level for each common signal, up to 8 RGB 7-segment digits, or 24 monochrome 7-segment digits, up to 8 14-segment or 16-segment digits, up to 192 LEDs in an 8x8 tri-color matrix, active high or active low commons, active high or active low segments, driver is multiplexed requiring no CPU overhead or interrupts, functions for numeric and string display using 7-, 14- and 16-segments and independent brightness level for each common signal [57].

15.83.1 LED Segment and Matrix Driver (LED_Driver) Functions 1.10

- **LED_Driver_Init( )**—Clears the displays and initializes the display arrays and registers.
- **LED_Driver_Enable( )**—Initializes the DMAs and enables the component.
- **LED_Driver_Start( ) LED_Driver_Stop( )**—Enables and starts the component.
- **LED_Driver_Stop( )**—Clears the display, disables the DMA and stops the component.
- **LED_Driver_SetDisplayRAM( )**—Writes a value directly into the display RAM at the specified position.
- **LED_Driver_SetRC( )**—Sets the bit in the display RAM in the specified row and column.
- **LED_Driver_ClearRC( )**—Clears the bit in the display RAM in the specified row and column.
- **LED_Driver_ToggleRC( )**—Toggles the bit in the display RAM in the specified row and column.
- **LED_Driver_GetRC( )**—Returns the bit value in the display RAM in the specified row and column.
- **LED_Driver_ClearDisplay( )**—Clears the display for the specified common to zero.
- **LED_Driver_ClearDisplayAll( )**—Clears the entire display to 0.
- **LED_Driver_Write7SegNumberDec( )**—Displays a 7-segment hexadecimal number up to 8 characters long, starting at the specified position and extending to a specified number of digits.
- **LED_Driver_Write7SegNumberHex( )**—Displays a 7-segment null terminated string starting at the specified position and ending at either the end of the string or the end of the display.
- **LED_Driver_WriteString7Seg( )**—Displays a 7-segment ASCII encoded character at the specified position.
- **LED_Driver_PutChar7Seg( )**—Displays a single 7-segment digit (0. . . 9) on the specified display. Displays a single 7-segment digit (0. . . F) on the specified display.
- **LED_Driver_Write7SegDigitDec( )**—Displays a single 7-segment digit (0. . . 9) on the specified display.
- **LED_Driver_Write7SegDigitHex( )**—Displays a single 7-segment digit (0. . . F) on the specified display.
- **LED_Driver_Write14SegNumberDec( )**—Displays a 14-segment signed integer up to 8 characters long, starting at the specified position and extending to a specified number of digits.
- **LED_Driver_Write14SegNumberHex( )**—Displays a 14-segment hexadecimal number up to 8 characters long, starting at the specified position and extending to a specified number of digits.

- **LED_Driver_WriteString14Seg( )**—Displays a 14-segment null terminated string starting at the specified position and ending at either the end of the string or the end of the display.
- **LED_Driver_PutChar14Seg( )**—Displays a 14-segment ASCII encoded character at the specified position.
- **LED_Driver_Write14SegDigitDec( )**—Displays a single 14-segment digit (0...9) on the specified display.
- **LED_Driver_Write14SegDigitHex( )**—Displays a single 14-segment digit (0...F) on the specified display.
- **LED_Driver_Write16SegNumberDec( )**—Displays a 16-segment signed integer up to 8 characters long, starting at the specified position and extending to a specified number of digits.
- **LED_Driver_Write16SegNumberHex( )**—Displays a 16-segment hexadecimal number up to 8 characters long, starting at the specified position and extending to a specified number of digits.
- **LED_Driver_WriteString16Seg( )**—Displays a 16-segment hexadecimal number up to 8 characters long, starting at the specified position and extending to a specified number of digits.
- **LED_Driver_PutChar16Seg( )**—Displays a 16-segment null terminated string starting at the specified position and ending at either the end of the string or the end of the display.
- **LED_Driver_Write16SegDigitDec( )**—Displays a 16-segment null terminated string starting at the specified position and ending at either the end of the string or the end of the display.
- **LED_Driver_Write16SegDigitHex( )**—Displays a 16-segment ASCII encoded character at the specified position.
- **LED_Driver_PutDecimalPoint( )**—Displays a single 16-segment digit (0...9) on the specified display. Displays a single 16-segment digit (0...F) on the specified display.
- **LED_Driver_GetDecimalPoint( )**—Sets or clears the decimal point at the specified position.
- **LED_Driver_EncodeNumber7Seg( )**—Returns zero if the decimal point is not set and one if the decimal point is set.
- **LED_Driver_EncodeChar7Seg( )**—Converts the lower 4 bits of the input into 7-segment data that will display the number in hex on a display.
- **LED_Driver_EncodeNumber14Seg( )**—Converts the ASCII encoded alphabet character input into the 7-segment data that will display the alphabet character on a display.
- **LED_Driver_EncodeChar14Seg( )**—Converts the lower 4 bits of the input into 14-segment data that will display the number in hex on a display.
- **LED_Driver_EncodeNumber16Seg( )**—Converts the ASCII encoded alphabet character input into the 14-segment data that will display the alphabet character on a display.
- **LED_Driver_EncodeChar16Seg( )**—Converts the ASCII encoded alphabet character input into the 14-segment data that will display the alphabet character on a display.
- **LED_Driver_SetBrightness( )**—Sets the desired brightness value (0 = display off; 255 = display at full brightness) for the chosen common.
- **LED_Driver_GetBrightness( )**—Returns the brightness value for the specified common.
- **LED_Driver_Sleep( )**—Stops the component and saves the user configuration.
- **LED_Driver_Wakeup( )**—Restores the user configuration and enables the component.

15.83.2 LED Segment and Matrix Driver (LED_Driver) Function Calls 1.10

- **void LED_Driver_Init(void)**—Clears the display and initializes the DMAs. Also initializes the brightness array if brightness control is enabled.
- **void LED_Driver_Enable(void)**—Enables the DMAs and enables the PWM if brightness control is enabled. Once these are complete, the component is enabled.

- **void LED_Driver_Start(void)**—Configures the hardware (DMA and optional PWM) and enables the LED display by calling LED_Driver_Init( ) and LED_Driver_Enable( ). If LED_Driver_Init( ) had been called before, then the LEDs will display whatever values that are currently in the display RAM. If it is the first call, then the display RAM will be cleared.
- **void LED_Driver_Stop(void)**—Clears the display RAM, disables all DMA channels and stops the PWM (if brightness enabled)
- **void LED_Driver_SetDisplayRAM(uint8 value, uint8 position)**—Writes 'value' directly into the display RAM. This function writes a single byte into the display RAM associated with a set of 8 segments designated by the "position" argument.
- **void LED_Driver_SetRC(uint8 row, uint8 column)**—Sets the bit in the display RAM corresponding to the LED in the designated row and column. Note that rows are the segments and columns are the commons.
- **void LED_Driver_ClearRC(uint8 row, uint8 column)**—Clears the bit in the display RAM corresponding to the LED in the designated row and column.
- **void LED_Driver_ToggleRC(uint8 row, uint8 column)**—Toggles the bit in the display RAM corresponding to the LED in the designated row and column.
- **uint8 LED_Driver_GetRC(uint8 row, uint8 column)**—Returns the bit value in the display RAM corresponding to the LED in the designated row and column.
- **void LED_Driver_ClearDisplay(uint8 position)**—Clears the display (disables all the LEDs) for a set of 8 segments designated by the "position" argument.
- **void LED_Driver_ClearDisplayAll(void)**—Clears the entire display by writing zeros to all the display RAM locations.
- **void LED_Driver_Write7SegNumberDec(int32 number, uint8 position, uint8 digits, uint8 alignment)**—Displays a 7-segment signed integer up to 8 characters long, starting at "position" and extending for "digits" characters. The negative sign will consume one digit if it is required. If the number exceeds the digits specified, the least significant digits will be displayed, e.g., if number is −1234, position is 0 and digits is 4, the result will be: −234. Note that the positions of the digits are continuous and it is up to the user to choose the correct position for the application. Also note that any digits that extend beyond the configured number of commons are discarded.
- **void LED_Driver_Write7SegNumberHex(uint32 number, uint8 position, uint8 digits, uint8 alignment)**—Displays a 7-segment hexadecimal number up to 8 characters long, starting at "position" and extending for "digits" characters. If the number exceeds the digits specified, the least significant digits will be displayed. For example, if number is 0xDEADBEEF, position is 0 and digits is 4, the result will be: BEEF. Note that the positions of the digits are continuous and it is up to the user to choose the correct position for the application. Also note that any digits that extend beyond the configured number of commons are discarded.
- **void LED_Driver_WriteString7Seg(char8 const character[], uint8 position)**—Displays a 7-segment null terminated string starting at "position" and ending at either the end of the string or the end of the configured number of commons. Non-displayable characters will produce a blank space. Note that the positions of the digits are continuous and it is up to the user to choose the correct position for the application.
- **void LED_Driver_WriteString7Seg(char8 const character[], uint8 position)**—Displays a 7-segment null terminated string starting at "position" and ending at either the end of the string or the end of the configured number of commons. See the Functional Non-displayable characters will produce a blank space.[34]

---

[34]The positions of the digits are continuous and it is up to the user to choose the correct position for the application.

- **void LED_Driver_PutChar7Seg(char8 character, uint8 position)**—Displays a 7-segment ASCII encoded character at "position". This function can display all alphanumeric characters. The function can also display "–", ".", "_", "", *and* "=". All unknown characters are displayed as a space.[35]
- **void LED_Driver_Write7SegDigDec(uint8 digit, uint8 position)**—Displays a single 7-segment digit on the specified display. The number in "digit" (0–9) is placed at "position."[36]
- **void LED_Driver_Write7SegDigHex(uint8 digit, uint8 position)**—Displays a single 7-segment digit on the specified display. The number in "digit" (0–F) is placed at "position".[37]
- **void LED_Driver_Write14SegNumberDec(int32 number, uint8 position, uint8 digits, uint8 alignment)**—Displays a 14-segment signed integer up to 8 characters long, starting at "position" and extending for "digits" characters. The negative sign will consume one digit if it is required. If the number exceeds the digits specified, the least significant digits will be displayed. For example, if number is −1234, position is 0 and digit is 4, the result will be: −234.
- **void LED_Driver_Write14SegNumberHex(uint32 number, uint8 position, uint8 digits, uint8 alignment)**—Displays a 14-segment hexadecimal number up to 8 characters long, starting at "position" and extending for "digits" characters. If the number exceeds the digits specified, the least significant digits will be displayed. For example, if number is 0xDEADBEEF, position is 0 and digits is 4, the result will be: BEEF.
- **void LED_Driver_WriteString14Seg(char8 const character[], uint8 position)**—Displays a 14-segment null terminated string starting at "position" and ending at either the end of the string or the end of the display. Non-displayable characters will produce a blank space.
- **void LED_Driver_PutChar14Seg(char8 character, uint8 position)**—Displays a 14-segment ASCII encoded character at "position". This function can display all alphanumeric characters. The function can also display "-", ".", "_", "", and "=". All unknown characters are displayed as a space.
- **void LED_Driver_Write14SegDigDec(uint8 digit, uint8 position)**—Displays a single 14-segment digit on the specified display. The number in "digit" (0–9) is placed at "position."
- **void LED_Driver_Write14SegDigHex(uint8 digit, uint8 position)**—Displays a single 14-segment digit on the specified display. The number in "digit" (0–F) is placed at "position."
- **void LED_Driver_Write16SegNumberDec(int32 number, uint8 position, uint8 digits, uint8 alignment)**—Displays a 16-segment signed integer up to 8 characters long, starting at "position" and extending for "digits" characters. The negative sign will consume one digit if it is required. If the number exceeds the digits specified, the least significant digits will be displayed. For example, if number is −1234, position is 0 and digits is 4, the result will be: −234.
- **void LED_Driver_Write16SegNumberHex(uint32 number, uint8 position, uint8 digits, uint8 alignment)**—Displays a 16-segment hexadecimal number up to 8 characters long, starting at "position" and extending for "digits" characters. If the number exceeds the digits specified, the least significant digits will be displayed. For example, if number is 0xDEADBEEF, position is 0 and digits is 4, the result will be: BEEF.
- **void LED_Driver_WriteString16Seg(char8 const character[], uint8 position)**—Displays a 16-segment null terminated string starting at "position" and ending at either the end of the string or the end of the display.[38]

---

[35]The positions of the digits are continuous and it is up to the user to choose the correct position for the application.

[36]The positions of the digits are continuous and it is up to the user to choose the correct position for the application.

[37]The positions of the digits are continuous and it is up to the user to choose the correct position for the application.

[38]Non-displayable characters will produce a blank space.

- **void LED_Driver_PutChar16Seg(char8 character, uint8 position)**—Displays a 16-segment ASCII encoded character at "position". This function can display all alphanumeric characters. The function can also display "-", ".", "_", "'", and "=".[39]
- **void LED_Driver_Write16SegDigDec(uint8 digit, uint8 position)**—Displays a single 16-segment digit on the specified display. The number in "digit" (0–9) is placed at "position."
- **void LED_Driver_Write16SegDigHex(uint8 digit, uint8 position)**—Displays a single 16-segment digit on the specified display. The number in "digit" (0–F) is placed at "position."
- **void LED_Driver_PutDecimalPoint(uint8 dp, uint8 position)**—Sets or clears the decimal point at the specified position.
- **uint8 LED_Driver_GetDecimalPoint(uint8 position)**—Returns zero if the decimal point is not set and one if the decimal point is set.
- **uint8 LED_Driver_EncodeNumber7Seg(uint8 number)**—Converts the lower 4 bits of the input into 7-segment data that will display the number in hex on a display. The returned data can be written directly into the display RAM to display the desired number. It is not necessary to use this function since higher level API are provided to both decode the value and write it to the display RAM.
- **uint8 LED_Driver_EncodeChar7Seg(char8 input)**—Converts the ASCII encoded alphabet character input into the 7-segment data that will display the alphabet character on a display. The returned data can be written directly into the display RAM to display the desired number. It is not necessary to use this function since higher level API are provided to both decode the value and write it to the display RAM.
- **uint16 LED_Driver_EncodeNumber14Seg(uint8 number)**—Converts the lower 4 bits of the input into 14-segment data that will display the number in hex on a display. The returned data can be written directly into the display RAM to display the desired number. It is not necessary to use this function since higher level API are provided to both decode the value and write it to the display RAM.
- **uint16 LED_Driver_EncodeChar14Seg(char8 input)**—Converts the ASCII encoded alphabet character input into the 14-segment data that will display the alphabet character on a display. The returned data can be written directly into the display RAM to display the desired number. It is not necessary to use this function since a higher level APIs are provided to both decode the value and write it to the display RAM.
- **uint16 LED_Driver_EncodeNumber16Seg(uint8 number)**—Converts the lower 4 bits of the input into 16-segment data that will display the number in hex on a display. The returned data can be written directly into the display RAM to display the desired number. It is not necessary to use this function since higher level API are provided to both decode the value and write it to the display RAM.
- **uint16 LED_Driver_EncodeChar16Seg(char8 input)**—Converts the ASCII encoded alphabet character input into the 16-segment data that will display the alphabet character on a display. The returned data can be written directly into the display RAM to display the desired number. It is not necessary to use this function since higher level APIs are provided to both decode the value and write it to the display RAM.
- **void LED_Driver_SetBrightness(uint8 bright, uint8 position)**—Sets the desired brightness value (0 = display off; 255 = display at full brightness) for the chosen display by applying a PWM duty cycle to the common when the display is active.
- **uint8 LED_Driver_GetBrightness(uint8 position)**—Returns the brightness value for the specific display location.

---

[39] All unknown characters are displayed as a space.

- **void LED_Driver_Sleep(void)**—Prepares the component for sleep. If the component is currently enabled it will be disabled and reenabled by LED_Driver_Wakeup( ).
- **void LED_Driver_Wakeup(void)**—Returns the component to its state before the call to LED_Driver_Sleep( ).

15.84 Character LCD 2.00

The Character LCD component contains a set of library routines that enable simple use of one, two, or four-line LCD modules that follow the Hitachi 44780 standard 4-bit interface. The component provides APIs to implement horizontal and vertical bar graphs, or you can create and display your own custom characters. Features include: implements the industry-standard Hitachi HD44780 LCD display driver chip protocol, requires only seven I/O pins on one I/O port, contains built-in character editor to create user-defined custom characters and supports horizontal and vertical bar graphs [16].

15.84.1 Character LCD Functions 2.00

- **LCD_Char_Start( )**—Starts the module and loads custom character set to LCD, if it was defined.
- **LCD_Char_Stop( )**—Turns off the LCD.
- **LCD_Char_DisplayOn( )**—Turns on the LCD module's display.
- **LCD_Char_DisplayOff( )**—Turns off the LCD module's display.
- **LCD_Char_PrintString( )**—Prints a null-terminated string to the screen, character by character.
- **LCD_Char_PutChar( )**—Sends a single character to the LCD module data register at the current position.
- **LCD_Char_Position( )**—Sets the cursor's position to match the row and column supplied.
- **LCD_Char_WriteData( )**—Writes a single byte of data to the LCD module data register.
- **LCD_Char_WriteControl( )**—Writes a single-byte instruction to the LCD module control register.
- **LCD_Char_ClearDisplay( )**—Clears the data from the LCD module's screen.
- **LCD_Char_IsReady( )**—Polls the LCD until the ready bit is set or a timeout occurs.
- **LCD_Char_Sleep( )**—Prepares component for entering sleep mode.
- **LCD_Char_Wakeup( )**—Restores components configuration and turns on the LCD.
- **LCD_Char_Init( )**—Performs initialization required for component's normal work.
- **LCD_Char_Enable( )**—Turns on the display.
- **LCD_Char_SaveConfig( )**—Empty API provided to store any required data prior entering to a Sleep mode.
- **LCD_Char_RestoreConfig( )**—Empty API provided to restore saved data after exiting a Sleep mode.

15.84.2 Character LCD Function Calls 2.00

- **void LCD_Char_Start(void)**—This function initializes the LCD hardware module as follows:
  1. Enables 4-bit interface
  2. Clears the display
  3. Enables auto cursor increment
  4. Resets the cursor to start position

It also loads a custom character set to LCD if it was defined in the customizer's GUI.

- **void LCD_Char_Stop(void)**—Turns off the display of the LCD screen.
- **void LCD_Char_DisplayOn(void)**—Turns the display on, without initializing it. It calls function LCD_Char_WriteControl( ) with the appropriate argument to activate the display.
- **void LCD_Char_DisplayOff(void)**—Turns the display off, but does not reset the LCD module in any way. It calls the function.
- **void LCD_Char_PrintString(char8 const string[])**—Writes a null-terminated string of characters to the screen beginning at the current cursor location.
- **void LCD_Char_DisplayOn(void)**—Turns the display on, without initializing it. It calls function LCD_Char_WriteControl( ) with the appropriate argument to activate the display.
- **void LCD_Char_PutChar(char8 character)**—Writes an individual character to the screen at the current cursor location.[40]
- **void LCD_Char_DisplayOn(void)**—Turns the display on, without initializing it. It calls function LCD_Char_WriteControl( ) with the appropriate argument to activate the display.
- **void LCD_Char_Position(uint8 row, uint8 column)**—Moves the cursor to the location specified by arguments row and column. uint8 row: The row number at which to position the cursor. Minimum value is zero.[41]
- **void LCD_Char_DisplayOn(void)**—Turns the display on, without initializing it. It calls function LCD_Char_WriteControl( ) with the appropriate argument to activate the display.
- **void LCD_Char_WriteData(uint8 dByte)**—Writes data to the LCD RAM in the current position. Upon write completion, the position is incremented or decremented depending on the entry mode specified.
- **void LCD_Char_WriteControl(uint8 cByte)**—Writes a command byte to the LCD module. Different LCD models can have their own commands.
- **void LCD_Char_ClearDisplay(void)**—Clears the contents of the screen and resets the cursor location to be row and column zero. It calls LCD_Char_WriteControl( ) with the appropriate argument to activate the display.
- **void LCD_Char_IsReady(void)**—Polls the LCD until the ready bit is set or a timeout occurs.[42]
- **void LCD_Char_Sleep(void)**—This is the preferred routine to prepare the component for sleep. The LCD_Char_Sleep( ) routine saves the current component state. Then it calls the LCD_Char_Stop( ) function and calls LCD_Char_SaveConfig( ) to save the hardware configuration. Call the LCD_Char_Sleep( ) function before calling the CyPmSleep( ) or the CyPmHibernate( ) function. Reinitialize the component after saving or restoring component pin states.
- **void LCD_Char_Wakeup(void)**—Restores component's configuration and turns on the LCD.
- **void LCD_Char_Init(void)**—Performs initialization required for the component's normal work.[43]
- **void LCD_Char_Enable(void)**—Turns on the display.
- **void LCD_Char_SaveConfig(void)**—Empties API provided to store any required data prior to entering Sleep mode.
- **void LCD_Char_RestoreConfig(void)**—Empties API provided to restore saved data after exiting Sleep mode.

---

[40]Used to display custom characters through their named values. (LCD_Char_CUSTOM_0 through LCD_Char_CUSTOM_7).

[41]uint8 column is the column number at which to position the cursor.

[42]This function Changes pins to HI-Z.

[43]LCD_Char_Init( ) also loads the custom character set if it was defined in the Configure dialog.

15.85 Character LCD with I2C Interface (I2C LCD) 1.20

The I2C LCD component drives an I2C interfaced 2 line by 16 character LCD. The I2C LCD component is a wrapper around an I2C Master component and makes use of an existing I2C Master component. If a project does not already have an I2C Master component, one is required in order to operate. When one of the API functions is called, that function calls one or more of the I2C Master functions in order to communicate with the LCD. Features include: communicate on a 2-wire I2C bus, API compatible with the current character LCD component, one component may drive one or more LCDs on the same I2C bus, can coexist on an existing I2C bus if the PSoC is the I2C master and support for the NXP PCF2119x command format. Features include: communicate on a 2-wire I2C bus, API compatible with the current character LCD component, one component may drive one or more LCDs on the same I2C bus, can coexist on an existing I2C bus if the PSoC is the I2C master and support for the NXP PCF2119x command format [17].

15.85.1 Character LCD with I2C Interface (I2C LCD) Functions 1.20

- **I2C_LCD_Start( )**—Starts the module and loads custom character set to LCD if it was defined.
- **I2C_LCD_Stop( )**—Turns off the LCD
- **I2C_LCD_Init( )**—Performs initialization required for component's normal work
- **I2C_LCD_Enable( )**—Turns on the display
- **I2C_LCD_DisplayOn( )**—Turns on the LCD module's display
- **I2C_LCD_DisplayOff( )**—Turns off the LCD module's display
- **I2C_LCD_PrintString( )**—Prints a null-terminated string to the screen, character by character
- **I2C_LCD_PutChar( )**—Sends a single character to the LCD module data register at the current position.
- **I2C_LCD_Position( )**—Sets the cursor's position to match the row and column supplied
- **I2C_LCD_WriteControl( )**—Writes a single-byte instruction to the LCD module control register
- **I2C_LCD_ClearDisplay( )**—Clears the data from the LCD module's screen This function allows the user to change the default I2C address of the LCD.
- **I2C_LCD_SetAddr( )**—This function allows the user to change the default I2C address of the LCD.
- **I2C_LCD_PrintInt8( )**—Prints a two-ASCII-character hex representation of the 8-bit value to the Character LCD module.
- **I2C_LCD_PrintInt16( )**—Prints a four-ASCII-character hex representation of the 16-bit value to the Character LCD module.
- **I2C_LCD_PrintNumber( )**—Prints the decimal value of a 16-bit value as left-justified ASCII characters
- **I2C_LCD_HandleOneByteCommand( )**—This command adds a support of sending custom commands with 1 byte parameter
- **I2C_LCD_HandleCustomCommand( )**—Performs sending of the command that has variable parameters.

15.85.2 Character LCD with I2C Interface (I2C LCD) Function Calls 1.20

- **void I2C_LCD_Start(void)**—When this function called first time it initializes the LCD hardware module as follows:
  1. Turns on the display;
  2. Enables auto cursor increment;
  3. Resets the cursor to start position;
  4. Clears the display;
  5. It also loads a custom character set to LCD if it was defined in the customizer's GUI. Resets the cursor to start position.

  All of the following calls to this function will just turn on the LCD module.[44] The I2C Master must be initialized and global interrupts must be enabled before calling this function and if the NXP-compatible LCD module is being used, then a 1 ms reset pulse prior to calling I2C_LCD_Start( ) is required.
- **void I2C_LCD_Stop(void)**—Turns off the display of the LCD screen but does not stop the I2C Master component.
- **void I2C_LCD_PrintString(char8 const string[])**—Writes a null-terminated string of characters to the screen beginning at the current cursor location.[45]
- **void I2C_LCD_PutChar(char8 character)**—Writes an individual character to the screen at the current cursor location. Used to display custom characters through their named values. (I2C_LCD_CUSTOM_0 through I2C_LCD_CUSTOM_7).[46]
- **void I2C_LCD_Position(uint8 row, uint8 column)**—Moves the cursor to the location specified by arguments row and column.
- **void I2C_LCD_WriteData(uint8 dByte)**—Writes data to the LCD RAM in the current position. Upon write completion, the position is incremented or decremented depending on the entry mode specified.
- **void I2C_LCD_WriteControl(uint8 cByte)**—Writes a command byte to the LCD module. Different LCD models can have their own commands. Review the specific LCD datasheet for commands valid for that model.
- **void I2C_LCD_ClearDisplay(void)**—Clears the contents of the screen and resets the cursor location to be row and column zero. It calls I2C_LCD_WriteControl( ) with the appropriate argument to activate the display.
- **void I2C_LCD_SetAddr (uint8 address)**—This function allows you to change the default I2C address of the LCD. This function is not used for designs with a single LCD. Systems that have two or more LCDs on a single I2C bus use this function to select the appropriate LCD.
- **void I2C_LCD_PrintInt8(uint8 value)**—Prints a two ASCII character representation of the 8-bit value to the character I2C LCD module.
- **void I2C_LCD_PrintInt16(uint16 value)**—Prints a four ASCII character representation of the 16-bit value to the Character I2C LCD module.
- **void I2C_LCD_PrintNumber(uint16 value)**—Prints the decimal value of a 16-bit value as left-justified ASCII characters.

---

[44]This function sends commands to the display using the I2C Master.

[45]Because of the character set that is hardcoded to the NXP PCF2119x LCD module, which is used in the PSoC 4 processor module, some of the characters can't be displayed.

[46]Because of the character set that is hardcoded to the NXP PCF2119x LCD module, which is used in the PSoC 4 processor module, some of the characters can't be displayed.

- **void I2C_LCD_HandleOneByteCommand(uint8 cmdId, uint8 cmdByte)**—This command adds a support of sending custom commands with 1 byte parameter.
- **void I2C_LCD_HandleCustomCommand(uint8 cmdId, uint8 dataLength, uint8 const cmd-Data[])**—Performs sending of the command that has variable parameters.

15.86 Graphic LCD Controller (GraphicLCDCtrl) 1.80

The Graphic LCD Controller (GraphicLCDCtrl) Component provides the interface to an LCD panel that has an LCD driver, but not an LCD controller. This type of panel does not include a frame buffer. The frame buffer must be provided externally. This Component also interfaces to an externally provided frame buffer implemented using a 16-bit-wide async SRAM device. This Component is designed to work with the SEGGER emWin graphics library. This library provides a full-featured set of graphics functions for drawing and rendering text and images.[47] Features include: fully programmable screen size support up to HVGA resolution with QVGA (320x240) @ 60 Hz 16 bpp, WQVGA (480x272) @ 60 Hz 16 bpp, HVGA (480x320) @ 60 Hz 16 bpp, supports virtual screen operation, can be used with SEGGER emWin graphics library, performs read and write transactions during the blanking intervals, generates continuous timing signals to the panel without CPU intervention, supports up to a 23-bit address and a 16-bit data async SRAM device used as externally provided frame buffer and generates a selectable interrupt pulse at the entry and exit of the horizontal and vertical blanking intervals [50].

15.86.1 Graphic LCD Controller (GraphicLCDCtrl) Functions 1.80

- **GraphicLCDCtrl_Init( )**—Initializes or restores the Component parameters to the settings provided wit the Component customizer.
- **GraphicLCDCtrl_Enable( )**—Enables the GraphicLCDCtrl.
- **GraphicLCDCtrl_Start( )**—Starts the GraphicLCDCtrl interface.
- **GraphicLCDCtrl_Stop( )**—Disables the GraphicLCDCtrl interface.
- **GraphicLCDCtrl_Write( )**—Initiates a write transaction to the frame buffer.
- **GraphicLCDCtrl_Read( )**—Initiates a read transaction from the frame buffer.
- **GraphicLCDCtrl_WriteFrameAddr( )**—Sets the starting frame buffer address used when refreshing the screen.
- **GraphicLCDCtrl_ReadFrameAddr( )**—Reads the starting frame buffer address used when refreshing the screen.
- **GraphicLCDCtrl_WriteLineIncr( )**—Sets the address spacing between adjacent lines.
- **GraphicLCDCtrl_ReadLineIncr( )**—Reads the address increment between lines.
- **GraphicLCDCtrl_Sleep( )**—Saves the configuration and disables the GraphicLCDCtrl.
- **GraphicLCDCtrl_Wakeup( )**—Restores the configuration and enables the GraphicLCDCtrl.
- **GraphicLCDCtrl_SaveConfig( )**—Saves the configuration of the GraphicLCDCtrl.
- **GraphicLCDCtrl_RestoreConfig( )**—Restores the configuration of the GraphicLCDCtrl.

---

[47]The emWin graphics library available for PSoC 3, PSoC 4 and PSoC 5LP devices: www.cypress.com/go/comp_emWin.

15.86.2 Graphic LCD Controller (GraphicLCDCtrl) Function Calls 1.80

- **void GraphicLCDCtrl _Init(void)**—This function initializes or restores the Component parameters to the settings provided with the Component customizer. The compile time configuration that defines timing generation is restored to the settings provided with the customizer. The run-time configuration for the frame buffer address is set to 0; for the line increment it is set to the display line size.[48]
- **void GraphicLCDCtrl_Enable(void)**—This function activates the hardware and begins Component operation. It is not necessary to call GraphicLCDCtrl_Enable( ) because the GraphicLCDCtrl_Start( ) routine calls this function, which is the preferred method to begin Component operation.
- **void GraphicLCDCtrl_Start(void)**—Configures the Component for operation, begins generation of the clock, timing signals, interrupt, and starts refreshing the screen from the frame buffer. Sets the frame buffer address to 0 and the number of entries between lines to the line width.
- **void GraphicLCDCtrl_Stop(void)**—Disables the GraphicLCDCtrl Component.
- **void GraphicLCDCtrl_Write(uint32 addr, uint16 wrData)**—Initiates a write transaction to the frame buffer using the address and data provided. The write is a posted write, so this function returns before the write has actually completed on the interface. If the command queue is full, this function does not return until space is available to queue this write request.
- **uint16 GraphicLCDCtrl_Read(uint32 addr)**—Initiates a read transaction from the frame buffer. The read executes after all currently posted writes have completed. The function waits until the read completes and then returns the read value.
- **void GraphicLCDCtrl_WriteFrameAddr(uint32 addr)**—Sets the starting frame buffer address used when refreshing the screen. This register is read during each vertical blanking interval. To implement an atomic update of this register it should be written during the active refresh region.
- **uint32 GraphicLCDCtrl_ReadFrameAddr(void)**—Reads the starting frame buffer address used when refreshing the screen.
- **void GraphicLCDCtrl_WriteLineIncr(uint32 incr)**—Sets the address spacing between adjacent lines. By default, this is the display size of a line. This setting can be used to align lines to a different word boundary or to implement a virtual line length that is larger than the display region.
- **uint32 GraphicLCDCtrl_ReadLineIncr(void)**—Reads the address increment between lines.
- **void GraphicLCDCtrl_Sleep(void)**—Disables block's operation and saves its configuration. Should be called prior to entering Sleep.
- **void GraphicLCDCtrl_Wakeup(void)**—Enables block's operation and restores its configuration. Should be called after awaking from Sleep.
- **void GraphicLCDCtrl_SaveConfig(void)**—This function saves the Component configuration and nonretention registers. It also saves the current Component parameter values, as defined in the Configure dialog or as modified by appropriate APIs. This function is called by the GraphicLCDCtr_Sleep( ) function.
- **void GraphicLCDCtrl_RestoreConfig(void)**—This function restores the Component configuration and nonretention registers. It also restores the Component parameter values to what they were before calling the GraphicLCDCtrl_Sleep( ) function.

---

[48]This function does not clear data from the FIFOs and does not reset Component hardware state machines.

15.87 Graphic LCD Interface (GraphicLCDIntf) 1.80

The Graphic LCD Interface (GraphicLCDIntf) Component [49] provides the interface to a graphic LCD controller and driver device. These devices are commonly integrated into an LCD panel. The interface to these devices is commonly referred to as an i8080 interface. This is a reference to the historic parallel bus interface protocol of the Intel 8080 microprocessor. This Component is designed to work with the SEGGER emWin graphics library. This library provides a full-featured set of graphics functions for drawing and rendering text and images.[49] Features include: 8- or 16-bit interface to Graphic LCD Controller, compatible with many graphic controller devices, can be used with EGGER emWin graphics library, performs read and write transactions, 2 to 255 cycles for read low pulse width, 1 to 255 cycles for read high pulse width, and implements typical i8080 interface [51].

15.87.1 Graphic LCD Interface (GraphicLCDIntf) Functions 1.80

- **GraphicLCDIntf_Start( )**—Starts the GraphicLCDIntf interface.
- **GraphicLCDIntf_Stop( )** —Disables the GraphicLCDIntf interface.
- **GraphicLCDIntf_Write8( )**—Initiates a write transaction on the 8-bit parallel interface.
- **GraphicLCDIntf_Write16( )**—Initiates a write transaction on the 16-bit parallel interface.
- **GraphicLCDIntf_WriteM8( )**—Initiates multiple write transactions on the 8-bit parallel interface.
- **GraphicLCDIntf_WriteM16( )**—Initiates multiple write transactions on the 16-bit parallel interface.
- **GraphicLCDIntf_Write8_A0( )**—Initiates a write transaction on the 8-bit parallel interface, d_c line low.
- **GraphicLCDIntf_Write16_A0( )**—Initiates a write transaction on the 16-bit parallel interface, d_c line low.
- **GraphicLCDIntf_Write8_A1( )**—Initiates a write transaction on the 8-bit parallel interface, d_c line high.
- **GraphicLCDIntf_Write16_A1( )**—Initiates a write transaction on the 16-bit parallel interface, d_c line high.
- **GraphicLCDIntf_WriteM8_A0( )**—Initiates multiple write transactions on the 8-bit parallel interface, d_c line low.
- **GraphicLCDIntf_WriteM16_A0( )**—Initiates multiple write transactions on the 16-bit parallel interface d_c line low.
- **GraphicLCDIntf_WriteM8_A1( )**—Initiates multiple write transactions on the 8-bit parallel interface, d_c line high.
- **GraphicLCDIntf_WriteM16_A1( )**—Initiates multiple write transactions on the 16-bit parallel interface d_c line high.
- **GraphicLCDIntf_Read8( )**—Initiates a read transaction on the 8-bit parallel interface.
- **GraphicLCDIntf_Read16( )**—Initiates a read transaction on the 16-bit parallel interface.
- **GraphicLCDIntf_ReadM8( )**—Initiates multiple read transactions on the 8-bit parallel interface.
- **GraphicLCDIntf_ReadM16( )**—Initiates multiple read transactions on the 16-bit parallel interface.
- **GraphicLCDIntf_Read8_A1( )**—Initiates a read transaction on the 8-bit parallel interface.

---

[49]The emWin graphics library is available for PSoC 3, PSoC 4, PSoC 5LP devices: www.cypress.com/go/comp_emWin.

- **GraphicLCDIntf_Read16_A1( )**—Initiates a read transaction on the 16-bit parallel interface.
- **GraphicLCDIntf_ReadM8_A1( )**—Initiates multiple read transactions on the 8-bit parallel interface, d_c line high.
- **GraphicLCDIntf_ReadM16_A1( )**—Initiates multiple read transactions on the 16-bit parallel interface, d_c line high.
- **GraphicLCDIntf_Sleep( )**—Saves the configuration and disables the GraphicLCDIntf.
- **GraphicLCDIntf_Wakeup( )**—Restores the configuration and enables the GraphicLCDIntf.
- **GraphicLCDIntf_Init( )**—Initializes or restores the default GraphicLCDIntf configuration.
- **GraphicLCDIntf_Enable( )**—Enables the GraphicLCDIntf.
- **GraphicLCDIntf_SaveConfig( )**—Saves the configuration and disables the GraphicLCDIntf.
- **GraphicLCDIntf_RestoreConfig( )**—Restores the configuration of the GraphicLCDIntf.

15.87.2 Graphic LCD Interface (GraphicLCDIntf) Function Calls 1.80

- **void GraphicLCDIntf_Start(void)**—This function enables Active mode power template bits or clock gating as appropriate. Configures the Component for operation.
- **void GraphicLCDIntf_Stop(void)**—This function disables Active mode power template bits or gates clocks as appropriate.
- **void GraphicLCDIntf_Write8(uint8 d_c, uint8 wrData)**—This function initiates a write transaction on the 8-bit parallel interface. The write is a posted write, so this function returns before the write has actually completed on the interface. If the command queue is full, this function does not return until space is available to queue this write request.
- **void GraphicLCDIntf_Write16(uint8 d_c, uint16 wrData)**—This function initiates a write transaction on the 16-bit parallel interface. The write is a posted write, so this function returns before the write has actually completed on the interface. If the command queue is full, this function does not return until space is available to queue this write request.
- **void GraphicLCDIntf_WriteM8(uint8 d_c, uint8 wrData[], uint16 num)**—This function initiates multiple write transactions on the 8-bit parallel interface. Writing of multiple bytes with one execution of GraphicLCDIntf_WriteM8, instead of multiple executions of GraphicLCD-Intf_Write8 increases the write performance on the interface.
- **void GraphicLCDIntf_WriteM16(uint8 d_c, uint16 wrData[], uint16 num)**—This function initiates multiple write transactions on the 16-bit parallel interface. Writing of multiple words with one execution of GraphicLCDIntf_WriteM16, instead of multiple executions of GraphicLCD-Intf_Write16 increases the write performance on the interface.
- **void GraphicLCDIntf_Write8_A0(uint8 wrData)**—This function initiates a command write transaction on the 8-bit parallel interface with the d_c pin set to 0. The write is a posted write, so this function will return before the write has actually completed on the interface. If the command queue is full, this function will not return until space is available to queue this write request.Parameters: wrData: Data sent on the do_lsb[7:0] pins.
- **void GraphicLCDIntf_Write16_A0(uint8 wrData)**—This function initiates a command write transaction on the 16-bit parallel interface with the d_c pin se to 0. The write is a posted write, so this function will return before the write has actually completed on the interface. If the command queue is full, this function will not return until space is available to queue this write request.
- **void GraphicLCDIntf_Write8_A1(uint8 wrData)**—This function initiates a data write transaction on the 8-bit parallel interface with the d_c pin set to 1. The write is a posted write, so this function will return before the write has actually completed on the interface. If the command queue is full, this function will not return until space is available to queue this write request.

- **void GraphicLCDIntf_Write16_A1(uint8 wrData)**—This function initiates a data write transaction on the 16-bit parallel interface with the d_c pin set to 1. The write is a posted write, so this function will return before the write has actually completed on the interface. If the command queue is full, this function will not return until space is available to queue this write request.
- **void GraphicLCDIntf_WriteM8_A0(uint8 wrData[], int num)**—This function initiates multiple data write transactions on the 8-bit parallel interface with the d_c pin set to 0. The write is a posted write, so this function will return before the write has actually completed on the interface. If the command queue is full, this function will not return until space is available to queue this write request.
- **void GraphicLCDIntf_WriteM16_A0(uint16 wrData[], int num)**—This function initiates multiple data write transactions on the 16-bit parallel interface with the d_c pin set to 0. The write is a posted write, so this function will return before the write has actually completed on the interface. If the command queue is full, this function will not return until space is available to queue this write request.
- **void GraphicLCDIntf_WriteM8_A1(uint8 wrData[], int num)**—This function initiates multiple data write transactions on the 8-bit parallel interface with the d_c pin set to 1. The write is a posted write, so this function will return before the write has actually completed on the interface. If the command queue is full, this function will not return until space is available to queue this write request.
- **void GraphicLCDIntf_WriteM16_A1(uint16 wrData[], int num)**—This function initiates multiple data write transactions on the 16-bit parallel interface with the d_c pin set to 1. The write is a posted write, so this function will return before the write has actually completed on the interface. If the command queue is full, this function will not return until space is available to queue this write request.
- **uint8 GraphicLCDIntf_Read8(uint8 d_c)**—This function initiates a read transaction on the 8-bit parallel interface. The read executes after all currently posted writes have completed. This function waits until the read completes and then returns the read value.
- **uint16 GraphicLCDIntf_Read16(uint8 d_c)**—This function initiates a read transaction on the 16-bit parallel interface. The read executes after all currently posted writes have completed. This function waits until the read completes and then returns the read value.
- **void GraphicLCDIntf_ReadM8 (uint8 d_c, uint8 rdData[], uint16 num)**—This function initiates a read transaction on the 8-bit parallel interface. The read executes after all currently posted writes have completed. This function waits until the read completes and then returns the read value.
- **void GraphicLCDIntf_ReadM16 (uint8 d_c, uint16 rdData[], uint16 num)**—This function initiates a read transaction on the 16-bit parallel interface. The read executes after all currently posted writes have completed. This function waits until the read completes and then returns the read value.
- **uint8 GraphicLCDIntf_Read8_A1(void)**—This function initiates a data read transaction on the 8-bit parallel interface with the d_c pin set to 1. The read will execute after all currently posted writes have completed. This function will wait until the read completes and then returns the read value.
- **uint16 GraphicLCDIntf_Read16_A1(void)**—This function initiates a data read transaction on the 16-bit parallel interface with the d_c pin set to 1. The read will execute after all currently posted writes have completed. This function will wait until the read completes and then returns the read value.
- **void GraphicLCDIntf_ReadM8_A1 (uint8 rdData[], uint16 num)**—This function initiates a read transaction on the 8-bit parallel interface with the d_c pin set to 1. The read executes after

all currently posted writes have completed. This function waits until the read completes and then returns the read value.

- **void GraphicLCDIntf_ReadM16_A1 (uint16 rdData[], uint16 num)**—This function initiates a read transaction on the 16-bit parallel interface with the d_c pin set to 1. The read executes after all currently posted writes have completed. This function waits until the read completes and then returns the read value.
- **void GraphicLCDIntf_Sleep(void)**—Stops the Component operation and saves the user configuration.
- **void GraphicLCDIntf_Wakeup(void)**—Restores the user configuration and restores Component state.
- **void GraphicLCDIntf_Init(void)**—This function initializes or restores the Component according to the customizer Configure dialog settings. It is not necessary to call GraphicLCDIntf_Init( ) because the GraphicLCDIntf_Start( ) routine calls this function and is the preferred method to begin Component operation. Only the static Component configuration that defines Read Low and High Pulse Widths will be restored to its initial values.
- **void GraphicLCDIntf_Enable(void)**—This function activates the hardware and begins Component operation. It is not necessary to call GraphicLCDIntf_Enable( ) because the GraphicLCDIntf_Start( ) routine calls this function, which is the preferred method to begin Component operation.
- **void GraphicLCDIntf_SaveConfig(void)**—This function saves the Component configuration and non-retention registers. This function is called by the GraphicLCDIntf_Sleep( ) function.
- **void GraphicLCDIntf_RestoreConfig(void)**—This function restores the configuration of GraphicLCDIntf non-retention registers. The API is called by GraphicLCDIntf_Wakeup function.

15.88 Static LCD (LCD_SegStatic) 2.30

The Static Segment LCD (LCD_SegStatic) component can directly drive 3.3-V and 5.0-V LCD glass. This component provides an easy method of configuring the PSoC device for your custom or standard glass. Each LCD pixel/symbol may be either on or off. The Static Segment LCD component also provides advanced support to simplify the following types of display structures within the glass: 7-Segment numeral, 14-Segment alphanumeric, 16-Segment alphanumeric, 1-255 element bar graph, 1-61 pixels or symbols and 10–150 Hz refresh rate. Features include User-defined pixel or symbol map with optional 7-segment, 14-segment, 16-segment and bar graph calculation routines and direct drive static (one common) LCDs [88].

15.88.1 LCD_SegStatic Function 2.30

- **LCD_SegStatic_Start( )**—Starts the LCD component and DMA channels. Initializes the frame buffer. Does not clear the frame buffer RAM if it was previously defined.
- **LCD_SegStatic_Stop( )**—Disables the LCD component and associated interrupts and DMA channels. Does not clear the frame buffer.
- **LCD_SegStatic_EnableInt( )**—Enables the LCD interrupts.
- **LCD_SegStatic_DisableInt( )**—Disables the LCD interrupt.
- **LCD_SegStatic_ClearDisplay( )**—Clears the display RAM of the frame buffer.
- **LCD_SegStatic_WritePixel( )**—Sets or clears a pixel based on PixelState. The pixel is addressed by a packed number.

- **LCD_SegStatic_ReadPixel( )**—Reads the state of a pixel in the frame buffer. The pixel is addressed by a packed number.
- **LCD_SegStatic_WriteInvertState( )**—Inverts the display based on an input parameter.
- **LCD_SegStatic_ReadInvertState( )**—Returns the current value of the display invert state: normal or inverted.
- **LCD_SegStatic_Sleep( )**—Stops the LCD and saves the user configuration.
- **LCD_SegStatic_Wakeup( )**—Restores and enables the user configuration.
- **LCD_SegStatic_Init( )**—Configures every-frame interrupt and initializes the frame buffer.
- **LCD_SegStatic_Enable( )**—Enables clock generation for the component.
- **LCD_SegStatic_SaveConfig( )**—Saves the LCD configuration.
- **LCD_SegStatic_RestoreConfig( )**—Restores the LCD configuration.

15.88.2 LCD_SegStatic Function Calls 2.30

- **uint8 LCD_SegStatic_Start(void)**—Starts the LCD component, DMA channels, frame buffer and hardware. Does not clear the frame buffer RAM.
- **void LCD_SegStatic_Stop(void)**—Disables the LCD component and associated interrupts and DMA channels. Automatically blanks the display to avoid damage from DC offsets. Does not clear the frame buffer.
- **void LCD_SegStatic_EnableInt(void)**—Enables the LCD interrupts. An interrupt occurs after every LCD update (TD completion).
- **void LCD_SegStatic_DisableInt(void)**—Disables the LCD interrupts.
- **void LCD_SegStatic_ClearDisplay(void)**—This function clears the display RAM of the page buffer.
- **uint8 LCD_SegStatic_WritePixel(uint16 pixelNumber, uint8 pixelState)**—This function sets or clears a pixel in the frame buffer based on the PixelState parameter. The pixel is addressed with a packed number.
- **uint8 LCD_SegStatic_ReadPixel(uint16 pixelNumber)**—This function reads a pixel's state in the frame buffer. The pixel is addressed by a packed number.
- **uint8 LCD_Seg_WriteInvertState(uint8 invertState)**—This function inverts the display based on an input parameter.
- **uint8 LCD_Seg_ReadInvertState(void)**—This function returns the current value of the display invert state: normal or inverted.
- **void LCD_SegStatic_Init(void)**—Initializes or restores the component according to the customizer Configure dialog settings. It is not necessary to call LCD_SegStatic_Init( ) because the LCD_SegStatic_Start( ) routine calls this function and is the preferred method to begin component operation. Configures every frame interrupt and initializes the frame buffer.
- **void LCD_SegStatic_Enable(void)**—Enables clock generation for the component.
- **void LCD_SegStatic_Sleep(void)**—This is the preferred routine to prepare the component for sleep. The LCD_Sleep( ) routine saves the current component state. Then it calls the LCD_SegStatic_Stop( ) function and calls LCD_SegStatic_SaveConfig( ) to save the hardware configuration. Call the LCD_SegStatic_Sleep( ) function before calling the CyPmSleep( ) or the CyPmHibernate( ) function.
- **void LCD_SegStatic_Wakeup(void)**—This is the preferred routine to restore the component to the state when LCD_SegStatic_Sleep( ) was called. The LCD_SegStatic_Wakeup( ) function calls the LCD_SegStatic_RestoreConfig( ) function to restore the configuration. If the component was

enabled before the LCD_SegStatic_Sleep( ) function was called, the LCD_SegStatic_Wakeup( ) function will also re-enable the component.

- **void LCD_SegStatic_SaveConfig(void)**—This function saves the component configuration and non-retention registers. It also saves the current component parameter values, as defined in the Configure dialog or as modified by appropriate APIs. This function is called by the LCD_SegStatic_Sleep( ) function.
- **void LCD_SegStatic_RestoreConfig(void)**—This function restores the component configuration and non-retention registers. It also restores the component parameter values to what they were before calling the LCD_SegStatic_Sleep( ) function.

15.88.3 Optional Helper APIs (LCD_SegStatic Functions)

- **LCD_SegStatic_Write7SegDigit_n**—Displays a hexadecimal digit on an array of 7-segment display elements.
- **LCD_SegStatic_Write7SegNumber_n**—Displays an integer value o—Displays an integer location on a linear or circular bar graph
- **LCD_SegStatic_PutChar14Seg_n**—Displays a character on an array of 14-segment alphanumeric character display elements.
- **LCD_SegStatic_WriteString14Seg_n**—Displays a null terminated character string on an array of 14-segment alphanumeric character display elements.
- **LCD_SegStatic_PutChar16Seg_n**—Displays a character on an array of 16-segment alphanumeric character display elements.
- **LCD_SegStatic_WriteString16Seg_n**—Displays a null terminated character string on an array of 16-segment alphanumeric character display elements.

15.88.4 Optional Helper APIs (LCD_SegStatic) Function Calls

- **void LCD_SegStatic_Write7SegDigit_n(uint8 digit, uint8 position)**—This function displays a hexadecimal digit on an array of 7-segment display elements. Digits can be hexadecimal values in the range of 0 to 9 and A to F. The customizer Display Helpers facility must be used to define the pixel set associated with the 7-segment display elements Multiple 7-segment display elements can be defined in the frame buffer and are addressed through the suffix (n) in the function name. This function is only included if a 7-segment display element is defined in the component customizer.
- **void LCD_SegStatic Write7SegNumber_n(uint16 value, uint8 position, uint8 mode)**—This function displays a 16-bit integer value on a one- to five-digit array of 7-segment display elements. The customizer Display Helpers facility must be used to define the pixel set associated with the 7-segment display elements. Multiple 7-segment display element groups can be defined in the frame buffer and are addressed through the suffix (n) in the function name. Sign conversion, sign display, decimal points other custom features must be handled by application-specific user code. This function is only included if a 7-segment display element is defined in the component customizer.
- **void LCD_SegStatic_WriteBargraph_n(uint16 location, int8 Mode)**—This function displays an 8-bit integer location on a 1- to 255-segment bar graph (numbered left to right).The bar graph may be any user-defined size between 1 and 255 segments. A bar graph may also be created in a circle to display rotary position. The customizer Display Helpers facility must be used to define the pixel set associated with the bar graph display elements Multiple bar graph displays can be created

in the frame buffer and are addressed through the suffix (n) in the function name. This function is only included if a bar graph display element is defined in the component customizer

- **void LCD_SegStatic_PutChar14Seg_n(uint8 character, uint8 position)**—This function displays an 8-bit character on an array of 14-segment alphanumeric character display elements. The customizer Display Helpers facility must be used to define the pixel set associated with the 14-segment display element. Multiple 14-segment alphanumeric display element groups can be defined in the frame buffer and are addressed through the suffix (n) in the function name. This function is only included if a 14-segment element is defined in the component customizer.
- **void LCD_SegStatic_WriteString14Seg_n(uint8 const character[], uint8 position)**—This function displays a null terminated character string on an array of 14-segment alphanumeric character display elements. The customizer Display Helpers facility must be used to define the pixel set associated with the 14-segment display elements Multiple 14-segment alphanumeric display element groups can be defined in the frame buffer and are addressed through the suffix (n) in the function name. This function is only included if a 14-segment display element is defined in the component customizer
- **void LCD_SegStatic_PutChar16Seg_n(uint8 character, uint8 position)**—This function displays an 8-bit character on an array of 16-segment alphanumeric character display elements. The customizer Display Helpers facility must be used to define the pixel set associated with the 16-segment display elements. Multiple 16-segment alphanumeric display element groups can be defined in the frame buffer and are addressed through the suffix (n) in the function name. This function is only included if a 16-segment display element is defined in the component customizer
- **void LCD_SegStatic PutChar16Seg_n(uint8 character, uint8 position)**—This function displays an 8-bit character on an array of 16-segment alphanumeric character display elements. The customizer Display Helpers facility must be used to define the pixel set associated with the 16-segment display elements. Multiple 16-segment alphanumeric display element groups can be defined in the frame buffer and are addressed through the suffix (n) in the function name. This function is only included if a 16-segment display element is defined in the component customizer
- **(void) LCD_SegStatic_WriteString16Seg_n(uint8 const character[], uint8 position)**—This function displays a null terminated character string on an array of 16-segment alphanumeric character display elements. The customizer Display Helpers facility must be used to define the pixel set associated with the 16-segment display elements. Multiple 16-segment alphanumeric display element groups can be defined in the frame buffer and are addressed through the suffix (n) in the function name. This function is only included if a 16-segment display element is defined in the component customizer.

15.88.5 Pins API (LCD_SegStatic) Functions

- **LCD_SegStatic_ComPort_SetDriveMode**—Sets the drive mode for the pin used by a common line of the Static Segment LCD component.
- **LCD_SegStatic_SegPort_SetDriveMode**—Sets the drive mode for all pins used by segment lines of the Static Segment LCD component.

15.88.6 Pins API (LCD_SegStatic) Function Calls

- **void LCD_SegStatic_ComPort_SetDriveMode(uint8 mode)**—Sets the drive mode for the pin used by a common line of the Static Segment LCD component.

- **LCD_SegStatic_SegPort_SetDriveMode(uint8 mode)**—Sets the drive

---

15.89 Resistive Touch (ResistiveTouch) 2.00

This resistive touchscreen component is used to interface with a 4-wire resistive touch screen. The component provides a method to integrate and configure the resistive touch elements of a touchscreen with the emWin Graphics library. It integrates hardware-dependent functions that are called by the touchscreen driver supplied with emWin when polling the touch panel. This component is designed to work with the SEGGER emWin graphics library.[50] This graphics library provides a full-featured set of graphics functions for drawing and rendering text and images. Features include: supports 4-wire resistive touchscreen interface, supports the Delta Sigma Converter for both the PSoC 3 and PSoC 5LP devices and supports the ADC Successive Approximation Register for PSoC 5LP devices [72].

15.89.1 Resistive Touch Functions 2.00

- **ResistiveTouch_Start( )**—Calls the ResistiveTouch_Init( ) and ResistiveTouch_Enable( ) APIs.
- **ResistiveTouch_Stop( )**—Stops the ADC and the AMux component.
- **ResistiveTouch_Init( )**—Calls the Init( ) functions of the ADC and AMux components.
- **ResistiveTouch_Enable( )**—Enables the component.
- **ResistiveTouch_ActivateY( )**—Configures the pins to measure the Y-axis.
- **ResistiveTouch_ActivateX( )**—Configures the pins to measure the X-axis.
- **ResistiveTouch_TouchDetect( )**—Detects a touch on the screen.
- **ResistiveTouch_Measure( )**—Returns the result of the ADC conversion.
- **ResistiveTouch_SaveConfig( )**—Saves the configuration of the ADC.
- **ResistiveTouch_Sleep( )**—Prepares the component for entering the low power mode.
- **ResistiveTouch_RestoreConfig( )**—Restores the configuration of the ADC.
- **ResistiveTouch_Wakeup( )**—Restores the component after waking up from the low power mode.

15.89.2 Resistive Touch Function Calls 2.00

- **void ResistiveTouch_Start(void)**—Sets the ResistiveTouch_initVar variable, calls the ResistiveTouch_Init( ) function and then calls the ResistiveTouch_Enable( ) function.
- **void ResistiveTouch_Stop(void)**—Calls the Stop( ) functions of the ADC and the AMux components.
- **void ResistiveTouch_Init(void)**—Calls the Init( ) functions of the ADC and AMux components.
- **void ResistiveTouch_Enable(void)**—Calls the Enable( ) function of the ADC component.
- **void ResistiveTouch_ActivateX(void)**—Configures the pins to measure the X-axis.
- **avoid ResistiveTouch_ActivateY(void)**—Configures the pins to measure the Y-axis.
- **int16 ResistiveTouch_Measure(void)**—Returns the result of the ADC conversion.
- **uint8 ResistiveTouch_TouchDetect(void)**—Detects a touch on the screen.
- **void ResistiveTouch_SaveConfig(void)**—Saves the configuration of the ADC.
- **void ResistiveTouch_RestoreConfig(void)**—Restores the configuration of the ADC.

---

[50]This graphics library is provided by Cypress to use with Cypress devices and is available on the Cypress website at www.cypress.com/go/comp_emWin.

- **void ResistiveTouch_Sleep(void)**—Prepares the component for entering the low power mode.
- **void ResistiveTouch_Wakeup(void)**—Restores the component after waking up from the low power mode.

15.90 Segment LCD (LCD_Seg) 3.40

The Segment LCD (LCD_Seg) component can directly drive a variety of LCD glass at different voltage levels with multiplex ratios up to 16x. This component provides an easy method of configuring the PSoC device to drive your custom or standard glass. Internal bias generation eliminates the need for any external hardware and allows for software-based contrast adjustment. Using the Boost Converter, the glass bias may be at a higher voltage than the PSoC supply voltage. This allows increased display flexibility in portable applications. Each LCD pixel/symbol may be either on or off. The Segment LCD component also provides advanced support to simplify the following types of display structures within the glass: 7-segment numerals, 14-segment alphanumeric, 16-segment alphanumeric, 5x7 and 5x8 dot matrix alphanumeric.[51] Features include: 2 to 768 pixels or symbols, 1/3, 1/4 and 1/5 bias supported, 10- to 150-Hz refresh rate, integrated bias generation between 2.0 V and 5.2 V with up to 128 digitally controlled bias levels for dynamic contrast control, supports both type A (standard) and type B (low power) waveforms, pixel state of the display may be inverted for negative image, 256 bytes of display memory (frame buffer), user-defined pixel or symbol map with optional 7-, 14-, or 16-segment character; and 5x7 or 5x8 dot matrix; and bar graph calculation routines [79].

15.90.1 Segment LCD (LCD_Seg) Functions 3.40

- **LCD_Seg_Start( )**—Sets the initVar variable, calls the LCD_Seg_Init( ) function and then calls the LCD_Seg_Enable( ) function.
- **LCD_Seg_Stop( )**—Disables the LCD component and associated interrupts and DMA channels.
- **LCD_Seg_EnableInt( )**—Enables the LCD interrupts. Not required if LCD_Seg_Start( ) is called
- **LCD_Seg_DisableInt( )**—Disables the LCD interrupt. Not required if LCD_Seg_Stop( ) is called
- **LCD_Seg_SetBias( )**—Sets the bias level for the LCD glass to one of up to 64 values.
- **LCD_Seg_WriteInvertState( )**—Inverts the display based on an input parameter.
- **LCD_Seg_ReadInvertState( )**—Returns the current value of the display invert state: normal or inverted
- **LCD_Seg_ClearDisplay( )**—Clears the display and associated frame buffer RAM.
- **LCD_Seg_WritePixel( )**—Sets or clears a pixel based on PixelState
- **LCD_Seg_ReadPixel( )**—Reads the state of a pixel in the frame buffer.
- **LCD_Seg_Sleep( )**—Stops the LCD and saves the user configuration.
- **LCD_Seg_Wakeup( )**—Restores and enables the user configuration.—Saves the LCD configuration.
- **LCD_Seg_RestoreConfig( )**—Restores the LCD configuration.
- **LCD_Seg_Init( )**—Initializes or restores the LCD per the Configure dialog settings.
- **LCD_Seg_Enable( )**—Enables the LCD.

---

[51] The same look-up table for 5x7 and 5x8 is used for both. All symbols in the look-up table are the size of 5x7 pixels.

15.90.2 Segment LCD (LCD_Seg) Function Calls 3.40

- **uint8 LCD_Seg_Start(void)**—Starts the LCD component and enables required interrupts, DMA channels, frame buffer and hardware. Does not clear the frame buffer RAM.
- **void LCD_Seg_Stop(void)**—Disables the LCD component and associated interrupts and DMA channels. Automatically blanks the display to avoid damage from DC offsets. Does not clear the frame buffer.
- **void LCD_Seg_EnableInt(void)**—Enables the LCD interrupts. An interrupt occurs after every LCD update (TD completion). If the PSoC 5LP device is used, this API also enables an LCD wakeup Interrupt. This function should always be called when the component's operation in sleep is desired.
- **void LCD_Seg_DisableInt(void)**—Disables "every subframe" and LCD wakeup interrupts.
- **uint8 LCD_Seg_SetBias(uint8 biasLevel)**—This function sets the bias level for the LCD glass to one of up to 64 values. The actual number of values is limited by the Analog supply voltage, Vdda. The bias voltage cannot exceed Vdda. Changing the bias level affects the LCD contrast.
- **uint8 LCD_Seg_WriteInvertState(uint8 invertState)**—This function inverts the display based on an input parameter. The inversion occurs in hardware and no change is required to the display RAM in the frame buffer.
- **uint8 LCD_Seg_ReadInvertState(void)**—This function returns the current value of the display invert state: normal or inverted.
- **void LCD_Seg_ClearDisplay(void)**—This function clears the display and the associated frame buffer RAM.
- **uint8 LCD_Seg_WritePixel(uint16 pixelNumber, uint8 pixelState)**—This function sets or clears a pixel based on the input parameter PixelState. The pixel is addressed by a packed number.
- **void LCD_Seg_ClearDisplay(void)**—This function clears the display and the associated frame buffer RAM. **uint8 LCD_Seg_WritePixel(uint16 pixelNumber, uint8 pixelState)**—This function sets or clears a pixel based on the input parameter PixelState. The pixel is addressed by a packed number. uint8 pixelState: The pixelNumber specified is set to this pixel state.
- **uint8 LCD_Seg_ReadPixel(uint16 pixelNumber)**—This function reads the state of a pixel in the frame buffer. The pixel is addressed by a packed number.
- **void LCD_Seg_Sleep(void)**—This is the preferred routine to prepare the component for sleep. The LCD_Seg_Sleep( ) routine saves the current component state. Then it calls the LCD_Seg_Stop( ) function and calls LCD_Seg_SaveConfig( ) to save the hardware configuration. Call the LCD_Seg_Sleep( ) function before calling the CyPmSleep( ) or the CyPmHibernate( ) function.
- **void LCD_Seg_Wakeup(void)**—This is the preferred routine to restore the component to the state when LCD_Seg_Sleep( ) was called. The LCD_Seg_Wakeup( ) function calls the LCD_Seg_RestoreConfig( ) function to restore the configuration. If the component was enabled before the LCD_Seg_Sleep( ) function was called, the LCD_Seg_Wakeup( ) function will also re-enable the component.
- **void LCD_Seg_Sleep(void)**—This is the preferred routine to prepare the component for sleep. The LCD_Seg_Sleep( ) routine saves the current component state. Then it calls the LCD_Seg_Stop( ) function and calls LCD_Seg_SaveConfig( ) to save the hardware configuration. Call the LCD_Seg_Sleep( ) function before calling the CyPmSleep( ) or the CyPmHibernate( ) function.
- **void LCD_Seg_Wakeup(void)**—This is the preferred routine to restore the component to the state when LCD_Seg_Sleep( ) was called. The LCD_Seg_Wakeup( ) function calls the LCD_Seg_RestoreConfig( ) function to restore the configuration. If the component was enabled

before the LCD_Seg_Sleep( ) function was called, the LCD_Seg_Wakeup( ) function will also re-enable the component. CYRET_LOCKED Some of DMA TDs or a channel already in use CYRET_SUCCESS Function completed successfully

- **void LCD_Seg_SaveConfig(void)**—This function saves the component configuration. This will save non-retention registers. This function will also save the current component parameter values, as defined in the Configure dialog or as modified by appropriate APIs. This function is called by the LCD_Seg_Sleep( ) function.
- **void LCD_Seg_RestoreConfig(void)**—This function restores the component configuration. This will restore non-retention registers. This function will also restore the component parameter values to what they were prior to calling the LCD_Seg_Sleep( ) function.
- **tvoid LCD_Seg_Init(void)**—Initializes or restores the component parameters per the Configure dialog settings. It is not necessary to call LCD_Seg_Init( ) because the LCD_Seg_Start( ) routine calls this function and is the preferred method to begin component operation. Configures and enables all required hardware blocks and clears the frame buffer.
- **void LCD_Seg_Enable(void)**—Enables power to the LCD fixed hardware and enables generation of UDB signals.

15.90.3 Segment LCD (LCD_Seg)—Optional Helper APIs Functions

- **LCD_Seg_Write7SegDigit_n**—Displays a hexadecimal digit on an array of 7-segment display elements.
- **LCD_Seg_Write7SegNumber_n**—Displays an integer value on a 1- to 5-digit array of 7-segment display elements.
- **LCD_Seg_WriteBargraph_n**—Displays an integer location on a linear or circular bar graph.
- **LCD_Seg_PutChar14Seg_n**—Displays a character on an array of 14-segment alphanumeric character display elements.
- **LCD_Seg_WriteString14Seg_n**—Displays a null terminated character string on an array of 14-segment alphanumeric character display elements.
- **LCD_Seg_PutChar16Seg_n**—Displays a character on an array of 16-segment alphanumeric character display elements.
- **LCD_Seg_WriteString16Seg_n**—Displays a null terminated character string on an array of 16-segment alphanumeric character display elements.
- **LCD_Seg_PutCharDotMatrix_n**—Displays a character on an array of dot-matrix alphanumeric character display elements.
- **LCD_Seg_WriteStringDotMatrix_n**—Displays a null terminated character string on an array of dot-matrix alphanumeric character display elements.

15.90.4 LCD_Seg—Optional Helper APIs Function Calls

- **void LCD_Seg_Write7SegDigit_n(uint8 digit, uint8 position)**—This function displays a hexadecimal digit on an array of 7-segment display elements. Digits can be hexadecimal values in the range of 0 to 9 and A to F. The customizer Display Helpers facility must be used to define the pixel set associated with the 7-segment display elements Multiple 7-segment display elements can be defined in the frame buffer and are addressed through the suffix (n) in the function name. This function is only included if a 7-segment display element is defined in the component customizer.
- **void LCD_Seg Write7SegNumber_n(uint16 value, uint8 position, uint8 mode)**—This function displays a 16-bit integer value on a 1- to 5-digit array of 7-segment display elements. The customizer Display Helpers facility must be used to define the pixel set associated with the 7-

segment display element(s). Multiple 7-segment display element groups can be defined in the frame buffer and are addressed through the suffix (n) in the function name. Sign conversion, sign display, decimal points and other custom features must be handled by application-specific user code. This function is only included if a 7-segment display element is defined in the component customizer.

- **LCD_Seg_WriteBargraph_n(uint8 location, uint8 mode)**—This function displays an 8-bit integer Location on a 1- to 255-segment bar graph (numbered left to right).The bar graph may be any user-defined size between 1 and 255 segments. A bar graph may also be created in a circle to display rotary position. The customizer Display Helpers facility must be used to define the pixel set associated with the bar graph display elements Multiple bar graph displays can be created in the frame buffer and are addressed through the suffix (n) in the function name. This function is only included if a bar graph display element is defined in the component customizer

- **void LCD_Seg_PutChar14Seg_n(uint8 character, uint8 position)**—This function displays an 8-bit character on an array of 14-segment alphanumeric character display elements. The customizer Display Helpers facility must be used to define the pixel set associated with the 14-segment display element. Multiple 14-segment alphanumeric display element groups can be defined in the frame buffer and are addressed through the suffix (n) in the function name. This function is only included if a 14-segment element is defined in the component customizer.

- **void LCD_Seg_WriteString14Seg_n(uint8 const character[], uint8 position)**—This function displays a null terminated character string on an array of 14-segment alphanumeric character display elements. The customizer Display Helpers facility must be used to define the pixel set associated with the 14 segment display elements. Multiple 14-segment alphanumeric display element groups can be defined in the frame buffer and are addressed through the suffix (n) in the function name. This function is only included if a 14-segment display element is defined in the component customizer

- **void LCD_Seg_PutChar16Seg_n(uint8 character, uint8 position)**—This function displays an 8-bit character on an array of 16-segment alphanumeric character display elements. The customizer Display Helpers facility must be used to define the pixel set associated with the 16-segment display element(s). Multiple 16-segment alphanumeric display element groups can be defined in the frame buffer and are addressed through the suffix (n) in the function name. This function is only included if a 16-segment display element is defined in the component customizer

- **void LCD_Seg_WriteString16Seg_n(uint8 const character[], uint8 position)**—This function displays a null terminated character string on an array of 16-segment alphanumeric character display elements. The customizer Display Helpers facility must be used to define the pixel set associated with the 16 segment display elements. Multiple 16-segment alphanumeric display element groups can be defined in the frame buffer and are addressed through the suffix (n) in the function name. This function is only included if a 16-segment display element is defined in the component customizer.

- **void LCD_Seg_PutCharDotMatrix_n(uint8 character, uint8 position)**—This function displays an 8-bit character on an array of dot-matrix alphanumeric character display elements. The customizer Display Helpers facility must be used to define the pixel set associated with the dot matrix display elements. Multiple dot-matrix alphanumeric display element groups can be defined in the frame buffer and are addressed through the suffix (n) in the function name. This function is only included if a dot-matrix display element is defined in the component customizer.

- **void LCD_Seg_WriteStringDotMatrix_n(uint8 const character[], uint8 position)**—This function displays a null terminated character string on an array of dot-matrix alphanumeric character display elements. The customizer Display Helpers facility must be used to define the pixel set associated with the dot-matrix display elements. Multiple dot-matrix alphanumeric display element groups can be defined in the frame buffer and are addressed through the suffix (n) in the function name. This function is only included if a dot- matrix display element is defined in the component customizer.

15.90.5 LCD_Seg—Pins Functions

- **LCD_Seg_ComPort_SetDriveMode**—Sets the drive mode for all pins used by common lines of the Segment LCD component
- **LCD_Seg_SegPort_SetDriveMode**—Sets the drive mode for all pins used by segment lines of the Segment LCD component.

15.90.6 LCD_Seg—Pins Function Calls

- **void LCD_Seg_ComPort_SetDriveMode(uint8 mode)**—Sets the drive mode for all pins used by common lines of the Segment LCD component.
- **LCD_Seg_SegPort_SetDriveMode(uint8 mode)**—Sets the drive mode for all pins used by segment lines of the Segment LCD component.

15.91 Pins 2.00

The Pins Component allows hardware resources to connect to a physical port-pin and access to external signals through an appropriately configured physical IO pin. It also allows electrical characteristics (e.g., Drive Mode) to be chosen for one or more pins; these characteristics are then used by PSoC Creator to automatically place and route the signals within the Component. Pins can be used with schematic wire connections, software, or both. To access a Pins Component from Component Application Programming Interfaces (APIs), the Component must be contiguous and non-spanning. This ensures that the pins are guaranteed to be mapped into a single physical port. Pins Components that span ports or are not contiguous can only be accessed from a schematic or with the global per-pin APIs. There are #defines created for each pin in the Pins Component to be used with global APIs. A Pins Component can be configured into many combinations of types. For convenience, the Component Catalog provides four preconfigured Pins Components: Analog, Digital Bidirectional, Digital Input and Digital Output. This component allows rapid setup of all pin parameters and drive modes, PSoC Creator to automatically place and route signals and interaction with one or more pins, simultaneously [64].

15.91.1 Pins Functions 2.00

- **uint8 Pin_Read(void)**—Reads the associated physical port (pin status register) and masks the required bits according to the width and bit position of the component instance.
- **void Pin_Write(uint8 value)**—Writes the value to the physical port (data output register), masking and shifting the bits appropriately.
- **uint8 Pin_ReadDataReg(void)**—Reads the associated physical port's data output register and masks the correct bits according to the width and bit position of the component instance.
- **void Pin_SetDriveMode(uint8 mode)**—Sets the drive mode for each of the Pins component's pins.

- **void Pin_SetInterruptMode(uint16 position, uint16 mode)**—Configures the interrupt mode for each of the Pins component's pins. Alternatively you may set the interrupt mode for all the pins specified in the Pins component.
- **uint8 Pin_ClearInterrupt(void)**—Clears any active interrupts attached with the component and returns the value of the interrupt status register allowing determination of which pins generated an interrupt event.

15.91.2 Pins Function Calls 2.00

- **uint8 Pin_Read (void)**—Reads the associated physical port (pin status register) and masks the required bits according to the width and bit position of the component instance. The pin's status register returns the current logic level present on the physical pin.
- **void Pin_Write (uint8 value)**—Writes the value to the physical port (data output register), masking and shifting the bits appropriately. The data output register controls the signal applied to the physical pin in conjunction with the drive mode parameter. This function avoids changing other bits in the port by using the appropriate method (read-modify-write or bit banding). Note This function should not be used on a hardware digital output pin as it is driven by the hardware signal attached to it.
- **uint8 Pin_ReadDataReg (void)**—Reads the associated physical port's data output register and masks the correct bits according to the width and bit position of the component instance. The data output register controls the signal applied to the physical pin in conjunction with the drive mode parameter. This is not the same as the preferred Pin_Read( ) API because the Pin_ReadDataReg( ) reads the data register instead of the status register. For output pins this is a useful function to determine the value just written to the pin.
- **void Pin_SetDriveMode (uint8 mode)**—Sets the drive mode for each of the Pins component's pins. Note This affects all pins in the Pins component instance. Use the Per-Pin APIs if you wish to control individual pin's drive modes. Note USBIOs have limited drive functionality. Refer to the Drive Mode parameter for more information.
- **void Pin_SetInterruptMode (uint16 position, uint16 mode)**—Configures the interrupt mode for each of the Pins component's pins. Alternatively you may set the interrupt mode for all the pins specified in the Pins component.[52]
- **uint8 Pin_ClearInterrupt (void)**—Clears any active interrupts attached with the component and returns the value of the interrupt status register allowing determination of which pins generated an interrupt event.

15.91.3 Pins—Power Management Functions 2.00

- **void Pin_Sleep(void)**—Stores the pin configuration and prepares the pin for entering chip deep-sleep/hibernate modes. This function must be called for SIO and USBIO pins. It is not essential if using GPIO or GPI_OVT pins.
- **void Pin_Wakeup(void)**—Restores the pin configuration that was saved during Pin_Sleep( ).

---

[52]The interrupt is port-wide and therefore any enabled pin interrupt may trigger it.

15.91.4 Pins—Power Management Functions 2.00

- **void Pin_Sleep (void)**—Stores the pin configuration and prepares the pin for entering chip deep-sleep/hibernate modes. This function applies only to SIO and USBIO pins. It should not be called for GPIO or GPIO_OVT pins. Note This function is available in PSoC 4 only.
- **void Pin_Wakeup (void)**—Restores the pin configuration that was saved during Pin_Sleep( ).This function applies only to SIO and USBIO pins. It should not be called for GPIO or GPIO_OVT pins. For USBIO pins, the wakeup is only triggered for falling edge interrupts. Note This function is available in PSoC 4 only.

15.92 Trim and Margin 3.00

The component adjustment and control of the output voltage of up to 24 DC-DC converters to meet system power supply requirements. Users of this component enter the power converter nominal output voltages, voltage trimming range, margin high and margin low settings into the intuitive, graphical configuration GUI and the component calculates all the required parameters for injecting a pulse width modulated signal into the feedback network of a power converter. The component will also assist the user to select appropriate external passive component values based on performance requirements. The provided firmware APIs enable users to manually trim the power converter output voltages to any desired level within the operational limits of the power converter. Real-time active trimming or margining is supported via as a continuously running background task with an update frequency controlled by the user. Features include: compatible with most adjustable DC-DC converters or regulators, including low-dropouts (LDOs), switchers and modules. Features include support for positive and negative feedback control loops, up to 24 DC-DC converters, 8- to 10-bit resolution PWM pseudo-DAC outputs, real-time, closed-loop active trimming when used in conjunction with the Power Monitor or ADC component and built-in support for margining [100].

15.92.1 Trim and Margin Functions 3.00

- **TrimMargin_Start( )**—Starts the component operation.
- **TrimMargin_Stop( )**—Disables the component.
- **TrimMargin_Init( )**—Initializes component's parameters.
- **TrimMargin_Enable( )**—Enables the generation of PWMs outputs.
- **TrimMargin_SetMarginHighVoltage( )**—Sets the margin high output voltage parameter.
- **TrimMargin_GetMarginHighVoltage( )**—Returns the margin high output voltage parameter.
- **TrimMargin_SetMarginLowVoltage( )**—Sets the margin low output voltage parameter.
- **TrimMargin_GetMarginLowVoltage( )**—Returns the margin low output voltage parameter.
- **TrimMargin_SetNominalVoltage( )**—Sets the nominal output voltage parameter.
- **TrimMargin_GetNominalVoltage( )**—Returns the nominal output voltage parameter.
- **TrimMargin_ActiveTrim( )**—Adjusts the PWM duty cycle of the specified power converter to get the power converter actual voltage output closer to the desired voltage output.
- **TrimMargin_SetDutyCycle( )**—Sets PWM duty cycle of the PWM associated with the specified power converter.
- **TrimMargin_GetDutyCycle( )**—Gets the current PWM duty cycle of the PWM associated with the specified power converter.

- **TrimMargin_GetAlertSource( )**—Returns a bit mask indicating which PWMs are generating an alert.
- **TrimMargin_MarginHigh( )**—Sets power converter output voltage to the Margin high voltage.
- **TrimMargin_MarginLow( )**—Sets power converter output voltage to the Margin low voltage.
- **TrimMargin_Nominal( )**—Sets power converter output voltage to the Nominal voltage.
- **TrimMargin_PreRun( )**—Sets the pre-charge PWM duty cycle required to achieve nominal voltage before the power converter is enabled.
- **TrimMargin_Startup( )**—Sets power converter output voltage to the Startup voltage.
- **TrimMargin_StartupPreRun( )**—Sets the pre-charge PWM duty cycle to achieve the Startup voltage before power converter is enabled.
- **TrimMargin_ConvertVoltageToDutyCycle( )**—Returns the PWM duty cycle required to achieve the desired voltage on the selected power converter.
- **TrimMargin_PreRun( )**—Sets the pre-charge PWM duty cycle required to achieve nominal voltage before the power converter is enabled.
- **TrimMargin_Startup( )**—Sets power converter output voltage to the Startup voltage.
- **TrimMargin_StartupPreRun( )**—Sets the pre-charge PWM duty cycle to achieve the Startup voltage before power converter is enabled.
- **TrimMargin_ConvertVoltageToDutyCycle( )**—Returns the PWM duty cycle required to achieve the desired voltage on the selected power converter.
- **TrimMargin_ConvertVoltageToPreRunDutyCycle( )**—Returns the pre-charge PWM duty cycle required to achieve the desired voltage on the selected power converter.
- **TrimMargin_SetTrimCycleCount( )**—Sets the internal trim cycle counter that affects how often the PWM duty cycle is updated when calling TrimMargin_ActiveTrim( ) API. Applicable for Incremental controller type only.
- **TrimMargin_ConvertVoltageToPreRunDutyCycle( )**—Returns the pre-charge PWM duty cycle required to achieve the desired voltage on the selected power converter.
- **TrimMargin_PreRun( )**—Sets the pre-charge PWM duty cycle required to achieve nominal voltage before the power converter is enabled.
- **TrimMargin_PreRun( )**—Sets the pre-charge PWM duty cycle required to achieve nominal voltage before the power converter is enabled.
- **TrimMargin_Startup( )**—Sets power converter output voltage to the Startup voltage.
- **TrimMargin_StartupPreRun( )**—Sets the pre-charge PWM duty cycle to achieve the Startup voltage before power converter is enabled.
- **TrimMargin_ConvertVoltageToDutyCycle( )**—Returns the PWM duty cycle required to achieve the desired voltage on the selected power converter.
- **TrimMargin_ConvertVoltageToPreRunDutyCycle( )**—Returns the pre-charge PWM duty cycle required to achieve the desired voltage on the selected power converter.
- **TrimMargin_SetTrimCycleCount( )**—Sets the internal trim cycle counter that affects how often the PWM duty cycle is updated when calling TrimMargin_ActiveTrim( ) API. Applicable for Incremental controller type only.
- **TrimMargin_Startup( )**—Sets power converter output voltage to the Startup voltage.
- **TrimMargin_StartupPreRun( )**—Sets the pre-charge PWM duty cycle to achieve the Startup voltage before power converter is enabled.
- **TrimMargin_ConvertVoltageToDutyCycle( )**—Returns the PWM duty cycle required to achieve the desired voltage on the selected power converter.
- **TrimMargin_ConvertVoltageToPreRunDutyCycle( )**—Returns the pre-charge PWM duty cycle required to achieve the desired voltage on the selected power converter.

- **TrimMargin_SetTrimCycleCount( )**—Sets the internal trim cycle counter that affects how often the PWM duty cycle is updated when calling TrimMargin_ActiveTrim( ) API. Applicable for Incremental controller type only.
- **TrimMargin_SetTrimCycleCount( )**—Sets the internal trim cycle counter that affects how often the PWM duty cycle is updated when calling TrimMargin_ActiveTrim( ) API. Applicable for Incremental controller type only.

15.92.2 Trim and Margin Function Calls 3.00

- **void TrimMargin_Start(void)**—Starts the component operation. Calls the TrimMargin_Init( ) API if the component has not been initialized before. Calls TrimMargin_Enable( ) API.
- **void TrimMargin_Stop(void)**—sables the component. Stops the PWMs.
- **void TrimMargin_Init(void)**—Initializes component's parameters to those set in the customizer. It is not necessary to call TrimMargin_Init( ) because the TrimMargin_Start( ) routine calls this function, which is the preferred method to begin the component operation. PWM duty cycles are set to pre-run values to achieve the startup voltage target assuming that the power converters are not yet turned on (disabled).
- **void TrimMargin_Enable(void)**—Enables PWMs outputs generation.
- **void TrimMargin_SetMarginHighVoltage(uint8 converterNum, uint16 marginHiVoltage)**— Sets the margin high output voltage of the specified power converter. This overrides the present vMarginHigh[x] setting and recalculates vMarginHighDutyCycle[x] to be ready for use by TrimMargin_MarginHigh( ). Note: calling this API does NOT cause any change in the PWM output duty cycle.
- **uint16 TrimMargin_GetMarginHighVoltage(uint8 converterNum)**—Returns the margin high output voltage of the specified power converter

15.93 Voltage Fault Detector (VFD) 3.00

The VFD component allows monitoring up to 32 voltage inputs with user-defined over-and-under voltage limits.Good/bad status result ("power good" or pgood[x]) for each voltage being monitored. The component operates entirely in hardware without any intervention from PSoC's CPU core resulting in known, fixed fault detection latency. Features include: capable of monitoring up to 32 voltage inputs, ser-defined over and under voltage limits, good/bad digital status output, programmable glitch filter length, operates entirely in hardware without any intervention from PSoC's CPU core resulting in known, fixed fault detection latency [105].

15.93.1 Voltage Fault Detector (VFD) Functions 3.00

- **VFD_Start( )**—Starts the component operation.
- **VFD_Stop( )**—Stops the component.
- **VFD_Init( )**—Initializes the component.
- **VFD_Enable( )**—Enables hardware blocks.
- **VFD_GetOVUVFaultStatus( )**—Returns over/under voltagefault status of each voltage input (Applicable if Compare type is set to OV/UV).

- **VFD_GetOVFaultStatus( )**—Returns over voltagefault status of each voltage input (Applicable if Compare type is set to OV).
- **VFD_GetUVFaultStatus( )**—Returns under voltagefault status of each voltage input (Applicable if Compare type is set to UV).
- **VFD_SetUVFaultThreshold( )**—Sets the under voltage fault threshold for the specified voltage input.
- **VFD_GetUVFaultThreshold( )**—Returns the under voltage fault threshold for the specified voltage input.
- **VFD_SetOVFaultThreshold( )**—Sets the over voltage fault threshold for the specified voltage input.
- **VFD_GetOVFaultThreshold( )**—Returns the under voltage fault threshold for the specified voltage input.
- **VFD_SetUVGlitchFilterLength( )**—Sets the glitch filter length.
- **VFD_GetUVGlitchFilterLength( )**—Returns the glitch filter length.
- **VFD_SetUVDac( ) Sets UV DAC**—Value of each channel.
- **VFD_GetUVDac( ) Gets UV DAC**—Value for the specified voltage input.
- **VFD_SetOVDac( ) Sets OV DAC**—Value of each channel..
- **VFD_GetOVDac( ) Gets OV DAC**—Value for the specified voltage input.
- **VFD_Pause( )**—Pauses the state machine and fault detection logic.
- **VFD_IsPaused( )**—Checks to see if the component is paused.
- **VFD_Resume( )**—Resumes the control state machine and fault detection logic.
- **VFD_SetUVDacDirect( )**—Allows manual control of the UV VDAC value.
- **VFD_GetUVDacDirect( )**—Returns current UV VDAC.
- **VFD_SetOVDacDirect( )**—Allows manual control of the OV VDAC.
- **VFD_GetOVDacDirect( )**—Returns current OV VDAC.
- **VFD_ComparatorCal( )**—Runs a calibration routine.
- **VFD_SetSpeed( )**—Allows setting speed mode for the VDAC(s) (if Internal Reference option is enabled) and Comparator(s).

15.93.2 Voltage Fault Detector (VFD) Function Calls 3.00

- **void VFD_Start(void)**—Calls the Init( ) API if the component has not been initialized before. Runs a calibration routine for comparators and then calls Enable( ) to begin the component operation.
- **void VFD_Stop(void)**—Stops the component. Stops DMA controller and resets TDs. Disconnects AMux channels.
- **void VFD_Init(void)**—Initializes or restores default VFD configuration provided with the customizer. Initializes internal DMA channels. It is not necessary to call VFD_Init( ) because the VFD_Start( ) routine calls this function, which is the preferred method to begin component operation.
- **void VFD_Enable(void)**—Enables hardware blocks, the DMA channels and the control state machine. It is not necessary to call VFD_Init( ) because the VFD_Start( ) routine calls this function, which is the preferred method to begin the component operation.
- **void VFD_GetOVUVFaultStatus(uint32 * ovStatus, uint32 * uvStatus)**—Assigns over/under voltage fault status of each voltage input to its parameters. Bits are sticky and cleared by calling this API. Applicable only if Compare type is set to OV/UV.

- **void VFD_GetOVFaultStatus(uint32 * ovStatus)**—Assigns over voltage fault status of each voltage input to its parameter. Bits are sticky and cleared by calling this API. Applicable only if Compare type is set to OV.
- **void VFD_GetUVFaultStatus(uint32 * uvStatus)**—Assigns under voltage fault status of each voltage input to its parameter. Applicable only if Compare type is set to UV.
- **cystatus VFD_SetUVFaultThreshold(uint8 voltageNum, uint16 uvFaultThreshold)**—Sets the under voltage fault threshold for the specified voltage input. The uvFaultThreshold parameter is converted to a VDAC value and gets written to an SRAM buffer for use by the DMA controller that drives the UV DAC. This API does not apply when the Enable external reference option is selected.
- **uint16 VFD_GetUVFaultThreshold(uint8 voltageNum)**—Returns the under voltage fault threshold for the specified voltage. This API does not apply when the Enable external reference option is selected.
- **cystatus VFD_SetOVFaultThreshold(uint8 voltageNum, uint16 ovFaultThreshold)**—Sets the over voltage fault threshold for the specified voltage input. The ovFaultThreshold parameter is converted to a VDAC value and gets written to an SRAM buffer for use by the DMA controller that drives the OV DAC. This API does not apply when the Enable external reference option is selected.
- **uint16 VFD_GetOVFaultThreshold(uint8 voltageNum)**—Returns the over voltage fault threshold for the specified voltage input. This API does not when the Enable external reference option is selected.
- **void VFD_SetGlitchFilterLength(uint8 filterLength)**—Sets the glitch filter length.
- **uint8 VFD_GetGlitchFilterLength(void)**—Returns the glitch filter length.
- **void VFD_SetUVDac(uint8 voltageNum, uint8 dacValue)**—Sets the UV DAC value for the specified voltage input. Calling this API does not change the UV VDAC setting immediately. Instead, the dacValue gets written to an SRAM buffer for use by the DMA controller that drives the UV DAC for the specified voltage input. This API does not apply when the Enable external reference option is selected.
- **uint8 VFD_GetUVDac(uint8 voltageNum)**—Returns the dacValue currently being used by the DMA controller that drives the UV DAC value for the specified voltage input. This API does not apply when the Enable external reference option is selected.
- **void VFD_SetOVDac(uint8 voltageNum, uint8 dacValue)**—Calling this API does not change the OV VDAC setting immediately. Instead, the dacValue gets written to an SRAM buffer for use by the DMA controller that drives the OV DAC for the specified voltage input. This API does not apply when the Enable external reference option is selected.
- **uint8 VFD_GetOVDac(uint8 voltageNum)**—Returns the dacValue currently being used by the DMA controller that drives the OV DAC value for the specified voltage input. This API does not apply when the Enable external reference option is selected.
- **void VFD_Pause(void)**—Pauses the controller state machine. The current PGOOD states are kept when the component is paused. Note that calling this API does not stop the component until it completes the current process cycle. Therefore, if the purpose of calling this API is specifically to change the VDAC settings (for calibration purposes for example), sufficient time should be allowed to let the component run to a completion before an attempt to access the VDACs directly. This can be checked by calling VFD_IsPaused( ).
- **bool VFD_IsPaused(void)**—Checks to see if the component is paused.
- **void VFD_Resume(void)**—Enables the clock to the comparator controller state machine.
- **void VFD_SetUVDacDirect(uint8 dacValue)**—Allows manual control of the UV VDAC value. The dacValue is written directly to the UV VDAC component. Useful for UV VDAC calibration.

Note that if the VFD component is running when this API is called, the state machine controller will override the UV VDAC value set by this API call. Call the Pause API to stop the state machine controller if manual UV VDAC control is desired. This API does not apply when the Enable external reference option is selected.

- **uint8 VFD_GetUVDacDirect(void)**—Returns current UV VDAC. The returned dacValue is read directly from the UV VDAC component. Useful for UV VDAC calibration. Note: if this API is called while the component is running, it isn't possible to know which voltage input the returned UV VDAC value is associated with. Call the Pause API to stop the state machine controller if manual UV VDAC control is desired. This API does not apply when the Enable external reference option is selected.

- **void VFD_SetOVDacDirect(uint8 dacValue)**—Allows manual control of the OV VDAC value. The dacValue is written directly to the OV VDAC component. Useful for OV VDAC calibration. Note that if the VFD component is running when this API is called, the state machine controller will override the OV VDAC value set by this API call. Call the Pause API to stop the state machine controller if manual OV VDAC control is desired. This API does not apply when the Enable external reference option is selected.

- **uint8 VFD_GetOVDacDirect(void)**—Returns current OV VDAC. The returned dacValue is read directly from the VDAC component. This is useful for OV VDAC calibration. Note If this API is called while the component is running, it is impossible to know which voltage input the returned OV VDAC value is associated with. Call the Pause API to stop the state machine controller if manual UV VDAC control is desired. This API does not apply when the Enable external reference option is selected.

- **void VFD_ComparatorCal(uint8 compType)**—Runs a calibration routine that measures the selected comparator's offset voltage by shorting its inputs together. It corrects for it by writing to the CMP block's trim register. A reference voltage in the range at which the comparator will be used must be applied to the negative input of the comparator while the offset calibration is performed. When Enable external reference is selected, the negative comparator inputs are connected to ov_ref/uv_ref inputs respectively. Therefore, the reference voltage must be provided externally to the VFD component.When internal reference is selected, the negative comparator inputs are connected to internal OV/UV DACs. In this case, call the SetOVDacDirect( ) / SetUVDacDirect( ) function to provide the reference voltage.

- **void VFD_SetSpeed(uint8 speedMode)**—Allows setting speed mode for the VDAC(s) (if Internal Reference option is enabled) and Comparator(s).

15.94 Voltage Sequencer 3.40

This component supports power-down sequencing of up to 32 power converters. Sequencing requirements are entered into the configuration GUI and result in the sequencing implementation being managed automatically. Support is provided for sequencing and monitoring of up to 32 power converter rails. Features include: supports power converter circuits with logic-level enable, inputs and logic-level power good (pgood) status outputs, autonomous (standalone) or host driven operation, sequence order, timing and inter-rail dependencies can be configured through an intuitive, easy-to-use graphical configuration GUI [107].

15.94.1 Voltage Sequencer Functions 3.40

- **Sequencer_Start( )**—Enables the component and places all power converter state machines into the appropriate state.
- **Sequencer_Stop( )**—Disables the component.
- **Sequencer_Init( )**—Initializes the component.
- **Sequencer_Enable( )**—Enables the component.
- **Sequencer_Pause( )**—Pauses the sequencer, preventing sequencer state machine state transitions
- **Sequencer_Play( )**—Resumes the sequencer, if previously paused.
- **Sequencer_SingleStep( )**—Puts the sequencer in single step mode.
- **Sequencer_EnableCalibrationState( )**—Enables the sequencer calibration state, preventing sequencer state machine state transitions and disables fault detection and processing.
- **Sequencer_DisableCalibrationState( )**—Resumes the sequencer state machine state transitions and enables fault detection and processing Voltage Sequencer.
- **Sequencer_ForceOn( )**—Forces the selected power converter to power up.
- **Sequencer_ForceAllOn( )**—Forces all power converters to power up.
- **Sequencer_ForceOff( )**—Forces the selected power converter to power down either immediately or after the TOFF delay.
- **Sequencer_ForceAllOff( )**—Forces all power converters to power down either immediately or after their TOFF delays.
- **Sequencer_GetState( )**—Returns the current state machine state for the selected power converter.
- **Sequencer_GetPgoodStatus( )**—Returns a bitmask that represents pgood[x] status for all power converters.
- **Sequencer_GetFaultStatus( )**—Returns a bitmask that represents which power converters have experienced a fault that caused de-assertion of their pgood[x] inputs.
- **Sequencer_GetCtlStatus( )**—Returns a bitmask that represents which ctl[x] inputs have caused one or more converters to shutdowwn.
- **Sequencer_GetWarnStatus( )**—Returns a bitmask that represents which power converters have experienced a power down warning caused by exceeding the TOFF_MAX_WARN timeout
- **Sequencer_EnFaults( )**—Enables/disables assertion of the fault output terminal.
- **Sequencer_EnWarnings( )**—Enables/disables assertion of the warn output terminal.

15.94.2 Voltage Sequencer Function Calls 3.40

- **void Sequencer_Start(void)**—Enables the component and places all power converter state machines into the appropriate state (OFF or PEND_ON). Calls the Init( ) API if the component has not been initialized before. Calls the Enable( ) API.
- **void Sequencer_Stop(void)**—Disables the component, preventing sequencer state machine state transitions, system timer updates and fault handling.
- **void Sequencer_Init(void)**—Initializes the component. Parameter settings are initialized based on parameters entered into the various Configure dialog tabs.
- **void Sequencer_Enable(void)**—Enables the component. Enables sequencer state machine state transitions, system timer updates and fault handling.
- **void Sequencer_Pause(void)**—Pauses the sequencer, preventing sequencer state machine state transitions, system timer updates and fault handling.
- **void Sequencer_Play(void)**—Resumes the sequencer if previously paused. Re-enables sequencer state machine state transitions, system timer updates and fault handling.

- **void Sequencer_SingleStep(void)**—Puts the sequencer in single step mode. If the sequencer was paused, it will resume normal operation. The sequencer will then run until there is a state transition on any rail. At that time, the sequencer will be paused automatically until either the Play( ) API or the SingleStep( ) is called again.
- **void Sequencer_EnableCalibrationState(void)**—Enables the sequencer calibration state, preventing sequencer state machine state transitions, system timer updates and fault handling. Stops the hardware fast shutdowwn block.
- **void Sequencer_DisableCalibrationState(void)**—Disables the sequencer calibration state. Re-enables sequencer state machine state transitions, system timer updates and fault handling. Enables the hardware fast shutdowwn block.
- **void Sequencer_ForceOn(uint8 converterNum)**—Forces the selected power converter to the PEND_ON state. All selected power up prerequisite conditions must be satisfied for the power converter to turn on. The re-sequence counter for that converter's state machine is re-initialized.
- **void Sequencer_ForceAllOn(void)**—Forces all power converters to the PEND_ON state. All selected power up pre-requisite conditions must be satisfied for the power converter to turn on. The re-sequence counter for that converter's state machines is re-initialized.
- **void Sequencer_ForceOff(uint8 converterNum, uint8 powerOffMode)**—Forces the selected power converter to power down either immediately or after the TOFF delay. All selected power down pre-requisite conditions must be satisfied for the power converter to turn off.
- **void Sequencer_ForceAllOff(uint8 powerOffMode)**—Forces all power converters to power down either immediately or after their TOFF delays. All selected power down pre-requisite conditions must be satisfied for the power converter to turn off.
- **uint8 Sequencer_GetState(uint8 converterNum)**—Returns the current state machine state for the selected power converter.
- **uint8/uint16/uint32 Sequencer_GetPgoodStatus(void)**—Returns a bitmask that represents pgood[x] status for all power converters.
- **uint8/uint16/uint32 Sequencer_GetFaultStatus(void)**—Returns a bitmask that represents which power converters have experienced a fault that caused de-assertion of their pgood[x] inputs. Bits are sticky until cleared by calling this API.
- **uint8 Sequencer_GetCtlStatus(void)**—Returns a bitmask that represents which ctl[x] inputs have caused one or more converters to shutdowwn. Bits are sticky until cleared by calling this API.
- **uint8/uint16/uint32 Sequencer_GetWarnStatus(void)**—Returns a bitmask that represents which power converters have experienced a power down warning caused by exceeding the TOFF_MAX_WARN timeout. Bits are sticky until cleared by calling this API.
- **void Sequencer_EnFaults(uint8 faultEnable)**—Enables/disables assertion of the fault output terminal. Faults are still processed by the state machine and fault status is still available through the GetFaultStatus( ) API.

15.94.3 Voltage Sequencer—Run-Time Configuration Functions 3.40

- **Sequencer_SetStsPgoodMask( )**—Specifies which pgood[x] inputs participate in the generation of the specified general purpose sequencer status output.
- **Sequencer_GetStsPgoodMask( )**—Returns which pgood[x] inputs participate in the generation of the specified general purpose sequencer status output.
- **Sequencer_SetStsPgoodPolarity( )**—Configures the logic conditions that will cause the selected general purpose sequencer status output to be asserted.

- **Sequencer_GetStsPgoodPolarity( )**—Returns the polarity of the pgood[x] inputs used in the AND expression for the selected general purpose sequencer status output.
- **Sequencer_SetPgoodOnThreshold( )**—Sets the power good voltage threshold for power on detection.
- **Sequencer_GetPgoodOnThreshold( )**—Returns the power good voltage threshold for power on detection.
- **Sequencer_SetPowerUpMode( )**—Sets the power up default state for the selected power converter.
- **Sequencer_GetPowerUpMode( )**—Returns the power up default state for the selected power converter.
- **Sequencer_SetPgoodOnPrereq( )**—Determines which pgood[x] inputs are power up prerequisites for the selected power converter state machine.
- **Sequencer_GetPgoodOnPrereq( )**—Returns which pgood[x] inputs are power up prerequisites for the selected power converter state machine.
- **Sequencer_SetPgoodOffPrereq( )**—Determines which pgood[x] inputs are power down prerequisites for the selected power converter state machine.
- **Sequencer_GetPgoodOffPrereq( )**—Returns which pgood[x] inputs are power down prerequisites for the selected power converter state machine.
- **Sequencer_SetTonDelay( )**—Sets the TON delay parameter for the selected power converter.
- **Sequencer_GetTonDelay( )**—Returns the TON delay parameter for the selected power converter.
- **Sequencer_SetTonMax( )**—Sets the TON_MAX parameter for the selected power converter.
- **Sequencer_GetTonMax( )**—Returns the TON_MAX parameter for the selected power converter.
- **Sequencer_SetPgoodOffThreshold( )**—Sets the power good voltage threshold for power down detection.
- **Sequencer_GetPgoodOffThreshold( )**—Returns the power good voltage threshold for power down detection.
- **Sequencer_SetCtlPrereq( )**—Sets which ctl[x] input is a prerequisite for a power converter.
- **Sequencer_GetCtlPrereq( )**—Returns which ctl[x] input is a prerequisite for a power converter.
- **Sequencer_SetCtlShutdownMask( )**—Determines which ctl[x] inputs will cause the selected power converter to shutdowwn when de-asserted.
- **Sequencer_GetCtlShutdownMask( )**—Returns which ctl[x] inputs will cause the selected power converter to shutdowwn when de-asserted.
- **Sequencer_SetPgoodShutdownMask( )**—Determines which other pgood[x] inputs will shutdowwn the selected power converter when de-asserted.
- **Sequencer_GetPgoodShutdownMask( )**—Returns which other pgood[x] inputs will shutdowwn the selected power converter when de-asserted.
- **Sequencer_SetToffDelay( )**—Sets the TOFF delay parameter for the selected power converter.
- **Sequencer_GetToffDelay( )**—Returns the TOFF delay parameter for the selected power converter.
- **Sequencer_SetToffMax( )**—Sets the TOFF_MAX_DELAY parameter for the selected power converter.
- **Sequencer_GetToffMax( )**—Returns the TOFF_MAX_DELAY parameter for the selected power converter.
- **Sequencer_SetSysStableTime( )**—Sets the global System Stable parameter for all power converter state machines.
- **Sequencer_GetSysStableTime( )**—Returns the global System Stable parameter for all power converter state machines.
- **Sequencer_SetReseqDelay( )**—Sets the global Re-sequence Delay parameter for all power converter state machines.

- **Sequencer_GetReseqDelay( )**—Returns the global Re-sequence Delay parameter for all power converter state machines.
- **Sequencer_SetTonMaxReseqCnt( )**—Sets the re-sequence count for TON_MAX fault condition.
- **Sequencer_GetTonMaxReseqCnt( )**—Returns the re-sequence count for TON_MAX fault conditions.
- **Sequencer_SetTonMaxFaultResp( )**—Sets the shutdown mode for a fault group when a TON_MAX fault condition occurs on the selected master converter
- **Sequencer_GetTonMaxFaultResp( )**—Returns the shutdown mode for a fault group when a TON_MAX fault condition occurs on the selected master converter
- **Sequencer_SetCtlReseqCnt( )**—Sets the re-sequence count for fault conditions due to de-asserted ctl[x] inputs
- **Sequencer_GetCtlReseqCnt( )**—Returns the re-sequence count for fault conditions due to de-asserted ctl[x] inputs
- **Sequencer_SetCtlFaultResp( )**—Sets the shutdown mode for a fault group in response to fault conditions due to de-asserted ctl[x] inputs
- **Sequencer_GetCtlFaultResp( )**—Returns the shutdown mode for a fault group in response to fault conditions due to de-asserted ctl[x] inputs.
- **Sequencer_SetFaultReseqSrc( )**—Sets the power converter fault re-sequence sources.
- **Sequencer_GetFaultReseqSrc( )**—Returns the power converter fault re-sequence sources.
- **Sequencer_SetPgoodReseqCnt( )**—Sets the re-sequence count for fault conditions due to de-asserted pgood[x] inputs.
- **Sequencer_GetPgoodReseqCnt( )**—Returns the re-sequence count for fault conditions due to de-asserted pgood[x] inputs.
- **Sequencer_SetPgoodFaultResp( )**—Sets the shutdown mode for a fault group due to de-asserted pgood[x] inputs.
- **Sequencer_GetPgoodFaultResp( )**—Returns the shutdown mode a fault group due to de-asserted pgood[x] inputs.
- **Sequencer_SetOvReseqCnt( )**—Sets the re-sequence count for over-voltage (OV) fault conditions.
- **Sequencer_GetOvReseqCnt( )**—Returns the re-sequence count for over-voltage (OV) fault conditions.
- **Sequencer_SetOvFaultResp( )**—Sets the shutdown mode for a fault group due to overvoltage (OV) fault conditions.
- **Sequencer_GetOvFaultResp( )**—Returns the shutdown mode for a fault group due to overvoltage (OV) fault conditions.
- **Sequencer_SetUvReseqCnt( )**—Sets the re-sequence count for under-voltage (UV) fault conditions.
- **Sequencer_GetUvReseqCnt( )**—Returns the re-sequence count for under-voltage (UV) fault conditions.
- **Sequencer_SetUvFaultResp( )**—Sets the shutdown mode for a fault group due to under-voltage (UV) fault conditions.
- **Sequencer_GetUvFaultResp( )**—Returns the shutdown mode for a fault group due to under-voltage (UV) fault conditions.
- **Sequencer_SetOcReseqCnt( )**—Sets the re-sequence count for over-current (OC) fault conditions.
- **Sequencer_GetOcReseqCnt( )**—Returns the re-sequence count for over-current (OC) fault conditions.

- **Sequencer_SetOcFaultResp( )**—Sets the shutdown mode for a fault group due to overcurrent (OC) fault conditions.
- **Sequencer_GetOcFaultResp( )**—Returns the shutdown mode for a fault group due to overcurrent (OC) fault conditions.
- **Sequencer_SetFaultMask( )**—Sets which power converters have fault detection enabled.
- **Sequencer_GetFaultMask( )**—Returns which power converters have fault detection enabled.
- **Sequencer_SetWarnMask( )**—Sets which power converters have warnings enabled.
- **Sequencer_GetWarnMask( )**—Returns which power converters have warnings enabled.

15.94.4 Voltage Sequencer—Run-Time Configuration Function Calls 3.40

- **void Sequencer_SetStsPgoodMask(uint8 stsNum, uint8/uint16/uint32 stsPgoodMask)**—Specifies which pgood[x] inputs participate in the generation of the specified general purpose sequencer control output (sts[x]).
- **uint8/uint16/uint32 Sequencer_GetStsPgoodMask(uint8 stsNum)**—Returns which pgood[x] inputs participate in the generation of the specified general purpose sequencer control output (sts[x]).
- **void Sequencer_SetStsPgoodPolarity(uint8 stsNum, uint8/uint16/uint32 pgoodPolarity)**—Configures the logic conditions that will cause the selected general purpose sequencer control output (sts[x]) to be asserted.
- **uint8/uint16/uint32 Sequencer_GetStsPgoodPolarity(uint8 stsNum)**—Returns the polarity of the pgood[x] inputs used in the AND expression for the selected general purpose sequencer control output (sts[x]).
- **void Sequencer_SetPgoodOnThreshold(uint8 converterNum, uint16 onThreshold)**—Sets the power good voltage threshold for power on detection.
- **uint16 Sequencer_GetPgoodOnThreshold(uint8 converterNum)**—Returns the power good voltage threshold for power on detection
- **void Sequencer_SetPowerUpMode(uint8 converterNum, uint8 powerUpMode)**—Sets the power up default state for the selected power converter.
- **uint8 Sequencer_GetPowerUpMode(uint8 converterNum)**—Returns the power up default state for the selected power converter.
- **void Sequencer_SetPgoodOnPrereq(uint8 converterNum, uint8/uint16/uint32 pgood-Mask)**—Determines which pgood[x] inputs are power up prerequisites for the selected power converter.
- **uint8/uint16/uint32 Sequencer_GetPgoodOnPrereq(uint8 converterNum)**—Returns which pgood[x] inputs are power up prerequisites for the selected power converter.
- **void Sequencer_SetPgoodOffPrereq(uint8 converterNum, uint8/uint16/uint32 pgood-Mask)**—Determines which pgood[x] inputs are power down prerequisites for the selected power converter.
- **uint8/uint16/uint32 Sequencer_GetPgoodOffPrereq(uint8 converterNum)**—Returns which pgood[x] inputs are power down prerequisites for the selected power converter.
- **void Sequencer_SetTonDelay(uint8 converterNum, uint16 tonDelay)**—Sets the TON delay parameter for the selected power converter. Defined as the time between all power converters' prerequisites becoming satisfied and the en[x] output being asserted.
- **uint16 Sequencer_GetTonDelay(uint8 converterNum)**—Returns the TON delay parameter for the selected power converter. Defined as the time between all power converter's prerequisites becoming satisfied and the en[x] output being asserted

- **void Sequencer_SetTonMax(uint8 converterNum, uint16 tonMax)**—Sets the TON_MAX timeout parameter for the selected power converter. Defined as the maximum time allowable between a power converter's en[x] being asserted and its pgood[x] being asserted. Failure to do so generates a fault condition.
- **uint16 Sequencer_GetTonMax(uint8 converterNum)**—Returns the TON_MAX timeout parameter for the selected power converter. Defined as the maximum time allowable between a power converter's en[x] being asserted and its pgood[x] being asserted. Failure to do so generates a fault condition.
- **void Sequencer_SetPgoodOffThreshold(uint8 converterNum, uint16 onThreshold)**—Sets the power good voltage threshold for power off detection
- **uint16 Sequencer_GetPgoodOffThreshold(uint8 converterNum)**—Returns the power good voltage threshold for er off detection
- **void Sequencer_SetCtlPrereq (uint8 converterNum, uint8 ctlPinMask)**—Sets which ctl[x] input is a pre-requisite for the selected power converter
- **uint8 Sequencer_GetCtlPrereq (uint8 converterNum)**—Returns which ctl[x] input is a pre-requisite for the selected power converter
- **Sequencer_SetCtlShutdownMask(uint8 converterNum, uint8 ctlPinMask)**—Determines which ctl[x] inputs will cause the selected power converter to shutdowwn when de-asserted
- **uint8 Sequencer_GetCtlShutdownMask(uint8 converterNum)**—Returns which ctl[x] inputs will cause the selected power converter to shutdowwn when de-asserted
- **void Sequencer_SetPgoodShutdownMask(uint8 converterNum, uint8/uint16/uint32 pgoodMask)**—Determines which converter's pgood[x] inputs will shutdown the selected power converter when de-asserted. Note that a converter's own pgood[x] input is automatically a fault source for that converter whether or not the corresponding bit in the pgoodMask is set or not.
- **uint8/uint16/uint32 Sequencer_GetPgoodShutdownMask (uint8 converterNum)**—Returns which converters pgood[x] inputs will shutdowwn the selected power converter when de-asserted. Note that a converter's own pgood[x] input is automatically a fault source for that converter and the corresponding mask bit is not returned.
- **void Sequencer_SetToffDelay(uint8 converterNum, uint16 toffDelay)**—Sets the TOFF delay parameter for the selected power converter. Defined as the time between making the decision to turn a power converter off and to actually de-asserting the en[x] output.
- **uint16 Sequencer_GetToffDelay(uint8 converterNum)**—Returns the TOFF delay parameter for the selected power converter. Defined as the time between making the decision to turn a power converter off and to actually de-asserting the en[x] output.
- **void Sequencer_SetToffMax(uint8 converterNum, uint16 toffMax)**—Sets the TOFF_MAX_DELAY timeout parameter for the selected power converter. Defined as the maximum time allowable between a power converter's en[x] being de-asserted and power converter actually turning off. Failure to do so generates a warning condition.
- **uint16 Sequencer_GetToffMax(uint8 converterNum)**—Returns the TOFF_MAX_DELAY timeout parameter for the selected power converter. Defined as the maximum time allowable between a power converter's en[x] being deasserted and power converter actually turning off. Failure to do so generates a warning condition.
- **void Sequencer_SetSysStableTime(uint16 stableTime)**—Sets the global TRESEQ_DELAY parameter for all power converters. Defined as the time between making the decision to re-sequence and beginning a new power up sequence.
- **uint16 Sequencer_GetSysStableTime(void)**—Sets the global TRESEQ_DELAY parameter for all power converters. Defined as the time between making the decision to re-sequence and

beginning a new power up sequence Return Value: uint16 stableTime units = 8 ms per LSB Valid Range=0–65535 (0–534.28 s).

- **void Sequencer_SetReseqDelay(uint16 reseqDelay)**—Sets the global TRESEQ_DELAY parameter for all powers. Defined as the time between making the decision to re-sequence and beginning a new power up sequence.
- **uint16 Sequencer_GetReseqDelay(void)**—Returns the global TRESEQ_DELAY parameter for all power converters. Defined as the time between making the decision to re-sequence and beginning a new power up sequence Return Value: uint16 reseqDelay units = 8 ms per LSB Valid Range=0–65535 (0–534.28 s).
- **void Sequencer_SetTonMaxReseqCnt(uint8 converterNum, uint8 ReseqCnt)**—Sets the re-sequence count for TON_MAX fault conditions.
- **uint8 Sequencer_GetTonMaxReseqCnt(uint8 converterNum)**—Returns the re-sequence count for TON_MAX fault conditions.
- **void Sequencer_SetTonMaxFaultResp(uint8 converterNum, uint8 faultResponse)**—Sets the shutdown mode for all associated fault groups when a TON_MAX fault condition occurs on the selected power converter Valid range: 1–32.
- **uint8 Sequencer_GetTonMaxFaultResp(uint8 converterNum)**—Returns the shutdown mode for all associated fault group when a TON_MAX fault condition occurs on the selected power converter.
- **void Sequencer_SetCtlReseqCnt(uint8 converterNum, uint8 reseqCnt)**—Sets the re-sequence count for fault conditions due to de-asserted ctl[x] inputs.
- **uint8 Sequencer_GetCtlReseqCnt(uint8 converterNum)**—Returns the re-sequence count for fault conditions due to de-asserted ctl[x] inputs.
- **void Sequencer_SetCtlFaultResp(uint8 converterNum, uint8 faultResponse)**—Sets the shutdown mode for the selected power converter and rails in associated fault groups in response to de-assertion of ctl[x] inputs.
- **uint8 Sequencer_GetCtlFaultResp(uint8 converterNum)**—Returns the shutdown mode for the selected power converter and rails in associated fault groups in response to de-assertion of ctl[x] inputs.
- **void Sequencer_SetFaultReseqSrc(uint8 converterNum, uint8 reseqSrc)**—Sets the power converter fault re-sequence sources.
- **uint8 Sequencer_GetFaultReseqSrc(uint8 converterNum)**—Returns the power converter fault re-sequence source.
- **void Sequencer_SetPgoodReseqCnt(uint8 converterNum, uint8 reseqCnt)**—Sets the re-sequence count for fault conditions due to a de-asserted pgood[x] input on the selected rail
- **uint8 Sequencer_GetPgoodReseqCnt(uint8 converterNum)**—Returns the re-sequence count for fault conditions due to a de-asserted pgood[x] input on the selected rail.
- **void Sequencer_SetPgoodFaultResp(uint8 converterNum, uint8 faultResponse)**—Sets the shutdown mode for the selected power converter and rails in associated fault groups in response to de-assertion of the selected power converter's pgood[x] input.
- **uint8 Sequencer_GetPgoodFaultResp(uint8 converterNum)**—Returns the shutdown mode for the selected power converter and rails in associated fault groups in response to de-assertion of the selected power converter's pgood[x] input.
- **void Sequencer_SetOvReseqCnt(uint8 converterNum, uint8 reseqCnt)**—Sets the re-sequence count for over-voltage (OV) fault conditions.
- **uint8 Sequencer_GetOvReseqCnt(uint8 converterNum)**—Sets the re-sequence count for over-voltage (OV) fault conditions uint8 reseqCnt 0=no re-sequencing, 31=infinite re-sequencing, 1–30=valid re-sequencing counts.

- **void Sequencer_SetOvFaultResp(uint8 converterNum, uint8 faultResponse)**—Sets the shutdown mode for all associated fault groups due to over-voltage (OV) fault conditions on the selected power converter.
- **uint8 Sequencer_GetOvFaultResp(uint8 converterNum)**—Returns the shutdown mode for all associated fault groups due to over-voltage (OV) fault conditions on the selected power converter.
- **void Sequencer_SetUvReseqCnt(uint8 converterNum, uint8 reseqCnt)**—Sets the re-sequence count for under-voltage (UV) fault conditions.
- **void Sequencer_SetUvFaultResp(uint8 converterNum, uint8 faultResponse)**—Sets the shutdown mode for all associated fault groups due to under-voltage (UV) fault conditions on the selected power converter.
- **uint8 Sequencer_GetUvFaultResp(uint8 converterNum)**—Returns the shutdown mode for all associated fault groups due to under-voltage (UV) fault conditions on the selected power converter.
- **void Sequencer_SetOcReseqCnt(uint8 converterNum, uint8 reseqCnt)**—Sets the re-sequence count for over-current (OC) fault conditions.
- **uint8 Sequencer_GetOcReseqCnt(uint8 converterNum)**—Returns the re-sequence count for over-current (OC) fault conditions.
- **void Sequencer_SetOcFaultResp(uint8 converterNum, uint8 faultResponse)**—Sets the shutdown mode for all associated fault groups due to over-current (OC) fault conditions on the selected power converter.
- **uint8 Sequencer_GetOcFaultResp(uint8 converterNum)**—Returns the shutdown mode for all associated fault groups due to over-current (OC) fault conditions on the selected power converter.
- **void Sequencer_SetFaultMask(uint8/uint16/uint32 faultMask)**—Sets which power converters have fault detection enabled.
- **uint8/uint16/uint32Sequencer_GetFaultMask(void)**—Returns which power converters have fault detection enabled.
- **void Sequencer_SetWarnMask(uint8/uint16/uint32 warnMask)**—Sets which power converters have warnings enabled.
- **uint8/uint16/uint32 Sequencer_GetWarnMask(void)**—Returns which power converters have warnings enabled.
- **uint8 Sequencer_GetUvReseqCnt(uint8 converterNum)**—Returns the re-sequence count for under-voltage (UV) fault conditions.

15.95 Boost Converter (BoostConv) 5.00

The Boost Converter (BoostConv) component allows you to configure and control the PSoC boost converter hardware block. The boost converter enables input voltages that are lower than the desired system voltage to be boosted to the desired system voltage level. The converter uses an external inductor to convert the input voltage to the desired output voltage. The BoostConv component is enabled by default at chip startup with an output voltage of 1.9 V. This allows the chip to start up in scenarios where the input voltage to the boost is below the minimum allowable voltage to power the chip. The configuration parameters defined in the component customizer (default VIN = 1.8 V, VOUT = 3.3 V, Switching Frequency = 400 kHz) will not take effect until the BoostConv_Start( ) API is called. The BoostConv component parameters can also be adjusted during run time using the provided APIs. The boost converter has two main operating modes: Active mode is the normal mode of operation where the boost regulator actively generates a regulated output voltage, Standby mode is a low-power mode of operation with PSoC 3 and Sleep mode which is a low-power mode of operation with PSoC 5LP. Features include: a selectable output voltage that is higher than the input voltage,

Input voltage range between 0.5 V and 3.6 V, boosted output voltage range between 1.8 V and 5.25 V, sources up to 75 mA depending on the selected input and output voltage parameter values and two modes of operation: Active and Standby for PSoC 3 or Sleep for PSoC 5LP [13].

15.95.1 Boost Converter (BoostConv) Functions 5.00

- **BoostConv_Start( )**—Starts the BoostConv component and puts the boost block into Active mode.
- **BoostConv_Stop( )**—Disables the BoostConv component. Turns off power to the boost converter circuitry.
- **BoostConv_EnableInt( )**—Enables the boost block undervoltage interrupt generation.
- **BoostConv_DisableInt( )**—Disables the boost block undervoltage interrupt generation.
- **BoostConv_SetMode( )** Sets the boost converter mode to Active or Standby (PSoC 3) / Sleep (PSoC 5LP).
- **BoostConv_SelVoltage( )**—Selects the target output voltage the boost converter will maintain.
- **BoostConv_ManualThump( )**—Forces a single pulse of the boost converter switch transistors.
- **BoostConv_ReadStatus( )**—Returns the boost block status register.
- **BoostConv_ReadIntStatus( )**—Returns the contents of the boost block interrupt status register. BoostConv_Init( ) Initializes BoostConv registers with initial values provided from customizer.
- **BoostConv_Enable( )**—This function enables the boost block (only valid when in Active mode). Component is enabled by default.
- **BoostConv_Disable( )**—Disables the boost block.
- **BoostConv_EnableAutoThump( )**—Enables automatic thump mode (only available when the boost block is in Standby mode and the switching frequency is set to 32 kHz). (PSoC 3 API)
- **BoostConv_DisableAutoThump( )**—Disables automatic thump mode. (PSoC 3 API)
- **BoostConv_SelExtClk( )**—Sets the source of 32-kHz frequency: the 32-kHz ECO or 32-kHz ILO. (PSoC 3 API)
- **BoostConv_SelFreq( )**—Sets the switching frequency to one of two possible values: 400 kHz (generated internal to the boost converter block) or 32 kHz (sourced external to the boost converter block from the chip ECO-32kHz or ILO-32kHz oscillator). The 32kHz frequency is only applicable for PSoC 3. (PSoC 3 API)

15.95.2 Boost Converter (BoostConv) Function Calls 5.00

- **void BoostConv_Start(void)**—Starts the BoostConv component and puts the boost block into Active mode. The component is in this state when the chip powers up. This is the preferred method to begin component operation.
- **BoostConv_Start( )**—sets the initVar variable, calls the BoostConv_Init( ) function and then calls the BoostConv_Enable( ) function.
- **void BoostConv_Stop(void)**—saves the boost converter target output voltage and mode. Disables the BoostConv component.
- **void BoostConv_EnableInt(void)**—enables the boost block output undervoltage interrupt generation.
- **void BoostConv_DisableInt(void)**—disables the boost block output undervoltage interrupt generation.
- **void BoostConv_SetMode(uint8 mode)**—sets the boost converter mode: Active and Standby for PSoC 3 or Sleep for PSoC 5LP.

- **void BoostConv_SelVoltage(uint8 voltage)**—selects the target output voltage the boost converter will maintain.
- **void BoostConv_ManualThump(void)**—forces a single pulse of the boost converter switch transistors.
- **uint8 BoostConv_ReadStatus(void)**—returns the contents of the boost block status register.
- **void BoostConv_ReadIntStatus(void)**—returns the contents of the boost block interrupt status register.
- **void BoostConv_Init(void)**—initializes or restores the component according to the customizer Configure dialog settings. It is not necessary to call BoostConv_Init( ) because the Boost-Conv_Start( ) API calls this function and is the preferred method to begin component operation.
- **void BoostConv_Enable(void)**—This function enables the boost block when in Active mode. The component is enabled by default. Activates the hardware and begins component operation. It is not necessary to call BoostConv_Enable( ) because the BoostConv_Start( ) API calls this function, which is the preferred method to begin component operation.
- **void BoostConv_Disable(void)**—This function disables the boost block. (PSoC 3 API)
- **void BoostConv_EnableAutoThump(void)**—This function enables automatic thump mode. The AutoThump mode is available only when the boost block is in the Standby mode. The switching frequency clock source for the boost block must be set to the 32-kHz external clock. In this mode, standby boost operation is accomplished by generating a boost switch pulse on each edge of the switching clock when the output voltage is below the selected value.
- **void BoostConv_SelFreq(uint8 frequency)**—This function sets the switching frequency to one of two possible values. 400kHz (which is generated internal to the Boost Converter block with a dedicated oscillator) or 32kHz (which comes from the chips ECO-32kHz or ILO-32kHz).The 32kHz frequency is only applicable for PSoC 3.

15.96 Bootloader and Bootloadable 1.60

The bootloader system manages the process of updating the device flash memory with new application code and/or data. PSoC Creator uses the Bootloader project, i.e. a Project with a Bootloader Component and communication Component and Bootloadable project which uses a Bootloadable Component to create the code.Features include: separate Bootloader and Bootloadable Components and a Flexible component configuration [14].

15.96.1 Bootloader Functions 1.60

- **Bootloader_Start**—This function is called to execute the following algorithm.
- **Bootloader_GetMetadata**—Returns the value of the specified field of the metadata section.
- **Bootloader_ValidateBootloadable**—Verifies validation of the specified application.
- **Bootloader_Exit**—Schedules the specified application and performs software reset to launch it.
- **Bootloader_Calc8BitSum**—Computes the 8 bit sum for the specified data.
- **Bootloader_InitCallback**—Initializes the callback functionality.
- **Bootloadable_Load**—Updates the metadata area for the Bootloader to be started on device reset and resets the device.
- **Bootloader_Initialize**—Called for in-application bootloading, to initialize bootloading.
- **Bootloader_HostLink**—Called for in-application bootloading, to process bootloader command from the host.

- **Bootloader_GetRunningAppStatus**—Returns the application number of the currently running application.
- **Bootloader_GetActiveAppStatus**—Returns the application number of the currently active application.
- **Bootloadable_GetActiveApplication**—Gets the application which will be loaded after a next reset event.
- **Bootloadable_SetActiveApplication**—Sets the application which will be loaded after a next reset event.

15.96.2 Bootloader Function Calls 1.60

- **void Bootloader_Start (void)**—
  1. Bootloadable/Combination applications for the Classic Dual-app Bootloader/Launch-only Bootloader (Launcher for short) respectively.
  2. For the Classic Single-app Bootloader: if the Bootloadable application is valid, then the flow switches to it after a software reset. Otherwise it stays in the Bootloader, waiting for a command(s) from the host.
  3. For the Classic Dual-app Bootloader: the flow acts according to the switching table (see in the cod below) and enabled/disabled options (for instance, auto-switching). NOTE If the valid Bootloadable application is identified, then the control is passed to it after a software reset. Otherwise it stays in the Classic Dual-app Bootloader waiting for a command(s) from the host.
  4. For the Launcher: the flow acts according to the switching table (see below) and enabled/disabled options. NOTE If the valid Combination application is identified, then the control is passed to it after a software reset. Otherwise it stays in the Launcher forever.
  5. Validate the Bootloader/Launcher application(s) (design-time configurable, Bootloader application validation option of the Component customizer).
  6. Run a communication subroutine (design-time configurable, the Wait for command option of the Component customizer). NOTE This is NOT applicable for the Launcher.
  7. Schedule the Bootloadable and reset the device. See Switching Logic Table for details.
- **uint32 Bootloader_GetMetadata (uint8 field, uint8 appId)**—Returns the value of the specified field of the metadata section.
- **cystatus Bootloader_ValidateBootloadable (uint8 appId)**—Performs the Bootloadable application validation by calculating the application image checksum and comparing it with the checksum value stored in the Bootloadable Application Checksum field of the metadata section. If the "Fast bootloadable application validation" option is enabled in the Component customizer and Bootloadable application successfully passes validation, the Bootloadable Application Verification Status field of the metadata section is updated. Refer to the "Metadata Layout" section for the details. If the "Fast bootloadable application validation" option is enabled and the Bootloadable Application Verification Status field of the metadata section claims that the Bootloadable application is valid, the function returns CYRET_SUCCESS without further checksum calculation.
- **cystatus Bootloader_ValidateBootloadable (uint8 appId)**—Performs the Bootloadable application validation by calculating the application image checksum and comparing it with the checksum value stored in the Bootloadable Application Checksum field of the metadata section. If the "Fast bootloadable application validation" option is enabled in the Component customizer and Bootloadable application successfully passes validation, the Bootloadable Application Verification Status field of the metadata section is updated. Refer to the "Metadata Layout" section for the details. If the "Fast bootloadable application validation" option is enabled and the Bootloadable Application

Verification Status field of the metadata section claims that the Bootloadable application is valid, the function returns CYRET_SUCCESS without further checksum calculation.

- **uint8 Bootloader_Calc8BitSum (uint32 baseAddr, uint32 start, uint32 size)**—This computes an 8-bit sum for the provided number of bytes contained in flash (if baseAddr equals CY_FLASH_BASE) or EEPROM (if baseAddr equals CY_EEPROM_BASE).
- **void Bootloader_InitCallback(Bootloader_callback_type userCallback)**—This function initializes the callback functionality.
- **void Bootloadable_Load (void)**—Schedules the Bootloader/Launcher to be launched and then performs a software reset to launch it.
- **void Bootloadable_Load (void)**—Schedules the Bootloader/Launcher to be launched and then performs a software reset to launch it.
- **void Bootloader_HostLink(uint8 timeOut)**—Causes the Bootloader to attempt to read data being transmitted by the host application. If data is sent from the host, this establishes the communication interface to process all requests. This function is public only for Launcher-Combination architecture. For Classic Bootloader it is static, meaning private.
- **uint8 Bootloader_GetRunningAppStatus (void)**—Used for dual-app or in-app bootloader. Returns the value of the global variable Bootloader_runningApp. This function should be called only after the Bootloader_Initialize( ) has been called once.
- **uint8 Bootloader_GetActiveAppStatus (void)**—Used for dual-app or in-app bootloader. Returns the value of the global variable Bootloader_activeApp. This function should be called only after the Bootloader_Initialize( ) has been called once.
- **uint8 Bootloadable_GetActiveApplication (void)**—Gets the application which will be loaded after a next reset event. NOTE Intended for the combination project type ONLY!
- **cystatus Bootloadable_SetActiveApplication (uint8 appId)**—Sets the application which will be loaded after a next reset event. Theory: This API sets in the Flash (metadata section) the given active application number.[53]

15.96.3 Bootloadable Function Calls 1.60

- **void Bootloadable_Load (void)**—Schedules the Bootloader/Launcher to be launched and then performs a software reset to launch it.
- **uint8 Bootloadable_GetActiveApplication (void)**—Gets the application which will be loaded after a next reset event. NOTE Intended for the combination project type ONLY!
- **cystatus Bootloadable_SetActiveApplication (uint8 appId)**—Sets the application which will be loaded after a next reset event. Theory: This API sets in the Flash (metadata section) the given active application number.[54] Both metadata sections are updated. For example, if the second application is to be set active, then in the metadata section for the first application there will be a "0" written, which means that it is not active, and for the second metadata section there will be a "1" written, which means that it is active.

---

[53] The active application number is not set directly, but the boolean mark instead means that the application is active or not for the relative metadata. Both metadata sections are updated. For example, if the second application is to be set active, then in the metadata section for the first application there will be a "0" written, which means that it is not active, and for the second metadata section there will be a "1" written, which means that it is active. This is intended for the combination project type only!

[54] The active application number is not set directly, but the boolean mark instead means that the application is active or not for the relative metadata.

- **void Bootloader_Initialize (void)**—Used for in-app bootloading. This function updates the global variable Bootloader_runningApp with a running application number. If the running application number is valid (0 or 1), this function also sets the global variable Bootloader_initVar that is used to determine if the Component can process bootloader commands or not. This function should be called once in the application project after a startup.
- **uint8 Bootloader_GetRunningAppStatus (void)**—Used for dual-app or in-app bootloader. Returns the value of the global variable Bootloader_runningApp. This function should be called only after the Bootloader_Initialize( ) has been called once.
- **uint8 Bootloader_GetActiveAppStatus (void)**—Used for dual-app or in-app bootloader. Returns the value of the global variable Bootloader_activeApp. This function should be called only after the Bootloader_Initialize( ) has been called once.
- **uint8 Bootloader_Calc8BitSum (uint32 baseAddr, uint32 start, uint32 size)**—This computes an 8-bit sum for the provided number of bytes contained in flash (if baseAddr equals CY_FLASH_BASE) or EEPROM (if baseAddr equals CY_EEPROM_BASE).

15.97 Clock 2.20

The Clock Component provides two key features: it allows you to create local clocks and it allows you to connect to system and design-wide clocks. All clocks are shown in the PSoC Creator Design-Wide Resources (DWR) Clock Editor. Clocks may be defined in several ways, e.g., as a frequency with an automatically selected source clock, a frequency with a user-selected source clock and a divider and user-selected source clock. If a frequency is specified, PSoC Creator automatically selects a divider that yields the most accurate resulting frequency. If allowed, PSoC Creator also examines all system and design-wide clocks and selects a source and divider pair that yields the most accurate resulting frequency. Features include: quickly defines new clocks, refers to system or design-wide clocks and configures the clock frequency tolerance [18].

15.97.1 Clock Functions 2.20

- **Clock_Start( )**—Enables the clock.
- **Clock_StartEx( )**—Starts the clock initially phase aligned to the specified clock.
- **Clock_Stop( )**—Disables the clock.
- **Clock_StopBlock( )**—Disables the clock and waits until the clock is disabled.[55]
- **Clock_StandbyPower( )**—Selects the power for standby (Alternate Active) operation mode.[56]
- **Clock_SetDivider( )**—Sets the divider of the clock and restarts the clock divider immediately.
- **Clock_SetDividerRegister( )**—Sets the divider of the clock and optionally restarts the clock divider immediately.
- **Clock_SetDividerValue( )**—Sets the divider of the clock and restarts the clock divider immediately.
- **Clock_GetDividerRegister( )**—Gets the clock divider register value.

---

[55]The use of an external bypass capacitor is recommended if the internal noise caused by digital switching exceeds an application's analog performance requirements. To use this option, configure either port pin P0[2] or P0[4] as an analog HI-Z pin and connect an external capacitor with a value between 0.01 μF and 10 μF.
[56]Ibid.

- **Clock_SetMode( )**—Sets flags that control the operating mode of the clock.[57]
- **Clock_SetModeRegister( )**—Sets flags that control the operating mode of the clock.[58]
- **Clock_GetModeRegister( )**—Gets the clock mode register value.[59]
- **Clock_ClearModeRegister( )**—Clears flags that control the operating mode of the clock.[60]
- **Clock_SetSource( )**—Sets the source of the clock.[61]
- **Clock_SetSourceRegister( )**—Sets the source of the clock.[62]
- **Clock_GetSourceRegister( )**—Gets the source of the clock.[63]
- **Clock_SetPhase( ) [1]** - Sets the phase delay of the analog clock (only generated for analog clocks).
- **Clock_SetPhaseRegister( )**—Sets the phase delay of the analog clock (only generated for analog clocks).[64]
- **Clock_SetPhaseValue( )**—Sets the phase delay of the analog clock (only generated for analog clocks).[65]
- **Clock_GetPhaseRegister( )**—Gets the phase delay of the analog clock (only generated for analog clocks).[66]
- **Clock_SetFractionalDividerRegister( )**—Sets the fractional divider of the clock and restarts the clock divider immediately. (Only applicable to PSoC 4 devices.)
- **Clock_GetFractionalDividerRegister( )**—Gets the fractional clock divider register value. (Only applicable to PSoC 4 devices.)

15.97.2 Clock Function Calls 2.20

- **void Clock_Start(void)**—Starts the clock. Note On startup, clocks may already be running if the "Start on Reset" option is enabled in the DWR Clock Editor.
- **void Clock_StartEx(uint32 alignClkDiv)**—Starts the clock, phase aligned to the specified clock divider. This API requires the target phase align clock to be already running. Therefore, the correct procedure is to start the target align clock and then call this API to align to that target clock. If the target phase align clock is stopped and restarted, then phase alignment will be lost and the API should be called again to realign. Note This API is only available on PSoC 4 devices with the Phase Align clock feature. On startup, clocks may already be running if the "Start on Reset" option is enabled in the DWR Clock Editor.
- **void Clock_Stop(void)**—Stops the clock and returns immediately. This API does not require the source clock to be running but may return before the hardware is actually disabled. If the settings of the clock are changed after calling this function, the clock may glitch when it is started. To avoid the clock glitch, use the Clock_StopBlock( ) function.
- **void Clock_StopBlock(void)**—Stops the clock and waits for the hardware to actually be disabled before returning. This ensures that the clock is never truncated (high part of the cycle will terminate

---

[57] Ibid.
[58] Ibid.
[59] Ibid.
[60] Ibid.
[61] Ibid.
[62] Ibid.
[63] Ibid.
[64] Ibid.
[65] Ibid.
[66] Ibid.

before the clock is disabled and the API returns). Note that the source clock must be running or this API will never return as a stopped clock cannot be disabled.

- **void Clock_StandbyPower(uint8 state)**—Selects the power for standby (Alternate Active) operation mode. Note: The Clock_Start API enables the clock in Alternate Active Mode and the Clock_Stop and ClockStopBlock APIs disable the clock in Alternate Active Mode. If the clock is enabled, but needs to be disabled in Alternate Active Mode, Clock_StandbyPower(0) should be called after Clock_Start( ). If the clock is disabled, but needs to be enabled in Alternate Active mode, Clock_StandbyPower(1) should be called after Clock_Stop( ).
- **void Clock_SetDivider(uint16 clkDivider)**—Modifies the clock divider and thus, the frequency. When the clock divider register is set to zero or changed from zero, the clock is temporarily disabled in order to change a mode bit. If the clock is enabled when Clock_SetDivider( ) is called, the source clock must be running. The current clock cycle will be truncated and the new divide value will take effect immediately. The difference between this and Clock_SetDividerValue is that this API must consider the +1 factor.
- **void Clock_SetDividerRegister(uint16 clkDivider, uint8 reset)**—Modifies the clock divider and thus, the frequency. When the clock divider register is set to zero or changed from zero, the clock is temporarily disabled in order to change a mode bit. If the clock is enabled when Clock_SetDivider( ) is called, then the source clock must be running.
- **void Clock_SetDividerValue(uint16 clkDivider)**—Modifies the clock divider and thus, the frequency. When the clock divider register is set to zero or changed from zero, the clock will be temporarily disabled in order to change the SSS mode bit. If the clock is enabled when Clock_SetDivider( ) is called, then the source clock must be running. The current clock cycle will be truncated and the new divide value will take effect immediately.
- **uint16 Clock_GetDividerRegister(void)**—Gets the clock divider register value.
- **void Clock_SetMode(uint8 clkMode)**—Sets flags that control the operating mode of the clock. This function only changes flags from 0 to 1; flags that are already 1 remain unchanged. To clear flags, use the Clock_ClearModeRegister( ) function. The clock must be disabled before changing the mode. This API provides the same functionality as the SetModeRegister API.
- **void Clock_SetModeRegister(uint8 clkMode)**—Same as Clock_SetMode( ). Sets flags that control the operating mode of the clock. This function only changes flags from 0 to 1; flags that are already 1 will remain unchanged. To clear flags, use the Clock_ClearModeRegister( ) function. The clock must be disabled before changing the mode. This API provides the same functionality as the SetMode API.
- **uint8 Clock_GetModeRegister(void)**—Gets the clock mode register value.
- **void Clock_ClearModeRegister(uint8 clkMode)**—Clears flags that control the operating mode of the clock. This function only changes flags from 1 to 0; flags that are already 0 will remain unchanged. The clock must be disabled before changing the mode.
- **void Clock_SetSource(uint8 clkSource)**—Sets the input source of the clock. The clock must be disabled before changing the source. The old and new clock sources must be running. This API provides the same functionality as the SetSourceRegister API.
- **Clock_SetSourceRegister(uint8 clkSource)**—Same as Clock_SetSource( ). Sets the input source of the clock. The clock must be disabled before changing the source. The old and new clock sources must be running. This API provides the same functionality as the SetSource API.
- **uint8 Clock_GetSourceRegister(void)**—Gets the input source of the clock.
- **void Clock_SetPhase(uint8 clkPhase)**—Sets the phase delay of the analog clock. This function is only available for analog clocks. The clock must be disabled before changing the phase delay to avoid glitches. This API provides the same functionality as the SetPhaseRegister API.

- **void Clock_SetPhaseRegister(uint8 clkPhase)**—Same as Clock_SetPhase( ). Sets the phase delay of the analog clock. This function is only available for analog clocks. The clock must be disabled before changing the phase delay to avoid glitches. This API provides the same functionality as the SetPhase API.
- **void Clock_SetPhaseValue(uint8 clkPhase)**—Sets the phase delay of the analog clock. This function is only available for analog clocks. The clock must be disabled before changing the phase delay to avoid glitches. Same as Clock_SetPhase( ), except Clock_SetPhaseValue( ) adds one to the value and then calls Clock_SetPhaseRegister( ) with it.
- **uint8 Clock_GetPhaseRegister(void)**—Gets the phase delay of the analog clock. This function is only available for analog clocks.
- **void Clock_SetFractionalDividerRegister(uint16 clkDivider, uint8 fracDivider)**—Modifies the clock divider and the fractional clock divider and thus, the frequency. The fractional divider does not work with integer divider values by 1.
- **uint8 Clock_GetFractionalDividerRegister (void)**—Gets the fractional clock divider register value.

15.97.3 UDB Clock Enable (UDBClkEn) 1.00

The universal digital block (UDB) Clock Enable (UDBClkEn) component supports precise control over clocking behavior. Features include: clock enable support and the ability to add synchronization on a clock when needed [101].

15.97.4 UDB Clock Enable (UDBClkEn) Functions 1.00

NONE.

15.98 Die Temperature 2.10

The Die Temperature (DieTemp) component provides an Application Programming Interface (API) to acquire the temperature of the die. The System Performance Controller (SPC) is used to get the die temperature. The API includes blocking and non-blocking calls. Features include: accuracy of $\pm 5\,^{\circ}$C range $-40\,^{\circ}$C to $+140\,^{\circ}$C (0xFFD8 to 0x008C) and blocking and non-blocking API [28].

15.98.1 Die Temperature Functions 2.10

- **DieTemp_Start( )**—Starts the SPC command to get the die temperature
- **DieTemp_Stop( )**—Stops the temperature reading
- **DieTemp_Query( )**—Queries the SPC to see if the temperature command is finished
- **DieTemp_GetTemp( )**—Sets up the command to get the temperature and blocks until finished

15.98.2 Die Temperature Function Calls 2.10

- **cystatus DieTemp_Start(void)**—Sends the command and parameters to the SPC to start a Die Temperature reading. This function returns before the SPC finishes. This function call must always be paired with a call to the DieTemp_Query( ) API to complete the Die Temperature reading.
- **void DieTemp_Stop(void)**—There is no need to stop or disable this component. This component is naturally a slave that sends request to SPC through SPC API of cy_boot and waits for data to be ready or in case
- **cystatus DieTemp_Query(int16 * temperature)**—Checks to see if the SPC command started by DieTemp_Start( ) has finished. If the command has not finished, the temperature value is not read and returned. The caller will need to poll this function while the return status remains CYRET_STARTED. This can be used only in conjunction with the DieTemp_Start( ) API to successfully get the correct Die Temperature. The Die Temperature reading returned on the first sequence of DieTemp_Start( ) followed by DieTemp_Query( ) can be unreliable, so you must do this sequence twice and use the value returned from the second sequence.
- **cystatus DieTemp_GetTemp(int16 * temperature)**—Sends the command and parameters to the SPC to start a Die Temperature reading and waits until it fails or completes. This is a blocking API. This function reads the Die Temperature twice and returns the second value to work around an issue in the silicon that causes the first value read to be unreliable [28].

15.99 Direct Memory Access (DMA) 1.70

The DMA component allows data transfers to and from memory, components and registers. The controller supports 8-, 16- and 32-bit wide data transfers and can be configured to transfer data between a source and destination that have different endianess. TDs can be chained together for complex operations. The DMA can be triggered based on a level or rising edge signal. See the Hardware Request parameter selection for more details. Features include: 24 channels, eight priority levels, 128 Transaction Descriptors (TDs), 8-, 16- and 32-bit data transfers, configurable source and destination addresses, support for endian compatibility, can generate an interrupt when data transfer is complete and a DMA Wizard to assist with application development. Features include: 24 channels, eight priority levels, 128 Transaction Descriptors (TDs), 8-, 16- and 32-bit data transfers, configurable source and destination addresses, support for endian compatibility, can generate an interrupt when data transfer is complete and a DMA Wizard to assist with application development [34].

15.99.1 Direct Memory Access Functions 1.70

- **DMA_DmaInitialize( )**—Allocates and initializes a DMA channel to be used by the caller.
- **DMA_DmaRelease( )**—Frees and disables the DMA channel associated with this instance of the component.

15.99.2 Direct Memory Access Function Calls 1.70

- **uint8 DMA_DmaInitialize(uint8 burstCount, uint8 requestPerBurst, uint16 upperSrcAddress, uint16 upperDestAddress)**—Allocates and initializes a DMA channel to be used by the caller.

- **void DMA_DmaRelease(void)**—Frees the channel associated with this instance of the component. The channel cannot be used again unless DMA_DmaInitialize( ) is called again.

15.100 DMA Library APIs (Shared by All DMA Instances) 1.70

15.100.1 DMA Controller Functions

- **void CyDmacConfigure(void)**—Creates a linked list of all the TDs to be allocated. This function is called by the startup code if any DMA components are placed onto design schematic; you do not normally need to call it. You could call this function if all the DMA channels are inactive.
- **uint8 CyDmacError(void)**—Returns errors from the last failed DMA transaction.
- **void CyDmacClearError(uint8 error)**—Clears the error bits in the error register of the DMAC.
- **uint32 CyDmacErrorAddress(void)**—If there are multiple errors, only the address of the first error is saved. When a CY_DMA_BUS_TIMEOUT, CY_DMA_UNPOP_ACC and CY_DMA_PERIPH_ERR occur, the address of the error is written to the error address register and can be read with this function.

15.100.2 Channel Specific Functions 1.70

- **CyDmaChAlloc( )**—Allocates a channel of the DMA to be used by the caller.
- **CyDmaChFree( )**—Frees a channel allocated by CyDmaChAlloc( ).
- **CyDmaChEnable( )**—Enables the DMA channel for execution.
- **CyDmaChDisable( )**—Disables the DMA channel.
- **CyDmaClearPendingDrq( )**—Clears a pending DMA data request.
- **CyDmaChPriority( )**—Sets the priority of a DMA channel.
- **CyDmaChSetExtendedAddress( )**—Sets the high 16 bits of the source and destination addresses.
- **CyDmaChSetInitialTd( )**—Set the initial TD for the channel.
- **CyDmaChSetRequest( )**—Requests to terminate a chain of TDs or one TD, or start the DMA.
- **CyDmaChGetRequest( )**—Checks to see if the CyDmaChSetRequest( ) request was satisfied.
- **CyDmaChStatus( )**—Determines the status of the current TD.
- **CyDmaChSetConfiguration( )**—Sets configuration information for the channel.
- **CyDmaChRoundRobin( )**—Enables/disables the Round-Robin scheduling enforcement algorithm

15.100.3 Channel Specific Function Calls 1.70

- **uint8 CyDmaChAlloc(void)**—Allocates a channel from the DMAC to be used in all functions that require a channel handle.
- **cystatus CyDmaChFree(uint8 chHandle)**—Frees a channel handle allocated by CyDmaChAlloc( ).
- **cystatus CyDmaChEnable(uint8 chHandle, uint8 preserveTds)**—Enables the DMA channel. A software or hardware request still must happen before the channel is executed. Note While the channel is enabled and is currently being serviced by the DMA controller, any other configuration information for the channel should NOT be altered to ensure graceful operation.

- **cystatus CyDmaChDisable(uint8 chHandle)**—Disables the DMA channel. Once this function is called, CyDmaChStatus( ) may be called to determine when the channel is disabled and which TDs were being executed. If the DMA channel is currently executing it will complete the current burst naturally. Note PSoC 3 has a known silicon problem that may not allow the DMA to terminate properly if a new request occurs during the small terminate processing period. Refer to the Component Errata section for more information.
- **cystatus CyDmaClearPendingDrq(uint8 chHandle)**—Clears pending DMA data request.
- **cystatus CyDmaChPriority(uint8 chHandle, uint8 priority)**—Sets the priority of a DMA channel. You can use this function when you want to change the priority at run time. If the priority remains the same for a DMA channel, then you can configure the priority in the .cydwr file.
- **cystatus CyDmaChSetExtendedAddress(uint8 chHandle, uint16 source, uint16 destination)**—Sets the high 16 bits of the source and destination addresses for the DMA channel (valid for all TDs in the chain).
- **cystatus CyDmaChSetInitialTd(uint8 chHandle, uint8 startTd)**—Sets the initial TD to be executed for the channel when the CyDmaChEnable( ) function is called.
- **cystatus CyDmaChSetRequest(uint8 chHandle, uint8 request)**—Allows the caller to terminate a chain of TDs, terminate one TD, or create a direct request to start the DMA channel.
- **cystatus CyDmaChGetRequest(uint8 chHandle)**—This function allows the caller of CyDmaCh-SetRequest( ) to determine if the request was completed.
- **cystatus CyDmaChStatus(uint8 chHandle, uint8 * currentTd, uint8 * state)**—Determines the status of the DMA channel.
- **cystatus CyDmaChSetConfiguration(uint8 chHandle, uint8 burstCount, uint8 requestPer-Burst, uint8 tdDone0, uint8 tdDone1, uint8 tdStop)**—Sets configuration information for the channel.
- **cystatus CyDmaChRoundRobin(uint8 chHandle, uint8 enableRR)**—Either enables or disables the Round-Robin scheduling enforcement algorithm. Within a priority level a Round-Robin fairness algorithm is enforced. The default configuration has round robin scheduling disabled.

15.100.4 Transaction Description Functions 1.70

- **CyDmaTdAllocate( )**—Allocates a TD from the free list for use.
- **CyDmaTdFree( )**—Returns a TD back to the free list.
- **CyDmaTdFreeCount( )**—Gets the number of free TDs available.
- **CyDmaTdSetConfiguration( )**—Sets the configuration for the TD.
- **CyDmaTdGetConfiguration( )**—Gets the configuration for the TD.
- **CyDmaTdSetAddress( )**—Sets the lower 16 bits of the source and destination addresses.
- **CyDmaTdGetAddress( )**—Gets the lower 16 bits of the source and destination addresses.

15.100.5 Transaction Description Function Calls 1.70

- **uint8 CyDmaTdAllocate(void)**—Allocates a TD for use with an allocated DMA channel.
- **void CyDmaTdFree(uint8 tdHandle)**—Returns a TD to the free list.
- **uint8 CyDmaTdFreeCount(void)**—Returns the number of free TDs available to be allocated.
- **cystatus CyDmaTdSetConfiguration(uint8 tdHandle, uint16 transferCount, uint8 nextTd, uint8 configuration)**—Configures the TD.

- **cystatus CyDmaTdGetConfiguration(uint8 tdHandle, uint16 * transferCount, uint8 *nextTd, uint8 * configuration)**—Retrieves the configuration of the TD. If a NULL pointer is passed as a parameter, that parameter is skipped. You may request only the values you are interested in.
- **cystatus CyDmaTdSetAddress(uint8 tdHandle, uint16 source, uint16 destination)**—Sets the lower 16 bits of the source and destination addresses for this TD only.
- **cystatus CyDmaTdGetAddress(uint8 tdHandle, uint16 * source, uint16 * destination)**—Retrieves the lower 16 bits of the source and/or destination addresses for this TD only. If NULL is passed for a pointer parameter, that value is skipped. You may request only the values of interest.

15.101 EEPROM 3.00

The EEPROM component provides a set of APIs to erase and write data to non-volatile on-chip EEPROM memory. The term write implies that it will erase and then program in one operation. An EEPROM memory in PSoC devices is organized in arrays. PSoC 3 and PSoC 5LP devices offer an EEPROM array of size 512 bytes, 1 KB or 2 KB depending on the device. EEPROM array can be divided into sectors that have up to 64 rows with a size of 16 bytes. The Application Programming Interface (API) routines allow you to modify a whole EEPROM row, individual EEPROM bytes, or erase a whole EEPROM sector in one operation. The EEPROM memory is not initialized by the EEPROM component: the initial state of the memory is defined in the device datasheet. The default values can be changed in the PSoC Creator EEPROM Editor. For more details, refer to the PSoC Creator Help. The EEPROM component is tightly coupled with various system elements contained within the cy_boot component. These elements are generated upon a successful build. Refer to the System Reference Guide for more information about the cy_boot component and its various elements. Features include: 512 B to 2 KB EEPROM memory, 1,000,000 cycles, 20-year retention, Read/Write 1 byte at a time, 16 bytes (a row) can be programmed at a time [38].

15.101.1 EEPROM Functions 3.00

- **EEPROM_Enable( )**—Enables EEPROM block operation.
- **EEPROM_Start( )**—Starts EEPROM.
- **EEPROM_Stop( )**—Stops and powers down EEPROM.
- **EEPROM_WriteByte( )**—Writes a byte of data to the EEPROM.
- **EEPROM_ReadByte( )**—Reads a byte of data from the EEPROM.
- **EEPROM_UpdateTemperature( )**—Updates store temperature value.
- **EEPROM_EraseSector( )**—Erases an EEPROM sector.
- **EEPROM_Write( )**—Blocks while writing a row to EEPROM.
- **EEPROM_StartWrite( )**—Starts writing a row of data to EEPROM.
- **EEPROM_Query( )**—Checks the state of a write to EEPROM.
- **EEPROM_ByteWritePos( )**—Writes a byte of data to EEPROM.

15.101.2 EEPROM Function Calls 3.00

- **void EEPROM_Enable(void)**—Enables EEPROM block operation.

- **void EEPROM_Start(void)**—Starts the EEPROM. This has to be called before using write/erase APIs and reading the EEPROM.
- **void EEPROM_Stop(void)**—Stops and powers down the EEPROM.
- **cystatus EEPROM_WriteByte(uint8 dataByte, uint16 address)**—Writes a byte of data to the EEPROM. This function blocks until the function is complete. For reliable write procedure to occur you should call EEPROM_UpdateTemperature( ) API if the temperature of the silicon has changed for more than 10 °C since component was started.
- **uint8 EEPROM_ReadByte(uint16 address)**—Reads a byte of data from the EEPROM. Although the data is present in one of the memory spaces, this provides an intuitive user interface, addressing EEPROM memory as a separate block with first EERPOM address 0x0000.
- **uint8 EEPROM_UpdateTemperature(void)**—Updates store temperature value. This should be called anytime the EEPROM is active and temperature may have changed by more than 10C.
- **cystatus EEPROM_EraseSector(uint8 sectorNumber)**—Erases a sector (64 rows) of memory by making the bits zero. This function blocks until the operation is complete. Using this API helps to erase EEPROM a sector at a time. This is faster than using individual writes but affects cycle recourse of the whole row.
- **cystatus EEPROM_Write(const uint8 *rowData, uint8 rowNumber)**—Writes a row (16 bytes) of data to the EEPROM. This function blocks until the function is complete. Compared to APIs that write one byte, this API allows you to write a whole row (16 bytes) at a same time.
- **cystatus EEPROM_StartWrite(const uint8 *rowData, uint8 rowNumber)**—Starts the write of a row (16 bytes) of data to the EEPROM. This function does not block. The function returns once the SPC has begun writing the data. This function must be used in combination with EEPROM_Query( ). EEPROM_Query( ) must be called until it returns a status other than CYRET_STARTED. That indicates the write has completed. Until EEPROM_Query( ) detects that the write is complete the SPC is marked as locked to prevent another SPC operation from being performed. For reliable write procedure to occur you should call EEPROM_UpdateTemperature( ) API if the temperature of the silicon has changed for more than 10 °C since component was started.
- **cystatus EEPROM_StartErase(uint8 sectorNumber)**—Starts the EEPROM sector erase. This function does not block. The function returns once the SPC has begun writing the data. This function must be used in combination with EEPROM_Query( ). EEPROM_Query( ) must be called until it returns a status other than CYRET_STARTED. That indicates the erase has completed. Until EEPROM_Query( ) detects that the erase is complete the SPC is marked as locked to prevent another SPC operation from being performed.
- **cystatus EEPROM_Query(void)**—Checks the status of an earlier call to EEPROM_StartWrite( ) or EEPROM_StartErase( ). This function must be called until it returns a value other than CYRET_STARTED. Once that occurs the write has been completed and the SPC is unlocked.
- **cystatus EEPROM_ByteWritePos(unit8 dataByte, uint8 rowNumber, uint8 byteNumber)**— Writes a byte of data to EEPROM. Compared to EEPROM_WriteByte( ) this API allows to write a specific byte in a specified EEPROM row. This is a blocking call. It will not return until the function succeeds or fails.

15.102 Emulated EEPROM 2.20

The Emulated EEPROM Component emulates an EEPROM device in the PSoC device flash memory. Features include: EEPROM-Like Non-Volatile Storage, easy-to-use Read and Write API functions, Optional Wear Leveling and optional Redundant EEPROM Copy Storage. Features include:

EEPROM-Like Non-Volatile Storage, easy-to-use Read and Write API Functions and EEPROM-Like Non-Volatile Storage [39].

15.102.1 Emulated EEPROM functions 2.20

This Component includes a set of Component-specific wrapper functions that provide simplified access to the basic Cy_Em_EEPROM operation. These functions are generated during the build process and are all prefixed with the name of the Component instance [39].

- **Em_EEPROM_1_Init( )**—Fills the start address of the EEPROM to the Component configuration structure and invokes Cy_Em_EEPROM_Init( ) function.
- **Em_EEPROM_1_Read( )**—Invokes the Cy_Em_EEPROM_Read( ) function.
- **Em_EEPROM_1_Write( )**—Invokes the Cy_Em_EEPROM_Write( ) function.
- **Em_EEPROM_1_Erase( )**—Invokes the Cy_Em_EEPROM_Erase( ) function.
- **Em_EEPROM_1_NumWrites( )**—Invokes the Cy_Em_EEPRO_NumWrites( ) function.

15.102.2 Emulated EEPROM 2.20 function Calls to Wrapper Functions

- **cy_en_em_eeprom_status_t EmEEPROM_1_Init (uint32 startAddress)**—Fills the start address of the EEPROM to the Component configuration structure and invokes Cy_Em_EEPROM_ Init( ) function.
- **cy_en_em_eeprom_status_t EmEEPROM_1_Write (uint32 addr, void * eepromData, uint32size)**—Invokes the Cy_Em_EEPROM_Write( ) function.
- **cy_en_em_eeprom_status_t EmEEPROM_1_Read (uint32 addr, void * eepromDdata, uint32 size)**—Invokes the Cy_Em_EEPROM_Read( ) function.
- **cy_en_em_eeprom_status_t EmEEPROM_1_Erase (void)**—Invokes the Cy_Em_EEPROM_ Erase( ) function.
- **uint32 EmEEPROM_1_NumWrites (void)**—Invokes the Cy_Em_EEPROM_NumWrites( ) function.

15.103 External Memory Interface 1.30

This component enables access by the CPU or DMA to memory ICs external to the PSoC 3/PSoC 5LP. It facilitates setup of the EMIF hardware, as well as, UDBs and GPIOs as required. The EMIF can control synchronous and asynchronous memories without the need to configure any UDBs in synchronous and asynchronous modes. In UDB mode, UDBs must be configured to generate external memory control signals. Features include: 8-, 16-, 24-bit address bus width 8-, 16-bit data bus width, supports external synchronous memory, supports external asynchronous memory, supports custom interface for memory, supports a range of speeds of external memories (from 5 to 200 ns), supports external memory power-down, sleep and wakeup modes [40].

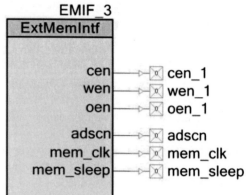

Asynchronous EMIF Macro   **Synchronous EMIF Macro**

15.103.1 External Memory Interface Functions 1.30

- **EMIF_Start( )**—Calls EMIF_Init( ) and EMIF_Enable( ).
- **EMIF_Stop( )**—Disables the EMIF block.
- **EMIF_Init( )**—Initializes or restores the EMIF configuration to the current customizer state
- **EMIF_Enable( )**—Enables the EMIF hardware block, associated I/O ports and pins
- **EMIF_ExtMemSleep( )**—Sets the external memory sleep signal high; note that depending on the type of external memory IC used, the signal may need to be inverted.
- **EMIF_ExtMemWakeup( )**—Sets the external memory sleep signal low; note that depending on the type of external memory IC used, the signal may need to be inverted.
- **EMIF_SaveConfig( )**—Saves the user configuration of the EMIF non-retention registers. This routine is called by EMIF_Sleep( ) to save the Component configuration before entering sleep.
- **EMIF_Sleep( )**—Stops the EMIF operation and saves the user configuration along with the enable state of the EMIF.
- **EMIF_RestoreConfig( )**—Restores the user configuration of the EMIF non-retention registers. This routine is called by EMIF_Wakeup( ) to restore the Component configuration when exiting sleep.
- **EMIF_Wakeup( )**—Restores the user configuration and restores the enable state.

15.103.2 External Memory Interface Function Calls 1.30

- **void EMIF_Start(void)**—This is the preferred method to begin Component operation. EMIF_Start( ) calls the EMIF_Init( ) function and then calls the EMIF_Enable( ) function.
- **void EMIF_Stop(void)**—Disables the EMIF block. Note Use CyPins_SetPinDriveMode function to change EMIF pins state to High-Z.
- **void EMIF_Init(void)** - Initializes or restores the Component according to the customizer Configure dialog settings. It is not necessary to call EMIF_Init( ) because the EMIF_Start( ) routine calls this function and is the preferred method to begin Component operation.
- **void EMIF_Enable(void)**—Activates the hardware and begins Component operation. It is not necessary to call EMIF_Enable( ) because the EMIF_Start( ) routine calls this function, which is the preferred method to begin Component operation.

- **void EMIF_ExtMemSleep(void)**—Sets the 'mem_pd' bit in the EMIF_PWR_DWN register. This sets the external memory sleep signal high. Depending on the type of external memory IC used, the signal may need to be inverted.
- **void EMIF_ExtMemWakeup(void)**—Resets the 'mem_pd' bit in the EMIF_PWR_DWN register. This sets the external memory sleep signal low. Depending on the type of external memory IC used, the signal may need to be inverted.
- **void EMIF_SaveConfig(void)**—This function saves the Component configuration. This will save non-retention registers. This function will also save the current Component parameter values, as defined in the Configure dialog or as modified by appropriate APIs. This function is called by the EMIF_Sleep( ) function.
- **void EMIF_Sleep(void)**—This is the preferred routine to prepare the Component for sleep. The EMIF_Sleep( ) routine saves the current Component state. Then it calls the EMIF_Stop( ) function and calls EMIF_SaveConfig( ) to save the hardware configuration. Call the EMIF_Sleep( ) function before calling the CyPmSleep( ) or the CyPmHibernate( ) function. Refer to the PSoC Creator System Reference Guide for more information about power management functions.
- **void EMIF_RestoreConfig(void)**—This function restores the Component configuration. This will restore non-retention registers. This function will also restore the Component parameter values to what they were prior to calling the EMIF_Sleep( ) function.
- **void EMIF_Wakeup(void)**—This is the preferred routine to restore the Component to the state when EMIF_Sleep( ) was called. The EMIF_Wakeup( ) function calls the EMIF_RestoreConfig( ) function to restore the configuration. If the Component was enabled before the EMIF_Sleep( ) function was called, the EMIF_Wakeup( ) function will also re-enable the Component.

15.104 Global Signal Reference (GSRef 2.10)

The Global Signal Reference Component allows access to device-specific global signals that must be used at a system level. These include interrupts from shared resources and system- wide interrupts. Features include: connections to device global signals and allows connections to shared resource interrupts. The Global Signal Reference Component can be used for accessing device global signals: watchdog interrupts and Time period interrupts, interrupts through wakeup sources such as, CTBm, LP comparator and I/O ports, status conditions such as PLL Lock, Low voltage detection, power management and interrupts for non-blocking Flash writes and error conditions such as XMHz Error and Cache Interrupt [48].

15.104.1 Global Signal Reference (GSRef Functions 2.10)

NONE.

15.105 ILO Trim 2.00

The ILO Trim component allows an application to determine the accuracy of the ILO. It provides a scaling function to allow the application to compensate for this inaccuracy. For PSoC 3 and PSoC 5LP devices, it can also directly improve the accuracy of the ILO by using a user-defined higher frequency, higher accuracy reference clock to count the number of ILO clock cycles. The derived information is then used to trim the ILO trim registers to incrementally approach the desired ILO frequency. The component supports both UDB and Fixed-Function implementations. The ILO consists of two low-speed oscillators (LSO): 100 kHz and 1 kHz. These are used to generate ILO clock frequencies of 1 kHz, 33 kHz, or 100 kHz. Post factory trim, the 1 kHz LSO has an accuracy of -50% to 100% and the 100 kHz LSO has an accuracy of -55% to 100% over the entire operating range of voltage and temperature. During the run-time trimming operation, the trim DACs and the bias block are adjusted using the ILO Trim registers. Features include: trims 1 kHz and 100 kHz ILO, UDB and Fixed-Function modes and a user-specified reference clock [53].

15.105.1 ILO Functions 2.00

- **ILO_Trim_Start( )**—Initializes and starts the component.
- **ILO_Trim_Stop( )**—Stops the component.
- **ILO_Trim_BeginTrimming( )**—Begins the implementation of the ILO trimming algorithm.
- **ILO_Trim_StopTrimming( )**—Disables the trimming algorithm.
- **ILO_Trim_CheckStatus( )**—Returns the current status of the ILO and the trimming algorithm.
- **ILO_Trim_CheckError( )**—Calculates ILO frequency error in parts per thousand.
- **ILO_Trim_Compensate( )**—Compensates for the ILO clock inaccuracy by converting from a desired nominal number of clock cycles to the effective number of ILO clock cycles required based on the current accuracy of the ILO.
- **ILO_Trim_RestoreTrim( )**—Restores the factory trim value.
- **ILO_Trim_GetTrim( )**—Returns the current ILO trim value.
- **ILO_Trim_SetTrim( )**—Returns the called number of words from the given address. If any address does not belong to the allocated register space, an error is returned.
- **ILO_Trim_Sleep( )**—Stops the component and saves the user configuration.
- **ILO_Trim_Wakeup( )**—Restores the user configuration and enables the component.
- **ILO_Trim_SaveConfig( )**—Saves the current user configuration of the component.
- **ILO_Trim_RestoreConfig( )**—Restores the current user configuration of the component.

15.105.2 ILO Function Calls 2.00

- **void ILO_Trim_Start(void)**—Starts the measurement of the ILO accuracy.
- **void ILO_Trim_Stop(void)**—Stops the measurement of the ILO accuracy. If trimming is currently active, then the trimming algorithm is also terminated at this point.
- **void ILO_Trim_BeginTrimming(void)**—Begins the implementation of the ILO trimming algorithm. The algorithm requires multiple ILO clock periods to converge with an accurately trimmed ILO clock frequency. This function is non-blocking and will return after configuring an interrupt process to implement the trimming algorithm. ILO_Trim_CheckStatus( ) and ILO_Trim_CheckError( ) can be used to determine the current status of the algorithm. Once the trimming algorithm converges on an accurately trimmed ILO frequency, the background trimming algorithm is disabled. Not (Supported for PSoC 4.)
- **void ILO_Trim_StopTrimming(void)**—Disables the trimming algorithm that was started by the ILO_Trim_BeginTrimming( ) function. This function is only used to terminate the trimming algorithm early. Normal operation of the trimming algorithm does not use this function since the algorithm will disable trimming once the ILO accuracy has been achieved.

- **uint8 ILO_Trim_CheckStatus(void)**—Returns the current status of the ILO and the trimming algorithm.[67]
- **int16 ILO_Trim_CheckError(void)**—Calculates ILO frequency error in parts per thousand. A positive number indicates that the ILO is running fast and a negative number indicates that the ILO is running slow. This error is relative to the error in the reference clock, so the absolute error will be higher and depend on the accuracy of the reference.[68]
- **uint16 ILO_Trim_Compensate(uint16 clocks)**—Compensates for the ILO clock inaccuracy by converting from a desired nominal number of clock cycles to the effective number of ILO clock cycles required based on the current accuracy of the ILO. The returned value can then be used instead of the nominal value when configuring timers that are based on the ILO. If the calculated result exceeds the capacity of the 16 bit return value, it will be saturated at the maximum 16-bit value. Note This function is an alternative to trimming the frequency of the ILO and should not be used in conjunction with the BeginTrimming( ) function. Note This function requires a full ILO cycle to occur prior to reading the ILO accuracy result. The application should insert a worst-case delay based on the chosen ILO frequency (for example, 32 kHz +100%) between ILO_Trim_Start( ) and calling this function.
- **void ILO_Trim_RestoreTrim(void)**—Restore the factory trim value. (Not Supported for PSoC 4.)
- **uint8 ILO_Trim_GetTrim(void)**—Returns the current ILO trim value. (Not Supported for PSoC 4.)
- **void ILO_Trim_SetTrim(uint8 trim)**—Sets the ILO trim value. (Not Supported for PSoC 4.)
- **void ILO_Trim_Sleep(void)**—Prepares the component for sleep. If the component is currently enabled it will be disabled and re-enabled by ILO_Trim_Wakeup( ).
- **void ILO_Trim_Wakeup(void)**—Returns the component to its state before the call to ILO_Trim _Sleep( ). In order to trim the ILO after waking from a low power mode, the ILO_Trim_BeginTrimming( ) function must be called.
- **void ILO_Trim_SaveConfig(void)**—Saves the configuration of the component. This routine is called by ILO_Trim_Sleep( ) to save the configuration.
- **void ILO_Trim_RestoreConfig(void)**—Restores the configuration of the component. This routine is called by ILO_Trim_Wakeup( ) to restore the configuration.

15.106 Interrupt 1.70

The Interrupt Component defines hardware triggered interrupts. It is an integral part of the Interrupt Design-Wide Resource system. There are three types of system interrupt waveforms that can be processed by the interrupt controller: Level, IRQ source is sticky and remains active until firmware clears the source of the request with an action (for example, clear on read). Most fixed-function peripherals have level-sensitive interrupts, including the UDB FIFOs and status registers, Pulse, ideally, a pulse IRQ is a single bus clock, which logs a pending action and ensures that the ISR action is only executed once. No firmware action to the peripheral is required and Edge, An arbitrary synchronous waveform is the input to an edge-detect circuit and the positive edge of that waveform

---

[67]The accuracy status is not reliable until two ILO clock cycles have occurred since the component was started.
[68]The accuracy error is not reliable until two ILO clock cycles have occurred since the component was started.

becomes a synchronous one-cycle pulse (Pulse mode).[69] Regardless of the InterruptType multiplexer selection, the interrupt controller is still able to process level, edge, or pulse waveforms. Features include: defines hardware-triggered interrupts, provides a software API to pend (hardware-connected) interrupts [55].

15.106.1 Interrupt Functions 1.70

- **ISR_Start( )**—Sets up the interrupt to function.
- **ISR_StartEx( )**—Sets up the interrupt to function and sets address as the ISR vector for the interrupt.
- **ISR_Stop( )**—Disables and un-configures the interrupt.
- **ISR_Interrupt( )**—The default interrupt handler for ISR.
- **ISR_SetVector( )**—Sets address as the new ISR vector for the Interrupt.
- **ISR_GetVector( )**—Gets the address of the current ISR vector for the interrupt.
- **ISR_SetPriority( )**—Sets the priority of the interrupt.
- **ISR_GetPriority( )**—Gets the priority of the interrupt.
- **ISR_Enable( )**—Enables the interrupt to the interrupt controller.
- **ISR_GetState( )**—Gets the state (enabled, disabled) of the interrupt.
- **ISR_Disable( )**—Disables the interrupt.
- **ISR_SetPending( )**—Causes the interrupt to enter the pending state, a software method of generating the interrupt.
- **ISR_ClearPending( )**—Clears a pending interrupt.

15.106.2 Interrupt Function Calls 1.70

- **void ISR_Start(void)**—Sets up the interrupt and enables it. This function disables the interrupt, sets the default interrupt vector, sets the priority from the value in the Design Wide Resources Interrupt Editor, then enables the interrupt in the interrupt controller.
- **void ISR_StartEx(cyisraddress address)**—Sets up the interrupt and enables it. This function disables the interrupt, sets the interrupt vector based on the address passed in, sets the priority from the value in the Design Wide Resources Interrupt Editor, then enables the interrupt in the interrupt controller. When defining ISR functions, the CY_ISR and CY_ISR_PROTO macros should be used to provide consistent definition across compilers:

```
Function definition example:
CY\_ISR(MyISR)
{
        /* ISR Code here */
}
Function prototype example:
CY\_ISR\_PROTO(MyISR);
```

- **void ISR_Stop(void)**—Disables and removes the interrupt.

---

[69]These interrupt waveform types are different from the settings made in the Configure dialog for the InterruptType parameter. The parameter only configures the multiplexer select lines. It processes the "IRQ" signal to be sent to the interrupt controller based on the multiplexer selection (Level, Edge).

- **void ISR_Interrupt(void)**—The default ISR for the Component. Add custom code between the START and END comments to keep the next version of this file from over-writing your code. Note You may use either the default ISR by using this API, or you may define your own separate ISR through ISR_StartEx( ).
- **void ISR_SetVector(cyisraddress address)**—Changes the ISR vector for the interrupt. Use this function to change the ISR vector to the address of a different interrupt service routine. Note that calling ISR_Start( ) overrides any effect this API would have had. To set the vector before the Component has been started, use ISR_StartEx( ) instead. When defining ISR functions, the CY_ISR and CY_ISR_PROTO macros should be used to provide consistent definition across compilers:

> Function definition example:
> CY_ISR(MyISR)
> {
>   /* ISR Code here */
> }
> Function prototype example:
> CY_ISR_PROTO(MyISR);

- **cyisraddress ISR_GetVector(void)**—Gets the address of the current ISR vector for the interrupt.
- **void ISR_SetPriority(uint8 priority)**—Sets the priority of the interrupt. Note Calling ISR_Start( ) or ISR_StartEx( ) overrides any effect this API would have had. This API should only be called after ISR_Start( ) or ISR_StartEx( ) has been called. To set the initial priority for the Component, use the Design-Wide Resources Interrupt Editor.
- **uint8 ISR_GetPriority(void)**—Gets the priority of the interrupt.
- **void ISR_Enable(void)**—Enables the interrupt in the interrupt controller. Do not call this function unless ISR_Start( ) has been called or the functionality of the ISR_Start( ) function, which sets the vector and the priority, has been called.
- **uint8 ISR_GetState(void)**—Gets the state (enabled, disabled) of the interrupt.
- **void ISR_Disable(void)**—Disables the interrupt in the interrupt controller.
- **void ISR_SetPending(void)**—Causes the interrupt to enter the pending state; a software API to generate the interrupt.
- **void ISR_ClearPending(void)**—Clears a pending interrupt in the interrupt controller. Note Some interrupt sources are clear-on-read and require the block interrupt/status register to be read/cleared with the appropriate block API (GPIO, UART [102] and so on). Otherwise, the ISR will to remain in a pending state even though the interrupt itself is cleared using this API.

15.107 Real Time Clock (RTC) 2.00

The Real-Time Clock (RTC) component provides accurate time and date information for the system. The time and date are updated every second based on a one pulse per second interrupt from a 32.768-kHz crystal. Clock accuracy is based on the crystal provided and is typically 20 ppm. The RTC keeps track of the second, minute, hour, day of the week, day of the month, day of the year, month and year. The day of the week is automatically calculated from the day, month and year. Daylight savings time may be optionally enabled and supports any start and end date, as well as a programmable saving time. The start and end dates may be absolute like 24 March or relative like the second Sunday in May. The Alarm provides match detection for a second, minute, hour, day of week, day of month, day of year, month and year. A mask selects what combination of time and date information will be used to generate the alarm. The alarm flexibility supports periodic alarms such as every twenty-third minute after the hour, or a single alarm such as 4:52 a.m. on September 14, 1941. User code stubs are provided for periodic code execution based on each of the primary time intervals. Timer intervals are provided at one second, one minute, one hour, one day, one week, one month and one year. Features include: multiple alarm options, multiple overflow options and daylight Savings Time (DST) option [73].

15.107.1 Real Time Clock (RTC) Functions 2.00

- **RTC_WriteAlarmDayOfYear( )**—Writes Alarm DayOfYear software register value
- **RTC_ReadSecond( )**—Reads Sec software register value
- **RTC_ReadMinute( )**—Reads Min software register value
- **RTC_ReadHour( )**—Reads Min software register value
- **RTC_ReadDayOfMonth( )**—Reads DayOfMonth software register value
- **RTC_ReadMonth( )**—Reads Month software register value
- **RTC_ReadYear( )**—Reads Year software register value
- **RTC_ReadAlarmSecond( )**—Reads Alarm Sec software register value
- **RTC_ReadAlarmMinute( )**—Reads Alarm Min software register value
- **RTC_ReadAlarmHour( )**—Reads Alarm Hour software register value
- **RTC_ReadAlarmDayOfMonth( )**—Reads Alarm DayOfMonth software register value
- **RTC_ReadAlarmMonth( )**—Reads Alarm Month software register value
- **RTC_ReadAlarmYear( )**—Reads Alarm Year software register value
- **RTC_ReadAlarmDayOfWeek( )**—Reads Alarm DayOfWeek software register value
- **RTC_ReadAlarmDayOfYear( )**—Reads Alarm DayOfYear software register value
- **RTC_WriteAlarmMask( )**—Writes the Alarm Mask software register with one bit per time/date entry
- **RTC_WriteIntervalMask( )**—Configures what interval handlers will be called from the RTC ISR
- **RTC_ReadStatus( )**—Reads the Status software register, which has flags for DST (DST), Leap Year (LY), AM/PM (AM_PM) and Alarm active (AA)
- **RTC_WriteDSTMode( )**—Writes the DST Mode software register
- **RTC_WriteDSTStartHour( )**—Writes the DST Start Hour software register
- **RTC_WriteDSTStartDayOfMonth( )**—Writes the DST Start DayOfMonth software register
- **RTC_WriteDSTStartMonth( )**—Writes the DST Start Month software register
- **RTC_WriteDSTStartDayOfWeek( )**—Writes the DST Start DayOfWeek software register
- **RTC_WriteDSTStartWeek( )**—Writes the DST Start Week software register
- **RTC_WriteDSTStopHour( )**—Writes the DST Stop Hour software register
- **RTC_WriteDSTStopDayOfMonth( )**—Writes the DST Stop DayOfMonth software register
- **RTC_WriteDSTStopMonth( )**—Writes the DST Stop Month software register
- **RTC_WriteDSTStopDayOfWeek( )**—Writes the DST Stop DayOfWeek software register
- **RTC_WriteDSTStopWeek( )**—Writes the DST Stop Week software register
- **RTC_WriteDSTOffset( )**—Writes the DST Offset register
- **RTC_Init( )**—Initializes and restores default configuration provided with the customizer
- **RTC_Enable( )**—Enables the interrupts, one pulse per second and interrupt generation on OPPS event

15.107.2 Real Time Clock (RTC) Functions 2.00

- **void RTC_Start(void)**—Enables the RTC component. This function configures the counter, sets up interrupts, does all required calculation and starts the counter.
- **void RTC_Stop(void)**—Stops RTC component operation.
- **void RTC_EnableInt(void)**—Enables interrupts from the RTC component.
- **void RTC_DisableInt(void)**—Disables interrupts from the RTC component. Time and date stop running.
- **RTC_TIME_DATE* RTC_ReadTime(void)**—Reads the current time and date.
- **void RTC_WriteTime(const RTC_TIME_DATE * timeDate)**—Writes the time and date values as current time and date. Only passes the Second, Minute, Hour, Month, Day of Month and Year.
- **void RTC_WriteSecond(uint8 second)**—Writes the Sec software register value.
- **void RTC_WriteMinute(uint8 minute)**—Writes the Min software register value.
- **void RTC_WriteHour(uint8 hour)**—Writes the Hour software register value.
- **RTC_TIME_DATE* RTC_ReadTime(void)**—Reads the current time and date.
- **void RTC_WriteTime(const RTC_TIME_DATE * timeDate)**—Writes the time and date values as current time and date. Only passes the Second, Minute, Hour, Month, Day of Month and Year.
- **void RTC_WriteSecond(uint8 second)**—Writes the Sec software register value.
- **void RTC_WriteMinute(uint8 minute)**—Writes the Min software register value.
- **void RTC_WriteHour(uint8 hour)**—Writes the Hour software register value.
- **void RTC_WriteAlarmHour(uint8 hour)**—Writes the Alarm Hour software register value.
- **void RTC_WriteAlarmDayOfMonth(uint8 dayOfMonth)**—Writes the Alarm DayOfMonth software register value.
- **void RTC_WriteAlarmMonth(uint8 month)**—Writes the Alarm Month software register value.
- **void RTC_WriteAlarmYear(uint16 year)**—Writes the Alarm Year software register value.
- **void RTC_WriteAlarmDayOfWeek(uint8 dayOfWeek)**—Writes the Alarm DayOfWeek software register value.
- **void RTC_WriteAlarmDayOfYear(uint16 goodyear)**—Writes the Alarm DayOfYear software register value.
- **uint8 RTC_ReadSecond(void)**—Reads the Sec software register value.
- **uint8 RTC_ReadMinute(void)**—Reads the Min software register value.
- **uint8 RTC_ReadHour(void)**—Reads the Min software register value.
- **uint8 RTC_ReadDayOfMonth(void)**—Reads the DayOfMonth software register value.
- **uint8 RTC_ReadMonth(void)**—This function reads the Month software register value.
- **uint16 RTC_ReadYear(void)**—Reads the Year software register value.
- **uint8 RTC_ReadAlarmSecond(void)**—Reads the Alarm Sec software register value.
- **uint8 RTC_ReadAlarmMinute(void)**—Reads the Alarm Min software register value.
- **uint8 RTC_ReadAlarmHour(void)**—Reads the Alarm Hour software register value.
- **uint8 RTC_ReadAlarmDayOfMonth(void)**—Reads the Alarm DayOfMonth software register value.
- **uint8 RTC_ReadAlarmMonth(void)**—Reads the Alarm Month software register value.
- **uint16 RTC_ReadAlarmYear(void)**—Reads the Alarm Year software register value.
- **uint8 RTC_ReadAlarmDayOfWeek(void)**—Reads the Alarm DayOfWeek software register value.
- **uint16 RTC_ReadAlarmDayOfYear(void)**—Reads the Alarm DayOfYear software register value.
- **void RTC_WriteAlarmMask(uint8 mask)**—Writes the Alarm Mask software register with one bit per time/date entry. Alarm true when all masked time/date values match Alarm values.

- **void RTC_WriteIntervalMask(uint8 mask)**—Configures what interval handlers will be called from the RTC ISR. See the Interrupt Service Routines section for information about how to use this functionality.
- **uint8 RTC_ReadStatus(void)**—Reads the Status software register, which has flags for DST (DST), Leap Year (LY), AM/PM (AM_PM) and Alarm active (AA).
- **void RTC_WriteDSTMode(uint8 mode)**—Writes the DST Mode software register That enables or disables DST changes and sets the date mo
- **void RTC_WriteDSTStartHour(uint8 hour)**—Writes the DST Start Hour software register. Used for absolute date entry. Only generated if DST is enabled.
- **void RTC_WriteDSTStartDayOfMonth(uint8 dayOfMonth)**—Writes the DST Start DayOf-Month software register. Used for absolute date entry. Only generated if DST is enabled.
- **void RTC_WriteDSTStartMonth(uint8 month)**—Writes the DST Start Month software register. Used for absolute date entry. Only generated if DST is enabled.
- **void RTC_WriteDSTStartDayOfWeek(uint8 dayOfWeek)**—Writes the DST Start DayOfWeek software register. Used for relative date entry. Only generated if DST is enabled.
- **void RTC_WriteDSTStartWeek(uint8 week)**—Writes the DST Start Week software register. Used for relative date entry. Only generated if DST is enabled.
- **void RTC_WriteDSTStopHour(uint8 hour)**—Writes the DST Stop Hour software register. Used for absolute date entry. Only generated if DST is enabled.
- **void RTC_WriteDSTStopDayOfMonth(uint8 dayOfMonth)**—Writes the DST Stop DayOf-Month software register. Used for absolute date entry. Only generated if DST is enabled.
- **void RTC_WriteDSTStopMonth(uint8 month)**—Writes the DST Stop Month software register. Used for absolute date entry. Only generated if DST is enabled.
- **void RTC_WriteDSTStopDayOfWeek(uint8 dayOfWeek)**—Writes the DST Stop DayOfWeek software register. Used for relative date entry. Only generated if DST is enabled.
- **void RTC_WriteDSTStopWeek(uint8 week)**—Writes the DST Stop Week software register. Used for relative date entry. Only generated if DST is enabled.
- **void RTC_WriteDSTOffset(uint8 offset)**—Writes the DST Offset register. Allows a configurable increment or decrement of time between 0 and 255 min. Increment occurs on DST start and decrement on DST stop. Only generated if DST is enabled.
- **void RTC_Init (void)**—Initializes or restores the component according to the customizer Configure dialog settings. It is not necessary to call RTC_Init( ) because the RTC_Start( ) API calls this function and is the preferred method to begin component operation.
- **void RTC_Enable(void)**—Enables the interrupts, one pulse per second and interrupt generation on OPPS event.

15.108 SleepTimer 3.20

The Sleep Timer component can be used to wake the device from Alternate Active and Sleep modes at a configurable interval. It can also be configured to issue an interrupt at a configurable interval. Features include: wakes up devices from low-power modes: Alternate Active and Sleep, contains configurable option for issuing interrupt, generates periodic interrupts while the device is in Active mode and supports twelve discrete intervals: 2, 4, 8, 16, 32, 64, 128, 256, 512, 1024, 2048 and 4096 ms [84].

15.108.1 SleepTimer Function Calls 3.20

- **SleepTimer_Start( )**—Starts Sleep Timer operation.
- **SleepTimer_Stop( )**—Stops Sleep Timer operation.
- **SleepTimer_EnableInt ( )**—Enables the Sleep Timer component to issue an interrupt on wakeup.
- **SleepTimer_DisableInt( )**—Disables the Sleep Timer component to issue an interrupt on wakeup.
- **SleepTimer_Subinterval( )**—Sets the interval for the Sleep Timer to wake up.
- **SleepTimer_Gestates( )**—Returns the value of the Power Manager Interrupt Status Register and clears all bits in this register.
- **SleepTimer_Init( )**—Initializes and restores the default configuration provided with the customizer.
- **SleepTimer_Enable( )**—Enables the 1-kHz ILO and the CTW counter.
- **SleepTimer_initVar**—Indicates whether the Sleep Timer has been initialized. The variable is initialized to 0 and set to 1 the first time SleepTimer_Start( ) is called. This allows the component to restart without reinitialization after the first call to the SleepTimer_Start( ) routine.—If reinitialization of the component is required, then the SleepTimer_Init( ) function can be called before the SleepTimer_Start( ) or SleepTimer_Enable( ) function.

15.108.2 Sleep Timer Function Calls 3.20

- **void SleepTimer_Start(void)**—This is the preferred method to begin component operation. SleepTimer_Start( ) sets the initVar variable, calls the SleepTimer_Init( ) function and then calls the SleepTimer_Enable( ) function. Enables the 1-kHz ILO clock and leaves it enabled after the Sleep Timer component is stopped.
- **void SleepTimer_Stop(void)**—Stops Sleep Timer operation and disables wakeup and interrupt. The device does not wake up when the CTW counter reaches terminal count, nor is an interrupt issued.
- **void SleepTimer_EnableInt(void)**—Enables the CTW terminal count interrupt.
- **void SleepTimer_DisableInt(void)**—Disables the CTW terminal count interrupt.
- **void SleepTimer_Subinterval(uint8 interval)**—Sets the CTW interval period. The first interval can range from 1 to (period + 1) milliseconds. Additional intervals occur at the nominal period. You can only change the interval value when CTW is disabled, which you can do by stopping the component.
- **uint8 SleepTimer_Gestates(void)**—Returns the state of the Sleep Timer's status register and clears the pending interrupt status bit. The application code must always call this function after wakeup to clear the ctw_int status bit. The code must call this function whether the Sleep Timer's interrupt is disabled or enabled.
- **void SleepTimer_Init(void)**—Initializes or restores the component according to the customizer Configure dialog settings. It is not necessary to call SleepTimer_Init( ) because the SleepTimer_Start( ) API calls this function and is the preferred method to begin component operation. Sets CTW interval period and enables or disables CTW interrupt (according to the customizer's settings).
- **void SleepTimer_Enable(void)**—Activates the 1-kHz ILO and the CTW and begins component operation. It is not necessary to call SleepTimer_Enable( ) because the SleepTimer_Start( ) API calls this function, which is the preferred method to begin component operation.

15.109 Fan Controller 4.10

The Fan Controller Component is a system-level solution that encapsulates all necessary hardware blocks including PWMs, or TCPWMs for PSoC 4, tachometer input capture timer, control registers, status registers and a DMA controller reducing development time and effort. The Component is customizable through a graphical user interface enabling designers to enter fan electromechanical parameters such as duty cycle-to-RPM mapping and physical fan bank organization. Performance parameters including PWM frequency and resolution as well as open or closed loop control methodology can be configured through the same user interface. Once the system parameters are entered, the Component delivers the most optimal implementation saving resources within PSoC to enable integration of other thermal management and system management functionality. Easy-to-use APIs are provided to enable firmware developers to get up and running quickly.[70] Supports up to 16 PWM-controlled, 4-wire brushless DC fans for PSoC 3/PSoC 5LP devices and up to 6 fans for PSoC 4, 25 kHz, 50 kHz or user-specified PWM frequencies, fan speeds up to 25,000 RPM, 4-pole and 6-pole motors, fan stall / rotor lock detection on all fans, firmware controlled or hardware controlled fan speed regulation for PSoC 3/PSoC 5LP, firmware-controlled fan speed regulation for PSoC 4, individual or banked PWM outputs with tachometer inputs and customizable alert pin for fan fault reporting [42].

15.109.1 Fan Controller Functions 4.10

- **FanController_Start( )**—Start the Component
- **FanController_Stop( )**—Stop the Component and disable hardware blocks
- **FanController_Init( )**—Initializes the Component
- **FanController_Enable( )**—Enables hardware blocks inside the Component
- **FanController_EnableAlert( )**—Enables alerts from the Component
- **FanController_DisableAlert( )**—Disables alerts from the Component
- **FanController_SetAlertMode( )**—Configures alert sources
- **FanController_GetAlertMode( )**—Returns currently enabled alert sources
- **FanController_SetAlertMask( )**—Enables masking of alerts from each fan
- **FanController_GetAlertMask( )**—Returns alert masking status of each fan
- **FanController_GetAlertSource( )**—Returns pending alert source(s)
- **FanController_GetFanStallStatus( )**—Returns a bit mask representing the stall status of each fan
- **FanController_GetFanSpeedStatus( )**—Returns a bit mask representing the speed regulation status of each fan in hardware control mode
- **FanController_SetDutyCycle( )**—Sets the PWM duty cycle for the specified fan or fan bank
- **FanController_GetDutyCycle( )**—Returns the PWM duty cycle for the specified fan or fan bank
- **FanController_SetDesiredSpeed( )**—Sets the desired fans speed for the specified fan in hardware control mode
- **FanController_GetDesiredSpeed( )**—Returns the desired fans speed for the specified fan in hardware control mode
- **FanController_GetActualSpeed( )**—Returns the actual speed for the specified fan

---

[70]Designs using the Fan Controller Component should not use PSoC's low power sleep or hibernate modes. Entering those modes prevents the Fan Controller Component from controlling and monitoring the fans.

- **FanController_OverrideAutomaticControl( )**—Enables firmware to override Automatic fan control
- **FanController_SetSaturation( )**—Changes the PID controller output saturation.
- **FanController_SetPID( )**—Changes the PID controller coefficients for the controlled fan.

15.109.2 Fan Controller Function Calls 4.10

- **FanController_SetPID( )**—Changes the PID controller coefficients for the controlled fan.
- **void FanController_Start(void)**—Enables the Component. Calls the Init( ) API if the Component has not been initialized before. Calls the Enable( ) API.
- **void FanController_Stop(void)**—Disables the Component. All PWM outputs will be driven to 100% duty cycle to ensure cooling continues while the Component is not operational. Note Due to PSoC 4 resource limitations, the PWM outputs will be cleared to low.
- **void FanController_Init(void)**—Initializes the Component.
- **void FanController_Enable(void)**—Enables hardware blocks within the Component.
- **void FanController_EnableAlert(void)**—Enables generation of the alert signal. Specifically which alert sources are enabled is configured using the FanController_SetAlertMode( ) and the FanController_SetAlertMask( ) APIs.
- **void FanController_DisableAlert(void)**—Disables generation of the alert signal.
- **void FanController_SetAlertMode(uint8 alertMode)**—Configures alert sources from the Component. Two alert sources are available: (1) Fan Stall or Rotor Lock, (2) Hardware control mode speed regulation failure.
- **uint8 FanController_GetAlertMode(void)**—Returns the alert sources that are enabled.
- **void FanController_SetAlertMask(uint16 alertMask)**—Enables or disables alerts from each fan through a mask. Masking applies to both fan stallalerts and speed regulation failure alerts.
- **uint16 FanController_GetAlertMask(void)**—Returns alert mask status from each fan. Masking applies to both fan stall alerts and speed regulation failure alerts.
- **uint8 FanController_GetAlertSource(void)**—Returns pending alert sources from the Component. This API can be used to poll the alert status of the Component. Alternatively, if the alert pin is used to generate interrupts to PSoC's CPU core, the interrupt service routine can use this API to determine the source of the alert. In either case, when this API returns a non-zero value, the FanController_GetFanStallStatus( ) and FanController_GetFanSpeedStatus( ) APIs can provide further information on which fan(s) has(have) a fault.
- **uint16 FanController_GetFanStallStatus(void)**—Returns the stall / rotor lock status of all fans.
- **uint16 FanController_GetFanSpeedStatus(void)**—Returns the hardware fan control mode speed regulation status of all fans. Speed regulation failures occur in two cases: (1) if the desired fan speed exceeds the current actual fan speed but the fan's duty cycle is already at 100%, (2) if the desired fan speed is below the current actual fan speed, but the fan's duty cycle is already at 0%.
- **void FanController_SetDutyCycle(uint8 fanOrBankNumber, uint16 dutyCycle)**—Sets the PWM duty cycle of the selected fan or fan bank in hundredths of a percent. In hardware fan control mode, if manual duty cycle control is desirable, call the FanController_OverrideHardwareControl( ) API prior to calling this API. Note Due to PSoC 4 resource limitations, setting 100% duty cycle will not generate a continuous high signal on respective "fan" output.
- **uint16 FanController_GetDutyCycle(uint8 fanOrBankNumber)**—Returns the current PWM duty cycle of the selected fan or fan bank in hundredths of a percent.

- **void FanController_SetDesiredSpeed(uint8 fanNumber, uint16 rpm)**—Sets the desired speed of the specified fan in revolutions per minute (RPM). In hardware fan control mode, the RPM parameter is passed to the control loop hardware as the new target fan speed for regulation. In firmware fan control mode, the RPM parameter is converted to a duty cycle based on the fan parameters entered into the Fans tab of the customizer and written to the appropriate PWM. This provides firmware with a method for initiating coarse level speed control. Fine level firmware speed control can then be achieved using the FanController_SetDutyCycle( ) API.
- **uint16 FanController_GetDesiredSpeed(uint8 fanNumber)**—Returns the currently desired speed for the selected fan.
- **uint16 FanController_GetActualSpeed(uint8 fanNumber)**—Returns the current actual speed for the selected fan. This function should be called the first time in the design only after the requested fan has made a full rotation. This can be ensured by calling the function after the end-of-cycle (eoc) pulse generation.
- **void FanController_OverrideAutomaticControl(uint8 override)**—Allows firmware to take over fan control in hardware fan control mode. Note that this API cannot be called in firmware fan control mode.
- **void FanController_SetSaturation(uint8 fanNum, uint16 satH, uint16 satL)**—Changes the PID controller output saturation. This bounds the output PWM to the fan and prevents what is known as integrator windup.
- **void FanController_SetPID (uint8 fanNum, uint16 kp, uint16 ki, uint16 kd)**—Changes the PID controller coefficients for the controlled fan. The coefficients are integers that are proportional to the gain.

15.110 RTD Calculator 1.20

The Resistance Temperature Detector (RTD) Calculator component generates a polynomial approximation for calculating the RTD Temperature in terms of RTD resistance for a PT100, PT500 or PT1000 RTD. Calculation error budget is user-selectable and determines the order of the polynomial that will be used for the calculation (from 1 to 5). A lower calculation error budget will result in a more computation intensive calculation. For example, a fifth order polynomial will give a more accurate temperature calculation than lower order polynomials, but will take more time for execution. After maximum and minimum temperatures and error budget are selected, the component generates the maximum temperature error and an error vs. temperature graph for all temperatures in the range, along with an estimate of the number of CPU cycles necessary for calculation using the selected polynomial. Selecting the lowest error budget will choose the highest degree polynomial. For the whole RTD temperature range, $-200\,°C$ to $850\,°C$, the component can provide a maximum error of $<0.01\,°C$ using a fifth order polynomial. Features include: calculation accuracy is $0.01\,°C$ for $-200\,°C$ to $850\,°C$ temperature range, a simple API function for resistance to temperature conversion and a graph of error vs temperature [74].

15.110.1 RTD Calculator Functions 1.20

- **int32 RTD_GetTemperature(uint32 res)**—Calculates the temperature from RTD resistance

15.110.2 RTD Calculator Function Calls 1.20

- **int32 RTD_GetTemperature(uint32 res)**—Calculates the temperature from RTD resistance.

15.111 Thermistor Calculator 1.20

The Thermistor Calculator calculates the temperature based on a provided voltage measured from a thermistor. The component is adaptable to most NTC thermistors. It calculates the Steinhart-Hart equation coefficients based on the temperature range and corresponding user-provided reference resistances. The API functions that use the generated coefficients to return the temperature value are based on measured voltage values. This component does not use an ADC or AMUX inside and thus requires those components to be placed separately in your projects. Features include: the component is adaptable for the majority of negative temperature coefficient (NTC) thermistors, Look-Up-Table (LUT) or equation implementation methods, selectable reference resistor, based on thermistor value, selectable temperature range and selectable calculation resolution for LUT method [93].

15.111.1 Thermistor Calculator Functions 1.20

- **uint32 Thermistor_GetResistance(int16 vReference, int16 vThermistor)**—The digital values of the voltages across the reference resistor and the thermistor are passed to this function as parameters. These can be considered as the inputs to the component. The function returns (outputs) the resistance, based on the voltage values.
- **int16 Thermistor_GetTemperature(uint32 resT)**—The value of the thermistor resistance is passed to this function as a parameter. The function returns (outputs) the temperature, based on the resistance value. The method used to calculate the temperature is dependent on whether Equation or LUT was selected.

15.111.2 Thermistor Calculator Function Calls 1.20

- **uint32 Thermistor_GetResistance(int16 vReference, int16 vThermistor)**—The digital values of the voltages across the reference resistor and the thermistor are passed to this function as parameters. These can be considered as the inputs to the component. The function returns (outputs) the resistance, based on the voltage values.
- **Thermistor_GetTemperature (uint32 resT)**—The value of the thermistor resistance is passed to this function as a parameter. The function returns (outputs) the temperature, based on the resistance value. The method used to calculate the temperature is dependent on whether Equation or LUT was selected.

15.112 Thermocouple Calculator 1.20

In thermocouple temperature measurement, the thermocouple temperature is calculated based on the measured thermo-emf voltage. The voltage to temperature conversion is characterized by the National Institute of Standards and Technology (NIST) and NIST provides tables and polynomial coefficients for thermo-emf to temperature conversion.[71] Thermocouple temperature measurement also involves measuring the thermocouple reference junction temperature and converting it into a voltage. The Thermocouple Calculator component simplifies the thermocouple temperature measurement process by providing APIs for thermo-emf to temperature conversion and vice versa for all thermocouple types mentioned above, using polynomials generated at compile time. The thermocouple component evaluates the polynomial efficiently to reduce computation time. Features include: Support for B, E, J, K, N, R, S and T Type Thermocouples, functions for thermo-emf to temperature and temperature to voltage conversions and displays a graph of Calculation Error Vs. Temperature [94].

15.112.1 Thermocouple Calculator Functions 1.20

- **int32   Thermocouple_GetTemperature(int32   voltage)**—Calculates   the   temperature   from thermo-emf in μV
- **int32 Thermocouple_GetVoltage(int32 temperature)**—Calculates the voltage given the temperature in 1/100ths degrees C. Used for calculating the cold junction compensation voltage based on the temperature at the cold junction.

15.112.2 Thermocouple Calculator Function Calls 1.20

- **int32   Thermocouple_GetTemperature(int32   voltage)**—Calculates   the   temperature   from thermo-emf in μV
- **int32 Thermocouple_GetVoltage(int32 temperature)**—Calculates the voltage given the temperature in 1/100ths degrees C. Used for calculating the cold junction compensation voltage based on the temperature at the cold junction.
- **int32   Thermocouple_GetTemperature(int32   voltage)**—Calculates   the   temperature   from thermo-emf in μV.
- **int32 Thermocouple_GetVoltage(int32 temperature)**—Calculates the voltage given the temperature in 1/100ths degrees C. Used for calculating the cold junction compensation voltage based on the temperature at the cold junction.

15.113 TMP05 Temp Sensor Interface 1.10

The TMP05 Temp Sensor Interface component is capable of interfacing with Analog Device's TMP05/06 digital temperature sensors in daisy chain mode only and can be configured to monitor the temperature readings in one of two ways [96]:

1. Continuously record temperatures, at a sample rate dictated by the temperature sensor(s)
2. One-shot mode triggers the temperature measurement at a rate you can control.

The first mode is intended for use in an environment where temperature variations are abrupt and need to be monitored frequently. The second option should be used when temperature measurements only need to be sampled once in a while or in applications where minimizing power consumption is

---

[71] The NIST tables and polynomial coefficients can be found in the following link: http://srdata.nist.gov/its90/download/download.html.

important. Since the component supports digital temperature sensors in daisy chain mode only, the device must be configured in daisy chain mode even when only a single device is connected. This component supports up to four TMP05 or TMP06 digital temperature sensors connected in daisy chain mode only, continuous and one-shot modes of operation, frequencies from 100 to 500 kHz and temperature ranging from $0°$ to $70°$, Celsius.

15.113.1 TMP05 Temp Sensor Interface Functions 1.10

- **TMP05_Start( )**—Starts the component
- **TMP05_Stop( )**—Stops the component
- **TMP05_Init( )**—Initializes the component
- **TMP05_Enable( )**—Enables the component
- **TMP05_Trigger( )**—Triggers the interfaced TMP05 sensors to start temperature measurement depending upon the mode of operation
- **TMP05_GetTemperature( )**—Calculates the temperature(s) in degrees Celsius
- **TMP05_SetMode( )**—Sets the operating mode of the component
- **TMP05_DiscoverSensors( )**—Automatically detects how many temperature sensors are daisy-chained to the component
- **TMP05_ConversionStatus( )**—Returns current state of temperature conversion (busy, completed or error)
- **TMP05_SaveConfig( )**—Saves the current state of the component before entering low power mode
- **TMP05_RestoreConfig( )**—Restores the previous state of the component after waking from low power mode
- **TMP05_Sleep( )**—Puts the component into low power mode
- **TMP05_Wakeup( )**—Wakes the component up from low power mode

15.113.2 TMP05 Temp Sensor Interface Function Calls 1.10

- **void TMP05_Start(void)**—Starts the component. Calls the TMP05_Init( ) API if the component has not been initialized before. Calls the enable API.
- **void TMP05_Stop (void)**—Disables and stops the component
- **void TMP05_Init(void)**—Initializes the component
- **void TMP05_Enable(void)**—Enables the component
- **void TMP05_Trigger (void)**—Provides a valid strobe/trigger output on the conv terminal.
- **int16 TMP05_GetTemperature (uint8 SensorNum)**—Calculates the temperature in degrees Celsius
- **void TMP05_SetMode (uint8 mode)**—Sets the operating mode of the component
- **uint8 TMP05_DiscoverSensors (void)**—This API is provided for applications that might have a variable number of temperature sensors connected. It automatically detects how many temperature sensors are daisy-chained to the component. The algorithm starts by checking to see if the number of sensors actually connected matches the NumSensors parameter setting in the Basic Tab of the component customizer. If not, it will retry assuming 1 less sensor is connected. This process will repeat until the actual number of sensors connected is known. Confirming whether or not a sensor is attached or not takes a few hundred milliseconds per sensor per iteration of the algorithm. To limit the sensing time, reduce the NumSensors parameter setting in the Basic Tab of the component customizer to the maximum number of possible sensors in the system.

- **uint8 TMP05_ConversionStatus (void)**—Enables firmware to synchronize with the hardware.
- **void TMP05_SaveConfig (void)**—Saves the user configuration of the TMP05 non-retention registers. This routine is called by TMP05_Sleep( ) to save the component configuration before entering sleep.
- **void TMP05_RestoreConfig (void)**—Restores the user configuration of the TMP05 non-retention registers. This routine is called by TMP05_Wakeup( ) to restore the component configuration when exiting sleep.
- **void TMP05_Sleep (void)**—Stops the TMP05 operation and saves the user configuration along with the enable state of the TMP05.
- **void TMP05_Wakeup (void)**—Restores the user configuration and restores the enable state.

15.114 LIN Slave 4.00

The LIN Slave Component[72] implements a LIN 2.2 slave node on PSoC 3, PSoC 4 and PSoC 5LP devices. Options for LIN 2.0, LIN 1.3 or SAE J2602-1 compliance are also available.[73] This Component consists of the hardware blocks necessary to communicate on the LIN bus and an API to allow the application code to easily interact with the LIN bus communication. The Component provides an API that conforms to the API specified by the LIN 2.2 Specification. This Component provides a good combination of flexibility and ease of use. A customizer for the Component is provided that allows you to easily configure all parameters of the LIN Slave. For PSoC 4 devices only, the LIN Slave Component is certified by the C&S group GmbH based on the standard protocol and data link layer conformance tests. A complete certification report can be made available on request. For PSoC 3 and PSoC 5LP devices, the LIN Slave Component is a prototype Component, because it is not certified for these devices. Features include: Full LIN 2.2, 2.1, or 2.0, Slave Node implementation, supports compliance with the LIN 1.3 specification and partial compliance with SAE J2602-1 specification, automatic baud rate synchronization, fully implements a Diagnostic Class I Slave Node, full transport layer support, automatic detection of bus inactivity, full error detection, automatic configuration services, customizer for fast and easy configuration, import of *.ncf/*.ldf files and *.ncf file export and an editor for *.ncf/*.ldf files with syntax checking [58].

15.114.1 LIN Slave Functions 4.00

- **l_bool_rd( )**—Reads and returns the current value of the signal for one-bit signals.
- **l_u8_rd( )**—Reads and returns the current value of the signal for signals of two to eight bits.
- **l_u16_rd( )**—Reads and returns the current value of the signal for signals of 9 to 16 bits.
- **l_bytes_rd( )**—Reads and returns the current values of the selected bytes in the signal.
- **l_bool_wr( )**—Sets the current value of the signal for one-bit signals to v.
- **l_u8_wr( )**—Sets the current value of the signal for signals of two to eight bits.
- **l_u16_wr( )**—Sets the current value of the signal for signals of 9 to 16 bits.
- **l_bytes_wr( )**—Sets the current values of the selected bytes in the signal.
- **l_u16_wr( )**—Sets the current value of the signal for signals of 9 to 16 bits.
- **l_bytes_wr( )**—Sets the current values of the selected bytes in the signal.

---

[72]This component is not supported in PSOC 3 and PSoC 5LP but is supported in PSoC 4.
[73]The J2602 protocol is not fully supported; SAE J2602-2 and SAE J2602-3 are not fully supported by the component.

15.114.2 LIN Slave Function Calls 4.00

- **l_bool_rd( )**—Reads and returns the current value of the signal for one-bit signals. If an invalid signal handle is passed into the function, no action is taken, function returns 0x00.
- **l_u8_rd( )**—Reads and returns the current value of the signal. If an invalid signal handle is passed into the function, no action is taken, function returns 0x00.
- **l_u16_rd( )**—Reads and returns the current value of the signal. If an invalid signal handle is passed into the function, no action is taken, function returns 0x00.
- **l_bytes_rd( )**—Reads and returns the current values of the selected bytes in the signal. The sum of the start and count parameters must never be greater than the length of the byte array. Note hat when the sum of start and count is greater than the length of the signal byte array, an accidental data is read. If an invalid signal handle is passed into the function, no action is taken. Assume that a byte array is 8 bytes long, numbered 0 to 7. Reading bytes from 2 to 6 from a user-selected array requires start to be 2 (skipping byte 0 and 1) and count to be 5. In this case, byte 2 is written to user_selected_array[0] and all consecutive bytes are written into user_selected_array in ascending order.
- **l_u8_wr( )**—Writes the value v to the signal. If an invalid signal handle is passed into the function, no action is taken.
- **l_bool_wr( )**—Writes the value v to the signal. If an invalid signal handle is passed into the function, no action is taken.
- **l_u16_wr( )**—Writes the value v to the signal. If an invalid signal handle is passed into the function, no action is taken.
- **l_bytes_wr( )**—Writes the current value of the selected bytes to the signal specified by the name sss. The sum of start and count must never be greater than the length of the byte array, although the device driver may choose not to enforce this in run time. Note that when the sum of start and count is greater than the length of the signal byte array an accidental memory area is affected. If an invalid signal handle is passed into the function, no action is taken. Assume that a byte array signal is 8 bytes long, numbered 0 to 7. Writing byte 3 and 4 of this array requires start to be 3 (skipping bytes 0, 1 and 2) and count to be 2. In this case, byte 3 of the byte array signal is written from user_selected_array[0] and byte 4 is written from user_selected_array[1].

References

1. 8-Bit Current Digital to Analog Converter (IDAC8) 2.0, Document Number: 001-84984 Rev. *D. Cypress Semiconductor (2017)
2. 8-Bit Voltage Digital to Analog Converter (VDAC8) 1.90, Document Number: 001-84983 Rev *D. Cypress Semiconductor (2017)
3. 8-Bit Waveform Generator (WaveDAC8) 2.10, Document Number: 002-10431 Rev. *A. Cypress Semiconductor (2017)
4. ADC Successive Approximation Register (ADC_SAR) 3.10, Document Number: 002-20501 Rev. **. 3.10. Cypress Semiconductor. Revised July 26, 2017
5. M. Ainsworth, Switch Debouncer and Glitch Filter with PSoC® 3, PSoC 4 and PSoC 5LP. AN60024. Document No. 001-60024 Rev. *P. Cypress Semiconductor (2016)
6. Analog Hardware Multiplexer (AMuxHw) 1.50, Document Number: 001-61006 Rev. *I. Cypress Semiconductor (2018)
7. Analog Multiplexer (AMux) 1.80, Document Number: 001-51245 Rev. *F. Cypress Semiconductor (2018)
8. Analog Multiplexer Sequencer (AMuxSeq) 1.80, Document Number: 001-87569 Rev. *F. Cypress Semiconductor (2018)
9. Analog Mux Constraint, Document Number: 001-79340 Rev. *C. Cypress Semiconductor (2017)
10. Analog Net Constraint 1.50, Document Number: 001-79342 Rev. *C. . Cypress Semiconductor (2017)

11. Analog Resource Reserve, Document Number: 001-63056 Rev. *H. Cypress Semiconductor (2017)
12. Basic Counter 1.0, Document Number: 001-84887 Rev. *C. Cypress Semiconductor (2017)
13. Boost Converter (BoostConv) 5.0, Document Number: 001-88522 Rev. *C. Cypress Semiconductor (2017)
14. Bootloader and Bootloadable 1.60, Document Number: 002-21055 Rev. *D. Cypress Semiconductor (2018)
15. Capacitive Sensing (CapSense® CSD) 3.50, Document Number: 001-96759 Rev. *B. Cypress Semiconductor (2017)
16. Character LCD 2.20, Document Number: 002-03681 Rev. *A. Cypress Semiconductor (2017)
17. Character LCD with I2C Interface (I2C LCD) 1.20, Document Number: 001-92579 Rev. *A. Cypress Semiconductor (2017)
18. Clock 2.20, Document Number: 001-90285 Rev. *C. Cypress Semiconductor (2018)
19. Comparator (Comp) 2.0, Document Number: 001-84985 Rev. *C. Cyptress Semiconductor (2017)
20. Control Register 1.80, Document Number: 001-96684 Rev. *B. Cypress Semiconductor (2017)
21. Controller Area Network (CAN) 3.0, Document Number: 001-96130 Rev. *E. Cypress Semiconductor (2017)
22. Counter 3.0, Document Number: 001-96201 Rev. *B. Cypress Semiconductor (2017)
23. Cyclic Redundancy Check (CRC) 2.50, Document Number: 002-20387 Rev. *A. Cypress Semiconductor (2017)
24. D Flip-Flop Enable 1.0, Document Number 001-84897 Rev. *B. Cypress Semiconductor (2017)
25. D Flip Flop 1.30, Document Number: 001-84971 Rev. *B. Cypress Semiconductor (2017)
26. Debouncer 1.0, Document Number: 001-82820 Rev. *C. Cypress Semiconductor (2017)
27. Delta-Sigma Analog to Digital Converter(ADC_DelSig) 3.30, Document Number: 002-22359 Rev. **. Cypress Semiconductor (2017)
28. Die Temperature (DieTemp) 2.10, Document Number: 002-22308 Rev. **. Cypress Semiconductor (2017)
29. Digital Comparator 1.0, Document Number: 001-84891 Rev. *B. Cypress Semiconductor (2017)
30. Digital Constant 1.0, Document Number: 001-84899 Rev. *D. Cypress Semiconductor (2017)
31. Digital Filter Block (DFB) Assembler 1.40, Document Number: 001-90472 Rev. *C. Cypress Semiconductor (2017)
32. Digital Logic Gates 1.0, Document Number: 001-50454 Rev. *F. Cypress Semiconductor (2017)
33. Digital Multiplexer and Demultiplexer 1.10, Document Number: 001-73370 Rev. *C. Cypress Semiconductor (2017)
34. Direct Memory Access (DMA) 1.70, Document Number: 001-84992 Rev. *E. Cypress Semiconductor (2017)
35. Dithered Voltage Digital to Analog Converter (DVDAC) 2.10, Document Number: 001-95076 Rev *A. Cypress Semiconductor (2017)
36. Down Counter 7-bit (Count7) 1.0, Document Number: 001-88468 Rev. *B (2017)
37. Edge Detector 1.0, Document Number: 001-84890 Rev. *A. Cypress Semiconductor (2017)
38. EEPROM 3.0, Document Number: 001-96734 Rev. *A. Cypress Semiconductor (2017)
39. Emulated EEPROM (Em_EEPROM) 2.20, Document Number: 002-24582 Rev. *A. Cypress Semiconductor (2018)
40. External Memory Interface (EMIF) 1.30, Document Number: 001-84998 Rev. *D. Cypress Semiconductor (2017)
41. EZI2C Slave 2.0, Document Number: 001-96746 Rev. *B. Cypress Semiconductor (2017)
42. Fan Controller 4.10, Document Number: 002-19745 Rev. *B. Cypress Semiconductor (2017)
43. File System Library (emFile) 1.20, Document Number: 001-85083 Rev. *H. Cypress Semiconductor (2017)
44. Filter 2.30, Document Number: 001-85031 Rev. *E. Cypress Semiconductor (2017)
45. Frequency Divider 1.0, Document Number: 001-84894 Rev. *D. Cypress Semiconductor (2017)
46. Full Speed USB (USBFS) 3.20, Document Number: 002-19744 Rev. *A. Cypress Semiconductor (2017)
47. Glitch Filter 2.0, Document Number: 001-82876 Rev. *C. Cypress Semiconductor (2017)
48. Global Signal Reference (GSRef) 2.10, Document Number: 002-17915 Rev. *B. Cypress Semiconductor (2017)
49. Graphic LCD Interface (GraphicLCDIntf) 1.80, Document Number: 002-24090 Rev. ** (2018)
50. Graphic LCD Controller (GraphicLCDCtrl) 1.80, Document Number: 002-24089 Rev. **. Cypress Semiconductor (2018)
51. Graphic LCD Interface (GraphicLCDIntf) 1.80, Document Number: 002-24090 Rev. **. Cypress Semiconductor (2018)
52. I2C Master/Multi-Master/Slave 3.50, Document Number: 001-97376 Rev. *C. Cypress Semiconductor (2107)
53. ILO Trim 2.0, Document Number: 001-95019 Rev. *F. Cypress Semiconductor (2017)
54. Inter-IC Sound Bus (I2S) 2.70, Document Number: 002-03587 Rev. *A. Cypress Semiconductor (2017)
55. Interrupt 1.70, Document Number: 001-85137 Rev. *G.. Cypress Semiconductor (2017)
56. Inverting Programmable Gain Amplifier (PGA_Inv) 2.0, Document Number: 001-84662 Rev. *C. Cypress Semiconductor (2017)
57. LED Segment and Matrix Driver (LED_Driver) 1.10, Document Number: 001-90024 Rev. *A. Cypress Semiconductor (2017)
58. LIN 5.0, Document Number: 002-26390 Rev. *A. Cypress Semiconductor (2019)
59. Lookup Table (LUT) 1.60, Document Number: 002-21309 Rev. **. Cypress Semiconductor (2017)

60. MDIO Interface 1.20, Document Number: 002-03587 Rev. *A. Cypress Semiconductor (2017)
61. Mixer 2, Document Number: 001-85078 Rev. *D. Cypress Semiconductor (2017)
62. Net Tie 1.50, Document Number: 001-79344 Rev. *C. Cypress Semiconductor (2017)
63. Operational Amplifier (Opamp) 1.90, Document Number: 001-84986 Rev. *D. Cypress Semiconductor (2017)
64. Pins 2.20, Document Number: 001-98278 Rev. *D. Cypress Semiconductor (2017)
65. Power Monitor 1.60, Document Number: 001-94248 Rev. *C. Cypress Semiconductor. 920170
66. Programmable Gain Amplifier (PGA) 2.0, Document Number: 001-84660 Rev. *D. Cypress Semiconductor (2017)
67. Precision Illumination Signal Modulation (PrISM) 2.20, Document Number: 001-84994 Rev. *C. Cypress Semiconductor (2017)
68. Pseudo Random Sequence (PRS) 2.40, Document Number: 001-88524 Rev. *B. Cypress Semiconductor (2017)
69. Pulse Converter 1.0, Document Number: 001-84896 Rev. *B. Cypress Semiconductor (2017)
70. Pulse Width Modulator (PWM) 3.30, Document Number: 001-97110 Rev. *D (2017)
71. Quadrature Decoder (QuadDec) 3.0, Document Number: 001-96233 Rev. *B (2017)
72. Resistive Touch (ResistiveTouch) 2.0, Document Number: 001-96234 Rev. *B. Cypress Semiconductor (2017)
73. Real-Time Clock (RTC) 2.0, Document Number: 001-88461 Rev. *C. Cypress Semiconductor (2017)
74. RTD Calculator 1.20, Document Number: 001-86910 Rev. *C. Ctpress Semiconductor (2017)
75. Sample/Track and Hold Component 1.40, Document Number: 001-85166 Rev. *D. Cypress Semiconductor (2017)
76. Scanning Comparator (ScanComp) 1.10, Document Number: 001-95018 Rev. *A. Cypress Semiconductor (2017)
77. SC/CT Comparator (SCCT_Comp) 1.0, Document Number: 001-88026 Rev. *C. Cypress Semiconductor (2017)
78. S/PDIF Transmitter (SPDIF_Tx) 1.20, Document Number: 001-85017 Rev. *D. Cypress Semiconductor (2017)
79. Segment LCD (LCD_Seg) 3.40, Document Number: 001-88604 Rev. *F. Cypress Semiconductor (2017)
80. Sequencing Successive Approximation ADC (ADC_SAR_Seq) 2.10, Document Number: 002-20508 Rev. **. Cypress Semiconductor (2017)
81. Serial Peripheral Interface (SPI) Master 2.50, Document Number: 001-96814 Rev. *D. Cypress Semiconductor (2017)
82. Serial Peripheral Interface (SPI) Slave, Document Number: 001-96790 Rev. *C. Cypress Semiconductor (2017)
83. Shift Register (ShiftReg) 2.30, Document Number: 001-87851 Rev. *B. Cypress Semiconductor (2017)
84. SleepTimer 3.20, Document Number: 001-85000 Rev. *C. Cypress Semiconductor. Semiconductor (2017)
85. SMBus and PMBus Slave 5.20, Document Number: 002-10834 Rev. *C. Cypress Semiconductor (2017)
86. Software Transmit UART 1.50, Document Number: 002-03685 Rev. *A. Cypress Semiconductor (2017)
87. SR Flip-Flop 1.0, Document Number: 001-84900 Rev. *B. Cypress Semiconductor (2017)
88. Static Segment LCD (LCD_SegStatic) 2.30, Document Number: 001-88605 Rev. *E. Cypress Semiconductor (2017)
89. Status Register 1.90, Document Number: 001-96683 Rev. *A. Cypress Semiconductor (2017)
90. Stay Awake 1.50, Document Number: 001-63288 Rev. *I. Cypress Semiconductor (2017)
91. Sync 1.0, Document Number: 001-65569 Rev. *E. Cypress Semiconductor (2017)
92. Terminal Reserve, Document Number: 001-63058 Rev. *H. Cypress Semiconductor (2017)
93. Thermistor Calculator 1.20, Document Number: 001-86908 Rev. *D. Cypress Semiconductor (2017)
94. Thermocouple Calculator 1.20, Document Number: 001-86911 Rev. *D. Cypress Semiconductor (2017)
95. Timer 2.80, Document Number: 002-19919 Rev. **. Cypress Semiconductor (2017)
96. TMP05 Temp Sensor Interface 1.10, Document Number: 001-86295 Rev. *. Cypress Semiconductor (2017)
97. Toggle Flip Flop 1.0, Document Number: 001-84903 Rev. *B. Cypress Semiconductor (2017)
98. Trans-Impedance Amplifier (TIA) 2.0, Document Number: 001-84989 Rev. *C. Cypress Semiconductor (2017)
99. Tri-State Buffer (Bufoe) 1.10, Document Number: 001-50451 Rev. *f (2017)
100. Trim and Margin 3.0, Document Number: 002-10607 Rev. *A. Cypress Semiconductor (2017)
101. UDB Clock Enable (UDBClkEn) 1.0, Document Number: 001-65568 Rev. *E. Cypress Semiconductor (2017)
102. Universal Asynchronous Receiver Transmitter (UART 2.50), Document Number: 001-97157 Rev. *D. Cypress Semiconductor (2017)
103. Vector CAN 1.10, Document Number: 001-85031 Rev. *E. Cypress Semiconductor (2017)
104. Virtual Mux 1.0, Document Number: 001-51245 Rev. *F. Cypress Semiconductor (2017)
105. Voltage Fault Detector (VFD) 3.0, Document Number: 001-97794 Rev. *B. Cypress Semiconductor (2017)
106. Voltage Reference (Vref) 1.70, Document Number: 002-10643 Rev. *A. Cypress Semiconductor (2017)
107. Voltage Sequencer 3.40, Document Number: 001-96670 Rev. *C. Cypress Semiconductor (2017)

Further Reading

1. Ainsworth, Mark. PSoC 3 8051 Code Optimization. Application Note: AN60630. Cypress Semiconductor Corporation. (2011)
2. Ainsworth, Mark. PSoC 3 to PSoC 5LP Migration Guide. AN77835. Document No. 001-77835Rev.*D1. Cypress Semiconductor. (2020)
3. Allen, Paul. Idea Man: A Memoir by the Cofounder of Microsoft. Portfolio; Reprint edition. (October 30, 2012)
4. Antoniou, A. Realization of Gyrators Using Operational Amplifiers and Their Use in RC-Active Network Synthesis. Proc. IEEE, 116(11), 1838–1850 (1969)
5. Anu, M D., Rastogi, Tush. PSoC 3/PSOC 5 *I2C* Bootloader. (AN60317). Document No. 001-60317 Rev. *L1. Cypress Semiconductor. (2020)
6. Appleton, E. V. Automatic synchronization of triode oscillators, Proc. Cambridge Phil. Soc., 21(Part III):231 (1922–1923)
7. Ashenden, Peter, J. The VHDL Cookbook. First Edition. http://tams-www.informatik.uni-hamburg.de/vhdl/doc/cookbook/VHDL-Cookbook.pdf (1990)
8. Ashenden, Peter J. The Designers Guide to VHDL. Third Edition. Elsevier, New York. (2008)
9. Ball, Roy and Pratt, Roger. Engineering Applications of Microcomputers Instrumentation and Control. Prentice Hall. (1984)
10. de Bellescize, Henri. "La réception Synchrone", L'Onde Electrique 11: 230–240. (June 1932)
11. Best, Roland E. Phase-Locked Loops, Theory Design, and Applications Fifth edition, McGraw Hill. (2003)
12. Birkner, John M., Chua, Hua-Thye, "Programmable array logic circuit," U. S. Patent 4124899 (Filed May 23, 1977. Issued November 7 (1978)
13. Birkner, John M., PAL Programmable Array Logic Handbook. Santa Clara: Monolithic Memories, (1978)
14. Birkner, John M. PAL Programmable Array Logic Handbook. Santa Clara: Monolithic Memories. (1978)
15. Birkner, John; Coli, Vincent, PAL Programmable Array Logic Handbook (2 ed.), Monolithic Memories, Inc. (1981)
16. Black, Harold S. Stabilized Feedback Amplifiers. Bell System Technical Journal, Vol. 13, No. 1, pp. 1-, January (1934)
17. Black, Harold S. Inventing the Negative Feedback Amplifier. IEEE Spectrum. (December, 1977)
18. Boehm, Barry. A Spiral Model of Software Development and Enhancement. Computer. (1988)
19. Borrie, John A. Modern Control Systems - A Manual of Design Methods. Prentice Hall. (1986)
20. Brennan, Paul V. Phase-locked loops: Principles & Practice. MacMillan. (1996)

© Springer Nature Switzerland AG 2021
E. H. Currie, *Mixed-Signal Embedded Systems Design*,
https://doi.org/10.1007/978-3-030-70312-7

21. Briggs, William L. and Henson, Van Emden. The An Owner's Manual for the Discrete Fourier Transform. SIAM, Society for Industrial and Applied Mathematics. Philadelphia. PA. (1995)

22. Butterworth, S. On the Theory of Filter Amplifiers. Experimental Wireless and the Wireless Engineer. pp 536–541, October (1930)

23. Cady, Walter Guyton. "Piezoelectricity, An Introduction to the Theory and Application of Electromechanical Phenomenon in Crystals. McGraw Hill Book Company, Inc. First Edition. (1946)

24. Cavlan, Napoleone. "Field Programmable logic array circuit" U. S. Patent 4,422,072 (Filed July 30, 1981. (Issued Dec. 20 (1983)

25. Chassaing, Rulph and Reay, Donald. Digital Signal Processing. Wiley Interscience. (2008)

26. P. Ciureanu and S. Middelhoek. 1992. Thin Film Resistive Sensors, New York: Institute of Physics Publishing. (1992)

27. Cline, R. "A Single-Chip Sequential Logic Element," IEEE International Solid Sate Circuits Conference, Digest of Technical Papers, 15–17, pp. 204–205, Feb (1978).

28. Coppens, A.B. Simple equations for the speed of sound in Neptunian waters. J. Acoust. Soc. Am. 69(3), pp 862–863. (1981) (1981).

29. Crawford, James A. Advanced Phase-Lock Techniques, Artech House, Boston, MA. (2008)

30. Crenshaw, Jack. A primer on Karnaugh Maps. Programmers Toolbox. EE Times Design. (2003)

31. E. Denton. Tiny Temperature Sensors for Portable Systems. National Semiconductor (2001)

32. Dijkstra Edsger, W. My recollections of operating system design. Operating Systems Review 39(2): pp 4–40. (2005)

33. Doboli, Alex N., Currie, Edward H. "Introduction to Mix-Signal, Embedded Design". Springer-Verlag. (2010)

34. Donglin, Lui; Xiabo, Hu; Lemmon, Sharon; Michael, D.; and Qiang, Ling. Firm Real-Time System Scheduling Based on a Novel QoS Constraint. IEEE Transactions of Computers, Vol. 55, No. 3, March (2006).

35. Dust, Todd and Reynolds, Greg. AN82156. Designing PSoC Creator Components with UDB Datapaths. www.cypress.com. Document No. 001-82156Rev. *I. Cypress Semiconductor. (2018)

36. Dust, Todd. PSoC 3 / PSoC 5LP - Temperature Measurement With Thermocouples. AN75511. Document No. 001-75511Rev.*F17. Cypress Semiconductor. (2017)

37. Anu M D., Tushar Rastogi. PSoC 3 and PSoC 5LP I2C Bootloader. Document No. 001-60317 Rev. *L1. AN60317. Cypress Semiconductor. (2020)

38. Egan, William F. Phase-Lock Basics, Second Edition, John Wiley and Sons, (2008).

39. Edwin Hewitt & Robert E. Hewi. The Gibbs-Wilbraham Phenomenon: An Episode in Fourier Analysis. Archive for History of Exact Sciences, Volume 21. Springer-Verlag. (1979)

40. Elliott Bros And Autonetics Fit Verdan Computers To Polaris Submarines. Electronics Weekly. 2 January 2018.

41. Evans, J.P. and Burns, G.W. "A Study of Stability of High Temperature Platinum Resistance Thermometers", in Temperature - Its Measurement and Control on Science and Industry, Reinhold, New York. (1962)

42. Fraden, J. AIP Handbook of Modern Sensors. Physics, Design and Application American Institute of Physics. (1993)

43. Gardner, Floyd M. Phaselock Techniques. John Wiley and Sons, Brisbane, 2nd edition. (1979)

44. Fernandez, Daniel; Garnier, A; Blanco, A.; Duran, A.; Jimenez-Jorquera, Cecilia; de Fuentes, Olimpia and Arias. Portable measurement system for FET type microsensors based on PSoC microcontroller. Journal of Physics Conference Series. (March 2013)

45. Gilbert, B. Translinear circuits: An historical overview. Analog Integrated Circuit Signal Processing 9, 95–118 (1996) https://doi.org/10.1007/BF00166408

46. Gilbert, Barrie. A High-Performance Monolithic Multiplier Using Active Feedback. IEEE Journal of Solid-State Circuits, Vol. SC-9, No. 6, December (1974)

47. Goldstein, Gordon; J, Neumann, Albrecht (April 1957). "COMPUTERS. U. S. A. Autonetics, RECOMP, Downey, Calif". Digital Computer Newsletter. VOLUME 9, NUMBER 2. 9: 2 via DTIC.

48. Goldstein, Gordon; J, Neumann, Albrecht. "COMPUTERS. U. S. A. Autonetics, RECOMP, Downey, Calif". Digital Computer Newsletter. VOLUME 9, NUMBER 2. 9: 2 via DTIC. (April 1957)

49. Gray, Paul R. and Meyer. Analysis and Design of Analog Integrated Circuits. John Wiley and Sons, Brisbane, 3rd edition. (1993)

50. Gray, Paul R., Hurst, Paul J., Lewis, Stephen H. and Meyer, Robert G. Analysis and Design of Analog Integrated Circuits. John Wiley and Sons, Brisbane, 4th edition. (2001)

51. Gupta, Sacchinb and Ntarajan, Lakshmi. Migrating from PSoC 3 to PSoC 5. Application Note: AN62083. Cypress Semiconductor Corporation. (2011)

52. Harbort, Bob and Brown, Bob. Karnaugh Maps. https://www.slideshare.net/hangkhong/karnaugh (2001).

53. Hastings, Mark. AN69133. PSoC 3/PSoC 5LP Easy Waveform Generation with the WaveDAC8 Component. Document No. 001-69133Rev. *F1.AN69133. Cypress Semiconductor. (2017)

54. Irvine, Robert G. Operational Amplifier Characteristics and Applications. Prentice Hall. (1994).

55. Janicke, J.M. The Magnetic Measurement Handbook, Magnetic Research Press, New Jersey. (1994)

56. Jayalalitha D.S. and Susan D. Grounded Simulated Inductor - A Review. Middle-East Journal of Scientific Research 15 (2): 278–286. (2013)

57. Johnson, J. B. Thermal Agitation of Electricity in Conductors. Phys. Rev. 32, 1928, p. 97. (1928)

58. Jung, Walt. OpAmp Applications Handbook. Newnes. (1994)

59. Kamen, E.W. and Heck, B.S. Fundamentals of Signals and Systems. 3rd Edition, Prentice-Hall. (2007)

60. Kannan, Vivek Shankar. PSoC 3 and PSoC 5 Interrupts. AN54460. Document No. 001-54460 Rev. *D 1. Cypress Semiconductor (2012).

61. Kannan, Vivek Shankar ; Chen, Julie. PSoC 3, PSoC 4,and PSoC 5LP Temperature Measurement with a Diode. AN60590. Document No. 001-60590 Rev. *K 1 Cypress Semiconductor. (2020)

62. Kathuria, Jaya and Keeser, Chris. Implementing State Machines with PSoC 3, PSoC 4, and PSoC 5LP. AN62510. Document No. 001-62510 Rev. *F. Cypress Semiconductor. (2017)

63. Keeser, Chris. PSoC Designer Boot Process, from Reset to Main. AN73617. Document No. 001-73617Rev. *C1. Cypress Semiconductor. (2017)

64. Kester, Walter. ADC Architecture III: Sigma-Delta Basics. TT-022 Tutorial. Analog Devices. (2008)

65. Kingsbury, Max. PSoC 3 Startup Procedure. Application Note AN60616. Cypress Semiconductor Corporation. (2011)

66. Kingsbury, Max. PSoC 3 and PSoC 5LP Clocking Resources. Application Note AN60631. Cypress Semiconductor Corporation. (2017)

67. Kingsbury, Max. PSoC 3 and PSoC 5LP External Crystal Oscillators. Application Note AN54439. Cypress Semiconductor Corporation (2019).

68. Klingman, Edwin E. Microprocessor Systems Design. Prentice Hall. (1977).

69. Konstas, Jason. PSoC 3 and LP Intelligent Fan Controller. Cpress Semiconductor. (2011)

70. H. Kreiger, H. VERDAN Technical Reference Manual. EM-1319-1. Autonetics A Division of North American Rockwell Corporation. (1959). Revised 13 June 1962

71. Krishswamy, Arvind and Gupta, Rajiv. Profile Guided Selection of ARM and Thumb Instructions. LCTES'02-Scopes'02, June 19–21. (2002)

72. Labrosse, Jean J. Embedded Systems Building Blocks, Second Edition. CMP Books. (2002)

73. Lancaster, Don. Active Filter Cookbook, Second Edition. Elsevier. (1998)

74. Lee, Edward A. The Problem with Threads. Technical Report UCB/EECS-2006-1. http://www.eecs.berkeley.edu/Pubs/TechRpts/2006/EECS-2006-1.html. (2006)

75. Lemieux, Joe. Introduction to ARM Thumb. Embedded Systems Design. September (2003).

76. Lenk, John D. Logic Designer's Manual, Reston Publishing. (1977)

77. Lenz, J.E. A Review of Magnetic Sensors, Proc IEEE, Vol. 78, No. 6:973–989. June 1990.

78. J.E. Lenz et al. A Highly Sensitive Magnetoresistive Sensor, Proc Solid State Sensor and Actuator Workshop. (1992)

79. Levine, John R. Linkers and Loaders. Morgan Kaufmann Publishers, California. (2000)

80. Long, Eric MARDAN Computer Photo, National Air and Space Museum, Smithsonian Institution (2015)

81. Lossio, Rodolfo. PSoC 3, PSoC 4, and PSoC 5LP Temperature Measurement with a TMP05/TMP06 Digital Sensor. AN65977 (2016).

82. Lui, Donglin; Hu, Xiabo Sharon; Lemmon D. Michael; and Ling, Qiang. Firm Real-Time System Scheduling Based on a Novel QoS Constraint. IEEE Transactions of Computers, Vol. 55, No. 3, March 2006.

83. Marrivagu, Vijay Kumar /Fernandez, Antonio Rohit De Lima. PSoC Creator -Implementing Programmable Logic Designs with Verilog. AN82250. Document NUMBER:001-82250Rev.*J1. Cypress Semiconductor. (2018)

84. Meador, Don. Analog Signal Processing with Laplace transforms and Active Filter Design. Delmar. (2002)

85. Megretsk, A. MULTIVARIABLE CONTROL SYSTEMS. Massachusetts Institute of Technology. Department of Electrical Engineering and Computer Science. April 3, 2004.

86. Meijering, E. H. W."A chronology of interpolation: From ancient astronomy to modern signal and image processing", Proc. IEEE, vol. 90, no. 3, pp. 319–342, Mar. 2002

87. Mendelson, Elliot. Schaum's Outline of Theory and Practice of Boolean Algebra. McGraw- Hill. (1970)

88. Meyers, C.H. "Coiled Filament Resistance Thermometers". NBS Journal of Research, Vol. 9. (1932)

89. Miller, John H. Dependence on the Input Impedance of a Three-Electrode Vacuum Tube Upon the Load in the Plate Circuit. Scientific Papers of the Bureau of Standards, 15(351) pp 367–385. (1920).

90. Mohan, Anup. SAR ADC in PSoC 3. AN60832. Cypress Semiconductor April (2010).

91. Murphy, Robert. USB 101: An Introduction to Universal Serial Bus 2.0. Cypress Application Note AN57294 (2011).

92. Murphy, Robert. PSoC 3 and PSoC 5LP–Introduction to Implementing USB Data Transfers. AN56377. Document No. 001-56377 Rev.*Cypress Semiconductor. (2017)

93. Nekoogar, Farzad and Moriarty, Gene. Digital Control Using Digital Signal Processing. Prentice Hall. (1999)

94. Nyquist, Harry. Certain topics in telegraph transmission theory. Trans. AIEE. 47 (2): 617–644. April 1928.

95. Nyquist, H. "Thermal Agitation of Electronic Charge in Conductors." Phys. Rev. 32, 1928, p. 110. (1928)

96. O'Donnell, C. F. Inertial Navigation Analysis and Design., pp. 139–176 and pp. 251–302. McGraw Hill Company. (1964)

97. Ohba, R. Intelligent Sensor Technology. John Wiley. (1992)

98. Orfinidis, Sophocles JH. Introduction to Signal Processing. Prentice Hall. (1996)

99. Paliy, Svyatoslav and Bilynskyy, Andrij. PSoC 1 - Interface to Four-Wire Resistive Touchscreen. Application Note AN2376. Cypress Semiconductor Corporation. (2011)

100. B.B. Pant. Fall 1987. "Magnetoresistive Sensors," Scientific Honeyweller, Vol. 8, No. 1:29–34.

101. Park, Sangil. Principals of Sigma-Delta Modulation for Analog-to-Digital Converters. Motorola. (1993)

102. Parr, E.A. The Logic Designer's Guidebook. McGraw-Hill. (1984)

103. Peckol, James K. Embedded Systems, A Contemporary Design Tool. John Wiley & Sons. (2008)

104. Pellerin, David and Holley, Michael. "Practical Design Using Programmable Logic". Prentice-Hall. (1991)

105. Phelan, Richard. Improving Arm Code Density and Performance (New Thumb Extensions to the Arm Architecture). Arm Limited, June 2003.

106. P., Phalguna. PSoC 3 and PSoC 5LP SPI Bootloader. AN84401. PSoC 3 and PSoC 5LP SPI Bootloader. Document No. 001-84401Rev.*D. Cypress Semiconductor. (2017)

107. Poincare, Henri. Science and Method, p 68. Translated by Maitland, Francis. Thomas Nelson and Sons, London, Dublin, New York. (1908)

108. Prandoni P. and Vetterli, M. Signal Processing for Communications. EPFL Press. (2008).

109. PSoC 3, PSoC 5 Architecture TRM (Technical Reference Manual). Document No. 001-50234 Rev D, Cypress Semiconductor. (2009)

110. Madaan, Pushek. Maintaining accuracy with small magnitude signals. EE Times Design (http://www.eetimes.com/design) (January 2011)

111. Ramsden, E. Measuring Magnetic Fields with Fluxgate Sensors. Sensors: 87–90. 1(1994.)

112. Reynolds, Greg. PSoC 3 and PSoC5LP Low-Power Modes and Power Reduction Techniques. AN77900. Document No. 001-77900 Rev.*G1. (2017)

113. Riewruja, V and Rerkratn, A. Analog multiplier using operational amplifiers. India Journal of Pure and Applied Physics. Vol. 48 January 2010, pp. 67–70.

114. Ripka, P. "Review of Fluxgate Sensors," Sensors and Actuators A, 33:129–141,(1996).

115. Rogatto, William D, Editor. The Infrared and Electro-Optical Systems Handbook, 3, pp 326–328, SPIE Optical Engineering Press, Bellingham, Washington. (1993)

116. Sadasivan, Shyam. An Introduction to the ARM Cortex-M3 Processor. ARM White Paper. 2006 https://class.ece.uw.edu/474/peckol/doc/StellarisDocumentation/IntroToCortex-M3.pdf

117. Sallen, R.P. A Practical Method of Designing RC Filters. IRE Transactions Circuit Theory, vol CT-2, pp 75–84, March 1955.

118. Schmitt, Otto H. A Thermionic Trigger. Journal of Scientific Instruments, Vol. XV, pp 24–26. (1938)

119. User's Reference Manual for emWin V5.14. SEGGER Microcontroller GmbH & Co. KG. (2012)

120. Sequine, Dennis. PSoC1 - Correlated Doubled Sampling for Thermocouple Measurement. AN2226, Cypress Semiconductor Corporation. (2011)

121. Shannon, Claude E., A Mathematical Theory of Communication. Bell System Technical Journal. 27 (3): 379–423. (1948)

122. Shannon, Claude E. Communication in the presence of noise. Proceedings of the Institute of Radio Engineers. 37 (1): 10–21. doi:10.1109/jrproc.1949.232969. S2CID 52873253. (1949)

123. Singh, Gaurav, Shetty, Shivprasad, Romit Pednekar. Implementation of a wireless sensor node using PSoC and CC2500 RF module. 2014 International Conference on Advances in Communication and Computing Technologies (ICACACT). (2014)

124. Smith, Carl H., Caruso, Michael J., and Schneider, Robert W. A New Perspective on Magnetic Sensing. www.sensorsmag.com. (1998)

125. Smith, Steven W. The Scientists and Engineers Guide to Digital Signal Processing. California Technical Publishing. (1997)

126. Ross Fosler/Srinivas NVNS. PSoC 3 and PSoC 5LP - Phase-Shift Full-Bridge Modulation and Control. AN76439. Document No. 001-76439 Rev. *A 1. (2017)

127. Stallings, William. Computer Organization and Architecture: Designing for Performance. Prentice-Hall. (1996)

128. Steinhart, J.S. and Hart, Stanley R. Calibration curves for thermistors, Deep Sea Research and Oceanographic Abstracts, Volume 15, Issue 4, Pages 497–503, August 1968,.

129. Sweet, Dan. Peak Detection with PSoC 3 and PSoC 5LP. AN60321. Document No. 001-60321Rev.*I. Cypress Semiconductor. (2015).

130. Sweet, Dan. Implementing Accurate Peak Detection. Cypress White Paper. Cypress Semiconductor. (2012).

131. Tanenbaum, Andrew S. Structured Computer Organization, Prentice-Hall. (2006)

132. Tellegen, B.D.H. The Gyrator. A New Electric Circuit Element. Philips Res. Rpt, 3,81–101 (Apr. 1948)

133. Van Dyke, Karl S. "The electric network equivalent of the piezoelectric quartz resonator". Phys. Rev., v. 25, p.895. (1925).

134. Van Dyke, Karl S. The Piezoelectric Quartz Resonator. http://www.minsocam.org/ammin/AM30/AM30_214.pdf

135. Van Ess, David. PSoC 1 Temperature Measurement With Thermistor. PSoC 1 Temperature Measurement With Thermistor. Document No. 001-40882 Rev. *E. Cypress Semiconductor.(2002)

136. Van Ess, David. Application Note: Comparator with Independently Programmable Hysteresis Thresholds (AN2310). Cypress Semiconductor, pp. 1–3, pp 1–2. (2005)

137. Van Ess, David, Learn Digital Design with PSoC, One Bit at a Time. CreateSpace Independent Publishing Platform. (2014)

138. Van Ess, D. E. Currie, and Doboli A. "Laboratory Manual for Introduction to Mixed-Signal Embedded Design", ISBN: 978-0-9814679-1-7. (2008).[1]

139. Warp™ Verilog Reference Guide. Document #001-483-52 Rev. *A. Cypress Semiconductor, San Jose, CA. (2009)

140. Wickert, Mark A. Phase Locked Loops with Applications. ECE 5675/4675 Lecture Notes Spring-Verlag. (2011))

141. Wilbraham, H. Cambridge and Dublin Math. Jour., 3 p198. (1848)

142. Wolaver, Dan H. Phase-Locked Loop Circuit Design. 1st Edition. Prentice Hall, New Jersey. (1991)

143. Yarlagadda, Archana. PSoC 3 and PSoC 5LP Correlated Double Sampling to Reduce Offset, Drift, and Low-Frequency Noise. AN66444. Document No. 001-66444 Rev. *D. Cypress Semiconductor. (2017)

144. http://en.wikipedia.org.

145. Yuill, Simon; Fuller, Mathew (ed.), Software Studies: A Lexicon, Cambridge, Massachusetts, London, England: The MIT Press. 2008

146. Noise Analysis in Operational Amplifier Circuits. App Note: SLVA043B. Texas Instruments. (2007)

147. A Primer on Jitter, Jitter Measurement and Phase-Locked Loops. AN687. Silicon Laboratories. Rev.0.1. (2012)

148. Digital Computers for Aircraft. Flight International. 85 (2867): 288. ISSN 0015-3710. Feb 1964.

---

[1] Converted to PSoC 3 by Anurag Umbarkar, Varun Subramanian and Alex Doboli.

149. The Amazing MARDAN - Accelerating Vector. https://acceleratingvector.com/2014/06/21/theamazing-mardan.
150. MARDAN Computer - Time and Navigation. https://timeandnavigation.si.edu/multimedia-asset/mardan-computer.
151. Recomp III Service Manual. A3958-501. Autonetics Division of North American Rockwell. 20 August 1959
152. MCS-51 Microcontroller Family User's Manual. Chapter 2, pp 28–75. Intel Corporation. (1993)
153. PSoC 3 Architecture Technical Reference Manual. Document No. 001-50235 Rev. *M. Cypress Semiconductor. April 8, 2020.
154. PSoC 5LP Architecture Technical Reference Manual. Document No. 001-78426 Rev. *G. Cypress Semiconductor. November 6, 2019.
155. CE210514-PSoC3, PSoC 4, and PSoC 5LP Temperature Sensing with a Thermistor. Document No. 002-10514Rev.*B. Cypress Semiconductor. (2018)
156. CE202479 -PSoC 4 Capacitive Liquid Level Sensing. Document No. 002-02479Rev. Cypress Semiconductor Corporation. (2015)
157. PSoC 4 Magsense. Inductive Sensing. Document Number: 002-24878 Rev.**. Cypress Semiconductor. Revised August 20, 2018.
158. https://www.planetanalog.com/design-considerations-the-analog-signal-chain-part-1-of-2/. January 30, 2011.
159. https://www.planetanalog.com/design-considerations-the-analog-signal-chain-part-2-of-2/#. February 4, 2011.
160. Complex Programmable Logic Devices (CPLD) Information. On GlobalSpec. N.p., n.d. Web. 6 Apr. 2013. www.globalspec.com/learnmore/analog_digital_ics/programmable
161. Xilinx "Programmable Logic Design – Quick Start Hand Book", 2nd edition, Jan 2002.

PSoC 3 Specification Summary

PSoC 3, a member of the CY8C36 family, consists of an MCU, memory, analog, and digital peripheral functions in a single chip. Supported signal functionality includes signal acquisition, processing, and control with high accuracy, high bandwidth, and high flexibility. Analog capability spans the range from thermocouples (near DC voltages) to ultrasonic signals. The CY8C36 family can handle dozens of data acquisition channels and analog inputs on every GPIO pin. It is a high-performance configurable digital system with optional interfaces such as USB, multi-master I2C, and CAN. In addition to communication interfaces, this family of systems on (a) chip has an easy to configure logic array, provides flexible routing to all I/O pins, and utilizes a high performance, single cycle, 8051 microprocessor core. System-level designs can be easily created using a rich library of pre-built components and Boolean primitives using PSoC Creator, a hierarchical schematic design entry tool.

Features

- Single Cycle 8051 CPU core
- DC to 67 MHz
- Multiply and divide instructions
- Flash program memory, up to 64K 10,000 write cycles, 20 year retention
- Up to 8KB of Flash ECC or configuration storage
- Up to 24 channels of DMA
  - Programmable chained descriptors and priorities
  - High bandwidth 32-bit transfer support
- Low voltage, ultra low power
  - Wide operating voltage range: 0.5–5.5 V
  - High efficiency boost regulator from 0.5 V input to 1.8–5.0 V output
  - 330 μA at 1 MHz, 1.2 mA at 6 MHz, 5.6 mA at 40 MHz
  - Low power modes including:
    * 200 nA hibernate mode with RAM retention and LVD
    * 1 μA sleep mode with real time clock and low voltage reset
  - Versatile I/O system
    * 28–72 I/O (62 GPIO, 8 SIO, 2 USBIO[1])
    * Any GPIO to any digital or analog peripheral routability
    * LCD direct drive from any GPIO, up to 46x16 segments[1]
    * 1.2–5.5 V I/O interface voltages, up to 4 domains
    * Maskable, independent IRQ on any pin or port
    * Schmitt trigger TTL inputs
    * All GPIO configurable as open drain high/low, pull up/down, High-Z, or strong output

© Springer Nature Switzerland AG 2021
E. H. Currie, *Mixed-Signal Embedded Systems Design*,
https://doi.org/10.1007/978-3-030-70312-7

   * Configurable GPIO pin state at power on reset (POR)
   * Configurable GPIO pin state at power on reset (POR)
   * 25 mA sink on SIO
 - Digital peripherals
   * 16–24 programmable PLD based Universal Digital Blocks
   * Full-speed (FS) USB 2.0 12 Mbps using internal oscillator[1]
   * Up to four 16-bit configurable timer, counter, and PWM blocks
   * Library of standard peripherals
   * 8, 16, 24, and 32-bit timers, counters, and PWMs
   * SPI, UART, I2C
   * Many others available in catalog
 - Library of advanced peripherals
 - Cyclic Redundancy Check (CRC)
 - Pseudo Random Sequence (PRS) generator
 - LIN Bus 2.0
 - Quadrature decoder
- Analog peripherals (1.71 V $\leq Vdda \leq$ 5.5 V)
 - 1.024 V $\pm$ 0.9% internal voltage reference across $-40\,°C$ to $+85\,°C$ (14 ppm/ °C)
 - Configurable Delta-Sigma ADC with 12-bit resolution
   * Programmable gain stage: $x0.25$ to $x16$
   * 12-bit mode, 192 ksps, 70 dB SNR, 1 bit INL/DNL
 - 67 MHz, 24-bit fixed point digital filter block (DFB) to implement FIR and IIR filters[1]
 - Up to four 8-bit, 8 Msps IDACs or 1 Msps VDACs
 - Four comparators with 75 ns response time
 - Up to four uncommitted OpAmps with 25 mA drive capability
 - Up to four configurable multifunction analog blocks. Example configurations are PGA, TIA, Mixer, and Sample and Hold
- Programming, debug, and trace
 - JTAG (4 wire), Serial-Wire Debug (SWD) (2 wire), and Single-Wire Viewer (SWV) interfaces
 - 8 address and 1 data breakpoint
 - 4 KB instruction trace buffer
 - Bootloader programming supportable through I2C, SPI,UART, USB, and other interfaces
- Precision, programmable clocking
 - 1–66 MHz internal 1% oscillator (over full temperature and voltage range) with PLL
 - 4–33 MHz crystal oscillator for crystal PPM accuracy
 - Internal PLL clock generation up to 67 MHz
 - 32.768 kHz watch crystal oscillator
 - Low power internal oscillator at 1, 100 kHz
- Temperature and packaging
 - $-40\,°C$ to $+85\,°C$ industrial temperature
 - 48-pin SSOP, 48-pin

PSoC 5LP Specification Summary

B

PSoC 5LP is part of the CY8C56LP family of Programmable System-on-Chip that was created by Cypress Semiconductor Corporation. It is a programmable, embedded system-on-chip, that integrates configurable analog/ digital peripherals, memory, and a microcontroller on a single chip. Its architecture utilizes a 32-bit Arm® Cortex®-M3 core and DMA controller and digital filter processor, capable of operating up to 80 MHz. In addition to operating at ultra-low power level over the industry's widest voltage range, PSoC 5LP's architecture supports a broad range of digital and analog peripherals as well as supporting custom functions. The system provides flexible routing of any of its myriad analog/digital peripheral function to any pin. The PSoC family of devices employs a highly configurable system-on-chip architecture for embedded control design. The integrated, configurable, analog/digital circuits are controlled by an on-chip microcontroller. A single PSoC device can integrate as many as 100 digital and analog peripheral functions, reducing design time, board space, power consumption, and system cost while improving system quality.

Features

- Operating Characteristics
  - Voltage range: 1.71–5.5 V, up to 6 power domains
  - Temperature range (ambient): −40 to 85 °C[1]
    Extended temperature parts: −40 to 105 °C
  - DC to 80-MHz operation
  - Power modes
    * Active mode 3.1 mA at 6 MHz, and 15.4 mA at 48 MHz
    * 300-nA hibernate mode with RAM retention
    * Boost regulator from 0.5-V input up to 5-V output
- Performance
  - 32-bit Arm Cortex-M3 CPU, 32 interrupt inputs
  - 24-channel direct memory access (DMA) controller
  - 24-bit 64-tap fixed-point digital filter processor (DFB)
- Memories
  - Up to 256 KB program flash, with cache and security feature
  - Up to 32 KB additional flash for error correcting code (ECC)
  - Up to 64 KB RAM
  - 2 KB EEPROM
- Digital Peripherals
  - Four 16-bit timer, counter, and PWM (TCPWM) blocks
  - I^2C, 1 Mbps bus speed

© Springer Nature Switzerland AG 2021
E. H. Currie, *Mixed-Signal Embedded Systems Design*,
https://doi.org/10.1007/978-3-030-70312-7

- Full CAN 2.0b, 16Rx, 8 Tx buffers
- 20–24 universal digital blocks (UDB), programmable to create any number of functions: 8-, 16-, 24-, and 32-bit timers, counters, and PWMs $\cdot I^2C$, UART, SPI, $I2S$, LIN 2.0 interfaces
- Cyclic redundancy check (CRC)
- Quadrature decoders
- Gate-level logic functions
- Programmable clocking
  - 3- to 74-MHz internal oscillator, 1% accuracy at 3 MHz a 4- to 25-MHz external crystal oscillator
  - Internal PLL clock generation up to 80 MHz
  - Low-power internal oscillator at 1, 33, and 100 kHz
  - 32.768-kHz external watch crystal oscillator
  - 12 clock dividers routable to any peripheral or 1/0
- Analog peripherals
  - Configurable 8- to 12-bit delta-sigma ADC
  - Up to two 12-bit SAR ADCs
  - Four 8-bit DACs
  - Four comparators
  - Four opamps
  - Four programmable analog blocks, to create:
  - Programmable gain amplifier (PGA)
  - Transimpedance amplifier (TIA) Mixer
  - Sample-and-hold circuit
  - CapSense $^\otimes$ support, up to 62 sensors
  - 1.024V $\pm$ 0.1% internal voltage reference
- Versatile I/O system
  - 48–72 I/O pins—up to 62 general-purpose I/Os (GPIOs)
  - Up to eight performance $1/O(SIO)$ pins
    * 25 mA current sink
    * Programmable input threshold and output high voltages
    * Can act as a general-purpose comparator
    * Hot swap capability and overvoltage tolerance
  - Two USBIO pins that can be used as GPIOs
  - Route any digital or analog peripheral to any GPIO
  - LCD direct drive from any GPIO, up to 46 $\times$ 16 segments
  - CapSense support from any GPIO
  - 1.2- to 5.5-V interface voltages, up to four power domains
- Programming, debug, and trace
  - JTAG (4-wire), serial wire debug (SWD) (2-wire), single wire viewer (SWV), and Traceport (5-wire) interfaces
  - Arm debug and trace modules embedded in the CPU core
  - Bootloader programming through I^2C, SPI, UART, USB, and other interfaces
- Package options:
  - 68-pin QFN, 100-pin TQFP, and 99-pin CSP
- Development support with free PSoC Creator tool
  - Schematic and firmware design support

- Over 100 PSoC Components^{TM} integrate multiple ICs and system interfaces into one PSoC. Components are free embedded ICs represented by icons. Drag and drop component icons' to design systems in PSoC Creator.
- Includes free GCC compiler, supports Keil/Arm MDK compiler
- Supports device programming and debugging

SFRPRT0DR = 0x80;

sfr SP = 0x81;
sfr DPL = 0x82;
sfr DPH = 0x83;
sfr DPL1 = 0x84;
sfr DPH1 = 0x85;
sfr DPS = 0x86;

sfr SFRPRT0PS = 0x89;
sfr SFRPRT0SEL = $0x8A$;
sfr SFRPRT1DR = $0x90$;
sfr SFRPRT1PS = $0x91$;

sfr DPX = $0x93$;
sfr DPX1 = $0x95$;

sfr SFRPRT2DR = $0x98$;
sfr SFRPRT2PS = $0x99$;
sfr SFRPRT2SEL = $0x9A$;

sfr P2AX = $0xA0$;
sfr $CPUCLK\_DIV$ = $0xA1$;
sfr SFRPRT1SEL = $0xA2$;
sfr IE = $0xA8$;
sbit EA = IE^7;
sfr SFRPRT3DR = $0xB0$;
sfr SFRPRT3PS = $0xB1$;
sfr SFRPRT3SEL = $0xB2$;
sfr SFRPRT4DR = $0xC0$;
sfr SFRPRT4PS = $0xC1$;
sfr SFRPRT4SEL = $0xC2$;
sfr SFRPRT5DR = $0xC8$;
sfr SFRPRT5PS = $0xC9$;
sfr SFRPRT5SEL = $0xCA$;

sfr PSW = $0xD0$;
sbit P = PSW^0;

© Springer Nature Switzerland AG 2021   833
E. H. Currie, *Mixed-Signal Embedded Systems Design*,
https://doi.org/10.1007/978-3-030-70312-7

sbit F1 = PSW^1;
sbit OV = PSW^2;
sbit RS0 = PSW^3;
sbit RS1 = PSW^4;
sbit F0 = PSW^5;
sbit AC = PSW^6;
sbit CY = PSW^7;
sfr SFRPRT6DR = $0xD8$;
sfr SFRPRT6PS = $0xD9$;
sfr SFRPRT6SEL = $0xDA$;
sfr ACC = $0xE0$;
sfr SFRPRT12DR = $0xE8$;
sfr SFRPRT12PS = $0xE9$;
sfr MXAX = $0xEA$;
sfr B = $0xF0$;

sfr SFRPRT12SEL = $0xF2$;
sfr SFRPRT15DR = $0xF8$;
sfr SFRPRT15PS = $0xF9$;
sfr SFRPRT15SEL = $0xFA$;

Mnemonics

ADC	Analog-to-Digital Converter (also A/D)
ALU	Arithmetic Logic Unit
AMD	Analog Modulator
ARM	Advanced RISC Machine
ATM	Automatic Thump Mode (Boost Converter)
CAN	Controller Area Network
CDAC	Current Digital-To-Analog Converter
CapSense	Capacitive Sensing
CLK	Clock
CLR	Clear
CMOS	Complementary Metal-Oxide Semiconductor
CPU	Central Processing Unit
CRC	Cyclic Redundancy Check
DAC	Digital to Analog Converter (also D/A)
DDA	Differential digital analyzer
DEC	Decimator
DFB	Digital Filter Block
DOC	Debug On Chip
DUT	Device Under Test
DSM	Delta-Sigma Modulator
EPROM	Electrically Programmable Read-Only Memory
EEPROM	Electrically Erasable/Programmable Read-Only Memory
FIFO	First In First Out
GP	General Purpose
GPIO	General purpose input/output
I2C	Inter-Integrated Circuit
ICO	Internal Crystal Oscillator
ILO	Internal Local Oscillator
IMO	Internal Main Oscillator
IPGA	Inverting Programmable Gain Amplifier
IAV	Interrupt Address Vector
INT	Interrupt
LCD	Liquid Crystal Display
LED	Light Emitting Diode
LUT	Look Up Table

© Springer Nature Switzerland AG 2021
E. H. Currie, *Mixed-Signal Embedded Systems Design*,
https://doi.org/10.1007/978-3-030-70312-7

MIPS	Millions of Instructions Per Second
MMIO	Memory Mapped Input Output
MUX	Multiplexer
NMI	Non Maskable Interrupt
NOP	No Operation
PFD	Phase frequency detector
PGA	Programmable Gain Amplifier
PLL	Phase-Locked Loop
PMIO	Port Mapped Input Output
PRT	Port (GPIO/SIO)
PWM	Pulse Width Modulator
RISC	Reduced Instruction Set Computer
PSC	Programmable signal change
SPISTK	SPI Stack
SFR	Special Function Register
SIO	Special Input Output
SWD	Serial Wire Debugging
TACH	tachometer
TMR	Timer
TST	Test
UART	Universal Asynchronous Receiver Transmitter
VDAC	Voltage Digital to Analog Converter
VCO	Voltage controlled oscillator
VLT	Low voltage reference
WDT	Watch Dog Timer

Glossary

Accumulator   In a CPU, a register in which intermediate results are stored. Without an accumulator, it would be necessary to write the result of each calculation (addition, subtraction, shift, and so on.) to main memory and read them back. Access to main memory is slower than access to the accumulator, which usually has direct paths to and from the arithmetic and logic unit (ALU).

Active High   A logic signal having its asserted state as the logic 1 state. A logic signal having the logic 1 state as the higher voltage of the two states.

Active Low

1. A logic signal having its asserted state as the logic 0 state.
2. A logic signal having its logic 1 state as the lower voltage of the two inverted logic states.

Address   The label or number identifying the memory location (RAM, ROM, or register) where a unit of information is stored.

Algorithm   A procedure for solving a mathematical problem in a finite number of steps that frequently involve repetition of an operation.

Ambient Temperature   The temperature of the air in a designated area, particularly the area surrounding the PSoC device.

Analog   (See analog signals.)

Analog Blocks   The basic programmable OpAmp circuits, i.e., SC (switched capacitor) and analog blocks CT (continuous-time) blocks. These blocks can be interconnected to provide ADCs, DACs, multi-pole filters, gain stages, etc.

Analog Output   An output that is capable of driving any voltage between the supply rails, instead of analog output just a logic 1 or logic 0.

Analog Signal   A signal represented in a continuous form with respect to continuous times, as analog signals contrasted with a digital signal represented in a discrete (discontinuous) form in a sequence of time.

Analog-to-Digital Converter (ADC)   A device that changes an analog signal to a digital signal of the corresponding magnitude. Typically, an ADC converts a voltage to a digital number. The digital-to-analog (DAC) converter performs the inverse operation.

AND   See Boolean Algebra.

Apodization   refers to the alteration of a signal to make it "smoother" and more tractable from a computational and mathematical perspective.

Application Program Interface (API)   A series of software routines that comprise an interface between a computer application and lower-level services and functions (e.g., user modules and programming interface libraries). APIs allow modules to serve as building blocks for programmers and thereby reduce the challenges of creating complex applications.

© Springer Nature Switzerland AG 2021   837
E. H. Currie, *Mixed-Signal Embedded Systems Design*,
https://doi.org/10.1007/978-3-030-70312-7

Array   Also referred to as a vector or list, is one of the simplest data structures in computer programming Arrays hold a fixed number of equally-sized data elements, generally of the same data type. Individual elements are accessed by index using a consecutive range of integers, as opposed to an associative array. Most high-level programming languages have arrays as a built-in data type. Some arrays are multi-dimensional, meaning they are indexed by a fixed number of integers; for example, by a group of two integers. One- and two-dimensional arrays are the most common. Also, an array can be a group of capacitors or resistors connected in some common form.

Assembly   A symbolic representation of the machine language of a specific processor. Assembly language is converted to machine code by an assembler. Usually, each line of assembly code produces one machine instruction, though the use of macros is common. Assembly languages are considered low level languages; where as C is considered a high-level language.

Asynchronous   a signal whose data is acknowledged or acted upon immediately, irrespective of any clock signal.

Attenuation   The decrease in intensity of a signal as a result of absorption of energy and of scattering out of the path to the detector, but not including the reduction due to geometric spreading. Attenuation is usually expressed in dB.

Bandgap Reference   A stable voltage reference design that matches the positive temperature coefficient of VT with the negative temperature coefficient of VBE, to produce a zero temperature coefficient (ideally) reference.

Bandwidth

1. The frequency range of a message or information processing system measured in Hertz.
2. The width of the spectral region over which an amplifier (or absorber) has substantial gain (or loss); it is sometimes represented more specifically as, for example, full width at half maximum.

Bias

1. A systematic deviation of a value from a reference value.
2. The amount by which the average of a set of values departs from a reference value.
3. The electrical, mechanical, magnetic, or other force (field) applied to a device to establish a reference level to operate the device.

Bias Current   The constant low level DC current that is used to produce stable operation in bias current amplifiers. This current can sometimes be changed to alter the bandwidth of an amplifier.

Binary   The name for the base 2 numbering system. The most common numbering system is the base 10 numbering system. The base of a numbering system indicates the number of values that may exist for a particular positioning within a number for that system. For example, in base 2, binary, each position may have one of two values (0 or 1). In the base 10, decimal, numbering system, each position may have one of ten values (0, 1, 2, 3, 4, 5, 6, 7, 8, and 9).

Bit   A single digit of a binary number. Therefore, a bit may have a value of "0" or "1". A group of 8 bits is called a byte. Because the PSoC's M8CP is an 8-bit microcontroller, PSoC's native data chunk size is a byte.

Bit Rate (BR)   The number of bits occurring per unit of time in a bitstream, usually expressed in bit rate (BR) bits per second (bps).

Block

1. A functional unit that performs a single function, e.g., an oscillator.

2. A functional unit that may be configured to perform one of several functions, such as a digital or analog PSoC block.

Boolean Algebra   In mathematics and computer science, Boolean algebras, or Boolean lattices, are algebraic structures that "capture the essence" of the logical operations AND, R, and NOT, as well as, the set-theoretic operations, i.e., union, intersection, and complement. Boolean algebra also defines a set of theorems that describe how Boolean equations can be manipulated. For example, these theorems are used to simplify Boolean equations, which will reduce the number of logic elements needed to implement the equation. The operators of Boolean algebra may be represented in various ways. Often they are simply written as AND, OR, and NOT. In describing circuits, NAND (NOT AND), NOR (NOT OR), XNOR (exclusive NOT OR), and XOR (exclusive OR) may also be used. Mathematicians often use + (for example, $A + B$) for OR and AND (for example, $A * B$) (since in some ways those operations are analogous to addition and multiplication in other algebraic structures) and represent NOT by a line drawn above the expression being negated (for example, $A, A, !A$).

Break-Before-Make   The elements involved go through a disconnected state entering (break) before the new connected state (make).

Buffer

1. A storage area for data that is used to compensate for a speed difference, when transferring data from one device to another. Usually refers to an area reserved for I/O operations, into which data is read, or from which data is written.
2. A portion of memory set aside to store data, often before it is sent to an external device or as it is received from an external device.
3. An amplifier used to lower the output impedance of a system.

Bus

1. A named connection of nets. Bundling nets together in a bus makes it easier to route nets with similar routing patterns.
2. A set of signals performing a common function and carrying similar data. It is typically represented using vector notation; for example, address[7:0].
3. One or more conductors that serve as a common connection for a group of related devices.

Byte   A digital storage unit consisting of 8 bits.

C   A high-level programming language.

Capacitance   A measure of the ability of two adjacent conductors, separated by an insulator, to hold a charge when a voltage differential is applied between them. Capacitance is measured in units of Farads.

Capture   To extract information automatically through the use of software or hardware, as opposed to hand-entering of data into a computer file.

Chaining   Connecting two or more8-bit digital blocks to form 16, 24, and 32-bit functions. Chaining allows certain signals such as Compare, Carry, Enable, Capture, and Gate to be produced from one block to another.

Checksum   The checksum of a set of data is generated by adding the value of each data word to a sum. The actual checksum can simply be the sum or a value that must be added to the sum to generate a predetermined value.

Clear   To force a bit/register to a value of logic "0".

Clock   The device that generates a periodic signal with a fixed frequency and duty cycle. A clock is sometimes used to synchronize different logic blocks.

Clock Generator   A circuit that is used to generate a clock signal.

CMOS   The logic gates constructed using CMOS transistors connected in a CMOS complementary manner. CMOS is an acronym for complementary metal-oxide semiconductor.

Comparator   An electronic circuit that produces an output voltage or current whenever two input levels simultaneously satisfy predetermined amplitude requirements.

Compiler   A program that translates a high-level language, such as C, into machine language.

Configuration   In a computer system, an arrangement of functional units according to their configuration nature, number, and chief characteristics. Configuration pertains to hardware, software, firmware, and documentation. The configuration will affect system performance.

Configuration Space   the PSoC register space accessed when the XIO bit, in the CPU_F configuration space register, is set to "1".

CPLD   Complex PLD consisting of multiple SPLDs.
FPGA Field-Programmable Gate Array a field programmable device capable of very complex logic functionality. Whereas CPLDs feature logic resources with a wide number of inputs (AND planes), FPGAs offer more narrow logic resources. FPGAs also offer a higher ratio of flip-flops to logic resources than do CPLDs.

Crowbar   A type of over-voltage protection that rapidly places a low resistance shunt (typically an SCR) from the signal to one of the power supply rails, when the output voltage exceeds a predetermined value.

Crystal Oscillator   An oscillator in which the frequency is controlled by a piezoelectric crystal. Typically a piezoelectric crystal is less sensitive to ambient temperature than other circuit components.

Cyclic Redundancy Check (CRC)   A calculation used to detect errors in data communications, typically performed cyclic redundancy using a linear feedback shift register. Similar calculations may be used for a variety check (CRC) of other purposes such as data compression.

Data Bus   A bi-directional set of signals used by a computer to convey information from a data bus memory location to the central processing unit and vice versa. More generally, a set of signals is used to convey data between digital functions.

Data Stream   A sequence of digitally encoded signals used to represent information in transmission.

Data Transmission   The sending of data from one place to another by means of signals over a channel.

Debugger   A hardware and software system that allows the user to analyze the operation of the system under development. A debugger usually allows the developer to step through the firmware one step at a time, set breakpoints, and analyze memory.

Dead Band   A period when neither of two or more signals are in their active state or in a dead-band transition.

Decimal   A base-10 numbering system, which uses the symbols 0, 1, 2, 3, 4, 5, 6, 7, 8, and 9 decimal (called digits) together with the decimal point and the sign symbols + (plus) and − (minus) to represent numbers.

Default Value   Pertaining to the pre-defined initial, original, or specific setting, condition, value, or default value action a system will assume, use, or take in the absence of instructions from the user.

Device   The device referred to in this manual is the PSoC chip, unless otherwise specified.

Die   An unpackaged integrated circuit (IC), normally cut from a wafer.

Digital   A signal or function, the amplitude of which is characterized by one of two discrete (digital) values: "0" or "1".

Digital Blocks   The 8-bit logic blocks that can act as a counter, timer, serial receiver, serial transmitter, CRC generator, pseudo-random number generator, or SPI.

Digital Logic   A methodology for dealing with expressions containing two-state variables that describe the behavior of a circuit or system.

Digital-to-Analog (DAC)   A device that changes a digital signal to an analog signal of corresponding magnitude. The analog-to-digital (ADC) converter performs the reverse operation.

Direct Access   The capability to obtain data from a storage device, or to enter data into a storage device, in a sequence independent of their relative positions by means of addresses that indicate the physical location of the data.

Duty Cycle   The relationship of a clock period high time to its low time, expressed as a percent.

Emulator   Duplicates (provides an emulation of) the functions of one system with a different system, so that the second system appears to behave like the first system.

External Reset (XRES)   An active high signal that is driven into the PSoC device. It causes all the operations of the CPU to stop, blocks interrupts to stop and return the system to a predefined state.

Falling Edge   A transition from a logic 1 to a logic 0. Also known as a negative edge.

Feedback   The return of a portion of the output, or processed portion of the output, of a (usually active) device to the input.

Filter   A device or process by which certain frequency components of a signal are attenuated.

Firmware   The software that is embedded in a hardware device and executed by the CPU.

Flag   The software may be executed by the end-user, but it may not be modified. Any of various types of indicators used for identification of a condition or event (for example, a character that signals the termination of a transmission).

Flash   An electrically programmable and erasable, volatile technology that provides users with the programmability and data storage of EPROMs, plus in-system erasability. Non-volatile means that the data is retained when power is off.

Flash Bank   A group of Flash ROM blocks where Flash block numbers always begin with "0" in an individual Flash bank. A Flash bank also has its own block-level protection information.

Flash Block   The smallest amount of Flash ROM space that may be programmed at one time and the smallest amount of Flash space that may be protected. A Flash block holds 64 bytes.

Flip-Flop   A device having two stable states and two input terminals (or types of input signals) each of which corresponds with one of the two states. The circuit remains in either state until it is made to change to the other state by application of the corresponding signal.

Frequency   The number of cycles or events per unit of time, for a periodic function.

Gain   The ratio of output current, voltage, or power to input current, voltage, or power, respectively. Gain is usually expressed in dB.

Gate

1. A device having one output channel and one or more input channels, such that the output channel state is completely determined by the input channel states, except during switching transients.
2. One of many types of combinational logic elements having at least two inputs (for example, AND, OR, NAND, and NOR (Boolean Algebra)).

Ground

1. The electrical neutral line having the same potential as the surrounding earth.
2. The negative side of the DC power supply.
3. The reference point for an electrical system.

4. The conducting paths between an electric circuit or equipment and the earth, or some conducting body serving in place of the earth.

Hardware   A comprehensive term for all of the physical parts of a computer or embedded system, as distinguished from the data it contains or operates on, and the software that provides instructions for the hardware to accomplish tasks.

Hardware Reset   A reset that is caused by a circuit, such as a POR, watchdog reset, or external reset. A hardware reset restores the state of the device as it was when it was first powered up. Therefore, all registers are set to the POR value as indicated in register tables throughout this document.

Harvard Architecture   separate memory areas are used for program instructions and data. Two or more internal data buses are employed to provide contemporaneous access data and instructions. The CPU fetches program instructions are fetched by the CPU on the program memory bus.

HCPLD   high-capacity PLD, e.g., FPGAs and CPLDs. Field-Programmable Device (FPD)—a type of programmable integrated circuit used for implementing digital hardware, where the chip can be configured by the end-user. Programming of such a device often involves placing the chip into a special programming unit, but some chips can also be configured in-system. Another name for FPDs is programmable logic devices (PLDs); although PLDs encompass the same types of chips as FPDs, we prefer the term FPD because historically the word PLD has referred to relatively simple types of devices.

Hexadecimal   A base 16 numeral system (often abbreviated and called hex), usually written using the symbols 0–9 and A–F. It is a useful system in computers because there is an easy mapping from four bits to a single hex digit. Thus, one can represent every byte as two consecutive hexadecimal digits. Compare the binary, hex, and decimal representations:

bin	hex	dec a
0000	0x0	0
0001	0x1	1
0010	0x2	2
...	...	...
1001	0x9	9
1010	0xA	10
1011	0xB	11
...	...	...
1111	0xF	15

So the decimal numeral 79 whose binary representation is $01001111b$ can be written as $4Fh$ in hexadecimal ($0x4F$).

High Time   The amount of time the signal has a value of "1" in one period, for a periodic digital high time signal.

I2C   A two-wire serial computer bus by Phillips Semiconductors. I2C is an Inter-Integrated Circuit. It is used to connect low-speed peripherals in an embedded system. The original system was created in the early 1980s as a battery control interface, but it was later used as a simple internal bus system for building control electronics. I2C uses only two bi-directional pins, clock, and data, both running at $+5$ V and pulled high with resistors. The bus operates at 100 kbits/second in standard mode and 400 kbits/second in fast mode. I2C is a trademark of Philips Semiconductors.

ICE   The in-circuit emulator that allows users to test the project in a hardware environment, while viewing the debugging device activity in a software environment (PSoC Designer).

Idle State   A condition that exists whenever user messages are not being transmitted, but the idle state service is immediately available for use.

Impedance

1. The response to the flow of current caused by resistive, capacitive, or inductive devices in a circuit.
2. The total passive opposition offered to the flow of electric current. Note the impedance is determined by the particular combination of resistance, inductive reactance, and capacitive reactance in a given circuit.

Input   A point that accepts data, in a device, process, or channel.

Input/Output   A device that introduces data into, or extracts data from, a system.

Instruction   An expression that specifies one operation and identifies its operands, if any, in an instruction programming language such as C or assembly.

Integrated Circuit (IC)   A device in which components such as resistors, capacitors, diodes, and transistors are formed on the surface of a single piece of semiconductor.

Interface   The means by which two systems or devices are connected and interact with each interface other.

Interrupt   A suspension of a process, such as the execution of a computer program, caused by an event external to that process and performed in such a way that the original process can be resumed.

Interrupt Service Routine   A block of code for which normal code execution is diverted to when the M8CP receives a hardware interrupt. Many interrupt sources may each exist with its own priority tine (ISR) and individual ISR code block. Each ISR code block ends with the RETI instruction, returning the device to the point in the program where it left normal program execution.

Jitter

1. A misplacement of the timing of a transition from its ideal position. A typical jitter form of corruption that occurs on serial data streams.
2. The abrupt and unwanted variations of one or more signal characteristics, such as the interval between successive pulses, the amplitude of successive cycles, or the frequency or phase of successive cycles.

Keeper   A circuit that holds a signal to the last driven value, even when the signal becomes un-driven.

Latency   The time or delay that it takes for a signal to pass through a given circuit or network.

Least Significant Bit (LSb)   The binary digit, or bit, in a binary number that represents the least significant least significant bit value (typically the right-hand bit). The bit versus byte distinction is made by using (LSb) a lower case for a bit in LSb.

Least Significant Bit (LSB)   The byte in a multi-byte word that represents the least significant values (typically the least significant byte the right-hand byte). The byte versus bit distinction is made by using an upper (LSB) case for byte in LSB.

Linear Feedback Shift Register (LFSR)   A shift register whose data input is generated as an XOR of two or more elements in the register chain.

Little-endian   the lower-order byte is stored at the lower address and the higher-order byte is stored at the upper address. (cf. Big-endian)

Load   The electrical demand of a process expressed as power (watts), current (amps), or resistance (Ohms).

Logic Block   A relatively small circuit block that is replicated in an array in an FPD. When a circuit is implemented in an FPD, it is first decomposed into smaller sub-circuits that can each be mapped into a logic block. The term logic block is mostly used in the context of FPGAs, but it could also refer to a block of circuitry in a CPLD.

Logic Capacity   The amount of digital logic that can be mapped into a single FPD. This is usually measured in units of the equivalent number of gates in a traditional gate array. In other words, the capacity of an FPD is measured by the size of the gate array that it is comparable to. In simpler terms, logic capacity can be thought of as the number of 2-input NAND gates.

Logic Density   the logic per unit area in an FPD.

Logic Function   A mathematical (Boolean) function that performs a digital operation on digital data and returns a digital value.

Look-Up Table (LUT)   A logic block that implements several logic functions. The logic function is selected look-up table (LUT) by means of select lines and is applied to the inputs of the block. For example, a 2-input LUT with 4 select lines can be used to perform any one of 16 logic functions on the two inputs resulting in a single logic output. The LUT is a combinational device; therefore, the input/output relationship is continuous, that is, not sampled.

Low Time   The amount of time the signal has a given value in one period, for a periodic digital signal.

Low Voltage Detect (LVD)   A circuit that senses Vdd and causes a system interrupt when Vdd falls below a selected threshold.

M8CP   An 8-bit, Harvard Architecture microprocessor. The microprocessor coordinates all activity inside a PSoC by interfacing to the Flash, SRAM, and register space.

Macro   A programming language macro is an abstraction, whereby a certain textual pattern is replaced according to a defined set of rules. The interpreter or compiler automatically replaces the macro instance with the macro contents when an instance of the macro is encountered. Therefore, if a macro is used 5 times and the macro definition required 10 bytes of code space, 50 bytes of code space will be needed in total.

Mask

1. To obscure, hide, or otherwise prevent information from being derived from a mask signal. It is usually the result of interaction with another signal, such as noise, static, jamming, or other forms of interference.
2. A pattern of bits that can be used to retain or suppress segments of another pattern of bits, in computing and data processing systems.

Master Device   A device that controls the timing for data exchanges between two devices. Or devices are cascaded in width, the master device is the one that controls the timing for data exchanges between the cascaded devices and an external interface. The controlled device is called the slave device.

Microcontroller   An integrated circuit chip that is designed primarily for control systems and products. In addition to a CPU, a microcontroller typically includes memory, timing circuits, and IO circuitry. The reason for this is to permit the realization of a controller with a minimal quantity of chips, thus achieving maximal possible miniaturization. This, in turn will reduce the volume and the cost of the controller. The microcontroller is normally not used for general-purpose computation as is a microprocessor.

Mixed-Signal The reference to a circuit containing both analog and digital techniques and components.

Mnemonic A tool intended to assist the memory. Mnemonics rely on not only repetition to remember facts, but also on creating associations between easy-to-remember constructs and lists of data. A two to four-character string representing a microprocessor instruction.

Mode A distinct method of operation for software or hardware. For example, the Digital modulation PSoC block may be in either counter mode or timer mode. A range of techniques for encoding information on a carrier signal, typically a sine wave signal. A device that performs modulation is known as a modulator.

Modulator A device that imposes a signal on a carrier.

MOS An acronym for metal-oxide semiconductor.

Most Significant bit (MSb) The binary digit, or bit, in a binary number that represents the most significant value (typically the left-hand bit). The bit versus byte distinction is made by using a lower case for bit in MSb.

Most Significant Byte (MSB) The byte in a multi-byte word that represents the most significant values (typically most significant byte the left-hand byte). The byte versus bit distinction is made by using an upper case for byte in MSB.

Multiplexer (Mux)

1. A logic function that uses a binary value, or address, to select between a number of inputs and conveys the data from the selected input to the output.
2. A technique that allows different input (or output) signals to use the same lines at different times, controlled by an external signal. Multiplexing is used to save on wiring and IO ports.

NAND See Boolean Algebra.

Negative Edge A transition from a logic 1 to a logic 0. Also known as a falling edge.

Net The routing between devices.

Net A signal that is routed throughout the microcontroller and is accessible by many blocks or systems.

Nibble A group of four bits, which is one-half of a byte.

Noise

1. A disturbance that affects a signal and that may distort the information carried by the signal.
2. The random variations of one or more characteristics of any entity such as voltage, current, or data.

NOR See Boolean Algebra.

NOT See Boolean Algebra.

OR See Boolean Algebra.

Oscillator A circuit that may be crystal controlled and is used to generate a clock frequency.

Output The electrical signal or signals which are produced by an analog or digital block.

Parallel The means of communication in which digital data is sent multiple bits at a time, with each simultaneous bit being sent over a separate line.

Parameter Characteristics for a given block that have either been characterized or may be defined by the designer.

Parameter Block A location in memory where parameters for the SSC instruction are placed prior to execution.

Parity   A technique for testing transmitting data. Typically, a binary digit is added to the data to make the sum of all the digits of the binary data either always even (even parity) or always odd (odd parity).

Path

1. The logical sequence of instructions executed by a computer.
2. The flow of an electrical signal through a circuit.

Pending Interrupts   An interrupt that has been triggered but has not been serviced, either because the processor is busy servicing another interrupt or global interrupts are disabled.

Phase   The relationship between two signals, usually the same frequency, that determines the delay between them. This delay between signals is either measured by time or angle (degrees).

Phase-Locked Loop (PLL)   An electronic circuit that controls an oscillator so that it maintains a constant phase angle relative to a reference signal.

Pin   A terminal on a hardware component. Also called a lead.

Pinouts   The pin number assignment: the relation between the logical inputs and outputs of the PSoC device and their physical counterparts in the printed circuit board (PCB) package. Pinouts will involve pin numbers as a link between schematic and PCB design (both being computer-generated files) and may also involve pin names.

Port   A group of input/output pins, usually eight.

Positive Edge   A transition from a logic 0 to a logic 1. Also known as a rising edge.

Posted Interrupts   An interrupt that has been detected by the hardware but may or may not be enabled by its mask bit. Posted interrupts that are not masked become pending interrupts.

Power-On Reset (POR)   A circuit that forces the PSoC device to reset when the voltage is below a preset level. This is one type of hardware reset.

Program Counter   The instruction pointer (also called the program counter) is a register in a computer processor that indicates where in memory the CPU is executing instructions. Depending on the details of the particular machine, it holds either the address of the instruction being executed, or the address of the next instruction to be executed.

Programmable Array Logic (PAL)   a small FPD that has a programmable AND-plane followed by a fixed OR-plane.

Programmable Logic Array (PLA)   a small FPD consisting of an AND-plane and an OR-plane, which are programmable.

Programmable Switch   a user-programmable switch that can connect a logic element to an interconnect wire, or one interconnect wire to another.

Protocol   A set of rules. Particularly the rules that govern networked communications.

PSoC   Cypress MicroSystems' Programmable System-on-Chip (PSoC) mixed-signal array. PSoC and Programmable System-on-Chip are trademarks of Cypress MicroSystems, Inc.

PSoC Blocks   See analog blocks and digital blocks.

PSoC Designer   The software for Cypress MicroSystems Programmable System-on-Chip technology.

Pulse   A rapid change in some characteristic of a signal (for example, phase or frequency), from a baseline value to a higher or lower value, followed by a rapid return to the baseline value.

Pulse Width Modulator (PWM)   An output in the form of a duty cycle which varies as a function of the applied measurand.

RAM   An acronym for Random Access Memory. A data-storage device from which data can be read out and new data can be written in.

Register   A storage device with a specific capacity, such as a bit or byte.

Reset   A means of bringing a system back to a known state. See hardware reset and software reset.

Resistance   The resistance to the flow of electric current measured in Ohms for a conductor.

Revision ID   A unique identifier of the PSoC device.

Ripple Divider   An asynchronous ripple counter constructed of flip-flops. The clock signal is fed to the first stage of the counter. An n-bit binary counter consisting of n flip-flops that can count in binary from 0 to 2^{n-1}.

Rising Edge   See positive edge.

ROM   An acronym for read-only memory. A data-storage device from which data can be read, but new data cannot be written in.

Routine   A block of code called by another block of code, that may have some general or frequent use.

Routing   Physically connecting objects in a design according to design rules set in the reference library.

RPM   revolutions per minute.

Runt Pulses   In digital circuits, narrow pulses that, due to non-zero rise and fall times of the signal, do not reach a valid high or low level. For example, a runt pulse may occur when switching between asynchronous clocks or as the result of a race condition in which a signal takes two separate paths through a circuit. These race conditions may have different delays and are then recombined to form a glitch or when the output of a flip-flop becomes metastable.

Sampling   The process of converting an analog signal into a series of digital values or reversed.

Schematic   A diagram, drawing, or sketch that details the elements of a system, such as the elements of an electrical circuit or the elements of a logic diagram for a computer.

Seed Value   An initial value loaded into a linear feedback shift register or random number generator.

Serial

1. Pertaining to a process in which all events occur one after the other.
2. Pertaining to the sequential or consecutive occurrence of two or more related activities in a single device or channel.

Set   To force a bit/register to a value of logic 1.

Settling Time   The time it takes for an output signal or value to stabilize after the input has changed from one value to another.

Shift   The movement of each bit in a word, one position to either the left or right. For example, if the hex value 0x24 is shifted one place to the left, it becomes 0x48. If the hex value 0x24 is shifted one place to the right, it becomes 0x12.

Shift Register   A memory storage device that sequentially shifts a word either left or right to output a stream of serial data.

Sign Bit   The most significant binary digit, or bit, of a signed binary number. If set to a logic 1, this bit represents a negative quantity.

Signal   A detectable transmitted energy that can be used to carry information. As applied to electronics, any transmitted electrical impulse.

Silicon ID   A unique identifier of the PSoC silicon.

Skew   The difference in arrival time of bits transmitted at the same time, in parallel transmission.

Slave Device   A device that allows another device to control the timing for data exchanges between two devices. Or when devices are cascaded in width, the slave device is the one that allows another device to control the timing of data exchanges between the cascaded devices and an external interface. The controlling device is called the master device.

Software   A set of computer programs, procedures, and associated documentation concerned with the operation of a data processing system (for example, compilers, library routines, manuals, and circuit diagrams). Software is often written first as source code, and then converted to a binary format that is specific to the device on which the code will be executed.

Software Reset   A partial reset executed by software to bring part of the system back to a known state. A software reset will restore the M8CP to a known state but not PSoC blocks, systems, peripherals, or registers. For a software reset, the CPU registers (CPU_A, CPU_F, CPU_PC, CPU_SP, and CPU_X) are set to $0x00$. Therefore, code execution will begin at Flash address $0x0000$.

SPLD   Simple PLD, typically a PAL or PLA

SRAM   An acronym for static random access memory. A memory device allowing users to store and retrieve data at a high rate of speed. The term static is used because, once a value has been loaded into an SRAM cell, it will remain unchanged until it is explicitly altered or until power is removed from the device.

SROM   An acronym for supervisory read-only memory. The SROM holds code that is used to boot the device, calibrate circuitry, and perform Flash operations. The functions of the SROM may be accessed in normal user code, operating from Flash.

Stack   A stack is a data structure that works on the principle of Last In First Out (LIFO). This means that the last item put on the stack is the first item that can be taken off.

Stack Pointer   A stack may be represented in a computer as inside blocks of memory cells, with the bottom at a fixed location and a variable stack pointer to the current top cell.

State Machine   The actual implementation (in hardware or software) of a function that can be considered to consist of a set of states through which it sequences.

Sticky   A bit in a register that maintains its value past the time of the event that caused its transition.

Stop Bit   A signal following a character or block that prepares the receiving device to receive the next character or block.

Switching   The controlling or routing of signals in circuits to execute logical or arithmetic operations, or to transmit data between specific points in a network.

Switch Phasing   The clock that controls a given switch, PHI1 or PHI2, in respect to the switch capacitor (SC) blocks. The PSoC SC blocks have two groups of switches. One group of these switches is normally closed during PHI1 and open during PHI2. The other group is open during PHI1 and closed during PHI2. These switches can be controlled in the normal operation, or in reverse mode if the PHI1 and PHI2 clocks are reversed.

Synchronous

1. A signal whose data is not acknowledged or acted upon until the next active edge of a clock signal.
2. A system whose operation is synchronized by a clock signal.

Tap   The connection between two blocks of a device created by connecting several blocks/components in a series, such as a shift register or resistive voltage divider.

Terminal Count   The state at which a counter is counted down to zero.

Threshold   The minimum value of a signal that can be detected by the system or sensor under threshold consideration.

Transistor   A transistor is a solid-state semiconductor device used for amplification and switching, and has three terminals: a small current or voltage applied to one terminal controls the current through the other two. It is the key component in all modern electronics. In digital circuits, transistors are used as very fast electrical switches, and arrangements of transistors can function as logic gates, RAM-type memory, and other devices. In analog circuits, transistors are essentially used as amplifiers.

Tri-state A function whose output can adopt three states: 0, 1, and Z (high-impedance). The function does not drive any value in the Z state and, in many respects, may be considered to be disconnected from the rest of the circuit, allowing another output to drive the same net.

UART A Universal Asynchronous Receiver-Transmitter translates between parallel bits of data and serial bits.

UDB Universal Digital Block.

User The person using the PSoC device and reading this textbook.

User Modules Pre-built, pre-tested hardware/firmware peripheral functions that take care of managing and configuring the lower level Analog and Digital PSoC Blocks. User Modules also provide high-level API (Application Programming Interface) for the peripheral function.

User Space The bank 0 space of the register map. The registers in this bank are more likely to be modified during normal program execution and not just during initialization. Registers in bank 1 are most likely to be modified only during the initialization phase of the program.

V_{dd} A name for a power net meaning "voltage drain". The most positive power supply. Usually 5 or 3.3 V.

Volatile Not guaranteed to stay the same value or level when not in scope.

V_{ss} A name for a power net meaning "voltage source". The most negative power supply signal.

von Neumann Architecture data and program instructions are stored in the same memory space. There is a single internal data bus that fetches Instructions and data are fetched over the same path.

Watchdog Timer A timer that must be serviced periodically. If it is not serviced, the CPU will reset after a specified period of time.

Waveform The representation of a signal as a plot of amplitude versus time.

XOR See Boolean Algebra.

Index

© Springer Nature Switzerland AG 2021
E. H. Currie, *Mixed-Signal Embedded Systems Design*,
https://doi.org/10.1007/978-3-030-70312-7

Printed in the United States
by Baker & Taylor Publisher Services